Atmospheric Aerosol Properties

Formation, Processes and Impacts

Kirill Ya. Kondratyev, Lev S. Ivlev,
Vladimir F. Krapivin and Costas A. Varotsos

Atmospheric Aerosol Properties

Formation, Processes and Impacts

Professor Kirill Ya. Kondratyev
Counsellor of the Russian Academy of Sciences
Scientific Research Centre for Ecological Safety
Nansen Foundation for Environment and Remote Sensing
St Petersburg
Russia

Professor Dr Vladimir F. Krapivin
Institute of Radioengineering and Electronics
Russian Academy of Sciences
Moscow
Russia

Professor Dr Lev S. Ivlev
St Petersburg State University
Institute of Physics
St Petersburg
Russia

Associate Professor Dr Costas A. Varotsos
University of Athens
Department of Physics
Laboratory of Meteorology
Athens
Greece

SPRINGER–PRAXIS BOOKS IN ENVIRONMENTAL SCIENCES
SUBJECT *ADVISORY EDITOR*: John Mason B.Sc., M.Sc., Ph.D.

ISBN 10: 3-540-26263-6 Springer-Verlag Berlin Heidelberg New York
ISBN 13: 978-3-540-26263-3

Springer is part of Springer-Science + Business Media (springeronline.com)

Bibliographic information published by Die Deutsche Bibliothek

Die Deutsche Bibliothek lists this publication in the Deutsche Nationalbibliografie; detailed bibliographic data are available from the Internet at http://dnb.ddb.de

Library of Congress Control Number: 2005928342

Printed in Germany

Cover design: Jim Wilkie
Project management: Originator Publishing Services, Gt Yarmouth, Norfolk, UK

Printed on acid-free paper

Contents

Preface xi

Acknowledgements xvii

Abbreviations xix

List of figures xxv

List of tables xxix

About the authors xxxiii

Part One The field observational experiments 1

1 Programmes of atmospheric aerosol experiments: The history of studies . 3

1.1 Complex Atmospheric Energetics Experiment 3

1.1.1 Basic objectives and peculiarities 3

1.1.2 CAENEX-70 6

1.1.3 CAENEX-71 7

1.1.4 CAENEX-72 7

1.2 The Complete Radiation Experiment 13

1.2.1 Spectral distribution of short-wave radiation flux divergence and the role of aerosol absorption in the cloud-free atmosphere 13

1.2.2 Complete Radiation Experiment (overcast cloudiness) . . 20

1.2.3 Atmospheric aerosol 24

1.2.4 Spectral transparency 30

1.2.5 The heat balance of the atmospheric boundary layer: interaction between radiative and turbulent heat exchange 32

1.3 Aerosol in the GATE region 34
1.4 Global Aerosol–Radiation Experiment (GAREX): scientific objectives of the programme 35
1.5 The Soviet–American balloon aerosol measurements. 41
1.5.1 Instrumentation 43
1.5.2 Discussion of the results. 43
1.5.3 The vertical profiles of long-wave radiation fluxes measured with the ARS radiosondes. 46
1.6 Aerosol studies in the 'Bering' experiment 48
1.7 Dust experiments in Tadjikistan 51
1.8 Results of complex aerosol–radiation studies. 54
1.9 Bibliography. 55

2 Field observational experiments in America and western Europe 61
2.1 Observation techniques 61
2.1.1 Remote sensing 61
2.1.2 *In situ* measurements 65
2.1.3 Aerosol observational network. 67
2.1.4 Regional studies 70
2.2 North America 73
2.2.1 A 'superexperiment' in the region of Atlanta 74
2.3 Central America 80
2.3.1 Mexico City. 80
2.3.2 Complex studies of aerosol in the lower atmosphere of Mexico 83
2.4 South America 105
2.5 Western Europe 109
2.5.1 LACE-1998 109
2.5.2 Other field experiments 113
2.6 High latitudes. 120
2.7 Conclusion. 123
2.8 Bibliography. 123

3 Field observational experiments in Eurasia and on the African continent 133
3.1 The Saharan dust experiment. 133
3.2 The field experiment to study the Asian aerosol properties: ACE-Asia 140
3.3 The region of the Indian Ocean 146
3.4 The horizontal distribution of the ion–chemical composition of atmospheric aerosol over the former USSR territory 153
3.5 Atmospheric aerosol properties over western Siberia: the 'aerosol of Siberia' 167
3.6 Bibliography. 175

Part Two Aerosol formation processes and evolution. Cloud cover dynamics 185

4 Aerosol formation processes 187
4.1 Various aerosol sources, formation processes, and types of aerosols 187
4.1.1 Formation processes and types of atmospheric aerosol . . 187
4.1.2 Secondary aerosols formed *in situ* 190
4.1.3 The ACE field experiment 191
4.1.4 Processes of aerosol formation (nucleation) 193
4.2 Background marine aerosol 207
4.3 Sulphur cycle and sulphate aerosol 212
4.4 Dust aerosol 218
4.4.1 Generation mechanisms 218
4.4.2 Dust aerosol from deserts 219
4.5 Carbon aerosol 222
4.5.1 General information 222
4.5.2 Soot 223
4.5.3 Organic carbon 226
4.5.4 Biogenic aerosol 227
4.6 Urban aerosol 228
4.7 Volcanic aerosol 231
4.8 High-latitude atmospheric aerosol 235
4.9 Global spatial–temporal variability of aerosol 240
4.10 Conclusion 254
4.11 Bibliography 254

5 Aerosol and chemical processes in the atmosphere 265
5.1 Interactions between aerosols and minor gas components 265
5.2 Photochemical processes with the participation of aerosol 278
5.3 Tropospheric ozone variations in the highly polluted atmosphere 283
5.4 Modelling atmospheric processes in the highly polluted atmosphere 288
5.4.1 General principles 288
5.4.2 Interaction between tropospheric aerosols and ozone . . . 290
5.5 Ozone latitudinal variations in the surface layer of the atmosphere 293
5.6 Biomass burning and its impact on tropospheric ozone and other MGCs 296
5.7 Conclusion 298
5.8 Bibliography 299

6 Interactions between aerosols and clouds 305
6.1 Introduction 305
6.2 The impact of cloudiness on solar radiation transfer 306
6.2.1 Repeatability of stratus cloudiness 307

6.2.2 Spectral radiation measurements under conditions of overcast cloudiness 309
6.3 The distribution of aerosol in the cloudy atmosphere 314
6.3.1 Vertical distribution of aerosol scattering under different synoptic conditions 316
6.4 The empirical diagnostics of aerosol–cloud interaction 320
6.5 Numerical simulation of the aerosol–cloud interaction 325
6.6 Conclusion. 337
6.7 Bibliography. 337

Part Three Numerical modelling of the processes and properties of atmospheric aerosol. 347

7 The optical properties of atmospheric aerosol and clouds. 349
7.1 Introduction 349
7.2 Statistical modelling of the optical characteristics of atmospheric aerosol 350
7.2.1 Variability of the optical characteristics of atmospheric hazes in the surface layer of the atmosphere. 350
7.2.2 The spectral dependences of aerosol optical characteristics 353
7.2.3 The vertical stratification of atmospheric aerosols 357
7.3 The basic parameters of the optical and microphysical models of atmospheric aerosols 362
7.3.1 The shape and morphological structure of particles 362
7.3.2 The homogeneous chemical composition of aerosol and the complex refraction index for particle substances. . . . 363
7.3.3 Number density and the function of particle size distribution 364
7.4 Studies of the morphological structure of atmospheric aerosols. . 366
7.5 Studies and modelling of the chemical composition of atmospheric aerosols. 369
7.5.1 The elemental composition 370
7.5.2 Modelling the chemical composition of aerosols 375
7.5.3 Modelling the complex refraction index of the atmospheric aerosol substance 375
7.5.4 Inverse problems of the atmospheric aerosols optics. . . . 377
7.5.5 The submicron fraction and mechanisms of its transformation 377
7.6 Absorption of solar and long-wave radiation by atmospheric aerosol 379
7.6.1 Specific features of spectral characteristics 379
7.6.2 The role of submicron and dust fractions of aerosol in radiation absorption 380

7.6.3 The impact of tropospheric aerosols on surface atmosphere albedo and radiative flux divergence in the atmosphere 381
7.7 Complex aerosol–radiation observations in the free atmosphere . 386
7.8 The dynamic and regional aerosol models 391
7.8.1 The mixing layer dynamics 392
7.8.2 The synoptic factor 392
7.9 The problem of 'excessive' absorption of short-wave radiation in clouds 394
7.9.1 The concepts of the effect of 'excessive' (anomalous) SWR absorption in the cloudy atmosphere 398
7.9.2 A comparison of results of measurements of the SWR absorption from the data of different aircraft experiments 399
7.9.3 The dependence of solar radiation absorption on the optical thickness of the cloud layer. 400
7.9.4 The dependence of solar radiation absorption on geographical latitude and solar zenith angle 401
7.10 The optical characteristics of stratified cloudiness. 404
7.10.1 The single-scattering albedo and the volume absorption coefficient 404
7.10.2 The optical thickness τ_0 and the volume scattering coefficient 405
7.10.3 The optical properties of aerosol in a cloudy medium – polluted clouds. 407
7.10.4 The coloration of clouds with aerosol 409
7.10.5 Empirical formulas to estimate the volume coefficients of scattering and absorption under conditions of multiple scattering of radiation in cloud layers 410
7.11 Bibliography. 415

8 Aerosol long-range transport and deposition 425
8.1 Introduction 425
8.2 The African aerosol 428
8.3 The Asian aerosol 437
8.4 Processes of aerosol deposition. 448
8.5 Numerical simulation of aerosol long-range transport 450
8.5.1 Relationships between the scales of atmospheric mixing processes and the choice of models. 450
8.5.2 Interrelationship between the types of models and aerosol characteristics. 452
8.5.3 Passive and active aerosol transport in the atmosphere. . 454
8.5.4 Types of aerosol models and their information base . . . 458
8.5.5 Modelling the wind field. 459
8.5.6 The Gauss-type models 460

8.5.7 Modelling the planetary boundary layer 465
8.5.8 The Euler-type models 466
8.6 An expert system for the physics of atmospheric pollution. 468
8.6.1 The expert system's structure. 468
8.6.2 Formation of database components 469
8.6.3 A subsystem for statistical solutions 473
8.6.4 A subsystem for control and visualization 474
8.7 Global sulphur cycle and its simulation modelling 475
8.7.1 Basic characteristics of the global sulphur cycle 475
8.7.2 The role of sulphur in the environment. 480
8.7.3 Global sulphur cycle model. 482
8.7.4 Modelling results 488
8.8 Conclusion. 497
8.9 Bibliography. 498

9 Aerosol radiative forcing and climate 507
9.1 Introduction. 507
9.2 Empirical diagnostics of the global climate. 513
9.2.1 Air temperature 515
9.2.2 Snow and ice cover. 516
9.2.3 Sea surface level and the ocean upper layer heat content 517
9.2.4 Other climatic parameters. 518
9.2.5 Concentrations of greenhouse gases and anthropogenic aerosol in the atmosphere. 519
9.2.6 Paleoclimatic information 521
9.3 Radiative forcing. 522
9.4 Results of numerical climate modelling and their reliability 525
9.5 Some aspects of aerosol radiative forcing retrieval techniques. . . 526
9.6 The IGAC programme, ACE-Asia, and Indoex projects 540
9.7 Some data of observations in western Europe 547
9.8 Numerical modeling of the 3-D distribution of aerosol and climate 549
9.9 Conclusion. 555
9.10 Bibliography. 555

Index 567

Preface

There is an increasing interest in studies of atmospheric aerosols in the context of their impact on the formation of the climate, chemical heterogeneous reactions in the atmosphere affecting in particular environmental quality, and problems of visibility and health issues. Following recent detailed overviews of these problems, a necessity has arisen to analyse the latest results concerning, first of all, three aspects:

(1) Aerosol properties from data of complex field observational experiments.
(2) Processes of aerosol formation and interaction of aerosol with clouds.
(3) Aerosol as a climate-forming atmospheric component (including the problem of long-range transport).

This book is dedicated mainly to the impact of aerosols on climatic processes in the atmosphere, from microphysical processes of formation and evolution of aerosol properties to the formation of macroscale cloud systems on aerosol particles.

Complex studies of the impact of aerosol (first of all, as a radiation factor) on climate change were undertaken in the early 1960s, when it became clear that solar radiation absorption by aerosols was approximately of the same order of magnitude as that of minor gas components (MGCs). At this time the former USSR undertook unique balloon actinometric and spectral measurements of solar radiation up to altitudes of about 30 km using aerosol impactor measurements sometimes followed by similar complex ground and satellite observations. The results obtained opened up the wide possibilities of analysing the role of aerosol in climate change on regional and global scales. The results of these studies are still of topical interest.

Beginning in 1960, national and international programmes have been devised to study the properties of atmospheric aerosol and its impact on radiation transfer. Such programmes include: the Complete Radiation Experiment (CRE), the Complex Atmospheric Energetics Experiment (CAENEX), the GARP Atlantic Tropical Experiment (GATE), the Global Aerosol–Radiation Experiment (GAREX), and the First Global GARP Experiment (FGGE). After the completion of these programmes, the role of satellite observations of atmospheric characteristics, especially cloud systems and aerosol formations, including their spatial–temporal variability, has strengthened.

The last few years have been marked by an undoubtedly growing interest in the problem of the complex studies of atmospheric aerosols in connection with the necessity to obtain reliable estimates of the impact of aerosols (both natural and anthropogenic) on the global climate. Field observations of various characteristics of aerosol have been made almost on all the continents and over the oceans. Satellite remote sensing has played a special role in providing information. All these aspects are discussed in the three chapters comprising the first part of this book.

It should be noted that the aspect of the climatic impact of aerosol connected with its effect on the processes of cloud formation has not been neglected. The third factor of climatic impact – changes in atmospheric chemistry as a result of heterogeneous chemical reactions – has not been taken into account in global climatological estimates, but its role raises no doubts. Of course, the problem of aerosol impact on climate change is of complicated interdisciplinary nature and requires the participation of specialists from different spheres of knowledge for its solution.

In this context, studies of the mechanisms of aerosol particle formation from MGCs during the course of gas-to-particle conversions and their growth on condensation nuclei as well as chemical reactions on particle surfaces with the subsequent formation of cloud elements are most important for understanding the climate-forming role of aerosol. Relevant problems have been discussed in the second part of this book.

One of the main uncertainties and difficulties in the assessment of the role of atmospheric aerosol in climate change is connected with the absence of adequate information about global spatial–temporal variability of the concentration of the different types of aerosols, the physical (especially optical) properties of aerosols, and their impact on the microphysical processes in clouds. Therefore, one of the objectives of this book is to analyse information concerning the physical properties and chemical composition of atmospheric aerosol, in order to simulate processes and characteristics of atmospheric aerosols. This is dealt with mainly in the third part of this book.

The results of numerical climate modelling show that apart from the internally induced variability of the climate system, the climatic implication of such factors as the anthropogenic growth of MGCs and aerosol concentration, volcanic eruptions, changes in extra-atmospheric insolation, and other issues, raises no doubts. From the viewpoint of the impact on rapid climate change, of special interest are ozone, water vapour, and aerosols – determining an urgency for their global modelling, since they are most responsible for the energy balance of different atmospheric layers and interactions between the atmosphere and the Earth's surface. On the other hand, the content of aerosols and MGCs in the atmosphere is rather variable and is affected both naturally and anthropogenically. Thus, there exists an interaction between aerosols, MGCs, and climate.

Estimates of contributions of various factors to climate change on different timescales (seasonal, interannual, decadal, centenarian), ozone content, UV solar radiation intensity, and atmospheric chemistry demonstrate an important role of aerosol in these processes. A very important aspect of the climate problem consists in the recognition of anthropogenically induced changes caused by increased emissions of aerosols and aerosol-forming gases. In this regard, highly uncertain quantitative estimates of anthropogenic impacts on global climate deserve special attention.

Since calculations made with the prescribed gradual increase of CO_2 concentrations in the atmosphere (as a rule, +1% per year) have led to an atmospheric overheating, the cooling factor due to the presence in the atmosphere of purely scattering sulphate aerosols was taken into account (or surface albedo decrease equivalent to its impact). Of course, this is a conditional approach, and the resulting agreement with observations is no more than an adjustment, though, of course, the problem of sulphate aerosols is very important. This problem, like the existing estimates of climate-forming contributions of other types of aerosol (dust, carbon, biogenic, etc.) has been discussed in detail mainly in the third part of this book.

The volcanic stratospheric aerosol that causes strong and long-period disturbances to the radiation regime and the resulting climate change, is one of the most substantial factors representing the impact of natural aerosol on radiation transfer and global climate. The presence of anthropogenic tropospheric aerosol determines, in particular, the increasing trend of atmospheric turbidity and aerosol haze formation in high latitudes of the northern hemisphere. In these cases of great importance is the mechanism of gas-to-particle aerosol formation (it prevails in the case of volcanic stratospheric aerosol), which demonstrates an interaction between the biogeochemical cycles of sulphur and nitrogen and the processes of atmospheric aerosol formation.

The global climate numerical simulation performed recently with consideration of not only an anthropogenically induced growth of greenhouse gas (GHG) concentration but also an increasing content of anthropogenic sulphate aerosol (the annual emissions of sulphur dioxide transformed into aerosol are equivalent to 7–80 Tg S) revealed a much more complicated pattern of climate formation than was previously supposed: the aerosol-induced climate cooling is most significantly compensated for by greenhouse warming, and the spatial–temporal variability of the aerosol content in the atmosphere determines the formation of a more variable climate.

The impact of desert dust aerosols on the regional and global climate requires more reliable estimates. The contribution of terpenes and other hydrocarbons emitted to the atmosphere by vegetation remains unknown (their photochemical oxidation produces submicron organic aerosols). For instance, observations on the east coast of the USA have shown that the contribution of biogenic compounds to the formation of radiative characteristics of the atmosphere in this industrial region is comparable with the contribution of anthropogenic sulphate aerosols.

The problem of the use of satellite remote sensing data for monitoring such natural and anthropogenic catastrophes as volcanic eruptions, forest fires, dust–sand storms, technogenic catastrophes in cities, and earthquakes is connected with dust aerosol observations. Specific clouds (of the type of smoke rings) form over seismicly active regions several days before underground tremors. Similar phenomena take place before volcanic eruptions. In particular, from the data of ground observations in Mexico, the concentration of aerosol increased several hours before an eruption.

A considerable number of pages of this book are dedicated to the problem of the indirect impact of aerosol on climate, in particular, through affecting the dynamics, microstructure, and optical properties of clouds manifested through increasing cloud albedo due to a growing concentration of small droplets. The effect of aerosol on the optical properties of clouds has been demonstrated by aircraft observations of short-wave radiation fluxes and flux divergences in the regions of the cities of Donetsk and Zaporozhye (the Ukraine) as well as over the Azov Sea (Russia). A strong absorption was detected in dirty clouds at the wavelength 0.5 μm (up to 0.15) compared with clean clouds ($\sim$0.03).

The results of the impact of aerosol on different processes, for instance, on radiation transfer and phase transitions of water, depend, as a rule, on the totality of the interactive chemical and physical processes; the dependence of chemical composition of aerosol on its size distribution almost always playing the substantial role. Therefore, an adequate description of the properties of real aerosol is only possible with the use of the results of a complex determination of its characteristics. One of the most widespread types of aerosol measurements is an estimation of mass concentrations, but this characteristic is less informative, since it says nothing about the sources, composition, and possible impacts.

The aerosol cycles are closely connected with the water cycle in the atmosphere as a result of a close interaction between aerosol and clouds: clouds and precipitation play an important role in the formation, transformation, and removal of aerosol from the atmosphere, but, on the other hand, aerosols affect substantially the microphysical processes in clouds and, hence, the processes of heat and mass-exchange within the atmosphere. It is quite natural that in most regions of the globe the aerosol lifetime in the lower and upper troposphere constitutes 1–2 weeks, and for water vapour it is about 1.5 weeks. The connection between clouds and aerosols testifies to the impossibility of an adequate understanding of the processes of formation and transformation of aerosol without having reliable ideas about the physics and chemistry of clouds. In this regard, of critical importance are studies of nucleation mechanisms.

A new important stage in the understanding and quantitative evaluation of the role of atmospheric aerosol as a climate-forming factor is the realization of the critical importance aerosols have regarding present climate change. The paradox is that with keen attention being paid to problems of global warming, the problem of the climate-forming contribution of atmospheric aerosols has not been adequately resolved. The remaining uncertainty of the estimates of the contribution of aerosols to the formation of the climate deprives such an important document as the Kyoto Protocol of any convincing scientific grounding.

A variety of aerosol characteristics and physico-chemical properties as well as spatial–temporal heterogeneity require the development of a modern complex system of aerosol monitoring, which would combine satellite operative measurements of global distributions with ground measurements of size distributions, chemical compositions, and optical properties of particles. In this context, of great importance are the numerical simulations of physical and chemical aerosol characteristics and processes.

General conclusions which can be drawn on the basis of the analysis of all the available information on atmospheric aerosols (including interactions with clouds) are reduced to the following:

(a) A conditional consideration of sulphate aerosol in climate models providing an illusive agreement with observational data is no more than an adjustment.
(b) An adequate simulation of the climate-forming contribution of aerosol can be reached only on the basis of its interactive consideration in climate models, bearing in mind the actual variety of the aerosol types.
(c) There is no doubt that the aerosol-induced radiative forcing (RF) is either of the same order of magnitude or greater than the greenhouse RF (depending on conditions).
(d) Observational data on aerosol properties are still fragmentary, which dictates the necessity to create a suitable global observation system with the use of both routine (ground) and satellite observations.

The authors thank scientists I. N. Melnikova and A. V. Vasilyev from the St Petersburg State University for some scientific material and comments presented here (Sections 6.2, 7.9, and 7.10). Section 7.1 was written with the participation of scientists G. I. Gorchakov and Yu. S. Lyubovtseva from the Institute of Atmosphere Physics.

Acknowledgements

The authors gratefully acknowledge scientists from the Institute of Atmosphere Physics (Russian Academy of Sciences, Moscow) and Institute of Atmosphere Optics (Siberian Branch of the Russian Academy of Sciences, Tomsk) for their cooperation during the production of this book. Warmest thanks are extended to G. I. Gorchakov, Yu. S. Lyubovtseva, B. D. Belan, A. I. Grishin, G. G. Matvienko, I. V. Samohvalov, and G. N. Tolmachev for their participation and consultations during the editing of the book. Scientists A. V. Vasiliev and I. N. Melnikova from St Petersburg State University provided useful information on aerosol properties. Their experimental data were used to assess different parameters of aerosols.

Abbreviations

ABL	atmospheric boundary layer
AFR	actinic flux of radiation
AI	aerosol index
AM	air mass
AMC	aerosol mass concentration
AO	Arctic oscillation
AOD	aerosol optical depth
AOF	aerosol organic fraction
AOM	aerosol organic matter
AOT	aerosol opitical thickness
AQM	air quality model
ARF	aerosol radiative forcing
AUAP	Analyses of the Urban Atmosphere Pollution
BC	black carbon
BHN	binary homogeneous nucleation
BL	buffer layer
BLAOT	boundary layer AOT
BSM	backscattering matrix
BTD	brightness temperature difference
CA	carbon aerosol
CACM	Caltech Atmospheric Chemistry Mechanism
CAENEX	Complex Atmospheric Energetics Experiment
CAM	Canadian aerosol model
CARE	clouds, aerosols and radiation explorer
CC	carbonate carbon
CCN	cloud concentration nucleii
CCNC	CCN counter
CF	contact freezing

CIMS	chemical ionization mass spectrometer
CN	condensation nucleii
CPC	condensation particles counter
CRE	Complete Radiation Experiment
CRF	cloud radiation forcing
CRI	complex refraction index
CRL	Communications Research Laboratory (Japan)
CTM	chemical transport model
DA	dust aerosol
DA	organic aerosol
DMS	dimethylsulphide
DO	Dansgaard–Oeschger
DOAS	differential optical adsorptive spectroscopy
DRH	deliquescence relative humidity
DSR	digital single-channel recorder
EC	elemental carbon
EE	enhancement effect
EM	emission factor
ENSO	El Niño–Southern Oscillation
EPA	Environmental Protection Agency
ERB	Earth's radiation budget
ESA	European Space Agency
ETR	European territory of Russia
FCCC	Framework Climate Change Convention
FOV	field of view
GAREX	Global Aerosol–Radiation Experiment
GARP	Global Atmospheric Research Programme
GATE	GARP Atlantic Tropical Experiment
GAW	Global Atmosphere Watch
GC–MS	gas chromatography–mass spectroscopy
GCM	general circulation model
GHG	greenhouse gases
GIS	geographical information system
GST	global surface temperature
HPLC	high-performance liquid chromatography
HULIS	'humus-like' substance
IC	ion chromatography
IGAC	International Global Atmospheric Chemistry
IMN	ion modulation nucleation
IOM	inorganic oxidised matter
IPCC	Intergovernmental Panel on Climate Change
ITBL	internal thermal boundary layer
ITCZ	inter-tropical convergence zone
KP	Kyoto Protocol

LBA	large-scale biosphere–atmosphere experiment in Amazonia
LMD	low-molecular dicarboxylic acid
LSAT	land surface air temperature
LUT	look up tables
LWC	liquid water content
LWR	long-wave radiation
MA	mineral aerosol
MABL	marine ABL
MCE	modified consumption efficiency
MCL	maximum (permissible) concentration level
MCS	mesoscale convective system
MD	mass dependent
MDA	monocarboxyl/dicarboxyl acid
MGC	minor gas component
MGSC	model of the global sulphur cycle
MI	mass independent
MPL	micropulse lidar
MSA	methylsulphide
MSD	macromolecular structure database
MSE	mass scattering efficiency
MSN	methanosulphonate
NA	nucleation aerosol
NAAQS	national atmospheric quality standard
NAM	northern hemishere annual mode
NASDA	Japanese National Space Agency
NAT	nocturnal air temperature
NCA	neutralizing capability of aerosol
NDVI	normalized difference vegetation index
NMHC	non-methane hydrocarbons
NMR	nuclear magnetic resource
NMVOC	non-methane VOC
NOAA	National Oceanic and Atmospheric Administration
NSS	non-sea salt
OC	organic carbon
ODF	ordinary differential equation
OM	organic matter
OMPS	ozone mapping and profiler suite
OOC	omitted organic components
OPAC	optical parameters of aerosols and clouds
OSR	outgoing short-wave radiation
PA	polycarboxyl acid
PAH	polycyclic aromatic hydrocarbons
PBL	planetary boundary layer
PIC	particulate inorganic carbon

PMF	positive Matrix Factorization
PNW	Pacific north-west
POA	primary organic carbon
POC	particulate organic carbon
POM	particles of organic matter
PSC	polar stratospheric clouds
PSD	particle size distribution
QBO	quasi-biannual oscillation
RF	radiative forcing
SA	sulphate aerosol
SADEX	Soviet–American Dust Experiment
SAL	Saharan aerosol layer
SAM	southern hemisphere annual mode
SAS	surface active substance
SAT	surface air temperature
SCE	snow cover extent
SEM	scanning electron microscope
SOA	secondary organic aerosols
SoCAB	South Coast Air Basin
SPM	'Summary for policy-makers'
SR	solar radiance
SRB	surface radiation budget
SSA	sea salt aerosol
SSA	single-scattering albedo
SST	sea surface temperature
STS	supercooled triple solution
SVOC	semivolatile oxidation compounds
SWR	short-wave radiation
SYNAER	synergic retrieval of aerosol properties
SZA	solar zenith angle
TC	total carbon
TCWV	total content of water vapour
TDMA	tandem differential mobility analyser
THN	triple homogeneous nucleation
THS	thermohaline circulation
TIR	Third IPCC Report
TOA	top of atmosphere
TOC	total organic carbon
TOVS	TIROS operational vertical sounding
TSC	thermodynamically stable cluster
TTO	total tropospheric ozone
UBTL	upper boundary turbulent layer
UFP	ultrafine particles
UTC	universal time constant
VISSR	visible and infrared spectral regions

VOC	volatile organic compounds
WMO	World Meteorological Organization
WSOC	water soluble organic compounds
XRF	X-ray flourescence

Figures

1.1	The spectral radiative flux divergence in the troposphere (0.3–8.4 μm) in the wavelength region 0.4–2.4 μm	14
1.2	Relative radiation flux divergence	15
1.3	Spectral distribution of the upward flux of short-wave radiation at different altitudes	17
1.4	Spectral distribution of the efficient flux of short-wave radiation at different altitudes	18
1.5	Spectral distributions of the short-wave radiation fluxes and albedo	22
1.6	Spectral distributions of albedo, short-wave balance, and radiation flux divergence (absorbed short-wave radiation) in the case of 1-layer stratified cloudiness	23
1.7	The diurnal variations of heat content, flux divergence due to absorption of short-wave and long-wave radiation, turbulent heat flux, and total flux divergence in the 0–3,000-m layer	33
1.8	The temporal change of wind speed at Shaartus station, at Esanbai station, and relative motion of soil in southern and northern wind directions	54
2.1	Temporal distributions of some elements at different observational sites	90
2.2	Averaged spectra of macroelements for samples of air, precipitation, and volcanic ash taken at different stages of measurements at several observation sites	92
3.1	Total concentrations of the groups of elements in various regions	155
3.2	The latitudinal change of the relative aerosol composition	155
3.3	The latitudinal change of concentrations of iron and elements correlating with it	161
3.4	The latitudinal change of concentration of aluminium and elements correlated with it	163
3.5	The latitudinal change of concentration of silver and elements correlated with it	164
3.6	The latitudinal change of concentration of manganese and elements correlated with it	166

3.7 The average multiyear number density of aerosol in the 0–3-km layer in May and December . . . 168
3.8 Total emissions of suspended matter in the USSR and the USA . . . 168
3.9 Concentration of suspended matter . . . 168
3.10 The annual mean number density of aerosol in the 0–3-km layer over western Siberia . . . 169
3.11 The temporal change of circulation indices . . . 171
3.12 The annual mean concentration of Na^+ and NH_4^+ ions in aerosol over western Siberia . . . 172
3.13 The change of aerosol number density over western Siberia . . . 172
3.14 The atmospheric aerosol composition over some regions of western Siberia . . . 173
3.15 The dynamics of the absolute composition of atmospheric aerosol in the south of the Novosibirsk region . . . 174

5.1 Dependence of the water–surface concentration of ozone on meteorological visibility . . . 287
5.2 Vertical profiles of the ozone mass concentration and coefficients of aerosol scattering . . . 291

6.1 Vertical profile of net, downward, and upward fluxes of solar radiation in cloud for three wavelengths . . . 313
6.2 The resultants of airborne sounding under overcast sky conditions . . . 313
6.3 Aerosol number density profiles under cloudy and cloud-free conditions . . . 314

7.1 Relative frequency of occurrence of different values of ε . . . 352
7.2 IR absorption spectra for different samples of natural aerosol . . . 355
7.3 Dependences of parameters of the microphysical optical model of aerosol on the attenuation coefficient . . . 357
7.4 Vertical profiles of the scattering coefficient from the data of twilight measurements . . . 359
7.5 Calculated and observed coefficients of aerosol attenuation for different situations in the atmosphere and comparison of calculated and observed estimates of the coefficients of aerosol attenuation for coastal regions in the Crimea . . . 376
7.6 The dynamics of altitudes of the mixing layer for Tomsk and Balkhash . . . 382
7.7 The vertical distribution of elements in aerosols over western Siberia and Kazakhstan . . . 393
7.8 The latitudinal dependence of the f_s parameters from and from aircraft measurements and their dependence on the solar zenith angle . . . 401
7.9 The global mean cloud amount averaged over the latitudinal belt, over the water surface, and land for 1971–1990 . . . 403
7.10 The spectral dependence of the single-scattering albedo and the optical thickness τ_0 from the results of measurements in the Arctic . . . 405
7.11 The volume coefficients of scattering and absorption obtained from observational data . . . 406
7.12 The volume coefficients of scattering and absorption . . . 413

8.1 The ESPhAP structure with functions of complex assessment of air pollution over a given territory . . . 469
8.2 Approximate scheme of ESPhAP functioning in the regime of dialogue with the user . . . 470
8.3 Procedure of topological referencing of the territory to the ESPhAP database components . . . 471
8.4 Basic components of ESPhAP functioning . . . 474
8.5 An approximate scheme of the operator–ESPhAP interaction in the regime of its adaptation to specific conditions of simulation experiments on the assessment of aerosol propagation over a given territory . . . 475
8.6 The conceptual scheme of the impact of anthropogenic sulphur emissions on the aquatic medium quality . . . 476
8.7 Sources of the transboundary transport of sulphur compounds in the territory of Norway and Sweden . . . 479
8.8 Relationship between molar parts of sulphite, bisulphate, and dissolved sulphur dioxide in the formation of pH level . . . 481
8.9 The scheme of sulphur fluxes in the environment considered in the MGSC. . . 482
8.10 Variations in the equilibrium latitudinal distribution of longitude–summed sulphur supplies Q in the environment to total sulphur reserve in the biosphere with initial conditions changing up to $\pm 70\%$ and parameters up to $\pm 25\%$ . . . 490
8.11 Dependence of dynamics of sulphur concentrations . . . 491
8.12 Dependence of average acidity of rain on anthropogenic sulphur fluxes . . . 491
8.13 Average production of H_2S in the oceans . . . 493
8.14 Dynamics of the average concentrations of SO_2, NO_2, and dust in the atmosphere over the territory of Vietnam . . . 494
8.15 Calculated dependence of the ratio of the current SO_2 concentration to its maximum value on the type of surface cover for the territory of South Vietnam 494
8.16 The spatial distribution of the impact of transboundary transports of sulphur over the territory of Vietnam . . . 495
8.17 Seasonal variability of the area-averaged sulphur content in the atmosphere over the territory of western Europe . . . 496
8.18 Distribution of SO_2 concentration in the atmosphere between some countries of western Europe calculated with respect to Russia . . . 497

Tables

1.1 Absorbed short-wave radiation 36
1.2 The share of different shapes of particles in samples taken over the Bering Sea 50

2.1 Correlation of temporal and spatial variations of element concentrations in air and precipitation samples 86
2.2 Coefficients of correlation for comparison of dependences of $C(t)$ for some elements during the wintertime series of measurements in 1994–1995 91
2.3 Correlation matrix of elemental spectra of $C(Z)$ of samples of precipitation and air from Colima as well as volcanic ash for the series of macroelements Al–Fe and macroelements Cr–Pb 97
2.4 Correlation coefficient 99
2.5 The diurnal change of the volume number density of aerosol particles size distribution in the surface layer for three series of measurements 102
2.6 The temporal change of deviations of volume concentrations of different fractions of aerosol at sudden increases of volcanic emissions from average values 104
2.7 The change of the sizes of particles in the surface layer of the atmosphere before and during the earthquake in Guadalajara on 9 October 1995 104

3.1 Average concentrations of elements and ions in the atmospheric aerosol composition over various regions 154
3.2 The relative elemental composition of aerosol in different regions 156
3.3 The enrichment coefficients for elements in aerosol of the regions of the former USSR 157
3.4 Correlations between elements 160
3.5 The level of correlations between elements 160
3.6 Correlations of iron with other elements 162
3.7 Correlations of aluminium with other elements 162
3.8 Correlations of silver with other elements 164
3.9 Correlations of manganese with other elements 165

4.1 Emissions of carbon particles due to vegetation burning 226
4.2 Factors of enrichment for various elements in volcanic aerosols of Augustin volcano 236
4.3 Factors of enrichment by Si for several elements in aerosol samples from the Popokatepetl volcanic eruption 238
4.4 Levels of emissions of various MGCs and aerosol 241

5.1 Global mean changes of concentrations of eight MGCs due to dust aerosol . . 272

6.1 The probability of the conservation of cloudiness, with the cloud amount equal to 1, above the European territory of the former USSR 308
6.2 The average altitudes of cloud tops and bottoms and cloud thickness 308
6.3 Recurrence of the extension of the frontal upper cloudiness above the European territory of the former USSR 309
6.4 Recurrence of cloud areas with different extensions of the frontal lower cloudiness 309
6.5 Results of the airborne radiative observation in a cloudy atmosphere 312
6.6 Number density of aerosol particles in different types of clouds 315
6.7 Basic statistical characteristics of the scattering coefficient 316
6.8 Basic statistical characteristics of the backscattering coefficient for the four types of atmospheric circulation 317

7.1 The spectral characteristics of haze 351
7.2 Spectral functions 352
7.3 The impact of relative humidity on the attenuation coefficient 353
7.4 Seasonal variations in P and τ 353
7.5 The vertical profile of σ_{mol} 362
7.6 The content of different shapes of particles 368
7.7 The content of particles of different morphological structure in the surface layer of the equatorial Atlantic 368
7.8 The content of elements in the Earth's crust 370
7.9 The global mean elemental composition of soils 371
7.10 The elemental composition of the basic types of atmospheric aerosols 373
7.11 The measured and calculated values of $b = f(\lambda)$ for $A = 0.9$ and 0.6 at $\cos\theta = 0.2$ 385
7.12 Averaged latitudinal variations of albedo for background aerosol 386
7.13 The aerosol correlation matrices 394
7.14 The size distribution of aerosol number concentration at different altitudes . . 395
7.15 The chemical composition of aerosol over the western Siberia and Kazakhstan 396
7.16 The chemical composition of aerosol in different air masses 397
7.17 Coefficients of absorption by atmospheric aerosol in a cloud layer 414

8.1 Scales of cartographic information presentation characteristic of the developed monitoring systems 450
8.2 A fragment of the scale for development of the structure of the atmospheric pollution dynamics model 453
8.3 The ESPhAP software 470
8.4 Classification of the atmospheric pollution sources 472

8.5 Components of the ESPhAP database 474
8.6 SO_2 emissions from anthropogenic sources in the USA and Canada 478
8.7 Main sources of formation of primary sulphates in the USA 479
8.8 Characteristics of the land and hydrospheric fluxes of sulphur shown in Figure 8.9 483
8.9 Some estimates of the sulphur reservoirs that can be used as initial data 485
8.10 Qualitative estimates of the GSM sulphur unit parameters used in numerical experiments 489
8.11 Results of numerical experiments 492
8.12 Estimates of intensities of atmospheric pollutant emissions in the territory of Vietnam 493

9.1 Annual average values of the total content of different types of aerosol in the atmosphere in the northern and southern hemispheres and globally 532
9.2 Estimates of the contribution of various factors to the ARF formation and ASWR during the period 5–15 April 2001 in eastern Asia 545

About the authors

Kirill Ya. Kondratyev received his Ph.D. in 1956 at the University of Leningrad where he also became a Professor in 1956 and was appointed Head of the Department of Atmospheric Physics in 1958. From 1978–1982 he was Head of the Department of the main Geophysical Observatory and from 1982–1991 Head of the Institute for Lake Research in Leningrad. Since 1992 he has been Councillor of the Russian Academy of Sciences. Kirill has published more than 100 books in the fields of atmospheric physics and chemistry, remote sensing, planetary atmospheres, and global change.

Vladimir F. Krapivin was educated at the Moscow State University as a mathematician. He received his Ph.D. in geophysics from the Moscow Institute of Oceanology in 1973. He became Professor of Radiophysics in 1987 and Head of the Applied Mathematics Department at the Moscow Institute of Radioengineering and Electronics in 1972. He is a full member of the Russian Academy of Natural Sciences and has specialized in investigating global environmental change by the application of modelling technology. Vladimir has published 14 books in the fields of ecoinformatics, the theory of games, and global modelling.

Lev S. Ivlev, Ph.D is a Professor and Head of the Laboratory of Aerosol Physics at Fok's Institute of Physics, St Petersburg State University. His scientific interests lie in the areas of aerosol investigation and ecology. Lev Ivlev developed direct methods and optical models to study atmospheric aerosols. He is the author of 11 books and 300 papers. At present Lev is studying the problems of the prognosis and control of atmospheric processes. He is member of the editorial board of the Russian journal "Ecological Chemistry".

Part One

The field observational experiments

1

Programmes of atmospheric aerosol experiments: The history of studies

The problem of the surface atmosphere system's energetics and consideration of diabatic factors is principal in the numerical modelling of atmospheric general circulation. An adequate climate theory can be developed only on the basis of correct simulation, taking into account different forms of heat and moisture exchange in the atmosphere. Digressing from the long-term internal fluctuations of the 'atmosphere–ocean–lithosphere–cryosphere–biosphere' system, the following factors remain that determine the present climate changes: (1) solar constant variations [48, 57, 77], (2) transformation of surface properties [51, 52, 71, 87], and (3) changes in the gas and aerosol (clouds including) composition of the atmosphere [32, 52, 88]. A study of the latter factor is reduced to the problem of the impact of variations of the atmospheric composition on radiation fluxes [54, 60, 62, 75, 80, 81]. In this case, basic objectives are connected with the studies of clouds, aerosol, and optically active gas components [8, 46, 61, 74, 75], as well as the impact on the intensity of water phase transformations in the atmosphere [31, 36, 40, 75] and on the dynamics of relevant processes [27, 40].

1.1 COMPLEX ATMOSPHERIC ENERGETICS EXPERIMENT

1.1.1 Basic objectives and peculiarities

In the process of realization of the Global Atmospheric Research Programme (GARP) [21, 72] an accomplishment of the Complex Atmospheric Energetics Experiment (CAENEX) has been very important as a first attempt to extensively study the atmospheric energetics and interaction between the atmosphere and surface, with the role of aerosol being taken into account [3, 14, 46]. The general objective of CAENEX was to study the transport of all categories of energy and all types of radiative heat flux divergence in the atmosphere, to analyse factors that determine these processes under different conditions, and on this basis, to work out recommendations on consideration of energy factors of the thermal regime and dynamics of the atmosphere. Naturally, this field experiment is rather complicated and can be realized only on a local basis through an organization of complex field studies in typical physico-geographical regions (desert, forest, steppe, water basins, etc.). Therefore, an experimental basis of CAENEX consisted of a number of successive expeditions between 1970–1975, aimed at obtaining a complete database that characterizes the energetics of a closed volume of the troposphere, with horizontal size ~200–300 km. Most detailed aircraft, balloon, actinometric measurements, and vertical microwave sensing of the atmosphere were carried out in the centre of this volume supplemented with simultaneous aircraft measurements, aerological soundings on the periphery as well as with results of satellite measurements (during the passage of meteorological satellites over the region under study). The choice of observations was determined by the requirement to close, with sufficient accuracy, the energy balance in different atmospheric layers and at the surface level.

According to the CAENEX substantiation given above, the basic sections of studies foresaw the solution of the following problems [3, 44]:

(1) Solar constant and its possible variations [57, 71, 77].
(2) A study of the observed laws and factors determining the radiation fluxes and heat flux divergences in the troposphere [16, 56]: (a) vertical profiles of radiation budget and its components; (b) spectral distribution of fluxes and flux divergences of short and long-wave radiation; (c) aerosol in the free atmosphere – vertical distribution of concentration, size distribution, chemical composition and optical characteristics [7, 38–40, 79]; (d) an accomplishment of the Complete Radiation Experiment (CRE) [15, 49].
(3) Approximate methods of phenomenological theory of radiation transfer, their comparisons, and test against the data of field experiments. Parameterization of the impact of aerosol (depending on its optical and microphysical characteristics) on radiation transfer [42, 43, 70].
(4) The impact of different types of clouds on radiation fluxes and flux divergences depending on the optical and microphysical properties of clouds. Analysis of the possibilities of using approximate methods of taking into account the effect of clouds [1, 2, 6, 20, 75].
(5) Statistical connections (from measurement data) of quantitative characteristics of the radiation field and parameters describing the atmospheric composition and cloud properties. On this basis, the development of techniques of statistical parameterization of radiation factors of the atmospheric general circulation [16, 20, 61].
(6) The surface heat balance: relationships between the components, the diurnal course, and the effect of surface non-horizontality. Study of the degree of surface horizontal heterogeneity from aircraft data of spectral and radiative surveys [9, 54, 70, 86, 87, 90].
(7) Fluxes and flux divergences of energy in the surface and boundary layers of the

atmosphere, connected with the transfer of heat, radiation, momentum, and moisture [10, 11, 15, 21]. Relationships between turbulent fluxes and the distribution of meteorological parameters. Analysis of approximate schemes of consideration of interaction between the surface and the atmosphere and their experimental testing [17, 19, 51].

(8) The heat balance of the Earth–atmosphere system; the heat balance system's components, their variability depending on the determining factors; the spatial–temporal meso and macrostructure of the fields of the heat balance system's components [47, 54].

(9) Study of the effects of various factors of atmosopheric energetics on the atmospheric thermal regime and dynamics. An experimental test of the consideration of energy factors in numerical modelling. The development of approximate schemes of parameterization of the considered energy factors using the observational data [75, 86].

To accomplish relevant observations, a 'quasi-stationary' test area was organized as well as episodic observations (lasting for 1.5–2 months) at different locations. The expeditions were equipped with IL-18 aircraft and helicopters MI-1 or MI-4, means of actinometric microwave sensing, radar and laser sounding, as well as measurements of radiative characteristics, and composition and structure of the atmosphere (in the 0–5-km layer) using free balloons and obtaining space-derived meteorological information.

An accomplishment of the CAENEX programme was its aim to make the maximum use of the existing meteorological and aerological networks in the regions under study. With this aim in view, observations were made at the stationary network for additional time periods and following an additional programme.

In connection with the necessity to exclude the effect of local conditions (which is important, in particular, for correct referencing of aircraft and balloon data to data of surface measurements), sufficiently extended (about 200–300 km) regions were chosen, uniform by surface character and located far from mountains and water basins (in the case of surface observations). Real scales of averaging the measured parameters varied from hundreds of metres to tens of kilometres. Sometimes, however, the requirement for an extended uniform surface was broken (e.g., Rustavi, Georgia and Baku, Azerbaidzhan) [65, 68].

During the period 1970–1975, in order to solve these problems, observations were carried out over surfaces with different surface types (land, water) and physical properties (albedo, thermal–physical characteristics, roughness, etc.) as well as under the conditions of a large city [4–6, 88]. The latter was determined by the desire to study special features of the urban climate and their global aspects connected with the urban production of pollutants [4].

In this period, in connection with preparations for the GARP Atlantic Tropical Experiment (GATE) and GARP, the emphasis was laid on the discussion of the 'radiation units' of these programmes, in particular, on the requirements of the observational system [4, 5]. The CAENEX programme was developed so that the aims of this programme could be reached and the programme could be sufficiently complete (within reasonable limits). It should be noted that the CAENEX programme followed the programme of the Complete Radiation Experiment (CRE) proposed by Kondratyev in 1968, which included the following measurements [15, 49, 87]:

- surface spectral and integral measurements of short and long-wave radiation fluxes, surface albedo, angular distribution of sky brightness, and atmospheric transparency accompanied by routine meteorological observations;
- aircraft and balloon measurements of radiation parameters mentioned above and standard meteorological parameters in the free atmosphere at altitudes up to 30 km (and within cloud layers);
- satellite spectral and integral measurements of the outgoing radiation and solar constant whilst simultaneously obtaining satellite information about cloud cover;
- measurements of ozone concentration and characteristics of aerosol particles near the surface, in the troposphere and lower stratosphere (analysis of data on aerosol to estimate the size distribution, chemical composition and optical properties of aerosol particles), and measurements of scattering phase function in the free atmosphere; and
- studies of microphysical characteristics of clouds with the use of direct (sampling) and indirect (optical, radar) techniques.

Thus the CRE foresaw the obtaining of complete (closed) information on the quantitative characteristics of the radiation field as well as parameters of the atmosphere and surface which determine the radiation transfer. The CRE, in fact, coincided with the radiation unit of the CAENEX programme.

Now let us consider briefly the basic observational system of CAENEX in expeditions carried out in different years.

1.1.2 CAENEX-70

The CAENEX-70 was carried out in the Kara-Kum desert in the region of Chardjou, 200 km north-east of Ashkhabad. The main objectives of this expedition were: to work out a technique of complex energy exchange experiments under conditions of a sufficiently uniform surface and, in fact, a single source of dust aerosol and trace gases; to work out a technique of airborne spectral radiation measurements and to obtain data on spectral radiative flux divergences for different atmospheric layers; to determine the nature of these influxes; to perform direct measurements of aerosol characteristics using a complex of ground and aircraft instruments, as well as laboratory studies of aerosol samples.

First test measurements of short-wave radiation fluxes with a spectrophotometer K-2 developed in the LSU Institute of Physics, and carried by the flying laboratory LI-2, were made in 1968 [82]. During the CAENEX-70 expedition, measurements were made from the flying laboratory IL-18 of the A. I. Voeikov Main Geophysical Observatory. The aircraft was also equipped with aerosol filters.

1.1.3 CAENEX-71

In June–July, 1971 a second CAENEX expedition was accomplished in north-western Kazakhstan, 80 km south of Uralsk (CAENEX-71). The programme of this expedition differed little from that of CAENEX-70. The difference consisted mainly in a more detailed study of turbulent heat fluxes. Both aircraft measurements of turbulent heat fluxes and momentum in the boundary layer and direct measurements of turbulent heat fluxes, moisture, and momentum in the surface layer were carried out. Substantial changes compared to CAENEX-70 were introduced into measurement techniques, in particular, to the scheme of aircraft sounding (IL-18). A great temporal variability of meteorological parameters in the boundary layer excluded the possibility of obtaining 'instant' (for a fixed time moment) vertical profiles of meteorological parameters and heat fluxes using such means of sounding as aircraft, helicopters, and balloons. In this connection, a task was set to study a detailed temporal course of all studied characteristics at fixed levels bearing in mind to interpolate the profiles of meteorological parameters and radiation fluxes for fixed time moments.

As a result of the CAENEX-71 expedition, extensive observational data were obtained which were published in a special issue of the Main Geophysical Observatory (St Petersburg), some papers of which are mentioned in this book. The complex material obtained has made it possible, for the first time, to consider the heat budget from data of observations carried out not only at the surface level but also in individual atmospheric layers. In particular, the heat budget was studied for the whole boundary layer whose altitude was determined from propagation of diurnal variations of temperature (for daytime – the layer 0–5 km).

1.1.4 CAENEX-72

The 'urban' CAENEX-72 carried out at the industrial centre of Zaporozhye (the Ukraine) differed substantially from the earlier expeditions. Therefore, we will consider in more detail the special objectives of this expedition and its observational programme [5].

Two aspects of the problem were considered: local (special features of the meteorological regime under urban conditions) and global (the impact of cities on planetary climate). The urban climate is, in a sense, a model of the impact of human activity on global climate due to such factors as gas and aerosol atmospheric pollutants, changes in the surface optical characteristics and geometry, additional sources of heat, and others. As for the scales of these effects, they are not confined to the size of a city – present trends of urbanization have led to, in some countries of western Europe, America and Asia, the appearance of mega-cities, conglomerates of adjacent cities covering vast territories of industrial regions. The scales of such mega-cities are so large that they are clearly traced, for instance, as strong thermal anomalies in IR images obtained from meteorological satellites [45, 55, 73].

The atmospheric pollution caused by large industrial centres and super-cities shows itself even on global scales and leads to an increase of concentration of carbon dioxide and other minor gas components (MGCs) (respectively, a decrease of atmospheric transparency). The atmospheric haze of the industrial origin is often observed in images of the Earth obtained from meteorological satellites. Photos of the Earth from manned spacecraft are a vivid source of information about atmospheric pollutions [4, 73]. The atmospheric dust load increase is confirmed by direct measurements of aerosol concentration in industrial regions. In particular, a distinct increase of the aerosol content has been recorded recently over the territories of the USA, India, and China [73].

Of course, the importance of studies of urban climatology goes beyond micro and meso-climatology. The objective of CAENEX-72 was, on the basis of complex observational data, to consider major features of the urban climate, to assess the factors of human impact on climate on global scales, and to reveal the aspects of urban climatology important from the viewpoint of simulating a global-scale human impact on the climate. The programme of the urban climatology study had the same objectives as CAENEX-72. Let us now consider some components of this programme.

1.1.4.1 The complex programme of the urban climate studies

The programme of observations covered both the urban territory and the atmospheric boundary layer, with due regard for the propagation of the effect of the city beyond its territory both vertically and horizontally [5]. Detailed observations of the meteorological regime and air pollution in the urban territory were made in the streets and squares as well as at different levels – on the ground and at roof top heights, which enabled one to study the transition from the zone of construction to the inflow of air masses. For this purpose, cars with telescopic masts were used. Results from these studies might serve as the basis for improving the allocation of industrial enterprises and residential areas and for solving other practical problems (e.g., protection of the atmosphere from pollution with toxic substances, consideration of wind load on constructions, solution of the problems of house heat exchange, and others).

The programme of studies included changes in atmospheric composition, meteorological regime (radiation, humidity, temperature, wind, etc.) foreseeing both experimental and theoretical developments. Let us dwell on the principles of this programme.

The atmospheric composition. One of the fundamental problems was to study, in detail, the special features of the urban atmosphere due to gas and aerosol pollution. For gas components, most important was the information about the spatial–temporal variability of concentrations of such components as sulphur dioxide, carbon monoxide, nitrogen oxides, and other admixtures in the surface and boundary layers, and in the lower troposphere. In studies of aerosol, along with estimation of weight concentration, measurements were organized of number density, size distribution, and chemical composition of particles (laboratory studies of samples supplemented this information with data on the complex refraction index of aerosol particles). From the viewpoint of global aspects of the problem, it was important to study the propagation of pollutants in the environment through advection, turbulent diffusion, and the accompanying processes of pollutant transformation (e.g., formation of solid aerosol particles from chemical reactions between gas components).

In this connection, it was very important to obtain comparative complex data on aerosols of the 'clear' and 'industrial' atmospheres whilst distinguishing between the aerosol components of natural and industrial origin. In addition to routine methods of chemical and physical analysis, as a means of control of the content of gas and aerosol pollutants, data were used, taken from ground and aircraft measurements, of atmospheric spectral transparency and laser sounding as well as results of observations of electrical conductivity of air masses – a sensitive characteristic of their dust load.

The atmospheric meteorological regime and composition. Of importance was to study the structure of the air flux at various stages of its passage over the city, with due regard for the special character of windward and leeward directions, as well as in the centre of the city. Changes in the temperature lapse rate near the surface due to asphalt cover of the urban streets and squares were determined, observations of the formation of elevated temperature inversions over the urban constructions were carried out, comparisons were made of wind direction and speed along the streets and in the blowing air flux with an estimated slowing down of air fluxes and assessed surface roughness. One of the objectives of observations was to establish the boundaries of the urban 'heat island' both horizontally and vertically, to analyse the dependence of these boundaries on the size and other characteristics of the city as well as on advection conditions.

A study of anomalies of the fields of temperature, wind, and air humidity under the urban conditions required a combined use of ground gradient measurements, measurements with tethered balloons and helicopters (boundary layer), as well as a means of aerological (laser included) and aircraft sounding. A rather efficient means of studying diffusion was the vertical smoke jet stereophotogrammetry. To estimate the thermal pollution together with ground measurements of temperature, aircraft calibrated IR images were used. The results of these measurements enabled one to obtain information on the vertical turbulent transfer of heat and momentum in the lower atmospheric layers.

Cloud conditions. The impact of pollution on conditions of the formation and characteristics of clouds was studied, in particular, spectral properties (including chemical composition and the characteristics of condensation nuclei) of clouds in industrial regions and the special radiative characteristics (albedo, in particular) of clouds in different spectral intervals.

Radiative regime. Specific features of radiation transfer in the polluted urban atmosphere determine the laws of the urban microclimate. In the absence of advection, the urban microclimate could serve as a good model of the possible impact of pollutions on the global climate ('urban greenhouse effect'). Therefore, of great interest for such modelling is a study of the polluted atmospheres radiative regime. This study, together with measurements of the atmospheric composition (both 'industrial' and 'clear'), enabled one to obtain data that characterizes a specific impact of aerosol on climate from the viewpoint of a relationship between observed values of surface albedo and coefficients of absorption and scattering by aerosol particles under different illumination conditions (sun elevation and cloud conditions). For this purpose, data were used from integral (actinometric) and spectral measurements of short-wave and long-wave radiation fluxes, scattering function, and integral and spectral atmospheric transparency. Of particular importance was an estimation of the surface albedo. Also, a decrease of UV radiation and visibility in the city was estimated.

Theoretical modelling. An important part of the considered complex programme was theoretical (numerical) simulation based on the use of observational data on both the microclimate of the city ('heat island', elevated inversions, etc.) and the potential changes of the global climate determined by pollution of the atmosphere causing a change in its optical and other parameters. Numerical modelling of some processes was successfully supplemented with results from laboratory simulations in wind tunnels.

The system and method of observations. An accomplishment of this programme required the organization of a large-scale expedition equipped with various technical means. Below are brief characteristics of the basic components of the CAENEX-72 expedition with the participation of the Main Geophysical Observatory, Leningrad State University, Odessa Hydrological Institute, Institute of Atmospheric Optics of the Siberian Branch of the USSR Academy of Sciences, and other organizations.

Observations of air pollution. To obtain data on the distribution of industrial impurities (dust, sulphur dioxide, nitrogen dioxide, carbon monoxide, etc.) in a city under different meteorological conditions, air samples were taken at four stationary and 12 route locations in different regions of the city. Observations at 12 route and two stationary locations were made between 06–22 hours, 3–4 times at route locations and every hour at stationary locations. Sometimes, at two stationary locations, hourly (24-hour) observations were carried out.

To assess the impact of meteorological factors on the distribution of industrial and natural aerosols, simultaneous measurements of aerosols were made at three points located at a distance of 5–10 km from each other: measurements were made at heights of 10, 50, 100, 200, 500, and 1,000 cm (4–5 daily series) as well as frequent measurements to study the diurnal course of the aerosol content at different heights (10–15 series of measurements).

Special attention was paid to measurements under conditions of strongly changing meteorological situations (precipitation, change of air masses). Several series of measurements were performed for the vertical profile of aerosol in the atmospheric boundary layer with the use of a telemast (150 m) and helicopter (2 km).

To study the spatial structure of the aerosol concentration field near the sources of industrial pollution and to determine the effect of industrial sources of aerosol on the formation of the urban aerosol cloud and the role of aerosol in changes of the atmospheric thermal regime, aircraft aerosol measurements in the free atmosphere were made [18, 31, 66]. The programme of measurements included measurements of particles' number density and size distribution, as well as sampling to determine the chemical composition of aerosols at different altitudes in the atmosphere. Soundings were made at altitudes of 200, 500, 1,000, 2,000, 4,000, and 7,200 m, and also at rapid descents and ascents – at lower and upper levels, 'platforms', of soundings. The 'platform' measurements lasted more than one minute, and for measurements of horizontal heterogeneity of aerosol lasted 10–15 minutes. Measurements were made with the use of impactors, filters, and photoelectric counters. The aircraft aerosol sounding was supplemented with laser sounding, which made it possible to obtain data on vertical profiles of concentration and size distribution of aerosols.

Surface meteorological observations. To assess special features of the meteorological regime in the region of Zaporozhye, a complex of observations was carried out both in and outside the city. Observations at the meteostation outside the city included: measurements of air temperature and humidity (the Assman psychrometers) at the heights 0.2, 0.5, and 2.0 m; air temperature and humidity (resistance thermometers) at the heights 0.25, 0.5, 1.0, 2.0, 4.0, 8, 12, and 17 m; the continuous record of wind direction pulsation (anemorhumbograph); actinometric and heat-balance observations; and visual observations of atmospheric phenomena (clouds, visibility, precipitation, rare phenomena).

Assman psychrometers, resistance thermometers, and contact anemometers were used for continuous observations. At the same time, visual observations were made, and in the daytime – observations of the heat balance. Round-the-clock wind recording was made with an anemorhumbograph M-12. Observations at the urban meteostations included: measurements of air temperature and humidity with the Assman psychrometers at the heights 0.2, 0.5, and 2.0 m; measurements of wind speed with contact anemometers at the heights 0.25, 0.5, 1.0, 2.0, and 4.0 m; and the continuous record of wind characteristics with an anemorhumbograph M-64. The schedule of observations was the same as observations made outside the city.

In addition to observations at meteorological stations at 16 locations in different regions of the city, observations were made of air temperature lapse rate in the 0.5–2.0-m layer, wind speed and direction at the 2-m level, as well as visual observations of atmospheric phenomena. These observations were made simultaneously with air sampling at stationary and route locations.

Aerological observations. To obtain data on the vertical profiles of temperature, wind speed and direction, and turbulent characteristics of air flows in the lower 1-km layer of the atmosphere over the city and within its environs, balloon soundings were carried out with the use of the MGO meteorograph and the gustiness-measuring instrument carried by the EmA-2 tethered balloon, as well as microwave sensing, and the double-theodolite and helicopter sounding (MI-1 helicopter) with the use of meteorograph A-10 and electrometeorograph. At the same heights, air sampling was carried out and sketches of the aerosol cloud were drawn.

Aircraft radiation measurements. Actinometric air-borne measurements enabled one to obtain data on the vertical profiles of radiation fluxes, albedo, and radiative flux divergence and its constituents in the free troposphere. With this aim in view, measurements were made of hemispherical upward and downward long-wave ($\lambda = 3$–$30\,\mu$m) and short-wave (0.3–$3\,\mu$m) radiation at different levels in the atmosphere: 200, 500, 1,000, 2,000, 4,000, and 7,200 m. The aircraft was at the 'platform' for 12–15 minutes. Simultaneously, spectral measurements were made of the downward and upward short-wave radiation fluxes (0.3–2.0 μm) at different levels [6, 11, 33, 34, 82]. As a rule, measurements were made at the 7,200 and 500-m levels (near noon).

Measurements of direct solar radiation spectra in the wavelength interval 1.8–8 μm at 'platforms' 500, 1,000, 2,000, 4,000, and 7,200 m, with the sun to the right of the flight route (one series of measurements lasted for 10 minutes), were aimed at obtaining data on spectral transparency of different atmospheric layers and estimating the content of pollutants with the use of spectroscopic techniques. To determine the radiative temperature of the surface, measurements of the surface thermal radiation were made in the 8–12-μm interval with the use of a wide-angle radiometer (15° viewing angle). These data were also used for control calibration of IR images of the surface obtained with the IR scanning radiometer (2.5–5.5-μm spectral interval). The scanning radiometer functioned at altitudes of 500, 1,000, 2,000, 4,000, and 7,200 m [66].

Airborne measurements were also made of the atmospheric meteorological parameters (temperature, humidity, pressure) at standard 'platforms' (200, 500, 100, 2,000, 4,000, and 7,200 m). At the same time, a description was made of the state of the atmosphere and surface from visual observations, as well as ground measurements of spectral fluxes of short-wave radiation and spectral transparency of the atmosphere.

Complex meteorological survey. To determine features peculiar to the thermal and wind regimes of the urban atmosphere and the formation of the urban 'heat island', 20 locations were chosen for observations. At each of them, frequent measurements were made of air temperature and humidity at the 0.5 and 1.3-m levels as well as wind speed and direction at a height of 2 m. Simultaneously, hourly meteorological and aerological observations were made at meteostations outside and inside the city as well as helicopter, balloon soundings, radiosonde launchings, and double-theodolite observations.

The second complex energy experiment under the conditions of an industrial city was accomplished in the summer of 1972 in the region of Rustavi (Georgia), in the mountains of Transcaucasia, where industrial emissions from the metallurgical plant were the main anthropogenic source of minor gases and aerosols, and the open surface of mountains and excavations the natural sources. The main results from this expedition have been published [68].

In 1987, in the region of Alma-Ata (Kazakhstan) a unique experiment (by experimental methods and the scope of the problems) AUAP (Analysis of the Urban Atmosphere Pollution) was accomplished. Note should be taken that in this city and its environs, during a long time period, various optical measurements were carried out both in the surface layer and in the free troposphere. Results of the combined use of data of optical, microphysical, and meteorological measurements, as well as the results of chemical analyses of air samples (in the interests of numerical simulation of the processes of pollution and radiation transfer in the complicated urban atmosphere), have been discussed in [29, 30].

1.2 THE COMPLETE RADIATION EXPERIMENT

1.2.1 Spectral distribution of short-wave radiation flux divergence and the role of aerosol absorption in the cloud-free atmosphere.

Within the CRE programme, during the CAENEX-72 expedition, observational data were obtained for the quantitative evaluation of the spectral distribution of the radiation flux divergence [49].

1.2.1.1 Spectral distribution of the radiation flux divergence

First, the spectral distribution of the radiation flux divergence in the free atmosphere was investigated in the wavelength interval 0.2–2.4 μm under desert conditions [59, 60, 62, 64, 70]. The observational complex included measurements of the vertical profiles of the upward and downward spectral short-wave radiation fluxes, the angular distribution of spectral brightness of the surface–atmosphere system, spectral brightness of the horizontal orthotropic reference plate irradiated from above, as well as auxiliary meteorological parameters and aerosol. Figure 1.1(a) shows the curve of spectral distribution of radiative flux divergence in the troposphere (the layer 0.3–8.4 km) in the interval 0.4–2.4 μm obtained on 25 October, 1970 at noon (solar zenith angle $z_{\Theta} = 55^{\circ}$) over the desert under conditions of a heavy haze. Figure 1.1(b) demonstrates the spectral distribution of radiative flux divergence (in %).

The data in Figure 1.1 point to the prevailing role of molecular absorption in the relative spectral radiative flux divergence in the troposphere. As seen from Figure 1.1(b), the relative aerosol absorption in the wavelength interval considered is <20% and changes approximately in proportion to λ^{-1}.

The dot-and-dash curve in Figure 1.1(a) was obtained by multiplying the ordinates of the smoothed curve of relative spectral radiative flux divergence (approximated by dependence of λ^{-1}) by the energy distribution in the spectrum of incoming radiation, which gives the spectral change of the absolute averaged radiative flux divergence in the troposphere. The area covered by this curve is equal to 0.095 cal cm^{-2} min^{-1} (4.8% of the solar constant) and characterizes the total absorption of radiation by aerosol in the troposphere in the wavelength interval under consideration. The area between this curve and the curve of the observed spectral radiative flux divergence equal to 0.075 cal cm^{-2} min^{-1} (3.8% of solar constant) determines the total molecular absorption. Thus, though molecular absorption plays the basic role in relative spectral radiative flux divergence, the energy distribution in the spectrum of the source (the sun) is such that the absolute radiative flux divergence due to radiation absorption by aerosol and gas components of the atmosphere are approximately equal. A comparison of the spectral distribution of radiative flux divergence with the dependence of optical parameters of aerosols on wavelength estimated from data of microphysical and chemical analyses of aerosol, shows that in the case considered, aerosol particles of limonite–haematite are the principal absorbing component of aerosol over the desert.

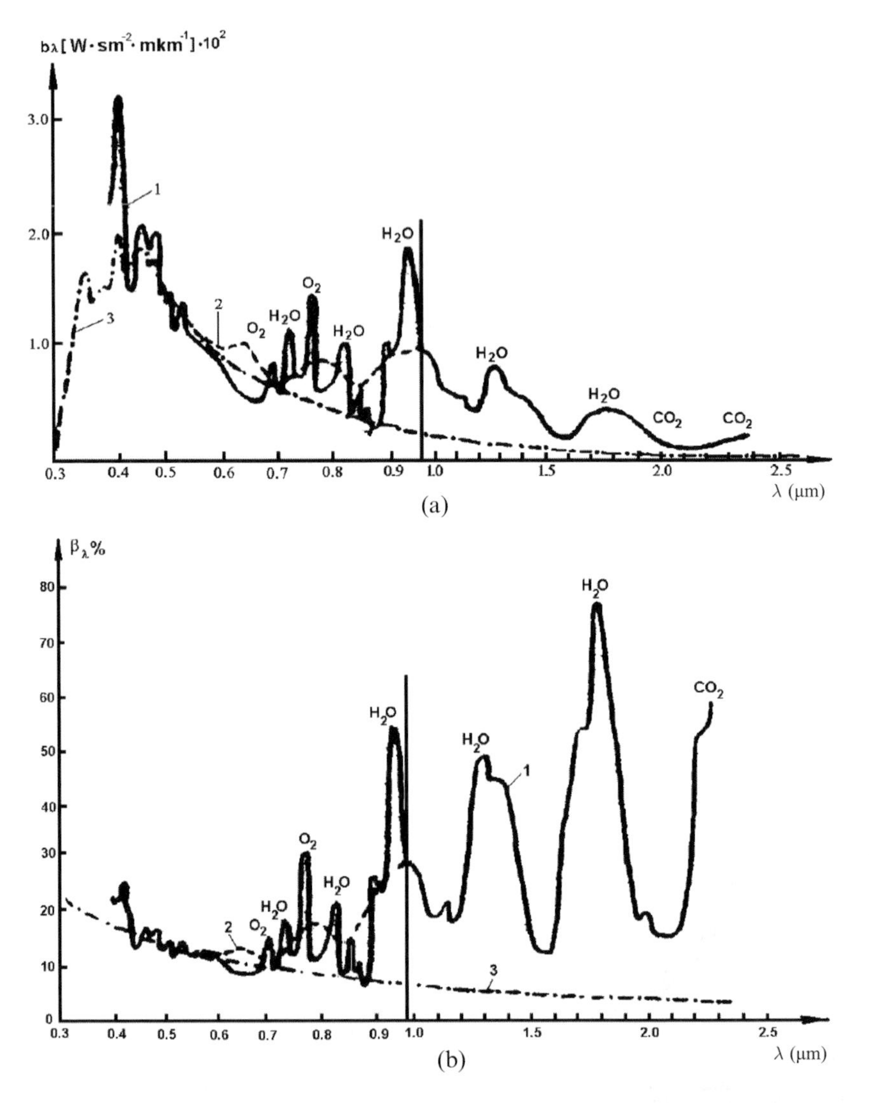

Figure 1.1. The spectral radiative flux divergence in the troposphere (0.3–8.4 μm) in the wavelength region 0.4–2.4 μm; (a) absolute; (b) relative. 1 – the measured spectral distribution of radiation flux divergence (K-2 spectrometer); 2 – the aerosol constituent of radiation flux divergence approximated by the dependence; 3 – results of radiation flux divergence calculations from data obtained with the SPI-2M spectrometer.

1.2.1.2 Variability of the radiative flux divergence

Similar measurements of spectral radiative flux divergence in the 0.86–4.2-km layer in the wavelength interval 0.4–0.9 μm were made on 17 June, 1970 over the Caspian Sea (Kara–Bogaz–Gol) at different solar zenith angles, and on 4 August, 1970 (over the desert) [59, 64, 70]. The latter data show that on this day the relative radiation

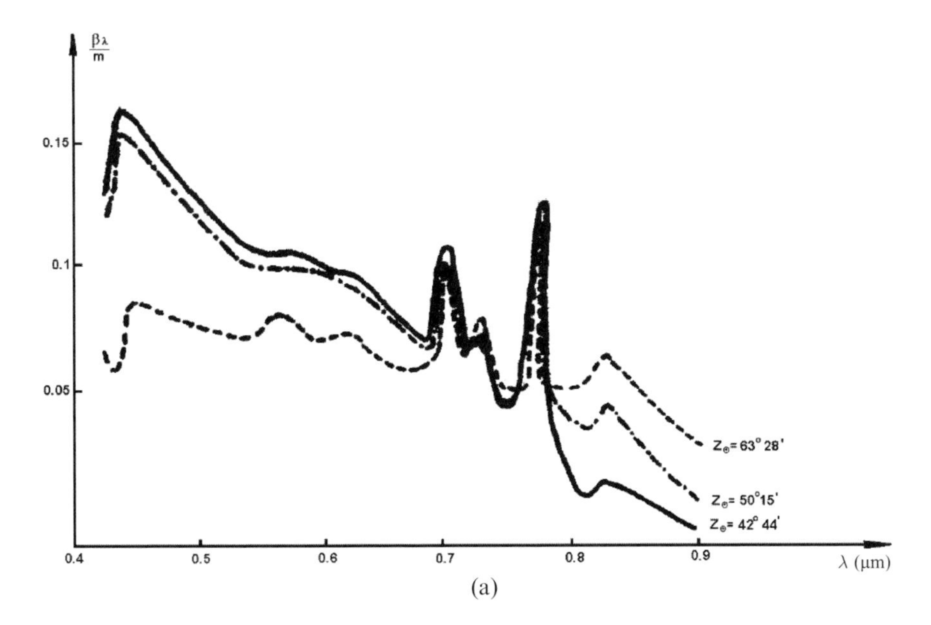

(a)

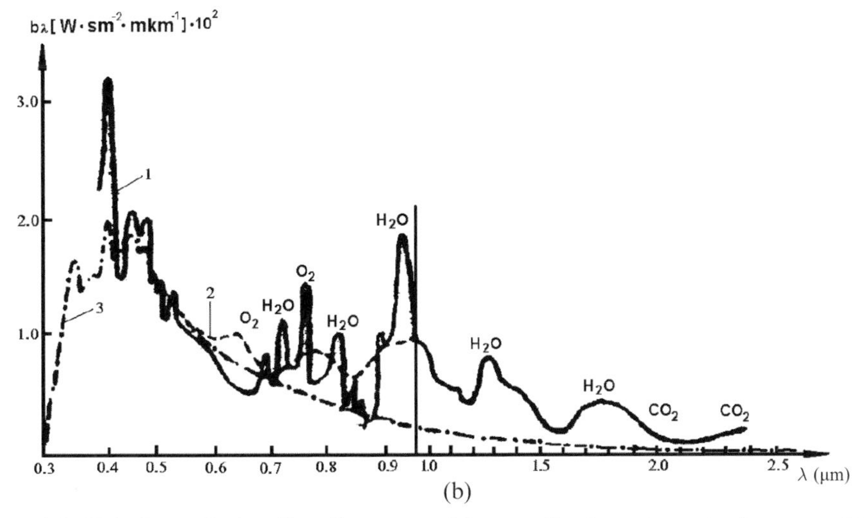

(b)

Figure 1.2. Relative radiation flux divergence: (a) normalized to atmospheric masses M (17 June, 1970), z_0 – solar zenith angle; (b) 1–25 October, 1970; 2–17 June, 1970; 3–4 August, 1970.

flux divergence in the troposphere was less than on other days by an order of magnitude (Figure 1.2). Since observations on 4 August, 1970 were made after a period of heavy rains, it means that on this day the absorbing aerosol particles in the atmosphere were practically absent.

The main result of studies of the spectral radiation flux divergence confirmed the fact that the 'residual' absorption of short-wave radiation detected earlier (see

[62, 87, 91]) is in fact the aerosol absorption which is rather selective. Its possible variations are determined, on the one hand, by changes of aerosol concentration in the atmosphere and, on the other hand, by variability of chemical composition and, hence, optical properties of aerosols.

1.2.1.3 Vertical profiles of spectral radiation fluxes and radiation flux divergence

Determination of radiative flux divergence from measurements of radiation fluxes is a complicated problem, since in this case it is necessary to calculate second differences. Naturally, this problem can be solved only with high-repeatability measurements. To obtain more reliable results, the values of radiative flux divergence considered above were determined for large atmospheric thicknesses. A search for techniques of data processing to determine the vertical profiles of a net flux ($F_\lambda\uparrow - F_\lambda\downarrow$), albedo $A_\lambda = F_\lambda\uparrow / F_\lambda\downarrow$), and radiation flux divergence ($\Delta(F_\lambda\downarrow - F_\lambda\uparrow / \Delta z$) have led to the use of the statistical method described below.

1.2.1.4 Methods of statistical processing of results of short-wave downward and upward radiation flux measurements

In calculations of the vertical profiles of various radiative characteristics, the following statistical methods of processing the observational results were used. First, the observed downward $F_\lambda\downarrow$ and upward $F_\lambda\uparrow$ fluxes of short-wave radiation were vertically smoothed, interpolated on equidistant values of an argument, and extrapolated beyond the interval of the argument, in which observations were actually made. For these operations to be made most successfully, pressure p but not height h in the atmosphere should be used as an argument. For the levels at which observations were made, pressure was calculated according to the standard model of the atmosphere. Pressure at the surface level was taken to be 1,000 hPa.

Smoothing, interpolation, and extrapolation of $F_\lambda\downarrow$ and $F_\lambda\uparrow$ were first made graphically. Then this operation was made by choosing the coefficients of polynomial not higher than the second power, namely:

$$F_\lambda\downarrow(p) = F_\lambda\downarrow(0) + a_\lambda p + b_\lambda p^2 \tag{1.1}$$

$$F_\lambda\uparrow(p) = F_\lambda\uparrow(0) + c_\lambda p + d_\lambda p^2 \tag{1.2}$$

To calculate the coefficients $F_\lambda\downarrow(0)$, a_λ, $F_\lambda\uparrow(0)$, c_λ, and d_λ by method of least squares, $F_\lambda\downarrow(p)$ and $F_\lambda\uparrow(p)$ should be measured, at least at four levels in the atmosphere.

Since at the first stage of studies the data were processed graphically (i.e., by hand), a wavelength step $\Delta\lambda$ was taken to be 10 nm. Thus, for each wavelength, graphs of dependences of $F_\lambda\downarrow(p)$ and $F_\lambda\uparrow(p)$ were drawn every $\Delta\lambda = 10$ nm. Smoothed curves were drawn across the respective points in the 0–1,000-hPa interval. Extreme pressure values for measurement points constituted approximately 400 hPa and 900 hPa. Thus, an extrapolation beyond the atmosphere was made from the 400-hPa level, and to the surface level – from 900 hPa. In both cases an extrapolation error, apparently, did not exceed 5–10%. Interpolated and extrapolated values of $F_\lambda\downarrow(p)$ and $F_\lambda\uparrow(p)$ were taken from the graphs for the interval $\Delta p = 100\,\text{hPa}$.

Though in the process of interpolation and extrapolation of $F_\lambda\downarrow(p)$ and $F_\lambda\uparrow(p)$ their dependences were vertically smoothed, both $F_\lambda\downarrow(p)$ and $F_\lambda\uparrow(p)$, as functions of wavelength, substantially varied due to random errors in the observations. These oscillations were smoothed with the use of the Whittaker probability operator with the coefficient of smoothing $\varepsilon = 10^{-2}$ in all the cases considered below (the results of this processing are illustrated below with calculations of the spectral albedo of the surface–atmosphere system as an example).

When smoothing the observational data with the Whittaker operator, the probability function of the combined sampling of the observed and smoothed values was introduced on the assumption of the normal law of distribution of differences between observed and smoothed values and differences of the nth order (in our case $n = 4$) between smoothed values. Then, from the maximum probability function condition, we obtain a system of linear algebraic equations to calculate smoothed values, which is solved using known methods.

1.2.1.5 Vertical profiles of radiative characteristics in the free atmosphere

Consider, as an illustration, some results of processing the data of observations performed on 25 October, 1970 (CAENEX-70, Kara-Kum Desert).

Figure 1.3 shows the dependence of $F_\lambda\uparrow(p)$ at different levels in the atmosphere.

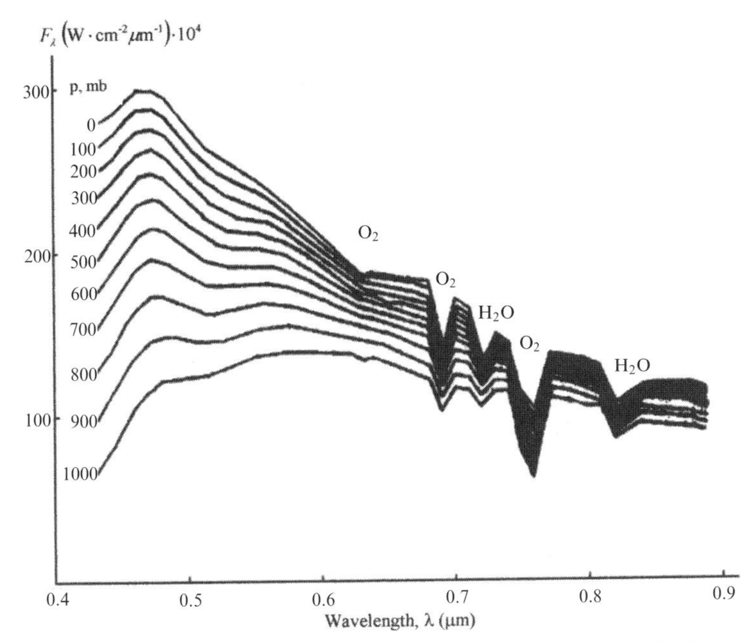

Figure 1.3. Spectral distribution of the upward flux of short-wave radiation at different altitudes.

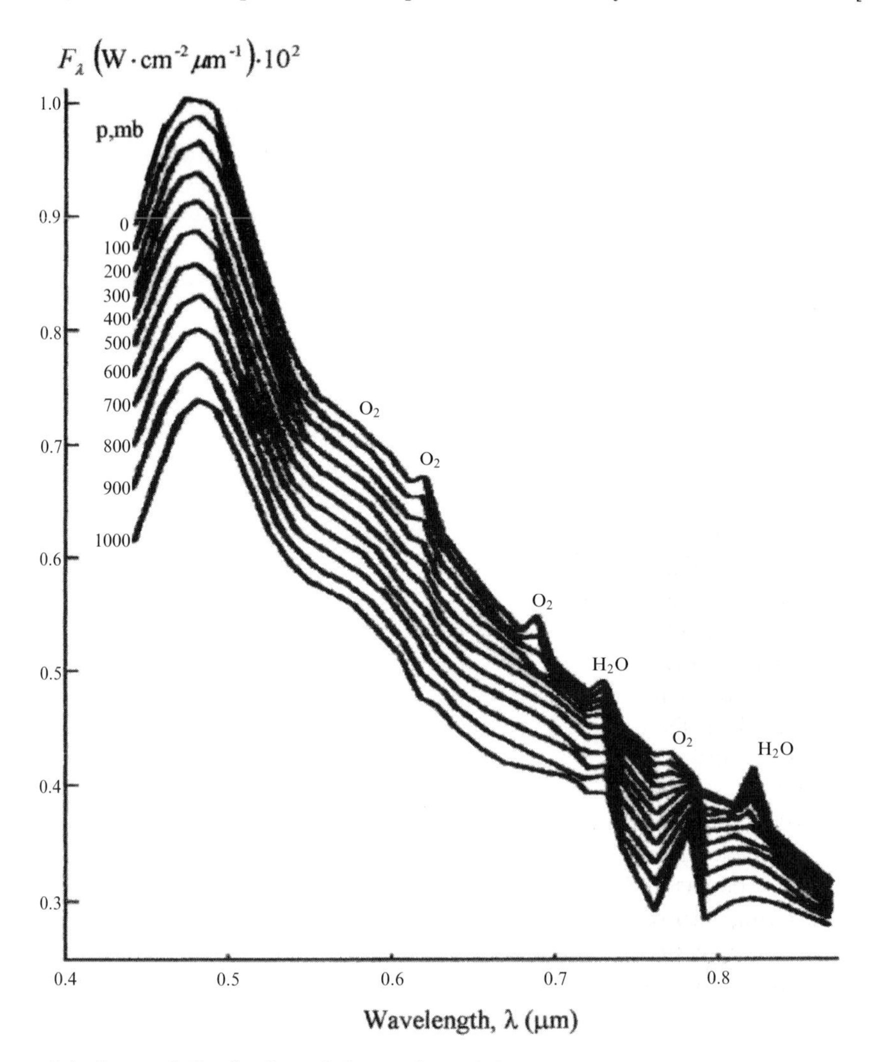

Figure 1.4. Spectral distribution of the net flux of short-wave radiation at different altitudes.

In the case considered, both upward and downward short-wave radiation fluxes grow with altitude. Naturally, the amplitude of the vertical variation of fluxes grows when the wavelength decreases. In the pure scattering atmosphere the net flux $B\lambda = F_\lambda\downarrow(p) + F_\lambda\uparrow(p)$ should remain constant. In the real atmosphere, in the presence of absorption, the net flux (short-wave balance) B_λ (Figure 1.4), like the fluxes, grows with altitude, being greater the shorter the wavelength.

The depth of absorption bands in $F_\lambda\downarrow(p)$ decreases, and in $F_\lambda\uparrow(p)$ increases with altitude. Naturally, in the case of the net flux B_λ, the contours of the bands are similar to those for $F_\lambda\downarrow(p)$ at $p \to 1,000\,\text{hPa}$ and $F_\lambda\uparrow(p)$ at $p \to 0$. Note, however, that all estimates for absorption bands are approximate.

Consider now the vertical transformation of the surface–atmosphere system's albedo $A_p(\lambda) = F_\lambda\uparrow(p)/F_\lambda\downarrow(p)$. The albedo of sand ($p = 1{,}000$ hPa) is characterized by standard variations typical of this class of surfaces: albedo grows smoothly with increasing wavelength. As the altitude increases, the surface–atmosphere system's albedo outside the absorption bands grows (growing greater the shorter the wavelength). This effect turns out to be so strong (at short wavelengths, the atmospheric haze brightness imposed on the radiation reflected by the surface is $\sim\lambda^{-4}$) that the sand surface albedo observed from space is almost independent of wavelength.

Of interest is a minimum albedo near $\lambda = 500$ nm due to an opposite wavelength dependence of the atmospheric haze brightness and albedo of sand. An albedo minimum at $\lambda = 600$ nm is of different nature and it is determined by the impact of radiation absorption in the Chappui band of ozone. Albedo changes in the near-IR region correspond to water vapour and oxygen absorption bands.

The data considered can serve as a vivid characteristic of the strong changes in the spectral variation in the surface–atmosphere albedo system determined by the effect of the intermediate atmospheric thickness. They demonstrate, in particular, the great importance of the correct determination of the transfer operator of the atmosphere when interpolating the data of spectral surface albedo measurements from space in order to solve the problem of identification of natural formations from specific features of their reflection spectra.

The vertical profile of albedo outside the absorption bands reveals the vertical variation of the surface–atmosphere system's albedo most strongly manifesting itself in the region of short wavelengths. In the Chappui absorption band, the albedo changes little with altitude, and in the absorption bands of water vapour and oxygen the albedo decreases with altitude. The latter fact is very important and refers to both the gas and aerosol components of the atmosphere. Calculations for purely scattering aerosol show that the presence of aerosol raises the Earth's albedo (hence, an increase of the aerosol content should lead to a planetary cooling). Bearing in mind that aerosols are absorbers, in this case the presence of aerosol can cause a decrease of albedo (and, hence, a warming). Such an important qualitative difference in the results of the climatic implication of aerosol has made the problem of studying these characteristics under real conditions extremely urgent. Outside the absorption bands of the atmospheric gas components, the radiative flux divergence takes place mainly due to absorption by aerosol.

The vertical profiles of B_λ outside the absorption bands characterize a decrease of radiative flux divergence with altitude due to an aerosol concentration decrease. In the Chappui bands of ozone, the B_λ practically does not change with altitude, since a decrease of aerosol concentration is compensated for by an increasing content of ozone in the layers $\Delta p = 100$ hPa. In all other bands, the B_λ increases with altitude. In the oxygen bands the flux divergence grows monotonically approaching the surface. In the water vapour bands the flux divergence grows sharply in the lower layers of the atmosphere, which can be explained by the presence of water vapour concentration in the lower atmospheric layers, especially in the case of observations in deserts.

1.2.2 Complete Radiation Experiment (overcast cloudiness)

Field experiments within the CAENEX programme accomplished in 1970–1971 were carried out under the simplest conditions: over a relatively homogeneous surface in the absence of clouds [15, 66]. However, an important component of the CAENEX programme was also a study of the impact of clouds on radiation fluxes and flux divergences. In this connection, special preparatory complex observations were undertaken in 1971–1972 in the cloudy atmosphere aimed at accomplishing the CRE under conditions of overcast cloudiness. Measurements of radiation fluxes, and microphysical and other parameters of clouds were made over the Black Sea. At the first stage, the dependence of spectral albedo for different clouds on their optical and microphysical characteristics was studied as well as special features of the spectral distribution of radiation flux divergence in the presence of overcast cloudiness.

As has been mentioned above, such studies can serve as the basis for experimental tests of theoretical schemes taking into account the radiative factors and working out techniques for their parameterization in the numerical simulation of atmospheric general circulation and climate theory. Another important aspect of such studies is connected with a study of the interactions between some physical characteristics of clouds, such as their microstructure and optical parameters. Finally, a more complete knowledge of the processes of radiation transfer in the cloudy atmosphere will be useful in the interpretation of radiative data about clouds of other planets, especially Venus and Mars.

In March–April, 1971, over the Black Sea, in the presence of overcast cloudiness, airborne complex radiative, microphysical, and meteorological measurements were made with two flying laboratories IL-18 and IL-14, the first being the radiation laboratory, the second – the microphysical laboratory.

The IL-18 aircraft carried two similar spectrometers with wavelength intervals 0.35–0.95 μm and 0.4–2.5 μm, aerial photography instrumentation, airborne thermo-hygrometer, transparency sensor, pyranometers, and pyrheliometers, as well as instruments to record the aircraft's orientation. The IL-14 aircraft was equipped with a complex of instruments for particle size distributions and aerological measurements.

Stratus clouds which are most homogeneous horizontally were an object of study. Measurements were made from two aircraft flying parallel routes. The IL-14 aircraft was to sound the chosen cloud layer with simultaneous measurements of droplet size distribution and water content every 100 m, whilst taking a continuous record of temperature, pressure, and humidity. This complex of measurements was made under the command of the IL-18 aircraft (via radio) to provide synchronicity of measurements. The IL-18 aircraft made radiative and other measurements beneath the bottom and above the top of clouds with a subsequent sounding of the atmosphere up to an altitude of 8,400 m. The duration of measurement 'platforms' was about 3 minutes. All measurements were made at about noon with the sun to the right of the aircraft to reduce, to a minimum, the errors of measurements caused by some non-horizontality of the spectrometer's receiving platform.

The use of this measurement technique enabled one to solve the main problem of the studies – to obtain simultaneous observational data on radiative characteristics of clouds and meteorological parameters of the atmosphere and clouds which determine the short-wave radiation transfer in the atmosphere in the presence of overcast horizontally homogeneous cloudiness.

Since the radiation field in the cloudy atmosphere is determined by numerous factors (which, in particular, hinders comparison of theory and observations), attempts have been undertaken, based on the results obtained: (a) to identify such conditions of observations, for instance, the time of their accomplishment, the type of clouds under study, the nature of the surface, spectral region, etc., most favourable from the viewpoint of reducing the number of parameters affecting the measured radiative characteristics of clouds; (b) to search for such cloud situations when radiation transformed by clouds depends mainly on one physical parameters (e.g., cloud optical thickness), other conditions being equal (sun elevation, average water content, particles' size distribution or droplets' modal radius) under similar synoptic conditions.

The main objective in processing the obtained material was to study the laws of variability of spectral reflection, transmission, and absorption of short-wave radiation by clouds. Let us consider some results of the complex measurements.

Figure 1.5 shows spectra of downward and upward radiation fluxes above cloud and beneath it, recorded from altitudes of 8,400 and 200 m, respectively, at a sun elevation $h_{\Theta} = 55^{\circ}$. Measurements were made over 1-layer, overcast stratocumulus cloudiness with an optical thickness $\tau \sim 9$, average water content $w = 0.4\,\mathrm{g\,m^{-3}}$, and modal radius of the averaged spectrum of cloud droplets' distribution $r_m = 6\,\mu\mathrm{m}$. Basic spectral features of radiation fluxes and cloud albedo in this case are determined by the effect of molecular absorption by liquid water, water vapour, and oxygen. Note that total albedos of clouds and the sea measured from altitudes 8,400 and 200 m, respectively, constitute 63.0 and 4.1%. A detailed analysis of measurement results makes it possible to analyse the dependence of spectral albedo on the cloud optical thickness, size distribution, phase state of the clouds, and albedo of the surface (the sea or cloud layer lying below).

Figure 1.6 demonstrates the results of measurements of albedo, spectral short-wave balance (the difference between upward and downward radiation fluxes), and radiative flux divergence for the whole cloud layer (absorbed short-wave radiation).

Figure 1.6(a) characterizes the data on spectral balances of short-wave radiation obtained in the presence of overcast stratus cloudiness with an optical thickness ~ 9, water content $0.4\,\mathrm{g\,cm^{-3}}$, modal radius of cloud droplets $6\,\mu\mathrm{m}$, and sun elevation $\sim 56^{\circ}$. Measurements of downward and upward radiation fluxes were made above the clouds from an altitude of 1,100 m and beneath clouds from 200 m. Naturally, the short-wave balance at 1,100 m – $B_{\lambda,H1}$ (curve 1) exceeds the balance at 200 m – $B_{\lambda,H2}$ (curve 2), which corresponds to the radiative warming of the atmospheric 0.2–1.1-km layer containing a cloud layer. The magnitude of radiative warming (radiative flux divergence due to radiation absorption) is determined by the difference between the balances $b_{\lambda} = B_{\lambda,H2} - B_{\lambda,H1}$. Curve 3 ($b_{\lambda}$) characterizes

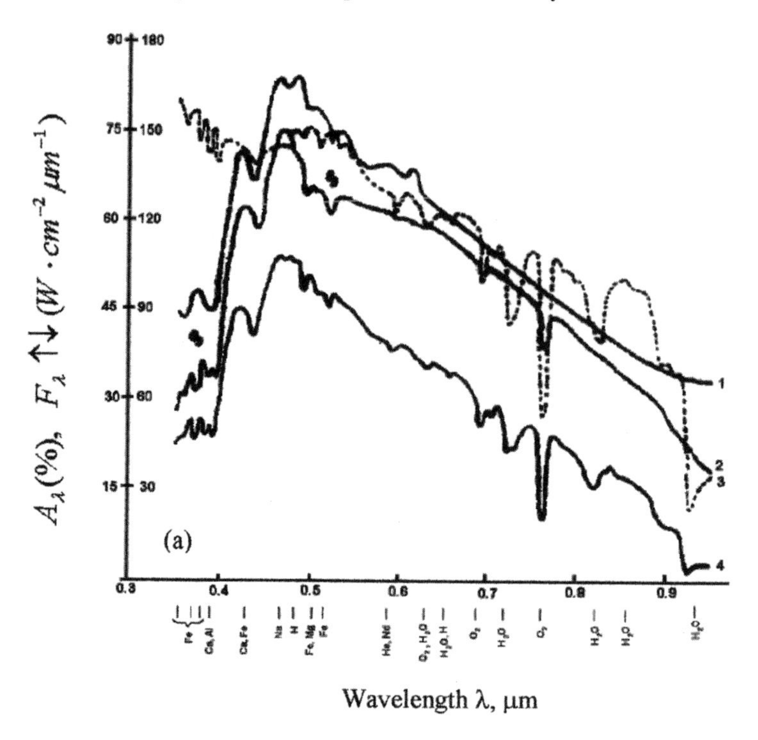

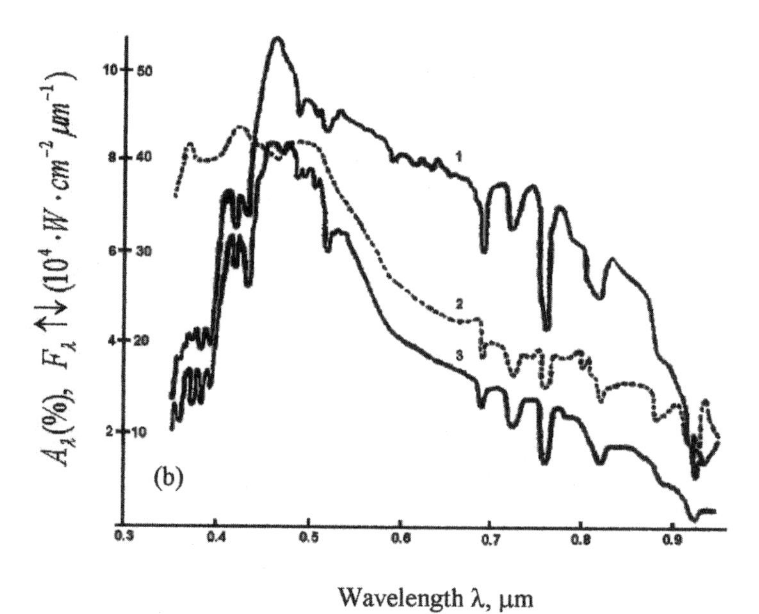

Figure 1.5. Spectral distributions of the short-wave radiation fluxes and albedo. (a) Spectra of downward and upward radiation fluxes above clouds: 1 – extra-atmospheric distribution of solar radiation; 2, 4 – spectra of total ($F_\gamma\downarrow$) and reflected-by-cloud ($F_\gamma\uparrow$) radiation fluxes, respectively; and 3 – dependence of cloud albedo A_λ on wavelength. The marks near the abscissa axis show the location of Fraunhofer lines in the solar spectrum and atmospheric absorption bands. (b) Spectra of radiation fluxes penetrating clouds and reflected from the sea: 1 – the downward radiation flux penetrating clouds ($F_\gamma, \downarrow$); 2 – spectral albedo of the sea $A'_\lambda(\%)$; and 3 – radiation flux reflected by the sea ($F_\gamma, \uparrow$).

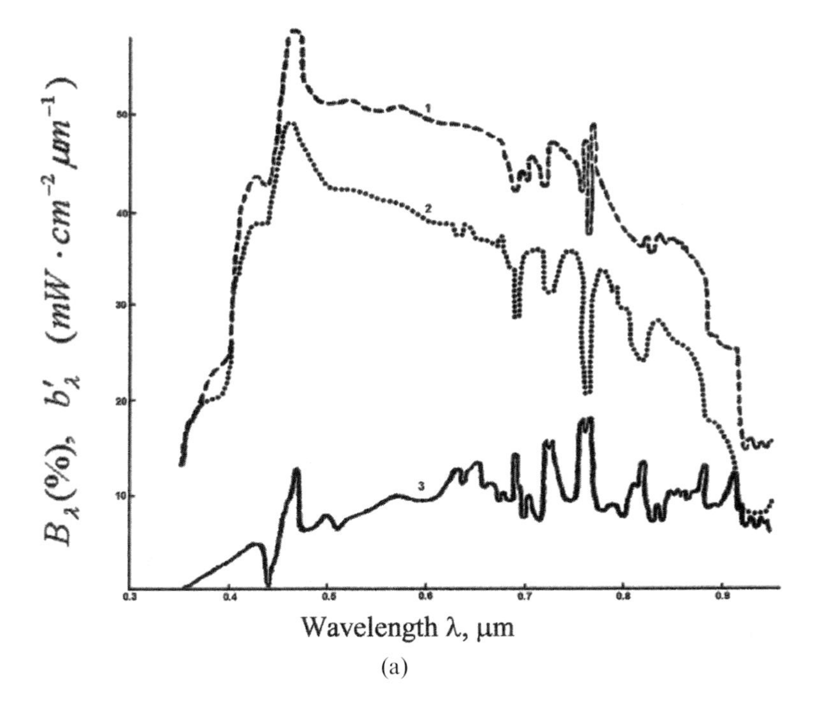

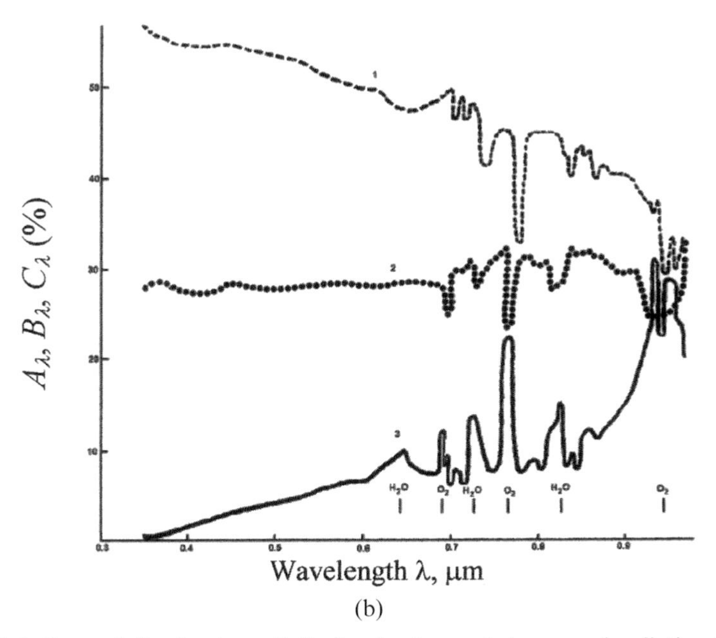

Figure 1.6. Spectral distributions of albedo, short-wave balance, and radiation flux divergence (absorbed short-wave radiation) in the case of 1-layer stratified cloudiness. Spectra of albedo (1), radiation penetrating clouds (2), and relative radiation flux divergence b'_λ (3).

the dependence on wavelength and radiative flux divergence in the spectral region under consideration.

The spectral curve of radiation flux divergence $b_\lambda^1 = b_\lambda / F_{\lambda.H2}\downarrow$ ($F_{\lambda,H2}\downarrow$ is the net radiation flux measured at an altitude of 1.1 km) demonstrates the wavelength dependence of radiation flux divergence determined by the combined effect of cloud droplets and dry aerosol as well as by molecular absorption. The relative spectral flux divergence in a cloud grows strongly in the bands of molecular absorption by water vapour, liquid water, and oxygen. So, for instance, the spectral flux divergence grows at wavelength 0.72 μm up to 16.6%, at 0.82 μm – up to 18.3%, and at 0.93 μm – up to 37.5%. As seen from calculations, radiation absorbed by clouds in the wavelength interval 0.35–0.95 μm constitutes 0.117 cal cm^{-2} min^{-1} or 7.2% of the incoming radiation at a given sun elevation. According to data from actinometric measurements, in the range 0.3–3 μm the total radiation flux divergence in the cloud layer constituted 0.08 cal cm^{-2} min^{-1} or 5% of the net radiation coming to the cloud top. From data of measurements [75], an average value of absorptance for stratocumulus and stratus cloud constitutes 7.4%, which agrees reasonably well with results of spectral and integral measurements. In general, the cloud spectral albedo decreases with growing wavelength, which is determined by the increasing role of absorption by cloud droplets. As seen from Figure 1.6(b) the spectral transmission is less selective. Integral values of albedo and transmission in this case constituted 59.1% and 36.5%, respectively.

Analysis of measurement data revealed a marked impact of the atmosphere above the cloud layer (atmospheric haze) on spectral albedo. A comparison of spectral albedo for the same cloud recorded from altitudes of 1,100 m (curve 1, Figure 1.6(b)) and 8,400 m (curve 3, Figure 1.5) shows that the atmospheric haze increases the cloud albedo in the region of shorter wavelengths because of a haze-induced increase of backscattering, and decreases the albedo in the long-wavelength region due to an increasing absorption by the atmosphere above the cloud layer.

A qualitative comparison of the observational data considered with similar results of calculations by Plass and Kattawar (see [75]) at $\lambda = 0.5$ μm for models of different forms of clouds as a function of cloud optical thickness, its droplet size distribution, surface albedo, as well as taking into account water vapour and water bands in the IR spectral region revealed a satisfactory agreement. Quantitative differences are connected, apparently, with the fact that parameters of calculations and observation conditions are not identical [75].

Results of the CRE accomplishment under overcast cloudiness conditions have opened up the possibilities of correct comparisons of theoretical calculations with observational data.

1.2.3 Atmospheric aerosol

When developing and accomplishing the CAENEX programme, special attention was paid to complex studies of radiation and aerosol, bearing in mind in the latter case the determination of vertical profiles of concentration, size distribution, chemical composition, and optical characteristics of aerosol [18, 40]. The aerosol 'unit' of the CAENEX programme was also important because the information about atmospheric aerosol published earlier was found to be rather inadequate (see, e.g., [1, 2, 87]).

1.2.3.1 Method of measurements

As has been mentioned above, *in situ* measurements of aerosol were made with filter and impactor aerosol sampling, and subsequent processing of these samples was undertaken with the use of optical and electron microscopes as well as running a chemical analysis for the content of individual elements.

During the first three experiments within the CAENEX programme, more than 600 samples of aerosol were taken from the surface layer with the use of membrane and quartz filters to be partially processed (to study the organic components). In addition, more than 30 filter samples in the free atmosphere at different altitudes and about 200 impactor samples were taken, which enabled one to obtain data on the vertical profiles of large aerosol particles at altitudes ranging from 300–8,400 m at different times of the day and night and under different meteorological conditions. A maximum volume of measurements was carried out during the 1972 expedition in the region of Zaporozhye (the Ukraine). Here, apart from filter and impactor samplings, measurements were made with the Scholtz counter, as well as gradient measurements of deposited dust, and measurements of the aerosol concentration with the electrophotochemical dust sensor. The filter gradient measurements at Zaporozhye were taken up to an altitude of 150 m [5, 31].

To obtain data on the content and the optical characteristics of aerosol in the atmosphere, simultaneously with microphysical and meteorological measurements, the following measurements were made: surface spectral measurements of atmospheric transparency in the UV, visible, and near-IR spectral regions; measurements of atmospheric transparency with the actinometer and Foltz filter radiometer; and aircraft measurements of spectral and total transparency of the atmosphere at altitudes up to 8.4 km. During the CAENEX-72 expedition, lidar soundings of the atmospheric aerosols were also made, comprising: measurements of polarization characteristics of scattered radiation; study of aerosol dynamics near its source; and measurements of the vertical structure of the coefficient of radiation backscattering by aerosol. Such a wide complex of aerosol studies has made it possible to reveal some features of the aerosol structure in the surface and boundary layers of the atmosphere and troposphere, to study the nature of aerosol layers absorbing the short-wave radiation, and to study the impact of natural and anthropogenic sources of aerosol on the formation of the spectrum of the aerosol particle sizes [3, 14].

1.2.3.2 Results of the measurements

The accomplished studies gave rich information on the spatial–temporal structure, chemical composition, and size distribution of aerosol. Unfortunately, this information was rather contradictory. In particular, data of different authors on global estimates of the power of various sources of atmospheric aerosols differ greatly [7,

38, 40, 45, 74, 79]. First of all, this refers to assessments of the contribution of such sources as sea surface, soil, and gas-to-particle aerosol formation. Results of the 'marine' CAENEX-72 [65] and GATE [1, 83] show that the contribution of marine aerosol to the total content of aerosol is small (about 10%), whereas the authors of [15] are of a different opinion, assuming that the share of marine aerosol is much larger (30%). In one monograph [74] the soil as a source of aerosol is given more importance than in [15, 86], where photochemical reactions play the leading role in aerosol formation, and first of all, in the formation of particles from organic gases. In [51] the contribution of photochemical aerosol to the total content of atmospheric aerosol is supposed to be very small (about 5%). Estimates of contributions from other sources of aerosol are similar (volcanic dust, 0.1–0.5%; forest fires, 2–5%; sulphur compounds, 9–25%; nitrogen compounds, 5–15%; anthropogenic particles, 5–10%). Note should be taken that these global mean assessments may not correspond to the role of various sources in different places and in different seasons. Such a conclusion follows, for instance, from the results of expeditions within the CAENEX programme accomplished in different climatic zones and during different seasons.

A very important conclusion was drawn concerning the spatial structure of the aerosol concentration field. The facts of the stratified vertical structure of aerosol has been reliably established; changes are wide ranging in the aerosol concentration, not only with altitude but also in horizontal direction (for particles with $r > 0.2\,\mu\text{m}$ from $N = 1\,\text{cm}^{-3}$ to $N = 1{,}000\,\text{cm}^{-3}$), as well as the dependence of aerosol concentration fields on latitude, season, time, and nature of the surface [14, 45, 61].

As a result of the first CAENEX-70 observational programme, due to the summertime aerosol measurements, the vertical distribution of short-wave radiation flux divergence in the free atmosphere and over desert was obtained for the first time. Analysis of these results is contained in [3, 5, 18, 31, 62, 65, 68, 74], whose basic brief conclusions are: the residual absorption of short-wave radiation detected earlier from data of aircraft pyranometric measurements is, in fact, the aerosol wavelength dependent absorption; estimates of the radiation flux divergence in the atmosphere due to absorption of short-wave radiation by aerosol and gas components may have the same order of magnitude.

The statistical processing of data showed that optical (absorptive) properties of aerosol particles in the free troposphere coincide with the properties of particles obtained from surface measurements, whereas in higher atmospheric layers (stratosphere) their properties differ substantially. The vertical profiles of various spectral radiative characteristics are mainly determined by the presence or absence of absorption at a given wavelength.

An estimation of a possible variability of the total and spectral short-wave radiation flux divergence in the troposphere [62, 67] revealed its dependence on the optical state of the atmosphere, solar zenith angle, and other factors. Therefore, it was decided to undertake several expeditions within the CAENEX programme over various surfaces: desert (CAENEX-70), steppe (CAENEX-71), industrial region (CAENEX-72), sea surface (CAENEX-73), and finally, tropical Atlantic Ocean (GATE radiation subprogramme) [1, 69].

When accomplishing each of these programmes, complex studies of specific features of radiative and aerosol characteristics of the atmosphere in these regions have been successfully carried out. To illustrate the peculiarities of the atmospheric radiative regime in different geographic regions, note, in particular, that observations during the GATE were often carried out under conditions of dust outbreaks from the Sahara, but the near-water-surface layer (up to 1 km in altitude) remained practically always free of aerosol particles. Thus, if under desert conditions a maximum of radiation flux divergence was observed in the atmospheric boundary layer, over the tropical Atlantic this maximum was observed in the layer located at a considerable altitude over the ocean.

In the 1971–1972 expeditions and during GATE, data were obtained on the radiation regime of the atmosphere in the presence of a homogeneous layer of stratus or stratocumulus clouds, as well as under conditions of broken cloudiness. These observations accompanied by measurements of the microphysical characteristics of clouds, made it possible to evaluate the effect of clouds on the radiation regime of clouds in the surface–atmosphere system, which, in turn, is expressed through a considerable decrease of the amount of radiation absorbed by the system (more than half as much).

Interesting results in these years were obtained during the accomplishment of various programmes of radiation studies in the USA [15, 16]. So, for instance, data of observations of the radiation flux vertical profiles at $\lambda = 0.38\,\mu m$, at solar zenith angle 55° (observations were made in 1970 from the Convair-990 aircraft), showed that radiation flux divergences usually observed in this case (Big Spring, Texas), in the visible spectral region (pyranometric measurements with the Schott filters), constituted about 0.022 0–0.000 5 in the units of solar constant (i.e., of the same order of magnitude as the estimates obtained during CAENEX-70 over the desert).

Despite certain progress in the accomplishment of the various programmes mentioned, many aspects of the problem remained far from being solved and in some cases the results obtained raised more questions than they provided answers. This testifies to the usefulness of combining some programmes into a broader single programme of global studies of radiative, microphysical, and chemical properties of aerosol in different geographical regions carried out on the basis of international cooperation.

The first two expeditions on the CAENEX programme were carried out in the absence of industrial sources of aerosol. The factor determining the aerosol production was the open surface sands of the Kara-Kum desert (CAENEX-70) as well as the agricultural lands and dried-up steppe of north-western Kazakhstan (CAENEX-71). For measurements in the region of Zaporozhye, the main factor was the presence of large industrial enterprises. This determined the chemical composition of aerosol both in the surface layer and in the troposphere, its mass concentration of the particles' size distribution, optical properties, and the dependence of the aerosol content on meteorological conditions. Here is a brief summary of the results obtained.

1.2.3.3 Aerosol size distribution

The data of the first two expeditions are characterized by a relatively high stability of the function of the particles' size distribution in the surface layer for a certain type of surface. This function was (very) approximately described by the Junge formula. Practically always in the surface layer there is a great amount of very large particles exceeding their concentration obtained with the Junge formula with the parameter $\nu < 3.5$. For the Kara-Kum desert conditions, instability of the content of particles with radius ranging between 0.25 and 0.35 μm can be considered typical. Aerosols in the region of the settlement of Repetek are characterized by a very steep function of distribution in the interval of large sizes of aerosol particles compared with the region of the settlement of Ankata (strong variability of particles formed in the desert).

The function of the aerosol particles size distribution for samples taken at Zaporozhye turned out to be less stable. The samples were characterized by the presence of a fine-size fraction and loose conglomerates of particles often exceeding several μm in size. In this case the shape of the distribution function depends comparatively little on the altitude at which measurements are made. At the same time, in the case of measurements in the Kara-Kum desert, the altitudinal dependence of the spectrum of particles' size distribution manifested itself distinctly.

For instance, at Ankata, at the surface level, there was a very high concentration of the finest particles (with $r < 0.3$ μm) and a very low concentration of gigantic particles with $r > 2$ μm, which is apparently connected with generation of particles by soils at the measurement site. At Repetek the generation of particles took place mainly far from the place of measurement, and therefore the function of distribution at heights $H > 50$ cm is not that steep.

For both points in the desert an altitudinal decrease of relative concentration of the largest aerosol particles ($r > 2.5$ μm) is practically always observed. The behaviour of the specific content of particles with $r \sim 0.25$ μm is unstable: as the altitude changes, both a strong increase and decrease of these particles' concentration can be observed at high altitudes. In the range of gigantic particles, the function of particles' size distribution deviates considerably from the Junge formula (an increased content). As a rule, particles were of an oval shape.

An electron microscopic analysis of aerosol particles reveals a reduced content (compared to Junge) of particles with $r < 0.1$ μm. The maximum for particles' size distribution is $r \approx 0.05$ μm. Data on tropospheric aerosol agree well with results of measurements obtained by Blifford [7].

1.2.3.4 Aerosol chemical composition

The data of the first two expeditions are characterized by a low content of organic components in aerosol. Though there was no quantitative analysis for organic matter, it can be considered that the amount of organic components did not exceed 5–10 μg m^{-3}. For Zaporozhye, this value is at a minimum, in this case the concentrations of compounds containing SO_4^{2-} are higher.

At Zaporozhye, there were especially high concentrations of compounds containing iron. For the surface layer at Repetek maximum values did not exceed $15\,\mu g\,m^{-3}$ (an average of about $5\,\mu g\,m^{-3}$), at Ankata $5\,\mu g\,m^{-3}$ (an average of about $1\,\mu g\,m^{-3}$), however, at Zaporozhye the concentration of this element always exceeded $30\,\mu g\,m^{-3}$. Concentrations of Pb, Mg, Ni, Ti, and Mn were also high.

The basic component of aerosol in the surface layer at Repetek was quartz particles. If the total weight concentration of aerosol averaged about $250\,\mu g\,m^{-3}$, the share of quartz particles constituted about $100\,\mu g\,m^{-3}$. The content of Fe, Ca, and Zn in aerosols was unstable. The basic component of aerosol in the region of Ankata was soil particles. Here, a higher content of Ca and Al was observed. In both regions a large amount (by mass) of the particles was of mineral origin. At Zaporozhye the aerosol component of industrial origin dominates. Clearly, organic aerosols were mainly of industrial origin.

1.2.3.5 Dependence on meteorological conditions

Different heights of aerosol sources (sand hill tops, steppe, plant chimneys) and different reliefs of the surface lead to different dependences of the aerosol content in the atmosphere on meteorological conditions. So, measurements at Repetek revealed a clearly expressed diurnal variation of aerosol number density at heights from 2–15 m, with a maximum in the evening and a minimum in the morning. The spectral density of particle size distribution depends little on the time of day. Only in the evening was there a small relative increase in the concentration of gigantic aerosol particles, probably, deposited onto the surface layer from higher atmospheric layers. The coagulation growth of dust particles is practically unobserved.

From measurements in the region of Ankata there were no distinct daily variations of number density. Here, of great importance was the wind speed that drove dust particles into the air. At the same time, the size distribution of particles had a distinct diurnal variation: during the first-half of day there was a strong decrease of relative concentration of particles in the range of radii 0.2–0.5 μm, by the evening, fractions of gigantic particles with $r > 2\,\mu m$ increased, and the share of particles with $r \sim (0.4\text{–}1.5\,\mu m)$ decreased. It should be noted that the diurnal variation of particle size distribution for different heights was different: there was a phase shift with altitude.

The diurnal variation of the spectrum of aerosol particle sizes took place not only in the surface layer but also within the whole atmosphere, which was distinctly seen not only from *in situ* measurements, but also from measurements of the diurnal variation of the atmospheric spectral transparency.

Analysis of the dependence of various characteristics of the surface-layer aerosol at Repetek on humidity, cloudiness, wind speed and direction, and precipitation, as well as the temperature regime of the surface layer showed that the effect of humidity (small changes in humidity) on the change of particle size distribution spectrum was negligible. There was a decreasing trend of aerosol number density at heights of 2–15 m, when the wind speed varied between 2 and $7\,m\,s^{-1}$. The main factor affecting the change of aerosol concentration was the temperature regime of the surface layer and its turbidity.

1.2.3.6 The vertical profile of concentration

Aircraft measurements over Repetek revealed the following characteristic features of the vertical profiles of large aerosol particle concentrations ($r > 0.2\,\mu m$): (1) a very high number density; (2) an exponential decrease of number density with altitude; and (3) a relative stability of the function of particle size distribution. This is explained by an almost complete absence of hygroscopic particles and a strong mixing of aerosol in the atmosphere over the desert. From measurements over Ankata, there was no evidence of an exponential decrease of aerosol number density with altitude. The vertical distribution of aerosol concentration in the troposphere for this region was of a stratified character. The stratification was particularly typical for the daytime. At night, the aerosol layers descended. Since the concentration of particles in the upper and middle troposphere in the evening and at night grows, one can assume that aerosol particles descend from higher atmospheric layers where the concentration of particles should be also high (the subtropopause layer). Though there is no direct connection between the vertical profile of aerosol and the vertical profile of relative humidity, there is a correlation between the altitudes of the prevailing residence of aerosol layers and layers of increased relative humidity.

1.2.4 Spectral transparency

Results of measurements of the diurnal variation of the atmospheric spectral transparency at Repetek (23–25 October, 1970) revealed a distinct trend of particle enlargement in the atmosphere by noon. Two weak maxima of aerosol attenuation for wavelength intervals 3,200–3,500 Å and 3,900–4,100 Å, probably determined by aerosol absorption bands, shift to the long-wave part of the spectrum. This is connected with a strong variability of the scale of scattering and absorption attenuation when the size of aerosol particles varies in the interval considered. The latter is characteristic, first of all, of hematite particles with an absorption band with a maximum at a wavelength of about 4,080 Å. Maximum factors of radiation attenuation on hematite particles at these wavelengths fall on the radii 0.08–1.1 μm and reach a value of 6–8. Therefore, only in the early morning, when a small-size fraction of particles prevails, is the aerosol attenuation of radiation in the hematite absorption band at a maximum, and then with the growth of particles this maximum levels out due to the growing attenuation of radiation by aerosol at adjacent wavelengths.

Laboratory experiments and chemical analyses of atmospheric aerosol samples enabled one to evaluate the optical constants of aerosol matter and the limits of variability of these constants depending on the chemical composition of aerosols and the relative humidity of the air [49, 52, 54]. Changes in the real part of the complex refraction index $m' = n - i\chi$ were mainly determined by the relative humidity of air. For the visible spectral region, the real part of the refraction index n varied from 1.65

to 1.33 when the relative humidity increased from zero to 98–100%. Higher values of the aerosol refraction index in the real atmosphere are practically unobserved, though for some particles (soot, iron oxides, etc.) they can reach 2.2–2.8. The imaginary part of refraction index χ in the short-wave region is determined by the presence in aerosol particles of such components as soot, iron oxides, some organic compounds, and sulphur compounds. The χ values for the visible spectral interval [42, 52, 54] at $\lambda = 0.4\,\mu m$ reach 0.005–0.015, which agrees well with the data of other authors [17, 22, 74, 89]. When the soot content grows (industrial aerosol), the χ can be even higher. Note should be taken that during measurements in Kara-Kum and in the GATE period a marked absorption of short-wave radiation by aerosol was observed due to iron oxides.

The complex refraction index and data on the tropospheric aerosol size distribution enabled one to calculate its optical parameters and to draw some conclusions about the aerosol optical properties which can be compared with data of optical measurements [39]. For the problem of calculating optical parameters it was most difficult to adequately choose the optical constants of aerosol matter. In this connection, the following characteristics were to be determined: (1) distribution of chemical compounds by aerosol size fractions; (2) relative content of chemical compounds in the substance of one particle; and (3) special features of the growth of particles when relative humidity increases.

On the assumption that all chemical compounds are distributed uniformly both by fractions and inside the particles, equivalent optical constants of aerosol matter were calculated with the optical constants of all (or most) components of the aerosol chemical composition known. Calculations were made for 'dry' aerosols in the surface layer and in the troposphere in the regions of the first two expeditions.

The magnitude of the imaginary part of the refraction index (the absorption index) corresponding to this model was determined by the presence of hematite (6%) and soot (0.1%), and in the visible it is $\chi = 0.005$–0.003. The real part of the refraction index in the same spectral interval is $n = 1.65$–0.2 and decreases substantially at $\lambda > 5\,\mu m$. A marked increase of χ at $\lambda = 2.7\,\mu m$ to $5.5\,\mu m$ is connected with the presence of bound water (6%) and sulphates (18%). This synthetic model of the complex refraction index for aerosol particles was a first approximation for the complex refraction index of atmospheric aerosol of soil origins published in the scientific literature [39].

1.2.4.1 Radiation absorption by aerosol

Of particular importance for calculations of the optical parameters of aerosol was to determine the imaginary part of the complex refraction index. Results of aircraft spectral measurements at Repetek confirmed the earlier supposition that the aerosol absorption in the short-wave spectral region is comparable with the total molecular absorption. A comparison of the spectral variation of the aerosol absorption function with that of the aerosol absorption coefficient for a dry model of aerosol, as well as analysis of the optical constants of some components of this model, have shown that in some cases one can use, in calculations, the optical constants of individual components. For instance, the difference of the short-wave radiation balances at altitudes 0.3 km and 8.4 km (absorbed radiation) obtained on 25 January, 1970 in the region of Repetek is characterized by a spectral change in the wavelength region 0.4–0.8 μm and is similar to the spectral dependence of the imaginary part of the hematite refraction index. As seen from calculations, hematite particles should, in this case, have a radius of ~0.1 μm. Some share of short-wave radiation absorption can be explained by the presence, in the atmosphere, of strongly absorbing submicron particles of soot, trivalent iron compounds, and particles of organic origin with strong absorption bands in the wavelength region 4,500 Å and 6,800 Å. The presence of these particles, which is not at variance with the data from the chemical analysis of the tropospheric aerosol, can explain the strong radiative heating of the atmosphere due to short-wave radiation absorption by aerosol particles. The monograph [39] contains a detailed discussion of the respective results.

The results of aerosol studies made it possible not only to obtain new data on the aerosol composition of the atmosphere and the laws of atmospheric aerosol behaviour depending on different meteorological parameters but also convincingly demonstrated that: (1) the solid fraction of aerosol plays a very important role in the atmospheric radiative regime, especially in short-wave radiation absorption; (2) the complex refraction index of aerosol obtained on the basis of numerical models differs strongly from the refraction index of water or from 1.5 – an estimate often used earlier which depends on relative humidity; and (3) condensation growth of particles is less significant than coagulation in the formation of the spectrum of particle sizes [40].

1.2.5 The heat balance of the atmospheric boundary layer: interaction between radiative and turbulent heat exchange

The existing schemes of parameterization of the surface–atmosphere interaction are known to be based on the supposition that the heat exchange between the atmosphere and surface is determined by turbulence and convection. Nevertheless, there are numerous data showing the substantial role of radiative heat exchange in the heat transport in the surface and atmospheric boundary layers (ABLs) [14, 15, 44]. One of the important problems of the CAENEX-71 expedition was the further investigation of this problem.

In the absence of advection, and over a homogeneous surface, a change in the heat content of a fixed atmospheric layer is determined by the difference between radiative and turbulent fluxes at the boundaries of the layer. The radiation flux divergence in the 0–3-km layer was determined from direct measurements. The turbulent heat flux at the top of the boundary layer can be considered to be zero and thus the heat flux divergence in the whole boundary layer due to the turbulent heat exchange is determined by turbulent heat flux measured at the surface level.

The closing of the heat balance is demonstrated in Figure 1.7, which shows the diurnal variations of heat content (ΔQ) in the 0–5,000-m layer, flux divergences due to short-wave (ΔS) and long-wave (ΔF) radiation, and turbulent heat flux near the

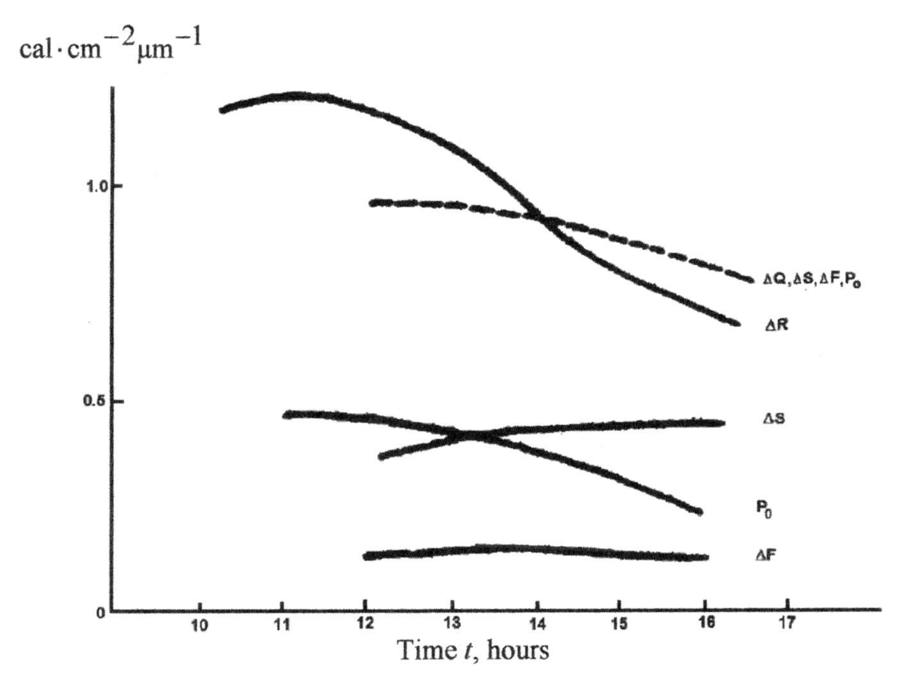

Figure 1.7. The diurnal variations of heat content (ΔQ), flux divergence due to absorption of short-wave (ΔS) and long-wave (ΔF) radiation, turbulent heat flux (ΔP), and total flux divergence ($\Delta Q + \Delta S + \Delta F + \Delta P$) in the 0–3,000-m layer.

surface ($P_0 = \Delta P$). According to estimates obtained from data of direct measurements with the additional use of synoptic maps, an adequate flux divergence did not exceed 0.1 cal cm^{-2} min^{-1}. As seen from Figure 1.7, the values of ΔQ and total flux divergence ΔR agree well, which testifies to a reliable closure of the heat balance in the boundary layer from data of direct measurements.

Analysis of the data obtained shows that:

(1) a considerable contribution to the heat balance in the ABL in the daytime is made by both short-wave and long-wave radiation;
(2) the contribution of radiative flux divergence is quantitatively comparable with that of the turbulent flux divergence (and sometimes $\Delta S + \Delta F > \Delta P$) for the whole boundary layer; and
(3) in the daytime, the boundary layer is heated due to long-wave radiative and turbulent heat exchange, and at night, a cooling takes place due to both these factors.

Thus in the case considered, both radiation and turbulence caused changes in the temperature of the same sign. It should be noted that such an idea about the role of turbulence and radiation in heat exchange (for the whole boundary layer) was more or less common. Therefore, it was interesting to study the impact of these factors on the thermal regime of not only the whole boundary layer but also individual, thinner layers – the surface layer in particular.

Investigations within the CAENEX programme made it possible to study special features of the diurnal variation of the heat balance in different atmospheric layers in the upper part of the troposphere due to the processes of radiative heat exchange. The ratio between ΔS and ΔF varied from day to day, but a substantial radiative heating of the surface layer was observed with a strongly heated soil surface exceeding in magnitude the observed changes in temperature. The data available show that whereas the whole boundary layer is heated due to radiative and turbulent heat exchange, in its lower part (in the layer about 100 m thick) in the daytime, a considerable heating due to radiation (first of all, long-wave) is compensated for by air cooling due to turbulent heat exchange (turbulent heat flux in this layer grows with altitude). At night (closer to sunrise) a surface layer cooling takes place due to turbulent heat exchange. These data testify to the important role of the interaction between the fields of radiation and motion in the surface and boundary layers of the atmosphere. Of course, this fact should be taken into account in working out the schemes of parameterization of the atmosphere–surface interaction.

1.3 AEROSOL IN THE GATE REGION

In the large-scale GATE-74 international field experiment, special attention was paid to studies of the structure and optical properties of atmospheric aerosol [1, 21, 26, 38, 50, 86]. Studies of aerosol size distribution and optical properties in the surface layer were carried out on several research vessels located at different points in the Atlantic Ocean, on the islands Barbados and Sal, from flying laboratories, as well as observations from satellites. The methods and programme of aerosol measurements were similar to those used in the CAENEX [3].

Special attention was paid to a study of the Saharan aerosol layer and optical properties of the Saharan dust. Most extensive studies in this direction were accomplished by a group of soviet participants in the GATE – within the GATE radiation subprogramme [51]. A considerable impact of the Saharan dust outbreaks on the atmosphere of the equatorial Atlantic Ocean was detected. The main component of aerosol in this region was non-soluble, mineral dust appearing in the deserts, arid, and semiarid regions of western Africa [1, 13, 15, 23, 38].

Dust storms raise dust up to 5–7 km where a large-scale anticyclonic vortex drives this dust across the Atlantic Ocean. This outflow is concentrated in the layer from 1–2 km to 5–7 km. Since the temperature of the lower boundary of the Saharan dust layer is 5–6° higher than that of the boundary layer located below, a strong inversion forms. The zone of dust outbreak is $\sim$10–15° latitude wide and it takes 5–7 days to cross the Atlantic Ocean in the trade winds. During the period of observations from the end of the 1960s a seasonal variation of the Saharan dust was detected with a distinct maximum in June–August and an increasing trend of concentration ($\sim$5-fold) in the 1970s determined by a multiyear drought in the Sahel region.

Aircraft measurements of the short-wave radiation fluxes have shown that the atmospheric dust loading can double the solar radiation levels absorbed by the atmosphere and lead to a strong attenuation of convection in the daytime (this conclusion is confirmed by observations pointing to a nocturnal maximum of convection in the tropics). An intensive absorption of solar radiation is, apparently, the cause of the inversion stability near the lower boundary of the Saharan dust layer remaining in the region of the Caribbean Sea and in other regions of the Atlantic Ocean. Therefore, clouds in the GATE region were characterized by the rare occurrence of well-developed cumulus clouds and non-coherence of their formation: clouds started to form simultaneously at different levels at altitudes ranging between 1.4 and 4 km. The bottom of well-developed (hot) cloud towers was often located at the 4-km level, which was the upper boundary of the haze layer. Much lower (at altitudes of about 0.6–0.8 km) there was a layer of weak trade winds with broken cumulus cloudiness. There was no interaction between these layers. Probably, the multilevel and chaotic character of the cumulus clouds formation is the result of interaction between the Saharan aerosol layer (SAL) and intra-tropical convergence zone. The SAL substantially changes the radiative regime of the atmosphere and the ocean surface, causing a considerable decrease in radiative air cooling.

Measurements of the atmospheric aerosol optical thickness in the periods of large-scale outbreaks of Saharan dust to the Atlantic Ocean basin have shown that the τ_a magnitude decreases in the westward direction, which indicates a gradual dust deposition. These data correlate well with the data of satellite observations, in particular, with the data from NOAA meteorological satellites.

1.4 GLOBAL AEROSOL–RADIATION EXPERIMENT (GAREX): SCIENTIFIC OBJECTIVES OF THE PROGRAMME

Consider schematically the interaction of radiation with the surface–atmosphere system. Assume in the case of the clear unpolluted atmosphere that about 83% of the solar energy reaches the surface (10% scatter to space, and about 7% is absorbed by atmospheric gases). At an average surface albedo $A = 0.2$, about 65% of solar energy is absorbed by the surface and 18% is reflected back to the atmosphere, of which 15% leaves the atmosphere and 3% is absorbed (secondary backscattering, absorption by the surface, etc., can be neglected in this approximate consideration). So, with the introduced characteristics of the surface–atmosphere system, 25% of the solar energy is reflected to space, 65% is absorbed by the surface, and 10% is absorbed by the atmosphere.

Absorption by aerosol particles in the atmosphere can both increase and decrease the system's albedo depending on the optical characteristics of aerosol particles and surface as well as the solar zenith angle. Consider the case of such an absorbing and scattering of aerosol particles which does not change the system's albedo. When the albedo remains constant, the absorbed energy is redistributed between the atmosphere and the surface: the atmosphere will absorb about 20%

Table 1.1. Absorbed short-wave radiation (in % with respect to insolation beyond the atmosphere).

	Atmospheric condition		
Absorbing medium	Clear, unpolluted	Clear with absorbing aerosol	Cloudy
Clouds	–	–	6
Atmosphere	10	20	11
Surface	65	55	18
Total	*75*	*75*	*35*

of the solar energy, and the share at the surface will constitute only 55%. Still greater changes in the radiative regime of the atmosphere will take place with the appearance of a cloud layer. In this case we assume that the system's albedo constitutes 0.65, with the above-cloud, dust-free layer absorbing about 5% of the solar energy. Assume that the clouds and subcloud layer absorb 6% each, then the share of the radiation absorbed by the surface will constitute only 18% of the incoming solar radiation. Thus, in this case the total absorbed radiation is substantially reduced and its more intensive redistribution takes place between the surface and the atmosphere in favour of the latter. The results of this schematic consideration are given in Table 1.1.

Naturally, the values given in Table 1.1 vary widely under real conditions depending on the specific composition and structure of the atmosphere, clouds and surface properties, as well as the solar zenith angle.

The need for an aerosol–radiation experiment was explained by the insufficient statistical information on the aerosol–radiation characteristics of the atmosphere, and the absence of observations for some typical conditions. In particular, there were almost no data on the spatial structure and properties of atmospheric aerosols in winter. Under conditions of strongly varying characteristics of atmospheric aerosols, an accumulation of statistical information may be insufficient. Under these conditions, of importance is the solution of the problem of simulating the processes of aerosol formation and transformation of its properties, including the optical characteristics.

In substantiation of an adequate model of atmospheric aerosol, the main difficulty consists not in calculations of the aerosol optical characteristics but in our insufficient knowledge about such properties of aerosol as particle size distribution, shape of particles, chemical composition, and internal structure of particles. In most cases an approximation of particle sphericity is quite satisfactory for calculations of the optical properties of real aerosols, since the energy characteristics of aerosol (coefficients of attenuation, scattering, and absorption of radiation) depend weakly on the shape of particles. It is more important to know the particle size distribution and distribution of the complex refraction index for the substance in a particle along its radius, especially near the particle's surface. It should be emphasized that an estimation of these two parameters of aerosol is needed but remain insufficient for calculation of their optical characteristics. These parameters are determined by physical and physico-chemical processes of aerosol formation and transformation in the atmosphere: coagulation and condensation mechanisms of growth, advective and convective transport of particles, capture of particles by droplets and other obstacles, deposition of particles in the atmosphere, turbulent diffusion, and other processes. Some of these can be studied only under natural conditions. Therefore, the problem appears of simultaneous complex studies of aerosol both in laboratories (chambers) and in the free atmosphere.

Different atmospheric processes transform in different ways the atmospheric aerosols of various chemical compositions. First of all, this refers to the processes of condensation and coagulation, which determine the distribution of a substance with different physical and physico-chemical properties within a particle. In this connection, it is necessary to study the laws of aerosol formation in the atmosphere from different sources. In particular, of importance are such problems as the role of the processes of aerosol formation from gas and sea sprays. It is also important that such aerosol most efficiently favours the formation of clouds. Of course, one can suppose that in the cases of formation of convective clouds and clouds formed via mixing of warm and cool air masses, there is no deficit of active condensation nuclei. In these cases there is a well-developed theory of cloud formation which does not take into account the specific properties of condensation nuclei of different origin [51]. However, the fact cannot be ignored that in some cases clouds are forming in the atmosphere (mainly stratus) due to the specific properties of condensation nuclei, which was understood long ago [75]. Such active condensation nuclei are known to form as a result of human activity. Therefore, in the regions of large industrial cities, clouds form and precipitation occurs much more often [75]. Since the share of anthropogenic aerosol continuously grows, studies of physical and physico-chemical properties of this type of aerosol ought to have been an integral part of the programme of the aerosol–radiation experiment.

Serious attention in the aerosol–radiation experiment was paid to aerosol data reliability. Unfortunately, data from aerosol measurements performed by different authors are still difficult to compare. An explanation is that measured quantities were not always evaluated unambiguously, and that the connection between different aerosol characteristics could be complicated and ambiguous. Therefore, it is necessary not only to specify relationships between various aerosol characteristics but also to make systematic, simultaneous measurements of these characteristics with different instruments. Of great importance are simultaneous studies of aerosols with methods of lidar sounding and *in situ* measurements using airborne and balloon-borne equipment. There is no doubt that lidar is the most valuable means of studying aerosol dynamics.

Bearing this in mind, the scientific objectives of GAREX were formulated as follows:

(1) Studies of radiative characteristics of the surface–atmosphere system (especially the system's albedo, distribution of absorbed radiation between the

atmosphere and the surface, radiative heat exchange, etc.) and their variability in different climatic zones.

(2) Studies of the atmospheric aerosol concentration field, size distribution, chemical composition, complex refraction index, and variability of these characteristics in different climatic zones.

(3) Continuation of developments connected with the assessment of the impact of aerosol on climate, and studies of the stratospheric composition (optically active MGCs and aerosol).

(4) Laboratory studies of the formation of the spectrum of aerosol particle sizes and their optical characteristics.

(5) Laboratory studies of photochemical processes of formation of the gas composition of the stratosphere and studies of the MGCs absorption spectra.

(6) Development and application of surface, aircraft, and satellite methods of remote sensing of aerosol for the most adequate determination of its properties.

(7) Development and application of direct and indirect methods of studying the microphysical and optical parameters of clouds (especially cirrus clouds).

(8) Comparison of the results of field and laboratory studies of various aerosol systems.

(9) Development of approximate methods of calculation of the radiative characteristics of the real atmosphere designed for parameterization of radiative processes in the numerical simulation of atmospheric general circulation and climate theory.

(10) Theoretical studies of radiation transfer in the real atmosphere based on the use of 'exact' methods and comparison of calculated results with the observational data.

The enumerated scientific problems were solved for various climatic and ecological conditions on the basis of the following coordinated (in some cases synchronous) theoretical, laboratory, surface, aircraft, and satellite studies.

1. Laboratory studies aimed at the simulation of:
   - 1.1. Physical and physico-chemical processes of atmospheric aerosol formation.
   - 1.2. Optical constants of atmospheric aerosol.
   - 1.3. Processes regulating the content and properties of aerosol in the atmosphere.
   - 1.4. Photochemical processes regulating the gas composition of the stratosphere.
   - 1.5. Absorption spectra for the MGCs of the stratosphere.
2. Surface complex experiments aimed at:
   - 2.1. Determination of the spectral optical thickness of the atmosphere.
   - 2.2. Evaluation of the spectral scattering function from sky brightness data.
   - 2.3. Determination of the aerosol content in the surface layer, its size distribution, and chemical composition.
   - 2.4. Assessment of the impact of meteorological conditions on the aerosol content in the surface layer.
   - 2.5. Remote sensing of clouds and aerosol.

3. Aircraft and balloon complex experiments aimed at:
   3.1. Determination of the vertical profiles of radiation spectral fluxes, balance, and flux divergence.
   3.2. Determination of the spectral optical thicknesses of different atmospheric layers.
   3.3. Estimation of the spectral scattering function for different atmospheric layers.
   3.4. Study of the vertical and horizontal structure of aerosol distribution, its size distribution, and chemical composition.
   3.5. Evaluation of the contribution from different sources of aerosol formation in the atmosphere.
   3.6. Measurements of the microphysical and optical parameters of clouds.
4. Theoretical studies involving:
   4.1. Calculations of vertical profiles, spectral fluxes, and balance and flux divergence of radiation with the use of approximate methods.
   4.2. Development of a theory of atmospheric aerosol formation and methods of parameterizing the aerosol effects in climate theory.
   4.3. Calculations of the optical characteristics of atmospheric aerosols.
   4.4. 'Exact' theoretical calculations of radiative characteristics of the real atmosphere.
   4.5. Development of the theory of radiation transfer in the presence of partial cloudiness with the aim of developing methods of parameterization of radiative processes.

The enumerated studies were accomplished within the following programmes of concrete observations and theoretical calculations.

(*A*) *Radiation studies.*
1. Surface.
   1.1. Study of the spectral optical thickness with the use of filter photometers and spectrometers.
   1.2. Estimation of the spectral scattering function with the use of filter photometers and spectrometers.
   1.3. Accomplishment of routine actinometric and, similar to aircraft, surface radiation measurements.
   1.4. Study of spectral and total downward and upward radiation fluxes at the surface level.
2. Aircraft.
   2.1. Measurements of spectral downward and upward radiation fluxes at different levels in the free and cloudy atmosphere: evaluation of spectral balances and spectral absolute and relative radiation flux divergences in the layers of different thicknesses.
   2.2. Determination of the vertical profile of the atmospheric spectral optical thickness.
   2.3. Determination of the vertical profile of the scattering function.

2.4. Measurements of total fluxes of downward and upward radiation in the free atmosphere: estimation of radiation balances and flux divergences for different layers of the atmosphere.

3. Satellite.
   3.1. Measurements of the surface–atmosphere system's radiation budget components.
   3.2. Sub-satellite synchronous field experiments providing simultaneous data on radiative characteristics of the surface, the atmosphere, and the surface–atmosphere system.
   3.3. Studies of the angular distribution of the outgoing radiation in different spectral regions.
   3.4. Measurements of total and spectral solar constant.
4. Theoretical studies.
   4.1. Numerical simulation of all radiative characteristics of the atmosphere for prescribed conditions and comparison of the calculated results with observed quantities.
   4.2. The same as 4.1. but using the Monte-Carlo method.
   4.3. The same as 4.1. but using 'exact' methods.

(*B*) *Aerosol studies.*

1. Laboratory studies.
   1.1. Complex refraction index of aerosol substance for typical samples and the most important chemical compounds.
   1.2. The effect of the distribution of substances with different refraction indices in a particle on its optical properties.
   1.3. Processes of gas-to-particle conversion.
   1.4. Processes of size distribution formation for aerosols of different origins and chemical compositions.
   1.5. The effect of changes in relative humidity on different physical characteristics of atmospheric aerosols.
   1.6. Comparison between various characteristics of aerosol obtained with the use of different instrumentation.
2. Surface.
   2.1. Measurements of concentration, size distribution, and chemical composition of aerosol and minor gases in different regions of the globe in order to determine their diurnal variation, dependence on meteorological parameters, and correlations between them as well as connections of the indicated aerosol parameters with the content of minor gases. For this purpose, both *in situ* and indirect methods were used (lidar sounding, measurements of spectral atmospheric transparency, and sky brightness).
   2.2. Complex and comparative measurements using instruments operating on the basis of different physical principles.
   2.3. Evaluation of aerosol parameters (transparency, scattering characteristics, etc.).

3. Aircraft – aircraft aerosol measurements similar to surface measurements but with the specific character of aircraft measurements taken into account. Of first priority was the study of the spatial distribution of the atmospheric aerosol concentration field with the use of *in situ* methods. Thus, for the programme of aircraft measurements, apart from those enumerated in the section of surface observations, the following measurements were included.
   3.1. Vertical profiles of atmospheric aerosol properties and their dependence on the specific feature of the surface, time of day, season, and cloud conditions.
   3.2. Horizontal heterogeneity of atmospheric aerosol distribution (determination of spatial and temporal characteristics of heterogeneities).
   3.3. Microphysical and optical characteristics of clouds.
4. Balloon – the programme of balloon measurements was similar to that of aircraft measurements, but special attention was paid to obtaining data on the characteristics of stratospheric aerosol as well as to study optically active MGCs. Such measurements, as far as possible, were accompanied by radiation measurements.
5. Satellite.
   5.1. Interpretation of satellite remote-sensing data for different spectral regions in order to study natural and anthropogenic aerosol.
   5.2. Application of obtained results to solve the problem of radiation–cloud interaction.

1.5 THE SOVIET–AMERICAN BALLOON AEROSOL MEASUREMENTS

The first stage of the Soviet–American field experiment to study atmospheric aerosol was accomplished in the period 27 July–11 August, 1975 [3]. It was carried out at the field station of the Central Aerological Observatory located in the vicinity of Rylsk. This point of sounding is in the central part of the European territory of Russia (ETR) in the valley of the Seima River. Climatically it is a part of the central, flat region of the ETR, near the southern boundary of this region. It is characterized by moderately warm summer seasons. The location of the sounding is in the zone of attenuation of cyclonic activity.

The synoptic situation during the period of the first stage of the aerosol–radiation experiment was characterized by interaction of two circulation processes. On the one hand, from the regions of the Barents and White Seas along the periphery of the high-pressure region over western Europe there was observed an outflow of cool polar air masses to the central regions of the ETR, and on the other hand – the motion of warm tropical air from the regions of Asia Minor across the Black Sea and Kuban′ to the Lower Volga. At the frontal interface between these two fluxes a cyclone formed, which, by 1 August reached the regions of the Volga – being already occluded. The presence of the cyclonic vortex was evident in the whole troposphere. The energy of this cyclone was supported by temperature contrasts between relatively cool air from the north-east and warmer air masses formed over the southern ETR. During the period 1–4 August, the cool air masses reached its central part, which determined a rapid filling of the cyclone. Due to the non-uniform growth of pressure, its centre gradually shifted toward the

Moscow region, and on 2–4 August the weather in the region of Rylsk was determined by the rear part of the filling cyclone and by the passing of secondary fronts.

On 4–5 August, the natural synoptic period changed, indicated by an attenuation of the Azores maximum, an enhancement of cyclonic activity over the northern seas, and further propagation of the high-pressure region to the east of the ETR. On 5 August the filling cyclone weakened considerably. The propagation of the high-pressure region to the north-western regions of the ETR led to a restoration of north-easterly winds in the free troposphere and to an enhancement of downward vertical air fluxes.

This development of synoptic processes during the period of undertaking the aerosol–radiation experiment determined the unstable weather in the region of Rylsk and favoured the development of cumulus cloudiness.

Here is a brief characteristic of the weather, synoptic situation, and vertical distribution of meteorological parameters for the period of balloon launchings. On 1 August, the point of sounding was on the western periphery of the filling cyclone, the near-surface centre of which was located over the Middle Volga region. Rylsk was located in the region of weak pressure gradients behind the slow cool front. At the moment of the deployment of the balloons the weather was almost cloudless (3/0, Ac, Ci), air temperature was 12.5°C, relative humidity 86%, and air pressure 985.9 hPa. During the flight, cirrus clouds grew to 9/0 due to the intermediate crest.

On 3 August, as on 1 August, the point of sounding was on the western periphery of the filling cyclone, which was stationary in the region of south-east of Moscow. The ridge located on 1 August over the territory of Poland and the former DDR became a surface anticyclone and moved to the south-east. The point of sounding was in the zone of weak pressure gradients.

Synoptically, 5 August could be considered a most favourable day. Compared with 3 August, the synoptic situation over the region of sounding changed a little. The cyclone existing at the surface on previous days became filled. The point of sounding was located on the south-eastern periphery of the anticyclone, pressure in its centre continued growing. There was clear, anticyclonic weather. The air temperature was 13.8°C, RH = 98%, and the pressure was 991.5 hPa.

Before launching a balloon, there was a radiative fog, which substantially reduced the atmospheric transparency both horizontally and vertically. During the flight the visibility varied between 6–11 km, the coefficient of total atmospheric transparency was 0.65–0.62. From radiosonde data, a surface inversion occurred. The tropopause was at an altitude of 10.6 km and its temperature was −55.0°C. During the flight the atmosphere was clear all the time.

On the whole, the aerosynoptic situation during the aerosol–radiation experiment was typical of the region of Rylsk in the summer period.

The second stage of the Soviet–American field experiment was carried out from 25 July to 22 August, 1976 in Laramie (Wyoming State). The region of the experiment was located at the western boundary of the Great Plains. The characteristic feature of this region is that it is protected from the impact of the Pacific Ocean by the Rocky Mountains . However, it is open in the north and south, and therefore the meridional air mass exchange takes place freely. Therefore, the weather in the region formed under the influence of air mass interaction both from the regions of Canada and from the Gulf of Mexico. There was a sufficiently warm summer even in the high-mountain part of the region. With the growing temperature contrasts and instability in late July–early August, thunderstorm activity sharply increased.

1.5.1 Instrumentation

The Soviet balloon measurements of aerosol at the first stage of the field experiments in 1975 were made with a 72-sample impactor and a one-time filter. The impactor was functioning during the whole ascent, the filter was continuously sampling aerosol from a certain altitude: in the first flight from the 150-hPa level, whilst in the second and third flights – from 200 hPa. At the second stage of the experiment the Soviet side used a similar impactor and a filter-trap with the 3-filter cassette which made it possible to sample aerosol from three layers of the atmosphere.

The balloon-borne impactor was one of the modifications of the instrument described earlier [40]. The impactor used in 1975 and 1976 differed from those used earlier in that its nozzle opening was round, not rectangular as in the earlier designs. With the rectangular nozzle impactor, the width of the dust path changed little with altitude, and the particle size distribution and their number density across the path were rather uniform [46]. In the version with the round nozzle opening and discrete operating regime, the area of the aerosol spot changed substantially with altitude and the particle distribution over the spot was not uniform. This created difficulties in processing the observational data.

The main instrument used successfully by the American side in balloon measurements of stratospheric aerosol at the first and second stages of the experiment was a dust sonde developed by the Wyoming State University [35, 38]. The sounded air is pumped at a rate of $0.75\,\mathrm{l\,min^{-1}}$ and passes the focus of the condenser lens and scattering volume $450\,\mathrm{cm^3}$, where the scattered light goes to the optical system. The optical system gathers the light scattered at an angle of 8–38° from the direction of forward scattering. At an angle of 25° efficiency is at a maximum. The light pulses go to the photomultipliers whose outputs are connected with a system of discriminators. Simultaneous pulses from both photomultipliers increases the signal/noise ratio. The sources of noise were mainly: (1) Reyleigh scattering from air molecules at low altitudes; (2) scattering from the walls of the scattering volume; and (3) oscillation of cosmic rays on the photomultiplier's glass at a high altitude. This background should be removed, since in the case of the finest particles (effective radius $>0.15\,\mu\mathrm{m}$), in the field of view of the instrument, only several photons are scattered.

1.5.2 Discussion of the results

1.5.2.1 The first stage of the experiment

From measurements with the balloon impactor for 1, 3, and 5 August, 1975, the vertical profiles of aerosol particle number density distinctly reveal the layered structure of aerosol at all altitudes. All three profiles are similar. Unfortunately, since the 3 August flight took place under conditions of well-developed cloudiness up to an altitude of 7–8 km, this led to the washing away of the impactor sample for practically the whole troposphere. The aerosol layers corresponded neither to aircraft data nor to data of the balloon dust sonde. Probably, there existed a systematic vertical shift between data of the dust sonde and the impactor caused by the specific character of sampling and processing of the data. However, the common characteristic features of the profiles from data of all types of measurements were preserved. On 1 August, the following altitudes in the troposphere with a maximum of aerosol number density could be detected: 0.2, 3.2, 4.8, 6.3, and 10 km. The optical microscope could not detect any aerosol particles in the lower stratosphere, because 'spots' of samples consisted of transparent particles, almost invisible on the substrate. In the stratosphere, there were two altitudes with an increased concentration of particles which were sufficiently contrasting and could be seen under the optical microscope: 18.6 and 22.6 km.

On 3 August, the altitudes of stratospheric layers were, respectively 20.3 and 24.3 km. On 5 August, from impactor data, the altitudes of aerosol layers were 0.5, 1.4, 2.35, 3.75, and 5.65 km. There was a layer of increased concentration of particles between altitudes of 10–12 km. In the stratosphere, mainly transparent particles were observed on this day, and above 19 km particles were practically absent.

A comparison of the aerosol vertical profiles obtained from impactor samples with the data of the filter sampler and the American dust sonde showed that the lower stratosphere contained particles undetected on the impactor substrate when seen under the optical microscope. One can assume that in this layer some particles are droplets of sulphutic or nitric acid. The electron microscope images of filter samples revealed a lot of homogeneous light spots with $r = 0.15$ and $0.24\,\mu m$, which probably formed with evaporation of liquid droplets on the substrate in a high vacuum. In higher atmospheric layers there were many particles with $r > 0.15\,\mu m$. Data on the size distribution for particles with $r > 0.2\,\mu m$ on 1 August reveal a maximum steepness of the decrease of particle concentration with growing size of particles. There were many particles that were dark-brown in colour and often of a regular spherical shape. The aerosol layers coincided vertically with the layers where the wind increased. These layers could be assumed to be particles of iron oxides. A qualitative agreement of the results of impactor measurements with those of measurements with the photoelectric counter could be considered satisfactory. One question remains unanswered – why at altitudes above 20 km did the impactor concentration of aerosol particles read much higher than the concentration measured with the photoelectric counter for the fraction with $r > 0.25\,\mu m$?

The results of the chemical analysis of filter samples for the content of various elements showed high values of mass concentrations of Mg and Fe. These data could not be taken as reliable, however, and probably they indicate a pollution of the filters with large-sized dust particles of soil origin. This pollution did not reflect upon the results of morphological and size-distribution analysis. In connection with the detection of a distinct layer of aerosol particles at altitudes 10–13 km a hypothesis was tested about the possible volcanic origin of the particles in this layer. One can suppose that the trajectories of particles emitted to the lower stratosphere by the volcano Tolbachik on 23 July, 1975 moved across Rylsk. The time taken for their appearance over Rylsk was 5–7 days at an average rate of particle motion of $100\,\mathrm{km\,hr^{-1}}$. Apparently, the Fuego eruption also affected the results. Therefore, the supposition of the volcanic origin of the aerosol layer observed at altitudes 10–13 km could be considered reliable.

1.5.2.2 The second stage of the experiment at the Wyoming State University

Measurements with the dust sonde were made in November, 1973. At the moment of launching one of the balloons carried two identical dust sondes. Layers at an altitude of 20 km were reliably detected. The altitudinal dependence of the particle size distribution can be demonstrated via the ratio of the number of particles with $r > 0.15\,\mu\mathrm{m}$ to the number of particles with $r > 0.25\,\mu\mathrm{m}$. For the particle size considered, this ratio is practically independent of the refraction index.

As a parameter for identification of long-term trends in the Junge layer, one can use a maximum ratio $N(r > 0.15)/N(r > 0.25)$. This value is not too sensitive to meteorological parameters and depends on the special features of the particles' source. Maximum values of the aerosol mixing ratio taken from some smoothed profiles, drawn as a function of time for all soundings in the summer of 1976 near Laramie, revealed a widely varying long-term trend. Apparently, they indicate the amplitude of natural zonal fluctuations in the Junge layer. However, values obtained in late 1971–early 1972 did not repeat during the subsequent two years, reflecting a short-term increase of aerosol concentration in the lower stratosphere. The decreasing trend of the mixing ratio observed in early 1973 characterized the intensity of the total aerosol content decrease. The sudden increase of the aerosol content in late 1974 was caused by the eruption of the Fuego volcano in Guatemala.

The cyclonic activity during the aerosol–radiation experiment led to a lowering of the tropopause altitude. Whilst on 31 August, 1976 it was at the 140-hPa level, on subsequent days the tropopause became layered and dropped to the 180-hPa level. This fact is especially interesting since the altitude of the tropopause affects substantially the vertical profile of aerosol concentration in the lower stratosphere and upper troposphere. The aerosol layers in this region of the atmosphere are partly determined by exchange processes in the zone of the tropopause and partly caused by volcanic activity [51, 55]. Also, a similarity of the vertical profiles of aerosol and ozone mixing ratio is observed and a decrease of air temperature lapse rate. This affects the long-wave radiation transfer in the atmosphere.

Determination of the contribution of any factor to the rate of the radiative temperature change would be much easier from data of spectral measurements. With actinometric (total) sounding it is especially important to use the results of simultaneous aerological measurements, data on the content of water vapour, aerosol, ozone, and other optically active gas components of the atmosphete (carbon dioxide, nitrogen oxides, etc.). Use of these data and results of numerical simulations, as applied to conditions of the experiments, enables one to estimate the effect of aerosols and other factors on the long-wave radiation transfer. We confine ourselves to a qualitative consideration of the data of aerosol–radiation measurements for 31 August, 1976. The rate of the radiative temperature change $(\partial T/\partial t)_R$ is expressed in the following way [58]:

$$\left(\frac{\partial T}{\partial t}\right)_R = -\frac{q}{c_p}\frac{\partial F}{\partial z} = \frac{\sigma_q}{c_p}\left[4\int_{T_z}^{T_\infty} \gamma T^3 dT - T_\infty^4 \gamma(u_\infty) + 4\int_{T_z}^{T_\theta} \gamma T^3 dT\right] \quad (1.3)$$

where T_z and T_∞ are temperatures at the altitude z and at the level above which the water vapour can be neglected; γ is the gradient of the vertical change of emissivity; σ is the Stefan-Boltzman constant; q is the specific humidity; c_p is the specific heat at a constant pressure; and u is the optical thickness.

Measurements of the vertical profiles of air temperature and fluxes of downward $F_\lambda\downarrow$, upward $F_\lambda\uparrow$, and net radiation $B_\lambda = (F_\lambda\uparrow - F_\lambda\downarrow)$ in the main part of the atmosphere revealed an increase of the net flux of thermal radiation with altitude and hence a radiative cooling. Only in the region of inversion and subinversion layers, in the region of tropopause where according to data from the American aerosol sonde the aerosol layers were located, a radiation warming was observed. At the 400–520-hPa level an aerosol layer was recorded with a concentration of particles $>0.15\,\mu m$ of about $3\,cm^{-3}$. Here, near the bottom of the temperature inversion, a radiative heating of $0.43°C\,hr^{-1}$ was observed and then a strong radiative cooling of $1.13°C\,hr^{-1}$. In the tropopause region, at the 70–80-hPa level, the concentration of aerosol particles reached $1.8\,cm^{-3}$. A sharp temperature increase at these altitudes, due to long-wave radiation absorption by ozone, caused a radiative heating of $0.27°C\,hr^{-1}$ in the lower part of the aerosol layer considered and to a radiative cooling of $0.06°C\,hr^{-1}$ in its upper part. Thus, the aerosol layer in the region of the temperature inversion and, in particular, in the zone of the tropopause can be compared with a cloud which, like a 'heat hole', has radiatively active boundary layers [56]. In the middle of this layer, the radiative change of temperature is close to zero, whereas the upper boundary corresponds to a heat sink, and the lower boundary – to the source of heating.

1.5.3 The vertical profiles of long-wave radiation fluxes measured with the ARS radiosondes

1.5.3.1 The first stage of the experiment

The Soviet–American aerosol–radiation experiment has opened up wide possibilities for comparing data of synchronous measurements of such parameters as concentration, size distribution, and chemical composition of aerosol particles as well as radiative characteristics of the atmosphere.

To study the radiative effects of the aerosol component of the atmosphere, launchings of actinometric radiosondes ARS-1 were included in the programme of the experiment – made at night before launchings of the USSR and USA aerosol instrumentation. Three launchings were made over Rylsk. Let us consider the results obtained.

The first launching was made on 1 August, 1975 at 02:00 Moscow time. The sky at the moment of the ARS launching was almost cloud-free, with only cirrus clouds (5–6 cloud amount) – a layer of radiative temperature inversion formed near the surface. In this layer, net radiation decreased from 0.070 to 0.021 $\text{cal cm}^{-2}\,\text{min}^{-1}$. Above the layer of inversion, net radiation F mainly increased. Since in this case the most interesting fact was the radiative effect of aerosol layers, we will not dwell upon the characteristics of the long-wave radiation budget constituents (i.e., on values of the upward and downward long-wave radiation fluxes). Note only that in all three cases, reliable data from measurements of radiation fluxes, temperature, humidity, and wind were obtained. Unfortunately, during radio-control of the first launching, at the 14.76 km altitude, the radiosonde transmitter malfunctioned. Therefore, here the radiation fluxes were estimated only up to this altitude. In all other cases the altitudes of the ceiling of sounding were 27.5 and 26.8 km.

In careful studies of the layers with different heating or cooling an interesting phenomenon was observed – the layers with heating or cooling near any altitude remained for several days. For instance, heatings in the 19.5-km layer and coolings in the 21.8–22.8-km layer, etc. In general, by the character of the vertical distribution of F a maximum of tropospheric dust load should be expected in the first and second cases and a minimum in the third case. These suppositions were confirmed by *in situ* aerosol measurements.

Most interesting is the complex analysis of results with the use of measurements of aerosol concentrations performed with aerosol instruments from both countries. Note one interesting fact – the data of the first launching of ARS were compared with the results of measurements with the USA dust radiosonde. This comparison made it possible to detect a good qualitative agreement between measurements of the vertical profiles of F and aerosol concentration. From data of the USA sonde, at an altitude of about 5 km, a small layer with an increased concentration of dust particles was observed. In this layer, the ARS 'sensed' a change in net radiation with the radiation flux divergence approaching zero, and a cooling at the level of the layer top was equal to $-0.1^{\circ}\text{C hr}^{-1}$ compared with $-0.01^{\circ}\text{C hr}^{-1}$ within the layer itself. A good qualitative agreement was detected near the tropopause.

1.5.3.2 The second stage of the experiment

The second stage of the Soviet–American aerosol–radiation experiment near Laramie foresaw nocturnal surface actinometric measurements of long-wave radiation fluxes before each sounding of the atmosphere. Measurements were made two hours after sunset and two hours before sunrise, to exclude the effect of short-wave radiation on measurement results. The actinometric soundings were aimed at:

(1) obtaining the vertical profiles of long-wave radiation fluxes, aerological parameters, and studies of their variability in time during the accomplishment of the experiment; and
(2) analysis of a possible impact of aerosol and optically active gas components on the rate of radiative temperature change using data from respective measurements of aerosol, condensation nuclei, and ozone.

The launchings of ARS were made in clear skies (2–3 cloud amount) with a light breeze. The results are similar to the measurements performed in the USA, but specific features of the long-wave radiation transfer mentioned above manifested themselves more clearly, namely, near the tropopause, in the zone of temperature inversion, the aerosol layer was similar to cloud by its absorptive and radiative properties.

From data of actinometric radiosoundings, the presence of thin layers of heating and cooling in the atmosphere in the absence of clouds was detected long ago. Results of this experiment suggest a supposition that the thin vertical structure of radiative properties of the atmosphere is determined by the presence in it of thin layers with an increased concentration of aerosol. A comparison of the aerosol vertical profiles obtained during the respective time periods shows that the aerosol layers cause a change in the sign of the radiative temperature variation. The latter fact can serve an indirect indicator to detect aerosols from data of actinometic sounding. The results obtained make it possible to conclude that substantial impact of aerosol layers on the longwave radiation transfer takes place [78].

1.6 AEROSOL STUDIES IN THE 'BERING' EXPERIMENT

The main goal of studies within the Soviet–American field experiment 'Bering' was to solve a number of problems of microwave remote sensing of the zones of precipitation, ice cover characteristics, determination of temperature, and state of the ocean surface with the use of instruments carried by aircraft and ships. Microwave remote sensing was accompanied by *in situ* measurements of meteorological and oceanological characteristics. The expedition was carried out from 15 February to 7 March, 1973 in the neutral waters of the Bering Sea. The flying laboratory IL-18, apart from microwave instruments, carried a complex of aerosol equipment and the K-3 spectrometer operating within the CAENEX programme.

Of the data obtained in the 'Bering' expedition, most interesting are the results of spectral measurements of short-wave radiative flux divergence in the presence of stratus clouds, confirmed later with similar measurements in the Arctic [6, 60] and by the results of *in situ* measurements of the vertical structure of aerosol over the sea surface [2, 37].

Especially distinct was the dependence of the morphological structure of particles on the altitude over the sea surface and ice concentration. In the morphological classification of aerosol, with the help of electron microscope analysis, it was established that tropospheric particles had a sufficiently dense structure and, mainly, a spherical or oval shape. Detailed studies of particle size distribution have led to the conclusion that one of the most important factors determining the morphological structure of particles was the level of relative humidity.

With the help of the optical microscope one could often observe the watering of some particles under conditions of comparatively low humidity ($f = 50$–60%). Under the electron microscope such particles are seen as spheres with a central nucleus or nuclei and a low density of matter on the periphery. Such particles, of course, are not of condensation but condensation–coagulation origin (i.e., are loose conglomerates of small particles formed by the coagulation of particles and filled with water or other liquid (a capillary effect). In the case of impactor capture of large particles, and their splitting, there was no watering of fragments even at large values of relative humidity ($f = 70\%$). This favours the initial coagulation production of large loose conglomerates. Filling of microcapillaries with water in such particles can take place at relative humidity $> 20\%$. The strength of water surface tension makes a particle's matter more dense. Therefore, an increased density of solid matter is observed at the particle–air interface. In transition of a particle to drier layers of the atmosphere a 'dry' coagulation occurs resulting in a looser layer. The electron microscope analysis of aerosol samples enabled one to detect particles repeatedly moving from layers with low air humidity to layers with high air humidity and back. Such particles have a multilayered structure and are observed, mainly, in the upper troposphere [37].

There is a direct dependence of the morphological structure and size of aerosol particles formed in the atmosphere as a result of reactions of sulphur-containing compounds with water vapour: at a low content of water vapour, smaller spherical particles form, and in the presence of metal-containing substances even crystal particles form, whereas with a high content of water vapour the transformation of liquid-droplet particles of large sizes takes place. In the first case, small particles coagulate creating loose aggregates of the fractal structure. Similar formations are observed in the case of condensation or sublimation of the products of burning and subsequent rapid coagulation of the formed particles at $N \geq 10^8\,\mathrm{cm}^{-3}$.

Fractal aggregates are specifically organized structures in which each selected element is similar to the system as a whole. As applied to the problem of atmospheric aerosols, fractals exist in the form of aggregates formed by numerous (10^2–10^3) particles with similar physico-chemical properties whose size is much smaller compared with the size of the system and differs little between individual (initial) particles – their mutual location within an aggregate is described by general statistical laws. Concrete values of the fractal structure parameters depend on the mechanism of aggregates formation. For such formations, the relationship $R = aN^{1/D}$ is valid where R is the size of a cluster. The range of possible D variations is limited. For instance, for the volume clusters $1 \leq D \leq 3$, the D value for smoke aggregates averages 1.78 and reveals an increasing trend due to an increased humidity. The size of initial particles depending on the source of smoke (as well as on temperature and other parameters) constitutes $a = 0.01$–$0.05\,\mu\mathrm{m}$, and the size of aggregates varies from several tenths to about $10\,\mu\mathrm{m}$. Bearing in mind the loose structure of such particles, the process of gravitational sedimentation for them is not very efficient. They are removed from the atmosphere mainly as a result of washing out or are driven by downward air fluxes.

The electron microscope analysis of aerosol samples taken in the atmosphere at different altitudes and under different conditions of the formation and evolution of aerosol confirmed the conclusions about the formation in the atmosphere of fractal structures in the process of their coagulation growth.

Table 1.2. The share of different shapes of particles in samples taken over the Bering Sea (%).

Number of samples	H (m)	Type of particles							
		1	2	3	4	5	6	7	8
4	1,000	18	65	–	8.0	–	7.5	–	1.5
10	4,000	41	30	2.0	2.5	10	3.7	7.0	3.8
3	9,000	12	40	–	15	9.0	13	7.0	4.0

To solve the problems of aerosol optics, a morphological classification of the types of atmospheric aerosols has been developed. With the observational data taken into account, eight morphological types of particles were proposed: 1 – dense spheres; 2 – loose spheres; 3 – particles with the shell of very small particles (with a 'fur-coat'); 4 – dense non-spherical (fragments of rocks and other objects); 5 – loose non-spherical (agglomerates); 6 – chain structures, including fractals; 7 – loose with dense nuclei (e.g., sulphate particles of the Junge layer); and 8 – crystals and particles with a dry shell [38].

Relative shares of particles of different types vary widely in the real atmosphere, depending both on the altitude of sampling and on the range of the sizes of the observed particles. For samples taken over the Bering Sea in March, 1973 the results obtained are demonstrated in Table 1.2.

With decreasing altitude the share of spherical particles grows due to increasing relative humidity (at altitudes 1 km – 83%, 4 km – 71%, 9 km – 52%). A maximum share of dense spheres (41%) is observed at an altitude of 4 km, where 'older' particles reside. The fractal structure and agglomerates are observed mainly at an altitude of 9 km, and the 'fragments' in the lower layer and at 9 km. The presence of fractals at an altitude of 1 km indicates a relatively recent formation of clusters. Spherical particles have the size $<0.2\,\mu m$, whereas the dust non-spherical particles have relative maxima of size distributions in different size ranges. Most of the particles of type 8 (crystals, tetrahedrons, plates, etc.), whilst modelling the optical characteristics of aerosols, can be considered as equivalent spheres, bearing in mind the chaotic orientation of such particles in a sufficiently turbulent medium.

Analysis of the elemental composition of aerosol over the Bering Sea revealed a marked dependence of the composition of particles on the prevailing direction of the motion of air masses: at altitudes above 1,000 m particles, mainly of volcanic origin, were observed in the cases of latitudinal winds and a sharp decrease of aerosol concentration took place in the case of longitudinal winds. At altitudes below 1,000 m the elemental composition of particles corresponded more to particles of marine origin. However, in all cases, a relatively high content of sulphate was observed in particles [37].

1.7 DUST EXPERIMENTS IN TADJIKISTAN

The aim of the expedition was to study: the conditions and mechanisms for dust formation in the desert; chemical and physical characteristics of dust; as well as to evaluate the effect of arid aerosol on local meteorological conditions and climate [41]. In planning and accomplishment of the field experiment the possibility was taken into account to evaluate the dynamics of the chain 'dust generation–transport–deposition in the Aral Sea basin'. The main goal of the experiment was to obtain information, as complete as possible, on the physical characteristics of dust, including its optical and radiative characteristics and chemical composition [29].

In September, 1989, in the territory of Tadjikistan, in accordance with the Soviet–American agreement on environmental protection, a complex field experiment was carried out to study dust lifting, and the chemical, morphological, and optical properties of dust aerosol and its impact on local meteorological conditions and the climate. Instruments were installed at several stations on 5 September and measurements were made from 6–27 September, 1989.

The necessity to study the aerosol of the soil erosion origin is explained by the fact that this type of aerosol is considered to be a very important aerosol component of the atmosphere and constitutes, from different estimates, between 50–80% of the total (by mass) amount of aerosol. Deserts covering about 30% of land are the main source of aerosol [17].

The USSR had gained considerable experience in accomplishing field experiments studying dust aerosol and its impact on radiation fluxes. They were carried out mainly in the Kara-Kum desert at the base of the Repetek sand-desert reserve (Turkmenistan). In the first experiment – the CRE – accomplished in 1970, important data were obtained on the profiles of solar and long-wave radiation fluxes [3, 49, 67]. More detailed studies of the optical characteristics of the dust loaded atmosphere were carried out within GAREX [1, 51, 52].

Great efforts to study the properties of dust aerosol of the Sahara desert were made by the scientists of the USA, France, and Germany [7, 9, 12, 13, 16, 17, 19, 22–27]. As a result, important observational data were obtained on chemical composition, optical, and radiative properties of dust aerosol. However, for correct consideration of arid aerosol in the problems of modelling the atmospheric general circulation and climate as well as regional weather, data were needed on the properties of arid aerosol from different geographical regions.

In this connection, one of the main problems of the 1989 Soviet–American field experiments consisted of establishing characteristic features of chemical and physical properties of dust in the zone of deserts in the south of Tadjikistan [29, 30, 41] approximately 500 km east of Repetek. The emphasis was laid on observational and theoretical studies of desert aerosol outbreaks to quantitatively describe the wind-driven dust transport from the source to the sink. For the first time, for a given region, conditions were studied of the lift-off of dust particles from different types of surface and their further transport. The chemical composition of aerosol was also studied as well as microphysical (particle shape and size distribution), optical (spectral change of the complex refraction index), and radiative (spectral transmission for solar and thermal radiation) characteristics; records were also made of meteorological conditions (pressure, temperature, wind, humidity, and visibility), their connection with atmospheric turbidity, and conditions for the formation of dust outbreaks. The ultimate aim was to determine changes in the thermal and radiative regimes of the surface and the atmosphere depending on the amount, vertical distribution, and properties of aerosols. The observational data obtained in combination with the earlier results made it possible to get an adequate characterization of the properties of arid aerosol over a large territory of central Asia.

The Soviet–American Dust Experiment (SADEX) included a complex of surface and aircraft observations. Surface observations were carried out from 5–26 September, 1989 at three observational sites. Site 1 was located in a semidesert region not far from the settlement Shaartus, 15 km from the border with Afghanistan, at an altitude of 380 m above sea level. Site 2 was located at the meteorological station in Esanbai (Tadjikistan). This station was located at an altitude of 564 m above sea level in the Kafirnigan River valley surrounded by hills up to 2 km high. Site 3 was located on the outskirts of Dushanbe, in the territory of the Physicotechnical Institute.

The sources of dust can be in this case the desert and semidesert sites near Shaartus as well as the desert regions of northern Iran and Afghanistan (the local people call the dust-driving wind 'afghanets'). Site 2 does not have local sources of dust and is located along the most probable trajectory of dust transport from the source to Dushanbe, since in this season, southern winds prevail, blowing up the valley of the Kafirnigan River. Aircraft measurements were made from the flying laboratory of the Institute of Atmospheric Optics of the Siberian Branch of the USSR Academy of Sciences in the zone of dust transport in the valley of the Kafirnigan River between observational sites 1 and 3. Measurements were made of the coefficients of aerosol attenuation and backscattering (lidar), particle size distribution, and meteorological parameters of the atmosphere in the altitude interval from 100 m to 6 km. Additional measurements of aerosol were also made at the meteorological station Altyn-Mazar located in the valley of the Vakhsh River at an altitude of 2,746 m. Separate measurements for background monitoring were made at the meteorological station at the Fedchenko glacier at an altitude of 4,100 m.

September was chosen because during this month dust storms or dust haze are most probable. On 16 September, at all three points there was observed a marked dust haze, and on 20–21 September a dust storm struck, the strongest in the 1980s according to estimates of local weather forecasters. At each point, measurements were made of chemical, microphysical, optical, and radiative characteristics of arid aerosols for different conditions: clear atmosphere, dust haze, and dust storm.

The observational data showed that by their mineralogical composition, arid aerosols are close to sand and consist (by mass) of 65% of quartz, 18% of field spar, 9% of apatite, 3% of organics, and 5% of other minerals. The particles volume distribution reveals approximately log-normal dependences of which three modes can be detected with efficient radii at about 0.5, 1, and 3 μm.

The complex refraction index (χ) was measured by two groups of specialists. From the data of both groups, the real part of the refraction index was 1.5. At the 600-nm wavelength, the imaginary part χ, from the American data, varied from 0.001–0.003; from the Soviet data, $\chi > 0.003$ and depended weakly on wavelength in the visible. A simple model was developed describing the transport and deposition of dust from the source at the southern boundary of the region considered, across the valley of the Kafirnigan River to Dushanbe. This model gave results comparable with data of *in situ* measurements [84].

To assess the role of dust in the radiative and thermal regimes of the atmosphere and the surface, it is important to know the ratio between optical thicknesses τ in the visible and thermal spectral intervals. From the observational data, the ratio $\tau(0.55\,\mu m)/\tau(10.2\,\mu m)$ varied from 1.4 to 3.2, with most probable values of 2.0–2.5. Marked decreases of diurnal temperatures were recorded, especially after 20 September.

The frequency of occurrence of dust storms in the south of Tadjikistan, north of Amu-Darya River, in the valley of the Kafirnigan River, is from 10–30 dust storms annually, the peak season being summer and early autumn. In the summer, winds blow mainly along river valleys (from the south). For the field experiment, the Kafirnigan valley was chosen along the route from Amu-Darya (the Afghanistan border) to Dushanbe. In this region the width of the river valley is 25 km, and in the vicinity of Dushanbe about 5 km. The desert region in which site 1 is located is the southern part of the valley with an area of about 700 km^2. The meteorological data presented by the Tajik Hydrometeorological Service was used to evaluate the parameters of the dust flow model when estimating the dust flows observed on the outskirts of Dushanbe.

The dust experiment in the desert zone of the Kafirnigan valley made it possible to study the causes of the appearance of the desert dust aerosol and consequences of its impact on climate. The first direction of studies was connected with the problems of revealing mechanisms of dust formation in the region of dust source, which supposes a study of biological, physical, and chemical processes [78, 82]. The second problem was to determine the mechanism of the long-range transport of dust and to assess dust flows from the source, fluxes to the surface, and intensity of dust wind driving. The third and very important problem was to describe the physical and optical properties of dust. The related main goal of the experiment was to assess the climatic impact of the desert dust aerosols. The fourth aim was to describe the chemical composition of dust, its transformation and interaction with other aerosol components and gases, in particular, the products of anthropogenic pollution. Finally, the fifth aim was to determine the large-scale structure of dust storms with the use of the dust storm models.

As has been mentioned, experiments under desert storm conditions were made from 16 to 20 September, 1989. The 16 September dust storm was much weaker than the storm on 20 September. The wind speed during the latter storm exceeded $20\,m\,s^{-1}$ in the desert (Shaartus station) and $7\,m\,s^{-1}$ in the valley (Esanbai station,

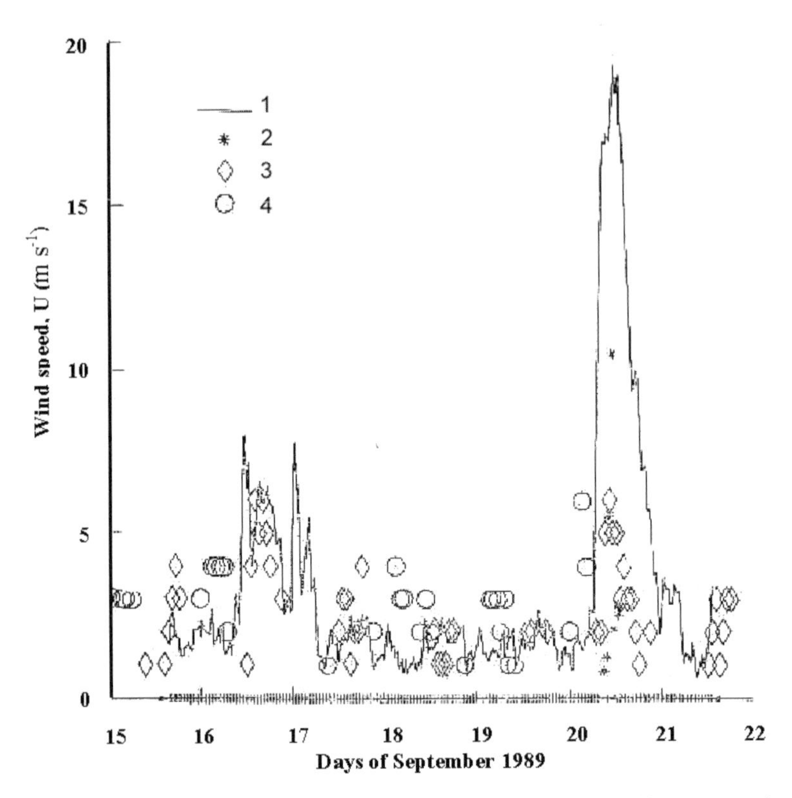

Figure 1.8. The temporal change of wind speed at a height of 3.3 m at Shaartus station (1), at a height of 10.6 m at Esanbai station (2), and relative motion of soil in southern (3) and northern (4) wind directions.

Figure 1.8). Station 1 was a weak source of dust during the 58-minute period on 29 September. In both dust episodes the station Shaartus was, on the whole, a dust sink even when during a short interval on 20 September it was, at the same time, a weak source. The experiment was successful from the viewpoint of obtaining data from all five directions of studies:

(1) study of mechanisms of dust formation in the region of the source;
(2) study of the laws of dust motion;
(3) description of physical and radiative characteristics of arid aerosols;
(4) description of transformations of the physico-chemical composition of dust aerosols; and
(5) determination of the large-scale structure of storms.

1.8 RESULTS OF COMPLEX AEROSOL–RADIATION STUDIES

Results of complex aerosol–radiation field experiments in the 1960s–1980s drastically changed the ideas about the role of aerosol in atmospheric processes. It was found that aerosol plays the most important role in the atmospheric energy transformation processes, especially from the viewpoint of the transfer of short-wave solar radiation. Aerosol contribution was detected and evaluated in the thermal radiation transport, especially in the wavelength interval 8–12 μm. The basic mechanisms of the formation of aerosols with different optical properties were studied in detail. Several field experiments were accomplished to study aerosols of soil (arid, in particular) origin. Experiments were carried out to evaluate the content and size distribution of background aerosol (an experiment at Ambastumani, Georgia), to study the stratospheric aerosol [1, 35, 63]. An analysis was made of impactor samples [18, 38, 40, 76, 92], balloon spectral observations of solar radiation attenuation in the atmosphere were made [57, 58], as well as complex studies of atmospheric pollution (including clouds) with industrial emissions [54, 60–62, 65, 67] and with products of volcanic eruptions [32, 38, 53]. A special direction of study was connected with the atmosphere and atmospheric processes over sea surfaces [2, 23, 65, 69].

It is generally acknowledged that as a radiation factor, clouds play the leading role in climatic processes, and the role of aerosol in cloud formation raises no doubts. In this sense, it can be more important than the direct impact of aerosol on the short-wave and thermal radiation transfer. Great efforts by scientific groups were directed at modelling the optical properties and characteristics of clouds as well as atmospheric aerosols [33, 34, 39, 61, 71, 74]. It has become clear that simulation of the optical characteristics from the data on size distribution and chemical composition of atmospheric aerosols is more useful than an accumulation of statistical data by *in situ* measurements. Accomplishment of such studies required new instruments and techniques. In these studies, spaceborne observational means were widely used [3, 47, 55, 73]. In this connection, of great importance are programmes of subsatellite field experiments enabling one to use the results of other complex, local studies on much larger spatial–temporal scales [14, 23, 67].

1.9 BIBLIOGRAPHY

1. Ackerman S. A. and Cox S. K. Aircraft observations of the shortwave fractional absorptance of non-homogeneous clouds. *J. Appl. Meteorol.*, 1981, **20**, 1510–1515.
2. Ackerman S. A. and Stephens G. L. The absorption of solar radiation by cloud droplets: An application of anomalous diffraction theory. *J. Atmos. Sci.*, 1987, **44**, 1574–1588.
3. *Atmospheric Aerosol and Its Impact on Radiation Transfer. On the Results of the Soviet–American Aerosol Experiment.* K. Ya. Kondratyev (ed.). Gidrometeoizdat, Leningrad, 1978, 120 pp. [in Russian].
4. Berlyand M. E. and Kondratyev K. Ya. *Cities and Climate of the Planet.* Gidrometeoizdat, Leningrad, 1972, 68 pp. [in Russian].
5. Berlyand M. E., Kondratyev K. Ya., and Vasilyev O. B. A complex study of the features of the meteorological regime of a large city with Zaporozhye as an example (CAENEX-72). *Meteorology and Hydrology*, 1974, **1**, 14–23 [in Russian].
6. Binenko V. I. and Kondratyev K. Ya. Spectral albedo of stratified cloudiness in the wavelength interval 0.35–0.95 μm. *Proc. of the Main Geophysical Observatory* (St Petersburg), 1973, **322**, 68–76 [in Russian].

7. Blifford J. H. (ed.) *Particulate Models: Their Validity and Application*. NCAR-TN Proc. 1968. Boulder, Fort Collins, Colorado, 184 pp.
8. Budyko M. I. *Climate and Impact on the Atmospheric Aerosol Layer*. Gidrometeoizdat, Leningrad, 1974, 42 pp. [in Russian].
9. Carisen T. N. and Benjamin S. G. Radiative heating rates for Saharan dust. *J. Atmos. Sci.*, 1980, **37**(1), 193–213.
10. Chapursky L. I., Chernenko A. P., and Andreeva N. I. Spectral radiative characteristics of the atmosphere during a dust storm. *Proc. of the Main Geophysical Observatory* (St Petersburg), 1975, **366**, 77–84 [in Russian].
11. Chapursky L. I. and Chernenko A. P. Spectral radiarion fluxes and flux divergences in the cloud-free atmosphere of the sea in the interval 0.4–2.5 μm. *Proc. of the Main Geophysical Observatory* (St Petersburg), 1975, **366**, 23–35 [in Russian].
12. Charlson R. J., Ahlquist N. C., and Selvidge H. Monitoring of atmospheric aerosol parameters with the integrating nephelometer. *J. Air Pollut. Contr. Assoc.*, 1969, **19**, 937–943.
13. Charney J. G. Dynamics of deserts and drought in the Sahel (Symon memorial lecture). *Quarterly J. Roy. Meteorol. Soc.*, 1975, **101**(428), 193–202.
14. Complex Energetics Experiment (Results of studies, 1970–1972). K. Ya. Kondratyev and L. P. Orlenko (eds). *Proc. of the Main Geophysical Observatory (Leningrad)*. Leningrad, Gidrometeoizdat, 1973, 84 pp. [in Russian].
15. *Complete Radiation Experiment*. K. Ya. Kondratyev and N. E. Ter-Markaryants (eds). Gidrometeoizdat, Leningrad, 1976, 240 pp. [in Russian].
16. Cox S. K.,Vonder Haar T. H., Hanson K. J., and Suomi V. E. Measurements of absorbed shortwave energy in a tropical atmosphere. *Solar Energy*, 1974, **14**(2), 169–174.
17. D'Almeida G. A. On the variability of desert aerosol radiative characteristics. *J. Geophys. Res.*, 1987, **92**(3), 3017–3026.
18. Dmokhovsky V. I., Ivlev L. S., and Ivanov V. N. Aircraft measurements of the vertical structure of atmospheric aerosol under the CAENEX programme. *Proc. of the Main Geophysical Observatory* (St Petersburg), 1972, **276**, 103–108 [in Russian].
19. Ellsaesser H. W., Mac Cracken M. C., Potter G. L., and Luther F. M. An additional model test of positive feedback from high desert albedo. *Quarterly J. Roy. Meteorol. Soc.*, 1976, **102**(2), 330–337.
20. Feigelson E. M. *Radiative Heat Exchange and Clouds*. Gidrometeoizdat, Leningrad, 1970, 230 pp. [in Russian].
21. *First Global GARP Experiment (Vol. 2): Polar Aerosol, Extended Cloudiness, and Radiation*. K. Ya. Kondratyev and V. I. Binenko (eds). Gidrometeoizdat, Leningrad, 1981, 89–91 [in Russian].
22. Fouquart Y. and Bonnel B. Observations of Saharan aerosols: Results of ECLATS field experiment/ Parts I and II. *J. Climate and Appl. Meteorol.*, 1987, **26**(1), 28–50.
23. Gille J. C. *A Program of Radiation Measurements during TROPEX*. NCAR, Boulder, Colorado, 1970, 48 pp.
24. Gillette D. A. On the production of soil wind erosion aerosols having the potential for long range transport. *J. Tech. Atmos.*, 1974, **8**, 735–744.
25. Gillette D. A. A model of transport deposition of desert dust in the Kafirnigan River Valley, from Shaartus to Dushanbe. In: K. Ya. Kondratyev (ed.), *The Soviet-American Experiment on Studies of Arid Aerosol*, 1992, pp. 17–24 [in Russian].
26. Gillette D. A. and Goodwin P. A. Microscale transport of sand-sized soil aggregates eroded by wind. *J. Geophys. Res.*, 1974, **79**, 4080–4084.

27. Gillette D. A. and Stockton P. The effect of non-erodible particles on wind erosion of credible surfaces. *J. Geophys. Res.*, 1989, **94**(12), 885–893.
28. *Global Atmospheric Aerosol Radiation Study (GAARS) Research Plan*. NCAR, Washington, 1973, 56 pp.
29. Golitsyn G. S. Complex Soviet–American dust experiment. In: K. Ya. Kondratyev (ed.), *The Soviet-American Experiment on Studies of Arid Aerosol*, St Petersburg, 1992, pp. 3–6 [in Russian].
30. Golitsyn G. S. and Shukurov A. Kh. Temperature effects of dust aerosol with dust storms in Tadjikistan as an example. *Proc. USSR Acad. Sci.*, 1987, **297**(6), 1334–1337 [in Russian].
31. Golovina E. G., Ivlev L. S., and Solomatin V. K. On the effect of humidity on the industrial aerosol structure. *Proc. of the Main Geophysical Observatory* (St Petersburg), 1974, **332**, 104–109 [in Russian].
32. Grigoryeva A. S. and Drozdov O. A. On the effect of volcanic eruptions on the NH precipitation. *Proc. of the Main Geophysical Observatory* (St Petersburg), 1975, **354**, 102–108 [in Russian].
33. Grishechkin V. S. and Melnikova I. N. Study of radiative flux divergences in stratus clouds in the Arctic regions. In: L. S. Ivlev (ed.), *Rational Use of Natural Resources*. Leningrad Polytechnic Institute Publ., Leningrad, 1989, pp. 60–67 [in Russian].
34. Grishechkin V. S., Shults E. O., and Melnikova I. N. Analysis of spectral radiative characteristics. *Problems of Atmospheric Physics*, LSU Publ., 1989, **19**, 32–42 [in Russian].
35. Hofmann D. J., Rosen J. M., and Pepin T. J. Global monitoring of stratospheric aerosol, ozone and water vapor. *Progress Report Contrib.* No. 00014-70-A-0266-0005. Univ. of Wyoming, June 1972, 32 pp.
36. Ivlev L. S., Odintsov O. A., and Popova S. I. Optical properties of the atmospheric aerosol. *Proc. Intern. Radiation Symposium, Sendai, Japan, 1972*, pp. 141–144.
37. Ivlev L. S., Dmokhovsky V. I., Ivanov V. A., and Solomatin V. K. Aerosol studies in the 'Bering' expedition. *Proc. of the Main Geophysical Observatory* (St Petersburg), 1975, **363**, 37–43 [in Russian].
38. Ivlev L. S. *Chemical Composition and Structure of Atmospheric Aerosols*. LSU Publ., Leningrad, 1982, 366 pp. [in Russian]
39. Ivlev L. S. and Andreev S. D. *Optical Properties of Atmospheric Aerosols*. LSU Publ., Leningrad, 1986, 359 pp. [in Russian].
40. Ivlev L. S. and Dovgaliuk Yu. A. *Physics of Atmospheric Aerosol Systems*. St Petersburg State University Publ., St Petersburg, 2000, 258 pp. [in Russian].
41. *Joint Soviet–American Experiment on Arid Aerosol*. G. S. Golitsyn, D. A. Gillette, T. Johnson (eds). Hydromet., St Petersburg, 1993, 220 pp.
42. Kondratyev K. Ya. *Radiation in the Atmosphere*. Academic Press, New York, 1969, 912 pp.
43. Kondratyev K. Ya. *Radiation in the Atmosphere*. WMO Monograph No. 309, Geneva, 1972, 214 pp.
44. Kondratyev K. Ya. *The Complete Atmospheric Energetics Experiment*. GARP Publ. Ser., 1972, No. 12, WMO, Geneva, 48 pp.
45. Kondratyev K. Ya. (ed.) *Studies of Natural Environment from Manned Orbital Stations*. Gidrometeoizdat, Leningrad, 1972, 400 pp. [in Russian].
46. Kondratyev K. Ya. *The Complex Energetics Experiment (CAENEX)*. Obninsk Information Centre, Obninsk, 1973, 79 pp. [in Russian].
47. Kondratyev K. Ya. Aerosol and climate. *Proc. of the Main Geophysical Observatory* (St Petersburg), Gidrometeoizdat, Leningrad, 1976, **381**, 3–84 [in Russian].

48. Kondratyev K. Ya. *New Results in Climate Theory*. Gidrometeoizdat, Leningrad, 1976, 64 pp. [in Russian].
49. Kondratyev K. Ya. *The Complete Radiation Experiment*. Gidrometeoizdat, Leningrad, 1976, 239 pp. [in Russian].
50. Kondratyev K. Ya., Barteneva O. D., Vasilyev O. B., Zhvalev V. F., Ivlev L. S., *et al.* Aerosol in the GATE region and its radiative properties. *Proc. of the Main Geophysical Observatory* (St Petersburg), 1976, **381**, 61–130 [in Russian].
51. Kondratyev K. Ya. *The Present Climate Changes and Their Determining Factors*. Progress in Sci. and Technol., Meteorology and Climatology, Moscow, ARISTI, 1977, Volume 4, 203 pp. [in Russian].
52. Kondratyev K. Ya. *Radiative Factors of Present Global Climate Changes*. Gidrometeoizdat, Leningrad, 1980, 280 pp. [in Russian].
53. Kondratyev K. Ya. *Stratosphere and Climate*. Progress in Sci. and Technol., Meteorology and Climatology, Moscow, ARISTI, 1981, Volume 6, 223 pp. [in Russian].
54. Kondratyev K. Ya. *The Earth's Radiation Budget, Aerosol, and Clouds*. Progress in Sci. and Technol., Meteorology and Climatology, Moscow, ARISTI, 1983, Volume 10, 315 pp. [in Russian].
55. Kondratyev K. Ya. *Satellite Climatology*. Gidrometeoizdat, Leningrad, 1983, 264 pp. [in Russian].
56. Kondratyev K. Ya. *Climatic Effects of Aerosols and Clouds*. Springer–Praxis, Chichester, UK, 1999, 264 pp.
57. Kondratyev K. Ya. and Nikolsky G. A. Variations of solar constant from data of balloon studies in 1962–1968. *Proc. of USSR Acad. Sci., Physics of the Atmosphere and Ocean*, 1970, **6**(4), 231–240 [in Russian].
58. Kondratyev K. Ya., Nikolsky G. A., Murcray D. G., Kosters J. J., and Gust P. R. The solar constant from data of balloon investigations in the USSR and the USA. *Space Research*, 1971, XI, **l**(1), 695–703.
59. Kondratyev K. Ya., Vasilyev O. B., Grishechkin V. S., and Ivlev L. S. Spectral radiative flux divergence and its variability in the troposphere. *Proc. of the Int. Rad. Symp., Sendai, Japan, 1972*, pp. 78–81.
60. Kondratyev K. Ya., Vasilyev O. B., and Grishechkin V. S. The shortwave spectral radiative flux divergence in the troposphere. *Annals of USSR Acad. Sci., Ser. Phys., Math.*, 1972, **207**(2), 334–336 [in Russian].
61. Kondratyev K. Ya., Vasilyev O. B., Ivlev L. S., Nikolsky G. A., and Smokty O. I. *The Effect of Aerosol on Radiation Transfer: Possible Climatic Consequences*. LSU Publ., Leningrad, 1973, 266 pp. [in Russian].
62. Kondratyev K. Ya., Vasilyev O. B., Grishechkin V. S., and Ivlev L. S. Spectral shortwave flux divergences in the troposphere and their possible variability. *Proc. of USSR Acad. Sci., Physics of the Atmosphere and Ocean*, 1974, **10**(5), 453–503 [in Russian].
63. Kondratyev K. Ya., Ivlev L. S., and Nikolsky G. A. Complex studies of stratospheric aerosol. *Meteorology and Hydrology*, 1974, **9**, 16–26 [in Russian].
64. Kondratyev K. Ya., Vasilyev O. B., Grishechkin V. S., and Ivlev L. S. Spectral radiative flux divergence and its variability in the troposphere in the 0.2–2.4 μm region. *Appl. Optics*, 1974, **13**(3), 478–486.
65. Kondratyev K. Ya., Vasilyev O. B., and Ivlev L. S. Complex observational studies over the Caspian Sea (CAENEX-73). *Meteorology and Hydrology*, 1975, **7**, 3–10 [in Russian].
66. Kondratyev K. Ya., Binenko V. I., Vasilyev O. B., and Grishechkin V. S. Spectral radiative characteristics of stratus clouds according to CAENEX and GATE data. *Proc. Symp. Rad. in Atm.*, Science Press, Garmisch-Partenkitchen, 1976, pp. 572–577.

67. Kondratyev K. Ya., Vasilyev O. B., Ivlev L. S. *The Global Aerosol–Radiation Experiment (GAREX)*. Information Centre, Obninsk, 1976, 28 pp. [in Russian].
68. Kondratyev K. Ya., Lominadze V. P., and Vasilyev O. B. A complex study of radiative and meteorological regimes in Rustavi (CAENEX-72). *Meteorology and Hydrology*, 1976, **3**, 3–14 [in Russian].
69. Kondratyev K. Ya., Barteneva O. D., Chapursky L. I., Chernenko A. P., Grishechkin V. S., Ivlev L. S., Ivanov V. A., and Vasilyev O. B. *Aerosol in the GATE area and its radiative properties*. Atmospheric Science Paper No. 247, Colorado State Univ., Fort Collins, 1976, 109 pp.
70. Kondratyev K. Ya., Welch R. M., Vasilyev O. B., Zhvalev V. F., Ivlev L. S., and Radionov V. F. *Comparison between the measured and calculated spectral characteristics of shortwave radiation in the free atmosphere over the desert (from the data of the CAENEX-70 expeditions)*. Atmospheric Science Paper No. 261, Colorado State Univ., Fort Collins, 1976, 79 pp.
71. Kondratyev K. Ya., Marchuk G. I., Buznikov A. A., Minin I. N., Mikhailov G. A., Nazarliev M. A., Orlov V. M., and Smoktiy O. I. *The Radiation Field of the Spherical Atmosphere*. LSU Publ., Leningrad, 1977, 214 pp. [in Russian].
72. Kondratyev K. Ya. and Zhvalev V. F. *The First GARP Global Experiment Aerosol and Climate*. Gidrometeoizdat, Leningrad, 1981, 167 pp. [in Russian].
73. Kondratyev K. Ya., Grigoryev Al. A., Pokrovsky O. M., and Shalina E. V. *Satellite Remote Sounding of Atmospheric Aerosol*. Gidrometeoizdat, Leningrad, 1983, 216 pp. [in Russian].
74. Kondratyev K. Ya., Moskalenko N. I., and Pozdnyakov D. V. *Atmospheric Aerosol*. Gidrometeoizdat, Leningrad, 1983, 224 pp. [in Russian].
75. Kondratyev K. Ya. and Binenko V. I. *Effect of Clouds on Radiation and Climate*. Gidrometeoizdat, Leningrad, 1984, 240 pp. [in Russian].
76. Lazrus A. L. and Gandrud B. W. Stratospheric sulfate aerosol. *J. Geophys. Res.*, 1976, **79**(24), 3424–3431.
77. Makarova E. A., Kharitonov A. V., and Kazachevskaya T. V. *The Solar Radiation Flux*. Nauka Publ., Moscow, 1991, 400 pp. [in Russian].
78. Marshall J. K. Drag measurements in roughness arrays of varying density and distribution. *Agricultural Meteorology*, 1971, **8**, 269–292.
79. Martell E. A. Tropospheric aerosol residence times: A critical review. *J. Tech. Atmos.*, 1974, **8**(3–4), 903–910.
80. Melnikova I. N. and Vasilyev A. V. *Shortwave Solar Radiation in the Earth's Atmosphere*. Springer Verlag, Berlin, 2004, 303 pp.
81. Mikhailov V. V. and Voitov V. P. An advanced model of the universal spectrometer to study the shortwave radiation fields in the atmosphere. *Problems of Atmospheric Physics*, 1966, **4**, 120–128 [in Russian].
82. Musick H. B. and Gillette D. Field evaluation of relationships between a vegetation structural parameters and sheltering against wind erosion. *Land Degrad., Rehabil.*, 1990, **2**, 87–94.
83. Nesmelova L. I., Rodimova O. B., and Tvorogov S. D. Absorption by water vapour in the near-IR region and some geophysical consequences. *Optics of the Atmosphere and Ocean*, 1997, **10**(2), 131–135 [in Russian].
84. Patterson E. M. Optical properties of dust collected in the dust storms on September 16 and 20, 1989 in SW Tadjikistan. In: K. Ya. Kondratyev (ed.), *Joint Soviet–American Experiment on Arid Aerosol*, Hydromet., St Petersburg, 1993, pp 162–165.

85. Patterson E. M. and Gillette D. A. Commonalities in measured size distributions for aerosols having a soil-derived component. *J. Gephys. Res.*, 1977, **82**, 2074–2082.
86. Petrosyants M. A. First results of the Soviet TROPEX-74 expedition. *Meteorology and Hydrology*, 1975, **3**, 3–18 [in Russian].
87. Sabatini R. R., Rabchevsky G. A., and Sissalla J. E. Nimbus earth resources observations. *Techn. Rep. No. 2, Contr. No. NAS-5-21617*, Allied Res. Assoc., Inc. Concord, Mass, 1971.
88. *Studies of Atmospheric Pollution over Alma-Ata. Part 1: The AUAP Experiment. Part 2: The TOPAZ System.* Gylym, Alma-Ata 1990, 185 pp. [in Russian].
89. Volz F. E. Infrared optical constants of aerosols at some locations. *Appl. Optics*, 1983, **22**, 3690–3700.
90. Yanishevsky Yu. D. *Actinometric Instruments and Methods of Observations*. Gidrometeoizdat, Leningrad, 1957, 415 pp. [in Russian].
91. Zdunkowsky M. and Korb W. G. An approximative method for the determination of shortwave radiative fluxes in scattering and absorbing media. *Contrib. to Atmospheric Physics*, 1974, **47**, 129–144.
92. Zuev V. E., Ivlev L. S., and Kondratyev K. Ya. New results from aerosol studies. *Proc. of USSR Academy of Sciences., Physics of the Atmosphere and Ocean*, 1972, **8**(11), 34–39 [in Russian].

2

Field observational experiments in America and western Europe

The aim of this chapter is to analyse the observational data on the structure, composition, and spatial distribution of aerosol referring to America, western Europe, and high latitudes. Before discussing these data, we shall discuss some general trends in the development of atmospheric aerosol studies. Apart from accomplishment of complex field observational experiments, these general trends include: (1) continued modification of observational techniques (both conventional and satellite) with emphasis on the remote sensing [7, 12, 15, 27, 30, 35, 37, 62, 65, 72, 97, 98, 105, 120, 123, 129, 130, 134]; (2) development of the global network to monitor aerosols (first of all, with the use of multichannel sun photometers and lidar sounding) [6–8, 16, 17, 27, 78, 84, 90, 105, 106, 111, 118]; and (3) substantiation of more adequate models of atmospheric aerosols [15, 35, 47–50, 56, 58, 61, 79, 86, 106, 126].

2.1 OBSERVATION TECHNIQUES

2.1.1 Remote sensing

In the development of observational means of importance are methods of lidar sounding of the atmosphere. According to the Third IPCC Report (Intergovernmental Panel on Climate Change), the global warming forecasts (even with the use of the most adequate climate models) are still rather uncertain [58, 61, 64]. In particular, there is still a great uncertainty in the estimation of radiative forcing (RF) determined by an inadequacy of satellite data on the Earth's radiation budget (ERB) and the absence of a sufficiently reliable understanding of the role of clouds, aerosol, and water vapour (as well as their interactions) in RF.

In this context, the Japan National Space Agency (NASDA) and Communication Research Laboratory (CRL) in cooperation with the European Space Agency (ESA) started an undertaking of phase A of the Earth CARE project. The main goals of this project are satellite observations of the 3-D fields of aerosol and cloud characteristics as well as ERB components bearing in mind subsequent studies of the impact of aerosol and clouds on the ERB and RF.

The equipment of the Earth CARE satellite described by [54] includes: (1) radar to sound clouds (Radar-CPR); (2) lidar to sound the atmosphere (ATLID); (3) multichannel video-radiometer (MSI); (4) broadband radiometer (BBR); and (5) Fourier spectrometer (FTS).

The 94-GHz Doppler radar has an antenna with $D = 2.5\,\mathrm{m}$ and a sensitivity of 38 dB. The vertical and horizontal resolutions constitute 100 m and 1 km, respectively. The spectral wavelength range of the Fourier spectrometer is 5.7–25 μm with a resolution $0.5\,\mathrm{cm}^{-1}$ and field of view (FOV) of $10 \times 10\,\mathrm{km}^2$. The Earth CARE satellite is planned to be launched in 2008 with 2-year lifetime.

Bravy *et al.* [17] analysed the possibilities of using the multifrequency chemical pulse DF laser (ten uniformly distributed channels in the wavelength interval 3.6–4.2 μm) for remote sensing of the chemical composition and size distribution of atmospheric aerosol, and briefly discussed the experience gained from laboratory testing of the instruments.

Solution of the problem of remote sensing of aerosol properties from the data of satellite spectral measurements of the outgoing short-wave radiation (OSR) requires a division of contributions to the OSR formation, determined by the surface and various optically active atmospheric components. Naturally, such a division is most easily made in the presence of a dark surface, that is, in the case of a water surface. Therefore, intense efforts have been undertaken to solve the problem of retrieval of the aerosol optical thickness (AOT) over the ocean from the data of OSR observations in the near-IR region. The use of similar retrieval algorithms has made it possible to determine the AOT over surfaces, both water and land, from the data of observations in the UV region.

With satellite data processing in the longer wavelengths region, the use of ATSR-2 satellite data for two viewing directions played an important role. In this connection, [26] discussed the results of testing the 'dual' algorithm from data of observations over the Indian Ocean (INDOEX) and South Africa (SAFARI), which were positive. Such algorithms have been also developed to process data of the AATSR and SCIAMACHY instruments (ENVISAT satellite) and OMI (EOS–Terra satellite).

Gassó and Hegg [34] proposed a technique to retrieve the aerosol mass concentration (AMC) and cloud concentration nucleii (CCN) number density counter (CCNC) from data of observations with the spaceborne moderate-resolution video-radiometer MODIS. The retrieval technique is based on the use of direct proportionality between AMC and cloud optical thickness with the use of retrieved data from MODIS for the effective radius of droplets r_{eff} and parameter η which determines the contribution of radiance that corresponds to the aerosol accumulation mode, to total radiance. The dependence of AMC on the relative humidity of the environment under conditions of the constant chemical composition of aerosol has also been taken into account. Parameterization of CCNC as a function of the retrieved volume of aerosol particles enables one to estimate the CCNC.

A comparison with data of *in situ* measurements has shown that the reliability of retrieval for the dry environmental conditions (in the presence of dust aerosol) increases with due regard to dependence of the proportionality constant on r_{eff} and η obtained for the same pixel. With a high humidity the use of the new technique is inefficient because of the difficulty to reliably take into account the RH vertical profile. Practically, the same results have been obtained with the use of a retrieval technique developed for processing the data of aircraft simulators, though the latter technique turned out to be more reliable.

Von Hoyningen-Huene *et al.* [120] proposed a retrieval technique for the atmospheric AOT over land surfaces from satellite data on OSR at nadir to process the data of SCIAMACHY and MERIS instrumentation carried by ENVISAT satellite. The technique is based on the use of tables of a priori data (look up tables, LUT) characterizing the dependence of AOT on the surface–atmosphere system's reflectance at wavelengths $<0.67\,\mu\text{m}$. Information about reflectance is obtained from the results of OSR calculations with the spectral surface albedo taken into account. The model of spectral albedo used was based on the data on bare soil surface and vegetation cover albedo obtained with due regard to satellite information about the normalized difference vegetation index (NDVI). The adequacy of the technique was tested using SeaWiFS data on OSR and data from the surface field observational experiment (LACE: the Lightweight Airborne Chromatograph Experiment) on atmospheric optical parameters. As applied to data for short-wave channels of SeaWiFS (wavelength interval 412–510 nm) the results of AOT retrieval and respective data of surface observations agreed with a difference not exceeding 20%.

Fussen *et al.* [33] substantiated a complex technique of atmospheric remote sensing based on the use of a combination of the retrieval operator applied in the case of high noise level in observational data and an optimal estimation technique. The success of the realization of such a complex technique depends on the adequacy of available a priori information about the state of the atmosphere. An example has been given [33] of processing the 'occultation' data for aerosol remote sensing which illustrates the robust nature of the proposed technique.

Holzer-Popp *et al.* [44] described a new technique of AOT retrieval called the synergic retrieval of aerosol properties (SYNAER). This technique is based on the use of data of simultaneous observations with ATSR-2 instruments (the radiometer scanning along the satellite's trajectory) and GOME spectrometer for the global monitoring of the total ozone content. Both these instruments were part of the complex scientific equipment carried by the European satellite ERS-2 for the remote sensing.

These observational data made it possible to retrieve AOT for the atmospheric boundary layer (BLAOT) and recognize the types of aerosol in the atmosphere over land and over the ocean. The second problem is solved by assessing the BLAOT share with respect to six representative components of a database on optical parameters of aerosols and clouds (OPAC) (the high spatial resolution of ATSR-2 data provides a reliable recognition of clouds).

The BLAOT retrieval technique is based on comparison of measured spectra of the outgoing radiation in the visible and near-IR regions (obtained from GOME data after filtering out the cloud contribution) with those calculated using various aerosol models. An agreement of spectra at a mean square minimization of differences determines the choice of an adequate aerosol model. This technique is planned to be used for processing the data of the SCIAMACHY video-spectrometer and advanced ATSR instrumentation to be carried by the ENVISAT satellite as well as GOME-2/AVHRR equipment developed for the METOP satellite.

One of the promising remote sensing techniques to retrieve the optical properties of atmospheric aerosol and clouds is the use of satellite data on the spectral distribution of OSR at a high spectral resolution in the A band of oxygen centred near the wavelength 762 nm. To apply this technique, the aircraft spectrometer LAABS was developed at the NASA Langley Centre with a wavelength range 12,980–13,160 cm^{-1} (~760–770 nm) and resolution of about 0.60 cm^{-1}. Pitts *et al.* [93] described in detail the results of laboratory studies of LAABS characteristics and the field tests which were carried out in July, 2001 during the flights in the region of Chesapeake Bay (USA). Preliminary results obtained during three flights turned out to be positive. Further tests are planned, followed by control *in situ* observations of the optical parameters of aerosol and clouds.

Yi *et al.* [130] described a retrieval algorithm for the single-scattering albedo (SSA) and the scattering function of atmospheric aerosol from data of surface observations of the angular distribution of sky brightness. The first stage of solving the problem is to substantiate (with the use of the theory of radiation field disturbances) the system of linear equations which make it possible to approximately express the sky brightness as a function of SSA and the coefficients of the Legendre polynomial, approximating the scattering function. This linear system is, however, poorly stipulated. Therefore, to obtain a stable solution, the process of regularization is carried out. The most important and difficult stage of regularization consists in the determination of the regularization parameter, especially when the process of retrieval is connected with iterations.

Yi *et al.* [130] tested the adequacy of this technique of retrieval of SSA and the scattering function of atmospheric aerosol from the data of observations of the angular distribution of the cloud-free sky brightness. With this aim in view, data were used of a conditional ('synthetic') experiment with prescribed different noise levels. Estimates have been obtained of the sensitivity of the retrieval results to variations in surface albedo and atmospheric optical thickness. This variability reflects on the SSA retrieval results. Also, the reliability of the retrieval was analysed with the use of data obtained at the network of aerosol observations.

Zhao *et al.* [133] analysed the significance of taking into account the non-sphericity of aerosol particles when retrieving the AOT τ from the data of the satellite AVHRR instrumentation in case of dust particles. The analysis was made by comparison with the use of surface observations of τ at two stations of the AERONET network in a desert. The consideration of non-sphericity ensured a decrease in retrieval errors of τ (especially random errors) with the use of the retrieval algorithm developed for AVHRR data processing.

Ellrod *et al.* [29] proposed a new technique to retrieve the content of volcanic ash suspended in the atmosphere from the data of meteorological satellites on the outgoing radiation at wavelengths 3.9, 9.7, and 12.0 μm. The technique is based on the consideration of the sum of two differences of brightness temperature (BTD) normalized in order to maximize the variability and its contrasts as indicators of the ash content. The following physical factors help to distinguish between the contributions to the formation of the outgoing radiation determined by ash and cloud: (1) differences in absorption determined by ash or sulphur dioxide at wavelengths 3.9, 10.7, and 12.0 μm; and (2) intense reflection of solar radiation by ash at 3.9 μm characterized by diurnal variations. The best results of retrieval were obtained in the daytime over all types of surface, and at night when the cloud of ash was located over the ocean or other water basins.

Alexandrov *et al.* [1] discussed the results of the statistical analysis of the spatial–temporal variability of AOT in the range of spatial scales from 0.2 to 1,200 km from data of the observational network obtained with the multichannel filter radiometer MFRSR in the southern sector of Great Plains (USA). Over a large territory in Oklahoma and Kansas, 21 radiometers were functioning. Surface observational data were compared with results of AOT retrieval from observations with the spaceborne MODIS instruments. Besides this, results of satellite observations were used to reveal a large-scale variability of the AOT. The results obtained make it possible to substantiate the presence of three major regimes of the AOT spatial variability. On small scales (0–30 km) the effect of the 3-D turbulent transport prevails. The effect of the 2-D turbulence on scales >30 km intensifies the AOT variability but it is reduced on scales >100 km.

The scientists of the Catalan Polytechnic University (Barcelona) developed a portable lidar on Nd:YAG with 3-D scanning to retrieve the aerosol properties in the free atmosphere. The lidar is simultaneously functioning at wavelengths 1,064 nm and 532 nm when scanning within zenith angle 15°–70° with a step of 5°. This instrument was used by [105] to retrieve the vertical profiles of various aerosol properties (first of all, the ratio of backscattering to extinction coefficients b/e) from data of observations in the region of Barcelona, where the effect of various sources of pollution and Saharan dust aerosol manifest itself all the year round.

The retrieval algorithm is based on the use of the variational method which provides a retrieval of the vertical profile of b/e and AOT without any a priori assumptions with respect to the type of aerosol. Reliable results of retrieval of the b coefficient were obtained for the 0.3–2.5-km layer with an average value of 0.030 sr^{-1}, whereas for higher latitudes, reliable results could not be obtained. The use of lidar profiles for small zenith angles made it possible to calculate the values of the Angström coefficient for backscattered radiation, which correlate well with the variational method. Difficulties in solving this problem for altitudes >1.6 km reflect, apparently, the effect of atmospheric heterogeneities connected with the strong spatial variability of aerosol properties.

2.1.2 *In situ* measurements

The considerable progress in remote sensing does not belittle the significance of *in situ* measurements of aerosol characteristics. This especially refers to the chemical composition of aerosol. In this connection, an adequate understanding of the nature and properties of atmospheric aerosols is seriously complicated by the absence of the required measuring instrument complex. Instruments to measure the mass, concentration, and number density as well as size distribution of aerosol have been developed, but the situation with measurements of aerosol chemical composition is poor. The method of chemical analysis of filter samples has been most widely used, though various difficulties occur with its application. During the last years, aerosol mass spectrometry, ensuring measurements of the aerosol chemical composition on real timescales, was widely used. This method is based on the use of aerosol samples, their subsequent evaporation, ionization of gas molecules, and the mass-spectroscopic analysis of ions. Powerful lasers are used for evaporation and ionization. However, numerous factors complicate the quantitative estimation with the use of the new technique. For instance, particles <200 nm are unreliably taken into account. Desorption and ionization are connected with sufficient losses of information in the case of complex organic compounds – determined by an intense molecular fragmentation.

Allan *et al.* [2] discussed the results of the use of new mass spectroscopic AMS equipment developed by the firm 'Aerodyne', in which the aerosol evaporates on a heated surface and then a quadruple mass spectrometry is applied with electron impactor ionization (EI). In this case, it is possible to obtain the quantitative information about the chemical composition and size of particles of volatile and semi-volatile small-sized aerosol substance with a high temporal resolution. The process of data processing consists of the use of the results of instrument calibration in order to proceed from the mass spectrum (MS) to mass concentration of the components under study ($\mu g\,m^{-3}$).

Illustrating the achievements of mass spectrometry, [3] discussed the results of measurements of the chemical composition of atmospheric aerosols with the use of the aerosol mass spectrometer from 'Aerodyne' obtained in two British cities – Edinburgh (October, 2000) and Manchester (July, 2001 and January, 2002). The results contain information about the size distribution of aerosol particles for nitrate, sulphate, and organic compounds. In all three cases, the effect of local transport has been recorded with the values of the mass modal aerodynamic diameter of particles ranging between 100 and 200 nm.

The main contribution to the formation of this mode of aerosol is made by hydrocarbon compounds which reveal only a weakly expressed indication of oxidation. Also, the mode of larger particles was observed, which consisted of inorganic compounds and oxidized organics. This mode was mainly determined by contributions of the sources located outside the cities and was characterized by the internally mixed particles. The mass modal aerodynamic diameter of particles varied between 200–500 nm in winter and 500–800 nm in summer. The summertime observations are characterized by an increase of mass concentration of aerosol, not followed by an increase in the number of particles. In the periods of low wind speeds, a change in the chemical composition of aerosols was observed as well as their ageing with time.

An improvement and wider use of various means of aerosol observations have made it possible to obtain more representative information and, in particular, to substantiate new aerosol models.

2.1.2.1 Aerosol models

Gong *et al.* [35] described the Canadian aerosol model (CAM) to be used in numerical models of climate and air quality. The model takes into account the main processes responsible for the formation of atmospheric aerosol: generation, hygroscopic growth, coagulation, nucleation, condensation, dry deposition, sedimentation, washing out in the subcloud atmospheric layer, and activation. The cloud module has been substantiated with explicit consideration of the microphysical processes reproducing the aerosol–cloud interactions and chemical transformation of sulphur compounds in the cloud-free atmosphere and in clouds. Gong *et al.* [35] carried out a numerical optimization of the algorithm determined by CAM to provide an effective solution to a complicated problem of describing the processes with multicomponent aerosols and their size distribution taken into account, which permits the use of CAM in global and regional numerical models. Aerosol particles are assumed to be internally mixed except for dust aerosol and black carbon which are considered to be externally mixed near their sources.

To test the adequacy of the proposed algorithm, [36] considered its application with due regard to emission to the atmosphere of anthropogenic and natural aerosols of two types: sea salt and sulphates. An analysis was made of the capabilities of two numerical methods of the algorithm realization: splitting and ordinary differential equations (ODE). A comparison of the results has shown that in calculations of mass concentration of total sulphates, the use of the splitting method provides more reliable data (by $\sim$15%) than does the ODE, but somewhat less reliable data ($<$1%) in the case of sea salt aerosol. It is also important, however, that the splitting method is 100-fold faster.

Gong *et al.* [36] studied the sensitivity of calculated aerosol size distribution to assess its precision. The 'diffuse' behaviour of each process is quantitatively characterized by the magnitude of difference between the modal radius and macromolecular structure database (MSD) for the log-normal curve of the size distribution in the cases of approximate and more accurate (with due regard to 96 'steps' of particles number density) presentations of the aerosol size distribution. The numerical modelling has shown that the size distributions of both number density and mass concentration turn out to be quite adequate considering 12 'steps' in many situations in the atmosphere, when the sink for condensed substance on the surface of aerosol particles is so large that nucleation of new particles is negligibly small. Quite reliable results of calculations of total mass concentration are obtained with the use of only 8 'steps'. However, to provide a realistic size distribution of nucleation mode, 18 'steps' or even more should be taken into account.

2.1.3 Aerosol observational network

Along with the data of satellite and aircraft observations, results of regular surface observations, especially at the global AERONET network, contribute substantially to raising the level of adequacy of the aerosol models.

With an appearance by the mid-1990s of high-quality instrumentation to measure direct solar radiation in order to retrieve the AOT of the atmosphere, the observational network has considerably broadened with the application of sun photometers (SPh). In this connection, of great importance became the intercomparisons of the sun photometers used at the existing network. At the observatory located near Bratt Lake (Canada) four types of sun photometers are functioning (for comparison purposes) used at three observational networks: Aerosols in Canada; Global Air Watch (GAW); Programme of UV-B radiation monitoring carried out under the auspices of the US Ministry of Agriculture. Since the instruments used at these and other networks are similar, results of SPh comparisons made in the summer of 2001 can be considered of general value. The results of comparisons performed by [78] have shown that the errors in the retrieved instant AOT values are within ± 0.01. During the 3-month comparisons the MSD of AOT values at 500 nm constituted 0.0069. Thus the sun photometers can be used for the global monitoring of AOT.

Continuing the developments based on the use of AERONET data, [106] studied the atmospheric aerosol properties over the ocean, which are characterized by a strong variability determined, first of all, by diverse types of aerosol including, for instance, urban/industrial aerosol, dust aerosol of deserts, products of biomass burning, and marine aerosol. Based on the use of data of aerosol observations at a robotic AERONET network on three islands (Bermuda in the Atlantic Ocean, Hawaii in the Pacific Ocean, and Kaashidhoo in the Indian Ocean), a model of the aerosol marine component has been substantiated [40], which can be used, in particular, to solve the problem of atmospheric correction. To retrieve the marine component of AOT at 500 nm, observations were made when AOT was <0.15, and the Angström parameter was below unity.

The results of determination of the size distribution of the aerosol marine component turned out to be very close for the three cases considered and is characterized by the bimodal mode with a small-size mode, to which the values of the effective radius of particles $r_{eff} \sim 0.11$–$0.14\,\mu m$ correspond, and a large-size mode, with $r_{eff} \sim 1.8$–$2.1\,\mu m$ (these values correspond to those published in literature). The independent-of-wavelength complex refraction index turned out to be equal to 1.37–$0.001i$ (this corresponds to the single-scattering albedo $\sim$0.98). The contributions of fine and coarse modes of aerosol to the AOT are, respectively, $\tau_{fine} \sim 0.03$–0.05 and $\tau_{coarse} \sim 0.05$–0.06. The Angström parameter varies between 0.8–1.0 (the wavelength interval 340–670 nm) and $\sim$0.4–0.11 (870–913 nm). Almost the same scattering functions are characteristic of all three cases.

The CERES instruments carried by the TRMM satellite are meant to measure the ERB components in three broad spectral intervals: 0.3–5.0, 0.3–1.00, and 8–12 μm. The TRMM scientific complex also included the scanning radiometer VIRS to measure the outgoing radiation in the visible and IR spectral regions. Zhao *et al.* [134] compared the TRMM/CERES-VIRS data and results of surface observations with sun photometers at the AERONET network to validate the AOT over the ocean retrieved from satellite data with due regard to possible errors determined by the presence of subgrid cloudiness (the VIRS data are used for this purpose) and the effect of wind near the ocean surface (roughness).

In the case of strong roughness there is a high positive correlation between wind speed near the ocean surface and τ, which is determined, apparently, not by the direct impact of sea surface roughness but by intensified input of marine aerosol to the atmosphere [40]. After minimization of the possible impact of the subgrid cloudiness and filtering out of cases with strong winds, the positive systematic difference between satellite and surface data could be reduced from 0.05 to 0.2 for the VIRS channel at 0.63 μm and from 0.05–0.03 (1.61 μm). Random errors also decreased from 0.09–0.06 (0.63 μm) and from 0.06–0.05 (1.61 μm). The remaining differences at small τ values are mainly determined by errors of calibration, radiometric noises, and inadequately prescribed aerosol parameters in the retrieval algorithm.

Subsequent developments will be aimed at assessing the errors of τ retrieval for different types of aerosol. Bearing in mind that the aerosol size distribution is bimodal, as a rule, the particle size distribution (PSD) can be used to distinguish between those AOT components (respectively, τ_f and τ_c) whose formation is determined by contributions of fine and coarse aerosols. The source of information for this distinction can be the spectral dependence of total optical thickness $\tau_a = \tau_f + \tau_c$. As O'Neill *et al.* [88] have shown, the adequacy of this method is provided by: (1) its physical substantiation; (2) an agreement of data on τ_c variability with data of cloud thin layer photography, whereas the reliability of τ_f is illustrated by observational data obtained under conditions of clear sky and haze; (3) a high correlation between τ_c and τ_f values and related values obtained by retrieval from data on sky brightness and general solar radiation attenuation. These data have made it possible to validate the algorithm of filtering out cloudiness.

Studies of the biogenic component of atmospheric aerosols have recently attracted serious attention. Rannik *et al.* [99] discussed the results of gradient (correlative) observations of the flux of atmospheric aerosol particles over the 40 year old Scots pine (*Pinus sylvestris*) forest performed at the station for measurements of the forest ecosystem–atmosphere interactions (the SMEAR programme) at Hyytiälä in southern Finland (51°61′N, 24°17′E, 181 m above sea level). The observational programme included measurements of aerosol size distribution ($D = 3$–500 nm), with instruments determining the differential mobility of particles mounted at a height of 2 m and the condensation particles counter (CPC) located at a height of 23 m (the forest height averaged 14 m). The CPC data were used to estimate the total number density of particles with sizes ranging between 10 and 500 nm.

Analysis of observational data has shown that the aerosol particles number density is more unstable than such scalar parameters as temperature, water vapour and carbon dioxide concentration. The results of observations obtained in the periods of steady-state conditions reflect the prevalence of the process of particles deposition on the forest canopy. The dependence of the rate of deposition on the particles' size was assessed [40] with the use of data on aerosol size distribution, and a semiempirical model of deposition was developed which could be used when the diameter of particles was <100 nm. However, in case of larger particles this model is unreliable.

2.1.4 Regional studies

In the context of assessment of the impact of black carbon aerosol on climate formation, of special interest are the measurements of light absorption by atmospheric aerosol. This aerosol appears as the product of incomplete burning of carbon-containing substances. While radiative forcing due to purely scattering sulphate and nitrate aerosols causes a cooling and determines thereby a partial (and sometimes complete) compensation of the warming caused by greenhouse gases, the presence of black carbon absorbing radiation in the visible spectral region favours an enhancement of radiative heating. One more effect consists in the impact of black carbon on cloud droplets size distribution and optical properties, which manifests itself through a decrease of cloud albedo.

Arnett *et al.* [8] carried out measurements of light absorption by black aerosol using two techniques:

(1) photoacoustic (ethalometer); and
(2) based on the use of aerosol filter samples (absorption photometer PSAP).

Measurements were made in the south of Texas (USA) in the 'Big Band' reserve within the field observational programme of studies of regional aerosol and visibility conditions. An analysis of measurement results revealed the presence of a satisfactory correlation between data of the ethalometer and PSAP. The absorption coefficient in the visible during one month of synchronous measurements did not exceed $2.1\,N\,m^{-1}$. Similar measurements with the use of the same two techniques were carried out in 2000 in the south of the Great Plains (USA). Again, a satisfactory agreement was observed between data obtained with the two techniques, but the absorption coefficient from the PSAP data was higher by a factor of 1.61. According to the results of photoacoustic measurements, the absorption coefficient decreases with the growing relative humidity $> 70\%$. This decrease can be connected with specific features of the measurement technique. Also, care should be taken with the use of PSAP under conditions of rapid changes of relative humidity.

With the use of the measuring complex carried by the NASA DC-8 flying laboratory, [25] performed measurements of concentration in the atmosphere of various minor gas components (MGCs) and condensation nuclei (CN) as well as aerosol to study the process of aerosol formation. An analysis of the observational data revealed a strong variability of CN concentration correlating with MGCs concentration whose changes were determined by the mixing of two air masses with different properties. The formation of CN took place in air masses 'flowing' from the zone of a storm, and the CN concentration before air mixing had been comparatively homogeneous along the flight route $\sim$200 km. An analysis was made [25] of the aerosol formation due to nucleation followed by the growth of the size and coagulation of particles under the assumption that the aerosol particles consisted of water and sulphuric acid, whose local formation was caused by SO_2 oxidation. It follows from the results that a 5-minute 'flash' of the nucleation process was followed by enhanced growth and coagulation during a period of about five hours. The observed values of both mass concentration and number density of aerosol correspond to the time constant characteristic of the storm dynamics. The final aerosol number density depended weakly on the SO_2 initial concentration.

At more than 60 stations of observations within the interdisciplinary programme IMPROVE of the environmental monitoring of national parks in the USA, a regular filter sampling of aerosol was carried out for the subsequent estimation of the content in the PM-2.5 fraction of total carbon (TC), organic carbon (OC), and elemental carbon (EC). Analysis of the results performed by Chow and Watson [24] has led to the conclusion that even in the presence of carbonates in samples they do not affect substantially the measured concentrations of TC, OC, and EC. This situation is explained by the fact that:

(1) with samples heated to a temperature of about 800°C carbonates are decomposed; and
(2) the maximum possible content of carbonates determined from the concentration of calcium (Ca) is almost always negligibly small compared with concentrations of TC, OC, and EC. At most locations the concentration of carbonate was $<100\,\mu g\,m^{-3}$ and sometimes it was below the detection threshold. A maximum measured concentration of carbonate reached $420\,\mu g\,m^{-3}$.

Blanchard *et al.* [14] considered the results of processing the filter samples of aerosol obtained at two locations in the region of Great Toronto (Canada) during several observational periods in 1998–1999 in order to study the organic fraction of atmospheric aerosol. The main purpose was to analyse differences in the composition of the aerosol organic fraction (AOF) from data of observations at urban and rural locations in different seasons. With the use of the thermal desorption methods, the transformation of aerosol samples into solution, derivatization, chromatography, and mass-spectrometry, data on the content in the atmospheric aerosol of organic and elemental carbon, alcanoic acids, n-alcanes and polycyclic aromatic hydrocarbons (PAHs) were obtained.

Results obtained in [14] turned out to be similar to data of earlier similar observations. The concentration of PM-2.5 particles at both locations was at a maximum in February due to weather conditions favouring the formation of aerosol particles. The ratio between the content of organic and elemental carbon was about the same as in the respective emissions except data for the rural location, where the higher potential of biogenic particle formation played a more substantial role as well as the appearance of secondary aerosols.

A comparison of the content of individual components of the chemical composition and total organic carbon has demonstrated that acids, alcanes, and PAHs constitute only a small part of the aerosol mass, and this indicates the necessity for further improvement of techniques of assessing the aerosol chemical composition. The results of preliminary processing of observational data suggest the conclusion that at the urban locations the content of alcanoic acids correlates with the level of emissions by diesel engines. Data on n-alcanes reflect the presence of prevalent contributions of anthropogenic sources at urban locations and biogenic sources at rural locations of observation. In the latter case the contribution of vegetation cover to biogenic emissions is clearly seen.

In August 2000, two broadband 13 and 16-channel radiometers were mounted on the roof of a building at Poker Flat (Alaska: 65.12°N, 147.43°E) for regular measurements of direct solar, scattered, and total radiation. The main goal of the observations was to analyse the effect of total ozone content variability, cloud amount, surface albedo, and aerosol in the atmosphere on the incoming UV solar radiation. In March–April, 2001 a special series of aerosol characteristic measurements was carried out with the use of instruments which determine the aerosol chemical composition depending on particle size. Wetzel *et al.* [126] discussed the results of observations obtained with four types of air masses entering the regions: three springtime air masses from the Gobi desert, polluted air masses from the Arctic and wet marine air, as well as late-summer air masses.

An analysis of the observational results has shown that the long-range transport of pollutants to the region of central Alaska caused only a small or moderate increase in the atmospheric AOT in the UV spectral region and, respectively, a limited decrease of atmospheric transparency at UV wavelengths. The strongest increase of extinction is caused by marine aerosol in the highly humid atmosphere. The results of calculations of the Angström coefficient, single-scattering albedo (from data on spectral AOT) absorption due to aerosol, and data of radiation flux observations have demonstrated specific features of all these characteristics for various types of aerosol. It means that even with a low concentration of aerosol, specific features of various air masses cause a substantial variability in the spectral and angular distributions of UV radiation intensity which should be taken into account when studying photochemical processes in the atmosphere and estimating the doses of biologically active UV solar radiation.

Natural and anthropogenic emissions of MGCs – aerosol precursors, and their subsequent transport to the upper troposphere (UT) – are the main source of aerosol in the UT. Within the INCA project of studies of interhemispherical differences between cirrus cloud properties due to anthropogenic emissions of MGCs, [83] carried out two series of aircraft measurements of the characteristics of atmospheric aerosol and cirrus clouds in the upper troposphere of the northern hemisphere and southern hemisphere mid-latitudes. Part of the relevant results were considered, including measurements of the vertical profiles of the number density of Aitken particles and particles of the accumulation mode, in order to analyse the interhemispherical differences of the results obtained. An analysis of the observational data has shown that the northern hemisphere particle number density is more variable than in the southern hemisphere, and it turns out to be 2–3 times higher, as a rule.

Feingold and Morley [30] demonstrated the possibilities of using the 1-channel lidar data on backscatter to assess an assimilation of water vapour by aerosol particles in the ABL, whilst being well mixed and covered by a cloud layer. Solution of this problem is important for the analysis of the impact of aerosol on the ERB formation. The zenith directed lidar data were used to retrieve the vertical profile of backscattering in the ABL beneath the stratocumulus clouds. An analysis was made of the dependence of backscattering on the aircraft measured relative humidity with values ranging between ~85% and 98.5%. To calculate an expected enhancement of backscattering due to water vapour assimilation, results of *in situ* measurements of the aerosol size distribution and composition were used.

A comparison of the observed and calculated enhancement of backscattering as a function of relative humidity revealed good agreement. Apparently, the aerosol backscattering at the sites of upward air motion is weaker than at the sites of downward motion (with similar relative humidity values). This situation agrees with ideas about the inertia of large hydrated particles with respect to their growth (or evaporation) on short timescales. Probably, systematic differences between the heights of cloud bottoms retrieved from lidar measurements in the regions of upward and downward motions also play some role. Calculations of the total scattering enhancement due to water vapour assimilation and of similar backscattering enhancement revealed an agreement between dependences on relative humidity within $\sim$20% at relative humidities $< \sim$95% but a considerable difference in the case of relative humidity $>$ 95%.

On the assumption of a cloud-free atmosphere, [123] substantiated the method of atmospheric correction of the results of retrieval of temperature and humidity vertical profiles with due regard to the presence in the atmosphere of mineral (dust) aerosol (DA) absorbing in the IR spectral region using the data from the instruments developed for the TIROS satellite for operational vertical sounding (TOVS) as part of the high-resolution IR sounder (HIRS) complex. The concentration and properties of DA are prescribed from the data for the GOCART model of global aerosol transport, taking into account chemical reactions accompanying the transport. The measure of the DA impact on the formation of the outgoing thermal radiation was the difference (O–P) between the observed brightness temperature (BT) and BT values calculated with prescribed vertical profiles of temperature and humidity every six hours.

It follows from these results that with an increase of DA concentration the difference O–P becomes negative. Thus, it means that the presence of aerosol leads to a decrease of the observed BT (i.e., aerosol affects markedly the BT). In the case of HIRS channels, which are sensitive to surface temperature as well as to temperature and humidity of the free atmosphere, a decrease of BT due to aerosol reached 0.5 K and more. Therefore, the scheme of radiation transfer parameterization was specified. This scheme was used in the retrieval algorithm to take into account the DA effect on the assumption that the aerosol consists of illite. The specification resulted in an increase of surface temperature (0.4 K) and the temperature of the lower troposphere in the regions of a moderate dust loading of the atmosphere in the tropical Atlantic Ocean.

2.2 NORTH AMERICA

On the North and South American continent a number of large-scale field observation experiments have been undertaken, an important component of which being studies of atmospheric aerosol. Naturally, in the USA emphasis was laid on assessment of anthropogenic impacts and, in this connection, on the field experiments in the regions of large cities.

2.2.1 A 'superexperiment' in the region of Atlanta

In the summer of 1999, in central Atlanta (Georgia, USA), in cooperation with the US Environmental Protection Agency (EPA), a 'supersite' was organized to study aerosol in the urban atmosphere within the SOS project on studies of oxidants in the south of the USA. The initial short-term goal was to compare instruments to measure the size distribution (mass), chemical composition, and physical properties of aerosol. The long-term problem was to study the physical and chemical processes taking place in the urban atmosphere in the presence of intense anthropogenic forcings.

To obtain information about aerosol, instruments of three types were used:

(1) filters with ensured integration of samples over their time of functioning in the discrete regime – like those used at the EPA network to monitor fine aerosol (PM-2.5);
(2) instrumentation for continuous and semicontinuous records of chemical composition of atmospheric MGCs and aerosol – still at the stage of development; and
(3) mass spectrometers to measure the MGCs concentration and the chemical composition of aerosol particles.

Solomon *et al.* [107, 108] discussed the results of comparisons of 12 different filter sensors operating in the discrete regime. Data were analysed for total aerosol mass as well as on the content of sulphate, nitrate, ammonium, organic and elemental carbon, and some trace elements. As a rule, the concentration of the studied atmospheric components substantially exceeded the detection limits, except nitrate and some trace elements (e.g., As and Pb). In most cases, differences were in evidence between various instruments, except data for sulphate and ammonium. These differences are determined by the effect of various factors including specific features of the instrumentation and methods of analysis of the chemical composition. The levels of differences are characterized by the following data: $\pm 20\%$ (aerosol mass); $\pm 10\%$ (sulphate); ± 30–35% (nitrate); ± 10–15% (ammonium); from $\pm 20\%$ to ± 35–40% (organic carbon); from $\pm 20\%$ to $\pm 200\%$ (elemental carbon); and ± 20–30% (trace elements).

Solomon *et al.* [107, 108] performed an overview of various instruments to study aerosol and MGCs applied within the EPA-supported programme on studies of oxidants in the atmosphere in the south of the USA (SOS). The supersite in Atlanta, designed for tests and comparisons of methods to measure aerosol characteristics, is located in the north-western sector of the city, in the territory used for the Aerosol Characterization Experiment (ACE) to observe atmospheric aerosols. The studied methods included various filters and impactors as well as instrumentation to measure the aerosol chemical compositions (sulphates, nitrates, ammonium, organic and elemental carbon, and trace elements) and MGC precursors of aerosols (chemical composition of aerosol and MGC was analysed with the use of four mass spectrometers). Apart from this, information was used about meteorological parameters, volatile organic compounds (VOC), oxidized VOC and NO_y, bearing in mind an analysis of conditions of the formation and accumulation of aerosol in Atlanta. Analysis of the results under discussion revealed some differences between data obtained with instruments of different types, but (with the use of an adequate method of data processing) provided rich information about the aerosol dynamics under the urban conditions of Atlanta.

During the period 6 August–1 September, 1999, at the Atlanta supersite, studies of aerosol in the urban atmosphere were carried out with the use of a laser desorption–ionization mass spectrometer ATOFMS, making it possible to measure the chemical composition of aerosol particles of sizes ranging from 0.2–2.5 μm. Analysis of the results obtained (the recorded mass spectra totaled 455,444) carried out by [75] revealed the presence of the following seven types of aerosol:

(1) Potassium containing.
(2) Potassium containing with an addition of secondary components (ammonium, nitrate, and sulphate).
(3) Carbonaceous (both elemental and organic).
(4) Carbonaceous with an addition of secondary components (ammonium, nitrate, and sulphate).
(5) Dust.
(6) Dust with an addition of secondary components (ammonium, nitrate, and sulphate).
(7) Secondary components (ammonium, nitrate, and sulphate). Basic characteristics of ion components of aerosol have been described in [75].

On the whole, the results obtained demonstrate a complexity and variety of the urban aerosol chemical composition.

A comparison of results of the aerosol size distribution measurements with the use of the mass spectrometer ATOFMS and laser particles counter carried out by [125] revealed substantial differences which depend on the chemical composition of aerosol particles. These differences were especially marked in the periods of an increased mass concentration of ammonium and sulphate. Aerosol particles not fixed in measurements referred to the category of submicron particles (0.35–0.54 μm) and were characterized by the data on scattering caused by them. A scaling procedure was proposed [125], which makes it possible to take into account systematic errors in the determination of the chemical composition of particles with the use of the ATOFMS instrumentation. The results obtained in this way are compared with data of measurements from other types of instruments.

Within the August 1999 field observational experiment on studies of oxidants at the Atlanta supersite, [100] performed measurements on the size resolved chemical composition of aerosol using unique desorption–ionization mass spectrometry. This technique provides the possibility of analysing the chemical composition of individual particles over real timescales with aerodynamic diameters of particles varying within 0.01–1.5 μm. With a high content of aerosol in the atmosphere over

Atlanta, an analysis was made of the chemical composition of particles with size 0.1–0.2 μm nearly once every second. The number of mass spectra recorded during one month (both in the daytime and at night) totaled about 16,000.

Analysis of the results obtained has led to the conclusion that carbon-containing compounds are the prevalent component of ultrafine (<100 nm) particles. A greater variability in the chemical composition is a characteristic feature of large particles with typical combinations of organic carbon and inorganic components (rocks, metals, etc.). A consideration of all the results obtained revealed the presence of 70 classes of aerosol chemical composition. Rhoads *et al.* [100] discussed the variability of averaged mass spectra and their dependence on meteorological parameters for basic and some secondary classes. The main classes of particle composition are determined as characterized by maxima in the spectra, which correspond to: organic carbon (~74% of particles), potassium (8%), iron (3%), calcium (2%), nitrate (2%), elemental carbon (1.5%), and sodium (1%). Many of these classes are characterized by a certain wind direction reflecting the location of an aerosol source.

With the use of the RSMS II mass spectrometer to measure the aerosol particles chemical composition during the period 23 August–18 September, 2000 in Houston (USA), measurements were made north of the main sources of industrial emissions to the atmosphere. Twenty-seven thousand mass spectra obtained by [91] from these observations were classified depending on specific maxima of spectra using the method of neural networks. Analysis was made of the correlation between the frequency of occurrence of each of the classes with wind direction as well as the time dependence of the occurrence. Some classes of particles were always present, whereas others – only from time to time or even for very short time intervals. Most frequent components of the particle compositions were potassium and silicon. Organics and heavy metals contributed less.

Establishment of harmful effects of the urban fine PM-2.5 aerosol on human health stimulated adoption by the US EPA in 1997 of a national standard of air quality (NAAQS) which determined the maximum permissible concentration limit (MCL) for PM-2.5 in the atmosphere. According to NAAQS, a short-term MCL (65 $\mu g\,m^{-3}$ averaged over 24 hr) and a long-term MCL (15 $\mu g\,m^{-3}$ for mean annual level of concentration) have been defined. In May, 1999, the adequacy of such standards was, however, questioned by the US Supreme Court.

In connection with this juridical incident, an observational network was organized in Atlanta (Georgia, USA) to monitor the mass concentration and chemical composition of fine aerosols. According to observational results obtained by [20], the annual mean concentration of aerosol varied (depending on observation location) within 19.3–21.2 $\mu g\,m^{-3}$, with maximum concentration levels 44.3–51.5 $\mu g\,m^{-3}$ (24 hours) and 73.3–87.9 $\mu g\,m^{-3}$ (1 hour). Considerable annual and diurnal changes of aerosol concentrations were observed. Measurements of the chemical composition for 75% of aerosol mass showed that carbon compounds, sulphate, ammonium, and nitrate were the prevalent components of the aerosol composition.

In the centre of Atlanta, on Jefferson Street, where the supersite was located for the monitoring of the urban atmosphere characteristics, during intense observations from 3 August–1 September, 1999, [74] performed semicontinuous measurements of the content of organics and elemental composition of atmospheric aerosol with the use of various types of instruments: thermal–optical analyser of carbon, aerosol carbon monitor, analyser of carbon transformed into gas, and ethalometer AE-16. Comparisons of results revealed differences between data on TC, OC, and EC constituting, respectively, 7, 13, and 26%. The correlation between paired data on OC was moderate ($R^2 = 54$–73%), and for EC paired – high ($R^2 = 74$–93%). Differences between measured values of OC turned out to be small compared with data for EC. On the whole, an agreement between the observational data discussed can be considered quite satisfactory, especially with due regard to the quality of similar results obtained earlier.

In August, 1999, in the period of intense observations at the Atlanta supersite to monitor the atmospheric chemistry, five new types of instruments were mounted for semicontinuous measurements of the composition of fine aerosol PM-2.5 (particles with $D < 2.5\,\mu m$). With intervals from 5 minutes to 1 hour, measurements were made of the chemical composition of aerosol samples using an ion chromatograph (IC) as well as processing filter samples (with automated removal of water).

According to data of one-month observations carried out by [124], values of nitrate concentration averaged over 15 minutes were small, varying within 0.1–$3.5\,\mu g\,m^{-3}$, with an average of $0.5\,\mu g\,m^{-3}$. Values of sulphate concentration averaged over 10 minutes changed from 0.3–$49\,\mu g\,m^{-3}$, with an average of $14\,\mu g\,m^{-3}$.

Comparisons of results obtained using different methods (including results of filter sample analysis averaged over 24 hours) revealed an agreement, within differences of 20–30%, for nitrate (this corresponds to a mass concentration of $\sim$0.1–$0.2\,\mu g\,m^{-3}$) and 10–15% for sulphate (1–$2\,\mu g\,m^{-3}$). In the case of both sulphate and nitrate there was no difference between data of IC and filter samples at the 95% confidence level. However, comparisons of semicontinuous IC data for nitrate revealed a much weaker variability compared with data of filter sample analysis averaged over 24 hours.

From the data on the concentration and chemical composition of PM-2.5 aerosol (aerosol particles with aerodynamic diameters $D_p < 2.5\,\mu m$) measured in the atmosphere over Atlanta, [23] analysed the radiative as well as physical and chemical properties of the urban aerosol. The obtained arithmetical mean values and MSD for the coefficient of scattering due to PM-2.5 particles σ_{sp} at 530 nm and the coefficient of absorption σ_{ap} (at the same wavelength) for a relative humidity $= 49 \pm 5\%$, constituted $121 \pm 48 \times 10^{-6}$ and $16 \pm 12 \times 10^{-6}\,m^{-1}$, respectively. Though the average coefficient of extinction σ_{ep} in Atlanta strongly exceeds the background level, it is comparable with values observed in non-urban regions in the centre of the south-eastern part of the USA, which reflects the presence here of a vast regional haze. The single scattering albedo ω_0 in Atlanta is 0.87 ± 0.08 which is only 10% less than that recorded in the non-urban regions with polluted atmospheres. This is explained, apparently, by the impact of mobile sources of pollutants contained in emissions of elemental carbon.

A dual structure of the diurnal change of aerosol properties was observed manifesting itself as an impact of mobile sources (in the morning 'rush' hours, when the aerosol concentration reaches a maximum, especially its soot component) and the effect of atmospheric mixing (responsible for the formation of the afternoon minimum of concentration). There is a high correlation ($R^2 = 0.80$–0.96) between σ_{sp} and PM-2.5, with the average value of the mass efficiency of absorption due to elemental carbon E_{sp} (ratios of σ_{sp} to mass concentration of PM-2.5) varying within 3.5–$4.4\,\mathrm{m}^2\,\mathrm{g}^{-1}$. There was also a correlation, though expressed more weakly, between σ_{sp} and PM-2.5 ($R^2 = 0.19$).

The use of four methods of assessing the mass efficiency of absorption due to elemental carbon gave E_{ap} values between 5.3–$18.3\,\mathrm{m}^2\,\mathrm{g}^{-1}$. The wide range of the optical parameters variability reflects a strong variability of the aerosol optical properties and the presence of considerable errors in the measurements of light absorption by aerosol, especially absorption due to EC. The best agreement of the E_{ap} value was obtained with the use of the observed (with the multistage impactor) mass distribution and the σ_{ap} value calculated with the Mie formulae (in this case $E_{ap} = 9.5 \pm 1.5\,\mathrm{m}^2\,\mathrm{g}^{-1}$) as well as mass distribution and measured σ_{ap} ($E_{ap} = 9.3 \pm 3.2\,\mathrm{m}^2\,\mathrm{g}^{-1}$). Calculations of the scattering coefficients with the use of the Mie formulae and the measured mass of the EC size distribution as input parameters gave average values of the extinction coefficient reflecting the prevalent contribution of submicron aerosol under urban conditions. The PM-2.5 value averaged $31 \pm 12\,\mu\mathrm{g}\,\mathrm{m}^{-3}$, and the AOT at 500 nm $\sigma_a = 0.44 \pm 0.22$. The aerosol RF at the atmospheric top level, obtained with due regard to this value of σ_a and other parameters, was $\Delta F = -11 \pm 6\,\mathrm{W}\,\mathrm{m}^{-2}$, which exceeds the level of respective global mean estimates by an order of magnitude.

Middlebrook *et al.* [81] compared the results of four mass spectrometers which took part in measurements of MGCs concentration and chemical composition of individual aerosol particles carried out at the Atlanta supersite to study the anthropogenic impacts on the chemical composition of MGCs and aerosols. Though all four mass spectrometers were of the same type, each of them had unique characteristics due to considerable constructive differences. Some constructive differences were even inherent in two laser desorption–ionization mass spectrometers RSMS-II, but this did not bring forth any substantial differences between results obtained with these instruments, though with respect to the data of two other mass spectrometers such differences did occur.

Jimenez *et al.* [51] described the method of sampling and analysis of atmospheric aerosol samples with the use of the 'Aerodyne' aerosol mass spectrometer (AMS) designed to measure the size distribution of the aerosol mass, total mass, and composition of submicron particles of volatile and semivolatile chemical compounds. The main AMS components are: aerodynamic lenses ensuring the focusing of particles into a narrow band; a heater for particle evaporation in a deep vacuum; and a quadruple mass spectrometer to analyse the composition of evaporated molecules. The time of a particle's run is an indicator of their size. The AMS instruments function in two regimes: (1) continuous mass spectrometric records without differentiation by size; and (2) determination of the aerosol size distribution for fixed values of the m/z ratio. Mass spectra of individual particles can also be recorded if their mass is sufficient to be measured.

Results of AMS testing carried out in August, 1999 at the Atlanta supersite and at a suburban location near Boston (USA) in September, 1999 have been discussed in [132]. In both these cases the main components of aerosol composition were sulphate and organic compounds with a small addition of nitrate, which agrees with the earlier observational results. Various components of the aerosol chemical composition were often characterized by different size distributions and specific evolutions in time. More than half the mass of sulphate constituted 2% of the total number of particles. Data on the trends of the mass concentration of sulphate and nitrate in Atlanta agree well with results obtained earlier with the use of the IC method. A distinct diurnal change is typical of aerosol nitrate. The observational results obtained have been generalized in the form of the urban aerosol model.

Based on the use of data from observations of the atmospheric aerosol composition obtained during the period of the field observational experiment at the Atlanta supersite (accomplished from 3 August to 1 September, 1999), [124] analysed a supposition about the presence of the thermodynamic equilibrium of nitrate (NO_3^-) and ammonium (NH_4^+) contained in fine aerosol (PM-2.5), with gas-phase nitric acid (HNO_3(g)) and ammonium (NH_3(g)). The equilibrium condition was assessed by calculation of equilibrium concentration of HNO_3(g) and NO_3(g) on the assumption of an inorganic composition of PM-2.5 (N^+, NH_4^+, Cl^-, NO_3^-, and SO_4^{2-}) as well as with prescribed observed temperature and relative humidity. Results of calculations of the equilibrium concentration were compared with respective values of the observed concentration of gas-phase components.

Data on the chemical composition of fine aerosol PM-2.5 are based on the use of PILS data averaged over 5-minute intervals, obtained by the scientists of the Georgia Institute of Technology (USA), and also information about the concentration of gas-phase components taken from other sources. To calculate the equilibrium concentration of gas components, the aerosol thermodinamic model was used. The combined database on the chemical composition of PM-2.5 as well as on the concentration of HNO_3(g) and NH_3(g) includes 272 5-minute intervals.

The initial analysis of the results obtained has led to a conclusion about the absence of a thermodynamic equilibrium in the presence of calculated values of NH_3(g) concentration, which are less than the measured ones (backscattering takes place in the case of HNO_3(g)). However, a relatively small decrease in the content of SO_4^{2-} in PM-2.5 (for the purpose of adjustment) provided an agreement between calculated and measured concentrations of NH_3(g) and HNO_3(g). Besides this, an exclusion of 31 series from 272 complexes of data, which were characterized by abnormally low observed concentrations of SO_4^{2-} or NH_3(g) confirmed the adequacy of the correction of the content of SO_4^{2-} mentioned above. Average relative values of corrections required to provide a thermodynamic equilibrium with respect to HMO_2 and NH_3 are equal to −14.1% and −13.7%, respectively. The level of these corrections substantially exceeds the possible impact of measurement errors.

Results discussed in [124] suggest the conclusion about the existence of a thermodynamic equilibrium for the chemical composition of fine aerosol PM-2.5 from observations made in Atlanta, assuming one of the following possibilities:

(1) PILS data on SO_4^{2-} concentration in PM-2.5 were systematically overestimated by about 15%;
(2) these data systematically underestimate the concentration of alkaline components by 15%; and/or
(3) the aerosol thermodynamic model systematically underestimates pH for PM-2.5 compared with that observed in Atlanta.

The main conclusion to be drawn from analysis of the data from the field experiment in Atlanta is that (as could be supposed) aerosol properties in the atmosphere of a large industrial city (located in southern latitudes) are diverse and variable. This seriously complicates consideration of the effect of aerosol as a climate-forming component. This conclusion is an illustration of the results of various local and regional studies of atmospheric aerosols in Central America considered below.

2.3 CENTRAL AMERICA

Observations over the territory of Mexico are of special interest because this territory is a unique combination of various regions with different aerosol types (desert, ocean, volcanic, and urban) as well as specific and variable meteorological conditions.

2.3.1 Mexico City

Mexico City, one of the largest megapolises of our planet, is also known as a city with the most heavily polluted atmosphere. Apart from developed industry and transport, this city is also of interest because of its geographical location. Mexico City is located in the central part of the continent at a height of ~1,800–2,000 m in a hollow surrounded by mountains, which causes weak and irregular transport of air masses and, respectively, an accumulation of pollutants from the urban atmosphere. Mexico City is a unique test area for scientific studies of heavily polluted urban atmospheres, therefore, a study of aerosol (and MGCs) in the atmosphere is of great value for aerosol monitoring and developing potential measures against their build up. Starting in the 1990s, complex studies of aerosols in Mexico City have been carried out at the Institute of Geophysics of the University of National Autonomous Mexico together with the Department of Atmospheric Physics of the Institute of Physics at St Petersburg State University.

To study aerosols, both indirect (optical) [46, 47, 50, 82, 116] and direct (*in situ*) methods were used to determine the aerosol particle number density and their size distribution [49, 50] as well as analysis of the chemical composition of aerosol samples by method of air filtering [18, 50, 70].

The field spectral optical measurements in the atmosphere of Mexico City were made in 1992 during the joint Russian–Mexican expedition [47, 49]. The spectral optical thickness of the atmosphere was measured with Bouger's long method using the spectrometer K-3 [112–114]. Note that at the latitude of Mexico City the sun rises very rapidly from the horizon to considerable altitudes (in about 1.5 hour) and practically vertically. Therefore, Bouger's long method gives rather stable results, and such traditional problems as the effect of changes in sun azimuth and atmospheric parameters during observations practically did not appear in these measurements. From experience taken from the 1992 expedition, one can draw a conclusion about the good quality of measurements from the optical thickness using Burger's long method in tropical latitudes (the long method is very easily realized, since, in contrast to the short method, it does not require an energy calibration of the instrumentation and preservation of its long stable functioning).

From results of these measurements using the standard technique of extracting the optical thickness constituent determined by molecular scattering, the spectral AOT of the atmosphere was calculated or rather the total optical thickness of aerosol and absorbing gases. In principle, concentrations and spectrum of aerosol sizes, as well as concentrations of MGCs, can be calculated from this value. However, the complex inverse problem is known to be very complicated, and its solution requires the use of various a priori information about the parameters to be estimated and, with insufficient information, can be ambiguous and unreliable. Hence, this complex inverse problem has some sense only after accumulation of a sufficient volume of information about microphysical and optical properties of aerosol in Mexico City.

With the use of a simpler approach, the purely aerosol component was selected from the total optical thickness by an approximation of the former using the simple analytical dependences on wavelength of the type of the known Angström formula $\tau(\lambda) = A\lambda^{-b}$ [6]. Some more similar approximations with three parameters have also been proposed. The dependences mentioned determined the spectral aerosol optical thickness. Analysis of the obtained coefficients revealed their distinct division into several groups, which should be interpreted as a prevalence of each group of aerosols with different optical properties. The aerosol constituent in Mexico City is characterized by a very strong temporal variability, which is seen visually: along with the usual heavily polluted atmosphere, sometimes there are days when the air becomes pure, and the snow-covered mountain tops located at a distance of 70 km from the city can be seen. Also it is of interest that the described division of spectral aerosol optical thickness and spectral optical thickness of molecular absorption has made it possible to identify the molecular absorption bands. Apart from the known bands of ozone, oxygen, and water vapour, the absorption band of NO_3 (622–644 nm) can also be observed in the spectra, which is probably formed in the polluted atmosphere of Mexico City as a result of photochemical reactions.

Photochemical processes play a significant role in the formation of aerosols, too [46]. These begin as the reactions of petroleum burning products in car engines. The main paths of aerosol removal from the atmosphere of Mexico City are by washing out during the rainy season (June–October). Gravitational deposition plays a weaker role, with a considerable share of deposited dust raised repeatedly to the atmosphere in the noon and afternoon hours.

From the data of neutron activation, mass spectrometry, and spectral analysis of filter samples of the Mexico City aerosols, estimates were obtained of the chemical composition of aerosols in the surface layer of the atmosphere [49]. Mineral components mainly of soil origin (dust) are the main components of aerosol substances. As compared with similar samples in non-industrial regions, concentrations of soot, sulphuric acid, sulphates, and organics are very high. The last three components are mainly products of photochemical reactions. The total mass concentration of aerosols (for particles with $D \leq 5\,\mu m$) in the surface layer of the atmosphere in Mexico City varies from 30–110 $\mu g\,m^{-3}$.

Data of measurements of the aerosol size distribution revealed two distinct maxima (modes). The first maximum in the interval of sizes 0.2–0.7 μm is determined by the photochemical processes discussed above. The second maximum in the range 1–10 μm is due to processes of soil surface dust cover. The first mode has a distinct diurnal change (with a maximum in the day–evening hours and a minimum at night), the second mode depends strongly on weather conditions.

For analysis of the results of measurements of microphysical parameters and their coordination with optical measurements, model calculations were made of the optical characteristics of the aerosol light scattering in the visible spectral region [46, 47]. Substantial differences in the chemical composition and parameters of the aerosols in Mexico City, in wet and dry seasons, were taken into account. Aerosol particles were modelled as 2-layer spheres with prescribed size distribution functions and a complex refraction index for the aerosol substance. These parameters and their variations were estimated from data of *in situ* microphysical measurements. For the consideration of the impact of possible non-sphericity, a model of cylindrical particles was used, the ratio between spheres and cylinders being taken from the analysis of aerosol samples.

Results of model calculations have demonstrated the important role of taking into account not only spectra of particle sizes but also non-sphericity of particles as well as their heterogeneous composition. Variations of the aerosol optical properties have been modelled, which correspond to *in situ* measured variations of microphysical parameters: the complex refraction index for aerosol substance and the function of particle size distribution. Variations of the complex refraction index affect weakly the optical properties in the visible spectral region, but variations of particle concentration in the two modes described above affect it strongly.

The next stage of modelling the optical properties of aerosol in Mexico City was to construct a statistical optico-microphysical model of the type described in [50, 56]. Such a model makes it possible to solve the problem, more adequately than the methods mentioned above, for instance, of selection of the spectral aerosol optical thickness from the total one and, respectively, to obtain from the optical measurements much more information about aerosols. An optico-microphysical model may be substantiated under conditions that *in situ* measurements of microphysical properties of atmospheric aerosols in Mexico City are available. Complex studies of aerosols in Mexico City have made it possible to obtain a considerable volume of information useful for the construction of an aerosol model.

Let us briefly discuss the important problem of developing methods of atmospheric pollution control in Mexico City. As practice shows, the administrative measures, such as limitation of urban traffic, turn out to be of low efficiency. One possible solution is to try to control the atmospheric processes themselves in the region of Mexico City.

The content of dust and poisonous gases in the lower atmosphere can be reduced in several ways: (1) the outflow of pollutants to higher layers of the atmosphere with their subsequent scattering in these layers; (2) the horizontal removal of pollutants via natural ventilation due to specially projected urban constructions; (3) collection of pollutants directly in the zone of maximum pollution; (4) initiation of reactions converting poisonous gases to aerosols with the subsequent removal of aerosol particles from the atmosphere, for instance, due to their accelerated condensation and coagulation growth; and (5) a washing out of aerosol particles and some gases via artificial precipitation. A combination of various means of cleaning the atmosphere is also possible. Clearly, for Mexico City the use of these methods requires serious economic assessment and the creation of a specialized system of complex monitoring and managing of the state of the environment [117], in particular, due to the large scales of the city, diverse conditions of formation and accumulation of pollutants, seismic perturbations of the city, and close location to the active volcano Popocatepetl. At present, an analysis has been made of the possibility of the impact on the environment of convective outflows of pollutants from the surface layer of the atmosphere and washing out by artificial precipitation [117].

2.3.2 Complex studies of aerosol in the lower atmosphere of Mexico

In 1994–1995, studies of concentration, chemical, and elemental composition of aerosols in the lower layers of the atmosphere in the western states of Mexico (Colima, Halisco, Micheocan) were carried out together with measurements in central Mexico of the aerosol components emitted by the volcano Popocatepetl in December 1994–March 1995. Measurements were made both in the wet and dry seasons as well as during the earthquake in October, 1995. Despite diverse conditions of generation and evolution of aerosols in the surface layer, the elemental composition and size distribution of aerosol particles followed certain regular features.

Aerosols were sampled on one-layer filters of the AFA type and 3-layer filters of the FPA-70 type. The elemental analysis of samples was made using the X-ray fluorescence technique with the X-ray spectrometer MECA 10-44 (XR-500). The chemical analysis of some samples for acid content was made using standard methods, and electron microscope analysis of some samples was undertaken with the electron microscope microanalyser EMMA-2. The choice of aerosol samples was accompanied by measurements of number densities and size distributions of particles with the help of the photoelectric counter A3-5M. At the same time, measurements were made of ozone, sulphur dioxide, nitrogen dioxide, as well as meteorological observations.

The first stage of observations of the elemental composition of aerosol and dust particles covered the period from mid-December, 1994 until early February, 1995 and was partially carried out over 24 hours. Sampling was made at the following locations: the city of Colima, the city centre and the university; Colima volcano; Popocatepetl volcano (Pueblo airport); and the city of Mansanilla. After the Popocatepetl eruption, the volcanic ashes were sampled and studied.

The second stage of observations was carried out in May, 1995 at the following locations: the city of Colima, the city university, and near the Colima volcano: at Playon and Refujio. Apart from the usual filters, blocks of 3-layer filters were used. At Playon, aerosol particles deposited onto polyethylene film were also sampled.

The third stage included air sampling in August–October, 1995 at Apo, Mansanilla, Colima (the university), and Guadalajara. In Mansanilla and Guadalajara observations were made around the clock, and in Colima – with short breaks. The elemental composition and the data on the mass distribution of elements was used to draw conclusions about the nature of the observed dust particles.

Apart from air sampling, precipitation was also sampled in Colima during the period from late May to early October, 1995; in the period of late May–early August continuously, and then selectively. The duration of sampling was usually one week. The volume of rain water samples depending on the rain rate varied from 1–20 l. In the elemental analysis, deposits on filters were studied by pumping rain water from the bottom of the settled volume (deposit), in some cases from the upper part of the same volume (suspension). Particles (and elements) of fine and coarse fractions contained in rain water were separated thereby. Besides this, samples were studied which deposited onto mylar film with 3–20 ml of settled water evaporated from the upper part of the volume, which made it possible to determine the elements which reside in water in molecular or ionic state. An analysis was also made of the sample taken in Mansanilla (the dried-up deposit on the polyethylene film) for the period June–August, 1995. At all stages, measurements were made with the photoelectric counter A3-5M.

2.3.2.1 The elemental composition of aerosol and dust particles

Chemicals contained in the air can be divided by concentration into two basic groups: macrocomponents (concentrations from 100,000–200,000 ng m^{-3} to several hundred ng m^{-3}) and microcomponents (from several hundred ng m^{-3} to 1–5 ng m^{-3}). The group of macrocomponents includes Al, Si, P, S, Cl, K, Ca, Ti, Fe which are considered natural lithogenous elements. Other elements – V, Cr, Mn, Cu, Zn, Ga, Se, Br, Rb, Sr, Y, Zr, W, Hg, Pb, Th – detected in air samples refer to the group of microcomponents, though the boundary between the groups is somewhat conditional. For instance, in some cases concentrations of zinc or lead reached 500–1,000 ng m^{-3}, and the concentration of iron decreased to 50 ng m^{-3}. The second group includes elements which can enter the atmosphere both naturally (Mn, Sr, etc.) and from anthropogenic sources. This division of elements into groups by the content in the air has been confirmed by the results of analysis of precipitation and volcanic ash samples, as well as by available geochemical data for the Earth's crustal rocks. For instance, in the case of ash, there are two ranges of elemental content: from tens to tenths of percent and from hundredths of percent and less. Variations of concentrations are determined by the content of chemicals in the Earth's crust and rocks (granite, basalt). Therefore, a selection of samples of air and precipitation from the group of natural lithogenous elements, on the basis of the criterion of greatest concentration, is a natural consequence of the geochemical composition. On the other hand, it indicates a natural lithogenous source bringing these elements to the atmosphere. This is also confirmed by a correlation analysis. First, the content of most elements of the lithogenous group in the atmosphere changes in time and space in a similar way (i.e., there is a common source of these elements). An exception is sulphur, the content of which in the atmosphere varies specifically. Second, elements of the lithogenous group are somewhat interrelated, this relation remaining stable for a long time (several months, at least) over a large area. This relationship is similar to the ratio of concentrations of the same elements in the samples of volcanic ash. Possible anthropogenic inputs practically cannot be noticed against the natural contents of elements, except in some cases.

A more complicated situation is observed with respect to microelements of the second group. Against a background of small concentrations of these elements, both in the Earth's crust and in the atmosphere, the supply of these elements from anthropogenic sources can change their content in the air substantially. To assess the degree of lithogenous or anthropogenic impact, a technique of estimating coefficients of elements enrichment in the atmosphere has been used.

2.3.2.2 *Temporal changes in concentrations of elements*

The use of correlation analysis has made it possible to evaluate the character of temporal changes in concentration of elements $C(t)$ (sometimes together with spatial variations) at some stages of observations between December, 1994–October, 1995 in the cities of Colima, Mansanilla, Guadalajara, and at other locations. The results of the analysis can be seen in Table 2.1, where for some observational locations the elements are divided into various groups. In each group the concentrations of elements $C(t)$ change similarly ($K_{corr} = 0.70$–1.00; the cases with $K_{corr} > 0.9$ are marked by an underlined element). Concentrations of macroelements Al, Si, P, K, Ca, Ti, and Fe in each series of measurements change in a similar way, which indicates a common source of their emission to the atmosphere. Considering a large absolute content of these elements in the atmosphere and a good correlation of the spatial–temporal dependences of their concentrations, the measured elements Al, Si, P, K, Ca, Ti, and Fe can be considered as elements of natural lithogenous origin. Almost at every location of observation, this group also includes the microelements Mn and Sr, sometimes with not very high coefficients of correlation. Variations of the content of these elements more or less correlates with changes in the $C(t)$ of macroelements of lithogenous origin. These elements constitute the most stable

Table 2.1. Correlation of temporal and spatial variations of element concentrations in air and precipitation samples (Mexico, 1994–1995): a selection of groups of elements.

Time and place of sampling	Al	Si	S	Cl	K	Ca	Ti	V	Cr	Mn	Fe	Ni	Cu	Zn	Se	Br	Rb	Sr	Zr	Pb
December 1994–	<u>Al-</u>	<u>Si-</u>		-	K-	<u>Ca-</u>	<u>Ti-</u>	-	Cr-	<u>Mn-</u>	<u>Fe-</u>	-	Cu-				-	Sr-	Zr	
February 1995,																				
Colima city,				Cl-							-	<u>Ni-</u>		-	<u>Se-</u>	<u>Br-</u>	Rb-			<u>Pb</u>
Popocatepetl									Cr-				-	Zn-				-	-	
volcano			S																Zr	
May 1995,	<u>Al-</u>	<u>Si-</u>		-	K-	<u>Ca-</u>	<u>Ti-</u>			-	<u>Fe-</u>	Ni-		-	Se-	Br-	-	<u>Sr-</u>	Zr-	Pb
Colima,									Cr-	Mn-						-	Rb			
Playon, Refujio			S										Cu	Zn						
August 1995,	Al-				-	Ca-	<u>Ti-</u>	-	Cr-	Mn-	<u>Fe-</u>		-	Zn						
Mansanilla		Si-	-	Cl-	K-											-	Rb-	Sr		
			S-				-	<u>V-</u>			-	<u>Ni-</u>		-	Se					
																			<u>Zr-</u>	<u>Pb</u>
													Cu			Br				
August–October 1995,	<u>Al-</u>	<u>Si-</u>	S-	<u>Cl-</u>	K-	<u>Ca-</u>	<u>Ti-</u>	-	Cr-	<u>Mn-</u>		-	Cu-	Zn-	-	Br-	-	<u>Sr</u>		
Colima City								V-			-	Ni-				-	Rb			
											-	Ni-								
																	-	<u>Zr-</u>	<u>Pb</u>	
											Fe									

October 1995,		$\underline{\text{Si}}$-		-	$\underline{\text{K}}$-	$\underline{\text{Ca}}$-	$\underline{\text{Ti}}$-		-	Mn-	$\underline{\text{Fe}}$-	-	Cu-	Zn-	-	Br-	-	Sr		
Guadalajara	Al-															-	Rb			
								V-			-	Ni								
																			Zr-	Pb
			S	Cl					Cr											
Rain water:	Al-			-	K-	$\underline{\text{Ca}}$-		-	Cr-	-	$\underline{\text{Fe}}$-	Ni-		-	Se-	-	Rb-	Sr		
May–August 1995,					$\underline{\text{K}}$-			-	$\underline{\text{Cr}}$-		-	$\underline{\text{Ni}}$-		-	Se-	Br-	Rb			
Colima City			$\underline{\text{S}}$-										-	$\underline{\text{Zn}}$						
		Si		Cl						Mn										
Water deposits:	Al-	$\underline{\text{Si}}$-		-	$\underline{\text{K}}$-	$\underline{\text{Ca}}$-	$\underline{\text{Ti}}$-	-	Cr-	$\underline{\text{Mn}}$-	$\underline{\text{Fe}}$-	Ni-	Cu-	Zn-			-	$\underline{\text{Sr}}$		
May–October 1995,									$\underline{\text{Cr}}$-		-	Ni-	$\underline{\text{Cu}}$-	$\underline{\text{Zn}}$-	Se-	-	Rb-		-	Pb
Colima City			S												-	Br				

Note:
(a) Each line contains elements combined by the character of changes in its concentrations into one group (combined with dashes).
(b) The last line, for each series of measurements (except the last series) contains elements whose variations do not correlate with changes in the content of other elements.
(c) Symbols of elements are underlined when their distributions of concentrations correlate with $K_{corr} > 0.9$.
(d) Elements are underlined if they are a part of the lithogenous group.

part of the lithogenous group (they are shown in bold type in the first line for each point of sampling) with some exceptions at several observation locations.

At different stages of observation the group of natural components was supplemented with other microelements with similar variations of $C(t)$. For instance, in measurements carried out in May in the cities of Colima, Playon, and Refujio they were Ni, Se, Zr, Pb and partly Br; in measurements in Mansanilla in August, 1995 – Zn, Cr; in Colima in August–October – Cr, Cu, Zn, Br and to a lesser degree Rb; in Guadalajara in October, 1995 – Cu, Zn, and Br. It is clear that for the enumerated microelements the process of transport to the atmosphere is determined by the impact of both natural and anthropogenic sources, and only concrete conditions at an observational location determine the dominant process. The 1994–1995 wintertime measurements revealed elements Cr, Cu, Y, and Zr, with variations of $C(t)$ corresponding to a certain degree to $C(t)$ variations for the elements of the lithogenous group, though they were subject to external forcings.

In different measurement series there were microelements for which the dependences of $C(t)$ did not correlate with changes in concentrations of natural lithogenous elements, whereas their variations were similar. In the series of wintertime measurements one isolated group included Ni, Ga, Se, Br, Rb, and Pb, and another – Cr, Zn, and Zr. In May it was Cr, Mn, Rb, and Th; in August in Mansanilla – Ni, Se, and V; in August–October in Colima – V, Ni, and Rb; in October, 1995 in Guadalajara – Ni, V, Al, and Rb. Since many of these microelements have been marked already as correlating (well or partially) with variations of natural components, apparently, at these locations and at a given time, their arrival to the atmosphere depends mainly not on lithogenous sources but on other sources.

For some observational locations individual elements, in particular, Cr, Ni and others, are indicated simultaneously in different lines. This means that the distributions of $C(t)$ for these elements correlate with temporal variations of concentrations of other elements from different groups, but the degree of correlation in this case is not high: $K_{corr} = 0.6$–0.8, though there are cases when $K_{corr} > 0.9$. In consideration of the results of the analysis of rain water and aqueous deposits, Cr, Ni as well as Cr, Cu, and Zn together with some other elements form separate groups ($K_{corr} > 0.9$), the concentrations of which change in time in a different way compared with lithogenous elements. At the same time, there is a weak correlation with the lithogenous group. Changes in concentrations of some elements at some observational locations are of an individual character (S, Cl, Cu, Zn, etc.), though at other locations these elements can be part of one or another group.

The behaviour of lead and zirconium should be discussed separately. During measurements in May and partly in winter (in dry season), the concentrations of these elements changed in time (and in space) like the variations of $C(t)$ for the elements of the lithogenous groups. In these cases the natural input of lead and zirconium to the atmosphere dominated, though apparently, the impact of other sources was observed.

However, in measurements carried out in August–October, 1995 (mainly in the wet season) in Colima, Mansanilla, and Guadalajara, lead and zirconium formed in pairs, for which the dependences of $C(t)$ changed synchronously but did not correlate with variations in lithogenous components as well as other elements (only for Ni and V in Colima were changes in $C(t)$ similar to changes in $C(t)$ for Pb and Zr). This phenomenon was followed by a considerable increase of the content of lead and zirconium in the atmosphere both in the daytime and at night (i.e., in this area (Colima–Mansanilla–Guadalajara) a powerful unknown source of these elements was in action.

The content of sulphur, by the magnitude of concentration referred to the group of macrocomponents, changes in a specific way, and most probably has a natural source of emissions to the atmosphere. Variations of $C(t)$ for sulphur is of an individual character, and from results of analysis of samples of the winter and May series of measurements, they do not correlate with variations of lithogenous macroelements. This situation is also observed in the case of measurements in Guadalajara in October, 1995, and only in Mansanilla in August, 1995 variations of the sulphur content in the atmosphere were similar to $C(t)$ for Ni, V and partly Se, but in Colima in August–September variations in $C(t)$ were similar for sulphur and lithogenous components. Thus a natural powerful source of input of sulphur to the atmosphere in a given region was, apparently, the December, 1994–January, 1995 eruption of the Popocatepetl volcano, the consequences of which were observed in the atmosphere until May, 1995 (i.e., until the end of the dry season). In contrast to emissions of lithogenous substance in this period, sulphur was emitted to the atmosphere in a gaseous phase (SO_2, etc.) – sulphur compounds then reacting with water vapour formed sulphuric acid. This affected the dynamics of the sulphur content in the atmosphere and caused a difference in the changes of $C(t)$ for sulphur and other elements of the lithogenous group. The same situation was also observed during the next dry season in early October, 1995 during measurements in Guadalajara, which resulted from an enhanced emission from the volcano Fuego de Colima in August, 1995. In the rainy season (late May–August, 1995) the volcanic sulphur was washed out from the atmosphere, and during measurements in August, sulphur was recorded which entered the atmosphere together with lithogenous elements from the soil surface (Colima) or from anthropogenic sources together with V, Ni, and Se (Mansanilla).

Figure 2.1 demonstrates the temporal distributions of $C(t)$ for some elements at several observational locations during the 1994–1995 wintertime series of measurements. For lithogenous elements Si, Ca, and Fe changes in $C(t)$ are synchronous over the whole interval of measurements and coincide even in small details. At the same time, the character of changes in the sulphur content in the atmosphere differs from $C(t)$ for lithogenous elements.

2.3.2.3 Analysis of the elemental composition of air samples

Studies of the character of changes in ratios between elements (changes in the composition of samples) by correlation analysis were carried out separately for macro and microelements in connection with a great quantitative difference between concentrations reaching a coefficient of 100,000. The macroelemental

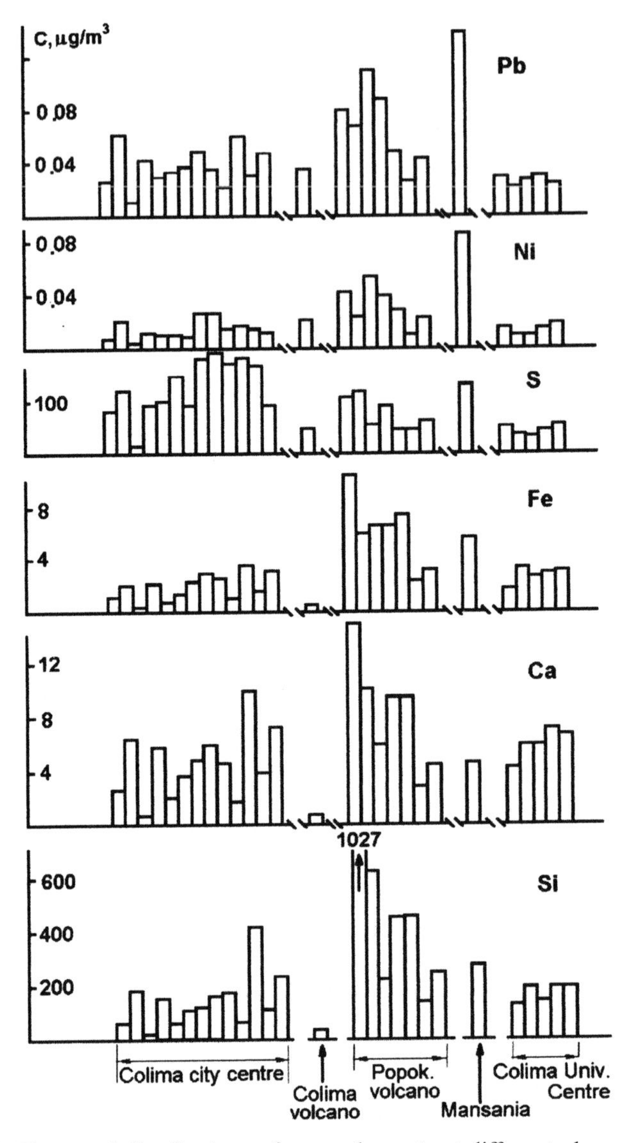

Figure 2.1. Temporal distributions of some elements at different observational sites.

composition of air samples is mainly characterized by the series of elements Al, Si, S, Cl, K, Ca, Ti, and Fe. Periods of continuous observations at most locations did not exceed 7–8 days and only near the Popocatepetl volcano were observations made with intervals during January, 1995.

Changes in concentrations of metals Ni and Pb in this series of measurements are similar, but not like the variations in concentrations of Si, Ca, Fe, and sulphur. Note that in December, 1994 (Colima city centre) changes in the content of lead in

Table 2.2. Coefficients of correlation for comparison of dependences of $C(t)$ for some elements during the wintertime series of measurements in 1994–1995.

Element	Si	Ca	Fe	S	Ni	Pb
Si	1	0.91	0.90	0.11	0.42	0.49
Ca		1	0.84	0.11	0.29	0.42
Fe			1	−0.01	0.65	0.68
S				1	0.18	0.20
Ni					1	0.93
Pb						1

the atmosphere did not correlate with $C(t)$ for the elements of the lithogenous group, but were closer, in particular, to $C(t)$ for Ni and other elements (Table 2.1) (i.e., the impact of the Popocatepetl eruption on the input of lead to the atmosphere compared with other sources was reduced substantially). The correlation coefficients for dependences of $C(t)$ shown in Figure 2.1 are given in Table 2.2 as quantitative characteristics of the performed comparison of the results.

2.3.2.4 Macroelements

An analysis of data at each stage of measurements has shown the following: during the wintertime stage (December, 1994–February, 1995) the existence of a certain stable concentration ratio is observed for macroelements in the air of Colima, to a lesser extent in Mansanilla, near Popocatepetl, and in samples of volcanic ash. This is seen from Figure 2.2 which demonstrates averaged elemental spectra (distributions of concentrations $C(Z)$) of air samples taken at two observational locations in Colima and near the Popocatepetl volcano. Distributions of $C(Z)$ for the database are characterized by the following order of elements with their concentration decreasing: Si, S, Al, Ca, Fe, and others. The correlation between distributions of $C(Z)$ for all samples is $K_{corr} = 0.8$–1.0. Nevertheless, in Colima (city centre) in December, 1994, in the correlation analysis of some samples, was exhibited short-term disturbances in this ratio of elements in the atmosphere, the $C(Z)$ for these samples correlated with $K_{corr} = 0.91$–1.00, and the correlation with other distributions of $C(Z)$, especially with results of measurements in January–February, 1995 was $K_{corr} < 0.7$, though in December, 1994 correlations were observed at $K_{corr} = 0.7$–0.8. This difference shows itself whilst comparing the averaged elemental spectrum for samples taken on 13–20 December, 1994 (Figure 2.2(c)) with spectra of samples taken during other wintertime measurements (Figure 2.2(b,d,e)): maximum concentration in these samples occurred not for silicon but sulphur, which was higher than in other samples at this stage, at the same time, the absolute content of all other macrocomponents was found to be below the values characteristic for this series of measurements. It can be assumed that the composition of macroelements in the air over this region of observations in January–February, 1995 was determined by the composition of volcanic ash emitted to the

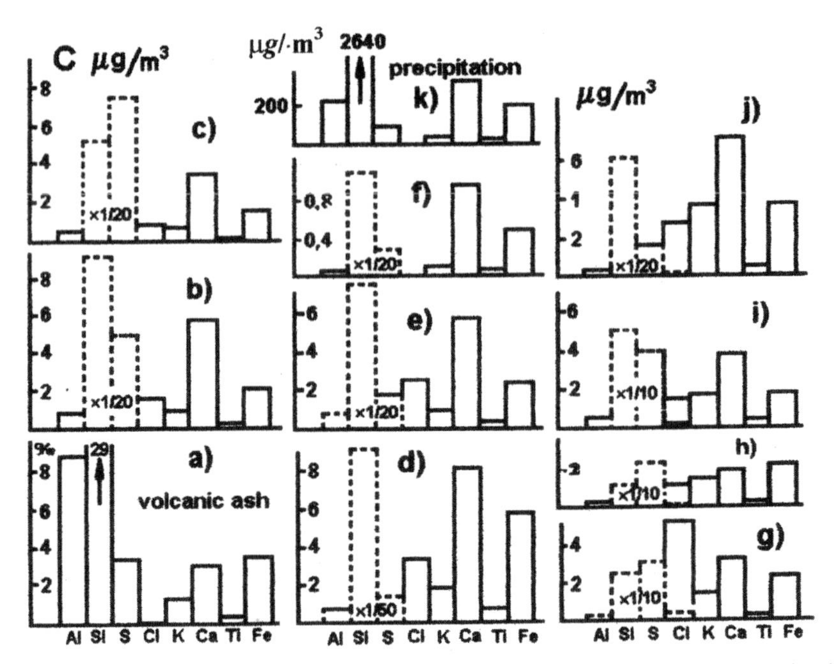

Figure 2.2. Averaged spectra of macroelements for samples of air, precipitation, and volcanic ash taken at different stages of measurements at several observation sites: (a) – samples of volcanic ash, December, 1994–January, 1995; (b) – air samples of volcanic composition, first stage of measurements, Colima (city centre), 14–15, 17–18, 19, 22 December, 1994; (c) – air sample of different composition in the same place, 13–14, 15–17, 18–19, 19–20 December, 1994; (d) – air, first stage, Popocatepetl volcano, 29 December, 1994–28 January, 1995; (e) – air, first stage, Colima (university), January–February, 1995; (f) – air, second stage, 9–20 May, 1995; (g) – air, third stage, Mansania, 10–14 August, 1995; (h) – air, third stage, Colima (university), 28 August–3 September, 1995; (i) – air, third stage, Guadalahara, 4–7 October, 1995; (j) – air, in the same place, 7–9 October, 1995; (k) – rain, May–October, 1995.

atmosphere during the Popocatepetl eruption. In December, 1994, when after the eruption of the Colima volcano the amount of ash in the atmosphere increased occasionally, there were cases of deviation of the macroelemental composition of samples from the total ratio of these elements in the air. The concentration of particles containing macroelements decreased, and sulphur compounds still remained in the atmosphere – their concentration even increasing due to volcanic gases emissions. The sample taken near the Colima volcano on 23 December, 1994 reflects the same situation (the time interval between eruptions): the macrocomposition of this sample coincides with the composition of samples taken in December, 1994 in Colima ($K_{corr} = 0.96$–0.99) but correlates less with compositions of other samples for this month ($K_{corr} = 0.73$–0.92) and with $C(Z)$ of samples for January–February, 1995 from other observational locations, including samples of volcanic ash ($K_{corr} < 0.5$–0.7). Probably, these cases of deviation of the composition of macroelements from the total ratio in samples correspond to the composition of macroelements in the pre-eruption atmosphere.

Measurements in May in Colima, Refujio, and Playon have shown that the macrocomposition of air samples at the first two locations (17–20 May, 1995) was practically the same. From samples taken in Playon, only the 9–10 May sample composition of macroelements is similar to that of samples taken in Colima and Refujio. Two samples taken on 10–11 May correlate with each other but differ strongly from all other samples at this stage of measurements. Hence, the stable ratio of the contents of macroelements recorded at the beginning of measurements was broken on 10–11 May, 1995. However, by 17–20 May, 1995 it was restored.

During the third stage of measurements a similarity was established between the macrocomponent composition of samples taken in the cities of Apo (13–16 July), Mansanilla (10–14 August), Colima (28 August–3 September and 12–13 October), and in the initial period of measurements in Guadalajara (4–7 October). In the period 13 July–7 October, 1995 the ratio of concentrations of elements in air samples was mainly preserved, not only at every point of sampling but also over the whole region under study (Figure 2.2(g,h,i)). Correlation between distributions of $C(Z)$ is characterized here by $K_{corr} > 0.8$, and in most cases by $K_{corr} > 0.9$.

Samples taken in Guadalajara on 8–9 October, 1995 correlate weakly in the composition of macrocomponents with air samples for previous days as well as with the main mass of samples from other locations – $K_{corr} < 0.6$ almost in all cases.

At the same time, these samples correlate in composition with the air samples of the third stage which at their observational points were an exception: with the sample taken on 15 July in Apo ($K_{corr} = 0.7$–0.9), with the sample taken on 13 August in Mansanilla (0.92–0.99), and with the sample taken on 12–13 October in Colima (0.87–0.98) after the earthquake on 9 October, 1995.

Thus, in most cases during each stage of measurements for all observational locations the stability of ratios has been established between concentrations of macroelements in air samples – though with some disturbances.

A question arises whether the concentration ratio for macroelements was preserved for the whole observational period, from December, 1994 to October, 1995. For this purpose, we compare changes in the composition of samples taken at one location, which will make it possible to exclude different anthropogenic impacts of each concrete locality. Such is the case with observations in Colima (university) and in Mansanilla, apart from measurements in August, 1995, samples were taken in January, 1995. At some stages of measurements we compared data from different locations with results obtained in Colima (university).

A comparison of compositions of macroelements in samples taken during the winter and May stages of measurements in Colima has shown that they are similar ($K_{corr} = 0.96$–1.00) and agree well with the composition of volcanic ash samples ($K_{corr} = 0.91$–0.97). In comparison with averaged elemental spectra obtained in Colima in January and May, 1995, the ratio of elements concentrations are similar in general, with a maximum of concentration falling on Si, and then, in decreasing order, on S, Al, Ca, Fe, and other components. Note that absolute values of concentrations in May were, on average, 5–10 times less than in winter. Bearing in mind that at each stage the constancy was established of the ratio of macrocomponents in samples for different observational locations, the macrocomposition of atmospheric aerosols can be considered similar in the whole region of measurements from December, 1994 until May, 1995 – being determined by volcanic emissions.

Results of the correlation analysis show (see Table 2.2 and Figure 2.2) that in August–September, 1995 in Colima the ratio of macroelements changed with respect to earlier measurements. This is confirmed by values of K_{corr} in comparisons of $C(Z)$ for samples taken in the period 28 August–3 September, with samples of the first and second stages of measurements – $K_{corr} = 0.14$–0.81 (in most cases <0.6) and in comparison with samples of volcanic ash – $K_{corr} = 0.05$–0.56. Maximum concentration in these measurements is observed for sulphur, and it is substantially less than in winter measurements. Then follows Si, also with a smaller concentration, and in decreasing order: Al, Ca, Fe as well as Cl, K, and Ti. The contents of Cl, K, Ca, Ti, and Fe in the atmosphere in winter and in summer are comparable. At the same time, the composition of samples from Colima is similar to that from Apo, Mansanilla, and Guadalajara at the beginning of measurements. Hence, the effect of volcanic ash on the ratio of macroelements in the air over the whole region of observations in August–early October, 1995 was strongly weakened. This is confirmed by the fact that sample no. 14 taken on 23 December, 1994 near the Colima volcano does not correlate with either the 'volcanic' composition of macrocomponents in ash or the atmosphere during this time period, but is similar to samples taken in Colima in August–September, when the effect of the Popocatepetl eruption products was markedly weaker (Table 2.2.) However, the 'volcanic' ratio of elements was restored in October, 1995, which is seen from sample no. 107 taken in Colima on 12–13 October, the composition of which correlates with the composition of samples taken in Guadalajara, including those with the disturbed composition compared with measurements in August–early October. The value of K_{corr} for compositions of this sample and samples of the first and second stages of measurements constitutes 0.85–0.96, and that of volcanic ash samples – 0.76–0.79. It means that the retrieval of the 'volcanic' composition of macroelements in Colima is similar to Guadalajara located far from the main observational location, and was apparently connected with the earthquake in early October. A similar composition of macrocomponents was observed in the atmosphere of Mansanilla in August, 1995 (the sample of 13 August, 1995). At this time, the Popocatepetl volcano was idle, but there was an enhanced emission of volcanic substances from the Colima volcano.

An additional argument in favour of this interpretation of changes in the elemental composition follows from a comparison of the sample taken in January 1995 in Mansanilla with samples taken there in August. This sample with the 'volcanic' ratio of macroelements correlates poorly in composition with the August samples. But between the compositions of this sample and the 13 August sample, which is abnormal in its series of measurements, $K_{corr} = 0.98$.

From late May to late July, 1995, there was a break in aerosol sampling. During this interval, rain water was sampled at Colima during late May–early October. Using these data, ratios were compared between macroelements in rain water samples and in air samples for the period of observations.

The macroelemental composition of rain water samples during the whole period of observations remained practically constant. A correlation between distributions of $C(Z)$ is characterized by $K_{corr} = 0.93$–1.00, with some deviation in the composition of the sample taken on 15–22 July – the correlation with compositions of other samples was at the level of 0.77–0.83. An exception was the sample taken on 3 September. Its composition differs from that of all other samples (K_{corr} is in the interval from 0.2 to negative values) by the absence of Si, S, Cl, Ti, and Fe.

The macroelemental composition of rain water samples (27 May–17 June) almost coincides with the composition of air samples taken in Colima in May ($K_{corr} = 0.99$) and with the composition of volcanic ash samples taken in December, 1994–January, 1995 ($K_{corr} = 0.97$). The same can be said about all other rain water samples ($K_{corr} = 0.93$–1.00) except the samples taken on 15–22 July ($K_{corr} = 0.83$–0.87) and 3 September, 1995 ($K_{corr} < 0$). However, the composition of all 12 samples of rain water with the 'volcanic' ratio of macroelements does not correlate ($K_{corr} < 0.6$) with the composition of air samples taken in Colima during 28 August–3 September, like the compositions of air samples taken in January and May as well as samples of volcanic ash. Moreover, the macroelemental composition of rain water samples does not coincide with samples of atmospheric air taken in the regions of Apo, Mansanilla, Colima, and Guadalajara in July–October, 1995, since ratios of macroelements in these samples mainly correlate with each other in all regions of observations. There is a contradiction between the 'volcanic' ratio of macroelements in rain water samples and their 'natural' ratio in air samples taken in Apo on 13–16 July, in Colima on 2–3 September, and in Guadalajara on 4–5 October, 1995.

In turn, the macrocomposition of air sample no. 107, taken in Colima on 12–13 October, correlated weakly with the database on the content of elements in air samples for July–September, but it was much closer to the 'volcanic' composition of air samples for January and May as well as to volcanic ash samples. The composition of this air sample approaches the 'volcanic' composition of rain water samples ($K_{corr} = 0.73$–0.86 except an abnormal sample taken on 3 September), which, once again, confirms the 'volcanic' character of the ratio of macroelements in the atmosphere of Colima on 12–13 October after the earthquake.

2.3.2.5 Microcomponents

The microelemental composition of air samples changes in time more chaotically than the ratio of macroelements. This is determined directly by small concentrations of elements in samples (mainly units and tens of $\text{ng}\,\text{m}^{-3}$) and the error of their estimation and impact of different sources of input of microelements to the atmosphere, especially under the urban conditions.

It was established from the results of the first stage of measurements that the microelemental composition of air samples (Cr, Mn, Ni, Cu, Zn, Ga, Se, Br, Rb, Sr, Zr, and Pb) over the whole territory correlates in many cases with the composition of volcanic ash samples ($K_{corr} = 0.61$–0.95). However, the microcomposition of the sample for 23 December, 1994 taken near the Colima volcano at a height of ~500 m, as in the case of macrocomponents, coincides neither with the composition of volcanic ashes nor with aerosol samples in this region (Table 2.3). Moreover, for each site of measurements, both similarities and differences of microcompositions of air samples were observed. ($K_{corr} = 0.2$–0.9) (i.e., the ratio of the microelement content in the atmosphere is less stable compared with the ratio of macroelements). In December, 1994, the correlation between microcompositions of samples in many cases was at the level of $K_{corr} \sim 0.80$–0.96, in January–February, 1995 such cases were few, and K_{corr} was mainly equal to 0.6–0.8 or less than 0.6. Respectively, the microelemental composition in the first samples often differed from the results obtained in the January–February samples.

For measurements of the second and third stages the ratio of microelements concentrations in the atmosphere also often deviates from a certain average state for a given observational site. This disturbance of the microelemental composition can coincide with a similar disturbance of the macroelemental composition. The correlation between distributions of $C(Z)$ is characterized by $K_{corr} = 0.4$–0.7 in May and 0.40–0.99 in August–October. Nevertheless, the microelemental composition of atmospheric aerosol samples at each stage of measurements did change, but on the whole, was preserved during all the observational period.

A consideration of changes in the microelemental composition of air samples taken in Colima over the period January–October, 1995 suggested the conclusion that during this period variations of the microelement ratio repeated, on the whole, the character of changes in the microelements, albeit less distinctly manifested. The correlation between $C(Z)$ for samples taken in January and May, 1995 was characterized by $K_{corr} = 0.5$–0.8, though smaller values were observed, too. A similarity between microcompositions of air samples and volcanic ash samples was at a level of $K_{corr} = 0.60$–0.83.

The ratio of microelement concentrations in the atmosphere of Colima in August–September, as in the case of macroelements, differed substantially from results of measurements in January and in May the correlation between $C(Z)$ was practically absent (Table 2.2), and only for one sample of the May measurements was the microcomposition similar to that of samples taken in the end of the rainy season. Also, there was no similarity between the microelemental compositions of these samples and samples of volcanic ash. However, in the sample for 12–13 October, as in comparisons of macroelemental compositions, the ratio between microelements was similar to results of measurements in January and May: the correlation with $C(Z)$ of air samples was mainly at $K_{corr} = 0.65$–0.79, and with $C(Z)$ of volcanic ash a maximum value of K_{corr} was 0.62, which was less than in the case of comparison of macroelemental compositions.

In the case of microelements there are at least three processes: the natural input of substances to the atmosphere, supply due to volcanic eruption, and from anthropogenic sources. In the cities these processes could overlap under certain conditions, in other cases one mechanism of input could prevail. Besides this, the microelements to be determined are mainly metals and, probably, they are removed from the atmosphere much more rapidly than macroelements. Therefore, while the macroelemental

Table 2.3. Correlation matrix (%) of elemental spectra of $C(Z)$ of samples of precipitation (May–October) and air (January, May, August–October) from Colima as well as volcanic ash for the series of macroelements Al–Fe (above, to the right of the unit diagonal) and macroelements Cr–Pb (bottom, on the right).

No. of sample	Precipitation												Popocatepetl Volcano			Air: January						May		August		Sept.			Oct.
	1	2	3	4	5	6	7	8	10	11	12	13	28	29	30	19	24	25	26	27	14	73	74	91	93	92	94	100	95	107
1	*1*	100	100	100	100	77	94	100	97	100	98	100	97	97	97	–	–	–	–	–	–	99	99	6	22	43	20	54	3	79
2	12	*1*	100	100	100	81	94	100	98	100	97	99	98	98	98	–	–	–	–	–	–	100	99	10	25	47	25	57	10	82
3	10	84	*1*	100	100	79	94	100	98	100	98	100	98	98	97	–	–	–	–	–	–	100	99	6	21	43	20	54	4	79
4	−33	3	11	*1*	100	80	94	100	98	100	98	100	98	98	98	–	–	–	–	–	–	99	99	6	21	44	21	54	6	79
5	80	31	46	−12	*1*	78	94	100	98	100	98	100	98	97	97	–	–	–	–	–	–	99	99	5	21	43	20	53	3	79
6	10	−01	−27	44	−08	*1*	83	78	82	80	78	77	84	86	87	–	–	–	–	–	–	83	84	41	49	69	55	77	49	86
7	37	54	65	25	53	−09	*1*	93	95	95	97	95	94	94	93	–	–	–	–	–	–	94	94	10	28	45	22	56	4	78
8	71	53	60	19	76	32	72	*1*	98	100	98	100	98	98	97	–	–	–	–	–	–	99	99	4	20	42	19	52	3	78
10	19	25	39	64	34	14	60	57	*1*	98	99	98	98	98	97	–	–	–	–	–	–	97	96	−03	11	36	12	47	−02	73
11	57	45	16	14	33	58	18	59	41	*1*	98	100	97	97	97	–	–	–	–	–	–	100	99	9	24	46	23	56	5	81
12	85	17	−02	−03	58	42	35	66	38	78	*1*	99	97	96	95	–	–	–	–	–	–	97	96	−03	13	35	11	46	−07	73
13	93	−07	−19	−37	61	30	5	51	8	63	83	*1*	97	97	96	–	–	–	–	–	–	99	99	5	21	43	19	53	1	79
28	69	−05	24	−11	76	−23	50	60	32	3	51	49	*1*	100	100	91	96	97	95	94	41	97	96	0	16	38	16	49	5	74
29	55	2	38	24	63	−07	65	71	69	17	44	36	83	*1*	100	92	97	97	96	96	46	97	97	6	21	43	22	53	11	77
30	69	−10	14	−31	73	−26	32	47	3	−05	47	52	95	63	*1*	93	97	97	96	96	48	97	96	7	22	44	23	54	13	78
19	–	–	–	–	–	–	–	–	–	–	–	–	66	70	57	*1*	97	98	98	99	73	96	97	44	55	74	56	81	41	96
24	–	–	–	–	–	–	–	–	–	–	–	–	75	61	73	68	*1*	100	100	99	54	100	100	21	35	56	35	65	19	87
25	–	–	–	–	–	–	–	–	–	–	–	–	31	31	23	47	70	*1*	100	100	57	100	100	24	38	59	38	67	22	88
26	–	–	–	–	–	–	–	–	–	–	–	–	72	62	71	74	91	65	*1*	100	59	99	100	27	41	62	41	70	24	90
27	–	–	–	–	–	–	–	–	–	–	–	–	82	60	83	52	66	52	76	*1*	64	99	99	33	46	66	46	73	30	92
14	–	–	–	–	–	–	–	–	–	–	–	–	0	14	−13	21	14	74	11	31	*1*	51	55	92	92	98	97	97	90	87
73	65	−35	−27	−32	37	0	−07	23	−12	20	58	68	69	43	77	28	78	41	78	76	−09	*1*	100	15	29	52	29	61	13	85
74	37	−13	21	7	29	−13	23	41	51	19	27	32	58	82	41	43	46	23	61	56	3	51	*1*	20	33	56	33	65	16	87
91	32	6	40	24	33	2	42	57	60	23	14	22	41	80	18	59	33	16	48	28	4	15	86	*1*	96	93	98	87	92	65
93	8	12	50	27	18	−17	36	40	58	9	−07	−02	30	70	7	43	11	7	33	27	10	3	84	95	*1*	94	97	91	87	74
92	4	36	52	−18	23	−27	−14	12	1	20	−17	2	−06	6	−07	23	−01	−00	34	14	−14	−06	34	39	48	*1*	97	99	85	89
94	−08	40	65	40	6	−02	36	45	45	15	−09	−22	22	52	4	12	−05	−10	12	21	−01	1	63	70	81	40	*1*	93	95	75
100	−02	−05	32	37	3	−08	27	29	63	5	−10	−08	20	67	−04	32	3	3	21	15	13	−03	83	91	95	38	71	*1*	80	94
95	−05	−07	30	37	1	−15	27	22	61	1	−13	−12	18	65	−06	33	3	11	22	19	23	−04	82	89	95	38	68	99	*1*	59
107	65	12	22	1	49	8	36	55	39	56	48	61	38	59	23	69	68	65	79	42	25	38	62	78	61	34	34	49	50	*1*

composition of samples is relatively constant, the composition of microelements changes rapidly and, again, restores due to a new input of substances from a certain source.

The correlation analysis of the microelemental composition of the settled rain water samples taken in the period 27 May–5 August has shown that the ratio of microelements in samples is preserved during the whole period indicated. For six samples, $K_{corr} = 0.84$–0.99, and the maximum concentration is of zinc. Two samples characterized by a lower content of zinc, compared with other elements, correlate with the rest at $K_{corr} = 0.57$–0.87. However, an analysis of rain water deposited onto filters both during the same time and during the whole sampling period 27 May–3 October demonstrates a contradictory pattern – only in some cases is there a similarity of microcompositions of samples characterized by $K_{corr} > 0.8$. Probably, the composition of these samples is affected not only by the content of elements in the atmosphere but also by the fractional distribution of substances in rain water samples as well as by the duration of sampling. Note that there is a similarity between microelement ratios for sample no. 1 for 27 May–17 June (the end of the dry season) and nos. 12 and 13 for 25 September and 2–3 October (the beginning of the next dry season) – a correlation between $C(Z)$ and these samples is 0.83–0.93. A comparison of microcompositions of rain water samples with air and volcanic ash samples has shown that a weak correlation between them ($K_{corr} = 0.6$–0.8) is observed only in some cases. For instance, the microcomposition of deposits from the first rain water sample taken on 27 May–17 June, on the verge of the dry and wet seasons, is similar to some extent to the composition of volcanic ash samples ($K_{corr} = 0.69$) as well as to the composition of one of the air samples taken in Colima on 19–20 May ($K_{corr} = 0.65$). The composition of the first sample of deposits together with a similar sample for 2–3 October is similar to atmospheric sample no. 107 for 12–13 October taken in Colima ($K_{corr} = 0.61$–0.65) which, in turn, is similar to samples of 'volcanic' character of the first and second stages of measurements.

It follows from the analysis of air samples that the ratio of macroelements of 'volcanic' nature in the atmosphere caused by the Popocatepetl eruption was formed over the whole region of observations in December 1994 and continued until May 1995 (i.e., until the end of the dry season). A comparison of compositions of air and rain water samples suggests the conclusion that this ratio of elements is preserved in the atmosphere until June (inclusive). Beginning in July (APO) and until early October this ratio was violated and, respectively, another ratio, also of natural character, was established. This can result from the process of washing out of volcanic substances from the atmosphere during the rainy season, which is testified by a decrease of concentrations of some elements, especially Al, Si, and S. Here a similarity was observed between macroelemental compositions of some samples taken in July–early October, 1995 with some samples for October, 1994, when the Popocatepetl volcano eruption began. However, on 8–13 October in Colima and Guadalajara, the ratio of macroelements of volcanic nature was observed, the same composition also being recorded in a sample taken in Mansanilla on 13 August. This character of changes in the atmospheric composition is also confirmed by the data of analysis of the microelemental composition of air and precipitation samples.

2.3.2.6 The elemental composition of fractions of aerosol and dust particles

A combined change of size distribution and elemental composition of aerosol samples for the period of 13–19 December, 1994 in Colima has been considered. For this purpose, for parallel measurements, comparisons were made of temporal dependences of the volume density of aerosol particles dV/dr obtained with the use of the photoelectric counter A3-5M for 12 ranges of particle sizes in the interval from 4 to 10 μm and larger, as well as temporal distributions of the element concentrations $C(t)$. The results of dV/dr measurements were averaged over the time interval corresponding to air filter sampling. A combined analysis did not reveal good correlations ($K_{corr} > 0.8$) between distributions of dV/dr and $C(t)$. Nevertheless, in some cases one can observe a weak correlation between such distributions with $K_{corr} = 0.6$–0.8 or slightly less (0.5–0.6) (see Table 2.4).

There were cases when $C(t)$ for any element overlapped a similar change of dV/dr for two or more measured fractions of aerosol particles. These values of K_{corr} do not allow one to draw any concrete conclusion about the fractional composition of particles containing certain elements, however, they can serve as additional information in the analysis of the results obtained in May–October, 1995 in experiments with packets of filters.

In particular, it should be pointed out that a group of natural lithogenous macroelements mainly correlates with variations of the content of large particles, on average, with $D = 1.5$–7 μm, and particles of calcium and titanium with $D = 4$–10 μm. Besides this, in the cases of sulphur, iron, and strontium, particles with $D = 0.7$–0.9 μm are observed. From the group of microelements, strontium, which is also considered to be lithogenous, refers to large particles with $D = 1.5$–4 μm, and lead refers to particles with $D = 4$–10 μm and larger.

Table 2.4. Correlation coefficient.

Al	and fractions with $d = 1.5$–7 μm	$K_{corr} = 0.50$–0.52
Si	and fractions with $d = 1.5$–4 μm	$K_{corr} = 0.50$–0.58
S	and fractions with $d = 0.7$–0.9 and 1.5–2 μm	$K_{corr} = 0.57$–0.59 and 0.73
Cl	and fractions with $d = 1.5$–4 μm	$K_{corr} = 0.55$–0.63
Ca	and fractions with $d = 4$–10 μm	$K_{corr} = 0.50$–0.52
Ti	and fractions with $d = 4$–15 μm	$K_{corr} = 0.53$–0.64
Cr	and fractions with $d = 0.6$–0.8 μm	$K_{corr} = 0.51$–0.52
Fe	and fractions with $d = 0.7$–0.8 and 1.5–2 μm	$K_{corr} = 0.56$ and 0.53
Ni	and fractions with $d = 0.6$–2 μm	$K_{corr} = 0.51$–0.76
Cu	and fractions with $d = 1.5$–2 μm	$K_{corr} = 0.66$
Ga	and fractions with $d = 0.7$–0.9 μm	$K_{corr} = 0.68$–0.73
Br	and fractions with $d = 0.6$–0.8 and 1.5–2 μm	$K_{corr} = 0.54$–0.56 and 0.83
Rb	and fractions with $d = 4$–7 μm	$K_{corr} = 0.59$
Sr	and fractions with $d = 0.7$–0.9 and 1.5–4 μm	$K_{corr} = 0.53$–0.58 and 0.56–0.71
Y	and fractions with $d = 0.7$–0.9 and 1.5–2 μm	$K_{corr} = 0.63$ and 0.86
Pb	and fractions with $d = 4$–15 μm	$K_{corr} = 0.61$–0.66

Samplings made in the first and second series of measurements, with the use of packets of filters, and an elemental analysis of samples, have given information for quantitative estimates of the size of particles.

This problem can be solved on a qualitative level using *in situ* data about the distributions of elements in the layers of the filters. In this packet the first layer catches large particles, the second layer catches particles moderate in size, and the third layer catches the remaining particles.

It follows from the observational data that in the case of macroelements of the lithogenous group, most of them are deposited on the first layer of the packet. For Al, 50–100% is deposited, Si – 70–100%, Cl – 80–100%, K – 40–100%, Ca, Ti – 60–100%, and Fe – 71–98%. Thus, even taking into account the possible passing of large particles through the first layer of the filter packet, one can state that the enumerated elements are contained in large particles several μm in size. This agrees with an earlier estimate of 2–7 μm. There are some cases for Al, K, Cl, and Ti, when their share on the first layer of the packet is <50%, and this can be connected with both measurement errors and specific features of the real measurement situation. But, on the other hand, the share of Si, Ca, and Fe on the first layer of the packet is always >60%.

In the case of sulphur, its share on the first layer of the packet constitutes, on the whole, 30–70% of the total amount in a sample, the share of the first layer in the August–October samples dominates. This agrees with the fact mentioned above about a correlation between sulphur and particles with $D = 1.5$–2 μm. At the same time, sulphur was also present among the particles with $D = 0.7$–0.9 μm (i.e., the share of sulphur on the first layer of the packet was less compared with other elements of the lithogenous group and comparable with the shares on the second and third layers).

Strontium was also present among the coarse particles ($D = 1.5$–4 μm), though it was associated with the fraction 0.7–0.9 μm. It follows from the observational results that the share of strontium on the first layer of the packet dominates most often, especially in samples taken in August–October, 1995, and constitutes 40–100%. However, there are samples where the shares of the second and third layers are either comparable with or exceed the share of the first layer of the packet.

An ambiguous situation was observed in the case of lead which was present earlier among particles with $D = 4$–10 μm and larger. There are few samples in which the share of lead in the first layer of the packet exceeds 50%, though there are some samples where its share constitutes 90–100%. Nevertheless, in most samples the share of lead in the first layer is <50% and even <10%, the share of the third layer being 80–97%. Hence, most often lead is transported by fine particles with $D < 1$ μm. In some samples the amount of lead in one of the layers of the packet of filters between 80–100% of the total mass of this element in a sample. In samples taken at different sites on 13–14 August and 1 September lead predominates in the first layer; but in the 2–3 September sample it predominates in the second layer; and in samples taken on 15–16 July, 12 August, 8 October, and partly 7 October and 8–9 October 1995 it predominates in the third layer. The same situation is observed in these samples in the case of zirconium with respect to the share of this element on the same layers of the packet (75–90%). Besides this, in these samples taken at four sites the concentration of lead and zirconium is very high. Considering the whole series of samples, in most cases zirconium prevails in mass on the first layer of the packet (i.e., it is connected with large particles).

As for other elements (Cr, Ni, Cu, Ga, Br, and Rb), earlier connected mainly with particles with $D < 1$–$2\,\mu m$, on the basis of a combined correlation analysis, the direct use of observational data on the mass distribution of elements from layers of the filters does not permit one to draw concrete conclusions on the size of the particles for individual elements. The elements Cr, Ni, and Cu dominate both in the first layer and in the second or third layers (or prevail in the second and third layers as a whole). This does not contradict the supposition about the size of particles for these elements ranging between 0.6 and $2\,\mu m$. The elements Zn, Ga, and Br are deposited most on the first layer of the packet. But while for Ga this results from serious observational errors (>50%), for Br it is connected with particles with $D = 1.5$–$2\,\mu m$ (and with $D = 0.6$–$0.8\,\mu m$). The situation with rubidium is uncertain: on the one hand, it deposits on large particles with $D = 4$–$7\,\mu m$, and on the other hand, the mass of rubidium often prevails on the first and second layers of the packet.

Proceeding from this analysis, it can be stated that macroelements of the lithogenous group Al, Si, Cl, K, Ca, Ti, and Fe deposit mainly on particles with $D = 2$–$7\,\mu m$, and sulphur is transported by smaller particles with $D = 1$–$2\,\mu m$. The microelements Zn, Sr, Zr, Br, and sometimes Pb are often connected with particles with $D > 1\,\mu m$, though in some cases one can suppose the presence of small particles with $D < 1\,\mu m$, especially for lead. It is difficult to connect the elements P, V, Cr, Mn, Ni, Cu, Ga, Se, Rb, W, Hg, and Th with particles of certain sizes.

2.3.2.7 *Measurements of aerosol size distribution*

The particle number density and size distribution were measured mainly with the use of the photoelectric counter A3-5M on the roof of the Atmospheric Research Centre at a height of 120 cm above the roof and about 5 m from the ground in rainy and dry seasons. Measurements at other sites gave similar results, therefore we shall dwell upon the results obtained at the Atmospheric Research Centre. Of principal interest are measurements during rain. Depending on precipitation character, various types of washing out of aerosol particles in the surface layer of the atmosphere were observed. Washing out is of selective character similar to that described earlier [50]. Particles with $D \geq 4\,\mu m$ are washed out most efficiently. However, quite unexpectedly their concentration is very rapidly restored to a certain magnitude. The only cause of this may be the activity of the Colima volcano. Probably, part of the volcanic aerosol substance has entered the atmosphere above the level of clouds and in their absence has diffused and deposited onto the lower layers. Different conditions for washing out and for the restoration of the spectrum of aerosol particle sizes does not make it possible to draw reliable conclusions from the observational data about an efficiency of washing out of particles of different sizes. Therefore, data were averaged over spectra of particle sizes for three series of measurements: 1 – May–June (192 spectra), 2 – July–August (264 spectra), and

Table 2.5. The diurnal change of the volume number density of aerosol particles size distribution V/r ($\mu m^3\,m^{-3}$) in the surface layer for three series of measurements (Colima City, Atmospheric Research Centre).

	Time (hr)											
D (μm)	01–02	03–04	05–06	07–08	09–10	11–12	13–14	15–16	17–18	19–20	21–22	23–24
10–15	6.24	5.10	8.03	11.0	9.68	10.4	11.9	9.57	9.13	4.62	3.99	4.7
	2.64	1.54	1.19	2.20	2.33	1.86	1.46	1.92	1.81	1.74	2.82	1.76
	0.22	0.044	0.15	0.53	0.92	1.34	0.93	0.77	0.64	0.55	0.46	0.37
7–10	4.60	2.68	3.64	5.16	6.84	7.15	7.09	8.52	6.27	9.14	22.0	16.9
	0.88	0.85	0.83	1.40	1.74	1.24	1.09	1.29	1.20	1.76	3.20	2.84
	0.28	0.15	0.22	0.44	0.74	0.69	0.90	0.95	0.78	0.37	0.56	0.47
4–7	4.10	3.56	5.17	8.16	7.76	9.84	10.1	8.34	5.43	6.72	11.6	10.1
	2.02	1.84	1.77	4.08	2.53	2.18	1.73	1.98	1.7	1.45	5.03	2.72
	0.65	0.31	0.31	0.57	1.13	1.05	0.94	0.91	0.76	0.82	0.92	0.84
2–4	4.10	3.54	28.4	40.4	7.47	6.48	6.30	7.26	5.42	6.04	5.76	4.63
	2.19	2.10	1.38	2.68	3.03	2.86	2.45	2.20	2.10	2.24	2.97	3.19
	0.72	0.60	0.58	0.85	1.84	1.51	1.79	1.36	0.91	1.03	0.90	0.81
1.5–2	2.01	1.83	1.05	2.08	2.96	3.32	9.88	5.62	4.72	2.88	5.19	5.46
	0.54	0.32	0.56	1.45	0.86	0.60	0.79	0.66	0.71	0.73	1.14	0.84
	0.34	0.23	0.22	0.47	1.07	0.65	0.81	0.58	0.49	0.47	0.43	0.42
1.0–1.5	0.86	0.70	0.55	0.68	0.98	8.82	24.7	7.03	2.48	1.86	0.83	0.56
	0.40	0.40	0.42	0.57	0.46	0.49	0.43	0.41	0.34	0.35	0.93	0.39
	0.92	0.77	0.66	0.29	0.35	0.59	3.32	2.66	0.98	0.93	1.25	1.10
0.9–1.0	1.83	2.19	3.78	3.74	6.04	46.2	164	129	19.0	10.6	16.6	9.97
	0.64	0.28	0.32	0.70	0.80	1.65	1.82	1.28	1.29	1.33	1.01	2.07
	7.18	9.66	10.8	6.58	1.38	3.68	6.66	9.20	9.35	1.74	4.14	5.48
0.8–0.9	3.35	3.60	9.79	5.95	10.7	45.8	141	106	30.4	10.4	2.78	37.9
	0.83	0.70	1.10	0.99	7.96	10	13.7	5.18	1.60	1.47	1.74	4.12
	11.2	12.5	13.7	13.6	12.8	33.8	91.6	60.8	26.2	2.58	7.68	9.90
0.7–0.8	6.66	9.12	17.0	38.6	10.7	45.8	157	170	67.0	0.48	1.96	36.7
	1.98	2.11	2.24	2.53	28.6	56.2	45.0	26.1	39.0	28.8	49.8	5.55
	84.9	70.4	59.4	40.0	49.3	74.0	87.4	88.6	90.0	95.9	112	97.7
0.6–0.7	18.6	40.6	37.6	55.0	43.6	59.0	109	62.8	30.4	6.18	5.94	37.8
	5.12	7.89	24.1	18.0	61.8	52.8	57.0	65.2	55.4	53.4	44.6	19.5
	72.3	70.8	72.0	66.8	74.4	66.4	55.0	57.6	62.2	67.2	75.3	73.8
0.5–0.6	51.2	39.9	32.9	38.2	53.5	63.6	27.6	11.8	12.0	21.2	27.6	37.8
	23.2	34.2	47.4	73.4	49.0	41.8	49.4	58.5	46.5	45.8	73.8	67.6
	37.8	43.0	45.2	52.1	27.4	25.8	24.4	43.2	54.2	31.5	30.2	33.5
0.4–0.5	50.2	37.0	27.5	19.6	62.6	58.5	0.05	12.9	62.0	92.6	83.0	56.2
	78.0	78.4	79.0	60.2	17.4	23.9	27.8	24.2	32.7	47.8	33.7	59.0
	7.70	8.75	10.4	11.8	16.0	5.10	4.80	7.75	11.0	3.80	4.50	6.05

Note: Line 1 – average values from measurements in May–June, 1995; Line 2 – July–early August; Line 3 – late August–September.

3– August–September (216 spectra). Data on obviously abnormal spectra of sizes were excluded from the analysis.

Table 2.5 contains data on the volume size distribution of particles $\Delta V/\Delta r$ ($\mu m^3\,m^{-3}$). Quite unexpected is an increase of particle concentration in the region $0.6 \leq D \leq 1.0\,\mu m$ for the third series of measurements (August–September). This is an obvious confirmation of an increase of intensity of the volcanic material emitted in August–September.

From 28 August, the photoelectric counter repeatedly recorded sudden increases of aerosol number density in the size range $0.6 \leq D \leq 7.0\,\mu m$. Fluctuations of particle concentrations continued for several hours, after which the particle size distributions returned to those close to the initial ones.

An analysis of fluctuations suggested the conclusion that their developments have some features in common. An obvious dominance of the concentration of particles in the range $0.6 \leq D \leq 1.0\,\mu m$ over average values is assumed to be the beginning of such a phenomena. After this, seven distinctly observed cases, for the period 28 August–3 September, were averaged over values $[\Delta V(r) - \Delta\bar{V}(r)]$ ($\mu m^3\,m^{-3}$), and the temporal evolution of these spectra was considered, beginning from 2 hours before a fluctuation initiation and up to 6 hours afterwards (Table 2.6).

An increase of concentration of particles with sizes $0.9 \leq D \leq 1.0$ and $2.0 \leq D \leq 0.4\,\mu m$ is observed two hours before the flash. After the beginning of the flash, the concentration of particles with $D \leq 1.0\,\mu m$ increased markedly, and then the concentration of particles with $D > (1.5\text{–}2)\,\mu m$ decreased to average and even below 4 hours after initiation. For each flash, the pattern can substantially differ from the average, however, the general pattern is preserved. The strongest changes of the spectrum are observed within the range $1.0 \leq D \leq 3.0\,\mu m$. This deformation of the spectrum can be reliably recorded both with the help of the photoelectric counter and with other optical methods, for instance, via the spectral aerosol scattering and attenuation in the near-IR region ($\lambda = 1.0\text{–}3.0\,\mu m$).

At the initial stages of the Earth's crust shifting, rocks are crushed, particularly of crystalline structure. This produces mainly charged aerosol particles of dominating sizes, determined by the size distribution of rocks, by the ratio of the possible charge, by the particle's mass, and by the electric field tension appearing. Table 2.7 reflects the dynamics of the change of the particles size spectrum and enhanced generation of particles with different sizes during the earthquake in Guadalajara on 9 October, 1995. The table shows the relative values of concentrations of particles of different fractions $\Delta N(r_i)$ in percentages normalized against average values $\Delta N(r_i)$ obtained during the previous four days (the meteorological conditions in this period were stable). Because of the absence of data from 1–8 a.m. on 9 October it is difficult to say anything about the period preceding the earthquake. One can note, however, that before the earthquake the concentration of particles started decreasing in the range $0.4 \leq D \leq 0.5\,\mu m$ and increasing in the range $0.8 \leq D \leq 7.0\,\mu m$. At the moment of the first shock, the generation of the largest particles increased, followed by increases in the case of particles in the two ranges $2.0 \leq D \leq 4.0\,\mu m$ and $1.0 \leq D \leq 1.5\,\mu m$ mentioned above.

The observational data obtained on the elemental composition, number density, and size distribution of aerosol in the surface layer of the atmosphere in the western states of Mexico suggested the following conclusions:

(1) There is a stable elemental composition of aerosols determined, first of all, by their volcanic origin.

Table 2.6. The temporal change of deviations of volume concentrations of different fractions of aerosol V_i ($\mu m^3\,m^{-3}$) at sudden increases of volcanic emissions from average values (28 August–4 September, 1995, Colima volcano).

	T (hr)								
d_i (μm)	−2	−1	0	+1	+2	+3	+4	+5	+6
0.4–0.5	+0.12	+0.04	−0.10	−0.16	−0.30	−0.20	−0.32	−0.30	−0.28
0.5–0.6	+0.75	+0.06	−0.43	−1.0	−1.4	−1.0	−1.5	−1.1	−0.38
0.6–0.7	−0.70	0	+1.0	0	+44.0	+0.34	0	+1.7	+1.0
0.7–0.8	−0.80	0	+1.2	0	+52.0	+0.40	0	+0.80	+1.2
0.8–0.9	−0.50	−0.50	+0.12	+6.0	+7.5	+7.2	+14.0	+11.0	+6.5
0.9–1.0	+0.65	+0.48	+1.3	+14.0	+8.5	+2.0	+1.8	+6.5	+5.4
1.0–1.5	−0.18	−0.22	−0.14	+20.0	+27.0	+3.0	+2.0	+10.0	−0.15
1.5–2.0	+0.08	−0.02	0	+1.4	+4.5	+0.80	+0.06	+1.2	−0.08
2.0–4.0	+0.43	0	+0.54	+0.43	+0.43	+1.0	+0.11	+0.11	−0.43
4.0–7.0	+0.20	0	+0.25	+0.20	+0.20	+0.47	+0.06	−0.03	−0.18
7.0–10.0	+0.18	+0.04	+0.08	+0.10	−0.15	−0.03	−0.06	−0.03	−0.18
10.0–15.0	+0.11	+0.07	+0.04	−0.03	−0.16	−0.03	−0.14	−0.13	−0.12

Table 2.7. The change of the sizes of particles in the surface layer of the atmosphere before and during the earthquake in Guadalajara on 9 October, 1995 (the ratio $Nr_i/\bar{N}r_i$ in (%)).

	8 October				9 October		
D (μm)	9 p.m.	10 p.m.	11 p.m.	12 p.m.	9:20 a.m.	10:10 a.m.	11 a.m.
>10	73	81	88	136	579	684	263
10–7	100	139	65	125	544	596	772
7–4	67	144	152	122	237	204	51
4–2	64	102	140	131	281	324	234
2–1.5	64	136	153	137	129	150	296
1.5–1.0	76	157	166	126	118	495	75
1.0–0.9	97	145	95	102	101	118	63
0.9–0.8	68	143	82	229	63	251	190
0.8–0.7	149	193	163	90	133	170	174
0.7–0.6	101	143	132	106	100	96	80
0.6–0.5	120	71	103	99	77	88	60
0.5–0.4	90	36	89	42	0	0	0
0.4	108	105	123	98	92	115	93

(2) The elemental composition of soil aerosol in the region under study is close to that of volcanic aerosol (i.e., it is formed by products of volcanic eruptions).
(3) The elemental composition of the products of Popocatepetl eruption is close to that of the products of eruption of the Fuego del Colima volcano.
(4) The products of the Popocatepetl eruption, despite their north-eastern transport immediately after the eruption in December, 1994 affected the elemental com-

position of precipitation in the western states of Mexico in the later period (May–July, 1995) at distances up to 600 km.

(5) During the rainy season from June–September, 1995, the elemental composition of aerosols changed and was similar to the elemental composition obtained at the beginning of Popocatepetl activity. From August–October, 1995 the 'volcanic' composition of atmospheric aerosol was restored due to a short-term eruption of the Colima volcano in August and the earthquake in the region of Guadalajara in October, 1995.

(6) The aerosols of volcanic origin exhibit a stable selectivity in the generation of particles of different sizes: maxima of intensities of particle generation (by mass) in the range of sizes $D \approx 1.0\,\mu m$ and $D \approx 2\text{–}3\,\mu m$. An analysis of the studies carried out in the Kamchatka region (Russia) confirmed this common feature of particle generation by volcanoes at a relatively low intensity of eruption.

Elements of volcanic origin, in their type of diurnal change of concentration, are divided into two groups. This phenomenon is explained by the existence of two sources of volcanic aerosols. The second source is not from the volcano crater but the soil surface on its slopes and surrounding areas. Aerosol substances result from mechanical deformations of the upper layers of the Earth's crust. From results of the elemental analysis, the power of both sources is approximately of the same order.

2.4 SOUTH AMERICA

An accomplishment of complex field observational experiments in South America was connected mainly with problems of the ecodynamics of the Amazonia region including, in particular, developments concerning atmospheric aerosol [38, 42, 68, 77].

The tropics whose area constitutes about half the Earth's surface, play a key role as a factor of chemical reactions and physical processes in the atmosphere. Vast regions of highly productive tropical land ecosystems are important sources of aerosol and MGCs affecting the formation of the radiation regime and oxidation power of the atmosphere. Vast regions covered with the largest wet tropical forests are located in the basin of the Amazon River, suffering from an enhancing anthropogenic impact. The biomass burning, which took place in the late 1970s–early 1980s in the tropics, resulted in emissions to the atmosphere of several key MGCs which seriously affected the chemical processes in the global atmosphere. So, for instance, the biomass burning is responsible for 10–30% of the global budget of carbon monoxide and has become a substantial source of MGCs such as NO_x, CH_4, COS, and CO_2.

In this context, within the programme of the field observational experiments studying the large-scale biosphere–atmosphere interaction (LBA) in Amazonia in March, 1998, [68] carried out aircraft measurements (ten flights in total) of the atmospheric aerosol characteristics over the north-eastern region of the wet tropical forests of Amazonia at altitudes from 0.2–12.6 km. An analysis of the observational results has shown that the aerosol number density in the free troposphere (normalized to standard values of temperature and humidity) is 2–15 times higher compared with that observed in the ABL, where the characteristic value of concentration is $500\,cm^{-3}$. Detection of such a contrast between ABL and free troposphere agrees with results obtained earlier. Calculations of backward trajectories of air masses have shown that even a week after the long-range transport of air masses in the middle troposphere newly formed ultrafine particles constitute more than one-third of the total aerosol number density.

The vertical profile of aerosol size distribution in the size range 0.001–3 μm determined from measurements in six sublayers of the troposphere had a C-shaped form with maximum concentration of the accumulation-mode aerosol in the ABL and in the free troposphere above 10 km. A minimum of aerosol number density was recorded in the 4–8-km layer. An increase in the concentration of the accumulation-mode aerosol particles at altitudes above 10 km can be explained by the impact of particle outflow from the penetrative convection cloudiness. The results obtained suggest the conclusion that convection plays the most important role in the evolution of aerosol and its chemical composition in the atmosphere over wet tropical forests. Further developments should foresee the following studies: (1) properties of cloudiness and its impact on the vertical transport of aerosol; (2) washing out and transformation of aerosol by clouds; and (3) formation of new particles in air fluxes from clouds.

Within the programme of the contribution of Europe to an accomplishment of the field observational experiment LBA-EUSTACH to study the atmosphere–biosphere interaction in the period of the season of biomass burning in 1999 and the transition from dry to wet seasons (September–October), [38] obtained aerosol samples in the representative region of pastures and in the primary wet tropical forest in Rondonia (Brazil). With the use of the methods of nuclear magnetic resonance spectroscopy (NMR) and gas chromatography–mass spectrometry (GC–MS), an analysis was made to recognize and evaluate the content of water-soluble organic components (WSOC) in aerosol.

The NMR data show that WSOCs are mainly aliphatic or oxidized aliphatic compounds (alcohols, carboxylic acids, etc.) with a small content of aromatic rings with carboxylic and phenolic groups. The most substantial component detected from the GC–MS data was levoglucosan (1,b-anhydro-β-D-glucose) – a well-known product of glucose burning, the concentration of which varied within 0.04–6.90 $\mu g\,m^{-3}$ and constituted 1–6% with respect to total carbon (TC) and 2–8% with respect to WSOC. Other gluco-anhydrites appearing with destruction of hemicellulose were detected in smaller amounts, in addition to acids, hydroxy-acids, oxy-acids, and poly-alcohols (in total, their content constituted 2–5% of TC and 306% of WSOC). The correlation of all these components with organic carbon (OC), black carbon (BC), and potassium reflects their origin as products of biomass burning.

As components of natural background aerosol, such compounds have been identified as various saccharic alcohols (manitol, arabitol, and erythritol) and saccharic compounds (glucose, fructose, mannose, galactose, sucrose, and trechalose). Apparently, these components appear due to microbes and biogenic substances contained in the atmosphere. The basic part of WSOC (86–91%) is not reflected in GC–MS data, being mainly compounds of high molecular weight.

Mayol-Bracero *et al.* [77] performed a chemical analysis of aerosol samples taken in October, 1999 in Rondonia (Brazilian Amazonia) for the content of carbon compounds. The observational period included a maximum of the season of biomass burning as well as a period of the transition from dry to wet season. Information about the carbon component of aerosol was obtained with the use of the method of thermal burning to determine the concentration of TC, BC, and OC. A considerable part of the BC fraction (usually about 50%) is water-soluble.

A detailed analysis has been carried out [77] of the water-soluble component of organic carbon (WSOC) in total carbon from the measured content of WSOC with the use of highly precise liquid chromatography (HPLC) to distinguish between water-soluble components whose concentration is determined with the use of the GC-MS method. According to the results obtained, the share of WSOC in OC varies within 45–75%. Such a high content of WSOC reflects the origin of this aerosol component as the product of smouldering during burning. The presence of GC–MS data enabled one to identify many various components including hydroxyl, carboxylated, and carbonyl groups. The share of WSOC determined from GC–MS data constituted about 10%. Three classes of compounds were selected from the UV data of HPLC: neutral compounds, monocarboxyl and dicarboxyl acids (MDA), polycarboxyl acids (PA). The total content of these three groups of compounds constituted ~70% of WSOC, with the highest (~50%) contents of MDA and PA.

A high correlation (the correlation coefficient $r^2 = 0.84$–0.89) between BC_{water} (dehydrated BC) and levoglucosan (both these compounds are indicators of biomass burning) and, on the other hand, water-soluble components (i.e., WSOC, N, MDA, and PA), as well as their high concentration during biomass burning, were convincing proof that biomass burning was the main source of WSOC. Of special interest is the fact that in the process of burning considerable amounts of PA form, and in this connection, probably, humus-like substances (due to their polyacid nature). Such substances, due to their high solubility in water and impact on the surface tension reduction, can play an important role as factors of formation of aerosol particles in biomass burning, which can function as CCN and, hence, affect substantially the cloud properties and climate. Further studies will be focused on the WSOC fraction and its impact on CCN formation.

Wet tropical forests located in the Amazon basin are the largest area of such forests on the globe and, probably, are the most significant continental source of emissions of biogenic MGCs and aerosols by vegetation. The same region gets intensively deforested due to forest fires. From the IPCC data, biomass burning is, on a global scale, a second important source of anthropogenic aerosols, about 80% of biomass burning falling on the tropics (and about one-third of this amount is the share of South America).

Conditions of an intense convection connected with the dynamics of the inter-tropical convergence zone (ITCZ) favour an intense transport of products of natural and pyrogenic emissions in the form of MGCs and aerosol to higher layers of the atmosphere and their long-range transport to higher latitudes. In this connection, within the large-scale field experiment on studies of the atmosphere–biosphere interaction in the Amazonia region as well as European developments to study MGCs and chemical processes in the atmosphere (LBA–EUSTACH), [41a] took samples of various fractions of atmospheric aerosol in the regions of wet tropical forests of Brazilian Amazonia during the periods April–May and September–October, 1999. These two periods include parts of the wet and dry seasons, respectively. The aerosol samples were taken in the daytime and at night at three levels (above, inside, and under the forest canopy) from a meteorological tower 54 m high located in the forest. The use of methods of X-ray emission and thermo-optical analyses has made it possible to study the aerosol composition for the content of 19 components. Also, a gravimetric analysis was carried out and the content of equivalent black carbon (BC_e) was estimated.

The measured average mass concentrations of particles with $D > 2\,\mu m$ were 2.2 and $3.5\,\mu g\,m^{-3}$ in wet and dry seasons, respectively. The aerosol components whose formation was connected with biomass burning and emissions of soil dust, reached maximum concentrations over the forest canopy in the daytime, whereas aerosols generated by the forest were concentrated mainly under the canopy at night. This variability can be explained by the impact of convective mixing in the daytime and the formation of a 'thin' boundary layer at night, as well as more favourable conditions for the formation of biogenic emissions of aerosol at night. The obtained values of mass coefficients of scattering (α_s) and absorption (α_a) point to a dominant contribution of fine aerosols to the formation of α_s, whereas the α_a value is determined nearly in the same way by the impact of fine and coarse aerosols during both seasons. Analysis of observational data has led to the conclusion that: it is not the elemental carbon contained in aerosol particles but other components that are responsible for a considerable part of the aerosol-induced absorption of radiation.

A number of local field experiments carried out in South America have also played a marked role in the studies of atmospheric aerosol.

An important contribution to the completion of the field observational experiment PRIDE in the period 26 June–24 July, 2000 in the region of Puerto Rico, in order to study the Saharan dust aerosol (DA) and its long-range transport, has been made by results of DA remote sensing with the moderate-resolution video-spectroradiometer MODIS on the Terra satellite. Comparisons of MODIS data and results of observations with the SPh as well as *in situ* surface and aircraft measurements of DA characteristics carried out by [72] constituted a substantial part of the PRIDE programme. Processing of MODIS data has made it possible to obtain results from retrieving the AOT over the ocean surface and the aerosol size distribution. A comparison of MODIS and SPh data on AOT at the wavelength $0.66\,\mu m$ revealed good agreement. However, the MODIS data turned out to be systematically overestimated at wavelengths 0.47 and $0.55\,\mu m$, but underestimated in the near-IR region ($0.87\,\mu m$). Such errors in remote sensing were explained by underestimation of the retrieved sizes of DA particles compared with data of both *in situ* and SPh measurements, which might be partially connected with the prescribed spherical shape of particles in the algorithm of the DA size distribution retrieval.

From data of measurements at two sites in Mexico City and at five urban sites in Germany, [13] analysed the ratio between concentrations of BC and carbon monoxide (CO) in the atmosphere. In all cases the coefficient of correlation was >0.90. An averaged slope of the straight lines of linear regression constituted $2.2\,\mu g\,mg^{-1}$ (Germany) and $1.1\,\mu g\,mg^{-1}$ (Mexico City). The most important factors determining this ratio were the ratio of diesel to petroleum fuel usage levels and the efficiency of these fuels when burning in engines. The results obtained testify to the possibility of using *in situ* measurements of CO concentration to estimate the BC content in the absence of related information.

2.5 WESTERN EUROPE

One of the most informative western European field observational experiments was accomplished in Germany in 1998.

2.5.1 LACE-1998

In the period from 13 July–12 August, 1998 in the region of Lindenberg Meteorological Observatory (Germany: 52.2°N, 14.1°E) a complex field observational experiment LACE-1998 was undertaken to study the properties of aerosol from data of surface and aircraft (3 aircraft) observations. The scientific complex included, in particular, one aircraft and four surface lidars. Observations were carried out under conditions of clear and polluted atmospheres to which corresponded a broad range of AOT values. Ansmann *et al.* [7] undertook a general overview of the applied complex of instruments illustrated with some observational results. Of special interest are data for 9–10 August, 1998, when in the free troposphere there was detected a layer of aerosol generated by forest fires in the west of Canada.

The aerosol-induced radiative forcing at the level of the tropopause, varying from $-4\,W\,m^{-2}$ (in the clear atmosphere) to $-16\,W\,m^{-2}$ (in the polluted atmosphere), was estimated from data of measurements and calculations. Processing of aircraft lidar data has shown that the SSA was always <0.9 (the polluted atmosphere) and was equal to 0.8–0.9 in the presence of an aerosol layer formed due to the forest fires in Canada. According to data of surface observations, the SSA averaged 0.8.

One of the important problems of LACE-1998 was the accomplishment of the 'closed' aerosol–radiation experiment. Obtaining the required complex of data on aerosol properties (size distribution and optical parameters) and radiation fluxes (both total and spectral) has made it possible to perform a correct comparison of results of observations and numerical modelling. Petzold *et al.* [90] described a detailed characteristic of specific features of the observed vertical profiles of size distribution, extinction coefficient, and the share of backscattered radiation, which should be taken into account in the climate numerical modelling. The AOT at 710 nm varied during observations from 0.18 to 0.06, and the aerosol-induced RF for the surface–atmosphere system changed from -33 ± 12 to $-12 \pm 5\,W\,m^{-2}$.

During LACE-1998, [18] performed surface observations of the aerosol microphysical and optical characteristics using various instruments (aerosol photometer, nephelometer, PSAP instruments to measure the radiation absorption by soot aerosol, integration plate, and lidar). The data of horizontally directed lidar were used to determine the extinction coefficients and their dependence on humidity. With this aim in view, the photometer data were also processed. Some results of observations have been given in [28]. The SSA values vary within 0.80–0.93.

To determine, on a real timescale, the chemical composition of aerosol particles of different sizes in the period of LACE-1998, [111] used the mobile laser mass spectrometer LAMPAS-2. The use of LAMPAS-2 enabled one to obtain information about the chemical composition of five fractions (by sizes) of aerosol with a time resolution 1 hour (the recorded mass spectra totaled several tens of thousands). The statistical processing of observational data resulted in substantiation of ten basic classes of particles differing in their chemical composition, including four most substantial classes determined by the presence of mineral, sea salt, and carbonaceous aerosols.

Schröder *et al.* [103] analysed the results of statistical processing of the data from aircraft (the *Falcon* flying laboratory) observations performed in July–August, 1998 in the region of Berlin (13.5°–14.5°W; 51.5°–52.7°N) within the Lindenberg field experiment LACE. An analysis of data on the vertical profiles of particle number density and related variability of nucleation and large-sized modes of particles has led to the conclusion that the upper troposphere is a powerful source of formation of new particles from gas-phase precursors. The observed extreme evolution of aerosol is considered to be determined by a succession of nucleation processes, formation of the accumulation mode, and the impact of aerosol on clouds. This interpretation supposes a close connection between the sources of aerosol (nucleation of particles from gas-phase precursors) and its sinks (the cloud-induced washing out of aerosol particles). A detailed consideration has been given to processes responsible for the transformation of the aerosol life cycles and an analysis has been made of the data on number density and the surface area of particles for different types of aerosols. The surface area of dry aerosol particles varies within 1–20 $\mu m^2\,cm^{-3}$ in the free troposphere and from 2–13 $\mu m^2\,cm^{-3}$ in the tropopause. Schemes of parameterization of aerosol size distribution and surface area of particles have been proposed which can be used in the atmospheric models. However, an adequate parameterization is only possible on the basis of accumulation of a larger (more representative) volumes of observational data.

During LACE-1998, [28] performed impactor measurements of atmospheric aerosol size distributions in the size range 0.1–25 μm, and using the method of X-ray fluorescence analysis, information was obtained about the elemental composition of aerosol – from potassium to lead ($11 \leq Z \leq 83$). The use of high-resolution scanning electron microscopy and X-ray microanalysis enabled one to study about 15.5 thousand particles – about 3.8 thousand particles were studied using a combined method of transmission electron microscopy and X-ray microanalysis.

A consideration of the data on morphology and the chemical composition of particles has made it possible to classify them into ten groups: ammonium sulphates, calcium sulphates, sea salts, metals oxides/hydrates, carbonates, silicates, soot, biological particles, mixed carbon–sulphate particles, and carbon-enriched residual particles. Using the data on size distribution and relative content of different groups of particles, average and size differentiated values of the complex refraction index (CRI) were found [28]. Depending on days of sampling, the real part of the CRI varied within 1.52–1.57. The variability of the imaginary part of the CRI (from 0.031–0.057) was due to variations of the content of soot and carbon–sulphate components. The obtained average values of the CRI agree well with results of photometric measurements of dry filter samples.

During the course of LACE-1998, [122] performed synchronous surface and aircraft (lidar) measurements of atmospheric aerosol characteristics. Two flying laboratories also carried instruments for *in situ* measurements of the optical and microphysical parameters of aerosol. Lidar data served as a source of information about the backscattering coefficients at 8 wavelengths in the interval 320–1,064 nm and about the extinction coefficient at 2–3 wavelengths in the interval 292–532 nm. The availability of such information has made it possible to retrieve the aerosol size distribution. The use of two retrieval algorithms has opened up the possibility to obtain data on the efficient radius, volume, surface area, number density, and complex refraction index of particles. Calculations with Mie formulas gave values of single scattering albedo. Results of retrieval were compared with data of *in situ* measurements of aerosol size distribution and coefficients of absorption of particles.

Two examples of observational data processing have been given in [122]. The 9–10 August, 1998 nocturnal measurements were made in the aerosol layer formed in the troposphere due to forest fires. The aerosol optical thickness at 550 nm constituted about 0.1. Comparisons revealed an excellent agreement between the data of *in situ* measurements and remote sensing. An efficient radius of particles in the central part of the plume of pollutants constituted about 0.25 μm, and the CRI values reached $(1.56\text{–}1.66) + (0.05\text{–}0.07)i$. At the wavelength 532 nm, very low SSA values were recorded (0.78–0.83). In the second case (11 August, 1998), conditions of industrial pollution of the atmosphere, typical of central Europe, were observed. The AOT at 550 nm constituted 0.35, and the efficient radius of particles varied within 0.1–0.2 μm. In contrast to the first case, the imaginary part of CRI turned out to be $<0.03i$. Respectively, the SSA changed from 0.87 to 0.95.

Wex *et al.* [127] compared the calculated values of coefficients of scattering, backscattering, and absorption of radiation for atmospheric aerosol particles with results of surface and aircraft observations, carried out during LACE-1998, to study the aerosol properties in the region of Lindenberg (Germany). In the cases of scattering and backscattering coefficients the difference between calculated and measured values did not exceed ±20%. The reliability of calculated values depended strongly on an adequately prescribed size distribution of aerosol. The measured values of the absorption coefficient turned out to be much lower than the calculated ones. The uncertainty of the latter was mainly determined by the share of the concentration of elemental carbon (EC) which depends on the size of particles. The use of the Monte Carlo method, with varying input parameters to calculate the variability of the considered coefficients, has led to the conclusion that the measured variability of the coefficients does not exceed the uncertainty of calculations (about ±20% for the coefficients of scattering and backscattering and ±30% in the case of absorption coefficients). Thus, to reduce the levels of uncertainties in the calculated coefficients, it is necessary to reduce the errors in measurements of aerosol size distribution. Also, methods of measurements of the coefficient of absorption by aerosol, and the share of mass concentration of EC, should be improved.

The interaction of tropospheric aerosol particles with water vapour affects the lifetime of particles in the troposphere, and mechanisms for their removal from the troposphere. The hygroscopic properties of aerosol particles depend on the chemical composition of the water soluble fraction of particles. These properties control the water content of particles and, eventually, the size of individual particles. Within the programme of the Lindenberg field experiment of LACE-1998, [19] performed measurements of the hygroscopic properties of atmospheric aerosol particles within the range of Aitken particles, as well as large and gigantic particles at a rural site 80 km south-east of Berlin. The hygroscopic properties of Aitken particles were determined for four classes of particle size (50, 100, 150, and 250 nm) with the use of the differential analyser of mobility in the relative humidity range 60–90%. At relative humidity = 90%, the aerosol particles could be classified into two groups ('more' and 'less' hygroscopic fractions), but at relative humidity = 60% this classification is impossible.

The measured average values of the factor of growth of 'more' hygroscopic particles constituted 1.43, 1.49, 1.56, and 1.63 for particles of the size classes mentioned above. Respectively, in the case of 'less' hygroscopic particles, the factor of growth constituting about 1.1 is practically independent of the size of particles. On the assumption that the aerosol particles consist of ammonium sulphate in the presence of an insoluble nucleus, the share of the volume of water soluble aerosol was estimated. According to estimates for 'more' hygroscopic particles, the shares of their volume are 0.47, 0.52, 0.59, and 0.68, whereas in the case of 'less' hygroscopic particles their size independent share of the volume constitutes about 0.1. At relative humidity = 60%, the measured average values of the factor of growth varied within 1.15–1.22, and the share of the volume of the water soluble fraction changed from 0.41–0.59.

According to the data of measurements for quasi-mono-disperse samples of particles in the range of diameters of 0.4–3.8 μm (if we differentiate with respect to seven levels of sizes), up to three classes of the share of volume for water soluble fractions were observed. In some cases, the presence of almost insoluble particles was observed, which corresponds to a less hygroscopic fraction with the share of water soluble particles of about 0.5–0.7. Usually, there is a third class of particles with the share of the volume of the water soluble fraction being about 0.85. In the range of particles with $D < 0.7\,\mu m$, almost completely water soluble particles of the third class dominate, whereas in the case of particles with $D > 0.7\,\mu m$, the distribution of the share of the volume of particles of all classes of sizes is sufficiently homogeneous. During LACE-1998, the share of the volume of large and gigantic particles did not vary substantially.

In November, 1997, in the region of Melpitz, and in July–August, 1998 near Lindenberg (Germany), complex field observation experiments (e.g., LACE-1998) were accomplished to study atmospheric aerosol and its impact on the short-wave radiation transfer. Using various surface and aircraft instrumentation, measurements were made of the size distribution and (depending on the size of particles) of the chemical composition of aerosol as well as optical parameters and radiation fluxes (both total and spectral in the short-wave interval).

To assess the quality of the information obtained, [86] compared different methods of measurements of the distribution of the mass of the particles by their sizes. The gravimetric mass of fine aerosol is interpreted as determined mainly by contributions of carbonaceous substances and ions, as well as water assimilated by hygroscopic components of aerosol. To characterize the coarse fraction of aerosol, whose contribution to total mass is less substantial, it is important to take into account the presence within it of an insoluble substance. The mass concentration calculated from data on aerosol size distribution agrees well with the gravimetric mass concentration. However, the calculated values exceed the measured ones, especially in the cases of the accumulation and Aitken modes. The mass concentration calculated from data on size distribution is especially sensitive to errors in the determination of the number density of different sizes of particles. The total error in calculated estimates constitutes about 10%.

For different types of air masses, estimates of log-normal distributions have been given in [86] concerning the distributions by number density, mass, and volume of particles. As a rule, the modal parameters for different air masses do not differ substantially. The highest values of mass concentration are connected with an increasing size of particles (the Aitken and accumulation modes), but not with the growth of the number of particles of a given mode. As a rule, the share of carbon mass grows as the size of particles decreases. The most significant feature of the annual variability is an increase of the content of sulphate. In the case of the content of nitrate, a strong dependence is observed on the direction of the coming air masses. Sulphate and nitrate are usually neutralized by ammonium.

2.5.2 Other field experiments

Apart from the large-scale experiment of LACE-1998, other numerous field studies of atmospheric aerosol have been accomplished in western Europe.

During four episodes of powerful emissions (plumes to heights above 500 m) of aerosol to the atmosphere from western Europe towards the Atlantic Ocean basin, taking place from 16 June–20 July, 1997, [84] performed lidar soundings to retrieve the vertical profiles of the physical characteristics of aerosol and SSA at the wavelength 532 nm. The results of observations with the 6-wavelength lidar mounted near Sagre, Portugal (37°N, 9°W), within the programme of the field observational experiment ACE-2, have been used to retrieve the vertical profiles of the backscattering coefficient at wavelengths 355, 400, 532, 710, 800, and 1,064 nm as well as the coefficient of extinction at wavelengths 355 and 532 nm.

According to the results obtained, an efficient radius of aerosol particles constituted $0.15 \pm 0.06\,\mu m$, the volume concentration of particles varied within 6–27 $\mu m^3\,cm^{-3}$, and the particle surface areas varied from 80–1,200 $\mu m^2\,cm^{-3}$. The average value of the imaginary part of the complex refraction index, independent of wavelength, reached only $0.009i \pm 0.07i$ (i.e., the radiation absorption by aerosol particles turned out to be negligibly small). The average value of the real part of the refraction index constituted 1.56 ± 0.07, and SSA at 532 nm was 0.95 ± 0.06. The results obtained show that the ammonium sulphate component prevails in the aerosol composition and the radiation-absorbing soot-like component is a negligible admixture. The correlation analysis revealed a high correlation between an efficient radius of particles and the Angström coefficient. Also, a correlation was observed between SSA, the ratio of coefficients of backscattering and extinction, and the imaginary part of the complex refraction index. A similarity to these results was obtained with data of the field observational experiment TARFOX, indicating that the processes of internal combustion, as a source of aerosol, taking place in Europe differs little from those taking place in North America. However, the difference from the INDOEX and SCAR-B data turned out to be substantial, which, in the first case, was explained by the formation of carbon compounds due to carbon burning, functioning of diesel engines, and in the second case by biomass burning.

Within the World Meteorologlcal Organisation (WMO)-supported aerosol programme of the Global Atmosphere Watch (GAW) at a station located at the Jungfraujoch mountain (2,580 m above sea level), [43] carried out continuous observations of the chemical aerosol composition starting from July, 1999, for $1\frac{1}{2}$ years. The filter sampling was made with due regard to two classes of particles: all particles and those with an aerodynamic diameter $<1\,\mu m$. The chromatographic analysis of the samples composition was made for the content of basic ions, whose mass concentration constituted $\sim$30% of the total mass concentration of aerosol. The mean global value of total ion mass concentration was $1.04\,\mu g\,m^{-3}$ [43]. Sulphate, ammonium, and nitrate were the main components of the fine fraction of aerosol, whereas the prevailing components of the water soluble coarse mode were calcium and nitrate. An analysis of individual particles of the coarse mode has shown that they are an internal mixture of calcium and nitrate. The total mass concentration of ions is characterized by a strong annual change with a concentration equal to 1.25, 1.62, 0.70, and $0.25\,\mu g\,m^{-3}$ in spring, summer, autumn, and winter, respectively. As for individual aerosol components, the concentration of sulphate, nitrate, and ammonium changed more strongly than that of calcium (the latter is determined, probably, by the presence of local sources of calcium and by the input of the Saharan aerosol which is independent of season).

Despite intense efforts in the field of studies of air quality, there are many unsolved problems, which are determined, first of all, by insufficiently complex developments. Therefore, [119] carried out the complex field observational experiment ESQUIF, with the main objective to study atmospheric pollutions in a large urban region of Paris, with the subsequent use of observational results to check an adequacy of the air quality models, reproducing the atmospheric pollution by photo-oxidants.

The main phases of ESQUIF included studies of: (1) the sensitivity of the processes of the surface-layer ozone formation to various factors (long-range transport, levels of the urban emissions, the ratio between anthropogenic and biogenic emissions); (2) the adequacy of data on emissions inventory (NO_x, CO, VOC, etc.); (3) the impact of aerosol and clouds on the rate of photolysis; (4) the role of photochemical processes in the episodes of the wintertime increase of NO_2 concentration; (5) the ways of improving the techniques of measurements and numerical modelling; and (6) the specific features of the spatial–temporal variability of aerosol in the region of Paris.

During 1998–2000, in different regions of Paris, observations were made consisting of 12 periods of intense observations, mainly in the summer. The observational conditions were characterized by a variety of meteorological situations, especially when the atmosphere of Paris and its environs became heavily polluted. The observational system included circular flights round the city together with surface observations using techniques of *in situ* measurements and remote sensing. The available complex data of observations enabled one to monitor the 3-D evolution of pollution events (on the whole, the atmosphere of Paris cannot be considered as heavily polluted).

A comparison of observational data has been made [119] for different intense observations from the viewpoint of revealing specific features of various events of atmospheric pollution, including, in particular, an assessment of the role of long-range transport. Also, a comparison was made of observational data and numerical modelling results (with the use of Monte Carlo and conjugate equation techniques). An important goal of this comparison was to assess the role of correct consideration of boundary conditions, the prescription of emissions polluting the atmosphere (biogenic including), and the correct simulation of the processes of photolysis.

An analysis of observational data obtained in the period of intense observations has shown, in particular, that events of pollution occur very rapidly in the stagnant atmosphere. Maximum concentrations of oxidants were observed at a high or low altitude of the ABL, even when there were no marked temperature inversions. Most substantial maxima of pollution resulted from local pollution under meteorological conditions favouring an enhancement of pollutants, but in some cases the long-range transport of pollutants contributed much.

Chemical analysis of the urban aerosol samples taken in Gent (Belgium) carried out by [69] with the use of GC–MS techniques revealed the presence of $\sim$100 different organic compounds. Samples taken in different seasons made it possible to monitor changes in the aerosol composition during a year. The composition of organic compounds reflected, first of all (both in winter and in summer), properties of the products of car engine emissions. The prevailing organic compounds included (during the whole year) n-alcanes and fatty acids, with the relative content of various organic components being characterized by a distinct annual change. The variability of the content of n-alcanes in summertime aerosols manifested itself through the impact of biogenic emissions by vegetation.

The characteristic feature of the annual change of the fatty acids content consisted of a lower concentration of unsaturated fatty acids in summer compared with winter, which could be connected with the summertime processes of more intense oxidation of unsaturated fatty acids. The content of dicarboxyl acids and the associated products which appear due to oxidation of hydrocarbons and fatty acids was at a maximum in summer. In some cases, such compounds could be detected only in summer, reaching a maximum of concentration on hot summer days at maximum air temperatures >25°C and an increased concentration of surface-layer ozone. Such compounds included recently detected derivatives of glutamic acid, 3-isopropyl, and 3-acetyl pentane-dionine acid which forms, probably, by the oxidation in the atmosphere of chemically active monoterpene or sesquiterpene-precursors which have not yet been identified. Some of the detected compounds are indicators of wood burning, including diterpenoic acids, products of lignin hydrolysis, and levoglucosan. Results of measurements of the content of diterpenoic acids and products of lignin pyrolysis indicate that the contribution of wood burning is greater in winter than in summer. The data obtained reflect the fact that burning of both solid (broad-leaved sort) and soft wood contributes to the formation of organic aerosol, but in winter the contribution of solid wood prevails. The wintertime increase in concentration is also characteristic of polyaromatic hydrocarbons.

De Tomasi *et al.* [27] discussed the results of retrieving the vertical profiles of water vapour and aerosol concentration from regular observations performed with the combined Rayleigh–Raman lidar based on the use of the excimer laser XeF (the wavelength 351 nm). This lidar mounted in the south of Italy (40°20′N, 18°06′E) was functioning within the programme of observations at the European network of aerosol lidar sounding. During several months, synchronous observations were made of the vertical profiles of aerosol backscattering $R(z)$, the aerosol coefficient of extinction $\alpha_{\lambda_0}^{\text{aer}}(z)$, and the mixing ratio of water vapour $w(z)$ to study the correlation between $R(z)$ and $\alpha_{\lambda_0}^{\text{aer}}(z)$ on the one hand, and $w(z)$, bearing in mind the laws of variability of aerosol optical properties. A strong correlation was observed between the spatial and temporal evolution of $R(z)$ and $w(z)$ in summer and in autumn. $R(z)$ increases with increasing w. The total content of aerosol and water vapour is less in autumn than in summer. An analysis of the calculated backward trajectories of air masses for four days has shown that the dependence of $\alpha_{\lambda_0}^{\text{aer}}(z)$ and $R(z)$ on w changes under the influence of variations in advection at the observation site. The contribution of aerosol to the formation of $R(z)$ and $\alpha_{\lambda_0}^{\text{aer}}(z)$ increased with air masses coming from the northern and eastern regions of Europe. Under these conditions, average values of lidar ratio S increased to 50–63 sr. With air masses from northern Africa, average values of S varied from 48–74 sr.

During the period from January–September, 2000 at a rural site in Hungary, [55] carried out samplings of fine aerosols ($D_p < 1.5\,\mu\text{m}$), an analysis of which showed that the total carbon content in aerosol varied from 5–13 $\mu\text{g}\,\text{m}^{-3}$ and from 3–6 $\mu\text{g}\,\text{m}^{-3}$ during the first three months, and the rest of the observational period, respectively. On average, the WSOC constituted 66% with respect to total carbon concentration independent of season. The variable part of the water soluble components (38–72% of WSOC depending on sample) were separated from inorganic ions and isolated in the form of purely organic compounds with the use of a method of filtering out the solid phase. This fraction of aerosol was characterized by the ratio of the contents of organic substances to organic carbon equal to 1.9, which remained practically constant during a year. The following molar ratio was obtained which characterized the elemental composition: $C:H:N:O \approx 24:23:1:14$ for the organic fraction and indicated the prevalence of oxidized fractional groups and a low ratio of concentration of hydrogen to carbon, which is seen from the presence of unsaturated or polyinterrelated structures.

These conclusions confirm the results of analysis with the use of data of UV, fluorescent, and Fourier spectrometry. Theoretical estimates of the ratio of mass-organic substances and organic carbon gave 2.3 for the non-isolated water soluble fraction and 2.1 – for the WSOC. To obtain data on the total organic carbon needed to close the balance of carbon, dehydration was followed by extracting with acetone and a solution of 0.01-M NaOH. Owing to this, the ratio was estimated between the total organic substance and total carbon constituting 1.9–2.0.

From the analysis of aerosol samples taken at the Observatory Puy de Dome (1,465 m above sea level) in 2000–2001, [104] studied the dependence of the chemical composition of aerosol on the size of particles. For samples obtained with the use of two types of cascade impactors, the chemical analysis was carried out for the content of ions of inorganic compounds (Na^+, NH_4^+, K^+, Mg^{2+}, Ca^{2+}, Cl^-, NO_3^-, and SO_4^{2-}) and organic compounds ($HCOO^-$, CH_3COO^-, and $C_2O_4^{2-}$) as well as OC and EC, non-soluble dust, and total mass of aerosol. In the presence of clouds, the samples also contained aerosols which were inside the clouds. Solid compounds remaining after evaporation of cloud droplets were also analysed.

The results obtained refer to air masses of three categories (depending on the content and composition of aerosol): background (BG), anthropogenic (ANT) and specific, including the transport of dust from the Sahara desert as well as the arrival of aerosol to the observational site from the polluted ABL. Depending on the presence or absence of large particles of sea salt aerosol, there were air masses affected (or not affected) by the ocean. For all three categories of air masses in the cases of submicron aerosol particles the balance of mass could be closed (with an error of 18.5%), which reflects the adequacy of the obtained information about aerosol properties in the free troposphere of western Europe. The total mass of aerosol at a relative humidity = 50% constituted approximately $2.7 \pm 0.6\,\mu g\,m^{-3}$ (BG); $5.3 \pm 1.0\,\mu g\,m^{-3}$ (ANT), and 15–$22\,\mu g\,m^{-3}$ (specific). An analysis of the aerosol size distribution revealed two submicron modes: Acc1 (average geometric diameter is $0.2 \pm 0.1\,\mu m$) and Acc2 ($0.5 \pm 0.2\,\mu m$), as well as one supermicron mode ($2 \pm 1\,\mu m$).

The aerosol properties are characterized by strong external mixing with respect to carbon-containing compounds (EC and OC) and ions, which are part of Acc1 and Acc2. The concentration of lightweight carboxylates and mineral dust did not exceed 4% with respect to the total content of these components in any case, except for the events of the Saharan dust transport, the contribution of which (in case of this transport) reached 26% with respect to total aerosol mass. The relative content of water soluble, inorganic carbon-containing compounds varied (depending on the type of air masses) within 25–70% with respect to total mass, respectively. The share of the OC fraction is greater in air masses with a low content of aerosol: 53%, 32%, and 22% for BG, ANT, and specific, respectively. On the contrary, the share of EC grows from 4% in BG to 10% in ANT, and 14% in specific. The content of the inorganic fraction is higher in specific (55%) and ANT (60%) than in BG (40%) due to an increase of the nitrate concentration and, to a lesser extent, of sulphate and ammonium.

Organic carbon observed in the atmosphere appears from several sources, including emissions of OC due to fossil fuel and biomass burning, as well as gas products of the processes of photochemical oxidation of both anthropogenic and biogenic origin. Since an important indicator of the processes of OC formation is the chemical composition of atmospheric aerosol, [4] performed an analysis of aerosol samples taken in an urban environment in Portugal (the cities of Lisbon and Aveiro) and in a Finnish forest. With the use of flame chromatography, extracts from the aerosol substance solution were divided into fractions characterizing the content of compounds such as aliphatic and aromatic compounds, carbonyls, alcohols, and acid fractions which were analysed with the use of the GC–MS techniques. To evaluate the content of carbonaceous components, water soluble fraction, and solution in samples, on the whole, an analyser of organic (black) carbon was used.

Analysis of the results obtained in [4] has shown that an extraction using dichlormethane ensures a solution of $<50\%$ of the mass of aerosol organic substance, but the use of water as a dissolving agent enables one to (totally) transform this into a solution of $\sim$70–90% organic carbon. Usually more than 70% of the extracted substances are organic compounds, among which organic acids and alcohols prevail, especially in the case of particles with $D < 0.49\,\mu m$. Analysis of the content of the aerosol organic components of anthropogenic and biogenic origin has demonstrated that in the latter case the main sources are vegetable wax and petrogenic sources, but products of oxidation of volatile organic compounds make some contribution, too, and the water soluble fraction includes oxycarboxyl and dicarboxyl acids as well as cellulose.

As for anthropogenic components of samples, non-polar fractions constitute up to 24% of extracted organic carbon in Lisbon (with a large share of petroleum products), whereas in the forest – only 8%. In samples from Aveiro and from Finland, oxidized organic compounds constitute 76–92% of the extracted carbon. Favourable conditions for photochemical processes during the period of observations result in a high level of concentration of secondary organic compounds characteristic of data obtained in Aveiro. The area of forest in Finland is characterized by a strong variability of water soluble components and a high level of concentration of such components.

In August 1998, using the instruments carried by the flying laboratory C-130 of the British Meteorological Service, [31, 32] performed measurements of the sulphur dioxide concentration and physical and chemical properties of aerosol in the free atmosphere over the Aegean Sea. The SO_2 mixing ratio in the atmosphere near Saloniki reached 18 ppb, and that of aerosol sulphate – up to 500 ppt. The highest concentration level was in the 1–2-km layer – near the surface the SO_2 concentration decreased to 1–4 ppb. The observational data revealed the southward transport of pollutants over the Aegean Sea to the island of Crete, where the recorded mixing ratio reached 5 ppb. A combined analysis of the results of measurements of SO_2 and sulphate concentration as well as aerosol size distribution has shown that air masses contain components of different ages varying from several hours to several days.

Pollution due to fossil fuel burning observed in the atmosphere over northern Greece was explained by the transport from eastern Europe (Bulgaria, Romania, Turkey). Besides, the observational data reflect a considerable contribution from forest fires. This especially concerns the layers of haze formed ten days earlier due to forest fires in the north-western regions of Canada transported across the Atlantic Ocean. At altitudes above 1 km there were layers of CO about 1–2 km thick, which had often a 2-layer vertical structure, with the second layer at altitudes >3.5 km. The estimates have shown that contributions of biomass burning to the formation of the CO layer over Greece exceeded about 4 times the contributions due to fossil fuel burning. The lower atmosphere is characterized by a high number density of N_p particles of the accumulation mode (0.1–1 μm) with the N_p/CO ratio of the order of 2–6 cm^{-3}/ppb, while in the upper part of the troposphere N_p decreases due to convection and washing out. The Aitken particles number density (5–100 nm) was underestimated both in the lower and upper layers. It means that such particles, initially present in the lower layers, at the point of their detection, grew to the sizes of accumulation-mode particles.

Over the site in north-eastern Greece (40°24′N, 23°57′E) where continuous surface spectral measurements of downward short-wave radiation fluxes (total, scattered and direct solar) were made, [32] carried out synchronous aircraft observations (during the descent of the flying laboratory C-130) of the vertical profile of the physical, chemical, and optical characteristics of atmospheric aerosols initially formed during biomass burning (during forest fires in the north-western region of Canada with the subsequent long-range transport across the Atlantic Ocean).

According to data of surface measurements, the AOT values at the 500-nm wavelength increased (up to 0.39) due to a haze layer formed at altitudes 1–3.5 km. The scattering coefficient due to dry aerosol particles in this layer constituted about 80 $M m^{-1}$ ($1 M = 10^{-6}$), and the absorption coefficient – 15 $M m^{-1}$, which corresponded to a SSA at 500 nm ~0.89 (in the case of dry particles). The share of black carbon was estimated at 6–9% with respect to the mass of the concentration-mode particles ($D < 1$ μm). The growth of the scattering coefficient due to increasing relative humidity reached ~40% when relative humidity grew from 30–60%. With the prescribed size distribution and optical properties (depending on altitude) of dry aerosol, calculations were made of spectral downward short-wave radiation fluxes at the surface level. A comparison of calculated and observed values revealed an agreement of the differences of not more than 10% in the cases of direct and scattered radiation at wavelengths 415, 501, and 615 nm. Calculations of radiative forcing at the levels of the surface and atmospheric top have shown that the impact of radiation absorption by aerosol manifests itself more strongly near the surface than at the atmospheric top level. The RF values over the sea calculated over the wavelength interval 280 nm–4 μm reached $-64 W m^{-2}$ near the surface and $-22 W m^{-2}$ at the top of the atmosphere. Thus the ratio of RF values near the surface and at the top of the atmosphere reached approximately 3, which agrees with the results obtained from the data of the field experiment INDOEX.

Johansen *et al.* [52] discussed the results of analysis of atmospheric aerosol samples for the content of labile Fe(II) and other trace metals. Samples were taken over the Arabian Sea from the German ship *Sonne* in March, 1997 within the JGOFS programme of the studies of atmosphere–ocean interaction. Earlier similar shipborne observations were accomplished during the other two seasons. Analysis of samples for the content of black iron was made immediately after sampling, and concentrations of other trace metals, anions, and cations were estimated in laboratory.

The main component of the aerosol mineral composition was determined by the average composition of rocks, and its mass concentration constituted $5.94 \pm 3.08\,mg\,m^{-3}$. An additional component of the type of clay enriched with water soluble Mg^{2+} and Ca^{2+} was recorded in fine aerosols in the air masses of Arabian origin. The total concentration of iron compounds varied within 3.9–17.2 $ng\,m^{-3}$ with an average of $9.8 \pm 3.4\,ng\,m^{-3}$, the share of Fe(II) concentrated in fine aerosol constituted $87.2 \pm 3.37\%$ (this is equivalent, on average, to $1.3 \pm 0.5\%$ with respect to total iron). The anthropogenic contribution to the aerosol composition reflected by the content of Pb and Zn, as well as some anions and cations, turned out to be greater (especially during first 10 days of the voyage) than that recorded earlier from the data of observations in the periods between monsoons and during the south-western monsoon in 1995.

2.6 HIGH LATITUDES

Studies of aerosol at high latitudes have attracted serious attention, and the respective results have been widely discussed in the literature [15, 58–60, 94, 95]. Therefore, we shall confine ourselves to some comments.

On the basis of aircraft impactor and filter measurements of the properties of atmospheric aerosols in the Arctic troposphere during the field observational experiment ASTAR, in the region of the Spitsbergen Islands (Svalbard) in March–April 2000, [42] analysed the mixing of aerosol particles. Processing the observational results has shown that the main components of both background aerosol and the Arctic haze are sulphate and soot, whereas the contribution of sea salt aerosol dominated only in the lower troposphere (at altitudes <3 km). Mineral/dust as well as unknown components were detected during the period of observations only as secondary elements.

Aircraft measurements of aerosol properties were carried out under different conditions, including the cases of the Arctic haze (23 March) and aerosol loaded atmosphere (20 March and 12 April). Maximum relative content of soot ($\sim$94.7%) was recorded in the free troposphere on 23 March, when the Arctic haze was at a maximum for the whole period of observations. During several days anthropogenic aerosol arrived at the region of the Svalbard island from the industrial regions of Russia. Analysis of aerosol samples has led to the conclusion that under conditions of Arctic haze and an aerosol loaded troposphere, particles dominated with externally mixed sulphate and soot, whereas in the cases of background aerosol, internal mixing prevailed, as a rule. On the other hand, most of the particles in the free troposphere contained sulphate, externally mixed with soot, and other aerosol components both in the presence of the Arctic haze and the background conditions. The sea salt aerosol dominated, however, in the lower layer of the troposphere (sometimes the concentration of Cl^- within it decreased), though some level of sea salt particles was also recorded in the middle and upper troposphere (at altitudes of 3–7 km).

The chemical composition of atmospheric aerosol in many regions of the globe (including urban conditions, marine, and polar atmospheres) is characterized by the presence of homologous series of low-molecular dicarboxylic acids (LMD). These are formed during fossil fuel and biomass burning as well as in the processes of photochemical oxidation of anthropogenic and biogenic organic precursors. At present, the LMD are considered as substantial components of the water soluble organic aerosols. Their important function is that they can function as cloud condensation nuclei and, thus, directly and indirectly affect the formation of radiative forcing. Recent observational data have shown that in the period of polar sunrise a strong (5–20-fold) increase of LMD concentration takes place in the atmosphere, which can be interpreted as a result of photochemical formation of dicarboxylic acids through oxidation of their anthropogenic precursors coming to the Arctic from middle and high latitudes. While the main precursors are VOCs, it is natural to suppose the possibility of an important role of the gas-phase processes in the LMD formation in submicron aerosol particles. Another possibility exists in the functioning of heterogeneous reactions as a mechanism for the LMD formation on particles of supermicron sizes. However, data of observations of the acid aerosol size distribution in the Arctic atmosphere are still absent.

To obtain such data, [85] took samples of fractions of fine ($D < 1\,\mu m$) and coarse ($>1\,\mu m$) aerosol during the period of polar sunrise at the Alert station (82°30′N, 62°21′W) from 29 March to 14 April, 1997, which was part of the programme of the field observational experiment PSE-1997. Gas chromatography techniques with the use of the ionization detector (GC-FID) and mass spectrometry (GC-MS) enabled one to perform an analysis of samples for the content of low-molecular (C_2–C_{11}) dicarboxylic acids.

Results of the analysis have shown that more than 80% of LMD is concentrated in the fine fraction of aerosol, which reflects the presence in the Arctic of gas-to-particle formation of aerosol. In both fractions of aerosol the contribution of oxalic acid prevails as an LMD component followed by succinic and malonic acids. On 5–7 April, a maximum of dicarboxylic acid concentration was recorded with shorter chains (C_2–C_5), whereas in the case of LMD with longer chains ($>C_6$) this maximum was not observed.

During the period of observations a considerable decrease in the concentration of surface ozone was noted – in the presence of a negative correlation between concentrations of ozone and (C_2–C_5) dicarboxylic acids in fine and coarse fractions of aerosol. In both these aerosol modes an increased concentration of filtered bromine was also observed. The presence of the maximum concentration of LMD in both fractions of aerosol during the ozone concentration decrease indicates the existence of heterogeneous reactions on the surface of large and, possibly, small particles. The formation of dicarboxylic acids could also be connected with oxidation of such compounds/precursors as glyoxalic, glyoxylic, and other w-oxycarboxylic acids, which contain compounds of the hydrated formaldehyde group taking an active part in the chemical processes of transformation of surface ozone and halogens in the marine atmospheric boundary layer of the Arctic.

During the periods 5–20 January, 2000 and 1–26 January, 2001, [66] carried out measurements of concentration and size distribution of atmospheric aerosol at the Antarctic station Aboa (73°03′S, 13°25′W) located at a distance of about 130 km from the coastline and at a height of 480 m above sea level. During these periods of the southern hemisphere summer, data were obtained on the total number density of aerosol particles with diameters between 3–800 nm, which reflect variations within 200–2,000 cm^{-3}. Estimates of concentration are higher in marine/coastal than continental air masses. The high level of concentration (>300–300 cm^{-3}) is assumed to be determined by the contribution of fine (<20 nm) aerosol particles of the nucleation mode.

It was found that the aerosol size distribution can be approximated as a superposition of 2–4 log-normal modes. An analysis of all size spectra revealed the presence of the accumulation mode, with maximum diameter within 70–150 nm, and the Aitken mode (30–50 nm). The average diameter of the accumulation mode particles is much lower in the case of continental (rather than marine) air masses, whereas the Aitken mode was characterized by almost similar values of particle diameters for both types of air masses. The nucleation mode was present in more than half the observed spectra of sizes, and sometimes even two nucleation modes were observed. Only in marine/coastal air masses was there recorded the formation of new particles with their subsequent growth to ~40 nm. In such cases the size of particles grew at a rate of 1–3 nm hr^{-1}, and the sizes exceeded, by several times, those corresponding to the Aitken mode. It means that climatically significant particles of the Aitken and accumulation modes can appear not only in the middle and upper troposphere, where conditions exist for their prolonged growth, but also in the ABL.

Results of observations discussed in [66] have demonstrated that newly formed particles contribute substantially to the formation of the total budget of aerosol particles in the Antarctic ABL and can be important as climatically substantial Aitken and accumulation modes of atmospheric aerosols. However, it remains unclear, whether new particles do actively form. In air masses formed near the centre of the continent, the nucleation mode was not observed. Also, it was very seldom observed in the remote marine ABL over the open ocean. On the other hand, the presence of the nucleation mode, as a rule, was recorded from data of ship observations when approaching the Antarctic coast. Thus, the formation of new particles is somewhat connected with the near-coastal processes. This specific feature may be due to either a special meteorological regime of the coastal zone or emissions of MGCs in the regions of pack ice. Clarification of these problems requires further studies.

2.7 CONCLUSION

The results of studies of the aerosol properties in different regions of America and western Europe, obtained during accomplishment of programmes of the field observation experiments discussed above, have demonstrated an exclusive diversity of physical properties and chemical composition of aerosol requiring further systematization of observational data in order to substantiate more adequate models of aerosol than existing ones. Only on the basis of the use of such models it is possible to obtain more reliable estimates of possible aerosol climatic impact.

2.8 BIBLIOGRAPHY

1. Alexandrov M. D., Lacis A. A., Carlson B. E., and Cairns B. Atmospheric aerosol and trace gases parameters derived from local MFRSR network: Multi-instrument data fusion in comparison with satellite retrievals. *Proc. of the International Society for Optical Engineering*, 2002, **4882**, 498–509.
2. Allan J. D., Jimenez J. L., Williams P. I., Alfarra M. R., Bower K. N., Jayne J. T., Coe H., and Worsnop D. R. Quantitative sampling using an Aerodyne aerosol mass spectrometer. 1. Techniques of data interpretation and error analysis. *J. Geophys. Res.*, 2003, **108**(D3), AAC1/1–AAC1/10.
3. Allan D. R., Alfarra M. R., Bower K. N., Williams P. I., Gallagher M. W., Jimenez J. L., McDonald A. G., Nemitz E., Canagaratna M. R., Jayne J. T., *et al.* Quantitative sampling using an Aerodyne aerosol mass spectrometer. 2. Measurements of fine particulate chemical composition in two U.K. cities. *J. Geophys. Res.*, 2003, **108**(D3), AAC2/1–AAC2/15.
4. Alves C., Carvalho A., and Pio C. Mass balance of organic carbon fractions in atmospheric aerosols. *J. Geophys. Res.,* 2002, **107**(D21), ICC7/1–ICC7/9.
5. Andreae M. O., Fishman J., and Lindesay J. The Southern Tropical Regional Experiment (STARE): Transport and Atmospheric chemistry near the Equator-Atlantic (TRACE-A) and Southern African Fire-Atmosphere Research Initiative (SAFARI): An introduction. *J. Geophys. Res.*, 1996, **101**(D19), 23,519–23,520.
6. Angstrom A. Techniques of determining turbidity of the atmosphere. *Tellus*, 1961, **13**, 214.
7. Ansmann A., Wandinger U., Wiedensohler A., and Leiterer U. Lindenberg Aerosol Characterization Experiment 1998 (LACE-98): Overview. *J. Geophys. Res.*, 2002, **107**(D21), LAC1/1–LAC1/12.
8. Arnott W. P., Moosmüller H., Sheridan P. J., Ogren J. A., Raspet R., Slaton W. V., Hand J. L., Kreidenweis S. M., and Collett J. L., Jr. Photoacoustic and filter-based ambient aerosol light absorption measurements: Instrument comparisons and the role of relative humidity. *J. Geophys. Res.*, 2003, **108**(D1), 15/1–15/11.
9. Arshinov M. Yu. and Belan B. D. The diurnal change of the fine aerosol concentration. *Optics of the Atmos. and Ocean,* 2000, **13**(11), 983–990.
10. Barth M. C. and Church A. T. Regional and global distributions and lifetime of sulfate aerosols from Mexico City and southeast China. *J. Geophys. Res.*, 1999, **104**(D23), 30231–30240.

11. Barth M. C., Rasch P. J., Kiehl J. T., Benkovitz C. M., and Schwartz S. E. Sulfur chemistry in the National Center for Atmospheric Research Community Climate Model: Description, evaluation, features, and sensitivity to aqueous chemistry. *J. Geophys. Res.*, 2000, **105**(Dl), 1387–1416.
12. Bates T. S., Huebert B. J., Gras J. L., Criffiths F. B., and Durkee P. A. International Global Atmospheric Chemistry (IGAC) Project's First Aerosol Characterization Experiment (ACE-1): Overview. *J. Geophys. Res.*, 1998, **103**(D13), 16297–16318.
13. Baumgardner D., Raga G., Peralta O., Rosas I., Castro T., Kuhlbusch T., John A., and Petzold A. Diagnosing black carbon trends in large urban areas using carbon monoxide measurements. *J. Geophys. Res.*, 2002, **107**(D21), ICC4/1–ICC4/9.
14. Blanchard P., Brook J. R., and Brazal P. Chemical characterization of the organic fraction of atmospheric aerosol at two sites in Ontario, Canada. *J. Geophys. Res.*, 2003, **108**(D21), ICC10/1–ICC10/8.
15. Bobylev L. P., Kondratyev K. Ya., and Johannessen O. M. (eds) *Arctic Environment Variability in the Context of Global Change*. Springer–Praxis, Chichester, UK, 2003, 471 pp.
16. Bokoye A. I., Royer A., O'Neill N. T., and McArthur L. J. B. A North American Arctic aerosol climatology using ground-based sunphotometry. *Arctic*, 2002, **55**(3), 215–228.
17. Bravy B., Vasiliev G., Agroskin V., and Papin V. Recognition of composition and of microphysical characteristics of aerosol clouds in multi-frequency sounding with DF laser based lidar system. *Proc. of the International Society for Optical Engineering*, 2002, **4882**, 394–399.
18. Bundke U., Hnel G., Horvath H., Kaller W., Seidl S., Wex H., Wiedensohler A., Wiegner M., and Freudenthaler V. Aerosol optical properties during the Lindenberg Aerosol Characterization Experiment (LACE-98). *J. Geophys. Res.*, 2002, **107**(D21), LAC5/1–LAC5/15.
19. Busch B., Kandler K., Schütz L., and Neusüss C. Hygroscopic properties and water-soluble volume fraction of atmospheric particles in the diameter range from 50 nm to 3.8 μm during LACE-98. *J. Geophys. Res.*, 2002, **108**(D21), LAC2/1–LAC2/11.
20. Butler A. J., Andrew M. S., and Russel A. G. Daily sampling of $PM_{2.5}$ in Atlanta: Results of the first year of the Assessment of Spatial Aerosol Composition in Atlanta study. *J. Geophys. Res.*, 2003, **108**(D7), SOS3/1–SOS3/11.
21. Buzoriuz G., Rannik U., Makela J. M., Keronen P., Vesala T., and Kulmala M. Vertical aerosol fluxes measured by the eddy covariance method and deposition of nucleation mode particles above a Scots pine forest in southern Finland. *J. Geophys. Res.*, 2000, **105**(D15), 19905–19916.
22. Cappellato R., Peters N. E., and Meyers T. P. Above-ground sulfur cycling in adjacent coniferous and decideous forest and watershed sulfur retention in the Georgia Piedmont, USA. *Water, Air, and Soil Pollut.*, 1998, **103**(1–4), 151–171.
23. Carrico C. M., Bergin M. H., Xu J., Baumann K., and Maring H. Urban aerosol radiative properties: Measurements during the 1999 Atlanta Supersite Experiment. *J. Geophys. Res.*, 2003, **108**(D7), SOS10/1-SOS10/17.
24. Chow J. C. and Watson J. G. $PM_{2.5}$ carbonate concentrations at regionally representative Interagency Monitoring of Protected Visual Environment sites. *J. Geophys. Res.*, 2002, **107**(D21), ICC6/1–ICC6/9.
25. Clement C. F., Ford I. J., Twohy C. H., Weinheimer A., and Campos T. Particle production in the outflow of a midlatitude storm. *J. Geophys. Res.*, 2002, **107**(D21), AAC5/1–AAC5/9.

26. De Leeuw G., Gonzalez C. R., Kusmierczyk-Michulec J., Decae R., and Veefkind P. Retrieval of aerosol optical depth from satellite measurements using single and dual view algorithms. *Proc. of the International Society for Optical Engineering*, 2002, **4882**, 275–283.
27. De Tomasi F. and Perrone M. R. Lidar measurements of tropospheric water vapor and aerosol profiles over southeastern Italy. *J. Geophys. Res.*, 2002, **107**(D21), AAC14/1–14/12.
28. Ebert M., Weinbinch S., Rausch A., Gorzawski G., Hoffmann P., Wex H., and Helas G. Complex refractive index of aerosols during LACE-98 as derived from the analysis of individual particles. *J. Geophys. Res.*, 2002, **107**(D21), LAC3/1–LAC3/15.
29. Ellrod G. P., Connell B. H., and Hillger D. W. Improved detection of airborne volcanic ash using multi-spectral infrared satellite data. *J. Geophys. Res.*, 2003, **108**(D12), AAC6/1–AAC6/13.
30. Feingold G. and Morley B. Aerosol hydroscopic properties as measured by lidar and comparison with in situ measurements. *J. Geophys. Res.*, 2003, **108**(D11), AAC1/1–AAC1/11.
31. Formenti P., Reiner T., Sprung D., Andreae M. O., Wendisch M., Wex H., Kindred D., Dewey K., Kent J., Tzortziou M., *et al.* STAARTE-MED 1998 summer airborne measurements over the Aegean Sea. 1. Aerosol particles and trace gases. *J. Geophys. Res.*, 2002, **107**(D21), AAC1/1–AAC1/15.
32. Formenti P., Boucher O., Reiner T., Sprung D., Andreae M. O., Wendisch M., Wex H., Kindred D., Tzortziou M., Vasaras A., *et al.* STAARTE-MED 1998 summer airborne measurements over the Aegean Sea. 2. Aerosol scattering and absorption, and radiative calculations. *J. Geophys. Res.*, 2002, **107**(D21), AAC2/1–AAC2/14.
33. Fussen D., Vanhellemont F., and Bingen C. Synthesis inverse mapping method applied to the retrieval of aerosol size distributions from extinction measurements. *J. Geophys. Res.*, 2003, **108**(D15), AAC1/1–AAC1/10.
34. Gassó S. and Hegg D. A. On the retrieval of columnar aerosol mass and CCN concentration by MODIS. *J. Geophys. Res.*, 2003, **108**(D1), 6/1–6/25.
35. Gong S. L., Barrie L. A., Blanchet J.-P., von Salzen K., Lohmann U., Lesing G., Spacek L., Zhang L. M., Girard E., Lin H., *et al.* Canadian Aerosol Module: A size-segregated simulation of atmospheric aerosol processes for climate and air quality models. 1. Module development. *J. Geophys. Res.*, 2003, **108**(D1), 3/1–3/16.
36. Gong S. L. and Barrie L. A. Simulating the impact of sea salt on global NSS sulphate aerosol. *J. Geophys. Res.*, 2003, **108**(D16), AAC4/1–AAC4/18.
37. Gorchakov G. I. and Shukurov K. A. Fluctuations of the sub-micron aerosol concentration under convection conditions. *Physics of the Atmos. and Ocean*, 2003, **39**(1), 85–97 [in Russian].
38. Graham B., Mayol-Bracero O., Guyon P., Roberts G. C., Decesary S., Facchini M. C., Artaxo P., Maenhaut W., Köll P., *et al.* Water-soluble organic compounds in biomass burning aerosols over Amazonia. 1. Characterization by NMR and GC–MS. *J. Geophys. Res.*, 2003, **108**(D15), LBA14/1–LBA14/16.
39. Grigoryev A. A. and Kondratyev K. Ya. *Ecological Disasters*. Sci. Centre, RAS, St Petersburg, 2001, 350 pp. [in Russian].
40. Guelle W., Balkanski Y. J., Schuiz M., Marticorena B., Bergametti G., Moulin C., Arimoto R., and Perry K. D. Modeling the atmospheric distribution of mineral aerosol: Comparison with ground measurements and satellite observations for yearly and synoptic timescales over the North Atlantic. *J. Geophys. Res.*, 2000, **105**(D2), 1997–2012.

41. Gullu G. H., Olmez I., Aygun S., and Tuncel B. Atmospheric trace element concentrations over the eastern Mediterranean Sea: Factors affecting temporal variability. *J. Geophys. Res.*, 1998, **103**(D17), 21943–21954.
41a. Guyon P., Graham B., Roberts G. C., Mayol-Bracero O. L., Maenhaut W., Artaxo P., and Andreae M. O. In-canopy gradients, composition, sources, and optical properties of aerosol over the Amazon forest. *J. Geophys. Res.*, 2003, **108**(D18), AAC9/1–AAC9/16.
42. Hara K., Yamagata S., Yamanouchi T., Sato K., Herber A., Iwasaka Y., Nagatani M., and Nakata H. Mixing states of individual aerosol particles in spring Arctic troposphere during ASTAR 2000 campaign. *J. Geophys. Res.*, 2003, **108**(D7), AAC2/1–AAC2/12.
43. Henning S., Weingartner E., Schwikowski M., Gäggeler H. W., Gehrig R., Hinz K.-P., Trimborn A., Spengler B., and Baltenspreger U. Seasonal variation of water-soluble ions of the aerosol at the high-alpine site Jungfrauhoch (3580 m asl). *J. Geophys. Res.*, 2003, **108**(D1), 8/1–8/10.
44. Holzer-Popp T., Schroedter M., and Gesell G. Retrieving aerosol optical depth and type in the boundary layer over land and ocean from simultaneous GOME spectrometer and ATSR-2 radiometer measurements. 1. Method description. *J. Geophys. Res.*, 2002, **107**(D21), AAC16/1–AAC16/17.
45. Ilyin A. P., Gromov A. A., and Popenko E. M. Toxotropic metal aerosols of the surface layer. *Optics of the Atmos. and Ocean*, 2000, **13**(11), 991–995.
46. Ivlev L. S., Leyva Contreras A., and Muhlia Velaskes A. On the role of photochemistry in processes of the atmospheric aerosol pollution. In: L. S. Ivlev (ed.), *Natural and Anthropogenic Aerosols*. St Petersburg State University Press, St Petersburg, 1998, 85 pp. [in Russian].
47. Ivlev L. S., Korostina O. M., Leyva A., and Muhlia A. Modelling the optical characteristics of surface aerosols in the region of Mexico City for dry and wet seasons. *Optics of the Atmos. and Ocean*, 1993, **6**(9), 1144–1150.
48. Ivlev L. S., Vasilyev A. V., Belan B. D., Panchenko M. V., and Terpugova S. A. Opticomicrophysical models of the urban aerosols. In: L. S. Ivlev (ed.), *Natural and Anthropogenic Aerosols*, St Petersburg State University Press, St Petersburg, 2001, pp. 161–170 [in Russian].
49. Ivlev L. S., Zhukov V. M., Korostina O. M., Leiva Contreras A., Mulia Velaskes A., and Bravo-Cabrera J. L. Specified optical characteristics of aerosol in the surface layer of Mexico City. *Optics of the Atmos. and Ocean*, 1994, **7**(9), 1202–1206.
50. Ivlev L. S. and Dovgaliuk Yu. A. *Physics of the Atmospheric Aerosol Systems*. St Petersburg State University Press, St Petersburg, 2000, 258 pp. [in Russian].
51. Jimenez J. L., Jayne J. T., Shi Q., Kolb C. E., Worsnop D. R., Yourshaw I., Seinfeld J. H., Flagan R. C., Zhang X., Smith K. A., *et al.* Ambient aerosol sampling using the Aerodyne Aerosol Mass-Spectrometer. *J. Geophys. Res.*, 2003, **108**(D7), SO13/1–SOS 13/13.
52. Johansen A. M., Siefert R. L., and Hoffmann M. R. Chemical composition of aerosols collected over the tropical North Atlantic Ocean. *J. Geophys. Res.*, 2000, **105**(D12), 15277–15312.
53. Khutorova O. G., Teptin G. M., and Latypov A. F. An empirical model of interaction of aerosol and chemical admixtures under the urban conditions. *Optics of the Atmos. and Ocean*, 2000, **13**(6–7), 678–680.
54. Kimura T., Kordo K., Kumagai H., Kuroiwa H., Ishida C., Oki R., Kyze A., Suzuki M., Okamoto H., Imasu R., *et al.* Earth CARE–Earth Clouds, Aerosols and Radiation Explorer: Its objectives and Japanese sensor designs. *Proc. of the International Society for Optical Engineering*, 2002, **4882**, 510–519.

55. Kiss G., Varga B., Galamos I., and Ganszky I. Characterization of water-soluble organic matter isolated from atmospheric fine aerosol. *J. Geophys. Res.*, 2002, **107**(D21), ICC1/1–ICC1/8.
56. Kondratyev K. Ya., Moskalenko N. I., and Pozdniakov D. V. *Atmospheric Aerosol.* Gidrometeoizdat, Leningrad, 1983, 224 pp. [in Russian].
57. Kondratyev K. Ya. Biogenic aerosol in the atmosphere. *Optics of the Atmos. and Ocean*, 2001, **14**(3), 171–179.
58. Kondratyev K. Ya., Krapivin V. F., and Phillips G. V. *Problems of the High-Latitude Environmental Pollution.* Sci. Centre, RAS, St Petersburg, 2002, 280 pp. [in Russian].
59. Kondratyev K. Ya. The Arctic environment. 1. Conceptual aspects. *Proc. Russian Geogr. Soc.*, 2002, **134**(5), 1–10 [in Russian].
60. Kondratyev K. Ya. The Arctic environment. 2. Prospects of studies. *Proc. Russian Geogr. Soc.*, 2002, **135**(6), 1–7 [in Russian].
61. Kondratyev K. Ya. Aerosol as a climate forming atmospheric component. 1 Physical properties and chemical composition. *Optics of the Atmos. and Ocean*, 2002, **15**(2), 123–146.
62. Kondratyev K. Ya. and Grigoryev Al. A. *Environmental Disasters: Natural and Anthropogenic.* Springer–Praxis, Chichester, UK, 2002, 484 pp.
63. Kondratyev K. Ya. Global climate change and the Kyoto Protocol. *Idöjárás*, **106**(2), 14413–14422.
64. Kondratyev K. Ya. Priorities of global climatology. *Proc. Russian Geogr. Soc.*, 2004, **136**(2), 1025 [in Russian].
65. Kondratyev K. Ya. From nano- to global scales: Properties, formation processes and consequences of the impact of atmospheric aerosol. 2. Field observational experiment. America, Western Europe, and high latitudes. *Optics of the Atmos. and Ocean*, 2005, **17**(9), 715–741.
66. Koponen I. K., Virkkula A., Hillamo R., Kerminen V.-M., and Külmälä M. Number size distributions and concentrations of the continental summer aerosols in Queen Maud Land, Antarctica. *J. Geophys. Res.*, 2003, **108**(D18), AAC8/1–AAC8/10.
67. Kozlov A. S., Ankilov A. N., Baklanov A. M., Vlasenko A. K., Eremenko S. I., and Malyshkin S. B. Study of mechanical processes of submicron aerosol formation. *Optics of the Atmos. and Ocean*, 2000, **13**(6–7), 664–666.
68. Krejci R., Ström J., de Reus M., Hoor P., Williams J., Fischeer H., and Hansson H.-C. Evolution of aerosol properties over the rain forest in Surinam, South America, observed from aircraft during the LBA-CLAIRE-98 experiment. *J. Geophys. Res.*, 2003, **108**(D18), AAC1/1–AAC1/17.
69. Kubátova A., Vermeylen R., Claes M., Cafmeyer J., and Maenhaut W. Organic compounds in urban aerosols from Gent, Belgium: Characterization, sources, and seasonal differences. *J. Geophys. Res.*, 2002, **107**(D21), ICC5/1–ICC5/12.
70. Kudriashov V. I. An analysis of the elemental composition of atmospheric aerosols using physical methods. In: L. S. Ivlev (ed.), *Physics and Chemistry of Atmospheric Aerosols.* St Petersburg State Univ. Press, St Petersburg, 1997, pp. 97–130 [in Russian].
71. Lavoue D., Liousse C., Cachier H., Stocks B., and Goldammer J. G. Modeling of carbonaceous particles emitted by boreal and temperate wildfires at northern latitudes. *J. Geophys. Res.*, 2000, **105**(D22), 26871–26890.
72. Levy R. C., Remer L. A., Tanré D., Kaufman Y. J., Ichoku C., Holben B. N., Livingston J. M., Russel P. B., and Mering H. Evaluation of the Moderate-Resolution Imaging Spectroradiometer (MODIS) retrievals of dust aerosol over the ocean during PRIDE. *J. Geophys. Res.*, 2003, **108**(D19), PRD10/1–PRD10/13.

73. Leyva A., Muhlia A., Valdes M., Hoblen B., Smirnov A., and Ivlev L. Photometric and meteorological characteristics of the Mexico City aerosol. Preliminary results. In: L. S. Ivlev (ed.), *Proc. Second Int. Conf. 'Natural and anthropogenic aerosols', St Petersburg, Petrodvorets, 27 September–1 October 1999*. St Peterburg State Univ. Press, St Petersburg, 2000, pp. 117–120.
74. Lim H.-J., Turpin B. J., Edgerton E., Hering S. V., Allen G., Maring H., and Solomon P. Semicontinuous aerosol carbon measurements: Comparison of Atlanta Supersite measurements. *J. Geophys. Res.*, 2003, **108**(D7), SOS7/1–SOS7/12.
75. Liu D.-Y., Wenzel R. J., and Prather K. A. Aerosol time-of-flight mass spectrometry during the Atlanta Supersite Experiment. 1. Measurements. *J. Geophys. Res.*, 2003, **108**(D7), SOS14/1–SOS14/16.
76. Maring H., Savoie D. L., Izaguirre M. A., McCormick C., Arimoto R., Prospero J. M., and Pilinis C. Aerosol physical and optical properties and their relationship to aerosol composition in the free troposphere at Izafia, Tenerife, Canary Islands, during July 1995. *J. Geophys. Res.*, 2000, **105**(Dl1), 14677–14700.
77. Mayol-Bracero O. L., Guyon P., Graham B., Roberts G., Andreae M. O., Decesary S., Facchini M. P., Fuzzi S., and Artaxo P. Water-soluble organic compounds in biomass burning aerosols over Amazonia. 2. Apportionment of the chemical composition and importance of the polyacidic fraction. *J. Geophys. Res.*, 2003, **108**(D20), LBA59/1–LBA59/15.
78. McArthur L. J. B., Halliwell D. H., Niebergall O. J., O'Neill N. T., Slusser J. R., and Wehrli C. Field comparison of network Sun photometers. *J. Geophys. Res.*, 2003, **108**(D1), 1/1–1/18.
79. Meszaros E. *Fundamentals of Atmospheric Aerosol Chemistry*. Akademiai Kiado, Budapest, 1999, 308 pp.
80. Meszaros E. and Molnar A. A brief history of aerosol research in Hungary. *Időjárás*, 2001, **105**(2), 63–80.
81. Middlebrook A. M., Murphy D. M., Lee S.-H., Thomson D. S., Prather K. A., Wenzel R. J., Liu D.-Y., Phares D. J., Rhoads K. P., Wexler A. S., *et al.* A comparison of particle mass spectrometers during the 1999 Atlanta Supersite Project. *J. Geophys. Res.*, 2003, **108**(D7), SOS12/1–SOS12/13.
82. Mikhailov V. V. and Voitov V. P. An improved model of the universal spectrometer to study the shortwave radiation fields in the atmosphere. *Problems of Atmospheric Physics*, 1966, **4**, 120–128 [in Russian].
83. Minikin A., Petzold A., Ström J., Krejci R., Seifert M., van Velthoven P., Schlager H., and Schumann U. Aircraft observations of the upper tropospheric fine particle aerosols in the Northern and Southern Hemispheres at midlatitudes. *Geophys. Res. Lett.*, 2003, **30**(10), 10/1–10/4.
84. Müller D., Ansmann A., Wagner F., Franke K., and Althausen D. European pollution outbreaks during ACE-2: Microphysical particle properties and single-scattering albedo inferred from multi-wavelength lidar observations. *J. Geophys. Res.*, 2002, **107**(D15), AAC3/1–AAC3/11.
85. Narukawa M., Kawamura K., Anlauf K. G., and Barrie L. A. Fine and coarse modes of dicarboxylic acids in the Arctic aerosols collected during the Polar Sunrise Experiment 1997. *J. Geophys. Res.*, 2003, **108**(D18), ACH3/1–ACH3/9.
86. Neusüss C., Wex H., Birmili W., Wiedensohler A., Koziar C., Busch B., Brüggemann E., Gnauk T., Ebert M., and Covert D. S. Characterization and parameterization of atmospheric particle number, mass, and chemical-size distributions in central Europe during LACE 98 and MINT. *J. Geophys. Res.*, 2002, **107**(D21), LAC9/1–LAC9/13.

87. Obolkin V. A., Potemkin V. L., and Khodzher T. V. Comparative data on aerosol chemistry in the continental and arctic regions of the Eastern Siberia. *Optics of the Atmos. and Ocean*, 1998, **11**(6), 632–635.
88. O'Neill N. T., Eck T. F., Smirnov A., Holben B. N., and Thulasiraman S. Spectral discrimination of coarse and fine mode optical depth. *J. Geophys. Res.*, 2003, **108**(D17), AAC8/1–AAC8/15.
89. Penenko V. V. and Panchenko M. V. Interdisciplinary studies of transport and transformation of admixtures in the atmosphere: Preliminary results and perspectives. *Optics of the Atmos. and Ocean*, 2000, **13**(6–7), 694–700.
90. Petzold A., Fiebig M., Flentje H., Keil A., Leiterer U., Schröder F., Stifter A., Wendisch M., and Wendling P. Vertical variability of aerosol properties observed at a continental site during the Lindenberg Aerosol Characterization Experiment (LACE-98). *J. Geophys. Res.*, 2002, **107**(D21), LAC10/1–LAC10/18.
91. Phares D. J., Rhoads K. P., Johnston M. V., and Wexler A. S. Size-resolved ultrafine particle composition analysis. 2. Houston. *J. Geophys. Res.*, 2003, **108**(D7), SOS8/1–SOS8/14.
92. Pirjola L., Laaksonen A., and Kulmala M. Sulfate aerosol formation in the Arctic Boundary Layer. *J. Geophys. Res.*, 1998, **103**(D7), 8309–8321.
93. Pitts M., Hansen G., and Lucker P. An airborne A-band spectrometer for remote sensing of aerosol and cloud properties. *Proc. of the International Society for Optical Engineering*, 2002, **4882**, 353–362.
94. Polissar A. V., Hopke P. K., Pantero P., Malm W. C., and Sisler J. F. Atmospheric aerosol over Alaska. 2. Elemental composition and sources. *J. Geophys. Res.*, 1998, **103**(D15), 19045–19057.
95. Polissar A. V., Hopke P. K., Malm W. C., and Sisler J. F. Atmospheric aerosol over Alaska. 1. Spatial and seasonal variability. *J. Geophys. Res.*, 1998, **103**(15), 19035–19044.
96. Raj P. E., Devara P. C. S., Pandithural G., Maneskumar R. S., and Dani K. K. Some atmospheric aerosol characteristics as determined from laser angular scattering measurements at a continental urban station. *Atmosfera*, 2004, **17**(1), 39–52.
97. Randall C. E., Bevilacqua R. M., Lumple J. D., Hoppel K. W., Rusch D. W., and Shettle E. P. Comparison of Polar Ozone and Aerosol Measurement (POAM) II and Stratospheric Aerosol and Gas Experiment (SAGE) II aerosol measurements from 1994 to 1996. *J. Geophys. Res.*, 2000, **105**(D3), 3929–3942.
98. Rankin A. M. and Wolff E. W. A year-long record of size-segregated aerosol composition at Halley, Antarctica. *J. Geophys. Res.*, 2003, **108**(D24), AAC9/1–AAC9/12.
99. Rannik Ü., Aalto P., Keronen P., Vesala T., and Külmälä M. Interpretation of aerosol particle fluxes over a pine forest: Dry deposition and random errors. *J. Geophys. Res.*, 2003, **108**(D17), AAC3/1–AAC3/11.
100. Rhoads K. P., Phares D. J., Wexler A. S., and Johnston M. V. Size-resolved ultrafine particle composition analysis. 1. Atlanta. *J. Geophys. Res.*, 2003, **108**(D7), SOS6/1–SOS6/13.
101. Romankevich E. A. and Vetrov A. A. *The Carbon Cycle in the Arctic Seas of Russia*. Nauka Publ., Moscow, 2001, 303 pp. [in Russian].
102. Savarino J. and Legrand M. High northern latitude forest fires and vegetation emissions over the last millenium inferred from the chemistry of a central Greenland ice core. *J. Geophys. Res.*, 1998, **103**(D7), 8267–8279.
103. Schröder F., Kärcher B., Fiebig M., and Petzold A. Aerosol states in the free troposphere at northern midlatitudes. *J. Geophys. Res.*, 2002, **107**(D21), LAC8/1–LAC8/8.

104. Sellegri K., Laj P., Peron F., Dupuy R., Legrand M., Peunkert S., Putraud J.-P., Cachier H., and Ghermandi G. Mass balance of free tropospheric aerosol at the Puy de Dôme (France) in winter. *J. Geophys. Res.*, 2003, **108**(D11), AAC2/1–AAC2/17.
105. Sicard M., Rocadenbosch F., López A. M., Comerón A., Rodriguez A., Muñoz C., and Garcia-Vizcaino D. Characterization of aerosol backscatter-to-extinction ratio from multi-wavelength and multi-angular lidar profiles. *Proc. of the International Society for Optical Engineering*, 2002, **4882**, 442–450.
106. Smirnov A., Holben B. N., Dubovik O., Frouin R., Eck T. F., and Slutsker I. Maritime component in aerosol optical models derived from Aerosol Robotic Network data. *J. Geophys. Res.*, 2003, **108**(D1), 14/1–14/11.
107. Solomon P. A., Cowling E. B., Weber R. Preface to special section: Southern Oxidants Study 1999 Atlanta Supersite Project (SOS3). *J. Geophys. Res.*, 2003, **108**(D7), SOS 0/1.
108. Solomon P. A., Chameides W., Weber R., Middlebrook A., Kiang C. S., Russel A. G., Butler A., Turpin B., Mikel D., Scheffe R., *et al.* Overview of the 1999 Atlanta Supersite Project. *J. Geophys. Res.*, 2003, **108**(D7), SOS1/1–SOS1/24.
109. Solomon P., Baumann K., Edgerton E., Tanner E., Eatough D., Modey W., Maring H., Savoie D., Natarajan S., Meyer M. B., *et al.* Comparison of integrated samplers for mass and composition during the 1999 Atlanta Supersite project. *J. Geophys. Res.*, 2003, **108**(D7), SOS11/1–SOS11/26.
110. Teinila K., Kerminen V.-M., and Hillamo R. A study of size-segregated aerosol chemistry in the Antarctic atmosphere. *J. Geophys. Res.*, 2000, **105**(D3), 3893–3905.
111. Trimborn A., Hinz K.-P., and Spengler B. Online analysis of atmospheric particles with a transportable laser mass spectrometer during LACE-98. *J. Geophys. Res.*, 2002, **107**(D21), LAC13/1–LAC13/10.
112. Vasilyev A. V., Melnikova I. N., Poberovskaya L. N., Tovstenko I. A. Spectral brightness coefficients for natural formations in the range 0.35–0.85 μm. 1. Instrumentation and processing technique of measurement results. *Studies of the Earth from Space*, 1997, **3**, 25–31.
113. Vasilyev A. V., Melnikova I. N., Poberovskaya L. N., and Tovstenko I. A. Spectral brightness coefficients for natural formations in the range 0.35–00.85 μm. II. Water surface. *Studies of the Earth from Space*, 1997, **4**, 43–51.
114. Vasilyev A. V., Melnikova I. N., Poberovskaya L. N., and Tovstenko I. A. Spectral brightness coefficients for natural formations in the range 0.35–0.85 μm. III. Land surface. *Studies of the Earth from Space*, 1997, **5**, 25–32.
115. Vasilyev A. V. and Ivlev L. S. An optical statistical aerosol model of the atmosphere for the Ladoga Lake region. *Optics of the Atmos. and Ocean*, 2000, **13**(2), 198–203.
116. Vasilyev O. B., Grishechkin V. S., and Kovalenko A. P. The spectral information-measurement system to study the shortwave radiation field in the atmosphere from surface and aircraft. In K. Ya. Kondratyev and V. A. Melentyev (eds), *Complex Remote Sensing of Lakes*. Nauka Publ., Leningrad, 1987, pp. 225–228 [in Russian].
117. Vasilyev S. L., Gudoshnikov Yu. N., and Ivlev L. S. Active impacts on atmospheric processes. Natural and anthropogenic aerosols. In: L. S. Ivlev (ed.), *Proc. Second Int. Conf. 'Natural and anthropogenic aerosols', St Petersburg, Petrodvorets, 27 September–1 October* 1999. St Petersburg State Univ. Press, Institute of Chemistry, St Petersburg, 2000, pp. 251–258 [in Russian].
118. Vasilyev S. L., Ivlev L. S., Krylov G. N. Complex monitoring and control of the environment. *Proc. of the Third Int. Conference 'Natural and anthropogenic aerosols', St Petersburg, 2001*, pp. 449–456 [in Russian].

119. Vautard R., Martin D., Beekmann M., Drobinski P., Friedrich R., Jaubertie A., Kley D., Lattuati M., Moral P., Neininger B., *et al.* Paris emission inventory diagnostics from ESQUIF airborne measurements and a chemistry transport model. *J. Geophys. Res.*, 2003, **108**(D17), ESQ7/1–ESQ7/21.
120. Von Hoyningen-Huene W., Freitag M., and Burrows J. B. Retrieval of aerosol optical thickness over land surfaces from top-of-atmosphere radiance. *J. Geophys. Res.*, 2003, **108**(D9), AAC2/1–AAC2/20.
121. Wahiin P. One year's continuous aerosol sampling at Summit in Central Greenland. In: E. W. Woiffand and R. C. Baler (eds), *Chemical Exchange Between the Atmosphere and Polar Snow* (NATO AST Series 1. 1996, **vol. 43**). Springer-Verlag, Berlin, pp. 131–143.
122. Wandinger U., Müller D., Böckmann C., Althausen D., Matthias V., Rösenberg J., Weiß V., Fiebig M., Wendisch M., Stohl A., *et al.* Optical and microphysical characterization of biomass-burning and industrial-pollution aerosols from multi-wavelength lidar and aircraft measurements. *J. Geophys. Res.*, 2002, **107**(D21), LAC7/1–LAC7/20.
123. Weaver C. J., Joiner J., and Ginoux P. Mineral aerosol contamination of TIROS Operational Vertical Sounder (TOVS) temperature and moisture retrievals. *J. Geophys. Res.*, 2003, **108**(D8), AAC5/1–AAC5/15.
124. Weber R., Orsini D., Duan Y., Baumann K., Kiang C. S., Chameides W., Lee Y. N., Brechtel F., Klotz P., Jongejan P., *et al.* Intercomparison of near real time monitors of $PM_{2.5}$ nitrate and sulfate at the U.S. Environmental Protection Agency Atlanta Supersite. *J. Geophys. Res.*, 2003, **108**(D7), SOS9/1–SOS9/13.
125. Wenzel R. J., Liu D.-Y., Edgerton E. S., and Prather K. A. Aerosol time-of-flight mass spectrometry during the Atlanta Supersite Experiment. 2. Scaling procedure. *J. Geophys. Res.*, 2003, **108**(D7), SOS15/1–SOS15/8.
126. Wetzel M. A., Shaw G. E., Slusser J. R., Borys R. D., and Cahill C. F. Physical, chemical, and ultraviolet radiative characteristics of aerosol in central Alaska. *J. Geophys. Res.*, 2003, **108**(D14), AAC9/1–AAC9/16.
127. Wex H., Neusüss C., Wendisch M., Stratmann F., Koziar C., Keil A., and Wiedensohler A. Particle scattering, backscattering, and absorption coefficients: An in situ closure and sensitivity study. *J. Geophys. Res.*, 2002, **107**(D21), LAC4/1–LAC4/18.
128. Wolff E. W. and Cachier H. Concentrations and seasonal cycle of black carbon in aerosol at a coastal Antarctic station. *J. Geophys. Res.*, 1998, **103**(D9), 11033–11042.
129. Yi Q., Box M. A., and Jupp D. L. B. Inversion of multi-angle sky radiance measurements for the retrieval of atmospheric optical properties. 1. Algorithm. *J. Geophys. Res.*, 2002, **107**(D22), AAC10/1–AAC10/10.
130. Yi Q., Jupp D. L. B., and Box M. A. Inversion of multi-angle sky radiance measurements for the retrieval of atmospheric optical properties. 2. Application. *J. Geophys. Res.*, 2002, **107**(D22), AAC11/1–AAC11/9.
131. Zakharenko V. S. and Moseichuk A. N. Chemical reactions in the troposphere. *Optics of the Atmos. and Ocean*, 2003, **16**(5–6), 447–453.
132. Zhang J., Chameides W. L., Weber R., Cass G., Orsini D., Edgerton E., Jongejan P., and Slanina J. An evaluation of the thermodynamic equilibrium assumption for fine particulate composition: Nitrate and ammonium during the 1999 Atlanta Supersite Experiment. *J. Geophys. Res.*, 2003, **108**(D7), SOS2/1–SOS2/11.
133. Zhao T. X.-P., Laszlo I., Dubovik O., Holben B. N., Sapper J., Tanré D., and Pietras C. A study of the effect of non-spherical dust particles on the AVHRR aerosol optical thickness retrievals. *Geophys. Res. Lett.*, 2003, **30**(3), 50/1–50/4.

134. Zhao T. X.-P., Laszlo I., Holben B. N., Pietras C., Voss R. J. Validation of two-channel VIRS retrievals of aerosol optical thickness over ocean and quantitative evaluation of the impact from potential subpixel cloud contamination and surface wind effect. *J. Geophys. Res.*, 2003, **108**(D3), AAC7/1–AAC7/12.

3

Field observational experiments in Eurasia and on the African continent

Complex aerosol studies over the territory of the former USSR were carried out starting from the late-1950s. Results of these studies have been discussed in several monographs and numerous papers [8, 43, 45, 55–58, 61–63, 78, 79, 83–87] as well as in Chapter 1 of this book. Of special interest are studies on biogenic components of aerosol [10–12, 92], specific features of the structure and properties of aerosol over vast basins of Lake Baikal, the White Sea, and the Atlantic Ocean [68, 78, 79, 83], aerosols formed on the dried-up bottom of the Aral Sea [45], as well as volcanic aerosol of Kamchatka [42, 44] and the urban aerosol of Eurasia. Some works on studies of the content and chemical composition of pollutants in glacier deposits in Altai and Tien Shan have made it possible to draw conclusions about the main sources and processes of transport of the aerosol dust component and about the interannual variability of aerosol properties in the region of northern Eurasia [8, 13, 25, 33, 43, 61–63, 69, 78, 79, 82–84, 87, 92, 102]. Below we shall discuss the results of complex field experiments accomplished in a number of regions during recent years.

3.1 THE SAHARAN DUST EXPERIMENT

The first complex observational experiments aimed at aerosol–radiation studies of dust aerosol of deserts were started in the region of Ashkhabad (Turkmenistan) within the CAENEX programme in October, 1970 and then grew in scale [53–55, 58]. One of the latest stages of similar developments is connected with an accomplishment of the field experiment SHADE – taking into account the key role of the Sahara desert as the most powerful source of natural dust aerosol (DA).

Haywood *et al.* [38] analysed the physical and optical properties of the Saharan DA using observational data from the flying laboratory C-130 of the British Meteorological Service, obtained within the field observational experiment SHADE. Four special flights were made from the Sal Island (the Cape Verde islands) on 21, 24, 25, and 28 September, 2000. The main sources of information were data of observations with the following instrumentation carried by C-130: (1) photometer PSAP to measure the short-wave radiation (SWR) absorption at the wavelength 0.567 μm; (2) nephelometer TSI-3563 (aerosol scattering at three wavelengths: 0.45, 0.55, and 0.70 μm); (3) spectrometer to measure the size of aerosol particles and cloud droplets in the interval of radii 1–23.5 μm; (4) broadband Eppley pyranometers for wavelength ranges 0.3–3.0 μm and 0.7–3.0 μm; (5) 16-channel filter radiometer (SAFIRE) for spectral SWR measurements; (6) and nadir-directed SWS spectrometer to measure in the intervals 0.30–0.95 μm and –0.95–1.70 μm (spectral resolution is 0.010 and 0.018 μm, respectively).

Processing of observational results has made it possible, in particular, to evaluate the aerosol optical thickness $\tau_{\text{aer}\,\lambda}$ and DA-induced direct radiative forcing (RF). An analysis of observational results revealed an overestimation of the DA-induced calculated SWR absorption with the prescribed standard values of the refraction index. According to the observational data considered, the estimate of the imaginary part of the complex refraction index (0.0015*i*) can be considered adequate at the wavelength 0.55 μm.

Various techniques to calculate $\tau_{\text{aer}\,\lambda}$ have been discussed in [38] and the respective errors have been estimated. The value of $\tau_{\text{aer}\,\lambda=0.55}$ under conditions of maximum dust loading of the atmosphere is 1.48 ± 0.05. Under these conditions, the instant value of RF is approximately $-129 \pm 5\,\text{W}\,\text{m}^{-2}$, which corresponds to an aerosol-induced increase of the surface–atmosphere system's albedo over the ocean by a factor of 2.7 ± 0.1. A comparison of the obtained RF values with aircraft observations within the CERES programme has demonstrated an agreement to within the coefficient 1.2. The results discussed indicate that in studies of the Earth's radiation budget, the RF due to the Saharan DA should be taken into account.

From observations with the use of the C-130 instrumentation carried out within the programme of the field observational experiment SHADE, the impact of the Saharan DA on the long-wave radiation (LWR) transfer in the atmosphere has also been studied [39]. A comparison of measurement results in the dust loaded and clean atmosphere (in clear sky) revealed, in the spectra of upward and downward LWR fluxes, a distinct signature of the DA impact on the LWR transfer. A comparison of measured and calculated spectral LWR fluxes with the required information prescribed from data of simultaneous observations enabled one to retrieve the refraction index for DA in the IR region, since the main factors affecting the radiation transfer are the refraction index, the DA content, and the height of the dust layer. The impact of aerosol size distribution is less substantial, considering the presence of particles with the radius >1 μm.

The impact of the DA layer manifested itself as a relative radiation-induced warming up to $0.5\,\text{K}\,\text{day}^{-1}$ under the layer, and a cooling (down to $0.5\,\text{K}\,\text{day}^{-1}$) above the aerosol layer. These radiative temperature changes are, by about an order of magnitude, less than the respective changes due to SWR absorption. In the field of LWR fluxes the presence of the dust layer is reflected as a decrease of the outgoing LWR ($6.5\,\text{W}\,\text{m}^{-2}$) at the atmospheric top level and an increase ($11.4\,\text{W}\,\text{m}^{-2}$) of the downward LWR flux at the surface level. The presence of dust has led to a decrease of the aircraft measured brightness temperature (from the data for the atmospheric transparency window) by about 2–4 K, which agrees with similar results of satellite measurements.

An estimation of the DA-induced globally averaged aerosol direct RF has led to a broad scattering of values ranging between +0.09 and $-0.46\,W\,m^{-2}$, which is explained by uncertain data on the chemical composition, size distribution, and content of dust aerosol. Important factors are that the diameter of DA particles can reach several μm, and the imaginary part of the complex refraction index is not zero (i.e., dust aerosol is an atmospheric component, which not only scatters but also absorbs SWR). It is also important that DA substantially affects the properties of clouds and the gas composition of the atmosphere. Apart from the sea salt aerosol, DA is the most significant component of atmospheric aerosols. An estimation of the DA content in the global atmosphere gave values ranging between 14 and 41 Tg with a distinct spatial–temporal variability of the DA content.

During accomplishment of the Saharan dust experiment (SHADE) in September, 2000 from the flying laboratory C-130, [27] obtained filter samples of aerosol in the atmosphere over the Atlantic Ocean in the region between the SAL Island and Senegal to study the aerosol chemical composition. Dust aerosol was detected in the 0.5–1-km layer. Within the main dust plume there were sublayers with different DA size distributions and of different origins. A negative correlation between concentrations of tropospheric ozone and dust indicates that dust layers have led to a removal of ozone. On days of a heavy dust loading of the atmosphere the DA concentration reached $54\,\mu g\,m^{-3}$ (including the submicron and supermicron aerosol fractions). An analysis of the fine DA fraction revealed the presence of the anthropogenic component of aerosol (mainly in the form of NH_4HSO_4) in the cases when air masses coming from western Europe were then transported to northern Africa.

Analysis of DA chemical composition showed that the ratio of the content of nitrate and non-marine sulphate constituted about 0.3. On this basis, one can draw the conclusion about the absence of a marked amount of components formed due to biomass burning (the latter is characterized by a higher content of nitrate). In some cases, traces of plumes of pollutants due to fossil fuel burning were observed, probably, coming from North America. The geochemical signatures of dust aerosol agree with those observed earlier. There was not, in particular, any enrichment of Si, Fe, and Ti compared with aerosol of soil origin, whereas in the case of elements such as Ca and S, the situation was opposite. Though Ca is present in African soils mainly as calcite, in the atmosphere it was also found as a calcium sulphate component.

Summarizing, [101] gave a brief characteristic of the programme, observational means, and results of the accomplishment of the field observational experiment SHADE, aimed at studies of the properties of the Saharan dust aerosol driven to the Atlantic Ocean basin from the region of northern Africa during periods of dust storms in the desert. The main observational means were complexes of various instruments carried by two flying laboratories in the region of Cape Verde during 19–29 September, 2000. As has been mentioned above, the on-board scientific instrumentation included: standard sensors to measure the meteorological parameters, the aerosol spectrometer of particle sizes PCASP, the optical spectrometer FFSSP to measure the forward scattered SWR (the FFSSP turned out to be functioning unreliably, however), ice particles sensor SID, instruments to take filter samples of aerosol, photometer to measure the soot-induced SWR absorption at 0.567 μm, and the nephelometer TSI-3563 to measure the aerosol scattering at 0.45, 0.55, and 0.70 μm. The Eppley pyranometers were used to measure the upward and downward SWR fluxes in the wavelength intervals 0.3–3.0 μm and 0.7–3.0 μm. The SWR intensity was measured with the scanning filter radiometer SAFIRE in 16 intervals of the wavelength range 0.55–2.1 μm, and data of SWR at a high spectral resolution ($\sim$0.5 cm^{-1}) were obtained with the Fourier spectrometer. Finally, the 2-range (0.30–0.95 and 0.95–1.70 μm) spectrometer for the SWR was the source of information about the spectral brightness of the surface–atmosphere system.

The scientific instrumentation of the French flying laboratory Mystere-20 (M-20) included standard meteorological sensors, Eppley pyranometers, 2-wavelegth (0.532 and 1.064 μm) lidar LEANDRE-1 and an aircraft version of the satellite instrumentation POLDER for polarization measurements at wavelengths 0.440 and 0.865 μm and to measure the system's brightness (0.440 and 0.865 μm). In the SHADE period on the Sal Island (14°43′N, 25°56′W) and at M'Bour (16°43′N, 22°56′W), sun photometers of the AERONET network were functioning. The M-20 observations were coordinated with remote sensing data from the Terra multi-channel spectral video-radiometer MODIS.

As has been mentioned above, studies of the aerosol chemical composition have shown that it does not contain products of biomass burning, but there is a small amount of anthropogenic aerosol. The typical aerosol size distribution near the Senegal coastline was characterized by two maxima near 0.40 μm and in the interval 1.5–2.0 μm. Non-sphericity of particles (in calculations of the SWR field) could be taken into account on the assumption of the spherical shape of particles. The estimates of the aerosol absorption turned out to be below those obtained earlier: the single-scattering albedo constituted 0.90 ± 0.02 in the 'blue' spectral region; 0.97 ± 0.02 at 0.670 μm, and was practically equal to zero in the near-IR spectral region.

The direct RF recorded on 25 September reached a maximum of $-130\ W\ m^{-2}$. Bearing in mind that the aerosol optical thickness (AOT) value at 0.55 μm is $\sim$1.5, it turns out that the RF value calculated per unit optical thickness constitutes $\sim -90\ W\ m^{-2}$. Near the surface, the RF is greater than at the atmospheric top level by a factor of 1.6 (this difference is less compared with that observed earlier from data of observations over the Indian Ocean). The DA impact on the RF formation in the IR region constitutes $\sim$10% with respect to the short-wave RF, and it is always positive. According to numerical modelling results, the aerosol RF in the regions 0°–30°N, 60°W – 40°E varies within -8 to $-10\ W\ m^{-2}$ in clear sky, constituting about $-6.0\ W\ m^{-2}$ in cloudy sky with contributions due to long-wave radiation equal, respectively, to $+1.0$–$1.2\ W\ m^{-2}$ and 0.7–0.9 $W\ m^{-2}$. An estimation of the global-scale contribution of the Saharan DA gave $-0.4\ W\ m^{-2}$.

In connection with DA studies, serious attention has been paid to an improvement of techniques to measure the aerosol characteristics. In particular, [64] discussed possibilities of improving the technique of retrieving the atmospheric aerosol properties from the joint data of active and passive satellite remote sensing bearing in mind prospects for an accomplishment of the satellite lidar sounding as well as the functioning of instruments such as the video-spectroradiometer MODIS and polarization radiometer POLDER. On the basis of the data of lidar sounding at wavelengths 0.532 and 1.064 μm, possibilities have been considered to retrieve the vertical profile of the backscattering/extinction ratio and then the altitudinal dependence of an efficient radius of aerosol particles (on the assumption of bimodality of its size distribution).

Analysis of sensitivity of retrieval results to input parameters has shown that the retrieval technique is rather robust with respect to noise levels and calibration errors. The reliability of the retrieval algorithm has been tested by processing the observational data obtained in September, 2000 over the Atlantic Ocean in the course of accomplishing the SHADE experiment. A comparison of the retrieved vertical extinction profile with the observational data has demonstrated a satisfactory agreement. Results of lidar observations on 25 September revealed the presence at altitudes of 2.2–4.5 km of an aerosol layer with an efficient radius of particles 1.19 ± 0.6 μm. There was also a transitional sublayer located over the marine ABL in which there were aerosol particles with far less effective radii than in the Saharan aerosol layer.

Beginning in the mid-1980s, with the support of the European Space Agency (ESA), a series of programmes of aircraft and surface observations have been accomplished aimed mainly at analysis of the prospects of remote sensing with the use of satellite video-spectroscopy. In particular, within the DAISEX programme, observations have been made with spectrometers in three test areas in south-eastern Spain (Barrax) and in the Upper Rhein valley (Colmar, France; Hartheim, Germany). The results obtained were very important in providing successful functioning of the video-spectrometer MERIS carried by the ENVISAT satellite.

Pedros *et al.* [81] discussed the results of observations carried out at Barrax with the use of two spectroradiometers Licor-1800 and one spectroradiometer Optronic OL-754 to retrieve spectral AOT of the atmosphere. The results considered referred to individual days of observations in 1998, 1999, and 2000. An analysis of the observational data in the visible spectral region led to the conclusion that in all cases the AOT formation was determined mainly by the contribution of the coarse fraction of aerosol. Results of calculations of air mass backward trajectories testify to the prevalence of marine aerosol in the lower layers of the troposphere in the presence of a small admixture of continental aerosols.

To obtain information about the chemical characteristics of final products of the Saharan dust particles' transformation into the atmospheric dust aerosol transported over long distances, [37] used samples of two types: (1) a fine fraction of sand particles from northern Africa; and (2) an aerosol phase of typical precipitation in the Sahara. In the second case, to determine the initial composition of particles captured by water droplets, a correction was applied considering the solution of part of the aerosol in cloud water.

Results of particles analysis for the content of Al, Fe, P, and Pb have led to the conclusion that (except Pb) the chemical composition of the long-range transported Saharan aerosol is more homogeneous than the composition of dust particles. According to data on the air mass backward trajectories, a higher homogeneity of the aerosol composition is partly determined by the fact that events of DA formation cover vast territories of the Sahara desert, and thus the aerosol composition, corresponds to an averaged composition of soil particles subject to erosion. Data on Pb reflect the presence of an anthropogenic component. A consideration of data on the Pb/Al concentration ratio for soil particles suggest the conclusion that the composition of the Saharan precipitation is affected, as a rule, by mixing with air masses flowing from western Europe.

The Saharan dust aerosol, as the final product of transformation of sand particles getting to the atmosphere, has the following typical composition (%): Al (7.09 ± 0.79); Fe (4.45 ± 0.49); P (0.082 ± 0.011); and Pb (24 ± 9 ppm). The concentration ratio of various minor gas components (MGCs) to Al or Fe should be considered as the most typical characteristics of DA composition. The important role of the Saharan dust is that it can be an important nutrient (P, Fe) for the Mediterranean Sea waters, constituting, in particular, about 30–40% of the phosphorus flux and $\sim$90% of the iron flux from the atmosphere in the western sector of the Mediterranean Sea. The DA can substantially affect the biogeochemical cycle of iron, being the main source of dissolved iron in this region. Observations in the regions adjacent to northern Africa became a significant source of information about the Saharan DA properties. So, for instance, an analysis of the data of observations in Leipzig (51.3°N, 12.4°E) revealed abnormally high lidar ratios of extinction to backscattering with the input of the Saharan dust particles. Initial observations of this dust outbreak recorded on 13–15 October, 2001 suggested the conclusion that high values of lidar ratio were determined by the presence of nonspherical particles of atmospheric aerosol. To more reliably interpret these observational data, [74] considered in detail the information about geometrical and optical characteristics of the DA plume. This has been done on the basis of the complex study of the Saharan aerosol plume first accomplished with the use of the 2-wavelength Raman lidar (Nd:YAG) laser with pulses repeated every 30 s and the sun photometer (wavelengths 340, 380, 440, 500, 670, 870, 940, and 1020 nm) located in Leipzig.

The considered observational site is part of the EARLINET network of lidar aerosol soundings. The automated observations of AOT and sky brightness were made with the sun photometer functioning within the worldwide AERONET network of robotic aerosol observations. Analysis of the observational data led to the conclusion that the dust plume reached an altitude of 6 km, and the AOT at 532 nm increased from 0.25 on 13 October, 2001 to a maximum of $\sim$0.83 on 14 October, 2001. According to lidar observations, the main contribution to the AOT formation at 532 nm was made by the dust layer at altitudes above 1,000 m.

From the data of the sun photometer in the wavelength interval 380–1,020 nm, the Angström coefficient constitutes about 0.45 with the input of the dust aerosol plume and decreases to a minimum of 0.14 during a maximum of dust outbreak. The retrieval of the vertical profile of the Angström coefficient from the data of lidar soundings at 355 and 532 nm revealed a strong variability in the coefficient when it is <-0.2 in the centre of the dust plume. The values of this coefficient averaged over the lower 1-km layer of the atmosphere changed from 1.0 in the beginning of atmospheric dust loading to 0.39 in the daytime on 14 October, 2001, when dust aerosol penetrated the ABL.

Comparisons of the AOT values averaged over the whole atmosphere and the Angström coefficient obtained from lidar and sun photometer data revealed good agreement. From the data of lidar observations at 532 nm the coefficient of depolarization due to dust aerosol increased to 25%. Analysis of the photometer data on scattering function at wavelengths 440, 670, 870, and 1,020 nm (for this purpose, the sky brightness measurement results were used) revealed the presence of non-spherical particles. It was the aerosol particles' non-sphericity that caused abnormally high values of the lidar extinction/backscattering ratio within 50–80 sr at 532 nm.

Using the electron microscopy method, [65] studied the aerosol samples taken in Punta del Hidalgo on the Canary Islands and in Sagra (Portugal) over the Atlantic Ocean in the period June–July, 1997 within the ACE-2 programme of the field experiment on the study of aerosol properties. Analysis of the chemical composition of individual aerosol particles has shown that the main types of aerosol are new ones as well as those which have partially or completely reacted with sea salts consisting of NaCl, mixed cations (Na, Mg, K, and Ca), sulphate Na_2SO_4 and $NaNO_3$, particles of industrial origin containing $(NH_4)_2SO_4$, soot, ashes, silicon, iron oxides, and $CaSO_4$, and minor admixtures in the form of mineral dust.

Since the sampling sites were located in different geographical regions, an analysis of the samples obtained has demonstrated the presence of different contributions of the sources of anthropogenic emissions located in Europe. In Sagra, depending on the sample, between 0–30% (0–50%) of salt particles did not participate (or partially participated) in chemical reactions, and 20–100% of particles turned out to be completely transformed into sulphates and nitrates as a result of reactions with polluting components from the European continent. In contrast to this, samples of sea salt aerosol taken in Punta del Hidalgo were weakly affected anthropogenically. Even in the polluted atmosphere, less than 5% of the total amount of particles was subject to this impact (through respective chemical reactions). This great difference in properties of aerosols sampled in Punta del Hidalgo and in Sagra reflects the difference in the processes of pollutants' solution, rates of chemical reactions, and a decrease in the amount of chemically active pollutants while they were transported over the ocean.

Israelevich *et al.* [46] proposed a technique to identify the sources of the input to the atmosphere of dust from deserts with the use of data averaged over long periods on local maxima of the aerosol index from satellite TOMS observations. An application of this technique ensures reliable information like that of the number of days with a dust-loaded atmosphere, but more simple and universal.

The results obtained in [46] enabled one, for the first time, to observe the springtime dust flows from the sources located near the sites with coordinates ~16°N, ~16°E, and ~19°N, 6°W, whose intensity exceeds the power of the dust aerosol sink due to deposition and transport. For this reason, the atmosphere over northern Africa turns out to be almost always dust loaded in the presence of a considerable amount of mineral dust aerosol in the atmosphere both in spring and in summer.

The source of dust in the basin of the Chad Lake (~16°N, 16°E) was most persistent and most active in April. A more variable source of dust (most intense in July) is the region around the point 19°N, 6°W. Due to the functioning of the so-called Sharav cyclones, dust is transported for long distances eastward and northward – toward the Mediterranean basin. The monitoring of the dust plume dynamics has shown that they first appear in the western sector of the Mediterranean Sea and then move eastward at an average rate of about 7–8° latitude per day. In spring, this motion continues to the western coast of the Mediterranean Sea. In summer, the dust plume does not move farther than 15°E.

Hoonaert *et al.* [41] studied the aerosol size distribution and chemical composition on the island of Tenerife in order to analyse the dependence of aerosol properties on the specific character of atmospheric circulation in the marine ABL (MABL) and in the free troposphere in June–July 1997. Processing the observational data has shown that the FT air masses can be classified (from the viewpoint of aerosol properties) as clear marine (Atlantic) containing the Saharan aerosol and pollution (the latter refers to air masses from western Europe). Approximately the same features have air masses in the MABL where, however, the contribution of the Saharan DA manifests itself weakly.

The characteristic feature of the chemical composition of aerosol in the free troposphere and MABL is its strong difference for fine and coarse fractions of aerosol. The fine aerosol in the free troposphere is rich in sulphur-depleted aluminosilicates (62%) in the cases of dust or sulphate aerosols, and under conditions of polluted air masses – carbonaceous particles (20%) and sulphur-enriched aluminosilicates (46%). The coarse aerosol is characterized by a strong decrease of sulphates (<20%) and carbonaceous components (10%) in samples of polluted air masses. The composition of fine aerosol in the MABL is completely determined by sulphur-enriched particles and, to a lesser degree, carbon and sea salt components. The coarse mode is characterized by the prevalence of sea salt aerosol (62%), whereas particles enriched with sulphur constitute only 5.3%.

Vakmirović *et al.* [106] analysed an episode of the input of the Saharan DA to the region of Belgrade on 14–17 April, 1994 and carried out a numerical simulation of the evolution of the aerosol chemical composition under the influence of dry and wet deposition.

3.2 THE FIELD EXPERIMENT TO STUDY THE ASIAN AEROSOL PROPERTIES: ACE-ASIA

Serious attention has been drawn to studies of DA entering the atmosphere in the deserts of China and then being subject to a long-range (west-to-east) transport toward the East China Sea and the Pacific Ocean [113].

During intense observations (March–May, 2001) within the programme of the field observational experiment ACE-Asia, at the China Digital Single-channel Recorder (DSR) network to monitor dust storms in China, atmospheric aerosol was regularly observed in order to determine its mass concentration and chemical composition. During this period, there were four dust storms. Zhang *et al.* [113–115] performed an analysis of observational data which showed that 45–82% of the DA mass was due to soil particles with an increased content of Ca and Fe (by 12% and 6%, respectively, for DA from the western and northern regions of China). Aerosols from the northern region were also characterized by a high concentration of Si (30%), but a low concentration of Fe (4%). In all cases the relative content of Al in DA constituted about 7%.

Calculations of backward trajectories of DA transport during dust storms revealed five main paths for its motion in the spring of 2001 – all these trajectories going through Beijing. Sandy soils of north-eastern China were one of the DA sources. The aerosol size distribution in the range of diameters 0.25–16 μm was characterized by a log-normal distribution with an average diameter of 4.5 μm and MSD ~1.5. Such characteristics are representative for ~69% of the total DA content. The shares of particles in the range of sizes 16 μm constituted about 1.7% and 30%, respectively.

Gong *et al.* [31] undertook a numerical simulation of the processes of formation and transport of soil dust aerosol with the use of the regional model NARCH based on the equation of DA mass balance determined by ratios between DA sources and sinks. Calculations were made for conditions of the field observational experiment ACE-Asia accomplished in the period March–May, 2001 aimed at study of the DA dynamics in south-eastern Asia. The a priori prescribed exhaustive meteorological information has made it possible to thoroughly simulate the processes of formation, transport (including changes of aerosol properties in the process of long-range transport), and dry and wet deposition of DA in the region of China with 12 categories of soil sources of aerosol taken into account. To calculate the DA vertical profile, its size distribution was prescribed from data of surface observations in deserts.

A comparison of numerical simulation results with data of surface and satellite observations for eastern Asia and North America confirmed the adequacy of the conclusions on the basis of numerical simulation results. According to the estimates obtained, in the deserts of eastern Asia during 1 March–31 May, 2001, about 252.8 Mt of soil aerosol was formed (and emitted to the atmosphere). About 60% of all DA entering the atmosphere during four large-scale dust storms had particle sizes $d < 40\,\mu m$. Information about the aerosol content above the 700-hPa level correlates well with estimates of the aerosol index obtained from data of the satellite TOMS instrumentation. Calculations of sensitivity of numerical simulation results to soil particles size distribution, water content in the atmosphere, and to meteorological conditions enabled one to determine optimal parameters and conditions typical (in the region considered) of the processes of formation and transport of aerosol.

During the period of the field observational experiment ACE-Asia (31 March–1 May, 2001), [68] carried out studies of atmospheric aerosol in the region of the

Pacific Ocean with improved methods of atmospheric aerosol sampling from aircraft (flying laboratory *Twin Otter*), using a thermo-optical analyser of carbon aerosol.

The aerosol samples were analysed to estimate the concentration of organic carbon (OC), elemental (EC), and carbonate (CC) carbon. During some flights, high concentrations of pollutants and/or mineral dust aerosol were observed.

To identify the nature of aerosol in different atmospheric layers from the Angström exponent ($\mathring{\alpha}$), data from the 3-channel nephelometer were used. The $\mathring{\alpha}$ values varied within 0.2–2, which corresponded to atmospheric layers containing dust or pollutants. The OC and EC concentrations at altitudes <3 km changed from 0.58–29 μgC m^{-3} and from 0.20–1.8 μgC m^{-3}, respectively. On the whole, the concentration of total carbon in different atmospheric layers in the ACE-Asia period turned out to be higher than that observed during ACE-2, TARFOX (an experiment on aerosol RF in the troposphere), and INDOEX (studies of the atmosphere over the Indian Ocean).

In some cases, mixed layers of dust and pollutants were observed. The presence of CC was found in samples with $\mathring{\alpha} = 1.6$, which indicated the presence in the samples of a considerable amount of dust aerosol with $\mathring{\alpha} > 0.2$. The linear regression of the aerosol absorption coefficient σ_{ap} (10^{-6} cm^{-1}) in the visible spectral region depending on EC was characterized by the correlation coefficient 0.50 (thus the σ_{ap} variability was determined by not only EC but also other factors). The mass aerosol coefficient of absorption E_{abs} (m^2 g^{-1}) changed by a factor of 8 (for different samples), with an average value of 11 ± 50 m^2 g^{-1}, which agrees well with data obtained earlier.

The key problems of observational data interpretation are connected with the necessity to answer the following questions: (1) do the observed levels of OC and EC concentration in the MABL affect the physico-chemical properties of cloud condensation nucleii (CCN) in the MABL; (2) does it affect the formation and lifetime of clouds in the MABL; and (3) to what extent does the OC chemical composition affect the level and physico-chemical properties of CCN. In the latter case an important unsolved problem consists in establishing relative contributions of carbonaceous aerosol produced by biomass or fossil fuel burning.

The DA emitted to the atmosphere in the spring in the deserts of north-western Asia and then wind-driven to the northern hemisphere Pacific Ocean basin has long since attracted attention in the context of its impact on climate and from the viewpoint of solving the problem of atmospheric correction. In the interests of solution of this problem, [66] analysed the results of simultaneous observations of the outgoing short-wave radiation (OSWR) from data from the satellite wide-angle instruments SeaWiFS and cloud-free sky brightness from data of observations on board the ship *Ronald H. Brown* in March–April, 2001 obtained within the ACE-Asia programme on studies of atmospheric aerosol of Asian origin. The AOT was retrieved from data on sky brightness using the optimization method. The measured AOT values were reproduced by iterative calculations with prescribed various characteristics of aerosol.

The results obtained by [66] agree well with data on AOT and *in situ* measurements of the ocean reflectance (albedo). An agreement of data from observations at three sites of the ocean has ensured a substantial decrease in the error of the ocean's albedo retrieval at 412 and 443 nm. The effect of DA was compared with the effect on the OSWR of fine absorbing aerosols. Taking one observation series as an example, it was shown that a combined consideration of DA and absorbing (soot) aerosol ensures a substantial decrease of AOT retrieval errors (from 44% to 13% at 865 nm). This example illustrates the necessity of taking into account the presence of absorbing aerosol in solving the problem of atmospheric correction to retrieve the optical properties of seawater.

The presence of increasing anthropogenic emissions of pollutants to the atmosphere over Asia, especially in the regions of rapid industrial development, has determined an urgency for studies of chemically and radiatively active MGCs and aerosol. The earlier studies revealed an emission-induced high concentration of atmospheric pollutants in a vast region of the Pacific Ocean in the northern hemisphere springtime. Prospero *et al.* [77] discussed the results of observations of aerosol concentration on the Midway Island located in the central sector of the northern hemisphere Pacific Ocean (28°13′N, 177°22′W) for the period 1981–2000. With the use of a relatively simple technique to separate the components, natural and anthropogenic shares of sulphate and nitrate aerosols have been determined.

It follows from the results obtained that during the period under consideration the concentration of anthropogenic sulphate and nitrate aerosol has almost doubled, which correlates with the growth of SO_2 emissions in China. This conclusion has been based, however, on the assumption that the springtime SO_4^{2-} surplus, with respect to concentration of dimethylsulphide (DMS) (determined from data on methylsulfonate MSA), is determined by the contribution of anthropogenic emissions. This assumption proceeds from the consideration of episodic nature of volcanic eruptions as the main source of natural emissions of SO_2. However, there are natural sources of NO_x emissions on the continents (e.g., lightning discharges, emissions by soil, biomass burning), but their contribution (compared with anthropogenic sources) can be considered small, especially in spring (the latter refers to biomass burning). Apparently, the contribution of the sources of considered MGCs outside Asia is negligible, their effect is manifested mainly in winter.

Data from observations for the late-1990s testify to the presence, beginning in the mid-1990s, of the trend of stabilizing or even decreasing concentrations of SO_4^{2-} and NO_3^-. It is important, however, that these data are less reliable because of (for different organizational reasons) the quality of observations (many aerosol samples had to be rejected). Nevertheless, the conclusion about the trend of stabilization or decrease of SO_4^{2-} concentration can be considered reliable, since in this period the scale of the use of fossil fuel in China stabilized and started decreasing. At the close of 1990, there was a marked decrease of emissions of chemically active nitrogen compounds. The observed stabilization or decrease of SO_4^{2-} and NO_3^- concentrations indicate that the earlier forecasts overestimate the concentration of these components in the troposphere, more so that the available data suggest the conclusion about a possible decrease of NO_x emissions in China. From the viewpoint of specification of information about the aerosol RF, of importance is a consideration of black carbon aerosol, whose content in emissions coming from China to the region of the Atlantic Ocean is substantial.

Widely practiced in rural regions of southern Asia, the use of wood fuel for house heating and cooking determines a large scale of biomass burning and emission of products in the form of smoke and various MGCs to the atmosphere. To study such aerosol products, [96] performed laboratory tests on burning various kinds of biofuel used in Bangladesh. The chemical composition of the products of burning was characterized by the presence of organic and elemental carbon, sulphate, nitrate, ammonium, and chloride ions – the content of potassium and natrium in different burned compounds differed little. The chemical composition of organics contained in smoke from different sources of biofuel is unique. Fingerprints of organic compounds contained in smoke aerosols differ substantially from respective wood fuels used in North America. Faecal stanols, including 5β–stipmastanol, corpostanol, and cholestanol are representative molecular markers of the products the burning of cow's manure. In addition to this, was information about methoxyphenols and vegetative sterols contains important information about biomass.

The lowering of the Aral Sea level and related processes of desertification are among the most serious natural disasters observed in the 20th century. During the last 40 years more than 36,000 km^2 of sea bottom have become exposed, which determined the appearance of a powerful source of Aeolian dust. The related enhancement of dust storms and subsequent deposition of dust aerosol are supposed to represent a serious danger for human health. In this context, [112] studied a connection between the atmospheric dust loading and respiratory disease among children in Kara-Kalpak Republic. The heaviest dust loading has been detected in the band of the formed sea bottom, which contributes most to the dust input to the atmosphere. Considerable sources of dust exist also in some regions of Kara-Kalpak located comparatively far from the Aral Sea, especially in early summer. A preliminary analysis of data on respiratory diseases among children has shown that children living in the north of Kara-Kalpak with intense deposition of aeolian dust suffer less from respiratory diseases. This negative correlation requires further verification, illustrating the complicated bonds between environmental conditions and human health.

Of course, dust aerosols are, though very important, only one of the components of atmospheric aerosol in Asia. From time to time, volcanic eruptions contribute substantially.

One of the most volcanically active global regions is the Japanese Archipelago. Here are 83 active volcanoes, including 12 volcanoes like Sakurajima, Asosan, Unsendake, and Miyakejima, which are especially active. An analysis of observational data performed by [29] has led to the conclusion that volcanic emissions of sulphur dioxide constitute about 20% of total emissions of SO_2, and in Japan their contribution to deposition of total sulphur constitutes $\sim$20%. In this context, the July, 2000 eruption of the stratovolcano Miyakejima (34.68°N, 139.53°E), located 200 km south of Tokyo, was unprecedented in scale and, in particular, affected the air quality in the region of the Tokyo megalopolis, where about 30 million people live.

The prolonged activity of the volcano has brought forth a considerable increase of SO_2 concentration in the atmosphere. The diurnal mean emissions of sulphur dioxide during the period from August, 2000–March, 2001 reached about $15 \times 10^9\ \mathrm{gS\,day^{-1}}$ (i.e., of the same level as that resulting from fossil fuel burning in eastern Asia). Estimates of total (dry and wet) post-eruption deposits of sulphur gave $\sim 4.26\ \mathrm{mgS\,m^{-2}\,day^{-1}}$, which by more than 3 times exceeded the magnitudes observed before the eruption ($1.41\ \mathrm{mgS\,m^{-2}\,day^{-1}}$). An increase of the contribution of wet deposition with enhanced precipitation reflected the considerable impact of the washing out of sulphur from the atmosphere and hence the removal of eruption products from the atmosphere.

In warm seasons characterized by prevailing subtropical anticyclones over the Pacific Ocean, episodes of high SO_2 concentration took place sometimes with some delay due to the northward transport of the volcanic plume. In cold seasons, most of the volcanic emissions were driven over the Pacific Ocean by westerlies blowing from the Asian continent. Though there was no direct correlation between dry deposition and volcanic emissions of SO_2, there was a positive correlation between wet deposition of non-sea salt (NSS) SO_4^{2-} and emissions of SO_2. One should think that the interaction between precipitating clouds and the plume of erupted products took place directly before the eruption of Miyakejima, and the role of the process of wet deposition in the removal of volcanic sulphur from the atmosphere was important beyond the region under consideration.

Gao *et al.* [30] analysed the chemical composition of 41 aerosol samples measured from aircraft over the ocean in the regions located north, south, and east of Japan. Analysis of samples was made first of all for the content of water soluble components, especially organic compounds. The presence of 21 identifiable components has been detected including inorganic anions, hydrocarbons, organic acids, and different metals. Three main sources of aerosol, identified from its specific chemical composition, were: anthropogenic emissions, biomass burning, and dust outbreaks.

As has been mentioned above, the dust aerosol appearing in the atmosphere during dust storms in the Sahel region of northern Africa and undergoing the long-range transport, often reached western Europe and even southern England. For instance, traces of such aerosol have been detected on the surface snow cover in the Alps. One of the representative signatures of an aerosol source location can be data on the isotopic composition of niobium contained in aerosol samples together with results of calculations of backward trajectories of the long-range transport of aerosol. So, for instance, analysis of 'red dust' samples obtained in France showed that northern Africa was the source of the aerosol.

An examination of the backward trajectory performed by [35] revealed, in one of the cases (6 March, 1990), that aerosol came from the region of China. The isotopic analysis of niobium revealed a close similarity with the composition of the Chinese loesses. Moreover, the results of numerical simulation of the global atmospheric circulation indicated that the dust plume preceding that observed had left China before 25 February, 1990, crossed the North American border in February–March, and reached the French Alps on 6 March, 1990. Thus the case considered revealed for the first time the possibility of transcontinental transport of dust aerosol across the Pacific Ocean and northern Atlantic in a latitudinal belt of western–eastern transport.

Within the programme of the field observational experiment ACE-Asia was aimed at study of atmospheric aerosols in eastern Asia. Wang *et al.* [108, 109] performed measurements of the aerosol size distribution and chemical composition using the following instrumentation carried by the aircraft *Twin Otter*: differential mobility analyzer (DMA), aerodynamic particles sensor (APS), impactors (MOUDI), and sensors for aerosol sampling. A detailed analysis has been made of the data of four flights to compare calculated (with prescribed observed aerosol parameters) and measured (with the use of the 14-channel sun photometer AATS-14) values of total solar radiation attenuation (extinction) in the cloud-free atmosphere.

The spatial distribution of aerosol was characterized by the presence of pollutants in the ABL and practical absence of mineral dust in the free troposphere. For data of the four flights mentioned above the best agreement takes place with the ratio of calculated to observed values of extinction equal to 0.97 ± 0.24 (0.96 ± 0.21) and 1.07 ± 0.08 (1.08 ± 0.08) at wavelengths 525 (1,059) nm, respectively, for the ABL and polluted layers in the ABL. In dust layers in the free troposphere the calculated aerosol extinction was, as a rule, less than that obtained from AATS-14 data, with the values of this ratio being 0.65 ± 0.06 (525 nm) and 0.66 ± 0.05 (1,059 nm). Probably, this difference is determined by the lack of account of the shape of mineral aerosol particles in numerical simulations as well as by the impact of horizontal heterogeneity of dust layers.

3.3 THE REGION OF THE INDIAN OCEAN

The main objective of the field observation experiment INDOEX, with a duration of intense observations between 11–25 February 1999, was to obtain complex information about chemical, microphysical, and optical properties of atmospheric aerosol, both of natural and anthropogenic origin, with the use of surface, aircraft, and ship observational means [72]. A comparison of measurement data with the use of various kinds of instruments, referring to different observational conditions, revealed the presence of a considerable variability of averaged values of such 'extensive' characteristics as mass, as well as aerosol-induced scattering, and absorption of radiation. On the other hand, such 'intensive' properties as mass scattering efficiency (MSE), single-scattering albedo (SSA), the share of backscattered radiation, the Angström coefficient, and the ratio of concentrations of various aerosol components are characterized by a considerably weaker variability. As a rule, concentration ratios are much more variable, however, than parameters of aerosol size distribution and optical properties.

Variability of aerosol characteristics revealed by comparisons of data for instruments installed on different carriers can be interpreted as really existing, and determined both by specific features of measurement techniques and different observational conditions. At a high level of aerosol haze concentration (respectively, at high values of the scattering coefficient) the MSE values turned out to be close (at relative humidity = 33%) to $3.8 \pm 0.3\,\mathrm{m^2\,g^{-1}}$ which determines a possibility to limit the variability of the properties of aerosol to this level. The most substantial variability of MSE was observed in the free atmosphere or in the regions of low aerosol concentrations. Results obtained by [10] suggest the conclusion that to realize a 'closure' of optical, chemical, and microphysical characteristics of aerosol with an error of not more than 20%, it is necessary to substantially improve the techniques of calibration and measurements.

An important aim of the field observational experiment INDOEX was to study the long-range transport and evolution of natural and anthropogenic aerosol emitted to the atmosphere in the region of the Indian subcontinent, as well as the interaction of aerosol with clouds, radiation, and climate. One of the means of observations to solve the problems mentioned was a system of balloon probes drifting horizontally at different altitudes. In the period of intense observations within the INDOEX programme (15 January–27 February), 17 balloon probes were launched from Goa (the western coast of India) which drifted at different altitudes in the lower troposphere. The monitoring of drifting balloon probes from satellites has shown that the drifting took place under the influence of the Madden–Julian oscillation, but the effect of coastal anticyclonic circulations was also manifested near the western coast of India, connected with orography and mesoscale circulation systems. The latter play an important role in the process of the long-range transport across the coastline of anthropogenic and mineral-polluting components. A comparison of the observed and precalculated trajectories and properties of air masses has shown that from the viewpoint of wind speed, air temperature, and humidity, an agreement is, on the whole, satisfactory. However, within the MABL, there are substantial differences between the results of calculations and observational data, including the temperature underestimation, which could be determined by the spatial smoothing of calculated values.

An important part of developments in the period of intense observations within the INDOEX programme, carried out in 1999, constituted ship observations of atmospheric aerosol properties. These observations were performed on board the ship *Ronald H. Brown* in the regions of the Indian Ocean north and south of the inter-tropical convergence zone (ITCZ) in the Arabian Sea as well as the Gulf of Bengal. An analysis of results of calculations of backward trajectories of air masses performed by [86] has led to detection of eight regions – sources of air masses (AM) – coming to the region, including: the northern hemisphere Indian Ocean; the eastern sector of the Indian subcontinent (here the AM trajectories went close to Calcutta, across southern India); the Indian subcontinent where the AM trajectories crossed central India; the Arabian Sea; the Arabian–Indian subcontinent with a typical combination of trajectories over Arabia in the lower troposphere and trajectories over India at higher levels; and the Arabian Sea/Indian coastline where the AM trajectories along the Indian coastline prevailed.

The studied properties of aerosol in the MABL included the chemical composition, size distribution, and coefficients of scattering and absorption. Besides this, there were also available data on the vertical profile of the backscattering coefficient and AOT. Quinn *et al.* [86] analysed the observed dependence of number density and mass concentration of various prevailing components of the aerosol composition, coefficients of scattering, absorption and attenuation (extinction), SSA, Angström coefficient, and AOT (at 500 nm) on specific features (origin) of air masses. All observations (except AOT) were made at relative humidity $= 55 \pm 3\%$. In the cases of AM from the southern hemisphere, an extinction due to particles $<1\,\mu m$ and $<10\,\mu m$ was mainly determined by the contribution of sea salt aerosols. The ratio of extinction coefficients for particles of these sizes was, in this case, at a minimum compared with all other cases, averaging, respectively, 28% and 40%, which was determined by the prevailing contribution of submicron sea salt aerosol to the formation of the aerosol mass. The AOT values were also at a minimum constituting, on an average, 0.06 ± 0.03.

The concentration of NSS sulphate aerosol in air masses from the Indian Ocean in the northern hemisphere is approximately six times greater than the respective values for the southern hemisphere air masses (with the comparable values of submicron sea salt aerosol concentration). In the case of submicron sulphate particles their contribution to the decrease of aerosol concentration reached 40%, the sea salt aerosol dominated in the presence of particles $>10\,\mu m$. The AOT values averaged 0.10 ± 0.03, and the SSA constituted about 0.89. The presence of the detectable concentration of black carbon (BC) equal to $0.14 \pm 0.05\,\mu g\,m^{-3}$ indicated the existence of long-range transport to the region of the ITCZ of aerosol from the Indian subcontinent. Two regions suffering the transport from Arabia at an altitude 500 m were characterized by higher concentrations of submicron NSS sulphate aerosol, particles of organic matter (POM), and inorganic oxidized matter (IOM) compared with marine regions. Concentrations of supermicron IOM and sea salt particles were comparable. The contribution of submicron particles of NSS sulphate aerosol dominated the decrease of aerosol concentration and remained considerable in the case of particles $>10\,\mu m$, when the impact of sea salt aerosol was at a maximum.

The average contribution of BC to the decrease of aerosol concentration due to submicron particles was, respectively, 8% and 12%, for air masses from marine and continental regions, whereas the SSA values were 0.93 ± 0.02 and 0.38 ± 0.07, and AOT 0.19 ± 0.12 and 0.38 ± 0.07 (higher values correspond to an air flux at an altitude of 2,500 m from India). The regions receiving air masses in the lower troposphere from the Indian subcontinent were characterized by maximum values of concentration of submicron NSS sulphate aerosol, POM, BC, and IOM. The concentration of supermicron NSS sulphate aerosol was below or comparable with the concentration of supermicron nitrate. The main contribution to extinction due to sub and supermicron particles was made by NSS sulphate aerosol, though the concentration of BC, HNO_3 and H_2SO_4 due to biomass burning was almost equivalent. The SSA value (0.85 ± 0.11) turned out to be the lowest. Average AOT values were at a maximum (0.3–0.4) in the regions of the long–range transport from the Indian subcontinent.

Since the carbon component constitutes one of the main fractions of the atmospheric aerosol composition, it is interesting to obtain information about the content in the aerosol of EC responsible for the aerosol absorbing properties and OC which contributes substantially to the aerosol-induced radiation absorption. This especially concerns the almost lack of data on the OC/EC ratio for different fractions of particles sizes. While the main source of EC is biomass burning, the sources of OC are various, including direct emissions to the atmosphere at incomplete burning of fossil fuel or biomass, gas phase reactions with participation of volatile organic compounds, etc. Neusüss *et al.* [77] discussed results of shipborne (*Ronald H. Brown*) observations of EC and OC concentrations in sub and supermicron fractions of aerosol in the atmosphere over Indian Ocean carried out within the INDOEX programme in February–March, 1999.

An analysis of data on measured total concentration has shown that it is low in air masses formed in the southern hemisphere. Under such conditions, only a small amount of OC was detected in the supermicron fraction of sea salt particles. In the atmosphere over the northern hemisphere Indian Ocean, where the continental air masses are affected most, the OC concentration in submicron and supermicron aerosol components turned out to be higher by an order of magnitude or more. EC was detected mainly in submicron particles, its content increased by more than two orders of magnitude, in the northern sector of the Indian Ocean compared with its southern sector. The mass share of carbon aerosol varied within 6–15% and 2–12% for sub and supermicron fractions, respectively, increasing with the growing level of air pollution.

The measured high values of aerosol absorption coefficient can be partly explained by a high degree of mixing with the radiation scattering component of aerosol. The efficiency of absorption turned out to be much higher at low levels of EC aerosol content. Various short-chain dicarboxylic and hydroxylated dicarboxylic acids have been detected as absorbing components. The first are concentrated mainly in supermicron particles, the second in submicron fractions of aerosol. Alkanes and polycyclic aromatic hydrocarbons are also detected in small amounts. This suggests the supposition that a considerable part of aerosol organic matter is of a secondary origin. This supposition confirms the presence of a high correlation between total concentration of carboxylic acids, concentration of OC, nitrate, and to a lesser degree, sulphate. However, the observed values of OC/EC concentration ratios also reflects the presence of primary OC. Very important are the observations of OC and EC sources in south-eastern Asia.

During the time of the field observational INDOEX programme there was a long-range transport of aerosols and MGCs from the Indian subcontinent to the region of the Indian Ocean towards the ITCZ. An analysis of calculated backward trajectories of air masses has made it possible to monitor the transformation of the chemical composition of aerosol and MGCs depending on their geographical origin. In connection with this, [36] performed a comparison of the observational data on temporal variability of the content of acetonitrile (a long-term indicator of biomass and biofuel burning), number density of submicron carbon-containing aerosol mass concentration of potassium (an indicator of the sources of fossil fuel burning), and submicron NSS potassium aerosol. These comparisons revealed the presence of high levels of correlation ($0.84 < R^2 < 0.92$) between the enumerated components. On this basis one can conclude, with a high probability, that the sources of these components are similar.

Data on aerosol and MGCs point to the impact of emissions from burning fossil fuel, biomass, and biofuel – the products of which then undertake long-range transport in the atmosphere. In air masses driven from the Indian subcontinent to the Indian Ocean basin, high levels of submicron NSS potassium were recorded as well as particles containing carbon and potassium and acetonitrile. This situation reflects the presence of a strong impact on the air composition of biomass and biofuel burning taking place during INDOEX (the contribution of such an impact to the formation of submicron carbon aerosol constitutes $74 \pm 9\%$). In contrast to this, air masses from the Arabian peninsula were characterized by a low content of the components mentioned above. In these cases, the impact of processes of fossil fuel burning prevailed, of which the contribution to the formation of submicron carbon aerosol was $63 \pm 9\%$. The results obtained indicate the urgency required for complex studies of the variability of the chemical composition of aerosol and MGCs in different regions.

In the INDOEX period, in the Indian Ocean region, aerosol soundings of the atmosphere were carried out with the use of micropulse lidars (MPL) carried by the ship *Ronald H. Brown* and also located on the Maldives. To measure the AOT of the atmosphere and to calculate the MPL, multichannel (wavelengths 380, 440, 500, 675, and 870 nm) sun photometers were used. Welton *et al.* [111] discussed the results of observations of the vertical profiles of aerosol optical properties from lidar soundings at 523 nm. According to ship observations, aerosol was located, as a rule, at altitudes below 4 km. At that time the altitude of the MABL was less than 1 km.

Results of aerosol soundings have been analysed together with data on calculated trajectories of air masses and aerological soundings. During the whole voyage the relative humidity varied from 30% near the ocean surface to 50% at the upper boundary of the aerosol layer. In the absence of aerosol input from the Indian subcontinent, the AOT values for marine aerosol constituted 0.05 ± 0.03, the ratio of extinction to backscattering $S = 33 \pm 6$ sr, and maximum value of the extinction coefficient (near the MABL upper boundary) was about $0.05\,km^{-1}$. These values agree with results obtained earlier. In the case of polluted air masses over the Indian Ocean, the AOT values exceeded 0.2; $S > 40$ sr and maximum coefficient of extinction reached $0.20\,km^{-1}$. All these characteristics are typical of continental aerosols.

Welton *et al.* [111] performed comparisons of the extinction coefficient near the ship (at a height of 75 m), obtained from MPL data, and results of observations of extinction at the ship's level from data on scattering observed with the nephelometer and of absorption observed with the photometer, which makes it possible to determine the radiation absorption by the soot component of aerosol. These comparisons have shown that the use of the algorithm of MPL data processing (with the prescribed constant value of S in the lower troposphere) makes it possible to obtain a reliable value of extinction near the ocean surface agreeing with the data of observations at the ship's level only under conditions when aerosol within the MABL is well mixed with aerosol located in higher atmospheric layers (above the MABL). An analysis of the vertical profiles of extinction obtained from MPL data has led to the conclusion that the vertical profile of extinction obtained earlier (1996), as typical of the region of the Maldives, cannot be considered representative for the whole region of the Indian Ocean. Therefore, a new model of the vertical profile of extinction has been proposed for the marine atmosphere over the Indian Ocean. This model characterizes conditions of the atmosphere polluted with continental aerosols.

In March 1999, in the period of intense observations within the field experiment INDOEX, [70, 71] performed aircraft measurements (flying laboratory C-130) of the content in fine aerosols ($D_p < 1.3\,\mu m$) of carbon components and water-soluble ion components at different altitudes in order to analyse the altitudinal dependence of the chemical composition of aerosol. In the region of the Indian Ocean, north of the equator, the polluted atmospheric layers were always present at altitudes up to 3.2 km. The total mass concentration of fine aerosol (the sum of masses of carbon and ion components) in these layers constituted $15.3 \pm 7.9\,\mu g\,m^{-3}$. The main components were POM (35%), SO_4^{2-} (34%), black carbon (BC, 14%), and NH_4^+ (11%).

The main difference between the chemical composition of MABL (altitude range 0–12 km) and, located above it, the residual continental ABL (1.2–3.2 km) was a higher content of SO_4^{2-} compared with POM within the MABL – probably explained by a faster transformation of SO_2 into SO_4^{2-} in the MABL. The results obtained have shown that carbon contributes most to the formation of aerosol mass, and this contribution increases with altitude. The low variability of the aerosol optical properties was detected in two layers. The regression analysis of the mass dependence of absorption coefficient at 565 nm (at BC concentration $>4.0\,\mu gC\,m^{-3}$) has led to the value of specific absorption cross section of $8.1 \pm 0.7\%$ during the whole period of observations. An abnormally considerable share of the BC content, as well as the presence of a high correlation between BC and absorption coefficient, indicated that BC was responsible for the strong absorption of light observed in the polluted layers during INDOEX. A high correlation between BC and total carbon (TC) ($R^2 = 0.86$) indicated mainly a primary origin. A high correlation was also recorded between the scattering coefficient at 550 nm and the mass of the fine fraction of aerosol. The specific cross section of scattering constituted $4.9 \pm 0.4\,m^2\,g^{-1}$. During the whole period of observations the concentration ratios BC/TC, BC/OC, SO_4^{2-}/BC, and K^+/BC were almost constant. It follows from these ratios that from 60–80% of aerosol residing in the polluted atmospheric layers in the period of INDOEX formed due to fossil fuel burning, and 20–40% was caused by biofuels burning.

During the period of INDOEX, [17] performed lidar soundings of the atmosphere in Goa (the western coast of India, 15.45°N, 73.08°E) from 1 to 15 March, 1999. The vertical profiles of backscattering against extinction ratio Φ_p were retrieved to study the laws of the vertical distribution and temporal evolution of the aerosol content in the atmosphere in the winter monsoon period (the micropulse lidar functioned at 532 nm). Considering the data from the sun photometer on AOT, the Φ_p value averages $0.03 \pm 0.010\,sr^{-1}$. The BC concentration turned out to be a representative tracer of the aerosol scattering coefficient α_s (from data of daytime observations), whereas the impact of relative humidity was of secondary importance.

A statistical ratio was obtained [17] between α_s and BC which was used to retrieve the aerosol properties at night, when the values of Φ_p are practically the same as in the daytime. The AOT value averaged over the observational period was 0.76 ± 0.15 at night and 0.55 ± 0.09 in the daytime. A considerable contribution to the AOT formation was made by the 0.7 km thick layer over the ocean surface. The second aerosol layer was detected above this layer (at altitudes 0.7–3.5 km) and was characterized by a substantial diurnal change of its vertical extent and optical thickness determined, apparently, by breeze variability. The optical thickness of the upper aerosol layer varied from 0.49 ± 0.14 at night to 0.25 ± 0.07 in the daytime.

During six days of INDOEX, [70] performed aircraft (C-130 flying laboratory based on the Maldives: 4.18°N, 73.53°E) *in situ* measurements and lidar Raman sounding of the vertical profiles of the aerosol extinction coefficient (σ_{ep}) and back-scattering coefficient (β_p). *In situ* measurements of σ_{ep} and β_p were made with the use of two integrating nephelometers and the photometer to measure the soot-induced absorption. The aerosol optical thickness of the atmosphere was also measured with a surface sun photometer.

The values of σ_{ep} and β_p, obtained from data of lidar sounding, turned out to be ~30% higher than those observed *in situ* (at a 95% confidence level). Possible causes of such differences are: (1) systematic errors of *in situ* data due to aerosol losses at the sampler's input; (2) underestimated impact of air humidity on extinction in processing the data of *in situ* measurements; (3) overestimated values of σ_{ep} and β_p obtained from data of lidar sounding explained by the impact of the non-observed cloudiness; and (4) errors in processing the observational results (especially lidar data). Incomplete spatial–temporal compatibility of *in situ* and lidar data could also play an important role.

During the INDOEX expedition in the region of the Indian Ocean (January–March, 1999), on board the ships *Ronald H. Brown* and *Sagar Kanya*, [34] obtained rainwater samples and performed a chemical analysis on them for the content of main ions and some trace metals. Using the results of calculations of backward trajectories of air masses for a period of 10 days, possible sources of air masses in Asia have been detected and a comparative analysis has been carried out of data on the chemical composition of rainwater and simultaneously obtained aerosol samples.

Though the observed concentration of NSS sulphates (NSS)-SO_4^{2-} ions, NO_3^-, NH_4^+, NSS-K^+, and NSS-Ca^{2+} in rainwater of a given region turned out to be 2–3 times lower than on the Indian subcontinent, it was determined that there was a considerable impact of continental sources of pollutants and dust aerosol from Asia. With the southward motion of air masses toward the ocean from the continent, the concentration of ions NSS-Ca^{2+} decreased, whereas the concentration of NSS-SO_4^{2-} grew. This agrees with a higher acidity of rainwater over the ocean (with a pH between 4.8–5.4) compared with the Indian subcontinent. The main cation was NH_4^+ (but not Ca^{2+}, as it was over land). There was a high correlation between Al and Fe concentrations and NSS-Ca^{2+} which indicated the continental origin (as the product of rocks) of these elements.

The ratio between concentrations of Na^+, Cl^-, and Br^- in rainwater was close to that for seawater which testifies to the absence of either an excess or a deficit of the halogen ions mentioned above. The ratios of concentrations of metal ions in rainwater and in the atmosphere were systematically higher in the case of the coarse mode of aerosol of salt origin (Na^+, Mg^{2+}, and Cl^-) compared with the fine mode (NH_4^+, NSS-K^+, and NSS-SO_4^{2-}). This testifies to the presence of an intracloud process of washing out of fine aerosols, whereas a substantial contribution to the removal of salt particles of coarse aerosols was made by rain droplets beneath clouds. Such components as NSS-Ca^{2+} and NO_3^- refer to the intermediate category and the process of their washing out under clouds is not so substantial as in the case of coarse sea salt aerosols.

The Asian–African wintertime monsoon is characterized by the presence of a stable north-western transport of air masses from the zone of the Indian subcontinent to the regions of the Arabian Sea and the equatorial Indian Ocean. Such a situation affects significantly the development of chemical processes on a global scale due to the transport of a large amount of anthropogenic atmospheric aerosols. To analyse the laws of this air mass transport, [59] calculated their trajectories for every day during the period from 1 February 1999, when the field observational experiment INDOEX was carried out. The numerical modelling for one of the episodes of transport, in the presence of an intense cyclone, has made it possible to monitor the evolution of the process of dust plume formation. The role of the weather-forming system in the transport of polluted air masses to Europe from the INDOEX region has been analysed in [59]. Such a transport can seriously affect the regime of precipitation and climate in the regions of southern Europe and the Mediterranean Sea.

3.4 THE HORIZONTAL DISTRIBUTION OF THE ION–CHEMICAL COMPOSITION OF ATMOSPHERIC AEROSOL OVER THE FORMER USSR TERRITORY

A general pattern of the content of elements in the composition of atmospheric aerosols is illustrated in Table 3.1, which contains aerosol chemical matrices for some regions of the former USSR [3–5, 9, 102]. These data were obtained by averaging over all the available material for each of the regions.

The zonal variation of total concentrations of the groups of elements is presented in Figure 3.1. It follows from the graphs that the total concentration of elements is at a minimum in the Urals, in the western and eastern Siberia, differing little, constituting 13.5, 13.3, and 14.2 $\mu g\,m^{-3}$. Maximum concentrations are observed in the regions of central Asia: Tadjikistan 31.8 $\mu g\,m^{-3}$ and Kirghizstan 28.2 $\mu g\,m^{-3}$. The rest of the regions occupy an intermediate place. Total concentrations of elements in the aerosols of these regions are almost similar and average 19 $\mu g\,m^{-3}$.

The concentration of the considered components of aerosols at Kamchatka was unexpectedly high – 22 $\mu g\,m^{-3}$ – an intermediate value between Uzbekistan and Kirghizstan. Kamchatka is characterized by a higher total concentration of ions than in other regions, which, with a high degree of accuracy, follow the latitudinal variation of total concentration of elements. A conclusion can be drawn that changes

Table 3.1. Average concentrations of elements and ions in the atmospheric aerosol composition over various regions (μm m^{-3}). Notations: ETR – European territory of Russia; Kz – Kazakhstan; Tk – Turkmenistan; Td – Tadjikistan; Uz – Uzbekistan; Kg – Kighistan; WS – West Siberia; ES – East Siberia; FE – Fast East; Km – Kamchatka.

Elem.	ETR	Ural	Kz	Tk	Td	Uz	Kg	WS	ES	FE	Km
Si	3.8950	2.0618	4.8914	1.3813	2.2378	1.4290	5.9067	2.6217	2.3619	0.6014	0.6749
Al	0.8654	0.6377	0.9748	2.4970	0.4218	0.8346	1.9614	0.7596	0.2703	0.5756	0.1319
Fe	0.4956	0.7334	0.4206	0.2588	1.5975	0.1283	0.8476	0.3379	0.2142	0.1562	1.9931
Mg	0.1606	0.1781	0.2472	0.4528	1.5818	0.2287	0.1274	0.1304	0.0896	0.3149	0.0906
Ca	0.3372	0.7110	0.3011	2.9871		3.0150	0.5022	0.1538	0.2093	0.2129	0.9086
Cu	0.0862	0.0640	0.0781	0.1029	nm	0.2200	0.2105	0.0762	0.0387	0.0775	0.0363
Ba			0.1150		nm		nm	0.0339			0.0870
	0.0659	0.0560	0.0008		nm			0.052	nm	nm	
Ti	0.0318	0.0934	0.0110	0.0302	0.1260	0.0169	0.0225	0.033	0.0283	0.0227	0.0855
Mn	0.0168	0.1100	0.0255	0.0654	0.1511	0.0215	0.0212	0.0123	0.0161	0.0162	0.0151
Cr	0.0470	0.0598	0.0593	0.1610	0.0453	0.0095	0.1200	0.0347	0.0388	0.0160	0.0580
Mo	0.0237	0.0021	0.0129	0.0063	0.0041	0.0036		0.004	0.0240		.0058
Ag	0.0027	0.0141	0.0043	0.0002			0.0049	0.0043	0.0014	nm	0.0154
Pb	0.0724	0.0483	0.0145	0.0035	0.0129	0.0190	0.0110	0.0176	0.0068	0.0185	0.0060
Ni	0.0307	0.0506	0.0457	0.0519	0.0534	0.0412	0.1410	0.0446	0.0551	0.0083	0.0364
Zn	0.0256	0.0492	0.0244	0.0549		2.7815	2.4600	0.0333	0.1418	0.0380	0.0810
B	0.0108	0.0110	0.0042				nm	0.0679	0.4327	nm	0.0672
V	0.0145		0.0302	0.0320	.0156	0.0038	nm	0.0056	0.0022	0.0016	0.0045
NO_3	4.9571	1.6867	3.1613	1.6335	9.8715	3.0365	9.6500	2.2814	1.3260	1.7433	
Na	0.8520	0.8100	0.4183	1.5807	1.2689	0.7823	0.0630	0.6803	0.4072	0.4767	0.9150
Cl	4.3102	2.8150	4.0318	2.4424	2.1174	3.5380	1.7250	1.7229	4.0518	5.9811	0.2484
SO_4	0.4398	1.0350	0.5532	1.1609	1.1187	0.5143		0.6628	0.4264	0.2650	0.2400
	0.1255	0.1175	0.0383	0.4243	0.4819	0.2179	0.0900	0.1128	0.1451	0.1010	0.8565
Br	1.3800	1.6908	1.8440	3.7679	7.2802	4.4711	2.4600	2.1817	2.8887	7.2267	13.770
NH_4	1.5749	0.4950	1.0096	0.8454	3.3764	2.4614	1.8800	1.2372	1.0262	0.9422	1.7313
Total	*19.821*	*13.530*	*18.317*	*19.940*	*31.762*	*23.774*	*28.204*	*13.302*	*14.203*	*18.796*	*22.058*

Note: nm = not measured.

in total concentration of elements in the latitudinal variations is determined mainly by changes of total concentration of ions in the atmospheric aerosol composition.

With the ratio of maximum to minimum total concentrations of ions taken into account, we note that the concentration changes almost by a factor of 3. Its minimum falls on the aerosol in the Urals (8.65 μg m^{-3}). A maximum is observed in the aerosol of Tadjikistan (25.5 μg m^{-3}). The total concentration of the group of terrigenous elements changes from region to region more substantially than that of ions and the ratio of maximum to minimum concentrations for this group is equal to 5 (Kirghizstan/Far East).

The most substantial changes in the latitudinal variation is characteristic of the total concentration of microelements – by a factor of 23.8. A minimum of this total concentration is observed in the Far East, with a maximum in Uzbekistan.

In Figure 3.2 the dashed curve indicates the trend obtained by averaging the main curve with the third-power polynomial.

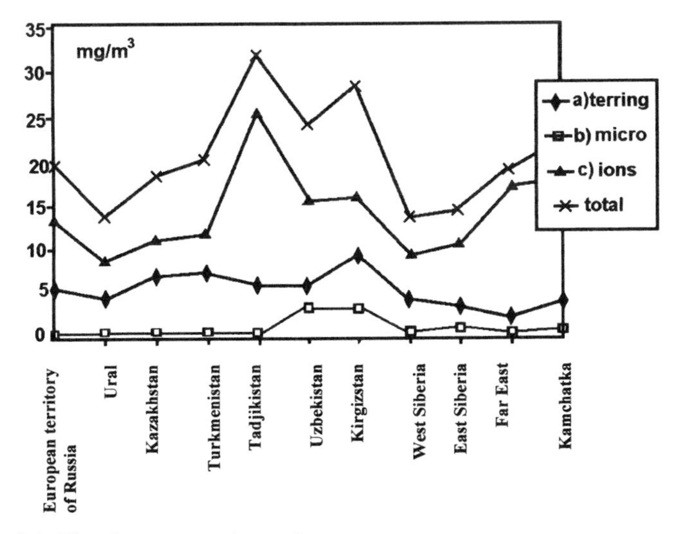

Figure 3.1. Total concentrations of the groups of elements in various regions.

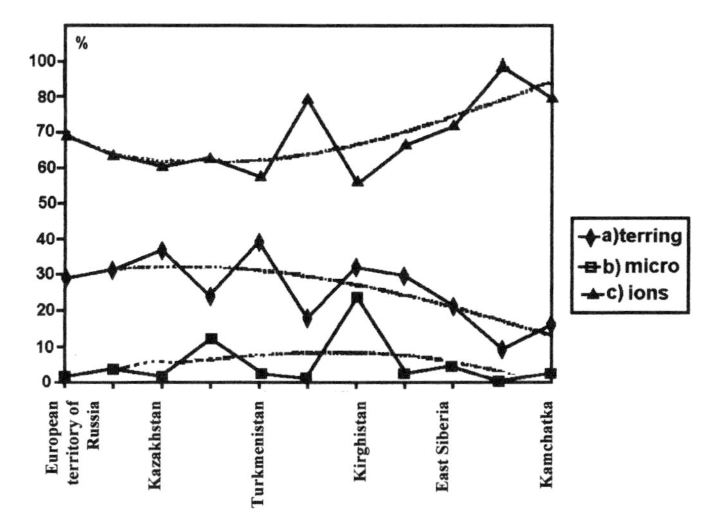

Figure 3.2. The latitudinal change of the relative aerosol composition.

Thus maximum values of total mass concentration of each of the three groups of elements are observed in the regions of central Asia (Tadjikistan, Uzbekistan, Kirghizstan). Minimum values of total concentration of terrigenous elements and microelements fall on Kamchatka, and minimum values of total concentration of ions are observed in the Urals.

Table 3.2 contains relative chemical aerosol matrices for each of the regions showing the percentage of elements in the atmospheric aerosol composition, and Figure 3.2 shows the latitudinal change of the relative share of these groups of elements.

Table 3.2. The relative (%) elemental composition of aerosol in different regions. Notation given in Table 3.1.

	ETR	Urals	Kz	Uz	Tk	Td	Kg	WS	ES	FE	Km
Si	19.65	15.24	26.7	6.01	7.29	7.05	20.96	19.71	16.63	3.2	3.06
Al	4.37	4.71	5.32	3.51	13.18	1.33	6.96	5.71	1.9	3.06	0.6
Fe	2.5	5.42	2.3	0.54	1.37	5.03	3.01	2.54	1.51	0.83	9.04
Mg	0.81	5.26	1.35	12.68	15.77		1.78	0.98	1.47	1.13	4.12
Ca	1.7	1.32	1.64	0.96	2.39	4.98	0.45	1.16	0.63	1.68	0.41
Cu	0.43	0.47	0.43	0.93	0.54		0.75	0.57	0.27	0.41	0.16
Ba			0.63					0.25			0.39
W	0.33	0.41						0.39			
Ti	0.16	0.69	0.06	0.07	0.16	0.4	0.08	0.25	0.2	0.12	0.39
Mn	0.08	0.81	0.14	0.09	0.35	0.48	0.08	0.09	0.11	0.09	0.07
Cr	0.24	0.44	0.32	0.04	0.85	0.14	0.43	0.26	0.27	0.09	0.26
Mo	0.12	0.02	0.07	0.02	0.03	0.01		0.03	0.17		0.03
Ag	0.01	0.1	0.02				0.02	0.03	0.01		0.07
Pb	0.37	0.36	0.08	0.08	0.02	0.04	0.04	0.13	0.05	0.1	0.03
Ni	0.15	0.37	0.25	0.17	0.27	0.17	0.5	0.34	0.39	0.04	0.17
Zn	0.13	0.36	0.13	11.7	0.29		0.73	0.25	1	0.2	0.37
B	0.05	0.08	0.02					0.51	3.05		0.3
V	0.07		0.16	0.02	0.17	0.05		0.04	0.02	0.01	0.02
NO_3	25.1	12.47	17.26	12.77	8.62	31.08	34.17	17.15	9.34	9.27	
Na	4.3	5.99	2.28	3.29	8.35	3.99	0.22	5.11	2.87	2.54	4.15

Terrigenous elements. The trendlines show that the share of terrigenous elements increases from the European territory of Russia (ETR) to Kazakhstan, and then decreases. In the regions of central Asia the percentage of terrigenous elements varies widely, with their maximum content recorded in Turkmenistan (40%). From western Siberia to the Far East, the share of terrigenous elements decreases monotonically reaching an absolute minimum of 9.9% at the Far East. At Kamchatka the relative content of terrigenous elements almost doubles compared with the Far East. The relative content of the group of microelements is characterized by a greater variability, especially in the regions of central Asia, with a maximum percentage of microelements observed in the aerosol of Kirghizstan (25%), and minimum percentage at the Far East – $<0.01\%$.

The group of ions. This group in all regions is relatively the largest. Its share in aerosol is at a minimum in Kirghizstan (56.2%), and at a maximum in the Far East (89%) and at Kamchatka (80.5%). The monotonic character of the curve of relative content of this group is broken in central Asia. In Tadjikistan, a partial maximum is observed comparable in magnitude with its share at Kamchatka. In other regions the share of the group of ions constitutes 60–72%. The composition of aerosol coming from different sources is different. In aerosols coming to the atmosphere from the surface of continents, the elements of the lithosphere should be present with their clarkes typical of the Earth's crust. This supposition is not valid for all components.

So, for instance, the content of silicon in aerosols of different regions varies from 3% at Kamchatka to almost 27% in Kazakhstan. The content of silicon in aerosol of Kazakhstan (26.7%) is practically similar to that found in the Earth's crust. Other regions are characterized by a decreased content of silicon in aerosols. The relative content of silicon in aerosols differs especially strongly from its percentage in the Earth's crust in the Far East and Kamchatka, as well as in Uzbekistan, Turkmenistan, and Tadjikistan. The deviation of the content of any element from its content in the Earth's crust can be estimated from the value of the enrichment coefficient which is calculated with:

$$K_x = (X/Fe)_{aer}/(X/Fe)_{E.c.} \tag{3.1}$$

where K_x is the enrichment coefficient for the element X; $(X/Fe)_{aer}$ and $(X/Fe)_{E.c.}$ are ratios of concentrations of elements X and Fe in aerosol and in the Earth's crust, respectively.

Elements whose coefficients of enrichment are close to unity, can be considered mainly terrigenous. Values of $K_x \geq 10$ can be explained thus: a substantial impact of other sources (anthropogenic activity or the ocean) on concentrations of respective elements in the atmosphere; various secondary transformations of aerosol in the atmosphere; and fractionating of aerosol particles through their removal from the atmosphere by moving air masses.

Table 3.3 gives the enrichment coefficient for the considered elements in aerosol in different regions calculated from equation (3.1).

Table 3.3. The enrichment coefficients for elements in aerosol of the regions of the former USSR (aluminium as a reference element). Notation given in Table 3.1.

	ETR	Urals	Kz	Uz	Tk	Td	Kg	WS	ES	FE	Km
Si	1.3	1.0	1.5	0.2	1.6	0.5	0.9	1.0	2.6	0.3	1.5
Al	0.9	0.4	1.2	3.3	4.8	0.1	1.2	1.1	0.6	1.8	0.03
Fe	1.1	2.3	0.9	0.2	7.6	0.3	0.9	0.9	1.6	0.5	30.2
Mg	0.7	1.1	0.0		15.0	0.1	0.3	0.7	1.3	0.2	2.8
Ca	1.0	3.0	0.8	3.2		9.6	0.7	0.5	2.1	1.0	18.4
Cu	79.7	80.3	64.1	33.0		210.9	85.9	80.3	111.9	107.7	20.2
Ba	87.0	100.4	0.9					78.2			
Ti	0.5	2.0	0.2	0.2	3.9	0.3	0.2	0.6	1.4	0.5	8.6
Mn	1.6	13.8	2.1	2.1	28.7	2.1	0.9	1.3	4.8	2.3	9.2
Cr	2.2	3.8	2.4	2.6	4.3	0.5	2.5	1.8	5.7	1.1	17.6
Mo	219.1	26.3	105.9	20.2	77.8	34.5	1998.6	42.1	710.3		351.8
Ag	2,496.0	17,688.6	3528.9	64.1	15.3			4,528.7	4.1		93,404.1
Pb	37.2	37.9	7.4	0.7	101.3	11.4	2.8	11.6	12.5	16.1	22.7
Ni	28.4	63.5	37.5	16.6		39.5	57.5	47.0	163.1	11.5	220.8
Zn	0.1	0.3	0.1	0.1		13.3	5.0	0.2	2.1	0.3	2.5
B	99.8	138.0	34.5					715.1	12,806		4,075.8
V	83.8		165.2	68.4	184.9	22.8		39.3	4.3	14.8	182.0

For elements Si, Al, Fe, Mg, Ca, Ti, Mn, Cr, and Zn, most values of the coefficient of enrichment are close to unity and, on the whole, do not exceed 10, which indicates their natural origin. The exception is iron, calcium, and chromium in the aerosol at Kamchatka, for which the coefficients of enrichment are estimated at 39.2, 18.4, and 17.6, respectively. The rest of the elements, such as copper, tungsten, molybdenum, silver, nickel, boron, and vanadium are characterized by high values of the coefficient of enrichment.

Kamchatka is a leader in the number of heavily enriched elements. In the aerosol of this region only six elements (Si, Al, Mg, Mn, Zn) have a coefficient of enrichment <10. An increased concentration of some elements in continental aerosol can be explained for several reasons. First, it is the composition of the initial material entering the troposphere as aerosol particles. If only small fractions of rocks were captured by wind, the composition of aerosol particles would be similar to that of the lithosphere. However, fresh rock fractions are rarely scattered – instead loose products of weathering and soils are scattered. In the upper horizon of weathering products the concentration of some elements grows due to their accumulation in vegetative residuals, humus, or on the soil surface. It has been established that concentrations of metals in continental dust and soil are close. Thus, particles entering the troposphere from the land surface can be initially enriched with some elements.

Mineral dust in the surface layer of the troposphere over ore deposits contains increased concentrations of metals due to scattering of loose products of weathering enriched with ore elements. As a rule, the enrichment with metals of loose products of weathering is considerably less than that detected in aerosol. The concentration of some heavy metals in aerosol exceeds, by hundreds of times, their clarke value for the lithosphere. Apparently, an enrichment of aerosol with metals takes place directly in the troposphere.

It has been established that chemical elements scattered in the troposphere differ in their phase (including the gas phase). Not only well sublimating elements (J, As, and Hg) are present in the gas phase but also heavy metals Zn, Cu, and Pb. The enumerated elements are closely connected with aerosol particles less than 0.5 μm. The main mass of such elements as Al, Fe, Sc, Ba, and some others is concentrated in relatively large particles. Calculations of the coefficients of enrichment of aerosol elements with respect to the average composition of the Earth's crust have shown that the concentration of elements present in the gas phase increases strongly in ultrafine fractions of aerosols. The relative concentration of elements, whose main mass is contained in particles >0.5–1.0 μm, changes weakly (see Tables 3.2 and 3.3). The relative content and the coefficient of enrichment of elements of the terrigenous group in the chemical matrix of aerosol change within one order of magnitude. The range of changes of percentages of microelements is much broader. So, the percentage of lead and chromium in aerosol from different regions changes almost 20 times, of zinc 90 times. The change of the enrichment coefficient for lead and zinc constitutes more than two orders of magnitude. Still wider are the range of changes of the enrichment coefficient for silver – more than three orders of magnitude.

Certain chemicals, including many heavy metals entering the troposphere in gas phase, are adsorbed by aerosol particles. Since the flux of molecules onto particles is proportional to the particles' area, and the coefficient of enrichment is inversely proportional to their mass, the effect of enrichment is most clearly manifested in the case of fine fractions. Determination of sources of the gas flux of heavy metals and related elements with variable valency is a complicated problem. For the region of Kamchatka, far from large industrial centres, without industries of their own, and deposits of ore, the presence and strong enrichment of heavy metals in aerosol can be explained either by their input to the atmosphere due to volcanic eruptions (i.e., in the gas phase) or by long-range transport by air masses from industrial regions.

Volcanic aerosols are enriched with heavy metals by hundreds and thousands of times compared with the Earth's crust – taking place as a result of adsorbing the metals erupted by volcanoes in the gas phase. Hence, the reasons for enrichment of elements in aerosols can be different, depending on physico-geographical conditions of the region.

Additional information about the origin of elements contained in atmospheric aerosol can be obtained using correlations between elements. By the magnitude of the coefficient of correlation between concentrations of elements, some elements can be chosen whose concentrations change synchronously either in space or in time.

Table 3.4 demonstrates the matrix of pair coefficients of correlation for concentrations of elements (by latitudinal changes) calculated from data in Table 3.2, from which data are excluded for the regions of central Asia – Turkmenistan, Tadjikistan, Uzbekistan, and Kirghizstan. The aerosol in these regions was sampled mainly at high latitudes [102]. Geographically, these regions are located far south of other regions, and their physico-geographical and orographical features make the consideration of data for these regions incorrect when the combined latitudinal change is being looked at. The number of statistically significant correlations between elements is given in Table 3.5.

Iron. Iron has a significant correlation with three elements – Ca, Ag, and Ti, and four ions – K^+, Cl^-, Br^-, and Na^+. The latitudinal change of iron and elements correlating with it is shown in Figure 3.3(a,b).

The concentration of iron increases from ETR to the Urals. From the Urals to the Far East it decreases monotonically, practically in a linear manner, and then increases to a maximum at Kamchatka. The highest coefficient for a pair correlation for iron is observed with the potassium ion – 0.945 (Table 3.6).

The latitudinal change of the concentration of potassium corresponds, on the whole, to the change of iron concentration. Differences are observed over the area of Kamchatka–eastern Siberia. Here the concentration of potassium increases.

Calcium. This element has almost the same correlation of its latitudinal changes as iron ($R = 0.9048$) and titanium ($R = 0.8992$), which corresponds to the character of the curves in Figure 3.3(a). A minimum of calcium concentration is observed in western Siberia, a maximum at Kamchatka. A maximum of titanium concentration is observed in the Urals, and a minimum in Kazakhstan. It is seen in Figure 3.3(b) that the concentration of bromine ions in aerosol increases monotonically from west to east. The coefficient of the iron–bromine correlation is 0.761.

Table 3.4. Correlations between elements.

	Si	Al	Fe	Mg	Ca	Cu	Ti	Mn	Cr	Mo	Ag	Pb	Ni
Si	1	0.758	−0.389	0.0554	−0.386	0.502	0.477	0.085	0.346	0.466	−0.684	0.293	0.366
Al		1	0.548	0.526	−0.422	0.919	−0.401	0.109	0.061	0.0262	−0.371	0.499	−0.044
Fe			1	−0.459	0.862	0.564	0.666	0.090	0.560	−0.406	0.749	−0.127	0.076
Mg				1	−0.294	0.655	−0.348	0.076	−0.355	−0.157	−0.0206	0.102	−0.61
Ca					1	−0.478	0.909	0.561	0.709	−0.565	0.953	0.164	0.216
Cu						1	0.436	0.015	−0.236	0.0327	−0.353	0.532	−0.396
Ti							1	0.747	0.584	−0.684	0.944	0.272	0.272
Mn								1	0.542	−0.582	0.657	0.488	0.430
Cr									1	−0.431	0.725	0.263	0.653
Mo									1	−0.781	0.0630	−0.202	
Ag											1	0.0810	0.040
Pb												1	−0.019
Ni													11

Zn	B	V	NO_3	Na	Cl	SO_4	K	Br	NH_4	
−0.447	−0.197	0.859	0.697	−0.241	0.208	0.186	−0.525	−0.691	0.0264	Si
−0.811	−0.576	0.723	0.725	−0.138	0.392	0.350	−0.730	−0.722	−0.221	Al
0.141	−0.274	−0.101	0.085	0.688	0.784	−0.122	0.927	0.731	0.411	Fe
−0.606	−0.626	0.307	0.0322	−0.408	0.713	−0.018	−0.500	−0.140	−0.384	Mg
0.0799	−0.383	−0.0648	−0.178	0.735	−0.628	0.300	0.656	0.448	−0.0596	Ca
−0.878	−0.627	0.469	0.703	−0.0791	0.506	0.162	−0.668	−0.548	−0.111	Cu
0.115	−0.298	−0.429	−0.365	0.758	−0.612	0.550	0.452	0.233	−0.292	Ti
−0.118	−0.352	0.876	−0.363	0.374	−0.120	0.898	−0.220	−0.361	−0.800	Mn
−0.355	−0.572	0.624	0.173	0.483	−0.546	0.506	0.260	−0.0958	−0.119	Cr
0.337	0.547	0.0730	0.470	−0.441	0.680	−0.594	−0.197	−0.189	0.406	Mo
−0.066	−0.448	−0.213	−0.428	0.673	−0.652	0.403	0.525	0.482	−0.276	Ag
−0.473	−0.502	0.241	0.566	0.542	0.195	0.482	−0.353	−0.522	−0.187	Pb
0.3906	0.481	0.242	−0.245	0.0527	−0.458	0.606	−0.086	−0.432	−0.325	Ni
1	0.914	−0.520	−0.541	−0.221	−0.148	−0.221	0.327	0.260	0.0456	Zn
	1	−0.573	−0.451	−0.569	0.208	−0.364	0.0051	0.0441	0.0504	B
		1	0.581	−0.166	0.131	0.441	−0.324	−0.443	−0.091	V
			1	0.320	0.137	−0.290	−0.224	−0.363	0.739	NO_3
				1	−0.633	0.281	0.510	0.231	0.224	Na
					1	−0.171	−0.698	−0.394	−0.298	Cl
						1	−0.432	−0.632	−0.775	SO_4
							1	0.892	0.601	K
								1	0.532	Br
									1	NH_4

Table 3.5. The level of correlations between elements.

Fe	Al	Ag	K	Ca	Ti	Cl	Si	NO_3	Br	Cu	Mn	Zn	V	Na	Mo	NH_4	Mg	Cr	Pb	B	SO_4	Ni
7	6	6	6	5	5	5	4	4	4	3	3	3	3	3	2	2	1	1	1	1	1	0

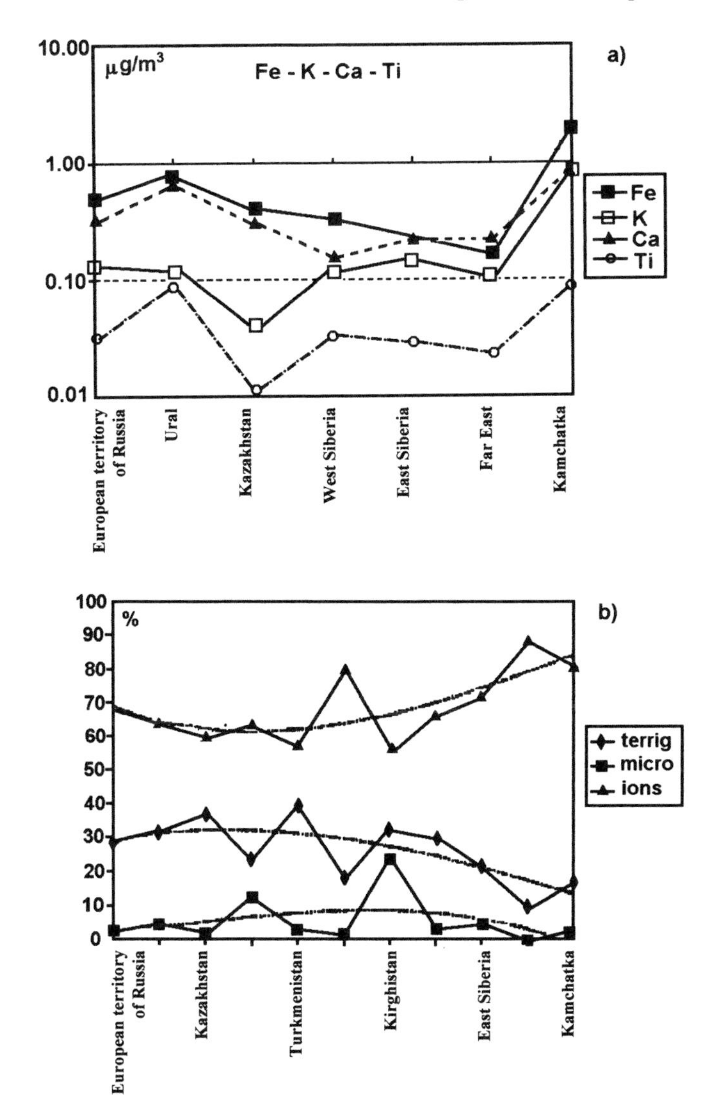

Figure 3.3. The latitudinal change of the concentration of iron and elements correlating with it.

Natrium. The coefficient of correlation of iron with the natrium ion is 0.6974. Maximum values of concentration of natrium ions are observed in aerosol over ETR and Kamchatka – natrium is traditionally considered an element of mainly marine origin, and the amount of marine air masses passing these regions is greater than in other regions. Close minimum concentrations are observed in Kazakhstan and in eastern Siberia.

Chlorine. Latitudinal changes of the concentration of chlorine ions from ETR to Kamchatka take place in opposite directions. This is confirmed by the value of the

Table 3.6. Correlations of iron with other elements.

	Fe	K	Ca	Ti	Br	Na	Cl	Ag
Fe	1	0.945	0.905	0.735	0.761	0.697	−0.785	0.821
K		1	0.766	0.605	0.893	−0.366	−0.721	−0.550
Ca			1	0.899	0.588	0.710	−0.641	0.949
Ti				1	0.411	0.749	−0.655	0.424
Br					1	−0.606	−0.429	−0.505
Na						1	0.228	0.572
Cl							1	−0.718
Ag								1

Table 3.7. Correlations of aluminium with other elements.

	Al	Cu	Zn	K	Br	NO_3	V
Al	1	0.921	−0.810	−0.731	−0.730	0.732	0.723
Cu		1	−0.882	−0.676	−0.568	0.696	0.469
Zn			1	0.322	0.253	−0.566	−0.520
K				1	0.893	−0.202	−0.324
Br					1	−0.441	−0.443
NO_3						1	0.581
V							1

coefficient of correlation between them being equal to −0.7832. Note that a similar pattern is observed for the natrium–chlorine pair. Minimum values of chloride concentrations are observed at Kamchatka, where their absolute content is lower, by an order of magnitude, compared with other regions.

Aluminium. The number of statistically significant correlations of aluminium with other elements is similar to that of iron. Table 3.7 shows the correlation matrix for aluminium and elements correlating with it.

The concentration of aluminium, by the latitudinal change, correlates positively with copper, silicon, vanadium, and ions of nitrates and bromides, and negatively with zinc and the potassium cation. As follows from Figure 3.4, in all regions, except eastern Siberia and Kamchatka, concentrations of aluminium are close – in the interval 0.6–1 $\mu g\, m^{-3}$. In eastern Siberia, the aluminium concentration constitutes 0.27 $\mu g\, m^{-3}$ and at Kamchatka 0.13 $\mu g\, m^{-3}$. An absolute minimum of aluminium concentration is observed at Kamchatka. The aluminium–silicon correlation, by the latitudinal change, is explained by the fact that they are the two most widely spread (in nature), typically terrigenous elements. The aluminium–copper correlation is difficult to explain when only considering their natural origin. Copper, in contrast to aluminium, can enter the atmosphere in large quantities as part of aerosol particles from anthropogenic origins in a gas phase. A close correlation is observed between aluminium and the NO_3^- ion.

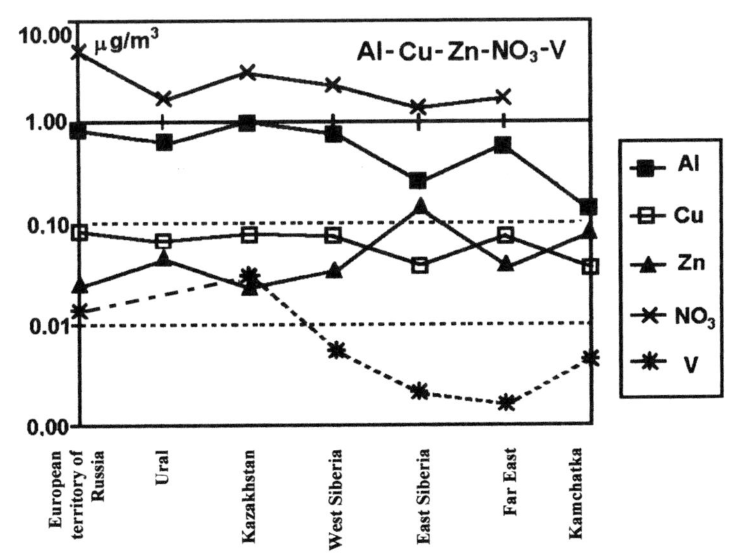

Figure 3.4. The latitudinal change in the concentration of aluminium and elements correlated with it.

Nitrates also correlate well with silicon latitudinally. The correlation of nitrates with terrigenous elements suggests the conclusion that nitrates comprise nitrogen compounds entering the atmosphere from the surface. Formation of nitrates in soils is the result of several processes following in a certain sequence. Nitrogen oxides interacting with water give nitric acid which transforms into nitrates in soils. It is difficult to explain the negative correlation between aluminium and zinc without additional studies. The latitudinal change in zinc concentration, despite its variability, exhibits a general trend of west-to-east increase. The concentration of zinc increases in the Urals, in eastern Siberia, and at Kamchatka. Zinc can be present in the atmosphere in the gas phase, emitted by higher plants, and can get to the atmosphere as a result of forest fires. All three mechanisms can function in the regions with an increased concentration of zinc, but in case of the gas phase zinc, a correlation of zinc with copper and lead should be observed. However, the correlation of copper and zinc is negative, and with lead it is absent altogether. But forest fires during sampling over eastern Siberia were observed over some large territories.

Silver. The correlation matrix for silver, and elements correlated with it, by latitudinal change, is given in Table 3.8. In the former USSR, silver is obtained from silver–lead ore whose deposits are located in the Urals, Altai, northern Caucasia, and Kazakhstan.

In this connection, one could expect a correlation with copper and lead, but it is absent. Due to its low chemical activity, silver does not form chemical compounds with the elements enumerated in Table 3.8. Only with the interaction of silver ions with chloride ions, is silver chloride formed, which decomposes giving metal silver. Table 3.8 demonstrates a negative correlation of silver with chlorides, but one can speak only hypothetically about silver chloride decomposition on this basis.

Table 3.8. Correlations of silver with other elements.

	Ag	Fe	Ca	Ti	Cr	Mo	Cl
Ag	1.00	0.821	0.949	0.931	0.678	0.738	−0.718
Fe		1.00	0.905	0.735	0.571	−0.431	−0.783
Ca			1.00	0.899	0.669	0.498	−0.641
Ti				1.00	0.511	−0.615	−0.655
Cr					1.00	0.340	−0.547
Mo						1.00	0.733
Cl							1.00

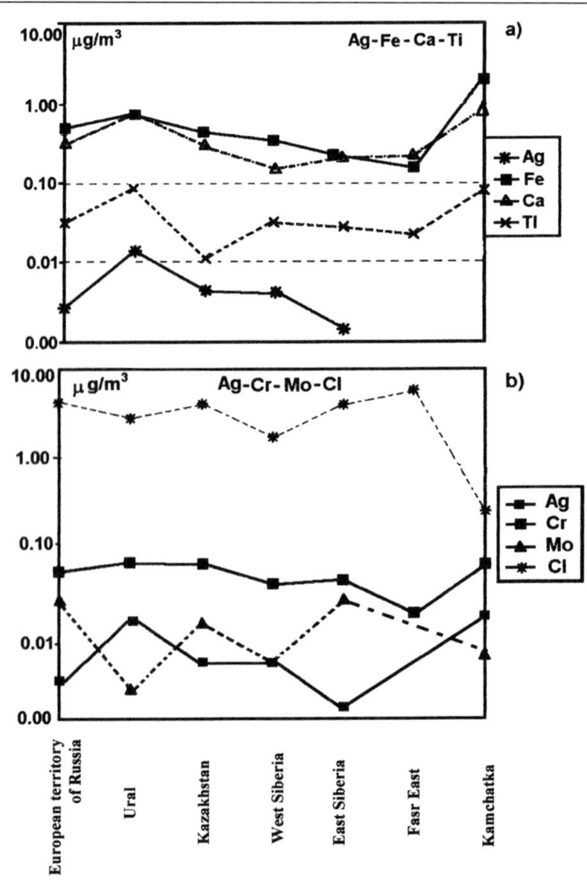

Figure 3.5. Latitudinal change of the concentration of silver and elements correlated with it.

The latitudinal change of silver and other elements enumerated in Table 3.8 is shown in Figure 3.5. Increased, practically equal, concentrations of silver are observed in the Urals and at Kamchatka. In Kazakhstan and western Siberia they are equal too, but lower by an order of magnitude. A minimum content of silver is observed in the western Siberian aerosol.

Table 3.9. Correlations of manganese with other elements.

	Mn	V	SO_4	NH_4
Mn	1	0.876	0.840	−0.715
V		1	0.441	−0.0909
SO_4			1	−0.681
NH_4				1

An absolute concentration of silver in the latitudinal change is two orders of magnitude lower compared to concentrations of terrigenous iron and calcium correlating with it. The silver clarke constitutes 10^{-5}% by mass.

Calcium. The next element, by a decreasing number of correlations with other elements, is calcium, which has been considered in the analysis of the latitudinal change of iron. The latitudinal change of calcium illustrated in Figure 3.5 shows that changes in its concentration (latitudinally) are within one order of magnitude. An increased content of calcium is observed in aerosol in the Urals and at Kamchatka. Natural compounds containing calcium are deposits of apatites and phosphorites in Apatity, silicates in Povolzhye and marble in the Urals. In other regions there are no deposits of calcium. Thus, the location of deposits of natural compounds of calcium explains, to some extent, the latitudinal change in its concentration.

Titanium. Apart from correlation with iron, calcium, and natrium, titanium exhibits a very close correlation with silver.

Manganese. Manganese, by its latitudinal change in concentration, correlates with the change of vanadium concentration as well as with anions of sulphates and the cation of ammonium. The respective correlation matrix is shown in Table 3.9. The correlation with representatives of the group of ions is negative. A specific feature of the latitudinal change of manganese concentration is its increase in the Urals by almost an order of magnitude. In other regions, concentrations of manganese are closely similar. The former USSR is the World's leader in ore deposits containing high contents of manganese. The latitudinal change of manganese concentration is characterized, to some extent, by its deposits.

Table 3.9 has iron omitted, but iron is always present in manganese ores. This fact indicates that manganese enters the atmosphere not only from the surface. Correlation with sulphates, which can form directly in the atmosphere, and with vanadium, favours the gas mechanism for input of manganese to the atmosphere.

Sulphate ion SO_4^{2-}. Formation of sulphate ions in the atmosphere takes place in the following scheme: sulphur dioxide SO_2, present in the atmosphere, reacts with atmospheric water, dissolves in it to give sulphurous acid:

$$SO_2 + H_2O = H_2SO_3$$

The sulphurous acid solution is slowly oxidized into sulphuric acid:

$$2H_2SO_3 + O_2 = 2H_2SO_4$$

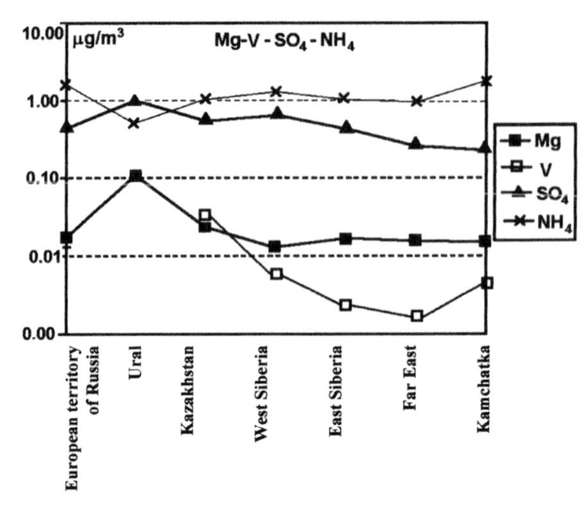

Figure 3.6. The latitudinal change in the concentration of manganese and elements correlated with it.

Sulphuric acid, being a strong electrolyte, dissociates in a diluted water solution:

$$H_2SO_4 = 2H^+ + SO_4^-$$

The ion combines with cations of metals, for instance, manganese, and the cation of ammonium. In the first case, sulphates of metals are formed, in the second case – ammonium sulphate $(NH_4)_2SO_4$.

This scheme gives the possibility of explaining a sharp increase of manganese concentration in aerosol over the Urals, where all conditions are present for the formation of manganese sulphate in the atmosphere: deposits of manganese, anthropogenic emissions of sulphur dioxide, and a sufficient amount of atmospheric moisture.

The presented scheme of sulphates formation is not a single mechanism explaining the presence of sulphates in aerosol particles. Sulphates are widely spread in nature. Most important of them are natrium sulphate (N_2SO_4) (in the form of crystalline hydrate – Glauber's salts), potassium sulphate (K_2SO_4), magnesium sulphate ($MgSO_4$) contained in seawater, and calcium sulphate ($CaSO_4$) found in nature in the form of the mineral gypsum. Thus, sulphates can reside in the atmosphere, they either form directly in the atmosphere or are of continental and marine origin. The latitudinal change of sulphate concentrations is shown in Figure 3.6, which demonstrates two maxima – the Urals and western Siberia. Most likely, this is the result of anthropogenic emissions of sulphur dioxide by metallurgical plants located in these regions. In an eastward direction we find that the concentration of sulphates decreases monotonically.

Ammonium (NH_4^+). Its correlation with sulphates is negative. A positive correlation, in its latitudinal change, is observed with manganese and nitrate ion NO_3^-. Most probably, these two ions are ammonium nitrate NH_4NO_3 which forms with the interaction of ammonia NH_3, always present in the atmosphere, with water, which gives an ammonium ion and nitrogen oxides. Ammonium ions become nitrified upon interaction with nitric acid, which eventually gives ammonium nitrate.

Minimum concentrations of ammonium ions and nitrate ions are observed over the Urals, maximum concentrations over the ETR and Kamchatka.

Elements and ions in atmospheric aerosols form groups of correlating elements by their latitudinal change. Maximum numbers of such correlations are exhibited by elements of terrigenous origin as well as ions of potassium, chlorine, and nitrate. The latitudinal changes in the total concentration of terrigenous elements reflects mainly the physico-geographical features of the regions. The latitudinal changes of total concentrations of microelements and ions do not have this property, which, on the whole, indicates their efficient scattering in the atmosphere.

A maximum contribution to the relative chemical composition of aerosol is made by the group of ions. Both the relative elemental composition of aerosol and the absolute content of the elements and ions comprising its composition exhibit maximum variability in the regions of central Asia.

Compared with the moderate content in the Earth's crust, it is microelements that get mainly enriched. The process of enrichment can take place both in soil layers and in the atmosphere. For some elements the degree of enrichment depends on the region.

A combined analysis of correlations, latitudinal change of concentrations, and chemical properties of elements gives the possibility of qualitatively identifying the source of input of one or another element to the composition of aerosol particles, and in some cases the chemical substance itself which is formed by these elements.

3.5 ATMOSPHERIC AEROSOL PROPERTIES OVER WESTERN SIBERIA: THE 'AEROSOL OF SIBERIA'

Prolonged measurements of aerosol concentration enable one to speak about long-period variations of its content in the air. This has been accomplished in [3], where, on the basis of aircraft sounding data, a multiyear trend of aerosol number density decrease in the 0–3-km layer over western Siberia has been detected. This trend is shown in Figure 3.7.

As seen from this figure, during the period 1983–1987 there was a distinct trend toward decreasing aerosol number density over western Siberia, a decrease in December by a factor of 4.5, and in May by a factor of 5.5.

Against a background of the enhanced anthropogenic pollution of the atmosphere (although the sounding was made outside the industrial centres) this result was quite unexpected. Therefore, other data had to be used. Since there was no regular vertical aerosol soundings over the territory of the former USSR, an assessment had to be made from data of surface measurements carried out at the network of air pollution monitoring stations.

From data published by different sources, Figures 3.8 and 3.9 give the changes in total aerosol emissions over the territory of the former USSR and USA, and changes

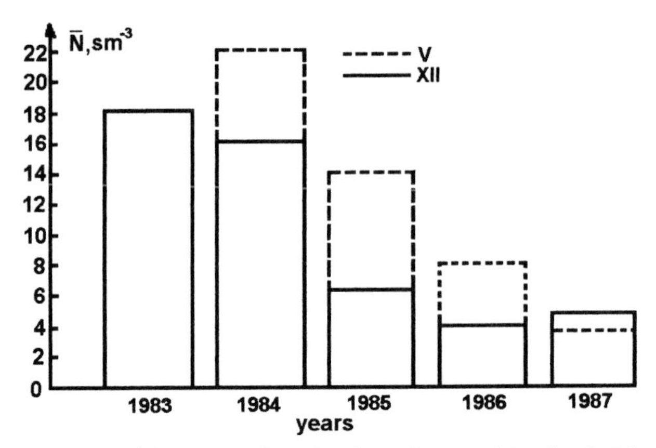

Figure 3.7. The average multiyear number density of aerosol in the 0–3-km layer in May (dashed line) and in December (solid line). May and December were chosen because for these months a good sampling was available.

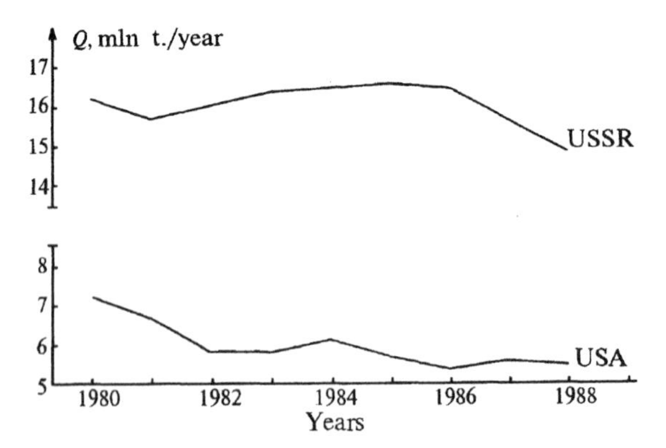

Figure 3.8. Total emissions of suspended matter in the USSR and the USA.

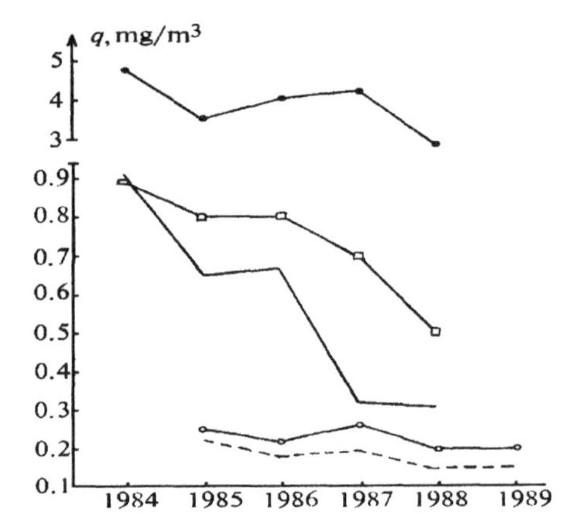

Figure 3.9. Concentration of suspended matter: – – – Karaganda; ——Novosibirsk; •–•– Omsk, –□–□– Kaliningrad, ○–○– Tolyatti.

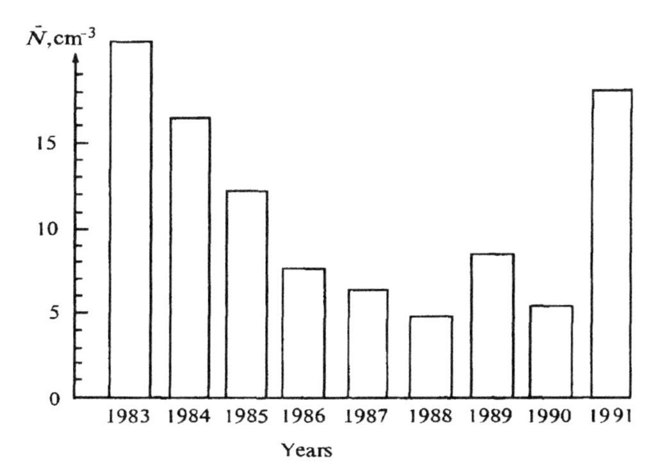

Figure 3.10. The annual mean number density of aerosol in the 0–3-km layer over western Siberia.

in aerosol concentrations in some cities. Bearing in mind that the contribution of these two countries constitutes a significant portion of the whole global emission and a decrease of concentrations demonstrated in Figure 3.7 is rather substantial, one can assume that data in Figure 3.8 excludes completely the version of the anthropogenic nature of the revealed trend.

At the same time, data in Figure 3.9 show that in some cities of the former USSR there was a trend of aerosol concentration confirmed by data in Figure 3.7. There was also a decreasing trend of suspended substances concentration in the air of the cities of Calcutta, Athens, Madrid, and Milan over a 10-year period (1975–1985). A comparison of results suggests the following conclusion: a decrease of aerosol concentration shown in Figure 3.7 is clearly not of anthropogenic origin, since the global volume of emissions barely changes in the 1980s.

Bearing in mind that the urban aerosol concentration is the sum of background concentration and created by local emissions, it may be assumed that in this case the background component decreases substantially due to natural processes, and the obtained result reflects the multiyear dynamics of the atmospheric aerosol.

The annual mean number densities of aerosol in the 0–3-km layer (Figure 3.10) decreased from 1983 to 1988 with some increase of concentration in 1989. In 1991 there was observed an increase in values close to those recorded in 1983. Thus, during 9 year's observations over western Siberia, during the period 1983–1991, there was a wavelike change of aerosol concentration in the lower atmosphere. Most likely, interannual variations of aerosol concentration are explained by natural causes, and the period of oscillations is close to the known 11-year cycle. An analysis was made of the possible reasons of long trends of aerosol number density. The following hypotheses were tested:

(1) As a result of nature protection activity, the concentration of anthropogenic aerosol decreased, which reflected upon its total concentration.

(2) Since the obtained series of measurements began in 1983, the revealed trend could be the long-term result of the El Chichon eruption (March–April 1982) – either direct or indirect.
(3) The trends of the aerosol component reflect the cyclicity of natural atmospheric processes which are characterized by a set of multiyear periods.

The test of the first hypothesis has shown that not more than 15% of the trend can be explained by a decrease of anthropogenic aerosol concentration, whereas the concentration changed several times. Analysis of cyclicity impact has led to the conclusion that it may be determined by a multiyear variability of general atmospheric circulation and, in particular, by transformation of the zonal component into the meridional one. A decrease of aerosol number density over the territory of western Siberia during the period 1984–1990 was followed by a decrease over this territory of the forms of the eastern circulation and a small increase of the western and meridional forms. Data for 1983 and 1991 deviate from this trend. Apparently, the reason is inadequate statistics for observational data. A similar conclusion can be drawn from the Katz circulation indices shown in Figure 3.11(b). As seen from the figure, a decrease of aerosol concentration took place against a background of general increase in intensity of the zonal form of circulation with relatively small variations of meridional intensity. A comparison of data in Figure 3.11(a,b) shows that the aerosol concentration trend in the mid-1980s was determined by circulation processes: increasing frequency of occurrence and intensity of the zonal western circulation with increasing frequency of occurrence of east–west circulation without substantial changes in its intensity.

This conclusion is also confirmed by the growth of the Blinova circulation index (Figure 3.11(c)) which is the ratio between the linear rate of air motion along the latitudinal belt and the distance to the Earth's axis. It follows from Figure 3.11(c) (inverse ordinate) that between 1984 and 1989 the index increased from 34 to 42 (i.e., the intensity of zonal circulation in this period increased, too). With an increasing intensity of the western zonal flux over the Urals, the probability of a high-pressure ridge formation increases, which plays a blocking role. Air masses come to western Siberia from the Arctic Ocean by ultrapolar trajectories.

As follows from Figure 3.11(b), against a background of increasing intensity of the western zonal circulation at the same time, the frequency of occurrence of blocking processes grew over the Urals from 16% in 1983 to 30% in 1988. The aerosol trend is determined by a superposition of two processes: increasing repeatability of mid-latitude air masses from the Atlantic Ocean by zonal trajectories and changing trajectories of Arctic masses which entered the territory of western Siberia not across the ETR but by ultrapolar (meridional) trajectories from the Kara Sea. Therefore, the Arctic air masses were cleaner. This conclusion is based not only on analysis of circulation indices but also on analysis of data on the aerosol chemical composition. During the period considered the content of NH_4^+ and Na^+, which refer to marine aerosol, increased substantially in the aerosol composition (Figure 3.12). An anthropogenic factor can be excluded from the growth of these components. Studies of aerosol number density in the surface layer from the data of

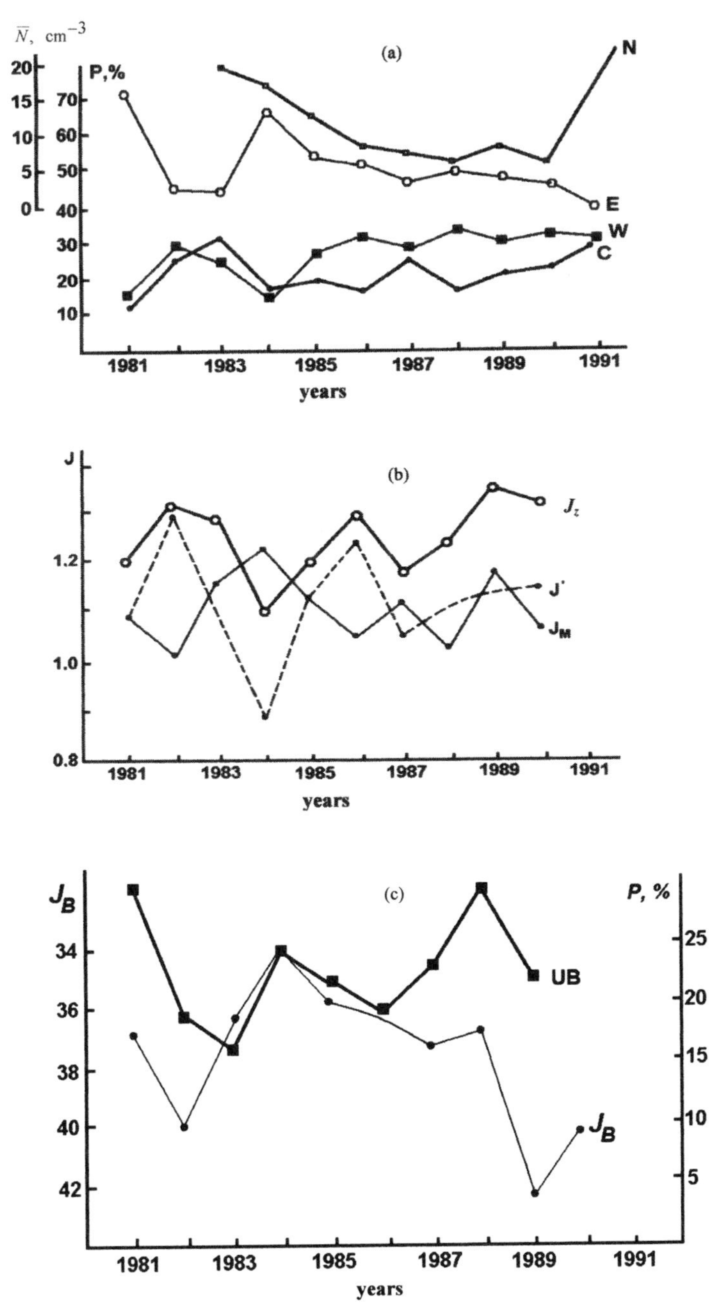

Figure 3.11. The temporal change of circulation indices. (a) Multiyear change of aerosol number density ($\bar{N}$) and repeatability of the western (W), eastern (E), and meridional (C) forms of circulation. (b) The Katz indices of circulation: J_z – zonal, J_m – meridional, and J' – their ratio. (c) The Blinova index of circulation (J_B) and frequency of occurrence of the Ural block (UB) over the territory of western Siberia.

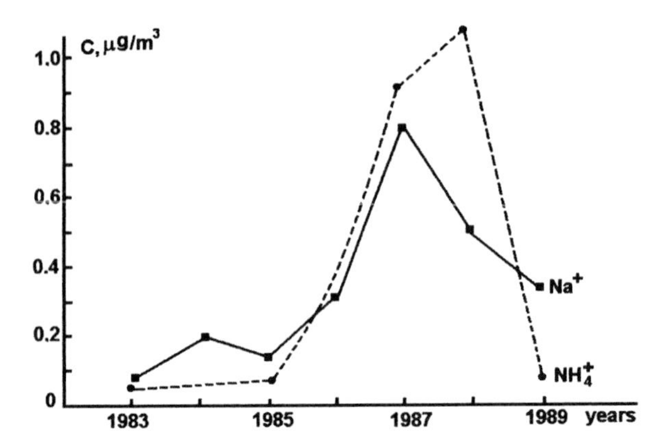

Figure 3.12. The annual mean concentration of Na^+ and NH_4^+ ions in aerosol over western Siberia.

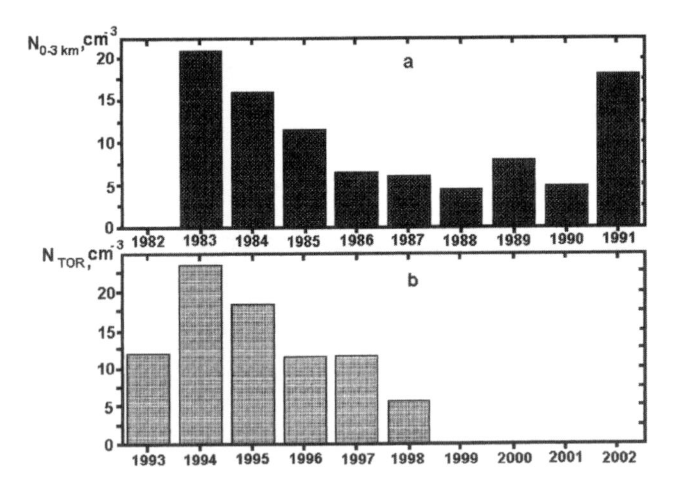

Figure 3.13. The change of aerosol number density over western Siberia: (a) aircraft data obtained in the 1980s; (b) data from TOR-stations in the 1990s.

TOR-stations [9] and aircraft soundings in the 1980s are given in Figure 3.13. There is an 11-year shift between these data. It is seen that in the early 1990s the aerosol concentration increased and then started decreasing with the trend close to that observed in the 1980s. This coincidence corresponds to a further increase in concentration in 2001–2002.

An analysis has been made [4] of data on aerosol samples taken over the forests in the regions of Plotnikovo (Bakcharskiy region, western Siberia, 24 July, 1997), Kamen-na-Obi (south of the Novosibirsk region, 25 September, 24 October, and 4 December, 1997), and oil–gas deposits: Olenye, Lakhtyniakhskoye, Ozernoye, Lomovoye, Vakhskoye, Strezhevskoye, Sovetskoye, Tatylginskoye, Pervomaiskoye

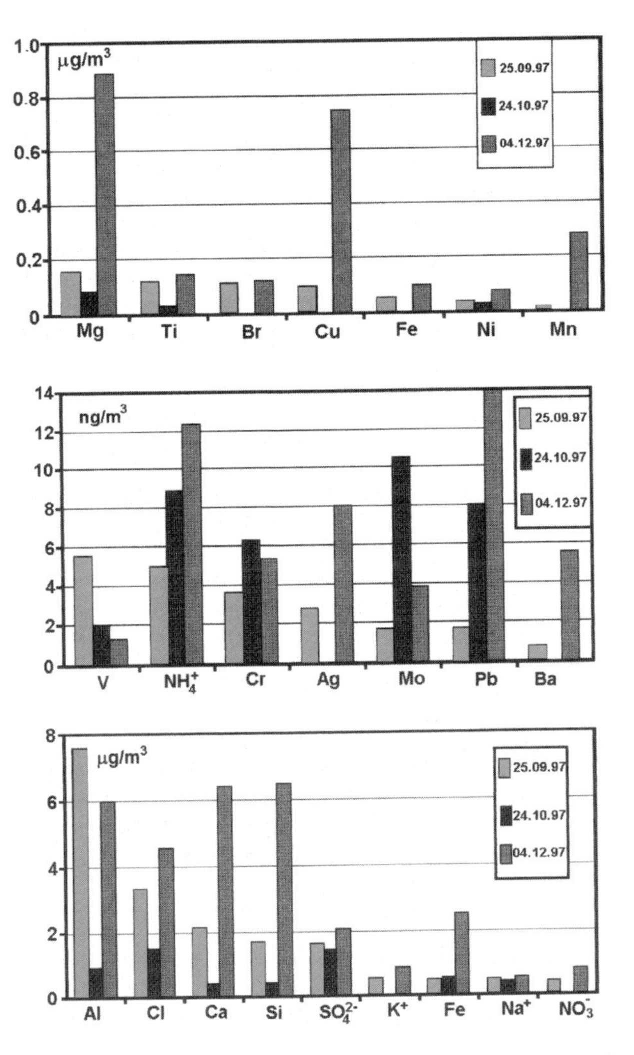

Figure 3.14. The atmospheric aerosol composition over some regions of western Siberia.

(28–29 October, 1997). Samples were taken in the second-half of 1997 at fixed altitudes from 500–7,000 m on filters AFA-20. When sampling, a wind direction was chosen so that the urban plumes remained outside the region of sampling.

Figure 3.14 shows histograms of the aerosol chemical composition, and Figure 3.15 shows the temporal dynamics of the main components of atmospheric aerosol for the region of Kamen-na-Obi, where a continuous 3-month series of measurements was obtained in the autumn–winter season of 1997. It follows from the analysis that there are several trajectories of air mass transport: distant – from the northern Atlantic across the regions of the Omskaya Oblast and Novosibirskaya Oblast; close – intrusions from the south, from the Aral Sea and northern

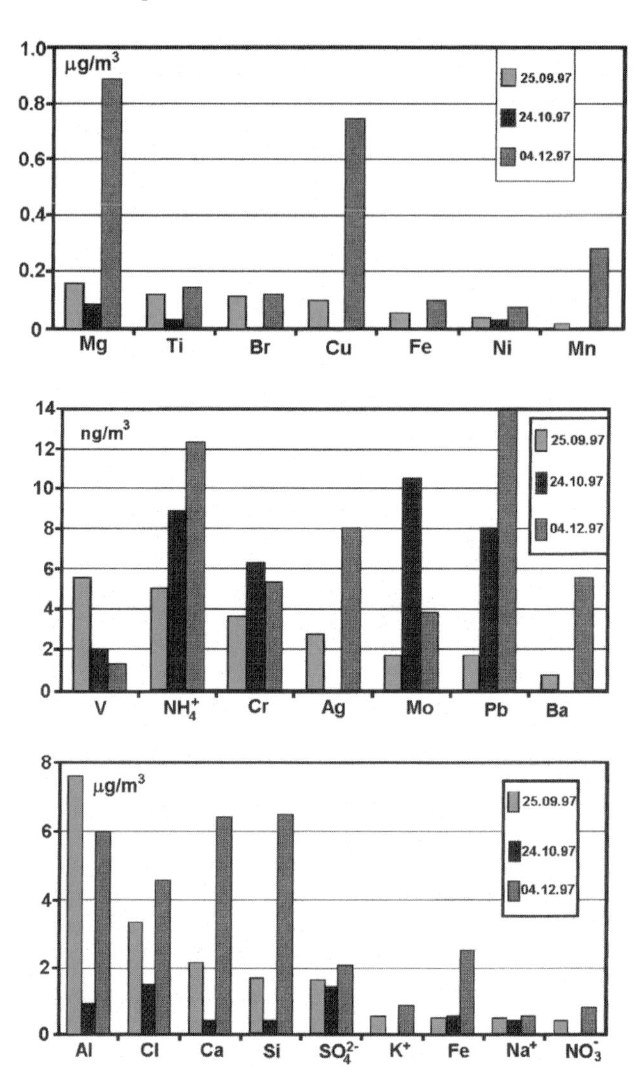

Figure 3.15. The dynamics of the absolute composition of atmospheric aerosol in the south of the Novosibirsk region (north of Kamen-on-Obi).

Kazakhstan; and one more branch of the air flux along the Urals and Lower Ob′ and Irtysh Rivers, across western Siberia overlapping the first trajectory, which comes from the Arctic.

Apparently, the air from Kazakhstan, where the wind speed on the preceding days and hours reached $20\,\mathrm{m\,s^{-1}}$, played the determining role in the aerosol formation, since even the Arctic cold front which had passed several hours before sounding could not remove the prevalence of the silicon component in the troposphere above 2,000 m (i.e., above the ABL). An abrupt increase in concentration in the surface layer, after the passing of the cold front, agrees well with the results of surface observations in Tomsk [9]. A considerable share of components following silicon in the histogram are of 'Kazakhstan–Aral' origin, since they all have a considerable secondary maxima, many of them, like silicon, at altitudes of 2,400–2,500 m (Al, Ca, Fe, Cu, B, Ti); a shifting of the peaks of ions of natrium, potassium, chloride, and nitrate above 3,000 m is apparently determined by their volatility and their marine and oceanic origin. On the whole, such interrelationships of elements and ions is characteristic of aerosol over the southern sector of western Siberia in the summer period, differing only in absolute concentrations. The autumn–winter sounding flights took place under conditions of an increasing impact of Arctic air masses. So, on 25 September the sampling was carried out more than 5 hours after the passing of the cold front and the change of moderate continental air masses from the semidesert regions of Kazakhstan for Arctic ones.

Values of tropospheric aerosol component concentrations sampled on 24 October (Figure 3.15) reflect best the composition of local aerosol, since at this moment the region of Kamen-na-Obi was for a long time to the rear of a powerful anticyclone (on the western side), which had just started disintegrating. This aerosol can be considered as the background one for this region. The same cannot be said about the aerosol sampled on 4 December, when this region was located on the north-eastern side of the anticyclone, and along the cold trough, the west–north-western outflow from the north of the Urals was clearly observed. This was confirmed by growing concentrations of many metals in aerosols. December is characterized by an increase of the mineral component of aerosol formed via dispersing (Al, Ca, Mg, Si, Cu).

An analysis has been made [63] of the magnitude and fluctuations of concentrations of organic carbon and lead in the surface layer of the atmosphere in different regions of western Siberia in order to determine the sources of aerosol particles and their role in the formation of the concentration field. For the northern territories of Siberia, the main contribution to the composition of organic carbon mass is made by long-range transport, and local sources play a substantial role, too. For the southern regions, the situation is different. It suggests the conclusion that in the southern regions a regional (or global-scale) background aerosol is formed, affecting most of the content of organic carbon in the atmosphere above the northern territories. For lead, a decreasing north–south gradient of concentration is observed, and variations of local sources turn out to be masked with a relatively high regional background. Among possible local sources, car engine exhausts – creating small particles with a short lifetime – are the only substantial source. The main source creating a high regional background is obviously located in the northern part of western Siberia – the Norilsk nickel works.

3.6 BIBLIOGRAPHY

1. Alfaro S. C., Gomes L., Rajot J. L., Lafon S., Gaudichet A., Chatenet B., Maille M., Cautenet G., Lasserre F., Cachier H., *et al.* Chemical and optical characterization of

aerosols measured in spring 2002 at the ACE-Asia supersite, Zhenbeitai, China. *J. Geophys. Res.*, 2003, **108**(D23), ACE9/1–ACE9/18.

2. Anderson T. L., Masonis S. J., Covert D. S., Ahlquist N. C., Howell S. G., Clarke A. D., and McNaughton C. S. Variability of aerosol optical properties derived from in situ aircraft measurements during ACE-Asia. *J. Geophys. Res.*, 2003, **108**(D23), ACE15/1– ACE15/19.
3. Arshinov M. Yu., Belan B. D., Loguntsev A. E., Simonenkov D. V., and Tolmachev G. N. An assessment of the urban anthropogenic contribution to air pollution: Development of technique with the city of Tomsk as an example. In: L. S. Ivlev (ed.), *Natural and Anthropogenic Aerosols*. St Petersburg State Univ. Press, St Petersburg, 2000, 273 pp. [in Russian].
4. Arshinov M. Yu., Belan B. D., Loguntsev A. E., Simonenkov D. V., and Tolmachev G. N. The ion-elemental composition of atmospheric aerosol over the Western Siberia. In: L. S. Ivlev (ed.), *Natural and Anthropogenic Aerosols*. St Petersburg State Univ. Press, St Petersburg, 2000, 273 pp. [in Russian].
5. Arshinova V. G., Belan B. D., and Vorontsova E. V. The aerosol dynamics in passing the atmospheric fronts. *Optics of the Atmos. and Ocean*, 1997, **10**(7), 813–819.
6. Augustine J. A., Cernwall C. R., Hodger G. B., Long C. N., Medina C. I., and De Luisi J. J. An automated method of MFRSR calibration for aerosol optical depth analysis with application to an Asian dust outbreak over the United States. *J. Appl. Meteorol.*, 2004, **42**(2), 266–278.
7. Bahreini R., Jimenez J. L., Wang J., Flagan R. C., Seinfeld J. H., Jayne J. T., and Worsnop D. R. Aircraft-based aerosol size and composition measurements during ACE-Asia using an Aerodyne aerosol mass spectrometer. *J. Geophys. Res.*, 2003, **108**(D23), ACE13/1–ACE13/22.
8. Belan B. D., Grishin A. I., Matvienko E. V., and Samokhvalov I. V. *The Spatial Valiability of Atmospheric Aerosol Characteristics*. Nauka Publ., Siberia Branch, Novosibirsk, 1989, 152 pp. [in Russian].
9. Belan B. D. and Tolmachev G. N. The temporal variability of aerosol in the troposphere over the Western Siberia. *Optics of the Atmos. and Ocean*, 1996, **9**(1), 99–105.
10. Borodulin A. I., Safatov A. S., Marchenko V. V., Shabanov A. N., Belan B. D., and Panchenko M. V. The altitudinal and seasonal variability of the concentration of the biogenic component of tropospheric aerosol over the southern Western Siberia. *Optics of the Atmos. and Ocean*, 2003, **16**(5–6), 422–425.
11. Borodulin A. I., Safatov A. S., Belan B. D., and Panchenko M. V. On the statistics of tropospheric bio-aerosol concentration over the southern Western Siberia. *Optics of the Atmos. and Ocean*, 2003, **16**(5–6), 519–522.
12. Borodulin A. I., Safatov A. S., Shabanov A. N., Khutorova O. G., Kutsenogiy K. P., and Makarov V. I. The periodic structure of the surface concentration fields of aerosols containing atmospheric albumen on the outskirts of Novosibirsk. *Optics of the Atmos. and Ocean*, 2004, **17**(5–6), 457–460.
13. Bukaty V. I., Samoilov A. S., and Sutorikhin I. A. Dynamics of the microphysical parameters of the surface aerosol in the city of Barnaul. *Optics of the Atmos. and Ocean*, 2003, **16**(5–6), 461–463.
14. Cantrell C., Carder K. L., and Gordon H. R. Columnar aerosol single-scattering albedo and phase function retrieved from sky radiance over the ocean: Measurements of Saharan dust. *J. Geophys. Res.*, 2003, **108**(D9), AAC10/1–AAC10/11.

15. Carrico C. M., Kus P., Rood M. J., Quinn P. K., and Bates T. S. Mixtures of pollution, dust, sea-salt, and volcanic aerosol during ACE-Asia: Radiative properties as a function of relative humidity. *J. Geophys. Res.*, 2003, **108**(D23), ACE18/1–ACE18/18.
16. Chatfield R. B., Guo Z., Sachse G. W., Blake D. R., and Blake N. J. The subtropical global plume in the Pacific Exploratory Mission–Tropics A (PEM–Tropics A), PEM–Tropics B, and the Global Atmospheric Sampling Program (GASP): How tropical emissions affect the remote Pacific. *J. Geophys. Res.*, 2003, **108**(D16), ACH1/1–ACH1/20.
17. Chazette P. The monsoon aerosol extinction properties at Goa during INDOEX as measured with lidar. *J. Geophys. Res.*, 2003, **108**(D6), ACH6/1–ACH6/12.
18. Chu D. A., Kaufman Y. J., Zibordi G., Chern J. D., Mao J., Li C., and Holben B. N. Global monitoring of air pollution over land from the Earth Observing System – Terra Moderate Resolution Imaging Spectroradiometer (MODIS). *J. Geophys. Res.*, 2003, **108**(D21), ACH4/1–ACH4/18.
19. Clarke A. D., Howell S., Quinn P. K., Bates T. S., Ogren J. A., Andrews E., Jefferson A., Massling A., Mayol-Bracero O., Maring H., *et al.* INDOEX aerosol: A comparison and summary of chemical, microphysical, and optical properties observed from land, ship, and aircraft. *J. Geophys. Res.*, 2002, **107**(D19), INX2 32/1–INX2 32/32.
20. Collins W. D., Rasch P. J., Eaton B. E., Khattatov B. V., and Lamarque J.-F. Simulating aerosols using a chemical transport model with assimilation of satellite aerosol retrievals: Methodology for INDOEX. *J. Geophys. Res.*, 2001, **106**(D7), 7313–7336.
21. Conant W. C. An observational approach for determining aerosol surface radiative forcing: Results from the first field phase of INDOEX. *J. Geophys. Res.*, 2000, **105**(D12), 15347–15360.
22. Dibb J. E., Talbot R. W., Seid G., Jordan C., Scheuer E., Atlas E., Blake N. J., and Blake D. R. Airborne sampling of aerosol particles: Comparison between surface sampling at Christmas Island and P-3 sampling during PEM–Tropics B. *J. Geophys. Res.*, 2003, **108**(D2), PEM2/1–PEM2/17.
23. Dibb J. E., Talbot R. W., Scheuer E. M., Seid G., Avery M. A., and Singh H. B. Aerosol chemical composition in Asian continental outflow during the TRACE-P campaign: Comparison with PEM-West B. *J. Geophys. Res.*, 2003, **108**(D21), GTE36/1–GTE36/13.
24. Duncan B. N., Martin R. V., Staudt A. C., Vevich R., and Logen J. A. Interannual and seasonal variability of biomass burning emissions constrained by satellite observations. *J. Geophys. Res.*, 2003, **108**(D2), ACH1/1–ACH1/22.
25. Emilenko A. S., Kopeikin V. M., and Van Geng Cheng. Variations in the content of the urban soot and submicron aerosols. In: G. S. Golitsin (ed.), *Physics of Atmospheric Aerosols*. Dialog-MSU, Moscow, 1999, pp. 160–169 [in Russian].
26. Ethé C., Basdevant C., Sadourny R., Appu K. S., Harenduprakash L., Sarode P. R., and Viswanathan G. Air mass motion, temperature, and humidity over the Arabian Sea and western Indian Ocean during the INDOEX intensive phase, as obtained from a set of super-pressure drifting balloons. *J. Geophys. Res.*, 2002, **107**(D19), INX2 22/1–INX2 22/19.
27. Formenti P., Elbert W., Maenhaut W., Haywood J., and Andreae O. M. Chemical composition of mineral dust aerosol during the Saharan Dust Experiment (SHADE) airborne campaign in the Cape Verde region, September 2000. *J. Geophys. Res.*, 2003, **108**(D18), SAH3/1–SAH3/15.
28. Franke K., Ansmann A., Müller D., Althausen D., Venkataraman C., Reddy M. S., Wagner F., and Sheele R. Optical properties of the Indo-Asian haze layer over the tropical Indian Ocean. *J. Geophys. Res.*, 2003, **108**(D2), AAC6/1–AAC6/17.

29. Fujita S.-I., Sakurai T., and Matsuda K. Wet and dry deposition of sulfur associated with the eruption of Miyakejima volcano, Japan. *J. Geophys. Res.*, 2003, **108**(D15), ACH3/1–ACH3/9.
30. Gao S., Hegg D. A., and Jonsson H. Aerosol chemistry, and light-scattering and hydroscopicity budgets during outflow from East Asia. *J. Atmos. Chem.*, 2003, **46**(1), 55–88.
31. Gong S. L., Zhang X. Y., Zhao T. L., McKendry I. G., Jaffe D. A., and Lu N. M. Characterization of soil dust aerosol in China and its transport and distribution during 2001 ACE-Asia. 2. Model simulation and validation. *J. Geophys. Res.*, 2003, **108**(D9), ACH4/1–ACH4/19.
32. Gorchakov G. I. and Ivlev L. S. Experimental observations of sand-dunes dusting in Kalmykia in the summer of 1997. In: L. S. Ivlev (ed.), *Natural and Anthropogenic Aerosols.* St Petersburg State Univ. Press, St Petersburg, 1998, pp. 401–407 [in Russian].
33. Gorchakov G. I., Kopeikin V. M., and Isakov A. A. On the spatial distribution of submicron and soot aerosol over Eurasia continent. In: L. S. Ivlev (ed.), *Natural and Anthropogenic Aerosols.* St Petersburg State Univ. Press, St Petersburg, 2001, pp. 52–53.
34. Granat L., Norman M., Leck C., Kulshrestha U. C., and Rodhe H. Wet scavenging of sulfur compounds and other constituents during the Indian Ocean Experiment (INDOEX). *J. Geophys. Res.*, 2002, **107**(D19), INX2 24/1–INX2 24/10.
35. Grousset F. E., Ginoux P., Bory A., and Biscaye P. E. Case study of a Chinese dust plume reaching the French Alps. *Geophys. Res. Lett.*, 2003, **30**(3), 10/1–10/4.
36. Guazzotti S. A., Suess D. T., Coffee K. R., Quinn P. K., Bates T. S., Wisthalet A., Hansel A., Ball W. R., Dickerson R. R., Nensüß C., *et al.* Characterization of carbonaceous aerosols outflow from India and Arabia: Biomass/biofuel burning and fossil fuel combustion. *J. Geophys. Res.*, 2003, **108**(D15), ACL13/1–ACL13/14.
37. Guieu C., Loÿe-Pilot M. D., Ridame C., and Thomas C. Chemical characterization of the Saharan dust end-member. Some biogeochemical implications for the western Mediterranean Sea. *J. Geophys. Res.*, 2002, **107**(D15), ACH5/1–ACH5/11.
38. Haywood J., Francis P., Osborne S., Glew M., Loel N., Highwood E., Tanré D., Myhre G., Formenti P., and Hirst E. Radiative properties and direct radiative effect of Saharan dust measured by the C-130 aircraft during SHADE. 1. Solar spectrum. *J. Geophys. Res.*, 2003, **108**(D18), SAH4/1–SAH4/16.
39. Highwood E. J., Haywood J. M., Silverstone M. D., Newman S. M., and Taylor J. P. Radiative properties and direct effect of Saharan dust measured by the C-130 aircraft during Saharan Dust Experiment (SHADE). 2. Terrestrial spectrum. *J. Geophys. Res.*, 2003, **108**(D18), SAH5/1–SAH5/13.
40. Höller R., Ito K., Tohno S., and Kasahara M. Wavelength-dependent aerosol single-scattering albedo: Measurements and model calculations for a coastal site near the Sea of Japan during ACE-Asia. *J. Geophys. Res.*, 2003, **108**(D23), ACE16/1–ACE16/15.
41. Hoornaert S., Godoi R. H. M., and van Grieken R. Single particle characterization of the aerosol in the marine boundary layer and free troposphere over Tenerife, NE Atlantic, during ACE-2. *J. Atmos. Chem.*, 2003, **46**, 271–293.
42. Ivlev L. S., Karpov G. A., Kist A. A., Kulmatov R. B., Abdullaev R., Semova A. Yu., and Fedchenko M. A. Study of the structure and chemical composition of aerosols in the surface layer of the atmosphere in the regions of volcanic activity at Kamchatka. *Volcanology and Seismology*, 1986, **1**, 32–42 [in Russian].
43. Ivlev L. S., Zhvalev V. F., Zhukovsky D. A., Ivanov V. A., and Prokofyev M. A. The chemical and mineralogical composition of surface aerosols in the region of Dushanbe. *J. Ecological Chemistry*, 1992, **1**, 87–93 [in Russian].

44. Ivlev L. S., Zhukov V. M., Kudriashov V. I., and Mikhailov E. F. In-situ measurements of volcanic matter in the lower atmosphere. *Optics of the Atmos. and Ocean*, 1993, **6**(10), 1249–1267.
45. Ivlev L. S., Ivanov V. A., Zhukov V. M., and Kudriashov V. I. Complex studies of atmospheric aerosols in the Aral region. *J. Ecol. Chem.*, 1994, **3**(2), 87–95 [in Russian].
46. Israelevich P. L., Levin Z., Joseph J. H., and Ganor E. Desert aerosol transport in the Mediterranean region as inferred from the TOMS aerosol index. *J. Geophys. Res.*, 2002, **107**(D21), AAC13/1–AAC13/13.
47. Iwasaka Y., Shi G.-Y., Yamada M., Matsuki A., Trochkine D., Kim Y. S., Zhang D., Nagatani T., Shibata T., Nagatani M., *et al.* Importance of dust particles in the free troposphere over the Taklamakan Desert: Electron microscopic experiments of particles collected with a balloon-borne particle impactor at Danhuang, China. *J. Geophys. Res.*, 2003, **108**(D23), ACE12/1–ACE12/10.
48. Iwasaka Y., Shibata T., Nagatani T., Shi G.-Y., Kim Y. S., Matsuki A., Trochkine D., Zhang D., Yamada M., Nagatani M., Nakata H., *et al.* Large depolarization ration of free tropospheric aerosols over the Taklamakan Desert revealed by lidar measurements: Possible diffusion and transport of dust particles. *J. Geophys. Res.*, 2003, **108**(D23), ACE 20/1–ACE 20/8.
49. Jordan C. E., Anderson B. E., Talbot R. W., Dibb J. E., Fuelberg H. E., Hudgins C. H., Kiley C. M., Russo R., Scheuer E., Seid G., *et al.* Chemical and physical properties of bulk aerosols within four sectors observed during TRACE-P. *J. Geophys. Res.*, 2003, **108**(D21), GTE34/1–GTE34/19.
50. Kaneyasu N. and Murayama S. High concentrations of black carbon over middle latitudes in the North Pacific Ocean. *J. Geophys. Res.*, 2000, **105**(D15), 19881–19890.
51. Kaufman Y. J., Tanré D., and Boucher O. A satellite view of aerosols in the climate system. *Nature*, 2002, **419**, 215–223.
52. Kawamura K., Umemoto N., and Mochida M. Water-soluble dicarboxylic acids in the tropospheric aerosols collected over east Asia and western North Pacific by ACE-Asia C-130 aircraft. *J. Geophys. Res.*, 2003, **108**(D23), ACE7/1–ACE7/7.
53. Kondratyev K. Ya., Orlenko L. R., Rabinovich Yu., Ter-Markaryantz N. E., and Shliakhov V. I. Complex Energetic Experiment (CENEX). *WMO Bulletin*, 1970, **XIX**(4), 59–60.
54. Kondratyev K. Ya. and Orlenko L. R. (eds). *The Complex Energetic Experiment (materials of the CENEX–70 expedition). Proc. of the Main Geophysical Observatory (St Petersburg)*. St Petersburg, 1972, 276 pp. [in Russian].
55. Kondratyev K. Ya. *Global Climate*. Nauka Publ., St Petersburg, 1992, 359 pp. [in Russian].
56. Kondratyev K. Ya. Aerosol as a climate-forming component of the atmosphere. 1. Physical properties and chemical composition. *Optics of the Atmos. and Ocean*, 2002, **15**(2), 123–146.
57. Kondratyev K. Ya. Atmospheric aerosol as a climate-forming component of the atmosphere. 1. Properties of various types of aerosols. *Optics of the Atmos. and Ocean*, 2004, **17**(1), 1–20.
58. Kondratyev K. Ya. From nano- to global scales: Properties, formation processes, and consequences of the impact of atmospheric aerosols. 1. Field observational experiments. Africa and Asia. *Optics of the Atmos. and Ocean*, 2005, **17**(9), 699–714.
59. Krichak S. O., Tsidulko M., and Alpert P. A study of an INDOEX period with aerosol transport to the eastern Mediterranean area. *J. Geophys. Res.*, 2002, **107**(D21), AAC18/1–AAC18/8.

60. Kudriashov V. I. and Ivlev L. S. An analysis of the elemental composition of the glaciers of Altai, Tien Shan, and Pamir. *Aerosols of Siberia* (the fourth meeting of the Working Group), Inst. Opt. Atmos., Siberian Branch of the RAS, Tomsk, 1997, pp. 90–91.
61. Kutsenogiy K. P. and Kutsenogiy P. K. Aerosols of the Siberia. Results of 7-year studies. *Siberian Ecological Journal*, 2000, **7**(1), 11–20 [in Russian].
62. Kutsenogiy K. P., Samsonov Yu. N., Churkina T. V., Ivanov A. V., and Ivanov V. A. The content of microelements in aerosol emissions during boreal forests fires in the Central Siberia. *Optics of the Atmos. and Ocean*, 2003, **16**(5–6), 461–465.
63. Kutsenogiy P. K. and Makarov V. I. An analysis of the sources of aerosol particles in the atmosphere over Siberia from the observational data on the amount and fluctuations of the organic carbon content. *Optics of the Atmos. and Ocean*, 2004, **17**(5–6), 405–410.
64. Lèon J.-F., Tanrè D., Pelón J., Kaufman Y. J., Haywood J. M., and Chatenet B. Profiling of a Saharan dust outbreak based on a synergy between active and passive remote sensing. *J. Geophys. Res.*, 2003, **108**(D18), SAH2/1–SAH2/18.
65. Li L. P., Fukushima H., Fronin R., Mitchell B. G., He M.-X., Uno U., Takamuro T., and Ohta S. Influence of submicron absorptive aerosol on Sea-viewing Wide Field of View Sensor (SeaWiFS) – derived marine reflectance during Aerosol Characterization Experiment (ACE)–Asia. *J. Geophys. Res.*, 2003, **108**(D15), AAC13/1– AAC13/11.
66. Li J., Anderson J. R., and Buseck P. R. TEM study of aerosol particles from clean and polluted marine boundary layers over the North Atlantic. *J. Geophys. Res.*, 2003, **108**(D6), ACH8/1–ACH8/14.
67. Luo B. P., Voigt C., Fueglistaler S., and Peter T. Extreme NAT supersaturations in mountain wave ice PSC's: A clue to NAT formation. *J. Geophys. Res.*, 2003, **108**(D15), AAC4/1–AAC4/10.
68. Mader B. T., Flagan R. C., and Seinfeld J. H. Airborne measurements of atmospheric carbonaceous aerosols during ACE–Asia. *J. Geophys. Res.*, 2002, **107**(D23), AAC13/1–AAC13/21.
69. Makienko E. V., Kabanov D. M., Rakhimov R. F., and Sakerin S. M. Specific micro-physical features of the aerosol component in different regions of the Atlantic. *Optics of the Atmos. and Ocean*, 2004, **17**(5–6), 437–443.
70. Masonis S. J., Franke K., Ansmann A., Müller D., Althausen D., Ogren J. A., Jefferson A., and Sheridan P. J. An intercomparison of aerosol light extinction and 180° backscatter as derived using in situ instruments and Raman lidar during the INDOEX field campaign. *J. Geophys. Res.*, 2002, **107**(D19), INX2 13/1–INX2 13/21.
71. Mayol-Bracero O. L., Gabriel R., Andreae M. O., Kirchstetter T. W., Novakov T., Ogren J., Sheridan P., and Streets D. G. Carbonaceous aerosols over the Indian Ocean during the Indian Ocean Experiment (INDOEX): Chemical characterization, optical properties, and probable sources. *J. Geophys. Res.*, 2002, **107**(D19), INX2 29/1–INX2 29/21.
72. Mitra A. P. Indian Ocean Experiment (INDOEX): An overview. *Indian J. Mar. Sci.* 2004, **33**(1), 30–39.
73. Moorthy K. K. and Satheesh S. K. Characteristics of aerosols over a remote island, Minicoy in the Arabian Sea: Optical properties and retrieved size characteristics. *Quart. J. Roy. Meteorol. Soc.*, 2000, Part A, **126**(562), 81–110.
74. Müller D., Mattis I., Wandinger U., Ausmann A., Althausen D., Dubovik O., Eckhardt S., and Stohl A. Saharan dust over a central European EARLINET–AERONET site: Combined observations with Raman lidar and Sun photometer. *J. Geophys. Res.*, 2003, **108**(D12), AAC1/1–AAC1/17.

75. Murayama T., Masonis S. J., Redemann J., Anderson T. L., Schmid B., Livingston J. M., Russel P. B., Huebert B., Howell S. G., McNaughton C. S., *et al.* An intercomparison of lidar-derived aerosol optical properties with airborne measurements near Tokyo during ACE–Asia. *J. Geophys. Res.*, 2003, **108**(D23), ACE 19/1–ACE 19/19.
76. Nakajima T. *Findings and Current Problems in the Asian Particle Environmental Change Studies: 2003.* JST/CREST/APEX 2003 Interim Report. Univ. of Tokyo, Tokyo, 2003. 240 pp.
77. Neusüß C., Gnauk T., Plewka A., Hermann H., and Quinn P. K. Carbonaceous aerosol over the Indian Ocean: OC/EC fractions and selected specifications from size-segregated onboard samples. *J. Geophys. Res.*, 2002, **107**(D19), INX2 30/1–INX2 30/13.
78. Panchenko M. V. A thematic issue 'Aerosols of the Siberia'. *Optics of the Atmos. and Ocean*, 2003, **16**(5–6), 405–559.
79. Panchenko M. V. Thematic issue 'Aerosols of the Siberia'. *Optics of the Atmos. and Ocean*, 2004, **17**(5–6), 375–534; 2005, **18**(5–6), 370–532.
80. Papaspiropoulos G., Martinsson B. G., Zahn A., Brenninkmeijer C. A. M., Hermann M., Heintzenberg J., Fischer H., and van Velthoven P. F. J. Aerosol elemental concentrations in the tropopause region from intercontinental flights with the Civil Aircraft for Regular Investigation of the Atmosphere Based on an Instrument Container (CARIBIC) platform. *J. Geophys. Res.*, 2002, **107**(D23), AAC3/1–AAC3/14.
81. Pedrós R., Martinez-Lozano J. A., Utrillas M. P., Gómez-Amo J. L., and Tenc F. Column integrated aerosol optical properties from ground-based spectroradiometer measurements in Barrax (Spain) during the Digital Airborne Imaging Spectrometer Experiment (DAISEX) campaign. *J. Geophys. Res.*, 2003, **108**(D18), AAC5/1–AAC5/17.
82. Penenko V. V. and Kurbatskaya L. I. Studies of the 'heat island' dynamics with the radiation–aerosol interaction taken into account. *Optics of the Atmos. and Ocean*, 1998, **11**(6), 581–585.
83. Piketh S. J., Annegam H. J., and Tyson P. D. Lower troposphere aerosol loading over South Africa: The relative contribution of aeolian dust, industrial emissions, and biomass burning. *J. Geophys. Res.*, 1999, **104**(D1), 1597–1607.
84. Popova S. A., Makarov V. I., and Kutsenogiy K. P. The spatial–temporal variability of the organic and inorganic carbon concentration in atmospheric aerosols over the Novosibirsk Oblast. *Optics of the Atmos. and Ocean*, 2004, **17**(5–6), 464–469.
85. Prospero J. M., Savoie D. L., and Arimoto R. Long-term record of nss-sulfate and nitrate in aerosols on Midway Island, 1981–2000: Evidence of increased (now decreasing?) anthropogenic emissions from Asia. *J. Geophys. Res.*, 2003, **108**(D1), 10/1–10/11.
86. Quinn P. K., Coffman D. J., Bates T. S., Miller T. L., Johnson J. E., Welton E. J., Neusüß C., Miller M., and Sheridan P. J. Aerosol optical properties during INDOEX 1999: Means, variability, and controlling factors. *J. Geophys. Res.*, 2003, **107**(D19), INX2 19/1–INX2 19/25.
87. Rakhimov R. F., Uzhegov V. N., Makienko E. V., and Pkhalagov Yu. A. The microphysical interpretation of the seasonal and diurnal variability of the spectral dependence of the aerosol attenuation coefficient along the surface trajectories. *Optics of the Atmos. and Ocean*, 2004, **17**(5–6), 386–404.
88. Rasch P. J., Collins W. D., and Eaton B. E. Understanding the Indian Ocean Experiment (INDOEX) aerosol distribution with an aerosol assimilation. *J. Geophys. Res.*, 2001, **106**(D7), 7337–7355.
89. Reid J. S., Jonsson H. H., Maring H. B., Smirnov A., Savoie D. L., Cliff S. S., Reid E. A., Livingston J. M., Meier M. M., Dubovik O., *et al.* Comparison of size and

morphological measurements of coarse mode dust particles from Africa. *J. Geophys. Res.*, 2003, **108**(D19), PRD9/1–PRD9/20.

90. Rodhe H., Ayers G., Peng L. C., and Bala M. R. Composition of Asian Deposition (CAD): A task within the IGAC DEBITS activity. *IGACtiv Newslett.*, 2003, **28**, 12–13.
91. Rosen J., Young S., Laby J., Kjome N., and Gras J. Springtime aerosol layers in the free troposphere over Australia: Mildura Aerosol Tropospheric Experiment (MATE 98). *J. Geophys. Res.*, 2000, **105**(D14), 17833–17842.
92. Safatov A. S., Andreeva I. S., Ankilova A. N., Baklanov A. M., Belan B. D., Borodulin A. I., Buriak G. A., Ivanova N. A., Kutsenogiy K. L., Makarov V. I., *et al.* The share of biogenic component in atmospheric aerosols in the south of the Western Siberia. *Optics of the Atmos. and Ocean*, 2003, **16**(5–6), 532–536.
93. Sano I., Mukai S., Okada Y., Kolben B. N., Ohta S., and Takamura T. Optical properties of aerosols during APEX and ACE-Asia experiments. *J. Geophys. Res.*, 2003, **108**(D23), ACE17/1–ACE17/9.
94. Satheesh S. K., Ramanathan V., Li-Jones X., Lobert J. M., Podgorny I. A., Prospero J. M., Holben B. N., and Loeb N. G. A model for the natural and antropogenic aerosols over the tropical Indian Ocean derived from Indian Ocean Experiment data. *J. Geophys. Res.*, 1999, **104**(D22), 27,421–27,440.
95. Schmid B., Hegg D. A., Wang J., Bates D., Redemann J., Russel P. B., Livingston J. M., Jonsson H. H., Welton E. J., Seinfeld J. H., *et al.* Column closure studies of lower tropospheric aerosol and water vapor during ACE-Asia using airborne Sun photometer and airborne in situ and ship-based lidar measurements. *J. Geophys. Res.*, 2003, **108**(D23), ACE24/1–ACE24/22.
96. Sheesley R. J., Shaner J. J., Chowdhury Z., and Cass G. R. Characterization of organic aerosols emitted from the combustion of biomass indigenous to South Asia. *J. Geophys. Res.*, 2003, **108**(D9), AAC8/1–AAC8/15.
97. Sinha P., Hobbs P. V., Yokelson R. J., Blake D. R., Gao S., and Kirchstetter T. W. Distributions of trace gases and aerosols during the dry biomass burning season in southern Africa. *J. Geophys. Res.*, 2003, **108**(D17), ACH4/1–ACH4/23.
98. Streets D. G., Bond T. C., Carmichael G. R., Fernandes S. D., Fu Q., He D., Klimont Z., Nelson S. M., Tsai N. Y., Wang M. Q., *et al.* An inventory of gaseous and primary aerosol emissions in Asia in the year 2000. *J. Geophys. Res.*, 2003, **108**(D21), GTE30/1–GTE30/23.
99. Sun D., Chen F. and Bloemendal J., Su R. Seasonal variability of modern dust over the Loess Plateau of China. *J. Geophys. Res.*, 2003, **108**(D21), AAC3/1–AAC3/10.
100. Takemura T., Okamoto H., Maruyama Y., Numaguti A., Higurashi A., and Nakajima T. Global three-dimensional simulation of aerosol optical thickness distribution of various origins. *J. Geophys. Res.*, 2000, **105**(D14), 17853–17874.
101. Tanré D., Haywood J., Pelon J., Léon J. E., Chatenet B., Formenti P., Francis P., Goloub P., Highwood E. J., and Myhre G. Measurement and modeling of the Saharan dust radiative impact: Overview of the Saharan Dust Experiment (SHADE). *J. Geophys. Res.*, 2003, **108**(D18), SAH1/1–SAH1/12.
102. Tolmachev G. N. The horizontal distribution of the ion-elemental composition of atmospheric aerosol over the USSR territory. *Proc. 2nd Int. Conf. 'Natural and Anthropogenic Aerosols', 27 September–1 October* 1999, *St Petersburg*. St Petersburg State Univ. Press, St Petersburg, 2000, 273 pp. [in Russian].
103. Trockine D., Iwasaka Y., Matsuki A., Yamada M., Kim Y.-S., Nagatani T., Zhang D., Shi G.-Y., and Shen Z. Mineral aerosol particles collected in Dunhuang, China, and their

comparison with chemically modified particles collected over Japan. *J. Geophys. Res.*, 2003, **108**(D23), ACE10/1–ACE10/11.

104. Uno I., Carmichel G. R., Streets D., Satake S., Takemura T., Woo J.-H., Uematsu M., and Ohta S. Analysis of surface black carbon distributions during ACE-Asia using a regional-scale aerosol model. *J. Geophys. Res.*, 2003, **108**(D23), ACE4/1–ACE4/11.
105. Uno I., Carmichel G. R., Streets D. G., Tang Y., Yienger J. J., Satake S., Wang Z., Woo J.-H., Guttikunda S., Uematsu M., *et al.* Regional chemical weather forecasting system CFORS: Model descriptions and analysis of surface observations at Japanese island station during the ACE-Asia experiment. *J. Geophys. Res.*, 2003, **108**(D23), ACE36/1–ACE36/17.
106. Vukmirović Z., Unkašević M., Lazic L., Tošić I., Rajšić S., and Tasić M. Analysis of the Saharan dust regional transport. *Meteorol. Atmos. Phys.*, 2004, **85**(4), 265–273.
107. Wang J., Flagan R. C., Seinfeld J. H., Jonsson H. H., Collins D. R., Russel P. B., Schmid B., Redemann J., Livingston J. M., Gao S., *et al.* Clear-column radiative closure during ACE-Asia: Comparison of multi-wavelength extinction derived from particle size and composition with results from Sun photometry. *J. Geophys. Res.*, 2002, **10**(D23), AAC7/1–AAC7/22.
108. Wang J., Christopher S. A., Reid J. A., Maring H., Savoie D., Holben B. N., Livingston J. M., Russel P. B., and Yang S.-K. GOES-8 retrieval of dust aerosol optical thickness over the Atlantic Ocean during PRIDE. *J. Geophys. Res.*, 2003, **108**(D19), PRD11/1–PRD11/15.
109. Wang J., Christopher S. A., Brechtel F., Kim J., Schmid B., Redemann J., Russel P. B., Quinn P., and Holben B. N. *J. Geophys. Res.*, 2003, **108**(D23), ACE25/1–ACE25/14.
110. Weber R. J., Lee S., Chen G., Wang B., Kapustin V., Moore K., Clarke A. D., Mauldin L., Kosciuch E., Cantrell C., *et al.* New particle formation in anthropogenic plumes advecting from Asia observed during TRACE–P. *J. Geophys. Res.*, 2003, **108**(D21), GTE35/1–GTE35/15.
111. Welton E. J., Voss K. J., Quinn P. K., Flatau P. J., Markowicz K., Campbell J. P., Spinhirne J. D., Gordon H. R., and Johnson J. E. Measurements of aerosol vertical profiles and optical properties during INDOEX 1999 using multi-pulse lidar. *J. Geophys. Res.*, 2002, **107**(D19), INX2 18/1–INX2 18/20.
112. Wiggs G. F. S., O'Hara S. L., Wegert J., van der Meerst J., Small I., and Hubbard R. The dynamics and characteristics of aeolian dust in dryland Central Asia: Possible impacts on human exposure and respiratory health in the Aral Sea basin. *Geogr. J.*, 2003, **169**(2), 142–158.
113. Zhang D., Iwasaka Y., Shi G., Zang J., Matsuki A., and Trochkine D. Mixture state and size of Asian dust particles collected at southwestern Japan in spring 2000. *J. Geophys. Res.*, 2003, **108**(D24), ACH9/1–ACH9/12.
114. Zhang X. Y., Gong S. L., Arimoto R., Shen Z. X., Mei F. M., Wang D., and Cheng Y. Characterization and temporal variation of Asian dust aerosol from a site in the Northern Chinese deserts. *J. Atmos. Chem.*, 2003, **44**, 241–257.
115. Zhang X. Y., Gong S. L., Shen Z. X., Mei F. M., Xi X. X., Liu L. C., Zhou Z. J., Wang D., Wang Y. Q., and Cheng Y. Characterization of soil dust aerosol in China and its transport and distribution during ACE-Asia. 1. Network observations. *J. Geophys. Res.*, 2003, **108**(D9), ACH3/1–ACH3/13.
116. Zhao T. L., Gong S. L., Zhang X. Y., and McKendry I. G. Modeled size-segregated wet and dry deposition budgets of soil dust aerosol during ACE-Asia 2001: Implications for trans-Pacific transport. *J. Geophys. Res.*, 2003, **108**(D23), ACE33/1–ACE33/9.

Part Two

Aerosol formation processes and evolution. Cloud cover dynamics

4

Aerosol formation processes

4.1 VARIOUS AEROSOL SOURCES, FORMATION PROCESSES, AND TYPES OF AEROSOLS

The regional-scale studies of atmospheric aerosols testify to their great variety, which may be illustrated, for instance, by the results of the accomplishment of the programme 'Aerosols of Siberia' published in the thematic issue of the journal *Optics of the Atmosphere and Ocean* [1, 17] and many other studies [72–77, 79–81]. The global aerosol is most variable and includes such types as sulphate aerosol, dust aerosol (especially connected with dust storms in deserts which is subject to a long-range transport in the atmosphere), organic (biogenic including) aerosol with its considerable absorptance in the visible spectral region, volcanic aerosol, aerosol generated in biomass burning, and others.

4.1.1 Formation processes and types of atmospheric aerosol

Atmospheric aerosols are the product of a complicated totality of chemical and physical processes. As a result of the complexity of these processes and a relatively short lifetime, the chemical composition and physical characteristics of aerosols are rather variable. The spatial–temporal variability of aerosol characteristics is so great, and the observational data are so fragmentary that it is difficult to reliably estimate the total content of various types of aerosols, and the available estimates of the power of global sources of natural and anthropogenic aerosols are rather approximate.

The division of aerosols by origin into anthropogenic and natural is sufficiently conditional. Anthropogenic sources are those determined by human activity: industrial wastes from chimneys, toxic exhausts from cars, fires, explosions, soil erosion in agriculture, and open mining. This gives an input of $(3\text{–}4) \times 10^{8}\,\mathrm{t\,yr^{-1}}$ of aerosols to the atmosphere. In the case of anthropogenic aerosols, the aerosols of internal atmospheric origin, *in situ*, or secondary aerosols play an important role. The concentration of aerosol smog due to photochemical reactions with exhaust gases in industrial centres reaches $200\,\mu g\,m^{-3}$, which is comparable with the consequences of dust storms.

Of course, the estimates of anthropogenic aerosols are more reliable than those of natural aerosols (especially in the regions of the World Ocean and continents which are difficult to access). Such a situation determines a low reliability of data on the relationships between natural and anthropogenic aerosols, though, of course, there is a global-scale impact of human activity on the cycles of sulphur and nitrogen.

The atmospheric aerosols, depending on their composition or sources, are classified into the following types of natural aerosols:

(1) products of sea spray evaporation;
(2) mineral dust wind-driven to the atmosphere;
(3) volcanic aerosol (both directly emitted to the atmosphere and formed due to gas-to-particle conversion);
(4) particles of biogenic origin (directly emitted to the atmosphere and formed as a result of condensation of volatile organic compounds, for instance, terpenes, as well as chemical reactions between these compounds);
(5) smokes from biota burning on land; and
(6) products of natural gas-to-particle conversion (e.g., sulphates resulting from reduced sulphur incoming from the ocean surface with emissions of dimethyl sulphide) [6, 89].

Important types of anthropogenic aerosols are also:

(1) direct industrial emissions of particles (particles of soot, smoke, road dust, etc.); and
(2) products of gas-to-particle conversion.

Besides these, one should distinguish between tropospheric and stratospheric (mainly volcanic) aerosols. Of interest are gas-to-particle conversions of aerosol formation due to the following processes:

(1) heterogeneous homomolecular nucleation (formation of a new stable liquid or solid fine particles from its gaseous phase in the presence of only one gas component);
(2) heterogeneous heteromolecular nucleation (a similar process to (1) but in the presence of two and more gases); and
(3) heterogeneous heteromolecular condensation (the growth of the existing particles due to gas adsorption).

According to [74], one should distinguish between primary and secondary sources of aerosols: soil dust, sea salts, industrial dust, and the primary anthropogenic aerosols, carbon aerosol (organic and black carbon), primary biogenic aerosol, sulphates, nitrates, and volcanic aerosol.

The photochemical and chemical reactions responsible for the initial transformation of a 'highly volatile' gas into a gas component which can serve the formation of aerosol are extremely complicated and still poorly studied. Apparently, the following processes are most substantial:

(1) reactions of sulphur dioxide with hydroxyl radicals which eventually lead to the formation of sulphuric acid molecules and sulphuric acid aerosol; and
(2) reactions of non-methane hydrocarbon compounds with ozone and/or hydroxyl radicals with the formation of aldehydes, spirits, and carboxyl and dicarboxyl acids (as a rule, secondary products of these reactions react with nitrogen oxides giving organic nitrates).

To understand the role of aerosol as an atmospheric component, of great importance is the information about the phase state of aerosols. Estimates of the NO_x content in the upper troposphere have shown, for instance, that from 20–60% of the loss rate of NO_x is determined by heterogeneous hydrolysis of N_2O_5, the efficiency of this process depending on the phase state of aerosols. Liquid aerosols cause a more intensive hydrolysis than solid aerosols.

Of great importance in atmospheric chemistry are ozone and hydroxyl radicals (HO and HO_2), which directly or indirectly are products of photochemical reactions (therefore, the processes of gas-to-particle conversion are usually characterized by strong diurnal variations).

Very important types of atmospheric aerosols include particles of organic compounds [135]. Various organic compounds participate in numerous reactions, so that studies of organic aerosols are very difficult. The concentration of soot particles in the atmosphere is rather considerable, reaching over the oceans at about $0.5\,\mu g\,m^{-3}$, on average, which is comparable with the concentration of mineral aerosols. Apparently, the sources of global organic aerosols are both natural and anthropogenic – about half the products of natural aerosols constituting the share of the ocean.

The results of aerosol impact on various processes (e.g., radiation transfer) depend, as a rule, on chemical and physical processes, with the dependence of composition on the size of aerosol particles almost always playing a substantial role. Therefore, an adequate description of the properties of real aerosols is only possible on the basis of results of a complex assessment of their characteristics. An estimation of the mass concentration is one of the most widespread types of aerosol measurements, but it is this characteristic that is least informative, since it says nothing about the sources, composition, and possible impacts.

The aerosol cycles are closely connected with the hydrological processes in the atmosphere due to a substantial interaction between aerosols and clouds: clouds and precipitation play an important role in the formation, transformation, and removal of aerosols from the atmosphere, but, on the other hand, aerosols strongly affect the microphysical processes in clouds. It is not accidental that in most global regions the aerosol lifetime in the lower and upper troposphere constitutes one–two weeks, and for water vapour it is about ten days. The relationship between clouds and aerosol testifies to the impossibility of sufficiently complete understanding of the processes of formation and transformation of aerosol without reliable ideas about the physics and chemistry of clouds.

There are justified apprehensions with respect to possible anthropogenic growth of aerosol content, which can affect the climate both through changing the Earth's radiation budget and through affecting the hydrological cycle. The strong spatial–temporal variability of the aerosol characteristics hinders an extraction of its anthropogenic component. Understanding may be possible through a knowledge of the causes of this variability, which would require completion of an extensive programme of complex studies of atmospheric aerosols.

4.1.2 Secondary aerosols formed *in situ*

Photochemical and chemical reactions in the atmosphere are responsible for the fine aerosol fraction, the so-called secondary aerosols. Aerosol particles are formed not only of the products of organic compounds but also of sulphur dioxide, hydrocarbon, carbonyl sulphide, dimethyl sulphide, ammonium, nitrogen oxides, and some other gases with oxidants of ozone type and various radicals, as well as with water vapour and aerosol particles playing the role of catalysts [9, 10, 101, 119].

Apart from volcanoes, sources of sulphur dioxide include industrial enterprises (the global level of emissions is about $10^8\,t\,yr^{-1}$) as well as anaerobic bacteria. The sources of hydrocarbon are mainly vegetation and products of organism decomposition in soil and water. It is difficult to estimate the hydrocarbon emitted to the atmosphere, but presumably it is about $10^8\,t\,yr^{-1}$. Recalculated for all sources it is about $4 \times 10^8\,t\,yr^{-1}$ for (SO_4^{-2}). The power of the sources of other sulphides is an order of magnitude lower. The power of the sources of nitrogen oxides and ammonium is estimated at $\sim 4 \times 10^7\,t\,yr^{-1}$. Besides this, it should be borne in mind that only part of all these gases ($\sim 10 \pm 25\%$) forms aerosol particles [116]. In connection with the transition to natural gas burning from coal burning, the industrial emissions of sulphur dioxide and nitrogen oxides have recently increased. A large amount of these gases are oxidized to give anhydrides of acids which dissolve in cloud droplets [15]. There are three basic, generally accepted mechanisms of the formation of aerosol particles from these gases in the atmosphere.

(1) Photochemical oxidation, heterogeneous reactions. The process takes place in arid regions and in high-tropospheric layers. The rate of conversion constitutes 0.03% SO_2 oxidized photochemically during one hour in pure air. The formation of aerosol particles with solar irradiation takes place in the presence of NO_2. The estimates of the NO_x content in the upper troposphere have shown that from 20–60% of the rate of NO_x losses is determined by heterogeneous hydrolysis of N_2O_5, the efficiency of this process depends on the phase state of the aerosol. Liquid aerosol particles cause a more intensive hydrolysis than solid ones.

(2) Catalytic oxidation in the presence of heavy metals. The rate of reaction depends strongly on the presence of suitable catalysts (ions of heavy metals)

and can be sufficiently high in heavily polluted air. With certain values of pH this process ceases. The reaction takes place both in dry air and in cloud droplets.

(3) The reaction of ammonium with sulphur dioxide in the presence of liquid water (cloud droplets). The rate of the formation of sulphate particles in the reaction with participation of SO_2 and NH_3 depends on the supply of NH_3. If the pH value is maintained sufficiently high, for instance, due to the supply of NH_3, then the reaction can continue. The mechanism of the formation of ammonium sulphate is efficient only in the presence of liquid water (i.e., in the region where clouds and fogs are present). Model calculations show that the rate of oxidation in cloud droplets is 12% per hour. Aircraft measurements of $(NH_4)_2SO_4$ particles show that maximum concentrations of sulphate particles are often observed beneath the lower cloud boundary. Particles of ammonium sulphate can remain suspended in the air after the evaporation of cloud droplets and fog. The initial nuclei of ammonium sulphate have radii of the order of 3×10^{-9} cm and are transformed into droplets of size 10^{-6} cm.

Apart from sulphuric and nitric acids and their salts, hydrochloric acid and its salts can also form in the atmosphere. Their formation results from the supply of gas-phase chlorine from the sea surface [13].

4.1.3 The ACE field experiment

To obtain more reliable information about aerosol and to specify its impact on the climate within the key IGBP project on the global chemistry of the atmosphere (IGAC), the programme of the First Aerosol Characteristics Experiment (ACE-1) has been developed to study the chemical and physical processes determining the southern hemisphere marine aerosol properties and to assess its impact on the climate [66, 99]. Solution of these problems required further development of global models of aerosol long-range transport, taking into account its chemical transformation, and an application of radiation transfer theory under conditions of both clear and cloudy atmospheres in order to calculate radiation fluxes and flux divergences as well as to analyse the interaction of aerosol with different components of the climate system.

It was also important to obtain adequate information on the characteristics which determine the impact of aerosols on radiation transfer, such as effectiveness of light scattering per unit mass (α_{sp}); the share of backscattered light (β); coefficient of phase function asymmetry (g); single scattering albedo (ω); and the dependence of aerosol scattering on relative humidity (f_{sp}(RH)). All these characteristics depend on chemistry, size distribution, morphology, and other properties of aerosols. In view of the important role of aerosols formed via gas-to-particle conversion, great importance is given to the information about their gas precursors.

To obtain reliable information on the climatic impact of aerosols is only possible through accomplishing complex programmes of field and laboratory observations together with numerical modelling. The ACE programme was planned as a series of the field observations in different representative global regions with the following three basic objectives: (1) to obtain data on the physical and chemical properties of the main types of aerosols (as well as on interactions between these properties), including information about aerosol characteristics which determine its impact on cloud formation (nucleation properties); (2) to study the physical and chemical processes governing its size distribution, chemistry, radiative, and nucleating characteristics; and (3) to check the reliability of various schemes of aerosol parameterization in regional and global climate models.

The first experiment (ACE-1) was carried out during the period 15 November–14 December, 1995 over the south-western Pacific (south of Australia). Forty-seven research organizations from 11 countries took part in ACE-1. Observations were made on board the NASA flying laboratory C-130, two research ships, and at island stations.

Bearing in mind the specific role of dimethyl sulphide emitted to the atmosphere by the ocean, characteristics of four water masses were investigated in the region of the experiment in order to analyse the impact of their physical, chemical, and biological properties on the content of various minor gas components in the atmosphere. In this connection, it is important that the ocean is a source of biogenic emissions to the atmosphere, such as dimethyl sulphide (DMS), hydrocarbon compounds, methyl nitrates, and methyl halides [24]. The ocean can also be a source of biogenic calcium carbonate promoting an enhancement of alkalinity of sea salt aerosols and ozone-induced oxidation of sulphur dioxide contained in the 'aerosol' water.

The earlier results show that in the region considered the sea salt aerosol is a prevailing component of aerosols (this refers to 90% of particles with diameters >130 nm and up to 70% of particles with diameters >80 nm). Half the amount of particles with diameters >160 nm were found to contain organic components connected with sea salt. This reflects the necessity to take into account the sea salt aerosol in climate models, since it controls not only solar radiation scattering but also the concentration of cloud condensation nuclei.

The results discussed by [11, 12] have shown that even in remote ocean regions, anthropogenic impacts on the atmosphere and aerosol properties show themselves markedly. So, for instance, at altitudes above 3 km there were layers containing old products of biomass burning. About 11–16% of sulphate aerosol particles with diameters >100 nm contained soot, probably resulting from biomass burning in South Africa.

The ACE-1 data have made it possible to assess the role of biogenic components in the formation and growth of new aerosol particles. There was also photochemical formation of new particles of sulphuric acid at the sites of 'emissions' from clouds. The results of numerical modelling show that about 30–50% of DMS is transformed into SO_2, the main sink for sulphur dioxide being its ozone-induced oxidation in the 'aerosol' water. Non-salt sulphates are contained mainly in supermicron salt particles constituting $35 \pm 10\%$ in summer and $58 \pm 22\%$ in winter.

4.1.4 Processes of aerosol formation (nucleation)

The diverse processes of aerosol formation determine the necessity to confine the analysis of these processes to a consideration of the most substantial and better studied processes [38, 39, 83, 86, 120]. In this context, the problem of particle nucleation is in focus. Külmälä [86] justly noted that in studies of the processes of nucleation much remains unclear from the viewpoint of the detemination of study priorities, but unquestionable progress has been recently achieved in the development of measurement techniques, which enables one to monitor the development of the processes of nucleation and the growth of particles through their whole range of sizes, starting from nanoparticles – not only near the Earth's surface but also in the free atmosphere (the lower stratosphere included). Lee *et al.* [91, 92] obtained new important data on an unexpectedly high concentration of ultrafine particles in the upper troposphere and lower stratosphere, revealing a special role, in this case, of ion-induced nucleation as the globally manifested mechanism of aerosol particle formation.

Studies of nucleation processes in coastal regions have been recently in focus. Earlier studies revealed the presence, in different regions (especially referring to boreal forests and the coastal bands of the sea bottom exposed during a falling tide), of biogenic sources of aerosol formation in the atmospheric boundary layer (ABL). Apparently, as a rule, nucleation events take place in the coastal bands of the seas and oceans. At least, this refers to western Europe, where nucleation aerosols (NA) were often observed on the coasts of Scotland, Ireland, England, and France. NA formation was also recorded in the southern hemisphere near the coastline of Antarctica and Tasmania.

Observations of the increased concentration of aerosol particles in the coastal ABL were started as far back as 1880 when J. Aitken found a considerable increase in aerosol number density near the western coast of Scotland, from 200–300 cm^{-3} to more than 10^4 cm^{-1}. The present studies reveal regular increases of NA concentration reaching sometimes 10^6 cm^{-3} particles or more, mainly in the size interval 3–10 nm. During the period 1998–1999, at the atmospheric research station at Mace Head (the western coast of Ireland), the field observational programme PARFORCE was accomplished with the aim of studying in detail the formation of new aerosol particles in the coastal ABL. For two years the formation of new aerosol particles was continually measured. During two special periods in September, 1998 and in June, 1999 the programme of observations was substantially enhanced.

An analysis of the results carried out by [106–110] has shown that new particles were formed approximately in 90% of all the days of the all-year-round observations and in all air masses under observation. As a rule, the duration of the episodes of NA formation constituted several hours, but sometimes it exceeded eight hours. Such episodes happened in the daytime during the falling tide, when the coastal band of the sea bottom was exposed. At this time, the aerosol number density often exceeded 10^6 cm^{-3} (under conditions of a pure cloud-free atmosphere), and the rate of the formation of particles with diameters $d > 3$ nm varied within 10^4–10^5 cm^{-3} s^{-1}, when the concentration of sulphuric acid vapour in the ABL exceeded 2×10^6 molecules/cm^3 without any correlation between maximum concentration of H_2SO_4, falling tides, and nucleation events.

According to the triple nucleation theory for the system H_2SO_4–H_2O–NH_3, the rate of nucleation can easily exceed $10^6\,cm^{-3}\,s^{-1}$, with the concentration of H_2SO_4 mentioned above. However, it follows from calculations of the rate of particle growth that this concentration of H_2SO_4 is insufficient to provide the growth of initial particles, with diameters of the order of 1 nm, to a detectable size of 3 nm. Analysis of the hygroscopically induced growth of particles to 8 nm has shown that in this case, particles should contain some amount of components less soluble than sulphate aerosols.

According to data on NA, in the absence of emissions of volatile organic compounds by coastal biota but in the presence of intensive biogenic emissions of halogen–hydrocarbon compounds and with a trace content of iodine in newly formed particles with diameters of 7 nm, iodine oxide is the most probable component favouring the growth of new aerosol particles (this supposition is confirmed by the results of laboratory experiments). However, it is still unclear whether the nucleation is a self-nucleation of iodine compounds and derivatives of halogen hydrocarbon compounds or whether at first stable clusters appear through a triple nucleation in the system H_2SO_4–H_2O–NH_3 with subsequent condensation growth of particles to detectable sizes via condensation of iodine compounds. It follows from aircraft observations that nucleation takes place along the whole coastline and that the coastal plume of biogenic aerosols can stretch for many hundreds of kilometres from the source of initial aerosols. In the process of evolution of the coastal plume the size of aerosol particles grows to 100 nm, at which point the optical activity of aerosol shows itself markedly. Estimates of the increase of cloud condensation nuclei number density suggest the possibility of its increase to 100%. Taking into account the probability of the formation of new aerosol particles due to biogenic sources located along the western coast of Europe and along other coastlines, it becomes clear that the aerosol appearing in the coastal band contributes much to the formation of natural background aerosols. Though these results obtained in the course of accomplishing the PARFORCE programme are very important, they still do not reveal the real mechanisms of nucleation and growth of aerosol particles in the coastal zones.

One of the most efficient means of studying the formation of new aerosol particles in the atmosphere is a many-stage condensation particles counter (CPC) with various thresholds of detection. So, for instance, the miniature stylus holder-3025 instruments enable one to measure the total number density of N_3 particles with diameters >3 nm (with a 50% relative humidity), whereas in the case of the miniature stylus holder-3010, the detection threshold constitutes 10 nm (N_{10} number density). The N_3–N_{10} difference determines the number density of NA particles with diameters ranging between 3 and 10 nm. The low inertia of the instruments (of the order of 1 Hz) gives the possibility to monitor rapid changes in aerosol properties as well as to distinguish between NA particles, the Aitken mode, and accumulation mode.

Hämeri *et al.* [49] discussed the results of observations using four counters to measure N_3 located at a distance of 100 m from each other in the form of a triangle (two counters at 3 m and two others at 10 m and 20 m). The counters at 10 m and 20 m provided the measurement thresholds 5 nm and 10 nm. Analysis of the results obtained during the formation of new particles has shown that the particle number density reached a maximum of $1.8 \times 10^5\ cm^{-3}$. The field calibration of the instruments suggested the conclusion that with the number density above $10^5\ cm^{-3}$ there is a 5–6-fold systematic underestimation of the concentration measured with CPC counters compared with the real concentration. A comparison with additional data of the counter on the total number density with those obtained with the super-sensitive differential analyser of particle mobility (DMPS) confirmed that during the 'flashes' of new particle formation their number density could reach $1.2 \times 10^6\ cm^{-3}$. The rate of N_3 particle formation was estimated at $>10^5\ cm^{-3}\ s^{-1}$. Analysis of the spatial inhomogeneity of the concentration distribution revealed numerous microplumes of an increased concentration on spatial scales of the order of 10–100 m. Far from the NA coastal source these microplumes merged into one coastal plume of nucleation aerosols.

The direct impact of atmospheric aerosols on radiation transfer in the atmosphere and indirect impact manifested through changes in the microphysical and radiative properties of clouds as a result of their mixing with aerosols are important climate-forming factors. In the context of the problem of aerosols' indirect impact on radiative forcing, of critical importance is an adequate understanding of the processes of aerosol–cloud interactions, which are connected with coagulation, condensation, chemical reactions in the water phase, and sedimentation of both aerosols and clouds. In this regard, of special interest are studies of the formation of new aerosol particles in the pure atmosphere. In this context, [109] discussed the results obtained within the field programme PARFORCE carried out from September, 1998–June, 1999 to observe the formation of the nucleation mode aerosol in the atmosphere on the coast of Mace Head (Ireland).

Nucleation events took place almost every day in the presence of a low tide and insolation. According to the data obtained in September, 1998, the average number density of nucleation mode particles constituted $8{,}600\ cm^{-3}$ ($2{,}200\ cm^{-3}$) in a pure (polluted) atmosphere, whereas in June, 1999 the respective values reached $27{,}000\ cm^{-3}$ ($3{,}300\ cm^{-3}$). During the periods of the most intensive formation of new particles of nucleation mode of 2–8 hours (4.5 hour, on the average), the level of number density increased to 5×10^5–$1 \times 10^6\ cm^{-3}$. The power of the sources of detectable particles ($d > 3$ nm) varied within 10^4–$10^6\ cm^{-3}\ s^{-1}$, and the initial rate of the new particles' growth (in the tide zone) changed from 0.1 to $0.35\ nm\ s^{-1}$. Original particles, 8 nm in size, then suffered a hygroscopically induced growth with growth factor 1.0–1.1 at relative humidity $\approx 90\%$. Small values of growth factor testify to the presence of a condensed gas-phase component with a very low solubility. However, it is still unclear whether the existence of such a gas component is accompanied by a homogeneous nucleation.

The measured concentration of sulphuric acid vapour and ammonia indicates that the formation of thermodynamically stable sulphate clusters via triple nucleation is most probable. In the pure atmosphere, a considerable level of particle formation ($>10^5$ cm^{-3}) was observed with the concentration of gas-phase sulphuric acid at 2×10^6 molecules cm^{-3}, but in the case of the polluted atmosphere the concentration was 1.2×10^8 molecules cm^{-3}. The observed level of sulphuric acid vapour concentration was sufficient to form particles of nucleation mode, but too low for the formation of new particles of a measurable size. For the formation of the observable particles it is necessary to have an additional biogenic source of minor gas component (MGC) with a low solubility. Iodine oxide is the most probable MGC of this type. It results from the photodissociation of CH_2I_2. Further studies are needed to clarify whether the observed concentration of H_2SO_4 vapour is sufficient to obtain the required amount of stable clusters or whether the data of observations of the formation of particles at Mace Head can be interpreted only as a result of nucleation resulting, in turn, from the condensation growth of iodine oxide.

According to nucleation theory, the gas-phase sulphuric acid is a direct precursor and the main source of submicron particles of atmospheric sulphate aerosols. Such particles with diameters 0.1–1 μm contribute substantially to the attenuation of solar radiation reaching the Earth's surface both directly and indirectly through the aerosol-induced increase of cloud albedo. In connection with the numerous unsolved problems connected with the *in situ* formation of new aerosol particles through the gas-phase transformation of H_2SO_4 and insufficient reliability of the respective estimates of changes in the aerosol-induced radiative forcing, [14] undertook measurements of the concentration of gas-phase sulphuric acid, methylsulphide acid (MSA), and hydroxyl radicals (OH) in the marine ABL (MABL) using the chemical ionization mass spectrometer (CIMS).

Observations were carried out within the field observational experiment PARFORCE accomplished off the coast of Mace Head (Ireland) in July, 1999. Average measured values of the concentration of the enumerated components in the marine background atmosphere constituted 1.5, 1.2, and 0.12×10^6 cm^{-3}, respectively. The presence in the MABL of H_2SO_4 was also recorded at night, which reflected the presence of non-photochemical sources of this acid. There was observed a high correlation between concentrations of OH and H_2SO_4 in the pure marine air, which can be explained by a rapid local formation of H_2SO_4 from its gas precursors. The results of calculations of the stationary level of H_2SO_4 concentration agree with the data of observations, on the assumption that either the adhesion coefficient for H_2SO_4 molecules is low (0.02–0.03), or there is an additional source of H_2SO_4 apart from the reaction of $SO_2 + OH$. As a rule, changes in the concentration of H_2SO_4 in the MABL correlate neither with tidal cycles nor with the concentration of ultrafine particles (UFP). However, on some days there was an anticorrelation between the levels of concentrations of H_2SO_4 and UFP, which reflected the contribution of H_2SO_4 to the formation of new aerosol particles and/or to the increase of their size. In the case of gas-phase MSA, there was a negative correlation with the dew point temperature, which can be interpreted as evidence of a high sensitivity of the equilibrium distribution of this compound between the gas phase and aerosol.

Analysis of the observational results under discussion brings forth serious doubts with respect to the possibility of using MSA as a conservative tracer to estimate the relative contribution of H_2SO_4 formation to DMS oxidation. As a rule, the MSA/H_2SO_4 concentration ratio varied in the marine air near the surface within 0.06–1.0. Consideration of the measured vertical profiles of OH concentration revealed considerable deviations from simultaneous variations of the ozone photolysis frequency. Also, the observational data revealed lower values of OH (by an order of magnitude) concentration compared with those obtained with the use of a simple photochemical box model. Such differences were most considerable in the periods of new particle formation taking place on sunny days near noon and at low tide. The results obtained suggest the conclusion that both the oxidizing capability and the process of new particle formation in the coastal MABL depend much on the course of chemical reactions with the participation of compounds unknown, so far.

The development of new types of highly sensitive and low-inertia instruments to study the properties of nucleation aerosols has opened up new possibilities to monitor the detailed dynamics of the properties of such aerosols. Väkelä *et al.* [134] discussed the results of measurements of NA characteristics carried out with the use of the ultrasensitive mobility analyser UF-TDMA and cloud condensation nucleii (CCN) counter within the PARFORCE programme. The results obtained include information on the chemical composition and mixing of aerosol particles. The UF-TDMA data were used to assess the hygroscopic growth of particles with 'dry' diameter within 8–20 nm, at relative humidity = 90%. Data of the CCN counter were the source of information about NA particle activation in the oversaturated atmosphere (with the 'dry' diameter from 15–150 nm)).

Analysis of the observational results has led to the conclusion that during episodes of the formation of new NA particles in pure marine air masses, small values of the growth factor (1.0–1.1) corresponded to newly formed NA particles (8–10 nm) determined by a low solubility of particles. On this basis, the conclusion can be drawn that nucleation mode particles consist of insoluble or weakly soluble components with the probable presence of a small amount of soluble components. However, the results of NA particle observations beyond the episodes of new particle formation have led to much higher values of growth factor (1.3–1.4). From observations of the aerosol size distribution, particles of size 20 nm belonged to the Aitken mode and exhibited hygroscopic properties similar to those of some salt particles (with growth factors of 1.4–1.5). For all particles measured with the UF-TDMA instruments (8, 10, and 20 nm) the growth factor in the pure atmosphere was close to that for $(NH_4)_2SO_4$. According to the results obtained, there was no connection between the Aitken mode particles and NA reflecting the different origins of these particles.

For several reasons, an estimation of the rate of hygroscopic growth of particles due to water vapour condensation is important. In particular, from the viewpoint of the impact of CCN on the formation of cloud droplets, the presence of an organic film on the surface of aerosol particles can be a strong limitation for the condensation growth. In this regard [26] undertook an overview of the developments dedicated to the studies of the impact of films on the water transfer, and carried out laboratory measurements on timescales of the condensation growth (τ_g) of atmospheric aerosol particles. Measurements were made in Mexico during a wet season (25–27 September, 2000) and dry season (16–17 November, 2000) for particles with diameters 50 and 100 nm. The results show that for most of the studied particles $\tau_g < 2$–3 s. However, greater values of τ_g correspond to a small amount of particles (0–2%). These few particles are characterized by the values of the accommodation coefficient within $(1–4) \times 10^{-5}$ which are at a minimum for aerosols studied in the laboratory. An analysis of observations revealed the presence of a strong dependence of the occurrence of such particles on season and daytime. It is supposed that such particles can be observed more frequently, which is important from the viewpoint of assessing the indirect climatic impact of aerosols.

Great attention has been recently drawn to studies of the processes of nucleation (i.e., the formation of new aerosol nanoparticles >3 nm, and their subsequent growth to 100 nm during 1–2 days in the ABL). As has been mentioned, most probable nucleation mechanisms are triple nucleation with the participation of H_2O, NH_3, and H_2SO_4 as well as ion-induced nucleation. Nucleation takes place in the atmosphere almost everywhere (at least, in the daylight) and leads to the formation of the reservoir of thermodynamically stable clusters of molecules (TSCs) with diameters >3 nm, which, under certain conditions, grow in size to a detection threshold. However, still it is unclear, which MGC is responsible for the TSC growth. A high correlation had been earlier observed between the intensity of short-wave solar radiation in the wavelength interval 300–340 nm and the number density of the nucleation mode particles 3–10 nm in size. Radiation in the interval of wavelengths shorter than 330 nm leads to the formation of hydroxyl OH through an appearance of excited oxygen atoms with the subsequent formation of condensed vapour. Therefore, it is necessary to take into account reactions of monoterpenes with O_3, OH, and NO_3 as mechanisms of particle growth.

Boy *et al.* [19] performed an analysis of the observational data obtained at the Hyytiälä station (Finland) in 2000–2001 within the SMEAR II programme to study the atmosphere–forest ecosystems interaction. These studies included two sections: (1) studies of the trends and correlations between the number density of particles with $d_p = 3$–6 nm and various parameters; and (2) calculations of various characteristics of aerosols. The developments have resulted, in particular, in detection of a correlation between solar irradiance and the level of condensation sink (in the diurnal course of these quantities) with the number density of the nucleation mode nanoparticles. There is also a correlation between mean diurnal (0.9–15 local time) values of these parameters (from the data for two years 2000–2001) and the concentration of small particles. In the case of particle flow, variability of wind speed with altitude and ozone concentration and correlations with the particle concentration manifest themselves weakly on certain days, but these parameters are characterized by a special annual course similar to solar radiation and condensation sink. Calculations of the aerosol characteristics variability with the prescribed observed values of concentration of monoterpenes and sulphur dioxide as well as comparisons with the data of observations of the aerosol number density has led to the conclusion that a consideration of the products of monoterpenes oxidation (carboxyl diacids, including, in particular, pinic acid) can explain 8–50% of the observed growth of aerosol particles. Estimates of the number density of particles with $d_p = 1$–3 nm have led to the conclusion that the real concentration of such particles can be 2–3 times greater than the observed concentration of particles with $d_p = 3$–6 nm.

The complexity of the gas-to-particle conversion processes of aerosol formation and the imperfection of measurement techniques are the main causes of the inadequacy of the existing ideas about these processes. For instance, there is still no sufficiently convincing theory, and the required information to explain the process of a multicomponent nucleation is absent. As for measurements, the problem is that products of nucleation are particles with diameters of the order of 1 nm, whereas the present instruments can detect only particles about 3 nm in size. Therefore, the events of nucleation can be recorded only some time after the formation of particles. The existence of a correlation between the concentration of sulphuric acid vapour and nanoparticles with diameters 3–4 nm has made it possible to assume a binary homogeneous nucleation of H_2SO_4–H_2O as a possible mechanism of the formation of aerosol particles. However, the observed formation of particles could not be simulated based on the classical theory of binary nucleation, which has prompted one to propose the triple nucleation hypothesis. Laboratory experiments have shown that an addition of NH_3 to the H_2SO_4–H_2O system determines an acceleration of nucleation.

As one of the versions, [138] proposed a mechanism of nucleation regulated by ions. Of great importance were studies of the processes of nucleation in the urban environment. One of the important problems of the field observational experiment TRACE-P, accomplished in the spring of 2001, consisted in studies of the formation of new particles in the region of pollutant plumes propagating from the territory of the Asian continent to the north-western sector of the Pacific Ocean, using the data of aircraft measurements. Observations in the latitudinal belt 20–45°N revealed a high level of concentration of particles with $d = 3$–4 nm in the most polluted part of the plume, both within it and near the upper boundary. Analysis of the observational results obtained during two flights has led to the conclusion that the number density of particles with $d = 3$–4 nm correlated, as a rule, with the concentration of sulphuric acid vapour. The H_2SO_4 concentration and specific surface area of particles in the region of their formation constituted, respectively, $6 \times 10^7\,cm^{-3}$ and $750\,\mu m^2\,cm^{-3}$. In contrast to anthropogenic plumes in the pure (background) atmosphere, a low number density of particles with $d = 3$–4 nm was observed. A similar situation was observed even in the plumes of volcanic emissions with the highest concentration of H_2SO_4 ($>10^8\,cm^{-3}$). Apparently, the main factor of nucleation was an increased (within 2–7 ppb) concentration of SO_2 and, probably, the presence of other gas components.

The formation of new aerosol particles in the atmosphere attracts serious attention in the present-day atmospheric sciences. The respective processes cause an increase of the number density of submicron aerosol particles, which is of great importance from the viewpoint of their impact on climate and human health. In this context, [70] discussed the results of continuous observations of aerosol characteristics during the period June, 2000–August, 2002 in the National Park

Pallas-Ounastunturi in the north of Finland, where there are no important sources of atmospheric pollution. Observations were made at two points located at different heights above sea level (340 and 560 m), one of them being in the forest and the other (upper) on a hilltop above the forest. The aerosol size distribution was measured within the particle size range 7–500 nm. In 2001, the aerosol number density averaged 700 cm^{-3} (the upper point) and 870 cm^{-3} (the lower point). The average diurnal values varied within 40–3,500 cm^{-3}.

At both points a similar annual course of the total number density of aerosols with a maximum in spring and summer and a minimum in winter was observed. The diurnal variation of total concentration was also similar for both points. Analysis of data on the concentration for individual modes of aerosol has led to the conclusion that during the whole period of observations there were 65 episodes of formation of new particles. Most often these events happened in April and May. The formation of new particles began at 08:25–15:50 (Universal Time Constant: UTC + 2 hours), and the estimated time of the nucleation process (the formation of aerosol particles 1 nm in size) varied between 04:50–14:20. The rate of the growth of aerosol particles was estimated at 1.4–8.2 nm hr^{-1}, and the rate of the formation of particles 7 nm in size varied within 0.06–0.40 particles cm^{-3} s^{-1}. The difference in the times of the beginning of new particle formation at these two points did not exceed 30 minutes and could not be explained by the impact of wind speed and direction – the turbulence-determined vertical motions of air masses are one of the possible causes. As for the prevailing wind direction, on the days of new particle formation it varied from westerly to northerly, testifying to the polar or Arctic origin of the air masses.

A considerable increase of different types of tropospheric aerosols in the atmosphere, including, first of all, submicron aerosol particles consisting of sulphates, nitrates, and secondary organic compounds affect markedly the formation of climate and acid rains, air quality, and human health. During recent years, serious attention has been paid to the processes of the formation of sulphate aerosol, especially its submicron fraction. As has been mentioned, the mechanism of binary nucleation of sulphuric acid and water molecules is most often supposed to be the main one. However, the field observational data show that the observed rate of nucleation often exceeds that calculated on the basis of such a hypothesis as well as with data of laboratory measurements taken into account. The cause of such differences can be the participation in the nucleation processes of a third component, most probably, ammonia.

To check this hypothesis, numerical modelling of the nucleation processes was carried out with the use of a new scheme of parameterization of the binary nucleation of water and sulphuric acid and triple nucleation with the participation of these two components and ammonia. A comparison with the results obtained with the earlier technique (a parameterization together with the data on aerosol dynamics) has led to the conclusion that new data on the total number density of aerosols reflect a difference between new and old data on the rate of nucleation reaching 1–2 orders of magnitude. A comparison of the scheme of parameterization of binary and triple nucleation has shown that at a temperature above 240 K the contribution of triple nucleation to the formation of new particles exceeds that of binary nucleation by ten orders of magnitude (i.e., in most cases, the process of triple nucleation turns out to be prevalent in the lower troposphere).

Korhonen *et al.* [82, 83] performed an analysis of the functioning of the aerosol dynamics models, with various sectoral size distributions taken into account, compared with the model that describes the processes on a molecular level. This analysis has led to the conclusion that from the viewpoint of the results of calculations of the total aerosol number density, the results of numerical modelling based on the consideration of the size distribution are similar to 'molecular' calculations with a more detailed description of the size distribution. In the presence of a strong condensation growth the problem arises of 'numerical diffusion'. As a rule, the 'microphysical' ('sectoral') approach was inadequate without a sufficiently detailed description of the size distribution. Very good results have been obtained with the use of the model of the multimodal monodisperse aerosol (at least, in calculations of total number density), if the background aerosol modes provided a reliable simulation of condensation sinks for the respective log-normal modes. Based on the obtained results, one can assume that the multimodal monodisperse model can serve as the basis for parameterizing the aerosol properties in large-scale numerical models of the atmosphere.

It should be noted that despite the intensive studies of the initial processes of the formation of atmospheric aerosols accomplished during the last decades, the mechanism of the formation of ultrafine aerosol particles manifesting itself everywhere, remains poorly studied. As has been mentioned, a supposition with respect to the major role of H_2SO_4 and H_2O in the binary homogeneous nucleation (BHN) is not always adequate when compared with reality. This has stimulated the consideration of the mechanism of triple homogeneous nucleation (THN) in the H_2SO_4–H_2O–NH_3 system and the ion modulated nucleation (IMN) in the H_2SO_4–H_2O–ions system. According to the THN theory, a strong increase in the rate of nucleation is connected with the ammonia-determined stabilization of aerosol nuclei due to their decreasing size. It follows from the IMN theory that around ions residing in the atmosphere charged molecular clusters form which are much more stable and able to grow in size much faster than neutral clusters – thus reaching a stable observable size. One of the main differences between the IMN theory and the classical theory of nucleation is that with a given concentration of $[H_2SO_4]$ the IMN theory makes it possible to simulate the kinetics of the clusters' evolution beginning with the stage of monomers, whereas the classical theory supposes the presence of an instant pseudo-stable concentration of aerosol nuclei (this supposition raises doubts). Though both these theories (THN and IMN) are based on the consideration of new components in the process of nucleation (NH_3 and ions), H_2SO_4 plays, as always, the key role.

Based on the use of the model of the processes' kinetics, [140] estimated the time t_0 needed to reach a pseudo-stable state of nucleation, when the concentration of H_2SO_4 vapour suddenly increased to a certain fixed level. It was shown that t_0 is approximately proportional to $[H_2SO_4]$. A supposition about an instant provision of the pseudo-stable concentration of clusters used in the classical theory cannot be used if t_0 exceeds the time interval needed to reach the stable state of $[H_2SO_4]$. In the lower layers of the marine atmosphere the results of calculations of the rate of nucleation on the basis of IMN theory are very sensitive to $[H_2SO_4]$, if the H_2SO_4 concentration is small (<7–$10^6\,cm^{-3}$), but such a sensitivity decreases substantially with the growth of $[H_2SO_4]$. This conclusion radically differs from the results corresponding to the classical theory and THN.

As one of the possible mechanisms of the formation of particles in the coastal regions, was proposed the photolysis of diodomethane (CH_2I_2) in the presence of O_3. In this connection, [65] discussed the results of laboratory chamber studies which revealed the presence of a fast homogeneous nucleation with the concentration of CH_2I_2 varying within three orders of magnitude up to the level of 15 ppt ($\sim 4 \times 10^8$ molecules cm^{-3}). Such a concentration corresponds to the observed concentration of gas-phase iodine compounds in the coastal regions.

After the initial 'flash' of nucleation in the chamber, the process began determined by prevailing condensation of an additional amount of vapour on the existing particles and coagulation of particles. Particles formed in the 'dry' air were fractal agglomerates with fractal mass size $D_f \sim 1.8$–2.5. A higher relative humidity (65%) does not change the regime of nucleation or growth of particles compared with dry air conditions but leads to the formation of more compact and dense particles ($D_f \sim 2.7$).

Bearing in mind the known gas-phase chemical reactions, OIO can be considered as the most probable gas-phase component responsible for the observed process of nucleation and growth of particles. The present ideas about chemical reactions with the participation of OIO are, however, not quite adequate. The mass spectrometric measurements of the aerosol chemical composition revealed that particles consist mainly of iodine oxides but contain also water and/or iodine hydroxy-acids. The chamber studies of the processes enabled one to quite realistically simulate the nucleation events observed in the coastal environment. A comparison of the processes taking place in the chamber and under real conditions permits one to assume that the main mechanism of the formation of new particles on the western coast of Ireland is photo-oxidation of CH_2I_2 with the possible participation of other organic iodine compounds.

The marine salt aerosol, which results from the sea roughness spray evaporation and from air bubbles bursting, plays an important role in the chemical processes taking place in the atmosphere, and as a climate-forming factor. Suffering further long-range transport, the marine salt aerosol reaches land where it can interact with the polluting components of the urban atmosphere. There is also an inverse process of the input of pollutants from coastal cities. Therefore, the problems of interaction between MSA and pollutants of the continental atmosphere attract serious attention. In this connection, [21] described the present models of the formation of secondary organic aerosols (SOA) through an absorption of semivolatile oxidation compounds (SVOC) by a mixture of organic aerosols and/or their solution in water aerosols.

In coastal regions, the newly formed sea salt aerosols absorb the products of oxidation of SVOCs, which leads to the formation of the surface organic layer. The numerical modelling of such a process has demonstrated the formation of the equilibrium between the gas phase, continental primary organic aerosols, an organic component of the sea salt aerosols, and the continental water aerosols, as well as an equilibrium between an organic component of marine salt aerosol and the nucleus of its particles of the sea salt brine. A dissociation of acid components in the water phase and the processes determining the pH level have been considered in [21]. The molecular composition is assumed:

(1) of six semivolatile products of the oxidation of organic compounds – components of secondary organic aerosols;
(2) of eight components of primary organic aerosols that characterize the composition of the urban aerosols;
(3) of five compounds of the organic surface layer of sea salt aerosol particles; and
(4) of four organic components of marine brine constituting the nuclei of marine salt aerosol particles.

An iterative method was used to solve the system of nonlinear equations to calculate the concentration of each of the indicated components in the aerosol and gas phases as well as pH values for the respective liquid phases.

The results obtained have shown that liquid aerosols and nuclei of the brine contained in marine salt aerosol are the most favourable medium to absorb the gas-phase compounds, the presence of surface organic layers favours the formation of the brine nuclei, and secondary organic acids contribute much to the oxidation of the sea salt brine nuclei contained in sea salt aerosol particles.

According to a number of studies mentioned above, nucleation of atmospheric aerosol particles in the triple system that is 'sulphuric acid vapour–ammonium–water vapour' is an effective mechanism for the formation of new particles in a forest, in the remote (background) troposphere, as well as in combustion engine emissions. The earlier developments have shown that though the concentration of the gas-phase sulphuric acid can be sufficient for the formation of new particles, it is not enough to ensure the growth of particles to an observable size or for the rapid growth of nuclei with their subsequent 'washing out' from the atmosphere before they reach an observable size (≤ 3 nm). Therefore, the atmospheric chemistry components should exist (apart from sulphuric acid) which favour the growth of nuclei to an observable size. Zhang and Wexler [141] noted that organic compounds' vapour can be such a component.

In this case, heterogeneous chemical reactions are the main process of particle growth, and the functioning of this process depends on three characteristic time scales:

(1) τ_{heter} – for heterogeneous reactions between sulphuric acid and organic compounds;
(2) τ_{coag} – for coagulation of new nuclei; and
(3) $\tau_{alkanes}$ – for alkanes, considering their role as a main sink in the atmosphere through reactions with the participation of the hydroxyl radical and surface ozone.

For an efficient growth of particles it is necessary that $\tau_{heter} < \tau_{coag}$ and τ_{sink} (characteristic time for alkanes sink). Otherwise, the concentration of alkanes or new nuclei will decrease before heterogeneous reactions can contribute to the growth of the nuclei size.

Though the most probable mechanism of the formation in the atmosphere (sometimes in the form of 'explosive' events) of nucleation aerosols 3 nm in size is the triple nucleation, the simultaneous functioning of several mechanisms is not excluded. Laakso *et al.* [87, 88] undertook a theoretical analysis of the possible role of the ion-induced nucleation and dynamics of the charged aerosol particles. The results suggested the conclusion that such a nucleation can cause the formation of a considerable amount of particles, if the number density of the earlier particles is sufficiently small. One more condition is that only positive or negative ions should determine the nucleation of a great number of particles of observable size. It has been shown that at a low temperature and high concentration of the earlier particles a continuous nucleation is possible of particles whose size cannot be detected with the existing instruments. In some cases the numerical modelling results agree well with the data of observations in the boreal forests. According to the obtained results, under certain conditions, the ion-induced nucleation determines changes in the distribution of the particles' electrical charge, which determines the possibility to observe the ion-induced nucleation in the real atmosphere.

An explanation of the mechanism of the 'explosive' nucleation of aerosols, with the help of the homogeneous triple nucleation with the participation of water vapour, sulphuric acid, and ammonia, leads to considerably higher calculated values of the rate of nucleation compared with those observed. In this connection, it was mentioned that such a mechanism can provide the formation and the subsequent growth of the TSCs to measurable sizes only in the presence of the 'explosive' increase of concentration of some organic vapour, which, however, does not happen. On the other hand, if we assume as the mechanism the ion-induced (or ion-governed) nucleation (i.e., nucleation takes place only on ions, and a strong acceleration of the particle growth is determined by the electrostatic interaction between the charged particles and condensed molecules of the dipole electrostatic nature), then this gives quite realistic results.

In this context, [88] analysed the impact of an enhanced condensation due to the Coulomb forces with the main objective of estimating the upper level of such an impact and its importance for the dynamics of the growth of nucleation mode particles. The factor of concentration enhancement constitutes, apparently, <5 in the case of the growth of charged particles with radii ranging between 1.5 and 10 nm. However, only a limited part of the particle population becomes charged, therefore, the total effect is characterized by coefficients <2. Taking into account this fact, and that the water vapour concentration constitutes about 1×10^7 molecules cm^{-3}, the rate of condensation growth is about 1 nm hr^{-1}. The effect of condensation enhancement on the observed total aerosol concentration, calculated with the use of the aerosol dynamics model under conditions of neutral nucleation, is negligible at low values of the enhancement efficiency (EE). In the case of great EE values the final result of the effect is a decreased concentration of aerosols. In the case of the ion-induced nucleation the character of the impact depends on the initial concentration of particles and the EE. In these cases the total aerosol concentration can increase by a factor of 3.

Regular episodes of 'flashes' of NA formation, 3–5 nm in size in the ABL, and the subsequent growth of particles to 100 nm during 1–2 days, were observed in a wide range of global regions, from subtropical Lapland to industrialized agricultural regions of Germany. Dal Maso *et al.* [31] undertook a numerical modelling of a process of formation and growth of new NA particles, taking into account the gas-phase compounds condensation and sinks due to the coagulation of particles. Observation data was used for the particles 3 nm in size.

Estimates of the condensation and coagulation-induced sinks obtained from the observational data on the atmospheric aerosol size distribution and its changes are aimed at simulating the formation and rate of particle growth during the periods of nucleation events in the ABL of boreal forests and coasts. Calculations have shown that typical values of the condensation sink constitute $(4–7) \times 10^{-3}\,s^{-1}$ in the forests and $2 \times 10^{-3}\,s^{-1}$ along the coastlines, whereas the nucleation sinks for particles with diameters of 1, 2, and 3 nm are as a rule less, by factors of 1.5–2, 5–7, and 11–15, respectively. The observed rate of particle growth is $2–10\,nm\,hr^{-1}$ (boreal forests) and $15–180\,nm\,hr^{-1}$ (coastlines) at concentrations of gas-phase components $2–13 \times 10^7\,cm^{-3}$ and $10^8\,cm^{-3}$. The calculated rate of gas component emissions reached $1–2 \times 10^5\,cm^{-3}\,s^{-1}$ (forests) and $2–5 \times 10^6\,cm^{-3}\,s^{-1}$ (coastlines), and the rate of formation of particles with diameters of 1 nm was $8–20\,cm^{-3}\,s^{-1}$ (forests) and $300–10{,}000\,cm^{-3}\,s^{-1}$ (coastline).

Under coastal environmental conditions, the large size aerosol substantially affects the level of condensation growth, but in the ABL over the forests only the effect of the Aitken and accumulation mode particles are important. The calculated values of the rate of nucleation and the number density of 1-nm particles turned out to be much higher in the coastal ABL than over the forest. The high rate of particle growth in the coastal air reflects the existence of a higher concentration of condensed gas-phase components under these conditions compared with the forest as well as more intensive emissions of gas-phase components to the atmosphere.

Though there are hundreds of organic compounds contained in atmospheric aerosols, the total share of their mass in the urban and rural aerosols constitutes less than 10%. The main components of organic continental aerosols are 'humus-like' substances (HULIS) whose contribution to the formation of water soluble organic aerosols under urban and rural conditions varies within 20–50%. This means that HULIS substantially affect the hygroscopic properties of aerosols, the formation of CCN, and, hence, the climate. The IR spectroscopic fingerprints of such aerosols indicate that the origin of the respective macromolecular components is connected with the processes of 'agricultural burning'.

However, as [95] noted, the high concentration of HULIS observed at rural locations in Hungary in aerosol samples with a low content of 'black carbon' reflects the fact that the origin of HULIS in this case could not be explained by biomass burning. In this connection, of importance is the fact that emissions of isoprene and terpenes are the most substantial source of chemically active organic compounds on land. While photo-oxidation of isoprene does not contribute much to the aerosol formation, the situation with terpenes is quite different. This makes them one possible important source of secondary aerosol formation. The secondary organic aerosol formation can be seriously affected by an appearance, in the atmosphere, of polymers due to heterogeneous reactions with the participation of emissions of isoprenoid or terpenoid in the presence of H_2SO_4 aerosol, which serves as a catalyst. Though such a competitive oxidant as ozone, or the impact of humidity, can lower the level of HULIS formation, this level remains high. The quantitative estimates have been obtained [95] confirming the importance of this mechanism of HULIS formation in the continental aerosol and of the aerosol particle growth.

At high levels of concentrations of sulphur dioxide and hydroxyl radicals (of which sulphuric acid vapour is formed) and high concentrations of water vapour and low temperatures, the formation of new particles can take place through the homogeneous bimolecular nucleation in the process of gas-phase reactions with the participation of the sulphuric acid and water vapour. New particles are supposed to form most often, in clear air, when sulphuric acid vapour gets easily condensed on existing particles. Though there are no conditions for homogeneous nucleation in the atmosphere, the observational data indicate that new particles form more often than predicted by the homogeneous nucleation model with the participation of only sulphuric acid vapour and water. This means that in the process of new particle nucleation, other components may be important, including ammonium and organic compounds' vapour.

In this connection, [133] studied the process of new particles formation in the air 'flowing' from the mesoscale convective system (MCS), observed in the mid-western USA. Aircraft measurements were made of condensation nuclei (CN), aerosol particles' surface area, as well as the concentration of water vapour and other MGCs. Analysis of measurement results revealed an increased concentration of CN above the clouds and in the direction leading from the cirrus cloud anvil with maximum number densities up to 45,000 cm^{-3}. According to data on volatility and results from an electron microscopic analysis, most of the particles constituted, apparently, small sulphate particles. This increase of the aerosol particles concentration took place in a region not less than 400 km in size.

It follows from the results of the multiparametric statistical analysis that the high CN concentration was connected with the effect of MCS elements and the input of background aerosols. Due to convection, the gas-phase aerosol precursors (as the product of gas-phase reactions) moved from the surface level to the zone of the stormy air masses where a very low air temperature favoured the nucleation process. The results of an approximate estimation have demonstrated that the penetrating convection systems contribute substantially (up to 20%) to the formation of background aerosols in the upper mid-latitude troposphere.

4.2 BACKGROUND MARINE AEROSOL

Emissions to the atmosphere of sea salts by the World Ocean contribute substantially to the formation of global aerosols. According to present estimates, such global-scale emissions exceed all other natural and anthropogenic emissions taken together, reaching 3,300 $Tg\,yr^{-1}$ in 2000 – determining the substantial role of sea salt aerosols as an optically active atmospheric component affecting climate formation. The approximate calculations have led to the conclusion that by the year 2100 the global emissions of sea salts may reach 5,880 $Tg\,yr^{-1}$ and contribute to global mean radiative forcing $-0.8\,W\,m^{-2}$.

The atmosphere in the remote regions of the World Ocean is a chemically dynamic system in which the sea salt aerosol forms (as a result of destruction of water droplets getting to the atmosphere due to sea roughness) and non-sea salt substances (NSS) as the product of gas-to-particle transformation of biogenic DMS emitted from the ocean to the atmosphere. To check a number of hypotheses with respect to the sources and properties of marine background aerosol, [58] analysed the results of aircraft measurements of NSS concentration, anions and cations of methanosulphonate (MSN), ammonium, potassium, nitrate, and other gaseous components, carried out during the period of the field observational experiment ACE-1 and at the surface station at Cape Grim.

The observational data processing revealed a high MSN concentration gradient in the free troposphere of the springtime southern hemisphere, whereas in the case of NSS this gradient was absent. Most of the studied components were characterized by high vertical concentration gradients with higher values of the concentration near biogenic sources in the ABL compared with the free atmosphere. The data of 'Lagrange' experiments and observations at Cape Grim revealed the photochemically induced formation of MSN in the daytime with the constant or decreasing concentration of MSN at night. In the daytime, an increase of NSS concentration was also observed. The MSN/NSS mixing ratio is characterized by a high latitudinal gradient with the values at the southernmost point close to those measured in the snow cover of the Antarctic plateau.

Though the formation of the finest aerosol particles (nucleation) in the non-polluted marine boundary layer is very important (especially from the viewpoint of the impact of such particles on cloud properties), the observation data on the nucleation process remain fragmentary. In this context, [107, 108] undertook observations of several nucleation episodes (the formation of ultrafine aerosol particles) at the coastal station Mace Head (Ireland) in 1996 with the use of two CN counters recording the number density of particles with the range of radii >1.5 nm and >5 nm. The difference between the counters' readings enabled one to estimate the concentration of particles within the range of radii 1.5–5 nm. Simultaneous measurements were made of the concentration of soot carbon. The results demonstrated the presence (in two cases) of clear air with soot carbon concentrations $<20\,ng\,m^{-3}$ and radon concentrations (to estimate the possible outflow of the continental aerosol) at insignificant levels less than 0.4 or 0.8 $pCi\,m^{-3}$ (typical of marine air masses in the North Atlantic Ocean).

Nucleation events happened often and were characterized by spatial–temporal scales ranging between tens and hundreds of metres and from seconds to minutes under two conditions: (1) a low tide, and (2) sufficiently intensive solar radiation. The first of these conditions determines the presence of an exposed sea bottom band and a source of gas-phase CN precursors. These nucleation precursors are supposed to have been volatile organic compounds and/or alkyl halide derivatives. As for solar radiation, its stimulation of the photochemical reactions of the formation of new particles plays an important role (observations did not reveal any nucleation process at night). The rate of nucleation was estimated at 10^3–$10^4\,cm^{-3}\,s^{-1}$, which means that the coastal zone is a substantial source of atmospheric condensation nuclei.

One of the remaining unclear aspects of the sea salt aerosol studies is the formation on the sea salt aerosol particles' surface of organic films (apparently, consisting mainly of biogenic fatty acids), whose presence can substantially change the chemical, physical, and optical properties of sea salt aerosol.

Tervahattu *et al.* [132] performed mass spectrometric (using the TOF-SIMS) studies of sea salt aerosol particles whose results indicated that palmitic acid is the dominating component of organic films, while other fatty acids play a negligible role. Analysis of images obtained in the laboratory by the TOF-SIMS technique has shown that the surface cover of palmitic acid is typical of small particles similar to sea salt aerosol. Thus, these results confirm the presence of biogenic fatty acids as an important ingredient of the surface organic film covering the sea salt aerosol particles. Such aerosols seriously affect the chemical composition of the marine air layer.

The gas-phase hydrogen chloride (HCl) in the marine air plays an important role in chemical reactions taking place in the MABL which was confirmed by results of direct measurements of concentration in the MABL of such components as molecular chlorine and hypochlorous acid (Cl_2, HOCl). Such components are photolabile and in the presence of insolation determine the formation of free chlorine (Cl) which rapidly reacts with any available volatile organic compound (VOC). Marine aerosol particles can be also a source of chemically active chlorine. As laboratory studies have shown, data on the uptake of gas-phase hypoiodine acid (HOI) on the surface of NaCl and NaBr particles and sea salts can be used to assess the expected release of chlorine atoms by marine aerosol particles appearing in the MABL due to reactions with the participation of iodine compounds.

McFiggans *et al.* [100] substantiated a theoretical model of the respective processes taking into account reactions, in which chemically active chlorine takes part, and two mechanisms of activation. As limiters of the numerical modelling results, the results were used from field measurements of the concentration of iodine oxide (IO) radicals and nitrate (NO_3) to analyse the importance of mechanisms of reactions regulated by iodine and nitrogen.

Calculations have shown that an assimilation of HOI and N_2O_5 in the MABL can result in a considerable release of Cl atoms by sea salt aerosol particles. The functioning of such an iodine-governed mechanism can be proved by 'flashes' of Cl formation with a concentration of chlorine atoms in the order of several thousand per cm^3, occurring after sunrise. The numerical modelling predicts a stable formation of Cl in the daytime with a concentration from hundreds to thousands per cm^3 over vast ocean basins. Under certain conditions, chemical reactions with the participation of bromine emissions by fresh sea salts can play the leading role in the formation of free chlorine.

In the course of studies of aerosol radiative effects in the MABL, a hot discussion took place with respect to contributions of sea salt and NSS components of sulphate aerosol (SA). Under MABL conditions, volatile sulphate aerosol particles prevail (<0.08 μm in size and with a small number density), though the most significant part of the aerosol mass is concentrated in a much smaller number of sea salt particles > 1 μm. The aerosol-induced radiative forcing (RF) on climate is formed, however, due to submicron particles of intermediate sizes (0.08–1 μm), since they most intensively scatter solar radiation and serve as CCN. Therefore, an assumption has been made that sulphate aerosol resulting from emissions by the ocean of dimethylsulphide and subsequent gas-phase reactions, can substantially affect the climate through this radiative forcing.

Marine aerosol particles <1 μm in size are usually supposed to consist of sulphates of non-marine origin, but recent studies have led to the conclusion (from indirect data) that, at least partially, such particles of the submicron size range contain marine sulphates. In this connection, [103] discussed the results of direct measurements of the chemical composition of aerosol particles during the ACE-1 field experiment, from which it follows that almost all particles larger than 0.13 μm in the MABL contain sea salts. Such marine aerosols were mainly responsible for solar radiation scattering and contained considerable amounts of CCN. The aerosol particles were, as a rule, an internal mixture of sea salts and sulphates in which it was difficult to distinguish between the two. Submicron particles also contained a small amount of organic components.

Studies of the MABL have shown that atmospheric aerosols and especially particles which serve as CCN, not only affect clouds (changing their size distribution and optical properties) but also suffer the inverse impact of clouds, which leads to changes in the concentration and other characteristics of CCN. Such an impact can be manifested through various mechanisms [103].

First of all, there are data showing that directly beneath the upper boundary of clouds the process of nucleation takes place, which determines a high concentration of new sulphate particles, apparently, as the product of the process of homogeneous heteromolecular nucleation in the regions of high relative humidity. This process ensures the growth of the aerosol particles number density, which eventually can play the major role as CCN reach the required size.

The second probable mechanism is an increase of the weight of droplets due to various chemical reactions in the water phase. In the case of complete evaporation of such droplets, aerosol particles appear which can serve as CCN. Of importance is the fact that in this case the mass concentration of particles increases without changes in their number density and, thus, CCN become larger and get easily activated (as CCN) in the oversaturated environment.

One more possible mechanism of the impact of clouds ('processing' of aerosols by clouds) is connected with the processes of collision–coalescence of particles. All three mechanisms mentioned can manifest themselves differently in various clouds. So far, there have been no attempts to analyse their comparative significance. In this connection, with some concrete situations as an example, [38, 39] analysed the importance of mechanisms of collision–coalescence and chemical processes in water, with the major aim of studying the processes in the MABL with a 300-m layer of stratocumulus clouds located in its upper part. This choice of object to study was determined by several factors: an important contribution of the MABL to the formation of the global radiation budget; the supposed sensitivity of MABL cloud properties to CCN concentration; and as a rule, the low concentration of CCN in this situation, favouring an enhancement of the impact of processes in clouds on CCN.

The results obtained in [38, 39] are based on the use of the 2-D version of the regional vortex-resolving model of atmospheric processes developed at the University of Colorado and used earlier to study the dynamics of stratocumulus clouds. The numerical modelling of the effect of the collision–coalescence processes leading to a decrease of droplets' number density and, hence, CCN, was made without consideration of wet deposition, and the obtained results were compared with the data characterizing the role of chemical processes in water. Microphysical processes such as droplets' activation, condensation (evaporation), collisions, coalescence, sedimentation, and regeneration of particles taking place with the complete evaporation of droplets have been considered in detail.

Calculations have shown that the processes of collision–coalescence can cause a considerable decrease in the droplet number density (down to 22% hr^{-1}) leading to a marked growth of particle radii (by about 7% hr^{-1}). For a more detailed analysis of the processes of collision–coalescence (especially the time taken for droplet sedimentation within the clouds) and an assessment of the intensity of transformation of sulphur dioxide into sulphates taking place in clouds, the particle trajectory has been calculated, which suggested the conclusion that the impact of both studied mechanisms on CCN manifesting itself through the growth of average sizes of particles (under certain conditions) is comparable. Estimates for stratocumulus clouds, far from the sources of pollution, indicate that chemical reactions in water can, in these cases, affect more strongly at a lower water content of the lower level clouds, whereas the mechanism of collision–coalescence can dominate in the case of a high water content or in a wide range of droplet sizes.

Many earlier numerical models of aerosol formation were suggested to simulate the results of laboratory experiments, and at a later stage of their development they considered multicomponent aerosols with the processes of gas-phase transformation taken into account. Such results served as the basis for 'box' models of the processes of aerosol formation in the [40, 41] for the numerical modelling of homogeneous nucleation of new particles and their subsequent growth to the size of CCN in the MABL.

Since these models cannot simulate the altitudinal changes in the process of aerosol formation, a 1-D sectional model has been proposed [40] which provides a simulation of temporal and altitudinal changes in aerosol properties within the MABL. An important feature of this model consists in its capability to adequately describe the transport of aerosol in the atmosphere in the presence of a humidity gradient. However, a limiting factor is the supposition of an instant equalization of the water vapour content in droplets with respect to the ambient atmosphere, which artificially limits the maximum size of particles with radii $<30\,\mu m$.

The model under discussion takes into account all substantial processes known that determine the formation of the four-component aerosol size distribution (sulphates, sea salts, insoluble continental components, water) within the MABL: the sea salt aerosol formation near the ocean surface; new particle nucleation; coagulation; condensation growth of gas phase particles; growth of sulphate particles in the process of their 'processing' by clouds; washing out of aerosols by rain; sedimentation onto the surface; turbulent mixing; gravitational sedimentation; and an exchange with the free troposphere. Processes of gas-to-particle conversion are taken into account, which describe the formation of sulphate aerosol as a result of oxidation of DMS (DMS photo-oxidation to SO_2) and sulphur dioxide (SO_2 oxidation to H_2SO_4) with the subsequent formation of particles (with a prescribed diurnal course of hydroxyl concentration). The proposed model can be realized as an interactive component of the meteorological mesoscale 3-D model of the MABL.

Fitzgerald *et al.* [41] discussed the results of numerical modelling of the processes of the formation and transformation of aerosols in the MABL under conditions of the remote atmosphere over the southern hemisphere ocean (near Tasmania) with the use of the 1-D model of the dynamics of the multicomponent aerosols in the MABL. The major aim of the numerical modelling was to analyse the processes responsible for the formation of the aerosol size distribution via the successive consideration of various factors determining the size distribution. The emphasis has been laid on a particularly interesting process of aerosol 'processing' by clouds and the aerosol exchange between the MABL and the free troposphere.

Calculations have shown that the effect of transformation of aerosol properties by clouds manifesting itself through an appearance of a characteristic dual maximum of the particle size distribution in the upper part of the MABL, at the sea surface level, demonstrate characteristic scales of aerosol lifetimes, which are much smaller than those corresponding to the interaction with the troposphere (in this case the typical speed of exchange constitutes about $0.6\,cm\,s^{-1}$). According to calculated data, the free troposphere can be a substantial source of particles for the MABL in the size range corresponding to minimum sizes of particles 'processed' by clouds, whose growth was determined by the conversion of dissolved SO_2 into sulphates in cloud droplets. The MABL–troposphere exchange contributes substantially to the formation of the aerosol size distribution in the MABL.

The model under consideration was also used to monitor the evolution of aerosol in air masses suffering an advection from the eastern coast of the USA in the middle sector of the Atlantic Ocean for time periods up to 10 days. In the process of this advection the size distribution gets transformed from continental to marine (characteristic of the remote sector of the ocean) during 6–8 days. There was no nucleation under typical meteorological conditions, but under less typical conditions, including a considerable washing out of aerosols (rain rate $5\,mm\,hr^{-1}$

during 12 hours), a decrease in temperature of 10°C (from 283 K near the surface down to 279 K at the height of 1,000 m), and a large flow of DMS (40 μmol/m^{2}day) considerable nucleation may take place. An assessment of sensitivity of the results to initial conditions has shown that it does not show itself during the whole period of numerical modelling (8–10 days).

4.3 SULPHUR CYCLE AND SULPHATE AEROSOL

The particular importance of the biogeochemical sulphur cycle in the context of natural and anthropogenic climate changes determines the key role of sulphur compounds as an atmospheric component. Sulphate aerosols, as always, attract serious attention. Aerosol as a climate-forming factor can be interactively considered in the numerical modelling of the global climate by taking into account the global biogeochemical cycle of sulphur. In this context, [18] discussed the evolution of the global sulphur cycle using data on SO_2 emissions in 1990 and the earlier published data on the trends of emissions in different countries. The results obtained enabled one to assess the sulphate aerosol-induced change in the global average aerosol radiative forcing (ARF), which gradually enhanced (due to both direct and indirect impacts of aerosols) for the period 1850–1990 from values close to zero (0 to $-0.17\,W\,m^{-2}$) to (-0.4 to $-1\,W\,m^{-2}$). In the case of direct impact the ARF intensity (defined as the ratio of ARF to the content of anthropogenic sulphate aerosol) remained constant in time during this period constituting $-150\,W\,g^{-1}$ of sulphate, but in the case of indirect impact it substantially decreased even with a growing content of sulphate. With the use of the model of the global sulphur cycle developed in the Laboratory of Dynamic Meteorology (France), calculations have been made [18] of the evolution of the sulphate aerosol content in the atmosphere and of respective ARF changes in the regions of the USA and western Europe, for the period 1980–1998. Results of calculations were compared with the observational data and revealed a comparative constancy of the ARF during this time period. The necessity of the numerical modelling and observations for other types of atmospheric aerosols has been emphasized.

In continuation of developments on the problems of sulphate aerosol, [52] analysed the information about the surface air quality from data of long-term observations at the network of complex monitoring of the atmosphere controlled by the U.S. National Oceanic and Atmospheric Administration (NOAA), together with the data on the chemical composition of precipitation obtained within the National Programme on studies of depositions from the atmosphere. The main goal of this development – based on the data of seven stations located east of the Mississippi River for the observational period from 1983–1986 to 1997 – is a substantiation of the quantitative indicators of chemical processes taking place in the lower atmosphere.

Analysis of the results obtained has shown that the level of the sulphur compounds content in the atmosphere reached a maximum in late 1980, and since then it has been gradually decreasing at a rate of about 2.8% yr^{-1}. During some period in 1990 it increased at a rate of about 5% yr^{-1}. The observational data did not reveal any stepwise change, which could be attributed to the beginning of taking measures with respect to emission trading and to the associated decrease of SO_2 emissions approximately in 1995. According to observations, these measures have only favoured the continuation of the earlier long-term trend of air quality improvement. The results of observations at some locations show that positive consequences of the total decrease of emissions do not necessarily show themselves at all locations, since in some cases the effect of local conditions may prevail. However, on the whole, the data on air quality correlates well with sulphur emissions from the data of the U.S. Environmental Protection Agency. A relative decrease of sulphur concentration in the planetary boundary layer east of the Mississippi River completely corresponds to a relative decrease of emissions by electric power stations. The regional air quality is mainly determined by contributions of emissions from large sources, but not by the total impact of all small sources.

Though in the northern hemisphere the anthropogenic emissions of sulphur compounds have long ago exceeded the natural sources, in the southern hemisphere the contribution of natural sources exceeds that of anthropogenic emissions by a factor of 2 or more, with the most substantial natural source being an input of DMS to the atmosphere from the ocean. In the Antarctic, sulphur is a widely spread element always detected in atmospheric aerosols, constituting up to 80–90% of the aerosol mass deposited on the Antarctic plateau. Even in winter the contribution of sulphur is next to Na and Cl, reflecting the impact of intensive intrusions of sea air (salt storms) during the time periods when other sources of sulphur are negligible or are absent completely.

Initially, it has been supposed that sulphur compounds get to the Antarctic as a result of either the transport from the stratosphere or volcanic activity. Further studies have led to the conclusion, however, that local volcanic sources can explain no more then 10% of the sulphur budget. Of secondary importance is also the input of sulphur from the stratosphere, except cases of powerful volcanic eruptions. The main natural source of sulphur for the Antarctic atmosphere is DMS emissions by the ocean. A correlation was found out between phytoplankton *Phaecyctis*, blooming in the seas surrounding the Antarctic, and numerous parameters connected with DMS, and it became clear that the annual course of the sulphur compounds content in the Antarctic troposphere reflects the seas' biodynamics.

These conclusions have prompted glaciologists to search for a possible impact of sulphate aerosol on climate in the past by analysing the cores of sea bottom sediments. For the purpose of further studies, the project SCATE has been formulated to study chemical processes determining DMS oxidation. Within this project, measurements of the concentration of such MGCs as methanosulphonic acid in the gas-phase MSA(g), dimethyl-sulphoxide DMSO(g), $SMSO_2$(g), and H_2SO_4(g) have been carried out for the first time at the Antarctic station Palmer during the period 18 January–25 February, 1994.

Analysis of the results obtained has led to the conclusion that the main source of DMSO is the reaction between DMS and OH, which is also responsible for the formation of MSA(g) after successive oxidation of DMSO and MSIA products. Probably, the OH/DMS reaction and the subsequent reaction, with OH/DMSO taken into account, is the main source of SO_2. There were no data which would serve as evidence that important sources of H_2SO_4 (apart from SO_2) are from any other intermediate compounds (e.g., SO_3). There is no doubt that at the Palmer station a considerable part of the cycle of DMS oxidation takes place above the ABL in the layer called the buffer layer (BL).

The main result of observations and numerical modelling within the SCATE project consists in establishing a close interaction between chemical processes and atmospheric dynamics. The observational data revealed frequent episodes of a rapid vertical transport from a very thin MABL to the above BL. Due to the long photochemical lifetime of DMS and the frequent phenomena of an intensive convection in the ABL, a considerable share of the ocean-emitted DMS comes to the BL remaining unoxidized. In the BL where the OH concentration is high and aerosols are weakly washed out, an accumulation of oxidized sulphur takes place reaching a high concentration. Some portion of air from the BL can sometimes return to the ABL affecting the SO_2, DMSO, and $DMSO_2$ contained in it. Besides, since SO_2 and DMSO are major precursors of H_2SO_4 and MSA, chemical processes in the BL, together with vertical transport, control the content of H_2SO_4 and MSA in the ABL.

It has been emphasized [52] that inadequate information on chemical processes in the Antarctic troposphere explains the preliminary character of the conclusions drawn. This is determined, first of all, by the absence of data on the SO_2 content in the ABL and total lack of results of direct measurements in the BL of both DMS and all oxidized sulphur compounds. Details of the vertical transport between the ABL and BL, as well as of episodic air intrusions from the BL to the ABL are still unclear. It is also unclear whether the Palmer station is sufficiently representative.

The aerosol sulphate is the main final form of sulphur compounds in the atmosphere resulting from sulphur oxidation. As a rule, reduced sulphur compounds, in the process of their chemical conversions, get oxidized (before transformation to sulphate) as SO_2. Sulphur dioxide oxidation is the most important anthropogenic source of sulphate in the free troposphere, the oxidation process taking place both in water (heterogeneous process) and gaseous (homogeneous reactions) phases. The homogeneous oxidation is mainly determined by reactions with the participation of hydroxyl:

$$SO_2 + OH^\bullet + M \rightarrow HOSO_2 + M$$
$$HOSO_2 + O_2 \rightarrow SO_3 + HO_2$$
$$SO_3 + H_2O + M \rightarrow H_2SO_4$$

As for the process of heterogeneous oxidation, various versions of transformation of S(IV) ($\equiv SO_2$, HSO_3^- and SO_3^{2-}) into S(VI) are possible – the most effective oxidants being H_2O_2, O_3, and O_2. Savarino *et al.* [126] have estimated the relative significance of various heterogeneous reactions of oxidation on the basis of analysis of the isotopic composition (^{16}O, ^{17}O, and ^{18}O) of sulphates resulting from these reactions. With this aim in view, laboratory studies have been carried out of sulphur oxidation due to H_2O_2, O_3, and O_2, catalyzed by Fe(III) and Mn(II). In the process of gas-phase oxidation, only the reaction $SO_2 + H_2O$ and its final phase $SO_3 + H_2O$ mentioned above have been considered.

According to the results obtained, none of the heterogeneous or homogeneous reactions of oxidation can explain the observed oxygen isotopic composition of sulphate independent of mass. Since H_2O_2 and O_3 have an isotopic signature independent of mass, possibilities have been analysed of applying this anomaly to sulphates. The isotopic analysis has shown that both atoms of oxygen in H_2O_2 are found in H_2SO_4 as the final product of oxidation, but in the case of O_3 only one atom of oxygen goes to sulphate. On this basis, one can assume that the mass-independent isotopic composition of oxygen in sulphate can be explained only by heterogeneous oxidation. This result enables one, in particular, to obtain quantitative estimates of contributions to sulphate formation due to both homogeneous reactions with the participation of OH and heterogeneous process with the participation of H_2O_2 and O_3. The results obtained can also be useful for checking the adequacy of the models of chemical processes in the atmosphere with long-range transport taken into account.

Particular attention has been paid to the problem of sulphate aerosol deposition on forests, which can be illustrated by the results of some investigations carried out in the USA [22] (see also [77]). After the Clean Air Act was adopted in the USA in 1970, in most states sulphur dioxide emissions decreased, except Georgia where SO_2 emissions almost doubled during the period 1975–1985 (by 88%) – now ranked seventh in the USA for sulphur emissions. In connection with the serious danger of these emissions for the ecosystems on land and in water basins, studies are necessary of the processes of dry and wet deposition of sulphur compounds (especially this refers to forests) and subsequent estimates of the impact of such depositions need to be made.

Being emitted to the atmosphere, sulphur dioxide can then deposit onto the forests both in gas phase and in the form of submicron ($(NH_4)SO_4$) and larger ($CaSO_4$) aerosol particles resulting from gas-to-particle conversion. The wet deposition of sulphur determining mainly acid rains, takes place in the form of SO_4^{2-} and may be reliably estimated, whereas in the case of dry deposition, it is difficult to do because of the complexity and variability of this process. It is essential, however, that the dry deposition of sulphur compounds on the forest ecosystems (especially in the form of gas-phase SO_2) constitutes a considerable part of deposition of acid components (very often $>50\%$). Since S sedimentation is not a limiting factor for the growth of forests, until recently it did not attract serious attention as in cases of other nutrients. However, the growth of sulphur compound emissions to the atmosphere required intensive studies. In this connection, [22] have undertaken a study of the processes of dry and wet S depositions in adjacent coniferous and broad-leaved forests at a key site in the watershed basin in the Panola mountains (Georgia state, USA).

According to the results obtained, the total (dry and wet) deposition for the period October, 1987–November, 1989 constituted 12.9 and 12.7 $kg\,ha^{-1}\,yr^{-1}$, respectively, for broad-leaved and coniferous forests. The contribution of dry deposition (mainly due to SO_2) reached $>40\%$. The annual average input of S under the forest canopy is 37% more in the broad-leaved forests than in the coniferous ones, which can be explained by the effect of a more substantial dry deposition. Though the input of S through deposition exceeds the needs of tree growth, more than 95% of deposited S remained within the watershed, this being mainly due to SO_4^{2-} adsorption by iron oxides and hydroxyls contained in soils. The content of S in white oak and in the stems of turpentine pines for the last 20 years increased by 200% (probably, due to increasing emissions).

The complexity of quantitative estimates of the impact of aerosol on climate dictates the necessity to obtain more reliable information on several poorly studied aspects of the problem. One of these aspects is the hygroscopic growth of aerosol particles. Using the H-TDMA differential analyser of the growth of particles as a function of relative humidity, [13] have carried out ship measurements of the hygroscopic properties of submicron aerosols in the MABL over the Pacific and other oceans of the southern hemisphere within the ACE-1 programme. Results of measurements characterize a change in the particle diameters with its gradual evolution from the dry state to relative humidities equal to 89–90%.

Initial values of the 'dry' diameter of particles constituted 35, 50, 75, and 150 (or 165) nm. Natural background aerosols in the MABL consisted of two types of particles with completely different hygroscopic properties: sea salt and NSS aerosol. Since the use of the H-TDMA instruments provides a reliable means of distinguishing between these two types of aerosol, it is possible to estimate their contribution to the formation of the observed mixed aerosol.

Observations over the ocean south of Australia revealed an increase of the particle diameter from 'dry' values 35, 50, and 150 nm, respectively, by factors of 1.62, 1.66, and 1.78, which considerably exceeded the growth of particles in the polluted continental air masses. In the cases of externally mixed salt particles the coefficient of growth was even greater: 2.12 and 2.14 for initial diameters of 50 and 150 nm, respectively. Observations over the Pacific Ocean (at relative humidity = 89%) gave values of the coefficient of growth of 1.56, 1.59, 1.61, and 1.63, respectively, for initial diameters 35, 50, 75, and 165 nm. Cases of particles' solution and relative humidity-hysteresis were observed only in the interval of relative humidity = 68–90% in air masses north of the South Pacific region and only for particles of the Aitken mode (the diameter of particles is within 20–80 nm). The formation of externally mixed salt particles could be connected with a high wind speed caused by either frontal motions or low-pressure systems. However, it should be mentioned that the number density of externally mixed particles with diameter 150 nm correlated weakly with local wind speed (apparently, this took place because of the long lifetime of such submicron particles). Particles with a considerably lower coefficient of growth compared with NSS particles (they were called less hygroscopic particles) were only observed in the periods of manifestation of the anthropogenic impact.

Initially, it was assumed that the dominant component of the tropospheric sulphate aerosol is H_2SO_4, but observations revealed also a considerable amount of ammonium ions. Thus, it can be concluded that ammonium sulphate and bisulphate, as well as letovicite, may be important components of aerosols. In this connection, [42] have performed laboratory studies of the phase state of ammonium sulphate $(NH_4)_2SO_4$ at low temperatures using data from Fourier spectroscopy. Whereas the water vapour deliquescence by aerosol from ammonium sulphate at a temperature above the ice eutectic point and from anhydrite ammonium sulphate at 254 K has been studied quite thoroughly, the phase properties at lower temperatures have been studied much less thoroughly. Therefore, the experiments discussed were conducted at temperatures from 166–235 K. The results obtained at a low relative humidity confirm the presence of the ferroelectric phase of ammonium sulphate at temperatures below 216 ± 8 K. The dependence on relative humidity indicates that at temperatures below the eutectic point phase transformation occurs: crystal $(NH_4)_2SO_4$ transforms into the metastable water solution (a similar process was called deliquescence). At a temperature above 203 K, deliquescence took place at relative humidity $= (88 \pm 8)\%$, which agrees well with the results of extrapolation of measurements at temperatures above the eutectic point, as well as with the theoretical conclusions.

In the experiments conducted at temperatures from 166–203 K, deliquescence was sometimes observed, and in other cases ice was formed directly from the gas phase, which is possibly indicative of selective heterogeneous nucleation. Formation of ice particles prevents the increase of relative humidity up to the level required for deliquescence, which explains the variability of the experimental results obtained at low temperatures.

The problem of the sulphate aerosol was mostly discussed in the context of its role as a factor of climate cooling. As was noted earlier, it is important to determine the radiative forcing taking into account both direct and indirect effects of aerosol on the climate. In the latter case, the effect takes place through changes of microphysical and optical characteristics of clouds as a result of aerosol particles functioning as condensation nuclei. This complicated problem includes various aspects, one of which is connected with the fact that chemical activity and efficiency of aerosol particles as cloud condensation nuclei depends on their phase state. While deliquescence (water uptake at the formation of solutions) and efflorescence (loss of water at the formation of crystals) in case of pure ammonium sulphate are well known, it is quite different with respect to organic components whose share in the aerosol composition may reach 50% and higher. Organic components, like dicarboxylic acids, are often water-soluble.

The transition of pure ammonium sulphate (AS) from solid to liquid states occurs at room temperature and deliquescence relative humidity (DRH) of 80% typical of the solution process – DRH in the presence of a SA mixture with another electrolyte is always lower than the DRH for pure AS. This lower DRH was called the eutonic DRH. Brooks *et al.* [20] discussed the results of laboratory measurements of eutonic DRH for the mixture of dicarboxylic acids (from C-2 to C-6) and ammonium sulphate at temperatures from +24°C to −10°C. These experiments confirmed the earlier conclusion about the decrease of DRH for the mixture as compared with pure ammonium sulphate and revealed the dependence of DRH on temperature. Thus, in such a case, aerosol particles can become liquid at lower values of relative humidity.

The initial phase of tropospheric aerosols substantially affects the development of the mechanism of ice particle nucleation, on which the properties of cirrus clouds depend.

In aerosol particles consisting largely of concentrated (in aqueous solution) mixtures of H_2SO_4, $(NH_4)_2SO_4$, and NH_4HSO_4, homogeneous nucleation of ice may occur, if liquid particles do not include any insoluble substance, or heterogeneous nucleation may occur on the surface of insoluble nuclei inside aerosol particles or in crystal salt nuclei. Using the technique of optical microscopy, [142] have performed a laboratory investigation of heterogeneous nucleation of ice on liquid $(NH_4)_2SO_4$–H_2O particles with mineral dust immersions of two types: kaolin and montmorillonite – having almost equal levels of efficiency in ice nucleation. The measurement results show that the freezing point differs by 8–20 K from that under conditions of homogeneous nucleation, and the difference increases when the weight concentration of admixtures in particles exceeds 27%. At temperatures ranging from 198–239 K, heterogeneous nucleation occurs in $(NH_4)_2SO_4$–H_2O particles with mineral dust admixtures only at the supersaturation level (with respect to ice) from 1.35–1.51. The results from [142] suggest that mineral dust particles serve as efficient nuclei for ice nucleation in $(NH_4)_2SO_4$–H_2O aerosol and thus favour the formation of ice clouds in the upper troposphere at a higher temperature and lower supersaturation level than under conditions of homogeneous nucleation.

4.4 DUST AEROSOL

4.4.1 Generation mechanisms

As was already mentioned, mineral particles, getting into the atmosphere over deserts as a result of saltation process, form the most significant component of the dust aerosol [72, 74]. The saltation process (at its initial stage) responsible for the entry of mineral dust aerosol into the atmosphere is supposedly connected with soil 'bursts' which produce aerosol particles. The intensity of this process is largely determined by the surface wind shear, and earlier this process was considered as oscillating. Grini *et al.* [47] proposed a new theoretical model of saltation which demonstrated that actual aerosol emissions are characterized by a monotonic dependence on the wind shear. Although oscillations of aerosol emissions are actually observed, they are connected with the specific size distribution of soil particles.

An important feature of the dust aerosol is its long-range transport. It is for this reason that the dust aerosol turns out to be a significant component of the tropospheric aerosol over the oceans during the events of dust storms in deserts.

Based on the combined use of the factor analysis model called Positive Matrix Factorization (PMF) and cluster analysis of back trajectories of the long-range aerosol transport on days of intense or weak transport, [91] have identified the types of aerosol observed at Brigantine station in the National Wildlife Refuge, NJ, USA, which is a good location to study the mid-Atlantic regional aerosol.

The most abundant component of aerosol chemistry is sulphate (characterized by the ammonium sulphate concentration) accounting for 49% of the annual mean fine aerosol mass (PM-2.5). The contribution of organic compounds constitutes 22% (this contribution is estimated as the organic carbon concentration multiplied by a factor of 1.4) and that of ammonium nitrate 10%. Apparently, in summer, the secondary organic aerosol can also play a substantial role.

The results obtained in [91] show that nine factors contributed to the PM-2.5 mass concentrations: coal combustion products (66% both in summer and in winter), sea salts (9%, fresh and 'old'), internal combustion engine emissions (8%), diesel oil (Zn–Pb) emissions (6%), waste burning (5%), oil burning (4%), and soil (2%). The aged sea salt concentrations were at a maximum in springtime, when the land breeze and sea breeze cycle is strongest. Comparison of backward air trajectories on high and low-concentration days suggested that the Brigantine station is surrounded by sources of oil burning, waste burning, and combustion engine emissions, whose total contribution to PM-2.5 mass concentration reaches 17%. In addition to SO_4^{2-}(43–49%), the contribution of organic carbon constitutes 22–33%, and that of NO_3^- – 10%.

4.4.2 Dust aerosol from deserts

Deserts, and first of all, the Sahara, are one of the most powerful global sources of dust aerosol input to the atmosphere [48, 72, 74]. Continuing from the earlier studies of the consequences of the Saharan dust transport across the Atlantic Ocean, [94] discussed the results of daily filter and impactor measurements of the variability of the concentration and size distribution of aerosol at Barbados Island (13°15′N, 59°30′W) during the period 5 April–3 May, 1994, when there were four episodes of Saharan dust transport reaching Barbados. Processing of the observational results revealed a high correlation between concentrations of NSS sulphate aerosol (nss SO_4^{2-}) and dust ($r^2 = 0.93$). Mass concentrations of aerosols varied, respectively, between 0.5–4.2 $\mu g\,m^{-3}$ and 0.9–257 $\mu g\,m^{-3}$. The interdiurnal variations of the aerosol size distribution of both types turned out to be negligible. However, the share of large particle concentrations (with aerodynamic diameter >1 μm) of NSS sulphate aerosol varied substantially (from 21–73%).

Maximum values of large particle concentrations of SO_4^{2-} were connected with the input of the Saharan dust (judging from the data of satellite observations), and minimum values with the intrusion of air masses from the central sector of the North Atlantic, where the dust concentration was <0.9 $\mu g\,m^{-3}$. The authors [94] supposed that high large particle concentration values for SO_4^{2-} resulted from heterogeneous reactions between SO_2 contained in the polluted air from western Europe and the Saharan dust, taking place in the atmosphere over northern Africa.

The interaction between pollutants and dust does not always result in a high concentration of large particle concentrations. Their low observed concentration points to the functioning of the process of SO_2 oxidation, with an SO_4^{2-} formation before the intrusion of the polluted air masses. On such days the prevailing source of NSS SO_4^{2-} is, apparently, oceanic DMS, and under these conditions, the large particle concentrations share remains close to 20%. Since the episodes of the Saharan dust outbreaks cause substantial changes in the size distribution of NSS SO_4^{2-}, it is necessary to take into account the effect of mineral dust on the NSS SO_4^{2-} size distribution when assessing the impact of NSS aerosols on radiation transfer in the atmosphere. The dust-induced removal from the atmosphere of the most radiatively efficient submicron sulphate aerosol fraction should lead to a considerable decrease of RF due to sulphate aerosol. Naturally, this situation is most substantial in the regions where episodes of dust outbreaks are most frequent (this is a considerable part of Asia and the Indian subcontinent). Also, of importance is the dust-induced removal of SO_2 from the atmosphere, which can manifest itself far from the sources of dust outbreaks, as it happens, for instance, with the Saharan dust transport across the Atlantic Ocean or across the Mediterranean Sea (in the direction of western Europe).

Deserts of north-western China cover territories of $1.3 \times 10^6\ km^2$, which is three times larger than the loess plateau ($0.4 \times 10^6\ km^2$), and are at high latitudes in the zone of west-to-east transport. This desert region is characterized by a dry climate and favourable conditions for the formation of the soil (dust) aerosol (DA), especially with a high repeatability of dust storms. The studies carried out during recent decades identified this region as the main source of DA to the North Pacific in spring and early summer. It was just in the period 1995–1996 when [96] collected about 180–190 aerosol samples from three observatories in the north-western China desert region (Minqin), coastal suburb area (Qingdao), and on an island in the Yellow Sea (Qianliyan).

Analysis of the samples for the concentrations of trace metals (Al, FE, Mn, Na, Cr, Zn, Cu, Co, Ni, Pb, V, Sr, and Cd) revealed a wide spatial and temporal variability and a pronounced annual trend (Qingdao and Qianliyan) of the trace metal levels, with concentrations generally 2–3 times higher in spring than in summer. However, diurnal variations of trace metal concentrations at Minqin are more significant than those in Qingdao and Qianliyan. Enrichment factors EF_{crust} at Minqin are relatively low compared with those at Qingdao and Qianliyan.

Metals such as Pb, Zn, and Cd contribute considerably to the anthropogenic source (R_p), with >98% found at Qingdao and Qianliyan, with EF_{crust} values 1–2 orders of magnitude higher than those at Minqin. Manganese, mainly from low-temperature weathering products, and contributions from the crustal source (R_c) is >70% at Qingdao and Qianliyan. The coastal atmosphere responds to the episodic dust storms in the desert region (Minqin) by exhibiting an increase in the level of aerosol content and concentrations of crustal dominated materials. The percentage contributions from crustal sources (R_c) of trace metals, such as Pb, Cd, and Zn, could increase 5–10 times during the passage of cold fronts than during calm weather periods over the Yellow Sea.

Analysis of aerosol samples from Zhenbeitai (38°17′N, 109°42′E), in order to determine the aerosol mass concentration during nine dust storm events in the spring of 2001, lasting for 26 days in total, has demonstrated that peak mass loadings greatly exceeded the average ($260\ \mu g\ m^{-3}$). About 82% of the total mass of aerosol particles was ascribed to Asian dust, of which Al, Ca, Fe, K, Mn, Si, and Ti amounted to 7%, 6%, 4%, 2%, 0.1%, 32%, and 1% by weight, respectively. The estimated dry deposition of Asian dust for the spring of 2001 was 189 g m^{-2}, of which 85% was due to dust storms. Only ~11% was due to local or background aerosol, whereas the share of long-range transport reached 89%.

Having carried out the scanning electron microscopy of aerosol samples collected in the Negev desert in Israel during the summer and winter campaign in 1996–1997, [128] found (through cluster analysis) the presence of 11 classes (groups) of particles. While the summer samples were rich in sulphates and mineral dust, the winter samples contained sea salts and industrial particles. The fine aerosol fraction with diameters below 1 μm was rich in secondary particles, which (in the case of sulphate particles) were mainly attributed to a long-range transport.

Colarco *et al.* [28] have developed a 3-D numerical model for simulating Saharan dust outbreaks and transport over the tropical North Atlantic Ocean. To control the adequacy of the numerical simulation results, the data used were obtained within the programme of the ACE-1 field experiment studying the atmospheric aerosol properties. The imaginary part of the complex refractive index of the aerosol substance (k) was retrieved from the data of the Total Ozone Mapping Spectrometer on the Earth Probe satellite (EP-TOMS) to map the total ozone content. The value of k was estimated from the observed ratio between the aerosol index (AI) and the aerosol optical depth (AOD), by comparing the observed values with the calculated ones.

This problem has been solved using the observational data for three sites [28]: Dakar, Sal, and Tenerife. At Sal and Tenerife the dust imaginary refractive index was $k = 0.0048$ (0.0024–0.0060) at 331 nm and $k = 0.004$ (0.002–0.005) at 360 nm. At Dakar the dust imaginary refractive index k was approximately 0.006 (0.024–0.0207) and 0.005 (0.0020–0.0175) at 331 and 360 nm, respectively. These values are considerably lower than the refractive index currently used in TOMS retrievals of AOD and single-scattering albedo (SSA). With regard to the retrieved refractive index values, the SSA was estimated by integrating over the particle size distribution. The SSA values (ϖ) at 331 nm at Dakar, Sal, and Tenerife turned out to be 0.81 (0.65–0.90), 0.84 (0.82–0.91), and 0.86 (0.83–0.89), respectively.

The presence of dust aerosol in the atmosphere significantly affects various processes occurring in the atmosphere and even in the ocean. Variations of the marine phytoplankton concentration due to dust aerosol exemplifies the impact on the ocean. The biomass of marine phytoplankton is characterized by a very intense variability: it is replaced, on average, every 1–2 weeks. Carbon compounds produced during photosynthesis, including particulate organic carbon (POC) and particulate inorganic carbon (PIC), as well as dissolved organic carbon are transported to deepwaters of 100 m (on a global scale) at a rate of about 10 Pg C yr^{-1}. This process, called the biological pump, plays an important role in the formation of the global carbon cycle.

The biological pump has been poorly studied, and this makes it necessary to obtain comprehensive information about the phytoplankton dynamics. For this purpose, starting on 10 April, 2001, two buoys were deployed in the sub-Arctic North Pacific (near 50°N, 145°E) which ensured automated regular records of carbon biomass variability from the surface down to 1,000 m. Analysis of the results of measurements for the first eight months carried out by [16] has made it possible to study the impact on phytoplankton dynamics of such events as hydrographic changes, numerous sea storms, and the April, 2001 dust storm. High-frequency observations of the upper ocean POC variability showed an almost doubled biomass in the mixed layer over a 2-week period after the passage of a dust cloud from the Gobi desert. This fact can be explained as a 'fertilization' of the ocean due to iron-containing dust particles, which stimulates the formation of POC.

4.5 CARBON AEROSOL

4.5.1 General information

A concern about the anthropogenic changes in the gas and aerosol composition of the atmosphere and their climatic implications has been caused, in particular, by emissions of carbon aerosol (CA) into the atmosphere due to fossil fuel burning. This aerosol consists of two fractions: organic carbon (OC) and chemically inactive black carbon (BC) – both with quite different optical properties. OC is characterized by the prevalence of scattering in the visible spectrum, while BC strongly absorbs short-wave solar radiation, and it is the only absorbing aerosol that can be found everywhere. Depending on the type of fuel and burning process, the BC/OC ratio can widely vary – such an important characteristic as radiative forcing strongly depends on this ratio.

There are three sources of CA in the atmosphere: natural primary and secondary emissions of organic gases by vegetation, fossil fuel burning, and biomass burning. The available data suggest the conclusion that the two latter sources (mainly anthropogenic) dominate in the global CA budget, with the basic regions of CA formation being the tropics (biomass burning) and the northern hemisphere mid-latitudes (fossil fuel burning).

The growing global-scale fossil fuel and wood burning results in the input to the atmosphere of a large amount of various optically active MGCs and aerosol particles. An important component of aerosol resulting from the processes of burning is the total carbon (TC) including black carbon (BC) and organic carbon (OC), or aerosol organic matter (AOM), if the effect of reactions with the participation of oxygen and hydrogen is taken into account (in this connection, the relationship has been assumed for mass transformation: POM = 1.3 POC). Both TC components are characterized by different optical and chemical properties. While the BC is an absorber, the POM is a scatterer, which is typical of sulphates.

The importance of biomass and fossil fuel burning as sources of greenhouse gases has also determined a great interest in estimates of the budgets of the products of burning getting to the atmosphere, such as DC and POC. On global scales, the average annual emissions of BC and POC as well as aerosol due to fossil fuel burning are comparable in quantity, the main component of emissions being POC, which requires special attention, in particular, to estimates of emissions to the atmosphere due to forest fires in middle and high latitudes.

Using the Geophysical Fluid Dynamics Laboratory SKYHI model of general atmospheric circulation (latitude–longitude resolution $\sim 3^\circ$ lat. $\times$ 3.6° long.), [30] have conducted a numerical simulation of the global distribution of CA for a detailed analysis of sensitivity of the CA content in the atmosphere to various parameters governing its distribution (in this connection, a detailed analysis has been made of the annual and interannual variability of the CA content). A comparison has been made between the calculated distribution of CA with that observed under different conditions (with special attention to data obtained at four stations: Mace Head, Mauna Loa, Sable Island, and Bondville).

A comparison of mean monthly values of the CA surface concentration from multiyear measurements revealed both similarities and differences. As a rule, CA concentrations observed in conditions of the remote (background) atmosphere exceed those calculated. On the whole, in most cases, both observed and calculated concentrations of OC and BC agree within a coefficient of 2. Of course, of importance (in the context of these differences) is an inadequacy of both numerical simulation and observational techniques. A comparison of data on the interannual variability of CA has demonstrated the necessity of future comparisons of the observed and calculated data over multiyear periods. Limited possibilities to compare data on the vertical profiles of aerosol concentrations have only suggested the conclusion that the calculated values of aerosol concentration in the free troposphere are overestimated.

The numerically simulated 'standard' global mean values of the total content of organic and 'black' aerosol in the atmosphere should be considered a lower limit of possible values, since they turned out to be systematically underestimated compared with those found earlier. Rain rate and its frequency of occurrence are the most important factors affecting the total 'black' aerosol content. Very important are also processes such as wet deposition, transformation from a hydrophobic to a hydrophilic state, and distribution of aerosol emissions between these states. Aerosols in remote regions are characterized by a comparatively universal sensitivity to the variability of various factors governing its content. The most important goals for future developments are as follows:

(1) modernization of the observation system to obtain more complete and representative data sets;
(2) studying of physical characteristics and transformation processes of aerosol as well as secondary OC;
(3) more comprehensive observations of rain rate and its frequency of occurrence; and
(4) an adequate numerical simulation of the tropospheric meteorology, including the wind and precipitation fields.

4.5.2 Soot

The formation in the global atmosphere of the CA anthropogenic component was estimated at 12–24 $\mathrm{Tg\,yr^{-1}}$, which testifies to the substantial impact of these aerosols on such processes as scattering and absorption of light in the atmosphere, chemical reactions, as well as the impact on human health. Since the extent of this impact depends on the level of CA particle assimilation of water, this determines the urgency of the problem of interaction of CA particles with water. For instance, it is well known that SO_2 oxidation regulated by soot carbon and photodegradation of poly-aromatic hydrocarbons bound with soot particles depends on the presence of water.

To further study the hydratation of soot particles, [27] performed laboratory measurements of the capability of various soot particles to uptake water (contained in TP-8 fuel for jet aircraft, kerosene, diesel oil, synthetic matter with metals and sulphates). Drawing isotherms of adsorption and absorption by soot (n-hexane model) has made it possible to determine the parameter of water desorption, for such a surface covering, for the threshold level of chemisorption and relative humidity (83%). The values obtained increase with the soot surface oxidation in the relative humidity interval of 35–85%, whereas the level of water assimilation at a higher relative humidity (up to 92%) depends on the composition of fuel and conditions of its burning. Soot containing metals and sulphur is characterized by a higher level of water assimilation. The obtained results were used to assess the effect of relative humidity on the kinetics of reactions of soot oxidation with the participation of O_3, NO_2, and other gas components. Posfai *et al.* [115] studied the aerosol particles containing soot aggregates in the marine atmosphere of both hemispheres, from very clear to heavily polluted atmospheres. Analysis of observational data has shown that even in the clear atmosphere of a remote region of the southern hemisphere from 10–45% of sulphate aerosol particles contained soot (such a component is often called 'black carbon' – BC), whose sources could be products of aircraft emissions or biomass burning.

The internally mixed soot and sulphates are considered to be a significant part of atmospheric aerosols on global scales. Such aerosols should be taken into account in estimation of both the direct and indirect (through changes in cloud properties) impacts of aerosol on climate.

It follows from laboratory measurements that hygroscopic properties of soot are determined by its composition. If organic substances on the surface of aerosol particles can enhance the water adsorption, such metals as Fe stimulate the catalytic activity of soot. It follows from the data obtained during the ASTEX/MAGE field experiment in the North Atlantic, that as a rule, soot is contained as sulphate particles. Similar results have also been obtained over the southern hemisphere oceans during the ACE-1 experiment. Possibly, the presence of soot particles, functioning as heterogeneous nuclei, enhance the formation of sulphate aerosol. Besides this, soot particles covered with a water film can serve as effective cloud condensation nuclei.

It is important that in the presence of aerosol particles of internally mixed sulphates and soot, the aerosol-induced direct RF substantially decreases. The estimates have shown that each percent of increase of soot/sulphates mass ratio leads to a positive RF at the upper boundary of the atmosphere equal to $0.034\,W\,m^{-2}$. However, this RF depends strongly on surface albedo, increasing when it increases.

The impact of soot particles on cloud optical properties is rather complicated. On the one hand, particles from biomass burning favour an increase of cloud albedo due to decreasing size of cloud droplets. On the other hand, the presence of strongly absorbing cloud droplets containing soot enhances the solar radiation absorption by clouds.

Since in the context of the problem of the aerosol climatic impact, biomass burning is especially important [121], this motivated the field experiment BIBEX with the main purpose of assessing the input of gases and aerosols to the atmosphere due to biomass burning as well as the impact of such emissions on the atmospheric chemical composition and climate. The respective international field experiments to achieve this aim have been accomplished, first of all, in the tropical Atlantic Ocean in the southern hemisphere (the Amazon River basin, the tropical Atlantic, South Africa), where the effect of fires manifests itself in the tropical forests and savanna. In particular, the America–Brazil programme TRACE-A has been accomplished, aimed at studies of biogenic emissions in Brazil, their subsequent long-range transport, and photochemical transformation. Based on the co-operation of the countries of Africa, Europe, and America, studies have been started of the emissions to the atmosphere due to forest fires and processes in soils in South Africa, which were initially called the SAFARI programme, but on agreement about coordination with development within the TRACE-A programme, they were called STARE (southern hemisphere tropical Atlantic regional experiment). These multidisciplinary field studies began in the second-half of 1992, and were based on the use of various observational means (surface, ship, aircraft, and satellite).

Lavoue *et al.* [90] obtained estimates which enabled one to draw the first maps of the distribution of monthly emissions to the atmosphere of carbon particles due to forest fires, bush fires, and grass cover fires. Along with this, statistical data have been given on the areas of burned out territories, the mass of burned organics, characteristics of the fires, and coefficients of emissions. Analysis of data for the period 1960–1997 revealed a considerable interannual variability. On the whole, the share of the vegetation-burning products getting to the atmosphere in middle and high latitudes constitutes about 4%, but in some years it can increase to 12%, with contributions of BC and POM constituting 9% and 20%, respectively. Absolute levels of the respective emissions due to fires in the boreal forests of North America (Canada and Alaska) vary within 4–122 $\mathrm{GgC\,yr^{-1}}$ (BC) and 0.07–2.4 $\mathrm{Tg\,yr^{-1}}$ (POM), whereas in the case of the Eurasian continent (Russia and northern Mongolia), rather uncertain variability intervals are possible equal to 16–474 $\mathrm{GgC\,yr^{-1}}$ (BC) and 0.3–9.4 $\mathrm{Tg\,yr^{-1}}$ (POM).

As for emissions due to forest fires in middle latitudes, for the USA (the continental part of the USA territory south of the frontier with Canada) and Europe, together they are much less, averaging 11 $\mathrm{GgC\,yr^{-1}}$ (BC) and 0.2 $\mathrm{Tg\,yr^{-1}}$ (POM). The scale of grass fires in Mongolia is substantial: 62 $\mathrm{GgC\,yr^{-1}}$ (BC) and 0.4 $\mathrm{Tg\,yr^{-1}}$ (POM). The annual mean BC emissions due to bush fires in Mongolia and California (in total) constitute 20 $\mathrm{GgC\,yr^{-1}}$, and the POM emissions – 0.1 $\mathrm{Tg\,yr^{-1}}$. Table 4.1 contains total data on emissions due to vegetation burning. Maps of geographical distribution of emissions have been drawn with a resolution 1° lat. $\times$ 1° long.

Table 4.1. Emissions of carbon particles due to vegetation burning.

Type of fires	Burned biomass	Black carbon (TgC)	Organic carbon (Tg)
Mid and high-latitude forests	66–700	0.07–0.54	1.01–10.70
Savannas, tropical forests, agricultural, and house fires	5,375	5.63	44.50
Global biomass burning	5,441–6075	5.70–6.17	45.51–55.20
Contribution of forests in middle and high latitudes	1.2–11.5%	1.3–8.8%	2.2–19.4%

4.5.3 Organic carbon

Organic carbon constitutes a significant share of atmospheric aerosol, and therefore it can play an important role as a climate-forming factor. Organic substances are universal components of atmospheric aerosol which either enters the atmosphere forming on aerosol surfaces during the course of fires or forms *in situ* in the atmosphere due to gas-to-particle conversion of the oxidation of VOCs.

The primary CA is emitted to the atmosphere mainly by combustion engines and biomass burning, whereas the secondary organic aerosol (OA) is the product of gas-to-particle conversion with the participation of gaseous organic compounds of both anthropogenic and natural origin. In the latter case, such compounds are mainly isoprene and terpenes, the products of their oxidation in the atmosphere (with the participation of ozone O_3 and OH hydroxyl radical in daytime and NO_3 nitrate radical at night) being one or more functional groups such as COOH, COH, C—O, $CONO_2$, etc.

Formation of OA due to gas-to-particle conversion takes place because the vapour pressure decrease favours the condensation formation and growth of aerosol particles and/or the solubility of the substance grows compared with that of the parent organic compounds.

Significant contribution of the secondary organic aerosol to the OA concentration has stimulated the development of techniques to assess the distribution of organic compounds between the aerosol and gas phases. In this connection, surrogate products of organic compounds oxidation had been divided into two groups of hydrophobic and hydrophilic compounds depending on their tendency to solve in organic or aqueous phase, respectively. The phase distribution of substitutes was supposed to take place through only one dominating mechanism determined by the molecular properties of molecules–substitutes. Classification into hydrophobic and hydrophilic substances is based on the structural and physical characteristics of compounds. As a rule, the secondary oxidation products, due to their low vapour pressure and high polarity, are similar to both organic and water phases of aerosol.

Griffin *et al.* [44–46] have developed a new, totally interactive model of the hydrophobic–hydrophilic distribution of organic components between aerosol and gas phase, which ensures a conservation of mass and equilibrium between: (1) organic aerosol and gas phases; (2) aqueous aerosol and gas phases; and (3) molecular and ionic forms of the rearranged components in the aqueous phase. Based on the use of this model in 0-D version and the California Institute of Technology (USA) 3-D model, taking into account chemical processes and transport, a numerical simulation was carried out of the process of distribution of components between the aerosol and gas phases. The numerical simulation revealed a shift in the distribution toward the phase of organic aerosol, and an increase of the amount of secondary organic aerosol compared with the earlier results.

An important feature of the organic aerosol is its hygroscopic properties: about 20–70% of organic matter of this aerosol is water soluble and can absorb water until an equilibrium with the atmospheric water vapour pressure is reached. Using the model of the monodisperse aerosol dynamics, [112] studied the process of condensation growth of the atmospheric aerosol particles formed of the water-insoluble vapourous organic compounds, taking into account the chemical reactions accompanying the process of condensation growth. Special attention has been paid to the variable power of the source of vapour and its saturating pressure on the surface of the particles. The initial size distribution of particles was assumed to be bimodal. Aerosol is assumed to be initially an internal mixture of particles consisting of sulphuric acid and insoluble organic matter at a ratio 1 : 1.

Calculations have shown that if the ratio of the vapour source power and the condensation sink (Q/CS) exceeds $10^8\,cm^{-3}$, and vapour is non-volatile or weakly volatile, then the 10-nm nuclei grow at a rate of 2.5–3 $nm\,hr^{-1}$ and can function as cloud condensation nuclei at a 1.6% oversaturation or lower during the 24 hours after the beginning of numerical simulation. As the Q/CS ratio grows, the level of the vapour saturating pressure should increase. If the vapour is highly volatile (and hence, the vapour saturating pressure is sufficiently high), then instead of the growth of the nucleation mode particles, their coagulation with larger particles takes place, and therefore the climatic impact is negligible.

4.5.4 Biogenic aerosol

As has been mentioned above, recent results of numerical simulation have indicated a significant direct and indirect effect of the NSS sulphate aerosol (NSS SO_4^{2-}) on climate because of its specific size distribution determining an intense scattering of the short-wave solar radiation. On the other hand, SA particles acting as CCN change the size distribution and optical properties (first of all, albedo) of clouds. The estimates have shown that the joint direct and indirect effect of SA can lower the global radiation by 1–2 $W\,m^{-2}$ or more, which is comparable with the anthropogenic enhancement of the atmospheric greenhouse effect (i.e., SA compensates the greenhouse climate warming).

The formation of NSS SO_4^{2-} aerosol over the ocean is determined by contributions from the following two sources: (1) biogenic sources of gaseous sulphur compounds (mostly dimethylsulphide, DMS); and (2) anthropogenic sources (mostly SO_2). Both of these sources vary strongly both in space and time.

The biogenic contribution is usually estimated from the concentration of methanesulphonate (MSA), which is one of the stable products of DMS oxidation by OH hydroxyl. The process of this oxidation remains unclear, and this makes it difficult to estimate the SO_4^{2-}/MSA concentration ratio. In order to obtain data on the biogenic and anthropogenic SA sources within the field observational Atmosphere/Ocean Chemistry Experiment (AEROCE), [127] have collected daily aerosol samples in the MABL at Barbados (13.17°N, 59.43°W), the Bermudas (32.27°N, 64.87°W), and at Mace Head (53.32°N, 9.85°W) as well as in the free troposphere over Isaña, the Canary Islands (28.30°N, 16.48°W, 2360 m above sea level).

Multiple regression analysis has been carried out on the relationship between the observed concentrations of MSA and Sb and/or NO_3 as independent variables characterizing the anthropogenic sources of aerosol, to assess the relative contributions of marine biogenic and anthropogenic sources to the formation of NSS SO_4^{2-} aerosol. From the data of two-year observations on the Bermudas and Barbados, 'marine' values of SO_4^{2-}/MSA (respectively, 19.6 ± 2.1 and 18.8 ± 2.2) are stable over a year and agree with those observed at American Samoa in the tropical Pacific (18.1 ± 0.9). From the data for Mace Head, where observations lasted one year, this ratio constituted only 3.01 ± 0.53 (apparently, this value is representative for the whole year, though the possibility of its increase in winter is not excluded). From the data of four-year observations at Mace Head the ratio of biogenic SO_4^{2-}/MSA could not be estimated because of its low level. The 'continental' values of SO_4^{2-}/Sb strongly vary in space. At the Bermudas, where the North American sources dominate, this ratio reaches 29,000, being twice as high as Mace Head (13,000), where the European sources dominate. Intermediate values are observed at Barbados (18,000) and Isaña (24,000), where both the European and North American sources contribute. The annual mean relative contributions of anthropogenic sources constitute: about 50% (Barbados), 70% (the Bermudas), 75–90% (Mace Head), and 90% (Isaña). With the biogenic SO_4^{2-}/MSA values for rainwater assumed to be the same as for aerosol, it turns out that relative biogenic and anthropogenic contributions to precipitation at Barbados and the Bermudas are about the same as for aerosols.

4.6 URBAN AEROSOL

This type of aerosol is quite special because of its highly complicated formation conditions in the polluted atmosphere. The gas-phase chemical reactions in the urban atmosphere on a regional scale determine the formation of oxidants such as ozone (O_3), the hydroxyl radical (OH), and the nitrate radical (NO_3), as well as their responsibility for the consumption of chemically active organic compounds and reactions between peroxide radicals and MGCs such as nitrogen oxides ($NO_x = NO + NO_2$).

An important product of the reactions discussed is the secondary organic aerosol (SOA), which is formed in two steps. First, a considerable amount of initial ('parent') organic compound is oxidized resulting in a MGC whose vapour pressure is below that of the initial compound. This MGC, with a low vapour pressure, can be a source of the aerosol phase formation (by condensation and through a homogeneous nucleation). As a rule, molecules of initial compounds which can form MGCs with a low vapour pressure and then SOA, should contain six or more atoms of oxygen.

The present knowledge of the gas-phase reactions in the atmosphere does not, so far, permit a detailed description of the mechanism of SOA formation. One of the reasons is that chemical processes with the participation of large organic molecules, which lead to the formation of semivolatile products, remain unknown and unstudied. In this connection, [44–46] proposed a new chemical mechanism called the Caltech Atmospheric Chemistry Mechanism (CACM) developed at the California Institute of Technology, which ensures the achievement of two goals:

(1) an adequate account of the chemical reaction that determines the formation of tropospheric ozone; and
(2) calculation of concentrations of secondary and tertiary semivolatile oxidation products which can be SOA constituents.

The mechanism suggested in [46] includes detailed chemical processes with the participation of organic compounds, which enables one to simulate the formation of multifunctional MGCs which are characterized by a low-level vapour pressure and are the sources of organic aerosol in the atmosphere. The CACM incorporates 191 MGCs, including:

(1) 120 thoroughly considered components (in this case the concentration of each MGC, including 15 inorganic, 71 chemically active organic, and 34 chemically inactive organic compounds, is calculated from the data on kinetics, emissions, and depositions);
(2) 67 pseudo-stable MGCs (2 organic and 65 inorganic); and
(3) MGCs with persistent concentrations.

The main goal of this numerical simulation was dual:

(1) to describe chemical reactions with the participation of inorganic and organic MGCs, responsible for the CACM mechanism; and
(2) to analyse the functioning of this mechanism from the viewpoint of simulating the gas-to-particle conversion and MGC concentration fields as applied to the conditions of the polluted atmosphere (examples only) in the region of the South Coast Air Basin (SoCAB) of California, during 27–29 August, 1987.

Calculated concentrations of the gas-phase compounds that served as SOA constituents, were compared with the SOA characteristics measured in this period. The comparison has demonstrated that the CACM-determined MGC concentration fields make it possible to almost adequately explain the observed levels of SOA concentration. The main source of uncertainty of the numerical simulation results is the inadequacy of data on emissions of numerous MGCs and their properties, such as rates of reactions and the temporal dynamics of their products. Of great importance are uncertainties in the chemical mechanism which determine a direct conversion of aldehydes into MGCs of the acid group.

Pun *et al.* [117] have developed a model to calculate the formation of the secondary organic aerosol based on the thermodynamically balanced distribution of molecular components between the gas-phase and SOA particles. The distribution of secondary hydrophobic products of chemical reactions between the absorbing organic aerosol, consisting of the primary organic aerosol (POA) and other secondary organic compounds, is supposedly determined by the coefficient of the balanced distribution, iteratively calculated for each secondary compound taken into account. An analysis of the hydrophobic compounds module is based on the results of studies of dissociation of octadecanoic acid into the POA surrogate components.

As expected, the precalculated amount of octadecanoic acid in the aerosol phase increases with increased amounts of the absorbing matter or total amount of the acid components. The secondary hydrophilic compounds are dissociated into the aqueous phase in accordance with Henry's law: the dissociated part of the mass of each component is determined by its respective (Henry's law) constant as well as by the constant of acid dissociation. The liquid water content (LWC) in aerosols is iteratively calculated as an intermediate value between data corresponding to the inorganic aerosol module and hydrophilic module (considering the dissociation into glyoxalic and malic acids). If the glyoxalic acid tends to remain in the gas phase, the malic acid is totally transformed into the aqueous phase, ions being the prevalent form.

As expected, the increasing relative humidity causes an increase in the amount of water bound with organic compounds (ΔLWC), and the lower level of pH for aerosol favours the formation of molecular solutions compared with ionized forms. The growing pH results in increasing values of Henry's law constants for acids, which leads to the growth of organic aerosol concentrations. An increased ΔLWC determines an additional conversion of inorganic compounds into aqueous phase.

Using a 3-D regional model to simulate changes in the gas and aerosol composition of the atmosphere, [46] considered the results of the respective numerical simulation for conditions of smog formation on the south coast of California (near Los Angeles) on 8 September, 1993. An analysis of the results was focused on the formation of the secondary organic aerosol. The results obtained show that in this case, the secondary hydrophobic products of oxidation of anthropogenic organic compounds have the greatest affect on SOA formation. The biogenic contribution to the total SOA increased in rural areas of the region as did the fraction of hydrophilic SOA (the latter reflects an increase of SOA oxidation taking place with SOA particle residence times in the atmosphere). Further developments foresee a detailed comparison of the numerical simulation results with observational data.

Within the framework of the Trans-Siberian Observations into the Chemistry of the Atmosphere (TROICA) project on studies of the MGCs content in the surface-air layer during June–July, 1999, [105] carried out measurements of the concentration of O_3, NO, NO_2, CO, CO_2, CH_4, ^{222}RN, J(NO_2) (the rate of NO_2 photodissociation), and black carbonaceous aerosol along the Trans-Siberian railway, at a distance of 16,000 km, from Kirov (~58°N, 49°E) to Khabarovsk (~48°N, 135°E). The diurnal variation of concentrations of the enumerated MGCs was estimated as well as the dependencies of the concentration on the types of soil and vegetation cover, in order to determine the flows of CO_2 and CH_4 for various ecosystems.

Maximum values of CH_4 flow from soil reached $(70 \pm 35)\,\mu mol\,m^{-2}\,hr^{-1}$ and were recorded in the arid areas of east Siberia. Emissions by wet tundra in the latitudinal belt 67°–77°N are similar to those observed in lower latitudes, but the Siberian boreal wetlands in the belt 50°–60°N make a very important contribution to the formation of the global budget of methane. The CO_2 flow is characterized by the variability trend which is opposite to that for CH_4. The mixing ratio values for the surface-air ozone varied from several $nmol\,mol^{-1}$ in the presence of inversions at night to more than $60\,nmol\,mol^{-1}$ in the daytime. These values turned out to be higher than those obtained during the TROICA-2 expedition in the summer of 1996. The CH_4 and CO concentrations were close to those observed in the TROICA-2 period. Increased CH_4 concentrations at an average mixing ratio of $(1.97 \pm 0.009)\,\mu mol\,mol^{-1}$, were observed over the western Siberian lowlands, decreasing to $(1.88 \pm 0.13)\,\mu mol\,mol^{-1}$ with transport toward east Siberia. In contrast, the background level of CO concentration over the western Siberian lowlands was, as a rule, below $140\,nmol\,mol^{-1}$, whereas east of Chita (~52°N, 113°E) the CO concentrations were high, in one case exceeding $2\,\mu mol\,mol^{-1}$ – caused by forest fires leading to considerable changes in atmospheric chemistry in some regions of Russia.

4.7 VOLCANIC AEROSOL

Volcanic eruptions are another significant source of many MGCs of the atmosphere and aerosol [76], and the mechanisms of aerosol formation as a secondary product of gas-to-particle conversion and other reactions are quite various. Sansone *et al.* [124] have carried out observations along the shoreline of the Kilauea volcano (Hawaii) and detected trace elements produced by lava–seawater interactions in the volcanic aerosol plume. Analysis of MGC concentrations measured in the plume, normalized to the Hawaiian basalt composition, showed a linear log–log (on a double logarithmic scale) relationship between concentration and emission coefficients (as indicators of volatility). Normalized aerosol concentrations correlate with respective values for dissolved fumarolic gas from the Kilauea volcano as well as for fumarolic condensates from Kudriavy and Merapi volcanoes, despite different mechanisms for elements volatization. Roughly estimated regional ocean surface deposition rates of Cu, Cd, Ni, Pb, Mn, Zn, Fe, and P were more than 50 times higher than the background rates. Thus, volcanoes may be an important source of both toxic and nutrient elements for the surrounding ocean. However, it is unlikely that this volcanism can exert a significant impact on the global ecosystem, even with massive lava outflows to the ocean.

With eruptive emissions the cloud of volcanic matter rises to a very high altitude (Krakatau up to 60 km, El Chichón up to 37 km) and deposits much later. For instance, the Krakatau cloud deposited after three years, and the El Chichón cloud was observed in Italy and Japan six months later and caused a temperature decrease in the northern hemisphere by -0.5° during the subsequent three years. For climatologically important eruptive emissions, the volcanic particles together with gases of volcanic origin can be supposed to rise to the atmosphere up to >20 km, on average, and the smallest particles can reside in the stratosphere for several years. Despite the episodic nature of eruptions, their mean annual power can be estimated at about 10^8 t. Besides this, sulphur dioxide emitted to the stratosphere ($\sim 10^7$ t yr^{-1}), carbonylsulphide, hydrogen chloride, nitrogen oxides, and water vapour chemically and photochemically react with various gas and aerosol atmospheric components, forming new aerosol particles, and initially, sulphuric acid and sulphates.

Volcanic eruptions are also powerful sources of sulphate aerosol. Thus, for example, the explosive eruption of Mt. Pinatubo (Philippines, 15.1°N, 120.4°E), the most powerful eruption of the 20th century which occurred on 14–16 June, 1991, emitted 14–26 Mt SO_2 into the stratosphere (up to 30 km). Within about 35 days, SO_2 quickly transformed into sulphuric acid aerosol H_2SO_4/H_2O with mass reaching about 30 Mt. The main part of the volcanic cloud rapidly moved westward, and in three weeks encircled the globe. In about two weeks a considerable part of the cloud crossed the equator and reached $\sim$10°S. For the first two months, the main part of the cloud was concentrated in the latitude belt between 20°S and 30°N, and thus it formed the tropical reservoir of aerosol matter strongly influenced by the quasi-biannual oscillation (QBO). This reservoir is stable or unstable depending on the QBO phase (east or west) (the eruption took place during the east QBO phase, and therefore the tropical maximum of the eruptive aerosol layer was stable; only after three–four months did the aerosol begin to spread to the mid-latitudes of the southern hemisphere). At altitudes below 20 km the aerosol moved relatively rapidly to the middle and high latitudes of the northern hemisphere. Stratospheric aerosol entered the troposphere through breaks in the tropopause and due to gravitational sedimentation. The global mass of the volcanic sulphate aerosol reached its maximum in October, 1991 and then it decreased approximately by e times per year.

Liu and Penner [97] have performed a numerical simulation of eruptive aerosol transformation on a global scale using the LLNL/IMPACT model developed at the Livermore National Laboratory (USA). This model takes into account both chemical processes and aerosol transport under meteorological conditions reproduced using a 46-layer model of the atmospheric general circulation (with a spatial resolution of 2° lat. × 2.5° long.) developed in the Goddard Space Flight Center, NASA (USA).

Analysis of the numerical simulation results showed that they reliably simulate the formation of the tropical reservoir of aerosol in the 20°S–30°N latitude belt after a few months of the eruption, and agree with the aerosol propagation characteristics obtained from the SAGE-II satellite data. The data on the global distribution of the eruptive sulphate aerosol were used to calculate the rate of homogeneous nucleation of ice particles from H_2SO_4/H_2O (number of frozen aerosol particles per cm^3 s^{-1}) in the lower stratosphere and upper troposphere. The results obtained were compared with similar data on the near-surface emissions of natural and anthropogenic sulphate aerosol.

The high values of the nucleation rate are characteristic (in the case of volcanic aerosol) of the layer near the equatorial tropopause (at the bottom of the main volcanic aerosol layer) during the first year after the eruption. For the eruptive aerosol, the nucleation rate is much higher than that for the sulphate aerosol from ground-based sources, but during the second year after the eruption, the nucleation rates become comparable. From the results discussed, it follows that the volcanic aerosol can significantly affect (through homogeneous nucleation) the conditions of formation of cirrus clouds and their global evolution. However, to obtain reliable estimates, it is necessary to perform a numerical simulation with microphysics and the dynamics of cirrus clouds taken into account.

During the period 1974–1981, measurements were repeatedly carried out of the surface layer aerosols near the active volcanoes of Kamchatka (Tolbachik, Kliuchevskoy, Gorely, Karymsky, Mutnovsky). A particle size distribution for $r \geq 0.2\,\mu m$ was determined with the use of a photoelectric counter A3–5M and for $r \leq 0.5\,\mu m$ using the electron microscopic analysis of impactor and filter aerosol samples [85].

Sufficiently numerous measurements of the chemical and elemental composition of volcanic aerosols testify to processes of their formation and further evolution: emission of products from crater wall abrasions and particles of crushed lava as well as water vapour, sulphur dioxide, hydrochloric acid, and volatile metals, further participate in various chemical reactions and become condensed. The chemical analysis of smoke and dust material of volcanoes reveals the prevalent content of silicon compounds (60–80%), sulphates (10–30%), tiffs (3–10%), aluminium compounds (0–20%), and iron (0–1%). However, a more detailed consideration of the results of chemical and elemental analyses reveals considerably different compositions of the emitted material from different volcanoes.

Powerful volcanic eruptions are characterized by changes in the chemical composition of aerosol particles with increasing altitude: a strong enrichment of some moderately volatile elements (arsenic, selenium, lead, cadmium, and zinc) in small particles as well as elements typical of magma (silicon, calcium, scandium, titanium, iron, zinc, and thorium) contained in larger particles. This can be interpreted only as the result of the fact that the source of the matter in the upper boundary of the plume is not of particles of the destroyed top of the volcano but represents hot emissions of magma. Also, changes in the chemical and elemental composition of volcanic material were observed at different periods of eruption [61]. When analysing the variability of the elemental composition of volcanic aerosols one can normalize the content of elements in them by reference material – the elemental composition of magma emitted lava and ashes of various rocks [75]. Analysis of the composition of rocks during some eruptions revealed considerable variations in their elemental composition. Most constant is the content of silicon dioxide. The following data on the content of silicon dioxide in different types of volcanic lava are typical: basalt–toleite (Kilauea volcano) 47–52%, andesite

(Fuego) 48–54%, phonolite–tephrite 50–55%, dacite (St. Helens) 65–70%, and rhyolite (Askya) 68%.

Changes in the Al content are much greater, especially in dunite lava, and Ca and Mg – in syenites, where marked losses of Fe, Na, K, Ti, Mn, Ni, and P are also observed. Still more substantial are variations in the content of elements in ash. For instance, the elemental analysis of ash of the Popokatepetl volcano eruption (21 December, 1994–28 January, 1995) showed that ash was heavily enriched, compared with the Earth's crust, with such easily sublimable elements as S, Br, Pb, Hg, Zn, Cu, whose factor of enrichment $\mathrm{EF}(x) = [x]_a/[x]$ exceeds 10 and changes depending on the time of ashfall.

The normalization of element content should be made by the elemental composition of the Earth's crust using silicon as a reference element. In this case, the error for most of the elements should not exceed 20–30%. A unified technique has been used to detect the spatial–temporal variability of the elemental composition of volcanic aerosols. Tables 4.2 and 4.3 (pp. 236–238) give the FE values for the elements of the eruptive aerosols of volcanoes Augustin and Popokatepetl. The concentration ratio of the elements changes markedly with altitude.

For Na, K, Mn, Ba, S, V, Sc, Hf, Yb, As, Eu, W, Se, and Au the EF increases with increasing altitude, and for Cl, Pb, Br, and Cd it decreases. Probably, this is evidence of a faster aerosol formation and the condensation growth of particles containing Cl, Pb, Br, and Cd compared with compounds containing other elements. Still more interesting is the temporal variability of the EF for some other elements. The EF values are relatively constant for elements such as Al, K, Ba, Cl, Th, and Sm. For Fe, Ca, Mg, V, Cr, La, Co, and Sc the EF increased markedly only during the first day, and for elements Ti, Mn, W, Cd, Zn, Cu, Pb, and Au the EF decreased regularly. For Zn, Cu, Cd, and Pb the EF decreased by more than an order of magnitude. Similar behaviour was observed for elements As, Sb, Se, and Br, with maximum EF values during the first day of measurements. For the second day it dropped sharply and then decreased regularly. Emissions from the Augustin volcano to the atmosphere were at a maximum on 2 February, 1976. The EF values increased for elements Na, Pb, Hg, and Ca, and a marked decrease was observed for S, Mg, Cs, Rb, Cr, and Co with $\mathrm{EF} < 1$ and for Cu, Ba, Sr, and V with $\mathrm{EF} > 1$. Quite special is the element S with an $\mathrm{EF} < 1$ during 2–4 February and $\mathrm{EF} = 49$ by 21 February. Due to the data obtained during the Popokatepetl eruption, a considerable part of terrigenous elements (Fe, Al, Ca, K, Ti, Cr, Mn, Ga, and Zr) had very low EF values.

One should emphasize that prolonged and heavy pollution of the atmosphere is caused by emissions of gas-phase sulphur-containing compounds (sulphur dioxide and carbonyl sulphide) which are transformed into sulphuric acid and sulphates [60]. In particular, the long effect of the Pinatubo eruption was mainly determined by the process of formation of H_2SO_4 aerosols in the lower stratosphere [76]. There is a statistically significant connection between volcanic activity and climate characteristics.

4.8 HIGH-LATITUDE ATMOSPHERIC AEROSOL

Polissar *et al.* [113, 114] performed an analysis of measurements of the aerosol chemical composition from data of seven locations in Alaska (within the National Park) for the period 1986–1995. Results of the elemental analysis of soot ('black' carbon, BC) concentration and total mass of aerosol have been obtained with the use of samples containing particles with a diameter $<2.5\,\mu m$. On the supposition that sulphates SO_4^{2-} are the source of all elemental sulphur, calculations have been made of an excess (non-marine) concentration of sulphates (XSO_4^{2-}) by extracting the share of the mass of sea salt sulphates SO_4^{2-}.

The mass concentration of XSO_4^{2-} varied within $0.01–3.9\,\mu g\,m^{-3}$, and that of BC varied from $0.01–5\,\mu g\,m^{-3}$. Almost at all locations the annual variation of XSO_4^{2-} concentration was at a maximum in winter–spring and at a minimum in summer. However, such a variation was absent in the case of BC and finely dispersed aerosols. The amplitude of the annual variation of XSO_4^{2-} increased in the north-eastern region of Alaska. The concentration of soot carbon in central Alaska was at a maximum in summer, which was explained by BC emissions to the atmosphere due to forest fires. Within the studied time interval there was observed a decreasing trend of Pb concentration in central Alaska. Analysis of the obtained results reveals the presence, in all cases, of aerosol pollutants due to a long-range transport and from regional and local sources. The annual variation of XSO_4^{2-} concentration is connected with the annual course of the long-range transport of sulphur compounds due to emissions in the middle latitudes, as well as with changing rates of photochemical formation of SO_4^{2-} in fine-particle aerosols (and fine-particle aerosol removal from the atmosphere). Analysis of the spatial distribution of XSO_4^{2-} revealed a strong negative gradient from the north-west to the south-east, which can be explained by the effect of long-range transport of anthropogenic aerosol from the sources located north or north-west of Alaska.

Studies of the nature, origin, and transport of the arctic aerosol began long ago. It has been established, in particular, that in winter and spring a considerable long-range transport to the Arctic of anthropogenic aerosol from various sources takes place. In the period 1986–1995, at seven locations in the National Park in Alaska, measurements were carried out of the composition of the fine-particle fraction ($<2.5\,\mu m$) of atmospheric aerosol [114]. The results obtained have been processed with the use of a new method of factor analysis – the so-called Positive Matrix Factorization (PMF). This method provides a more reliable consideration of data which are absent or are below the detection threshold.

Analysis of data using the PMF method has made it possible to detect components emitted by six sources. All data show the presence of the soil component of aerosol with maximum number density in summer and a minimum in winter. At five locations a salt component was found. At all (except one) locations the presence of elemental carbon, H^+, and K was recorded, whose sources were, apparently, from forest fires. Universal components of the aerosol composition are S, BC–Na–S, and

Table 4.2. Factors of enrichment for various elements in volcanic aerosols of Augustin volcano, 1–21 February, 1976.

	1 February 1976		2 February 1976		3 February 1976	4 February 1976	20 February 1976	21 February 1976
Element	0–8 km	8–32 km	0–8 km	8–32 km	Troposphere			
Al	1.08	1.11	1.15	1.83	1.25	1.32	1.3	1.13
Fe	1.31	0.85	0.92	1.68	0.92	0.89	0.80	0.80
Ca	1.51	1.66	0.83	1.27	1.13	1.08	0.95	0.95
Na	1.18	1.43	1.61	2.70	1.74	1.74	1.2	1.2
K	0.44	0.51	0.41	0.78	0.61	0.61	0.54	0.51
Mg	1.55	0.81	0.41	0.54	0.46	0.49	0.38	–
Ti	1.51	1.43	0.83	1.08	0.75	0.63	0.68	0.54
Mn	0.98	1.43	0.94	1.59	0.97	0.91	0.86	0.77
Ba	2.2	3.2	1.67	3.4	2.6	2.2	2.3	1.92
S	1.31	3.1	0.121	0.063	0.050	0.058	49	49
Sr	<7.6	<3.3	0.66	2.6	1.42	0.79	–	–
Cl	7,900	4,100	350	5.3	20	62	–	–
V	1.46	1.75	1.13	1.31	0.94	0.95	0.84	0.84
Rb	<0.35	0.31	0.16	0.34	0.19	0.28	–	–
Cr	<15	<13	0.167	4.2	0.132	0.166	0.120	0.14
Zn	9.4	6.4	<7.4	12	5.2	2.8	–	–

Ce	0.66	0.71	0.61	0.72	0.98	0.82	0.87	0.54
Cu	29	16	4.7	7.7	3.7	1.72	3.4	1.5
La	1.64	0.60	0.58	0.88	0.69	0.58	0.61	0.51
Co	1.0	1.07	0.55	0.99	0.52	0.55	0.45	0.50
Sc	1.45	1.64	0.92	1.66	0.95	0.03	0.82	0.93
Pb	52	23	56	24	9.6	7.9	–	–
Th	0.49	0.44	0.59	0.98	0.67	0.60	–	–
Sn	0.68	0.83	0.79	0.29	1.00	0.73	–	–
Cs	0.18	0.16	0.12	0.23	0.95	0.55	–	–
Hf	1.28	1.49	1.79	3.2	2.1	1.92	1.82	1.63
Yb	0.89	1.32	0.91	1.75	1.50	1.20	–	–
As	370	460	36	80	28	17	25	14
Eu	0.2	0.84	0.90	1.62	1.12	1.06	1.02	0.92
W	5.9	8.3	1.79	2.8	<0.60	<0.73	–	–
Sb	304	260	33	66	32	17	27	14
Br	38,000	13,000	3,500	1,400	197	180	–	–
Cd	290	166	65	22	39	27	35	24
Hg	>55,000	>76,000	>56,000	>6,200	<2,500	>1,500	–	–
Se	3,600	4,900	320	630	240	122	214	110
Au	47	96	8.5	16	6.5	5.9	6.3	5.1
M $\mu g/m^3$	20	9.77	2,100	250	620	145	–	–

Table 4.3. Factors of enrichment EF(x) by Si for several elements in aerosol samples from the Popokatepetl volcanic eruption.

Element	29 December 1994	06 January 1995	14 January 1995	21 January 1995	27 January 1995	27–28 January 2001	28 January 1995
Fe	0.24	0.228	0.72	0.36	0.40	0.40	0.32
Al	1.04	1.20	1.64	1.36	1.12	1.24	1.60
Ca	0.48	0.52	0.84	0.68	0.68	0.72	0.72
S	1.92	340	432	364	180	92	436
P	4	7	21	7	4	6	6.8
Cl	28.8	44	144	100	26.4	52	68
K	0.16	0.20	0.20	0.12	0.32	0.28	0.28
Ti	0.36	0.24	0.48	0.52	0.48	0.48	0.44
Cs	0.72	0.44	3.00	2.16	0.60	1.24	0.76
Mn	0.12	0.22	0.60	0.15	0.27	0.32	0.24
Ni	0.37	0.33	2.24	0.80	0.56	0.92	0.80
Cu	8.4	2.08	18.0	3.16	0.96	1.36	2.40
Zn	1.12	0.64	7.2	2.32	1.56	2.64	2.44
Ga	0.23	0.19	2.44	1.56	0.52	0.44	1.20
S	13.6	18.0	104	28	6.4	1.8	44
Br	10.4	12.4	48	19.8	22.4	12.8	9.2
Rb	0.18	0.22	2.16	1.32	0.48	0.4	1.08
Sr	0.40	0.40	0.44	0.92	0.56	0.68	0.44
Y	0.54	1.08	296	3.44	0.72	1.20	3.52
Zn	0.34	0.27	1.52	0.72	0.72	0.72	0.52
Hg	180	260	740	360	180	240	330
Pb	5.2	7.6	36	13.6	7.2	12.8	11.6

Zn–Cu, and at three locations Pb and Br. Elements such as S, Pb, and BC–Na–S are characterized by the annual variation with a maximum in winter–spring and a minimum in summer – the amplitude of the annual variation increasing with latitude. Special features of the annual course and the elemental composition of these components testify to their anthropogenic origin from remote sources. The four main aerosol components are: anthropogenic aerosol transported from remote sources (Arctic haze), sea salt aerosol, local aerosol of soil origin, and aerosol with a high BC content due to forest fires.

Kerminen *et al.* [69] performed an analysis of the dependence of the physico-chemical properties of aerosol samples (changes of aerosol chemistry depending on particle size) gathered in the central part of the Greenland glacier in 1993–1995. The Aitken mode ($d < 0.1\,\mu m$), one or two modes of accumulation aerosol (0.1–1 μm), and supermicron particles have been studied. Particles of the Aitken mode were characterized by a relatively small total mass (<5%), but the largest number density. As a rule, the accumulation mode is characterized by two overlapping modes with the minimum diameter of particles being 0.4 μm. A possible reason for this bimodality was the 'processing' of particles by fog, which often took place above the glacier on summer nights.

The accumulation mode particles consisted mainly of sulphate, ammonium, methane, sulphonic acid (MSA), and dicarboxylic acid, these particles representing, apparently, an internal mixture of submicron particles. The mixing ratio of MSA and sulphates varied, depending on particle size, within the range of the accumulation mode. Therefore, different methods of particle deposition on the snow surface should lead to changes in the MSA/sulphates ratio.

Supermicron particles usually contained sulphates (<20%) and nitrates (>95%). Probably, sulphates and nitrates are products of reactions of SO_2 and HNO_3 with particles of rocks. Supermicron sulphate particles have smaller sizes (<3 μm) compared with nitrate particles (2–3 μm). Aerosol contains only a small amount of semivolatile acid compounds, except the cases when plumes of biomass burning products reach the glacier. Such plumes are characterized by an increased content of low-volatile compounds, and they may contain submicron particles of ammonium nitrate and ammonium fermate.

From June 1993 to June 1994 a continuous record was carried out at the Summit of the Greenland plateau (72°35′N, 37°38′W, 3,220 m a.s.l.) of the atmospheric aerosol elemental composition (fractions of small and large particles) using the x-ray instruments PIXE [69]. In many aerosol samples the concentration of Al, Si, S, Cl, K, Ca, Ti, Mn, Fe, Zn, Br, Sr, and Pb exceeded the detection threshold, the source of these elements being plateau rocks, sea salts, and anthropogenic components (in the case of S and Br, other sources were possible). The first of the enumerated sources manifested itself at a maximum in spring, whereas the power of the anthropogenic source was substantial all the year round, but especially in winter and spring. The presence of sea salts in aerosols was difficult to detect, but most markedly they manifested themselves in winter.

The obtained results [69] have been compared with the data of observations in the southern part of the Greenland ice sheet (point Dai-3) and at the station 'North' located on the north-eastern coast of Greenland. At all these locations the anthropogenic component concentration was at a maximum in winter and spring, when fuel burning was most intensive, in contrast to the Arctic, where a vividly expressed maximum of the anthropogenic component was connected with the formation of the Arctic haze in early spring.

The data on the chemical composition of solid deposits accumulated in the past by polar ice sheets are sources of information about the chemistry of the pre-industrial atmosphere and its changes on different timescales. Analysis of ice cores enabled one, for instance, to obtain information about sea salt and dust atmospheric aerosols, as well as sulphate aerosol – the product of natural gas-to-particle conversion in the atmosphere. Interesting data have been obtained on ammonium and light carboxylic acids, which has made it possible to simulate the evolution of biogenic emissions. Analysis of Greenland ice cores characterizing the climate change for the last 15,000 years showed that in the period of the last ice age the values of background concentrations of ammonium and formic acid vapour had been much lower than the present ones. Apparently, it had been caused by a decrease of biogenic emissions to the atmosphere by soil and vegetation, since a considerable part of North America had been covered with glaciers. On the other hand, the monitoring of the content of ammonium and formic acid in the atmosphere revealed substantial sporadic variations due to biomass burning in high latitudes. The chemical composition of the layers is characterized by a mixture of ammonium, formic acid, and nitrate with the $NH_4^+/(HCOO^- + NO_3^-)$ polar ratio close to unity. The difference in the chemical composition can be explained either by different forest fire conditions (the ratio between burning and smouldering) or by different meteorological conditions transporting the plume to Greenland.

Three periods of an enhanced input of biomass burning products have been recorded: 1200–1350, 1830–1930, and, to a lesser extent, 1500–1600. The period 1200–1350 coincides with the phase of a warm and dry climate typical of the Middle Age warming. After the decrease of the input of products of burning during the Little Ice Age (1600–1850), an increase occurred on the verge of the last two centuries but a decrease in the 20th century. A gradual decrease was observed of the formic acid concentration, which can, probably, be explained by degradation of boreal vegetation in North America.

The results of measurements of BC concentration carried out with the use of an ethalometer at the Antarctic plateau in 1992–1993 have been discussed [70]. Analysis of the results revealed some episodes of an increased BC concentration due to local sources. Such data were excluded from consideration. Processing of the filtered data revealed a clear annual variation of BC concentration with monthly mean values ranging within $0.3–2\,ng\,m^{-3}$ (these values exceed somewhat those observed at the South Pole). A maximum in the annual course is observed in summer in the presence of a secondary (and general) maximum in October (the same variability was also observed at the South Pole, and it is similar to the annual variation of dust aerosols from the data for the coastal station Neumayer). Apparently, such an annual course is determined by the process of biomass burning in the tropics subject to a strong modulation due to processes of BC transport in the region of the Antarctic (the respective mechanisms have not been simulated reliably, so far).

The BC concentration turned out to be too small to affect the snow cover albedo in the Antarctic. The similarity of observational data at stations on the plateau and at the South Pole testifies to the fact that analysis of data on BC in ice cores should reliably reflect the paleoevolution of BC in the Antarctic and, in particular, serve as an indicator of biomass burning in the southern hemisphere.

4.9 GLOBAL SPATIAL–TEMPORAL VARIABILITY OF AEROSOL

The observational data discussed earlier illustrate the great efforts made recently to obtain more reliable information about the global spatial–temporal variability of aerosol properties [4, 68, 71–76, 106, 131]. One of the vivid examples of such efforts is the field observational experiment APEX started in 1999 to study the properties and climatic impact of aerosol [104] (this programme is carried out under the direction of Professor T. Nakajima, Director of the Centre for Climate System Research at the

University of Tokyo). An intensive development of aerosol studies dictates the necessity to mention briefly the results of the latest studies before discussing the basic theme of this overview.

Streets *et al.* [129] have made a detailed summary of MGCs and aerosol emissions to the atmosphere in Asia from the observational data for 2000. Estimates of anthropogenic emissions have been obtained with all major sources taken into account (including biomass burning) in 64 regions of Asia. The total levels of emissions of various MGCs ($Tg\,yr^{-1}$) constituted (the respective component is given in brackets): 34.3 (SO_2), 26.8 (NO_x), 9,870 (CO_2), 279 (O), 107 (CH_4), 52.2 (NMVOC), 2.54 (aerosol black carbon BC), 10.4 (organic carbon OC), and 27.5 (NH_3). Apart from this, 19 subcategories of non-methane volatile organic components (NMVOC) have been considered and classified in accordance with their functional special features and chemical activity. This suggests the conclusion that in this summary, major sources and types of anthropogenic emissions of MGCs and aerosols in the regions of the field observational experiments TRACE-P and ACE-Asia have been taken into account.

China contributes most to the formation of the concentration field of anthropogenic MGCs and aerosols, with the following levels of emissions ($Tg\,yr^{-1}$): 20.4 (SO_2), 11.4 (NO_x), 3,820 (CO_2), 116 (CO), 38.4 (CH_4), 17.4 (NMVOC), 1.05 (BC), 3.4 (OC), and 13.6 (NH_3). Maps have been drawn [129] of emissions with different spatial resolution, from 1° lat. $\times$ 1° long. and higher. Data of Table 4.4 characterize the levels of emissions of various compounds. A comparison of the data of the summary under discussion and observations in the period of TRACE-P (February–April, 2001) revealed a sufficiently close agreement of average values. The errors in the considered inventory (at a 95% statistical significance) vary from very small ($\pm 16\%$ for SO_2) to very high ($\pm 450\%$ for OC).

Alfaro *et al.* [2] discussed the results of observations of the chemical composition and optical properties of aerosol carried out at Zhenbeitai located near Yulin city (38°17′N, 109°43′E, province Shaanxi, China) in April, 2002. A year before,

Table 4.4. Levels of emissions of various MGCs and aerosol (PM-10, PM-2.5, and sea salt).

Component	Natural sources	Internal burning	Open burning	All sources
SO_2	2.0	5.6	0.2	7.8
NO_x	0.005	4.0	1.1	5.1
CO_2	–	1,453.0	459.6	1,912.6
CO	–	34.7	27.8	62.5
CH_4	–	16.5	1.3	17.8
NMVOC	18.3	6.7	5.0	30.0
BC	–	0.3	0.2	0.5
OC	–	1.2	1.4	2.6
NH_3	–	4.7	0.4	5.1
PM_{10}	107.7	–	–	116.0
$PM_{2.5}$	26.2	–	–	31.2
Sea-salt	41.6	–	–	41.6

Zhenbeitai was one of the main continental locations of observations within the international programme ACE-Asia to study the aerosol properties in Asia. In spring, this observational location south-west of the desert Mu Us, turns out to be a 'crossroad', across which the long-range transport of dust aerosol (DA) takes place from major DA sources located in the deserts of China.

Instruments used to measure the aerosol size distribution could also measure the number density, mass concentration, elemental composition, and scattering properties of PM-9 aerosols (particles with $d < 9\,\mu m$). The sun photometer served to measure the aerosol optical depth (AOD) of the atmosphere and single-scattering albedo. During the observational period under consideration, several dust outbreaks took place, in one case the mass concentration of PM-9 aerosols reached $4{,}650\,\mu g\,m^{-3}$, and the coefficient of aerosol scattering constituted 2,800 M $(10^{-6})\,m^{-1}$. Between the events of dust outbreaks, local anthropogenic aerosol contributed much (sometimes most). In such periods the aerosol mass concentration was much lower ($<100\,\mu g\,m^{-3}$), with the BC mass concentration within 0.9–$6.7\,\mu g\,m^{-3}$ and aerosol scattering coefficient within 7–800 Mm^{-1}.

The chemical composition of aerosol in the periods of dust outbreaks remained practically constant. The measured values of concentration ratios Fe/Al (0.63 ± 0.04) and Mg/Al (0.32 ± 0.03) agree with the presence of the DA sources supposed to be in the region of the north-western deserts of China. The validity of this supposition is also confirmed by the results of calculations of the 'back' trajectories of air masses. Dust particles reaching the observational point contained anthropogenic carbon components captured in the process of long-range transport. The number density of such aerosols, the product of the long-range transport, was comparatively constant and characterized by prevailing particles with diameters 1–2 μm, on average. The mass scattering coefficient constituted $1.05 \pm 0.13\,m^2\,g^{-1}$ and the Angström exponent was close to 0.19. The single scattering albedo increases with increasing wavelength from 0.89 (at 441 nm) to 0.95 (at 873 nm). Relatively high values of the single-scattering albedo indicated that the aerosol was always a mixture of mineral, natural, and anthropogenic aerosols, the aerosol not always being a strong absorber even in the periods of the most powerful dust storms.

Iwasaka *et al.* [64] have analysed the results of lidar soundings of the vertical profile of aerosol concentration in the troposphere in Dunhuang (China, 40°00′N, 94°30′E) over the desert of Taklamakan in the summer of 2002. Data on the vertical profile of the backscattering ratio $(BSC_1 + BSC_2)/BSC_1$, where $BSC_{1,2}$ are backscattering coefficients due to molecular and aerosol scattering at 532 nm, testify to the fact that aerosol was concentrated in the lower 6-km layer of the atmosphere, above which this ratio rapidly decreased from 2–5 to 1. The ratio of depolarization $P_1/(P_1 + P_2)$ varied similarly ($P_{1,2}$ are orthogonal components of the polarization degree) in the presence of a sharp change near 6 km. The electron microscopic analysis of aerosol samples taken with the use of the balloon impactor has shown that high values of the depolarization factor are determined by the effect of dust aerosol particles of an irregular shape. Such non-spherical particles rise from the surface to the free troposphere with prevailing west–east transport, reaching altitudes of about 6 km.

Jordan *et al.* [67] performed an analysis of the chemical composition and physical properties of atmospheric aerosols from samples taken from the flying laboratory DC-8 at different altitudes in the lower troposphere in the spring of 2001 during the period of accomplishing the field experiment TRACE-P. Depending on the type of calculated 'back' trajectories of air masses (i.e., on the direction of the long-range transport), the data of observations under discussion are classified into four groups, corresponding to: the long-range transport from the west (WSW); the effect of regional circulation over the western Pacific and south-eastern Asia (SE Asia); the transport of polluted air from the north of the Asian continent, whose special feature is the presence, in the lower part of the ABL, of a considerable amount of sea salt aerosol (NNW); and the input of mainly dust aerosol (Channel).

The WSW conditions were characterized by low values of the aerosol mixing ratio in middle and high latitudes, with the main component of aerosol being water-soluble NSS inorganic compounds. Small values of the mixing ratio are also typical of both the SE Asia conditions in the lower troposphere and prevailing of NSS aerosols, but with a considerable contribution of a soot component. The NNW conditions are characterized by the highest level of sea salt aerosol concentration with an approximately similar contribution to the aerosol mass at altitudes <2 km of sea salt, NSS, and dust components. In the sector Channel at altitudes <2 km, the share of water-soluble ions and the soot component was most significant. The main mass of aerosol from Asia is concentrated in this sector in low and middle latitudes due to prevailing dust aerosol (compared with other regions).

Under Channel and NNW conditions, the aerosol number density is mainly determined by fine-particle components, whereas the volume of particles by sea salt and dust aerosols. At low altitudes in the sector Channel a maximum of the condensation nuclei (CN) number density was observed, as well as increased (compared with other regions) values of the scattering coefficients. In middle latitudes at altitudes 2–7 km, in the regions of the polluted atmosphere (Channel, NNW), low values of CN combined with a large share ($\geq 65\%$) of non-volatile particles prevail – explained by the effect of a washing out from the atmosphere of hygroscopic CN particles. Low values of the single-scattering albedo in the SE Asia sector reflect the effect of an enhanced content of the soot component of aerosols on their optical properties.

As a result of the progressive dehydration of internal regions of the Asian continent taking place due to a rise of the Tibetan plateau in the end of the Caenozoic era, the input to the atmosphere of dust aerosol from sandy deserts increased. This was then transported over long distances, with the major centre of DA accumulation being the loess plateau (LP). Stratographic studies showed that this accumulation began not less than 7,000–8,000 years ago in the main part of LP and 22,000 years ago in its western region. According to available paleoclimatic data, the process of accumulation that took place during the last 2.5,000 years was characterized by the presence of intervals of an intensive DA deposition and weak chemical weathering (paedogenesis) in the periods of glaciation, which alternated with a moderate DA deposition and strong paedogenesis during interglacial periods.

The dependence of DA deposition on the dynamics of paleomonsoons determines the possibility of DA retrieval from stratographic data.

Sun *et al.* [130] undertook an analysis of the DA samples taken at LP locations during the periods 1995–1996 and 2000. The DA is characterized by the presence of mineral particles together with a large amount of organic matter, especially in summer. The magnetic susceptibility of samples turned out to be higher, as a rule, than in the case of glacial loess. The annual variation, with a maximum of DA concentration in summer and a minimum in winter, was evident. Leaving out an account of the effect of pollution on the present DA, one can think that the high magnetic susceptibility of Palaeozol of the Quaternary period had been determined either by the direct impact of climate or by further enhancement of post-depositional paedogenesis. Special features of the annual course of DA flow suggest the conclusion that the total annual DA deposition is mainly determined by continuous year-round sedimentation, but not by contributions of dust storms during limited time intervals. The DA flow is characterized by a distinct annual course with a maximum in spring or early summer, especially in the northern part of LP.

The particle number density is characterized by asymmetry similar to that observed in the case of the Quaternary loess. The main feature of the annual course of the DA size distribution is an enhanced share of large-size particles in spring and summer, and a decreased share in autumn and winter. At different locations of aerosol sampling the DA size distribution depends strongly on altitude and varies non-linearly with increasing altitude. Data on the vertical profiles of DA concentration, trends, and annual change of DA flow as well as DA size distribution indicate that the long-range transport of the present DA (like the Quaternary loess) took place mainly in the lower troposphere during the whole year, with the DA deposition intensified in spring and early summer. At present, the DA contains an anthropogenic component.

The tropical Indian Ocean is, apparently, the single region on the globe where powerful sources of anthropogenic pollution in the northern hemisphere make contact with the clear atmosphere of the southern hemisphere when crossing the equator with the monsoons. Therefore, the Indian Ocean sector located west and south of the Indian subcontinent (10°N – 11°S, 55°E–75°E) was chosen as the main region for the field observational experiment INDOEX – its aim to study the Indo-Asian aerosol haze propagating annually (in the period November–April). In the course of accomplishing the phase of intensive field observations in January–March, 1999, various means of satellite, aircraft, ship, and surface observations were used.

In this connection, [43] discussed the results of aerosol soundings of the atmosphere with the use of the unique aerosol 6-channel Raman lidar carried out in the Maldives from February, 1999 to March, 2000. The obtained vertical profiles of the aerosol coefficient of extinction, at wavelengths 335 and 532 nm, refer to conditions of a heavily polluted atmosphere. The average value of the aerosol optical depth (AOD) at 532 nm constituted about 0.3, and maximum values reached 0.7. Above the polluted marine atmospheric boundary layer (MABL), haze plumes were observed at altitudes up to 4 km. On average, the contribution of aerosol layers in the free troposphere to the formation of AOD varied within 30–60%. The typical range of variations of the volume aerosol coefficient of extinction at 532 nm in the 'above' aerosol layers constituted 25–175 $M m^{-1}$.

Franke *et al.* [55] described the properties of aerosol plumes observed in air masses from SE Asia and northern and southern India. The ratio between aerosol coefficient of extinction and backscattering (lidar ratio) varied, as a rule, within 30–100 sr, and in the case of strongly absorbing particles from northern India it constituted 50–80 sr. Higher values (~by 20 sr) of lidar ratio for aerosols from northern India compared with European aerosols, agree on the supposition that Indian aerosol is characterized by a high (up to 20%) content of black carbon. In the case of particles from SE Asia and northern and southern India, the Angström exponent for wavelengths 355, 400, 532 nm varies within 1–1.6, 0.8–1.4, and 0.6–1.0. Lower values of the exponent for aerosol from India are determined, apparently, by a substantial contribution of the products of biomass burning to the formation of aerosol in India. Analysis of correlations between lidar ratio, aerosol coefficient of extinction, Angström exponent, and relative humidity has shown that in most cases this correlation is either weak or totally absent.

Within the programme of the period of intensive observations, constituting part of the field observational experiment ACE-Asia to study the aerosol characteristics, on 23 April, 2001 near Tokyo, [102] undertook simultaneous surface lidar soundings of the atmosphere, and the results were compared with the data of aircraft measurements. During the aircraft measurements, at altitudes up to 6 km over the Gulf of Sagali, south-west of Tokyo, four surface lidars were operated. The airborne complex included a sun photometer and sensors for direct (*in situ*) measurements of the aerosol optical properties, size distribution, and ion composition, as well as CO_2 concentration.

According to the data of three polarization lidars, the observed moderate concentration of the Asian DA in the free troposphere was recorded at altitudes up to 8 km. There was a good agreement between the data of different lidars and the nephelometer on the scattering coefficient at 532 nm. The comparison revealed a high stability and mesoscale homogeneity of the aerosol layer at altitudes of 1.6–3.5 km. There was a satisfactory agreement between the values of the extinction coefficient ($\sigma_a \sim 0.03\ km^{-1}$), obtained from data of the airborne sun photometer, and the results of lidar soundings over the ABL. The aerosol depolarization ratio (σ_a) changed substantially with altitude, and a negative correlation was observed between σ_a and the backscattering coefficient β_a at altitudes below 1.3 km.

It follows from the results of aircraft measurements of the optical parameters, that depend on the particle size (the share of scattering due to small-sized aerosols), and of the ion composition of aerosols, that the major factors of the observed σ_a variability are values of mixing ratios for particles of the accumulation mode and large-size (dust) aerosol. During the period of these comparisons, the atmosphere was characterized by the presence of three specific layers: (1) the ABL (from the surface to 1.2–1.5 km) where fine-particle (mainly sulphate) aerosols prevailed, with small values of σ_a (<10%); (2) the intermediate layer (between the upper ABL boundary and the level 3.5 km) where particles of fine-size and dust aerosol were moderately externally mixed up, with small values of σ_a; and (3) the upper layer (above 3.5 km) where the dust aerosol particles prevailed, determining a high level of σ_a (30%). From the data of aircraft measurements west of Japan, a substantial layer of dust aerosol was observed at altitudes 4.5–6.5 km, where the lidar ratio constituted 50.4 ± 9.4 sr. This result agrees well with results of nocturnal observations with Raman lidar made later in the same dust layer, when it moved over Tokyo – the lidar ratio was 46.5 ± 10.5 sr.

A rapid industrial development and population growth in the east Asian region has resulted in a strong increase of anthropogenic pollution of the atmosphere from industrial and agricultural sources, supplemented with powerful emissions to the atmosphere of dust aerosol during dust outbreaks in north-western China, especially in spring. The tropospheric pollution due to various MGCs as well as anthropogenic, dust, volcanic, and marine aerosols, gets transformed and transported over long distances to the western sector of the Pacific Ocean. Carrico *et al.* [23] analysed the results of observations of the MGCs and aerosol concentrations in the troposphere over the Pacific Ocean carried out on board the ship *Ronald H. Brown* during the period 15 March–2 April, 2001.

The data of nephelometer measurements enabled one to estimate the coefficients of scattering (σ_{sp}) and backscattering (σ_{bsp}) for particles with the aerodynamic diameter D_p within 1–10 μm at different wavelengths depending on relative humidity. During the first ten days of the voyage (starting near Hawaii under conditions of prevailing unpolluted air masses), the coefficient of scattering at 550 nm was $\sigma_{sp} = 23 \pm 13 \times 10^{-6}\,(\mathrm{M})\,\mathrm{m}^{-1}$ and the absorption coefficient was $\sigma_{ap} = 0.5 \pm 0.3\,\mathrm{M\,m}^{-1}$ (relative humidity = 19% and 55%, respectively, and $D_p < 10\,\mu\mathrm{m}$). Marine aerosol was characterized by strong hygroscopicity and exhibited distinct signs of solubility and hysteresis. Approaching the coast of east Asia, the atmospheric pollution increased, which was reflected in the values: $\sigma_{sp} = 64 \pm 30\,\mathrm{M\,m}^{-1}$, $\sigma_{ap} = 6.6 \pm 4.4\,\mathrm{M\,m}^{-1}$ and in a large contribution of submicron aerosol to the formation of σ_{sp}. In such cases the aerosol was less hygroscopic and was characterized by a wide range of variability.

A strong scattering was also typical of volcanic aerosol: $\sigma_{sp} = 114 \pm 66\,\mathrm{M\,m}^{-1}$ and $\sigma_{ap} = 11.7 \pm 5.6\,\mathrm{M\,m}^{-1}$ together with the substantial role of the submicron aerosol fraction. This aerosol was strongly hygroscopic. The most intensive scattering is typical of dust aerosol: $\sigma_{sp} = 181 \pm 82\,\mathrm{M\,m}^{-1}$, $\sigma_{ap} = 12.1 \pm 6.4\,\mathrm{M\,m}^{-1}$, with a considerable contribution to the formation of the optical properties both of super and submicron particles.

The most important features of aerosols observed over the Pacific Ocean in the northern hemisphere consisted in the prevalent scattering due to aerosol properties, such as solubility, hysteresis, and the existence of metastable droplets. Average values of relative humidity, at which both solution and crystallization took place, constituted, respectively, $77 \pm 2\%$ and $42 \pm 3\%$. Average ratio $\mathrm{R} = \sigma_{sp}$(relative humidity for the environmental atmospheric conditions)/ σ_{sp}(relative humidity = 19%) varied from 1.25 (dust aerosol) to 2.88 (volcanic aerosol). Values of the backscattering coefficient varied from 0.077 (marine aerosol) to 0.111 (dust aerosol), while the single-scattering albedo varied from 0.94 ± 0.03 (dust and pollutants) to 0.99 ± 0.01 (marine aerosol). The Angström exponent depended weakly on relative humidity but differed strongly for different types of aerosol, varying from 0.16 ± 0.60 (marine aerosol) to 1.49 ± 0.29 (volcanic aerosol).

The dominating factor of sensitivity of the aerosol-induced radiative forcing to chemical composition of aerosol is, apparently, the dependence of single-scattering albedo on the content of aerosol components absorbing the short-wave radiation (first of all, soot component). In this connection, during three weeks in March–April, 2001, in Yasaka (Japan: 35.37°N, 135.81°E), located on the coast of the Sea of Japan, [54] undertook studies of the dependence of the aerosol optical properties on the size and composition of aerosol particles within the field observational programme ACE-Asia. On the other hand, the optical properties of aerosol were calculated from its prescribed chemical composition and number density, including its absorbing component, to further compare the results with the observational data. Continuous measurements of the single-scattering and backscattering coefficients, depending on the size of particles, were made at wavelengths 450, 550, and 700 nm. Measurements of the aerosol size distribution were made with cascade impactors three times during the whole period of measurements, each session of these measurements continued for 3–4 days. During one of these sessions (20–25 March, 2001) a yellow dust aerosol appeared on the Asian continent, whereas in other cases the effect of dust aerosol was negligible.

According to the results obtained, the dust aerosol was accompanied by the fine-particle fraction of the accumulation mode. The effect of DA on the optical and chemical properties of aerosol was analysed and compared with observational data. Before comparing the calculated and observed optical properties of aerosol, the component-wise control was carried out of the estimates of total mass concentration of aerosol and of the effect of errors due to incomplete consideration of the aerosol components on the results of calculations of the optical characteristics (most substantial errors were caused by neglecting the volatile and absorbing components of aerosol). On the whole, the agreement between calculated and observed values of the coefficients of scattering and absorption turned out to be satisfactory, whereas in the case of the backscattering coefficient, a substantial difference was observed, especially in the case of the large-size mode of aerosol. There was a distinct difference in the dependence of single-scattering albedo on wavelength in the presence of DA in the atmosphere and without its presence. In the first case, when the aerosol mass concentration was at a maximum, the single-scattering albedo grew with increasing wavelength, whereas during the third period of observations (with a minimum aerosol content) the single-scattering albedo decreased with increasing wavelength.

Bahreini *et al.* [8] considered the results of observations of the size distribution and chemical composition of atmospheric aerosol carried out during the period of the field observational experiment ACE-Asia with the use of the aerosol mass spectrometer (AMS) Aerodyne carried by the flying laboratory Twin Otter. The flights made during the period 31 March–1 May, 2001 covered the region 127°E–135°E, 32°N–38°N, with reliable data obtained during 15 of the 19 flights. According to AMS data, during this period of observations, in the atmosphere between the ABL and altitude of about 3.7 km, there were distinct layers of submicron aerosol consisting mainly of sulphate, ammonium, and organic compounds. Between these layers, there were layers with sufficiently lower concentrations of aerosol. In some aerosol layers, the mass concentration of sulphate and organic matter reached, respectively, $10\,\mu g\,m^{-3}$ and $13\,\mu g\,m^{-3}$.

Analysis of 'back' trajectories of air masses has shown that the layers of aerosol pollution have formed in the urban and industrial regions of China and Korea. The size distribution of fine-particle aerosol (weighted-by-mass size distribution of particles) varied little (from day to day and from layer to layer), with typical values of particles having aerodynamic diameters within 400–500 nm and distribution widths (by half-value with respect to maximum) of about 450 nm. In the absence of dust outbreaks from the Asian continent, there was a correlation of the non-refracting aerosol mass concentration and its total concentration.

Aircraft measurements of the aerosol radiative characteristics (from the flying laboratory C-130) made during the period 28 March–6 May, 2001, with the use of nephelometers and sensors PSAP to determine the aerosol-induced solar radiation absorption, enabled one to obtain extensive information about the optical properties of atmospheric aerosol. An analysis of the observational results performed by [5] revealed the presence of both the prevailing modes of fine-particle aerosol, connected with atmospheric pollution, and large-size mineral dust aerosol. The fine-particle aerosol turned out to be moderately absorbing, with an average single-scattering albedo at 550 nm and a low relative humidity $\omega = 0.88 \pm 0.03$ (at a 95% level of statistical significance). This component of aerosol is moderately hygroscopic and at relative humidities within 40–85% the scattering enhances with increasing relative humidity, whereas the large-size component is a very weak absorber ($\omega = 0.96 \pm 0.01$), and it is almost non-hygroscopic. The results considered can be used to substantiate the optical models of the Asian aerosol and to estimate the aerosol radiative forcing.

Zhang *et al.* [141] performed an analysis of aerosol samples collected in Kumamoto, a coastal city in the south-west of Japan (32°48′N, 130°45′E) during the periods of three dust storms in the spring of 2000 on the Asian continent. Studies were carried out of chemical composition, size distribution, and mixing of aerosol particles, including such components as sea salt, sulphate, and nitrate using electron microscopy and x-ray analysis. About 60–85% of DA particles were internally mixed with sea salt. Analysis of this situation in the context of weather conditions has made it possible to suppose that such particles resulted from collision and coagulation of particles of original DA and sea salt.

Data on weight ratios of mineral components and sea salt in some particles have shown that the composition of mixed particles is characterized by a prevailing mineral component, sea salt, or both. In all cases considered the aerosol size distribution turned out to be similar to the particle diameter, ranging between 1 and 8 μm (number density is at a maximum at $D \sim 3\,\mu m$). Particles outside the range 1–8 μm were almost absent. A combination of DA and sea salt resulted in a DA size increase. Apparently, particles with $D > 3\,\mu m$ are rapidly removed from the atmosphere – accompanied by a decreasing number density for such particles.

Data on particles' chemical compositions indicate that 91% or more contain sulphate, and 27% or less contain nitrate. A comparison of data on relative weight content of natrium, sulphur, and chlorine in mixed and sea-salt particles confirmed the earlier established fact that the presence of mineral substance favours the formation of aerosol sulphate and nitrate and limits a decrease of chlorine content in the sea salt components of the mixed particles. It is emphasized that these results have been obtained in the absence of 'processing' of aerosol by clouds which determine a strong change in aerosol properties.

Within the programme of the field observational experiments of APEX-E2 on the studies of variability of aerosol properties in the environment and also ACE-Asia (the Asian regional experiment to study the aerosol characteristics over the East China Sea and Japan), [123] carried out studies of the optical properties of atmospheric aerosol using sun multichannel photometers CE-318-1 and -2 and a polarimeter PSR-1000 at three observational locations: Amami-Oshima Island in the East China Sea (28.37°N, 128.50°E), Noto (37.33°N, 137.13°E), and Shirahama (33.68°N, 135.35°E).

From the data of radiometers, the AOD and Angström exponent values were found, and with the use of all the results of measurements aerosol size distribution, complex refraction index (CRI), and some other characteristics were identified. An analysis of the observational results to recognize the sources of aerosol has been made with the use of results of calculations of 'back' trajectories on the basis of the HYSLIT-4 model. The obtained results reflect an expediency of dividing data on AOD, size distribution, and CRI into two categories: (1) wintertime background aerosol; and (2) aerosol of soil origin appearing in spring during dust outbreaks in China. Data on size distributions testify to prevalence of large-size aerosol.

Within the programme of the Global Tropospheric Experiment (GTE), the fourth field observational experiment PEM-Tropics was accomplished in March–April, 1999, aimed at studies of the gas and aerosol composition of the troposphere using instruments carried by the flying laboratories DC-8 and P-3. During the P-3 flights from Wallops Island and surface measurements from the tower on Christmas Island with the use of filters, aerosol samples have been accumulated to study the aerosol associated final products of the sulphur cycle – non-marine sulphates NSS SO_4^{2-} and methanesulphonate MS^- – to test the adequacy of the model of the sulphur cycle in the MABL.

An analysis of the observational results performed by [33–35] has shown that the concentration of most of the aerosol connected ions decreases rapidly with altitude over the ocean surface, which hinders a comparison between results of measurements at the tower (at a height of 30 m) and at a minimum altitude of the P-3 flights (150 m). Theoretically, the mixing ratio of sea salt aerosol particles should decrease exponentially with altitude. The observed vertical gradients of Na^+ and Mg^+ concentrations calculated from the tower and P-3 aircraft (at altitudes <1 km) data confirmed the validity of this conclusion, though an exponential approximation leads to a 25% underestimation of mixing ratios near the surface.

Analysis of samples taken from the tower using a cascade impactor has shown that more than 99% of the mass of Na^+ and Mg^+ is in the share of the supermicron fraction of aerosol, including 65% of particles in the size range 1–6 μm and 20%

>9 μm. These results show that the airborne sensors should be very efficient in sampling the aerosol particles <6 μm. Also, the observational data show that the concentration of NSS SO_4^{2-} and NH_4^+ prevailing in the accumulation mode tends to decrease in the 150–1,000-m layer, but often it turned out that it was higher at a minimum flight altitude compared with the tower. The reasons are still unclear. Probably, in this case the concentration of NSS SO_4^{2-} and NH_4^+ decreases near the ocean surface as a result of their condensation on the surface of large particles which then get rapidly involved in the process of dry sedimentation.

Using measurement data for 414 filter samples obtained in the spring of 2001 from the flying laboratory DC-8, [34] analysed the concentration of the atmospheric aerosol associated ^7Be and ^{210}Pb, dissolved as ions and radioactive tracers. A comparison of the results for the coastal zone of the Pacific Ocean (Hong Kong and Japan), and a remote region of the ocean, revealed a considerable increase in concentration of NO_3^-, SO_4^{2-}, CO_4^{2-}, NH_4^+, K^+, Mg^{2+} and Ca^{2+} in the coastal band. Here the ABL and the lower troposphere were strongly affected by air masses from the continent, which most distinctly manifested itself in an increasing concentration of Ca^{2+} (dust aerosol tracer) and NO_3^- (indicator of HNO_3 uptake by dust aerosol). A comparison of data obtained in the TRACE-P period and during the field experiment PEM-West B, accomplished earlier (1994) in the same region, has demonstrated almost a doubling of concentrations of most of the ions mentioned above in the TRACE-P period. This comparison also revealed an increase in concentrations of Ca^{2+} and NO_3^-, which exceeded earlier values by almost 7 times (in the ABL) and 3 times (in the lower troposphere). A most reliable explanation of the causes of such a considerable increase in concentrations of these ions is that aircraft measurements, carried out in the TRACE-P period, were mostly concentrated within the region of intrusion of polluted air masses from the continent, compared with the PEM-West B period. Thus, there are no grounds to assume that the atmosphere became heavily polluted in the period 1994–2001.

Anomalies of the isotopic composition of sulphur independent of mass (MI), detected recently in samples of Pre-Cambrian rocks, were explained by the effect of the photochemical processes in the atmosphere, and were used to determine the level of oxygen concentration in the early atmosphere of the Earth. In this connection, [122] considered first the data on the MI isotopic composition of sulphur in the present atmosphere, obtained from analysis of sulphate aerosol samples taken in the northern hemisphere. Analysis of paleo and present-day data did not show any correlation between the MI isotopic composition of sulphur and signatures of oxygen, as was observed earlier. It turns out that soil sulphate from the dry valleys of the Antarctic is of atmospheric origin and is characterized by the MI oxygen signature – with respect to sulphur it is mass dependent (MD). Therefore, it is necessary to take into account another process than that proposed earlier to explain the MI oxygen signature in sulphate. Possible reasons of this anomaly, and the importance of the obtained results, have been discussed [122] from the viewpoint of global climate, Archaean era geology, and the composition of the early atmosphere of the Earth.

The origin of the volcanic sulphate aerosol is usually interpreted as connected with the gas-phase mechanism of its formation in the process of slow oxidation of sulphur dioxide. Allen *et al.* [3] considered the results of new observations of Masaya volcano emissions in Nicaragua, from which it follows that sulphate aerosol can also be directly emitted to the atmosphere. Simultaneous samples of aerosols, as well as gas-phase compounds SO_4^{2-}, Cl, and F, were obtained at the edge of the volcano crater in May, 2001. Concentrations of ions SO, Cl^-, and F^- within the plume of volcanic emissions averaged, respectively, 83, 1.2, and 0.37 $\mu g\,m^{-3}$ (in the case of the fine dispersed aerosol fraction <2.5 μm), and 16, 2.5, and 0.56 $\mu g\,m^{-3}$ (large dispersed fraction >2.5 μm).

At pH < 1.0, the fine-particle aerosol turned out to be strongly acid. Sulphate was mainly concentrated in small particles (fine-particle aerosols constituted about 80% of the mass). The basic mass of sulphate was emitted to the atmosphere directly through magma vents. The aerosol acidity was mainly determined by the presence of sulphuric acid and, to a lesser extent, of hydrofluoric acid, with concentration ratios $[H^+]/[SO_4^{2-}]$ within 0.5–0.8 and 0.3–3 for fine-particle and large-particle aerosol, respectively. The mass ratios of gas and aerosol phases of various components averaged, respectively, 95–1178, 37–659, and 43–259.

From the data of measurements with a cascade impactor at the Antarctic station Halley, located on the coast of the Weddell Sea, [118] analysed the dependence of the atmospheric aerosol composition on the particle size. According to the results obtained during the whole year (at 2-week intervals), the main component of aerosol is sea salt, about 60% of salt being from the sources located on the sea ice surface ('spots' of brine on the surface and 'frost flowers'), but not on the open water of the ocean. Sea salt particles are characterized by a low concentration of chloride in summer compared with natrium, which agrees with the idea that in interactions between sea salt particles and minor gas acid components, a loss of HCl, but an increase of HCl content occurs in large particles in winter due to fractionating during the formation of new sea ice cover.

The concentration of NSS SO_4^{2-} reaches a maximum in summer, being concentrated mainly in fine-particle aerosols, which reflects their gas-to-particle conversion. The distribution of methanesulphonic acid (MSA) follows variations of NSS sulphates. A maximum content of MSA is reached in summer mainly due to submicron aerosols. Apparently, the basic source of NSS SO_4^{2-} and MSA is the gas-phase oxidation of DMS. The mass concentration ratio MSA/NSS SO_4^{2-} increases with the increasing size of particles.

As a rule, a deficit of NSS SO_4^{2-} happens in winter, especially in the case of large-size aerosol. From this it follows that in winter the basic source of sea salts is the sea ice cover surface, not the open water surface. Aerosol nitrates are characterized by a maximum content in spring and summer, which demonstrates their connection with sea salt particles. The size distribution of nitrates is similar to that of natrium, and they appear, probably, due to a reaction of nitric acid vapour with particles of sea salt aerosol (this process contributes most, apparently, to the decrease of the chloride content).

Naturally, the strong spatial–temporal variability of aerosol properties necessitates the use of not only the usual but also satellite observational means. Hence, the long history of the development of spaceborne remote sensing can be explained [71, 73].

Observations of the evolution of biomass burning on regional and global scales are one of the successful examples of using the spaceborne aerosol monitoring. Depending on such factors as variability of precipitation and decisions on extents of forest cutting, a strong intergenetic variability of the scales of biomass burning took place for both natural (forest fires) and anthropogenic reasons. Among events of dramatic scale were forest fires in Indonesia (1997–1998) and Mexico (1998), connected with the El Niño–Southern Oscillation (ENSO) event, which caused droughts. Though such events take place mainly in the tropics, large-scale forest fires also occur in middle and high latitudes of the northern hemisphere (e.g., forest fires in western Europe in the summer of 2003).

Duncan *et al.* [37] proposed a technique to estimate the seasonal and interannual variability of biomass burning, to obtain information needed for global models of long-range transport and transformation of MGCs in the atmosphere. From the data of the spaceborne radiometer ATSR, scanning along the satellite's trajectory over the period of four years, as well as the AVHRR data for 1–2 years, information has been obtained about the annual variation of the number of forest fires. The aerosol index (AI) was calculated from observational results, obtained with a TOMS spectrometer mapping the total ozone content, which enable one to assess the conditions of biomass burning in six regions: SE Asia, Indonesia and Malaysia, Brazil, Central America and Mexico, Canada and Alaska, and the Asian part of Russia.

This technique has been used to analyse the averaged annual course and interannual variability of emissions of carbon monoxide due to biomass burning. The results obtained have not revealed any long-term trend of emissions for two decades (1979–2000), but displayed a distinct interannual variability. The level of CO emissions in these regions varied from 20 (the Asian part of Russia and China) to 170 $Tg\,yr^{-1}$ (Indonesia and Malaysia in 1997–1998) with average global (total) emissions of 437 $Tg\,yr^{-1}$ (annual values varied between 429 and 565 $Tg\,yr^{-1}$).

Wang *et al.* [137] analysed the combined (both in space and in time) results of retrieval of the atmospheric AOD at 0.67 μm, using the following sources of information: (1) surface observations with sun photometers at the 12-station network AERONET; (2) a 6-channel sun photometer AATS-6 carried by the flying laboratory C-130; (3) a 14-channel airborne (Twin Otter) sun photometer AATS-14; (4) a shipborne (*Ronald H. Brown*) sun photometer; and (5) a scanning radiometer for visible and IR spectral regions (VISSR) carried by the geostationary satellite GMS-5. Apart from this, data of surface observations of the aerosol size distribution at the Gosan station (Cheju Island, South Korea) were used. The emphasis has been laid on analysis of GMS-5 data every half-hour over 30 days (April, 2001) and a comparison of observational data with results of AOD calculations based on the use of the DISORT technique.

The results under discussion were obtained during the period of accomplishment of the programme of the field observational experiment ACE-Asia – to study aerosol characteristics and they refer to the western region of the Pacific Ocean (20°–45°N, 110°–150°E). A comparison of satellite and surface data on the AOD revealed a good agreement with the coefficient of linear correlation R for four surface locations equal to 0.86, 0.85, 0.86, and 0.87. The same good agreement was obtained when comparing with aircraft ($R = 0.87$) and ship ($R = 0.98$) data. An average error of AOD retrieval constituted about 0.08 with a maximum of 0.15, which is connected, mainly, with errors of calibration (±0.05), uncertainly of prescribed surface albedo (from ±0.01 to ±0.03), and the imaginary part of the complex refraction index (±0.05). Analysis of monthly mean distribution of AOD (from GMS-5 data) revealed the presence of a long-range transport of aerosol with maximum AOD values near the Asian coast and minimum values over the open ocean. An important advantage of the data of geostationary satellites is that they make it possible to obtain information about the diurnal variation of the AOD.

An inclusion into the scientific instrumentation complex of satellites *Terra* and *Aqua* of a moderate-resolution video-spectroradiometer MODIS (a 7-channel scanning radiometer for the wavelength region 0.47–2.1 μm provides a spatial resolution within 250–500 m) has opened up new possibilities to study atmospheric aerosols. The MODIS data enable one to retrieve not only the total aerosol content but also the share of its fine-size fraction. Prospects for MODIS instruments are also determined by the broad possibilities of atmospheric pollution monitoring on local, regional, and global scales. Such possibilities have been illustrated by [25] with results of data processing for the regions of northern Italy, Los Angeles, and Beijing, to retrieve the aerosol optical depth τ_a, bearing in mind an analysis of conditions of local and regional pollution of the atmosphere, taking into account the errors of retrieval $\Delta\tau_a = \pm 0.05 \pm 0.2\tau_a$.

Under conditions of a stagnated atmosphere (and, hence, aerosol accumulation), τ_a can reach values >1 (at 0.55 μm) before wind-driven aerosol removal from the atmosphere or removal by wet deposition. According to results of observations in Italy, the correlation between diurnal mean values of τ_a from the AERONET data (sun photometers) and mass concentration of PM-10 particles with $D > 10$ μm reached 0.82. An estimation of PM-10 from data of satellite observations is only possible in the presence of detailed information on the vertical profile of aerosol concentration. Prospects for obtaining such information (and, hence, data on aerosol concentration near the surface) are connected with the use of results from spaceborne lidar sounding (a laser altimeter GLAS was launched in 2003 and a 'cloud-aerosol' lidar CALIPSO had a planned launch in 2004).

To compare information about total aerosol content for two different regions of the globe, [25] analysed data for two of the most populated regions (eastern China and India) and for two of the most industrially developed regions (the eastern part of the USA/Canada and western Europe). A consideration of the time series of monthly mean values of τ_a for the period July, 2000–May, 2001 revealed a strong annual variation with maxima in spring/summer and minima in winter. A distinct difference between data for the eastern part of the USA/Canada and western Europe, and on the other hand eastern China and India, is explained by the fact that in the latter case the τ_a values turned out to be greater by 50% or 2–3 times than in the first case. The observed increase of the total aerosol content has been determined by smoke from forest fires in the states of Montana/Idaho (USA) subject to a long-range transport to the eastern part of the USA in late August, 2000, as well as by a long-range transport of DA to east China from Taklamakan and the Gobi deserts and smoke aerosol from South-East Asia and from south China in February–April, 2001.

Summarizing the results of analysis of observations of atmospheric aerosol properties, one should emphasize the fact that further developments and realization of a global observational system are needed, using both the usual observational means and satellite observational means as well as regular accomplishment of a coordinated totality of the problem oriented complex field observational experiments.

4.10 CONCLUSION

The basic conclusion to be drawn from analysis of the present ideas about mechanisms of formation, and properties of atmospheric aerosols is that these ideas are still far from being adequate. This situation is especially substantial in the context of consideration of aerosol as one of the most important climate-forming factors. On the one hand, it is obvious that an interactive consideration of aerosol as a climate-forming atmospheric component is needed, and on the other hand, there is no doubt that the still highly uncertain data on the global spatial–temporal variability of aerosol properties and mechanisms of formation do not permit an adequate parameterization of the aerosol dynamics in climate models.

4.11 BIBLIOGRAPHY

1. Panchenko M. V. (ed.). Aerosols of Siberia (thematic issue). *Optics of the Atmos. and Ocean*, 2003, **16**(5–6), 405–559.
2. Alfaro S. C., Gomes L., Rajot J. L., Lafon S., Gaudichet A., Chatenet B., Maille M., Cautenet G., Lasserre F., Cachier H., and Zhang X. Y. Chemical and optical characterization of aerosols measured in spring 2002 at the ACE-Asia supersite, Zhenbeitai, China. *J. Geophys. Res.*, 2003, **108**(D23), ACE9/1–ACE9/18.
3. Allen A. G., Oppenheimer C., Ferm M., Baxter P. J., Horrocks L. A., Galle B., McGonigle A. J. S., and Duffell H. J. Primary sulfate aerosol and associated emissions from Masaya Volcano, Nicaragua. *J. Geophys. Res.*, 2002, **107**(D23), ACH5/1–ACH5/8.
4. Anastasio C. and Martin S. T. Atmospheric nanoparticles. *Rev. Miner. and Geochem.*, 2001, **44**, 293–349.
5. Anderson T. L., Masonis S. J., Covert D. S., Ahlquist N. C., Howell S. G., Clarke A. D., and McNaughton C. S. Variability of aerosol optical properties derived from in situ aircraft measurements during ACE-Asia. *J. Geophys. Res.*, 2003, **108**(D23), ACE15/1-ACE15/19.

6. Andreae M. O., Fishman J., and Lindesay J. The Southern Tropical Regional Experiment (STARE): Transport and atmospheric chemistry near the Equator–Atlantic (TRACE-A) and Southern African Fire-Atmosphere Research Initiative (SAFARI): An introduction. *J. Geophys. Res.*, 1996, **101**(D19), 23519–23520.
7. Ankilov A. N., Baklanov A. M., Vlasenko A. L., Kozlov A. S., and Malyshkin S. B. Estimation of the concentration of aerosol-forming substances in the atmosphere. *Optics of the Atmos. and Ocean*, 2000, **13**(6–7), 644–648.
8. Bahreini R., Jimenez J. L., Wang J., Flagan R. C., Seinfeld J. H., Jayne J. T., and Worsnop D. R. Aircraft-based aerosol size and composition measurements during ACE-Asia using an Aerodyne aerosol mass spectrometer. *J. Geophys. Res.*, 2003, **108**(D23), ACE13/1–ACE13/22.
9. Barth M. C. and Church A. T. Regional and global distributions and lifetime of sulfate aerosols from Mexico City and southeast China. *J. Geophys. Res.*, 1999, **104**(D23), 30231–30240.
10. Barth M. C., Rasch P. J., Kiehl J. T., Benkovitz C. M., and Schwartz S. E. Sulfur chemistry in the National Center for Atmospheric Research Community Climate Model: Description, evaluation, features, and sensitivity to aqueous chemistry. *J. Geophys. Res.*, 2000, **105**(D5), 5123–5129.
11. Bates T. S. First Aerosol Characterization Experiment (ACE-1). Part 2. Preface. *J. Geophys. Res.*, 1999, **104**(D17), 21645–21647.
12. Bates T. S., Huebert B. J., Gras J. L., Criffiths F. B., and Durkee P. A. International Global Atmospheric Chemistry (IGAC) project's First Aerosol Characterization Experiment (ACE-1): Overview. *J. Geophys. Res.*, 1998, **103**(D13), 16297–16318.
13. Berg O. H., Swietlicki E., and Krejci R. Hydroscopic growth of aerosol particles in the marine boundary layer over the Pacific and Southern Oceans during the First Aerosol Characterization Experiment (ACE-1). *J. Geophys. Res.*, 1998, **103**(D13), 16535–17545.
14. Berresheim H., Elste T., Tremmel H. G., Allen A. G., Hansson H.-C., Rosman K., Dal Maso M., Mäkelä J. M., Külmälä M., and O'Dowd C. D. Gas–aerosol relationships of H_2SO_4, MSA, and OH: Observations in the coastal marine boundary layer at Mace Head, Ireland. *J. Geophys. Res.*, 2002, **107**(D19), PAR5/1–PAR5/12.
15. Berresheim H., Huey J. W., Thorn R. P., Eisele F. L., Tanner D. J., Jefferson A. Measurements of dimethyl sulfide, dimethyl sulfoxide, dimethyl sulfone, and aerosol ions at Palmer Station, Antarctics. *J. Geophys. Res.*, 1998, **103**(D1), 1629–1637.
16. Bishop J. K. B., Davis R. E., and Sherman J. T. Robotic observations of dust storm enhancement of carbon biomass in the North Pacific. *Science*, 2002, **298**(N5594), 817–821.
17. Borodulin A. I., Safatov A. S., Marchenko V. V., Shabanov A. N., Belan B. D., and Panchenko M. V. The vertical and seasonal variability of the concentration of the tropospheric aerosol biogenic component in the southern part of West Europe. *Optics of the Atmos. and Ocean*, 2003, **16**(5–6), 422–425.
18. Boucher O. and Pham M. History of sulfate aerosol radiative forcings. *Geophys. Res. Lett.*, 2002, **29**(9), 2211–2214.
19. Boy M., Rannik Ü., Lehtinen K. E. J., Tarvainen V., Hakola H., and Külmälä M. Nucleation events in the continental boundary layer: Long-term statistical analyses of aerosol relevant characteristics. *J. Geophys. Res.*, 2003, **108**(D21), AAC5/1–AAC5/13.
20. Brooks S. D., Wise M. E., Cushing M., and Talbart M. A. Deliquescence behavior of organic/ammonium sulfate aerosol. *Geophys. Res. Lett.*, 2002, **29**(19), 23/5–23/4.

21. Cai X. and Griffin R. J. Modeling the formation of secondary organic aerosol in coastal areas: Role of the sea-salt aerosol organic layer. *J. Geophys. Res.*, 2003, **108**(D15), AAC3/1–AAC3/14.
22. Cappellato R., Peters N. E., and Meyers T. P. Above-ground sulful cycling in adjacent coniferous and deciduous forest and watershed sulfur retention in the Georgia Piedmont, USA. *Water, Air, and Soil Pollut.*, 1998, **103**(1–4), 151–171.
23. Carrico C. M., Kus P., Rood M. J., Quinn P. K., and Bates T. S. Mixtures of pollution, dust, sea-salt, and volcanic aerosol during ACE-Asia: Radiative properties as a function of relative humidity. *J. Geophys. Res.*, 2003, **108**(D23), ACE18/1–ACE18/18.
24. Chin M., Savoie D. L., Huebert B. J., Bandy A. R., Thornton D. C., Bates T. S., Quinn P. K., Saltzman E. S., and De Bruyn W. J. Atmospheric sulfur cycle simulated in the global model GOCART: Comparison with field observations and regional budgets. *J. Geophys. Res.*, 2000, **105**(D20), 24689–24712.
25. Chu D. A., Kaufman Y. J., Zibordi G., Chern J. D., Mao J., Li C., and Holben B. N. Global monitoring of air pollution over land from the Earth Observing System – Terra Moderate Resolution Imaging Spectroradiometer (MODIS). *J. Geophys. Res.*, 2003, **108**(D21), ACH4/1–ACH4/18.
26. Chuang P. Y. Measurement of the time-scale of hydroscopic growth for atmospheric aerosols. *J. Geophys. Res.*, 2003, **108**(D9), AAC5/1–AAC5/13.
27. Chugthai A. R., Miller M. I., Smith D. M., and Pitts J. R. Carbonaceous particle hydratation III. *J. Atmos. Chem.*, 1999, **34**(2), 259–279.
28. Colarco P. R., Toon O. B., Torres O., and Rasch P. J. Determining the UV imaginary index of refraction of Saharan dust particles from Total Ozone Mapping Spectrometer data using a three-dimensional model of dust transport. *J. Geophys. Res.*, 2002, **107**(D16), AAC4/1–AAC4/18.
29. Collins W. D., Rasch P. J., Eaton B. E., Khattatov B. V., and Lamarque J.-F. Simulating aerosols using achemical transport model with assimilation of satellite aerosol retrievals: Methodology for INDOEX. *J. Geophys. Res.*, 2001, **106**(D7), 7313–7336.
30. Cooke W. F., Ramasvamy V., and Kasibhatla P. A general circulation model study of the global carbonaceous aerosol distribution. *J. Geophys. Res.*, 2002, **107**(D16), ACH2/1–ACH2/32.
31. Dal Maso M., Külmälä M., Lehtinen K. E. J., Mäkelä J. M., Aalto P., and O'Dowd C. D. Condensation and coagulation sinks and formation of nucleation mode particles in coastal and boreal forest boundary layers. *J. Geophys. Res.*, 2002, **107**(D19), PAR2/1–PAR2/10.
32. Davis D., Chen G., Kasibhatla P., Jefferson A., Tanner D., Eisele F., Lenshow D., Neff W., and Berresheim H. DMS oxidation in the Antarctic marine boundary layer: Comparison of model simulations and field observations of DMS, DMSO, $DMSO_2$, H_2SO_4(g), MSA(g), and MSA(p). *J. Geophys. Res.*, 1998, **103**(D1), 1657–1678.
33. Dibb J. E., Talbot R. W., Scheuer E. M., Blake D. R., Blake N. S., Gregory G. L., Sachse G. W., and Thomson D. C. Aerosol chemical composition and distribution during the Pacific Exploratory Mission (PEM) Tropics. *J. Geophys. Res.*, 1999, **104**(D5), 5785–5800.
34. Dibb J. E., Talbot R. W., Scheuer E. M., Seid G., Avery M. A., and Singh H. B. Aerosol chemical composition in Asian continental outflow during the TRACE-P campaign: Comparison with PEM-West B. *J. Geophys. Res.*, 2003, **108**(D21), GTE36/1–GTE36/13.
35. Dibb J. E., Talbot R. W., Seid G., Jordan C., Scheuet E., Atlas E., Blake N. J., and Blake D. R. Airborne sampling of aerosol particles: Comparison between surface

sampling at Christmas Island and P-3 sampling during PEM-Tropica B. *J. Geophys. Res.*, 2003, **108**(D2), PEM2/1–PEM2/17.

36. Donchenko V. K. and Ivlev L. S. On an identification of aerosols of different origin. In: L. S. Ivlev (ed.), *Natural and Anthropogenic Aerosols*. St Petersburg Univ. Press, St Petersburg, 2001, pp. 41–51 [in Russian].
37. Duncan B. N., Martin R. V., Staudt A. C., Vevich R., and Logen J. A. Interannual and seasonal variability of biomass burning emissions constrained by satellite observations. *J. Geophys. Res.*, 2003, **108**(D2), ACH1/1–ACH1/22.
38. Feingold G. and Kreidenweis S. Does cloud processing of aerosol enhance droplet concentrations? *J. Geophys. Res.*, 2000, **105**(D19), 24351–24362.
39. Feingold G., Kreidenweis S. M., Stevens B., and Cotton W. R. Numerical simulations of strato-cumulus processing of cloud condensation nuclei through collision–coalescence. *J. Geophys. Res.*, 1996, **101**(D16), 21391–21402.
40. Fitzgerald J. W., Hoppel W. A., and Gelbard F. A one-dimensional sectional model to simulate multicomponent aerosol dynamics in the marine boundary layer. 1. Model description. *J. Geophys. Res.*, 1998, **103**(D13), 16085–16102.
41. Fitzgerald J. W., Marti J. S., Hoppel W. A., Frick G. M., and Gelbard F. A one-dimensional sectional model to simulate multicomponent aerosol dynamics in the marine boundary layer. 2. Model application. *J. Geophys. Res.*, 1998, **103**(D13), 16103–16117.
42. Fortin T. J., Shilling J. E., and Tolbert M. A. Infrared spectroscopic study of the low-temperature phase behavior of ammonium sulfate. *J. Geophys. Res.*, 2002, **107**(D10), AAC4/1–AAC4/10.
43. Franke K., Ansmann A., Müller A., Althaen D., Hataraman C., Reddy M. S., Wagner F., and Sheele R. Optical properties of the Indonesian haze. *J. Geophys. Res.*, 2003, **108**(D2), AAC6/1–AAC6/5.
44. Griffin R. J., Dabdub D., and Seinfeld J. H. Secondary organic aerosol. 1. Atmospheric chemical mechanism for production of molecular constituents. *J. Geophys. Res.*, 2002, **107**(D17), AAC3/1–AAC3/26.
45. Griffin R. J., Dabdub D., Kleeman M. J., Fraser M. P., Cass G. R., and Seinfeld J. H. Secondary organic aerosol. 3. Urban/regional scale model of size and composition-resolved aerosol. *J. Geophys. Res.*, 2002, **107**(D17), AAC5/1–AAC5/14.
46. Griffin R. J., Nguyen K., Dabdub D., and Seinfeld J. H. A coupled hydrophobic–hydrophilic model for predicting secondary organic aerosol formation. *J. Atmos. Chem.*, 2003, **44**(2), 171–190.
47. Grini A., Zender C. S., and Colarco P. R. Saltation sandblasting behavior during mineral dust aerosol production. *Geophys. Res. Lett.*, 2002, **29**(18), 15/1–15/4.
48. Guelle W., Balkanski Y. J., Schulz M., Marticorena B., Bergametti G., Moulin C., Arimoto R., and Perry K. D. Modeling the atmospheric distribution of mineral aerosol: Comparison with ground measurements and satellite observations for yearly and synoptic timescales over the North Atlantic. *J. Geophys. Res.*, 2000, **105**(D2), 1997–2012.
49. Hämeri K., O'Dowd C. D., and Hoell C. Evaluating measurements of new particle concentrations, source rates, and spatial scales during coastal nucleation events using condensation particle counters. *J. Geophys. Res.*, 2002, **107**(D19), PAR6/1–PAR6/11.
50. Hanson D. R. and Eisele F. L. Measurement of prenucleation molecular clusters in the NH_3, H_2SO_4, H_2O system. *J. Geophys. Res.*, 2002, **107**(D12), AAC10/1–AAC10/18.
51. Harrison R. M., Grenfell J. L., Savage N., Alien A., Clemitshaw K. C., Penkett S., Hewitt C. N., and Davison B. Observations of new particle production in the atmosphere

of a moderately polluted site in eastern England. *J. Geophys. Res.*, 2000, **105**(D14), 17819–17832.

52. Hicks B. B., Artz R. S., Meyers T. P., and Hosker R. P., Jr. Trends in the eastern U.S. sulfur air quality from the Atmospheric Integrated Research Monitoring Network. *J. Geophys. Res.*, 2002, **107**(D12), ACH6/1–ACH6/12.
53. Hidy G. N. and Burton C. S. Atmospheric aerosol formation by chemical reaction. *Intern. Chem. Kinetics*, 1975, **7**(7), 509–541.
54. Höller R., Ito K., Tohno S., and Kasahara M. Wavelength-dependent aerosol single-scattering albedo: Measurements and model calculations for a coastal site near the Sea of Japan during ACE-Asia. *J. Geophys. Res.*, 2003, **108**(D23), ACE16/1–ACE16/15.
55. Huebert B. J., Howell S. G., Zhuang L., Heath J. A., Litchy M. R., Wylie D. J., Kreidler-Moss J. L., Coppicus S., and Pfeiffer J. E. Filter and impactor measurements of anions and cations during the First Aerosol Characterization Experiment (ACE-1). *J. Geophys. Res.*, 1998, **103**(D13), 16493–16509.
56. Ivlev L. S. *Chemical Composition and Structure of Atmospheric Aerosols*. Leningrad State Univ. Press, Leningrad, 1982, 368 pp. [in Russian].
57. Ivlev L. S. Special features of the volcanic aerosol size distribution. *Optics of the Atmos. and Ocean*, 1996, **9**(8), 1039–1057.
58. Ivlev L. S., Galindo J., and Kudriashov V. I. Estudio de aerosoles y cenizas dispersados durante la eruption de Volcan Popokatepetl del 21 de Diciembre 1994: Resultados preliminares. In: J. Galindo (ed.), *Report Centro Universitario de Investigaciones en Ciencias de la Tierra*. Universidad de Colima, Colima, Mexico, 1996, pp. 257–284.
59. Ivlev L. S., Kudriashov V. I., Arias M. E., and Vargas A. O. A complex study of the optical–meteorological parameters of the atmosphere near the Colima volcano (Mexico). Part 1. Dry season. Part 2. Wet season. *Optics of the Atmos. and Ocean*, 1998, **11**(7), 748–767; **11**(8), 884–898 [in Russian].
60. Ivlev L. S., Sirota V. G., and Khvorostovsky S. N. An impact of volcanic sulphur dioxide oxidation on the content of H_2SO_4 and ozone in the stratosphere. *Optics of the Atmos. and Ocean*, 1990, **3**(1), 37–43.
61. Ivlev L. S., Kudriashov V. I., and Edwards A. Study of the size distribution and elemental composition of aerosols near the Paricutin volcano (Mexico) in the period of rains. *Proc. Russian Geogr. Soc.*, 1998, **130**(2), 38–43 [in Russian].
62. Ivlev L. S. On relationship between volcanic activity and climate characteristics. In: L. S. Ivlev (ed.), *Natural and Anthropogenic Aerosols*, St Petersburg State Univ. Press, St Petersburg, 1997, pp. 64–72 [in Russian].
63. Ivlev L. S., Vlasenko S. S., Kudriashov V. I., and Terekhin N. Yu. Some results of measurements of the content of aerosols and aerosol-forming gases in the surface layer of St. Petersburg and south-eastern sector of the Gulf of Finland. In: L. S. Ivlev (ed.), *Natural and Anthropogenic Aerosols*, St Petersburg State Univ. Press, St Petersburg, 2001, pp. 75–95 [in Russian].
64. Iwasaka Y., Shi G.-Y., Yamada M., Matsuki A., Trochkine D., Kim Y. S., Zhang D., Nagatani T., Shibata T., Nagatani M., *et al.* Importance of dust particles in the free troposphere over the Taklamakan Desert: Electron microscopic measurements of particles collected with a balloon-borne particle impactor at Danhuang, China. *J. Geophys. Res.*, 2003, **108**(D23), ACE12/1–ACE12/10.
65. Jimenez J. L., Bahreini R., Cocker D. R. III, Zhuang H., Varutbangkul V., Flagan R. C., Seinfeld J. H., O'Dowd C. D., and Hoffmann T. New particle formation from photooxidation of diiodomethane (CH_2I_2). *J. Geophys. Res.*, 2003, **108**(D10), AAC5/1–AAC5/25.

66. Johansen A. M., Siefert R. L., and Hoffmann M. R. Chemical composition of aerosols collected over the tropical North Atlantic Ocean. *J. Geophys. Res.*, 2000, **105**(D12), 15277–15312.
67. Jordan C. E., Anderson B. E., Talbot R. W., Dibb J. E., Fuelberg H. E., Hudgins C. H., Kiley C. M., Russo R., Scheuer E., Seid G., *et al.* Chemical and physical properties of bulk aerosols within four sectors observed during TRACE-P. *J. Geophys. Res.*, 2003, **108**(D21), GTE34/1–GTE34/19.
68. Kaufman Y. J., Tanré D., and Boucher O. A satellite view of aerosols in the climate system. *Nature*, 2002, **419**, 215–223.
69. Kerminen V.-M., Hillamo R. E., Mäkelä T., Saffrezo J.-L., and Maenhut W. The physico-chemical structure of the Greenland summer aerosol and its relation to atmospheric processes. *J. Geophys. Res.*, 1998, **103**(D5), 5661–5670.
70. Komppula M., Lihavainen H., Hatakka J., Paatero J., Aalto P., Külmälä M., and Viisanen Y. Observations of new particle formation and size distributions at two different heights and surrounding in sub-Arctic area in northern Finland. *J. Geophys. Res.*, 2002, **108**(D9), AAC12/1–AAC12/11.
71. Kondratyev K. Ya. (ed.) *Studies of the Environment from the Manned Orbital Stations*. Gidrometeoizdat, Leningrad, 1972, 400 pp. [in Russian].
72. Kondratyev K. Ya. and Pozdniakov D. V. *Aerosol Models of the Atmosphere*. Nauka Publ., Moscow, 1981, 224 pp. [in Russian].
73. Kondratyev K. Ya., Grigoryev Al. A., Pokrovsky O. M., and Shalina E. V. *Satellite Remote Sensing of Atmospheric Aerosol*. Gidrometeoizdat, Leningrad, 1983, 216 pp. [in Russian].
74. Kondratyev K. Ya., Moskalenko N. I., and Pozdniakov D. V. *Atmospheric Aerosol*. Gidrometeoizdat, Leningrad, 1983, 224 pp. [in Russian].
75. Kondratyev K. Ya., Ivlev L. S., and Galindo I. Application of the 'enrichment factor' notion in studies of the volcanic eruption products. *Proc. RAS*, 1995, **394**(6), 581–583 [in Russian].
76. Kondratyev K. Ya. and Galindo I. *Volcanic Activity and Climate*. A. Deepak Publ. Co., Hampton, VA, 1997, 382 pp.
77. Kondratyev K. Ya. Biogenic aerosol in the atmosphere. *Optics of the Atmos. and Ocean*, 2001, **14**(3), 171–179.
78. Kondratyev K. Ya., Krapivin V. F., and Savinukh V. P. *Perspectives of Civilization Development. Multi-dimensional Analysis*. Logos Publ., Moscow, 2003, 573 pp. [in Russian].
79. Kondratyev K. Ya. Aerosol as a climate-forming component of the atmosphere. 1. Properties of aerosol of different types. *Optics of the Atmos. and Ocean*, 2004, **17**(1), 1–20.
80. Kondratyev K. Ya. Aerosol as a climate-forming component of the atmosphere. 2. Remote sensing of the global spatial–temporal variability of aerosol and its climatic impact. *Optics of the Atmos. and Ocean*, 2004, **17**(1), 23–34.
81. Kondratyev K. Ya. From nano- to global scales: Properties, formation processes, and implications of atmospheric aerosol 3. Processes of formation (nucleation) of aerosol. *Optics of the Atmos. and Ocean*, 2004, **17**(10), 1–21.
82. Korhonen H., Lehtinen K. E. J., Pirjola L., Napari I., Vehkamäki H., Noppel M., and Külmälä M. Simulation of atmospheric nucleation mode: A comparison of nucleation models and size distribution representations. *J. Geophys. Res.*, 2003, **108**(D15), AAC12/1–AAC112/8.

83. Korhonen H., Napari I., Timmrech C., Vehkamäki H., Pirjola L., Lehtinen K. E. J., Lauri A., and Külmälä M. Heterogeneous nucleation as a potential sulphate-coating mechanism of atmospheric mineral dust particles and implications of coated dust on new particle formation. *J. Geophys. Res.*, 2003, **108**(D17), AAC4/1–AAC4/9.
84. Kozlov A. S., Ankilov A. N., Baklanov A. M., Vlasenko A. L., Eremenko S. I., and Malyshkin S. B. A study of mechanic processes of the sub-micron aerosol formation. *Optics of the Atmos. and Ocean*, 2000, **13**(6–7), 664–666.
85. Kudriashov V. I. and Ivlev L. S. Analysis of the elemental composition of atmospheric aerosol in the region of volcanoes Colima and Popokatepetl (Mexico) in 1994–1995. In: L. S. Ivlev (ed.), *Proc. Int. Conf. 'Natural and Anthropogenic Aerosol', 29 September–4 October 1997, St Petersburg*. St Petersburg State Univ. Press, St Petersburg, 1998, pp. 457–479 [in Russian].
86. Külmälä M. How particles nucleate and grow. *Science*, 2003, **302**(5647), 1000–1001.
87. Laakso L., Mäkelä J. M., Pirjola L., and Külmälä M. Model studies on ion-induced nucleation in the atmosphere. *J. Geophys. Res.*, 2002, **107**(D20), AAC5/1–AAC5/19.
88. Laakso L., Külmälä M., and Lehtinen K. E. J. Effect of condensation rate enhancement factor on 3-nm (diameter) particle formation in binary ion-induced and homogeneous nucleation. *J. Geophys. Res.*, 2003, **108**(D18), ACH2/1–ACH2/6.
89. Laskin A., Caspar D. J., Wang W., Hunt S. W., Cowin J. P., Colson S. D., and Finlayson-Pitts B. J. Reactions at interfaces as a source of sulfate formation in sea-salt particles. *Science*, 2003, **301**(5631), 340–344.
90. Lavoue D., Liousse C., Cachier H., Stocks B., and Goldammer J. G. Modeling of carbonaceous particles emitted by boreal and temperate wildfires at northern latitudes. *J. Geophys. Res.*, 2000, **105**(D22), 26871–26890.
91. Lee J. H., Yoshida Y., Turpin B. J., Hopke P. K., Poirot R. L., Lioy P. J., and Oxley J. C. Identification of sources contributing to Mid-Atlantic regional aeroso. *J. Air and Waste Manag. Assoc.*, 2002, **52**, 1186–1205.
92. Lee S.-H., Reaves J. M., Wilson J. C., Hunton D. E., Viggiano A. A., Miller T. M., Ballenthin J. O., and Lait L. P. Particle formation by ion nucleation in the upper troposphere and lower stratosphere. *Science*, 2003, **301**(5641), 1886–1889.
93. Leiva A. Contreras, Ivlev L. S., Vasilyev A. V., and Vasilyev S. L. Complex studies of aerosol in Mexico City. In: L. S. Ivlev (ed.), *Proc. Third Int. Conf 'Natural and Anthropogenic Aerosols', Sankt-Petersburg*. St Petersburg State Univ. Press, St Petersburg, 2001, pp. 72–75 [in Russian].
94. Li-Jones K. and Prospero J. M. Variation in the size distribution of non-sea-salt sulfate aerosol in the marine boundary layer at Barbados: Impact of African dust. *J. Geophys. Res.*, 1998, **103**(D13), 16073–16084.
95. Limbeck A., Külmälä M., and Puxbaum H. Secondary organic aerosol formation in the atmosphere via heterogeneous resction of gaseous isoprene on acidic particles. *Geophys. Res. Lett.*, 2003, **30**(19), ASC6/1–ASC6/4.
96. Liu C. L., Zhang J., and Shen Z. B. Spatial and temporal variability of trace metals in aerosol from desert region of China and the Yellow Sea. *J. Geophys. Res.*, 2002, **107**(D14), ACH17/1–ACH17/17.
97. Liu X. and Penner J. E. Effect of Mount Pinatubo H_2SO_4/H_2O aerosol on ice nucleation in the upper troposphere using a global chemistry and transport model. *J. Geophys. Res.*, 2002, **107**(D12), AAC2/1–AAC2/18.
98. Maenhaut W., Schwarz J., Cafmeyer J., and Chi X. Aerosol chemical mass closure during the EUTRAC-2 Aerosol Intercomparison 2000. *Nucl. Instrum. and Meth. Phys. Res.* B, 2002, **189**, 233–237.

99. Mari C., Suhre K., Rosset R., Bates T. S., Huebert B. J., Bandy A. R., Thornton D. C., and Businger S. One-dimensional modeling of sulfur species during the First Aerosol Characterization Experiment (ACE-1), Lagrangian B. *J. Geophys. Res.*, 1999, **104**(D17), 21733–21749.
100. McFiggans G., Cox R. A., Mössinger J. C., Allan B. J., and Plane J. M. C. Active chlorine release from marine aerosols: Roles for reactive iodine and nitrogen species. *J. Geophys. Res.*, 2002, **107**(D15), ACH10/1–ACH10/13.
101. Meszaros E. *Fundamentals of Atmospheric Aerosol Chemistry*. Akademiai Kiado, Budapest, 1999, 308 pp.
102. Murayama T., Masonis S. J., Redemann J., Anderson T. L., Schmid B., Livingston J. M., Russel P. B., Huebert B., Howell S. G., McNaughton C. S., Clarke A., *et al.* An intercomparison of lidar-derived aerosol optical properties with airborne measurements near Tokyo during ACE-Asia. *J. Geophys. Res.*, 2003, **108**(D23), ACE19/1–ACE19/19.
103. Murphy D. M., Anderson J. R., Quinn P. K., McInnes L. M., Brechtel F. J., Kreidenweis S. M., Middlebrook A. M., Posfai M., Thomson D. S., and Buseck P. R. Influence of sea-salt on aerosol radiative properties in the Southern Ocean marine boundary layer. *Nature*, 1998, **392**(6671), 62–65.
104. Nakajima T. *Findings and Current Problems in the Asian Particle Environmental Change Studies:* 2003. JST/CREST/APEX 2003 Interim Report, Tokyo, 2003, 240 pp.
105. Oberlander E. A., Brenninkmeijer C. A. M., Crutzen P. J., Elansky N. F., Golitsyn G. S., Granberg I. G., Scharffe D. H., Hofmann R., Belikov I. B., Paretzke H. G., *et al.* Trace gas measurements along the Trans-Siberian railroad: The TROICA-5 expedition. *J. Geophys. Res.*, 2002, **197**(D14), ACH13/1–ACH13/15.
106. O'Dowd C. D. On the spatial extent and evolution of coastal aerosol plumes. *J. Geophys. Res.*, 2003, **108**(D15), PAR10/1–PAR10/13.
107. O'Dowd C. D., Becker E., Mäkelä J. M., and Külmälä M. Aerosol physico-chemical characteristics over a boreal forest determined by volatility analysis. *Boreal Envieon. Res.*, 2000, **5**(4), 337–348.
108. O'Dowd C. D., Geever M., Hill M. K., Smith M. H., and Jennings S. G. New particle formation: Nucleation rates and spatial scales in the clean marine coastal environment. *Geophys. Res. Lett.*, 1998, **25**(10), 1661–1664.
109. O'Dowd C. D., Hämeri K., Mäkelä J., Pirjola L., Külmälä M., Jennings S. G., Berresheim H., Hansson H.-C., de Leeuw G., Kunz G. J., *et al.* A dedicated study of New Particle Formation and fate in the Coastal Environment (PARFORCE): Overview of objectives and achievements. *J. Geophys. Res.*, 2002, **107**(D19), PAR1/1–PAR1/16.
110. O'Dowd C. D., Hämeri K., Mäkelä J., Väkeva M., Aalto P., de Leeuw G., Kunz G. J., Becker E., Hansson H.-C., Allen A. C., *et al.* Coastal new particle formation: Environmental conditions and aerosol photochemical characteristics during nucleation bursts. *J. Geophys. Res.*, 2002, **107**(D19), PAR12/1–PAR12/17.
111. Pinto J. P., Turco R. P., and Toon O. B. Self-limiting physical and chemical effects in volcanic eruption clouds. *J. Geophys. Res.*, 1989, **94**(D8), 11165–11174.
112. Pirjola L., Laaksonen A., and Külmälä M. Sulfate aerosol formation in the Arctic Boundary Layer. *J. Geophys. Res.*, 1998, **103**(D7), 8309–8321.
113. Polissar A. V., Hopke P. K., Malm W. C., and Sisler J. F. Atmospheric aerosol over Alaska. 1. Spatial and seasonal variability. *J. Geophys. Res.*, 1998, **103**(D15), 19035–19044.
114. Polissar A. V., Hopke P. K., Pantero P., Malm W. C., and Sisler J. F. Atmospheric aerosol over Alaska. 2. Elemental composition and sources. *J. Geophys. Res.*, 1998, **103**(D15), 19045–19057.

115. Posfai M., Anderson J. R., Buseck P. R., and Sievering H. Soot and sulfate aerosol particles in the remote marine troposphere. *J. Geophys. Res.*, 1999, **104**(D17), 21685–21694.
116. Pryde L. T. *Chemistry of the Air Environment*. Cummings Publ. Co., Menlo Park, California, 1973, 169 pp.
117. Pun B. K., Griffin R. J., Seigneur C., and Seinfeld H. Secondary organic aerosol. 2. Thermodynamic model for gas/particle partitioning of molecular constituents. *J. Geophys. Res.*, 2002, **107**(D17), AAC4/1–AAC4/15.
118. Rankin A. M. and Wolff E. W. A year-long record of size-segregated aerosol composition at Halley, Antarctica. *J. Geophys. Res.*, 2003, **108**(D24), AAC9/1–AAC9/12.
119. Restad K., Isaksen I. S. A., and Berntsen T. K. Global distribution of sulphate in the troposphere. A three–dimensional model study. *Atmos. Environ.*, 1998, **32**(20), 3593–3609.
120. Reus M., Strom J., Curtius J., Pirjola L., Vignati E., Arnold F., Hansson H. C., Külmälä M., Leieveld J., and Raes F. Aerosol production and growth in the upper free troposphere. *J. Geophys. Res.*, 2000, **105**(D20), 24751–24762.
121. Roberts G. C., Nenes A., Seinfeld J. H., and Andreae M. O. Impact of biomass burning on cloud properties in the Amazon Basin. *J. Geophys. Res.*, 2003, **108**(D2), AAC9/1–AAC9/19.
122. Romero A. B. And Thiemens M. H. Mass-independent sulfur isotopic compositions in present-day sulfate aerosols. *J. Geophys. Res.*, 2003, **108**(D16), AAC8/1–AAC8/7.
123. Sano I., Mukai S., Okada Y., Holben B. N., Ohta S., and Takamura T. Optical properties of aerosols during APEX and ACE-Asia experiments. *J. Geophys. Res.*, 2003, **108**(D23), ACE17/1–ACE17/9.
124. Sansone F. J., Benitez-Nelson C. R., Resing J. A., de Carlo E. H., Vink S. M., Heath J. A., and Huebert B. J. Geochemistry of atmospheric aerosols generated from lava-seawater interactions. *Geophys. Res. Lett.*, 2002, **29**(9), 49/1–49/4.
125. Savarino J. and Legrand M. High northern latitude forest fires and vegetation emissions over the last millennium inferred from the chemistry of a central Greenland ice core. *J. Geophys. Res.*, 1998, **103**(D7), 8267–8279.
126. Savarino J., Lee C. C. W., and Thiemens M. H. Laboratory oxygen isotopic study of sulfur (IV) oxidation: Origin of the mass-independent oxygen isotopic anomaly in atmospheric sulfates and sulfate mineral deposits on Earth. *J. Geophys. Res.*, 2000, **105**(D23), 29079–29088.
127. Savoie D. L., Arimoto R., Keene W. C., Prospero J. M., Duce R. A., and Galloway J. N. Marine biogenic and anthropogenic contributions to non-sea-salt sulfate in the marine boundary layer over the North Atlantic Ocean. *J. Geophys. Res.*, 2002, **107**(D18), 3/1–3/21.
128. Sobanska S., Coeur C., Maenhaut W., Adams F. SEM-EDX characterization of tropospheric aerosols in the Negev Desert (Israel). *J. Atmos. Chem.*, 2003, **44**, 299–322.
129. Streets D. G., Bond T. C., Carmichael G. R., Fernandes S. D., Fu Q., He D., Klimont Z., Nelson S. M., Tsai N. Y., Wang M. Q., Woo J.-H., *et al.* An inventory of gaseous and primary aerosol emissions in Asia in the year 2000. *J. Geophys. Res.*, 2003, **108**(D21), GTE30/1–GTE30/23.
130. Sun D., Chen F., Bloemendal J., and Su R. Seasonal variability of modern dust over the Loess Plateau of China. *J. Geophys. Res.*, 2003, **108**(D21), AAC3/1–AAC3/10.
131. Tegen I., Hollrig P., Chin M., Fung I., Jacob D., and Penner J. Contribution of different aerosol species to the global aerosol extinction optical thickness: Estimates from model results. *J. Geophys. Res.*, 1997, **102**(D20), 23895–23915.

132. Tervahattu H., Juhanoja J., and Kupiainen K. Identification of an organic coating on marine aerosol particles by TOF-SIMS. *J. Geophys. Res.*, 2002, **107**(D16), ACH18/1–ACH18/7.
133. Twohy C. H., Clement C. F., Gandrud B. W., Weinheimer A. J., Campos T. L., Baumgardner D., Brune W. H., Falona I., Sachse G. W., Vay S. A., *et al.* Deep convection as a source of new particles in the midlatitude upper troposphere. *J. Geophys. Res.*, 2002, **107**(D21), AAC6/1–AAC6/10.
134. Väkevä M., Hämeri K., and Aalto P. P. Hygroscopic properties of nuclearion mode and Aitken mode particles during nucleation bursts and in background air on the west coast of Ireland. *J. Geophys. Res.*, 2002, **107**(D19), PAR9/1–PAR9/11.
135. Virkula A., Dingenen R. V., Raes P., and Hjorth J. Hydroscopic properties of aerosol formed by oxidation of limonene, α-pinene, and β-pinene. *J. Geophys. Res.*, 1999, **104**(D3), 3569–3580.
136. Wahiin P. One year's continuous aerosol sampling at Summit in Central Greenland. In: E. W. Woiffand and R. C. Baler (eds), *Chemical Exchange between the Atmosphere and Polar Snow* (NATO AST Series 1. 1996, **vol. 43**.) Springer-Verlag, Berlin, Heidelberg, pp. 131–143.
137. Wang T.-J., Min J.-Z., Xu Y.-F., and Lam K.-S. Seasonal variations of anthropogenic sulfate aerosol and direct radiative forcing over China. *Meteorol. Atmos. Phys.*, 2003, **84**(3–4), 185–198.
138. Weber R. J., Lee S., Chen G., Wang B., Kapustin V., Moore K., Clarke A. D., Mauldin L., Kosciuch E., Cantrell C., *et al.* New particle formation in anthropogenic plumes advecting from Asia observed during TRACE-P. *J. Geophys. Res.*, 2003, **108**(D21), GTE35/1–GTE35/15.
139. Wolff E. W. and Cachier H. Concentrations and seasonal cycle of black carbon in aerosol at a coastal Antarctic station. *J. Geophys. Res.*, 1998, **103**(D9), 11033–11042.
140. Yu F. Nucleation rate of particles in the lower atmosphere: Estimated time needed to reach pseudo-steady state and sensitivity to H_2SO_4 gas concentration. *Geophys. Res. Lett.*, 2003, **30**(10), 33/1–33/4.
141. Zhang K. M. and Wexler A. S. A hypothesis for growth of fresh atmospheric nuclei. *J. Geophys. Res.*, 2002, **107**(D21), AAC15/1–AAC15/6.
142. Zuberi B., Bertram A. K., Cassa C. A., Molina L. T., and Molina R. J. Heterogeneous nucleation of ice in $(NH_4)_2SO_4$–H_2O particles with mineral dust immersions. *Geophys. Res. Lett.*, 2002, **29**(0), 142/1–142/4.

5

Aerosol and chemical processes in the atmosphere

The participation of aerosol particles in the heterogeneous chemical reactions with various minor gas components (MGCs) plays an important role in climate forming processes. Also, the presence of aerosols affects significantly the course of photochemical processes in the atmosphere. In this connection, of particular importance is the simultaneous reliable information about the concentration and properties of both aerosols and MGCs. The complexity and variety of the chemical composition of aerosols and MGCs hinder the solution of this problem. On the basis of the published material we will discuss the observational data and the possibilities of the numerical modelling of aerosol–MGC interaction.

5.1 INTERACTIONS BETWEEN AEROSOLS AND MINOR GAS COMPONENTS

Factors determining the content of ozone and aerosols in the troposphere are emissions into the atmosphere of these components and their precursors as well as chemical reactions and meteorological processes. In this connection, [7] have developed a coupled numerical model of interactions in the troposphere between chemical processes, aerosol dynamics, and climate on the basis of the Goddard Institute for Space Studies atmospheric general circulation model (GISS GCM-II). The model provides a detailed simulation of chemical processes with participation of tropospheric ozone, nitrogen oxides (NO_x), and hydrocarbons, as well as pre-calculated changes in the sulphates–nitrates–ammonium system, black carbon (BC), primary organic carbon and secondary organic carbon aerosols. The effect of all types of aerosol on the rates of photolysis, heterogeneous reactions with the participation of N_2O_5, NO_3, NO_2, and HO_2 on the wet surfaces of aerosol particles, as well as assimilation of SO_2, HNO_3, and O_3 by mineral dust particles, have been taken into account.

Though this coupled model does not imply a prognostic approach to the consideration of mineral aerosol (MA), its effect on photolysis and heterogeneous processes has been taken into account via prescribing the 3-D fields of parameters which determine the MA concentration. Distributions of ammonium and nitrate between the gas phase and aerosol are determined by conditions of thermodynamic equilibrium, and the process of formation of secondary organic aerosol is determined by its balanced distribution and parameters prescribed from observational data. A consideration of bilateral interactivity between aerosol and chemical processes provides the stability of the MGC concentration fields when simulating the dynamics and mass of aerosol determined by heterogeneous processes as well as when calculating the rate of MGC photolysis. A consideration of changes in NH_3 concentration, both through assimilation by mineral dust particles and through HNO_3 'washing-out' on ice particles, has made it possible to reach a closer agreement between the calculated concentration of gas-phase HNO_3 and observational data than in earlier global models of the transport with chemical reactions taken into account, especially in the middle and upper troposphere.

The results of interactive numerical modelling reveal a non-linear dependence of the global content of MGC and aerosol in the troposphere on varying emissions of NO_x, NH_3, and sulphur compounds. The coupled model has ensured a simulation of the process of formation of sulphate and nitrate aerosols on mineral dust particles. Not far from the regions of dust sources, more than 50% of the total sulphate near the Earth's surface forms in this way. On a global scale, the formation of nitrate aerosols on dust particles exceeds the scale of formation of such aerosols due to ammonium nitrate.

The numerical modelling has shown that the effect of aerosol (on a global scale) on chemical processes with the participation of MGC due to changes in the rate of photolysis is weak. However, heterogeneous processes are important both for MGCs and for aerosols. Though the total area of MA surfaces constitutes only a small part with respect to the area of the surfaces of all global aerosol particles, the MA particles play an important role in assimilation of MGCs, such as O_3, SO_2, and NH_3. However, the estimates of contributions of heterogeneous reactions on the aerosol particle surfaces (MA, in particular) remain rather approximate. Interactions between chemical processes with the participation of MGCs and aerosol are also important in the respect that the formation of sulphates inside clouds depends on changes in O_3 concentration and/or changes in pH of cloud droplets during aerosol formation. On the other hand, the formation of sulphate and nitrate aerosols connected with MA particles depends on the alkalinity of MA particles and on the presence of gas-phase HNO_3 and SO_2. Such interactive processes lead to a non-linear dependence of the content of MGCs and aerosol on emissions of ammonium and sulphur compounds.

Jacobson [34] proposed a new technique for the numerical modelling of the nucleation processes of coagulation, condensation, solution, and reversible chemical reactions with the particle size distribution and time dependence of the processes taken into account. Calculations made for a wide range of variations of aerosol characteristics (number density, molar and bulk concentrations, density of soluble and insoluble fractions, refraction index) and relative humidity suggested the following conclusions:

(1) The process of coagulation determines the internally mixed nature of particles initially of different composition through the entire range of particle sizes.
(2) The process of coagulation is more efficient in the case of large particles compared with small ones.
(3) Coagulation leads to a stronger internal mixing of large particles compared with small ones.
(4) In the presence of a mixture of aerosols with different size distributions, due to coagulation the unified size distribution is formed as in the case of coagulation under conditions of a single size distribution – if in both cases the sum of initial size distributions is the same.
(5) In the course of 'competition' for water vapour between the processes of homogeneous nucleation, the relative role of condensation grows with increasing number density of background particles.
(6) In the absence of any continuous source of new particles, the processes of coagulation, condensation, solution, hydration, and chemical reactions provide an internal mixing of particles during about half a day, under conditions of a moderately polluted atmosphere.
(7) Condensation favours, to a greater extent, a partial filming of large particles rather than small ones.
(8) The real part of the complex refraction index for particles containing electrolytes is greater in the case of low relative humidities.
(9) The difference between refraction indices for all particles and dissolved particles grows with decreasing relative humidities.
(10) As a rule, the refraction index for a solution increases with decreasing size of particles.

The earlier numerical simulation of the processes of formation and transformation of the global dust aerosols (DA) has made it possible to approximately simulate the DA transport and its subsequent deposition onto the surface under the present climatic conditions. None of the available models was able to reliably simulate the power of the sources of DA emissions in Sahara and Asia. All the models overestimated, in particular, the DA emissions and its transport from Australia.

It is important that dust emissions from the surface to the atmosphere are strongly controlled by vegetation cover properties. For instance, under conditions of semiarid climates, a rapid growth of grass cover (after rainfall) leads to a reduction of dust emission over time periods of several days to weeks, whereas the presence of bushes determines a substantial decrease of emissions, even when plants are without leaves. Even in the regions with scarce vegetation the surface is not everywhere an intensive source of dust. There are intensive DA emissions in the regions where recent geomorphological evolution has resulted in the formation of small-grained material with vast weakly roughened surface areas (this refers, for instance, to the surfaces of dried paleolake bottoms).

In connection with these facts, [66] proposed a new scheme of dust emissions which takes into account the presence of surface areas contributing most to the formation of DA emissions as well as the annual course of the vegetation cover characteristics. The new model more reliably simulates DA emissions and its long-range transport – illustrated by results of numerical modelling for the period 1982–1993. Special attention in the model has been paid to a consideration of contributions to DA emissions due to dried paleolake bottoms and vegetation cover. Calculations of DA emissions were made with the wind field observed between 1987 and 1990. Processes of dry and wet DA deposition were taken into account as well as below-cloud scavenging of aerosol. The reliability of numerical modelling results on variability scales from diurnal to interannual has been demonstrated by comparing the calculated values of DA content in the atmosphere with observed values of the aerosol index from TOMS data to obtain the total ozone content.

The calculated values of DA deposition agree with data of ship observations. To obtain realistic estimates of the aerosol optical depth, data are needed on emissions of submicron aerosols for the regions of the prevailing emissions. Numerical experiments on sensitivity suggested the conclusion that the intensity of DA emission sources in Asia is especially sensitive to the annual course of vegetation cover characteristics. Evaluation of the global level of DA emissions gave 800 Mt yr^{-1}, but with enhanced emissions of DA in the regions of Sahara, China, Australia, and Arabia this quantity reaches 1,700 Mt yr^{-1}. Therefore, the global emissions should be assumed to range between 800 and 1,700 Mt yr^{-1} (instead of the earlier extreme values of 600 and 3,000 Mt yr^{-1}).

It is well known that one of the most powerful sources of emissions into the atmosphere of dust aerosol (called 'koza', which means 'yellow sand') are the deserts in China. The long-range transport of this aerosol to the Pacific Ocean and transformation of its properties in the process of transport, affect substantially various processes taking place in the atmosphere, especially from the viewpoint of aerosol particle functioning as cloud condensation nuclei. Trockine *et al.* [69] performed an analysis of aerosol particles collected in Dunhuang (north-western China) during observations in 2001–2002, within the programme of the field observational experiment ACE-Asia. Analysis of the composition of particles, mainly of mineral origin, was carried out using the scanning electron microscope equipped with an X-ray analyser.

About 46–47% of DA particles had an increased content of Si (mainly due to quartz and aluminosilicates) and 13–14% of Ca (mainly as tiff $CaCO_3$, dolomite $CaMg(CO_3)_2$, and gypsum $CaSO_4 \times {}_2H_2O$). Particles enriched with iron (nitrogen oxides) constituted 3–10%. The mineral component dominated not only in China but also in the atmosphere over Japan, but aerosols in these cases were quite different due to chemical transformations during the process of transport. About 40–45% of particles turned out to be intermixed with sulphates during their transport in the troposphere. It is important that properties of aerosols contained in samples collected in the free troposphere over Japan differed from the properties of particles in surface samples collected in Nagasaki, Nagoya, and Fukuoke (Japan). The sizes of mineral particles intermixed with sea salt and sulphates observed in the free troposphere were considerably smaller than in the surface samples. This fact should be taken into account when studying the effect of DA particles on the biogeochemical cycles and climate.

To assess the effect of DA from the Asian continent on air quality in North America, [49] applied a relatively new method of factor analysis consisting of using Positive Matrix Factorization (PMF), with the aim of determining the chemical composition of PM-2.5 aerosol particles from observations at two alpine stations located in the National Park – Crater Lake (42.89°N, 122.14°W, 1,981 m a.s.l.) and Lassen volcano (40.54°N, 121.58°W, 1,798 m a.s.l.). These observations were accomplished within the IMPROVE programme of interdepartmental monitoring of the protected territories in the western USA. The PMF method is advantageous in that it is possible to assess the reliability of the results (point-by-point) obtained through determining the weights of various factors considering the observation errors, omitted data, and available data below the detection threshold. An application of multiple linear regression (with the contribution of each considered factor taken into account) to process data on fine-aerosol concentrations has provided a normalization of the factors under consideration.

From seven sources of aerosol discovered at two observational points under consideration, six sources were characterized by a close chemical composition of aerosols and a similar annual course. Dust aerosols from the Asian continent contained Al, Ca, Fe, NO_3, S, K, and Ti with a strong annual course. The concentration of secondary sulphates with a high content of S had a strong annual course correlating with the incoming Asian DA. The presence of smoke aerosol was parameterized as organic carbon (OC), elemental carbon (EC), and K. Besides this, sea salts with high concentrations of Na, S, and NO_3 were taken into account. Of importance are nitrates with a prevailing concentration of NO_3 and combustion engine emissions characterized by high concentrations of OC, C, and dust components. For the Crater Lake station, emissions containing Cu and Zn were also considered. On the whole, the sources of emissions at both stations have similar vertical profiles of emission chemistry and similar annual courses. The results obtained show that the PMF method is rather efficient as a means of detection of various sources of pollution from data on aerosols.

During an accomplishment of the field observational experiments within the programmes ACE-Asia (studies of aerosol characteristics in Asia) and TRACE-P (studies of the long-range transport and chemical evolution of aerosol in the atmosphere over the north-western Pacific), [41] performed airborne measurements (flying laboratories C-130 and P-3B, respectively) of concentrations of inorganic ionic components of fine aerosols including NH_4^+, SO_4^{2-}, NO_3^-, Ca^{2+}, K^+, Mg^{2+}, Na^+, and Cl^- using a new method of sampling and ionic chromatography (PILS-IC) with a 4-min. resolution and detection threshold $<0.05\,\mu g\,m^{-3}$. Maximum observed values of total concentration of ions constituted $27\,\mu g\,m^{-3}$ (C-130) and $84\,\mu g\,m^{-3}$ (P-3B).

In the period of ACE-Asia the dominating components of the ionic composition of aerosols were NH_4^+ and SO_4^{2-} with the contribution of Ca^{2+} due to DA comparable (by mixing ratios) with SO_4^{2-}. Concentrations of ions Na^+ and Cl^- connected with sea salt aerosols were about the same as of the ion K^+, a tracer of biomass burning. In the TRACE-P period, the ion NH_4^+ dominated, and then followed, in decreasing order of concentration: SO_4^{2-}, Cl^-, Na^+, NO_3^-, Ca^{2+}, and K^+. The presence of the ion Cl^- was connected not only with the contribution of sea salt aerosols but also with the urban emissions determined, probably, by biofuels burning. The Mg^{2+} concentration is similarly determined by contributions of sea salt and dust aerosols.

During both field airborne experiments a strong correlation was observed between concentrations of NH_4^+ , SO_4^{2-}, NO_3^-, and CO reflecting a prevalent contribution of emissions from combustion engines to the formation of concentrations of these components, as well as the fact that there was a sufficient amount of such alkaline components as NH_3 and others to neutralize H_2SO_4. The ratio of $[NH_4^+]$ concentration to $([NO_3^-]+2[SO_4^{2-}])$ constituted about 0.70 in the presence of substantial deviations from this level only in the zones of volcanic plumes (there was a correlation between SO_4^{2-} and SO_2). The charge balance of ions had different signs with an ~30% amplitude of variations permitting an estimation of the lower level of ion concentrations not measured, so far. Simultaneous increased concentrations of NO_3^- and Ca^{2-} were observed, as a rule, in the polluted atmosphere, which demonstrates the importance of the process of HNO_3 uptake by dust aerosol.

Such low-molecular dicarboxylic acids as oxalic, malonic, and succinic, were discovered in atmospheric aerosols both in the continental and marine atmosphere. Analysis of the chemical composition of aerosol samples collected [38] on board the flying laboratory C-130 over the western Pacific during April to May, 2001 revealed for the first time a homologous series of dicarboxylic acids C_2–C_5 in the polluted air incoming from the Asian continent. The content of oxalic acid (C_2) was the greatest, then followed malonic (C_3) and succinic (C_4) acids. The total concentration of diacids C_2–C_5 (44–870 ng m^{-3}, with 310 ng m^{-3} on average) turned out to be close to that observed near the Earth's surface in the urban atmosphere of Tokyo.

There was a positive correlation ($r^2 = 0.70$) of the concentration of oxalic acid with that of the total organic carbon (TOC). For other diacids this correlation was weaker. The obtained results show that the water-soluble dicarboxylic acids and TOC had common sources of emissions located on the land surface of the Asian continent and/or from the processes of photochemical oxidation of anthropogenic organic compounds in the atmosphere. Dioxicarbon compounds constitute 0.2–3.3% (1.8, on average) with respect to TOC. Water-soluble dicarboxylic acids can play an important role as factors determining the chemical and physical properties of organic aerosol in the polluted atmosphere over eastern Asia and the western Pacific.

The difference between chemical compositions of sea salt aerosol and seawater has been widely discussed. One of the characteristic features of this difference consists of the decreasing content of chloride in sea salt aerosol with respect to natrium as compared with seawater. With the prescribed deficit of chloride in the air over the ocean surface within 5–30 nmol m^{-3}, the HCl concentration ranged between 110 and 670 ppt (i.e., 200–1,100 ng m^{-3}). The deficit of chloride was estimated from the ratio of concentrations [Cl]/[Na] in aerosol samples on filters, and in this case one can suppose the presence of acid induced remobilization of chloride distorting the aerosol composition.

Analysis of impactor samples of aerosol, with differentiation by size of particles, carried out by [36] in January, 1998 in coastal Antarctica revealed a 70% loss of chloride (on average), but data on HCl for these conditions are absent. In this connection, in 1991 year-round observations were started in Dumont d'Urville (60.40°S, 140.01°E, coastal Antarctica) of the chemical composition of bulk and fractionated aerosol. Analysis of the chemical composition of sea salt aerosol revealed a ~10% decrease in the ratio of concentrations [Cl]/[Na] in summer. The loss of chloride mass reaches a maximum when the diameter of particles is of the order of 1–3 μm, the reason for chloride loss often being the nitrate anion. The summertime ratio $[SO_4^{2-}]/[Na^+]$ for submicron particles exceeds that for seawater due to the effect of biogenic sulphate, and in the case of coarse aerosol it is determined by ornithogeneous (contained in guano enriched soils) sulphate and heterogeneous assimilation of SO_2 (or H_2SO_4). The HCl mixing ratio varies from $41 \pm 28\,\mathrm{ng\,m^{-3}}$ in winter to $130 \pm 110\,\mathrm{ng\,m^{-3}}$ in summer, being close to the level of chloride loss by the bulk sea salt aerosol. In winter, the sea salt aerosol particles are characterized by mass concentration ratios $[Cl^-]/[Na^+]$ and $[SO_4^{2-}]/[Na^+]$, respectively, 1.9 ± 0.1 and 0.13 ± 0.04.

A decrease of the sulphate content with respect to natrium due to deposition of mirabilite resulting from seawater freezing takes place from May to October. An increase of temperature and/or decrease of sea ice cover extent taking place in March – April, limit the scale of this phenomenon. For submicron sulphate, the mass concentration ranges between 14 and $50\,\mathrm{ng\,m^{-3}}$, which confirms the presence of NSS SO_4^{2-} in the wintertime coastal Antarctica, with the highest concentration of non-salt sulphate recorded in the wintertime periods of 1992–1994 due to the Pinatubo eruption. During other seasons the NSS SO_4^{2-} concentration varied within $15–30\,\mathrm{ng\,m^{-3}}$, the origin of this non-salt sulphate remaining unclear. Its possible sources could be the quasicontinuous emissions by Erebus volcano or local oxidation of dimethylsulphide (DMS) in winter, as well as the long-range transport of by-products of DMS oxidation.

Bian and Zender [8] studied the effect of dust (mineral) aerosol on chemical processes in the atmosphere connected with photolysis and heterogeneous reactions with participation of MGCs on the surface of DA particles. To obtain the respective quantitative estimates, a numerical modelling has been accomplished with the use of the University of California global CTM model of chemical processes, with long-range transport taken into account. From the data for the DEAD model, the time-varying size distribution of aerosol has been prescribed. Table 5.1 contains the results of calculations made with a separate consideration of photolysis or heterogeneous processes as well as with an interactive consideration of both factors for northern and southern hemispheres and for the whole globe. The coefficient of interactivity λ characterizing a non-linear interaction of photolysis and heterogeneous reactions is estimated from the relationship $\lambda = \Delta_{p+H}/(\Delta_p + \Delta_H)$.

If λ is close to unity, it means that Δ_p and Δ_H contributions are additive (i.e., non-linearity does not show). As seen from Table 5.1, for most of the MGCs considered, the global mean changes in concentrations do not exceed several %, except for OH hydroxyl (~11.1%).

Table 5.1. Global mean changes (%) of concentrations of eight MGCs due to dust aerosol.

	Photolysis (Δ_p)				Heterogeneous reactions (Δ_H)				Interactive calculations (Δ_{p+H})				Non-linear interaction (λ)	
	NH	SH	Globe	NATA	NH	SH	Globe	NATA	NH	SH	Globe	NATA	Globe	NATA
O_3	0.2	0.3	0.2	0.9	−1.5	−0.3	−0.9	−5.0	−1.3	−0.0	−0.7	−3.8	1.00	0.93
OH	−4.0	−0.8	−2.4	−15.0	−16.4	−2.9	−9.6	−64.0	−18.5	−3.6	−11.1	−66.8	0.93	0.85
HNO_3	0.4	0.3	0.3	0.8	−6.1	−1.5	−3.8	−28.3	−5.8	−1.2	−3.5	−27.7	1.00	1.01
HO_2	−1.0	0.2	−0.4	−6.0	−9.1	−1.1	−5.1	−43.5	−9.6	−0.9	−5.2	−45.3	0.95	0.92
NO_3	1.9	0.8	1.3	5.4	−10.2	−1.5	−5.9	−47.2	−8.7	−0.8	−4.7	−44.2	1.02	1.06
NO_2	2.1	0.7	1.4	9.8	−0.5	−0.2	−0.3	−6.9	1.6	0.5	1.1	3.1	1.00	1.07
N_2O_5	3.3	1.2	2.2	12.0	−3.4	−0.8	−2.1	−19.6	−0.3	0.4	0.0	−9.4	–	1.24
H_2O_2	0.3	0.7	0.5	−0.6	−0.4	0.1	−0.2	−2.2	−0.2	0.8	0.3	−3.0	1.00	1.07

Note: NATA – regions of North Africa and Tropical Atlantic; NH – northern hemisphere; SH – southern hemisphere.

As was expected, maximum changes occur in the regions located in the direction of prevailing winds from DA sources (North Africa, tropical Atlantic, Arab peninsula). The DA induced 5-fold changes in concentrations of O_3 and OH in the northern hemisphere exceed those observed in the southern hemisphere. On a regional scale, the contribution of photolysis prevails in a few regions in low and middle latitudes, whereas the effect of heterogeneous processes dominates in the rest of the global atmosphere. On global scales, the interactivity of the two studied factors is weakly manifested (on average), but it shows markedly in the regions of the dust loaded atmosphere, where the local impact of interactivity consideration in the case of tropospheric ozone can exceed 20%.

Photolysis and heterogeneous reactions produce effects of the opposite sign from the viewpoint of impact on concentrations of ozone and odd nitrogen, which determines a weak total effect. Both these factors lead, however, to a decrease of OH and HO_2 concentrations. The global mean changes due to DA are −0.7% (tropospheric ozone), −11.1% (OH), −5.2% (HO_2), and −3.5% (HNO_3). Near the DA sources there appears a strong annual course of MGC concentrations. For instance, in the direction of DA from North Africa and the tropical Atlantic, a decrease of OH concentration (−66.8%) exceeds the global mean value by 6 times.

It is of interest that the level of the global mean effect of photolysis on ozone in the southern hemisphere is higher than in the northern hemisphere, where concentrations of DA and MGCs – precursors of ozone – are much higher. In the polar regions the prevailing contribution to changes in ozone concentration is made by the long-range transport of O_3, and therefore the impact of local DA is negligible. Changes in the concentration of tropospheric ozone due to photolysis depend not only on the vertical distribution of DA but also on the presence of O_3 precursors. The impacts of heterogeneous reactions on O_3 depends on the vertical distribution of DA, which is mainly governed by the temperature dependence of the rate of heterogeneous uptake.

Bian *et al.* [8, 9] emphasized the preliminary character of the results obtained and the necessity to collect adequate observational data (to check numerical modelling adequacy).

In the context of parameterization of the aerosol–MGC interactions, [23] performed laboratory studies of HNO_3 uptake on the water–ice film covering the walls of the reactor at a temperature corresponding to the upper troposphere conditions (210–235 K). Measurements were made in the 'ice' interval of the phase diagram HNO_3—H_2O. Within the range of partial HNO_3 pressure equal to $(0.3–2.0) \times 10^6$ torr there was a continuous uptake of HNO_3 at a temperature below 215 K, whereas at a temperature above 215 K the process of uptake depended on time. If we assume that the covering of the surface with the ice film can be characterized by the Langmuir isotherm for dissociative adsorption, then the HNO_3 adsorption enthalpy on the ice surface constitutes (54.0 ± 2.6) kJoule mol^{-1}.

With a constant partial HNO_3 pressure, maximum values of the coefficient of uptake γ varied, depending on temperature, from 0.03 at 215 K to 0.006 at 235 K. At 218 K changes in HNO_3 partial pressure did not affect significantly the coefficient of uptake. The uptake on the ice surface of HCl vapour, in which HNO_2 was preliminarily added, turned out to be a reversible process, and the coadsorption of HNO_3 and HCl indicates that HNO_3 molecules determined a removal of HCl molecules from surface areas. Analysis of HNO_3 uptake by the HCl^- containing surface has shown that HNO_3 molecules cause the transport of about 10^{33} HCl molecules km^{-3}. In this context, the efficiency of cirrus clouds for HNO_3 scavenging has been considered, as well as the significance of this process for activation of reactions with participation of chlorine in the troposphere.

Every year the troposphere receives about 1,000–3,000 Tg of mineral (dust) aerosol, wind-driven from the surface. According to available estimates, the present enhancement of the area of arid regions will lead to an increase of the DA content in the atmosphere. Since small particles of mineral aerosol have a relatively long time of residence in the atmosphere, they are transported for long distances, during which they can participate in heterogeneous chemical reactions with various MGCs in the atmosphere. These reactions can result in changes of the gas-phase chemical balance and physico-chemical properties of aerosol particles. Changes in aerosol physical characteristics such as size and shape, chemical composition, and hygroscopicity affect their optical properties and, hence, specific character of the climatic impact of aerosols.

Chemically active calcium carbonate is a widespread component of dust aerosol. Since nitric acid vapour is one of the MGCs in the atmosphere, the following reactions are possible:

$$CaCO_3 + 2HNO_3 \rightarrow Ca(NO_3)_2 + H_2CO_3 \rightarrow Ca(NO_3)_2 + CO_2 + H_2O \qquad (5.1)$$

Available data show that chemical activity of HNO_3 intensifies in the presence of water vapour, and solid particles of calcium carbonate are transformed into water droplets of calcium nitrate in the course of reaction (5.1), since solid calcium nitrate $Ca(NO_3)_2$ (s) transforms into liquid $Ca(NO_3)_2$ (aq). Transformation of solid particles into liquid droplets through heterogeneous chemical reactions is important in the problem of aerosol climatic impact. Laboratory studies using the scanning electron microscope (SEM) and dispersive x-ray analysis carried out by [43] have shown that individual particles of calcium carbonate react with gas-phase nitric acid at a temperature of 293 K. The SEM images show that in the course of the reactions mentioned, solid particles of $CaCO_3$ transform into liquid spherical droplets. The process takes place in two steps consisting of the transformation of calcium carbonate into calcium nitrate with subsequent dilution of calcium nitrate. Changes in the phase state of particles and a considerable chemical activity of nitric acid and $CaCO_3$ at a low relative humidity are direct results of the process of their dilution at a low relative humidity.

Photolysis of nitrous acid (HONO) taking place in the atmosphere is one of the important sources of OH hydroxyl under conditions of a polluted atmosphere. It has been established that HONO forms through transformation of nitrogen oxides NO_x on the surface of atmospheric aerosol particles, but the mechanism of this transformation has been poorly studied. To study this mechanism of heterogeneous chemical transformation, in the summer of 2001 [70] performed measurements of HONO and

NO_2 concentrations in the centre of Phoenix (Arizona, USA) using the technique of differential optical adsorptive spectroscopy (DOAS) and a long (about 3.3 km) route. Analysis of measurement results has led to the conclusion that the concentrations ratio [HONO]/[NO_2] seldom exceeds 3%. However, a substantial increase in this ratio was observed in the periods of two nocturnal dust storms. Unprecedented high values of the concentration ratio reaching ~19% suggest the presence of a very efficient process of heterogeneous transformation of NO_2 into HONO on the surface of mineral dust aerosol particles. It means that under conditions of a polluted atmosphere and dust storms, the transformation of NO_2 into HONO intensifies and, hence, the formation of OH intensifies, too.

Jordan *et al.* [35] discussed the results of analysis of chemical composition of aerosol samples collected from the flying laboratory DC-8 in the atmosphere over the western Pacific during completion of the field observational experiment TRACE-P. The major goal of the analysis was to study the uptake of NO_3^- and SO_4^{2-} on the surface of DA particles, mainly from the deserts of China, by comparing data for the dust loaded and clear atmospheres. The dust loaded air was characterized also by a higher concentration of such anthropogenic MGCs as HNO_3, SO_2, and CO, which exceeded that corresponding with clear atmospheric conditions by factors of 2.7, 6.2, and 1.4, respectively. Such a situation favoured an intensification of uptake of NO_3^- and NSS SO_4^{2-} on coarse DA particles, which determined an increase of mixing ratios for such particles, on average, by factors of 5.7 and 2.6, respectively.

The amount of NSS SO_4^{2-} ions in the troposphere was sufficient not only to uptake all available NH_4^+ but also to react with $CaCO_3$. This excludes a possibility of an increased concentration of NO_3^- in fine-mode NH_4NO_3. The aerosol ions $NO_3^-(p - NO_3^-)$ averaged ~54% with respect to total $NO_3^-(t - NO_3^-)$ reaching a maximum of 72% in the centre of the dust plume coming from the continent. In the dust-free part of the atmosphere the contribution of p-NO_3^- into t-NO_3^- constituted only 37% (the probable reason for this situation may be an abundance of sea salt aerosol). In the regions where the impact of DA and sea salt aerosols was at a minimum, the share of p-NO_3^- in total t-NO_3^- was less than 15%.

The results obtained in [35] testify to a possibility of a substantial *in situ* uptake of acidic gases on alkaline DA particles, and as a result, nitrogen was removed from the atmosphere, which can be important on regional scales. The presence of ions NO_3^- and SO_4^{2-} on coarse aerosol particles can intensify acid deposition and lead to harmful impacts on land ecosystems. In the case of marine ecosystems, an enhanced deposition of NO_3^- can lead to a deficit of this nutrient for coastal biota. Naturally, such effects can be more substantial under conditions of further industrialization of eastern Asia.

Using a box model, [54] performed a numerical modelling of possible forcings on chemical processes in the middle troposphere, determined by heterogeneous chemical reactions on the surface of cirrus clouds ice particles in their uptake of chemically active and inert MGCs. The emphasis has been laid on the process of deoxygenation due to heterogeneous reactions with the participation of HNO_3 with an adsorption by particles of gas-phase HNO_3, as well as denitrification determined by gravitational sedimentation of ice particles.

Results of the numerical modelling show that chemical processes in cirrus clouds can cause a strong local decrease of the content of HNO_3 and NO_x in the upper troposphere. A decrease of NO concentration results in a decreasing ratio of concentrations OH/HO_2 and concentration of OH. Estimates have been obtained [54] of the resulting decrease of the content of tropospheric ozone. Calculations have been made of the sensitivity of such a decrease to the efficiency of HNO_3 adsorption, the rate of particles sedimentation, and NO concentration in the troposphere.

Numerical sensitivity experiments have shown that the impact of heterogeneous reactions taking place in clouds is mainly determined by inert (not followed by chemical reactions) uptake of HNO_3 on cloud ice particles, which strongly depends on HNO_3 adsorption efficiency, and by subsequent sedimentation of particles. With the supposed enhanced uptake of HNO_3, the decrease of the ozone mixing ratio can reach 14%. Also, it follows from the results that the effect of the processes in cirrus clouds on tropospheric ozone is sensitive to the level of NO_x concentration in the troposphere. Under these conditions, chlorine activation, connected with heterogeneous chemical reactions on the surface of ice particles, exhibits only a secondary impact on chemical processes responsible for changes in the content of tropospheric ozone. The impact of uptake of OH, HO_2, and H_2O_2 by the ice particles shows markedly only in the period of a very intensive formation of cirrus clouds. So far, the absence of required observational data has not permitted the validation of the results under discussion.

At present, it is common knowledge that under conditions of heavy smog the main part of aerosol organic carbon forms due to secondary processes of the gas-to-particle conversion with an important contribution of biogenic precursors. Since the processes of nucleation are determined by contributions of several components, there is no need (for particle formation) for a high level of oversaturation with respect to any individual component. It means, in particular, that particles can form at a relatively low concentration of biogenic precursors of aerosols. Even if conditions for homogenous nucleation are unfavourable, ultrafine particles can form due to ionic nucleation. The resulting organic aerosol can further transform through reactions with the participation of free radicals and hydroxyl. Though humus constitutes about 70% of all organic matter in global continental and marine reservoirs, so far there has not been established any connection between these gigantic reservoirs and the composition of fine organic aerosols. In this connection, [17] proposed a new hypothesis according to which substances like humus are always present in fine aerosols. The formation of such substances is connected with the presence of a huge reservoir of organic matter in soils which can serve as a source of organic aerosol formation in the atmosphere due to processes of evaporation, condensation, and polymerization (in the aerosol phase) of low-molecular components of soil humus.

The hypothesis proposed [17] is based on results of a comparative study of the chemical composition of organic components contained in fine aerosols as well as in natural humus and fulvoacids. Though the order of magnitude of the net flux from the soil of all products of soil decomposition constitutes only $ng\,m^{-2}s^{-1}$, it is enough to maintain the mass concentration comparable with that corresponding to fine organic aerosols. This conclusion means that the proposed mechanism can contribute much to the formation of fine organic aerosols on land. A certain contribution to the formation of humic polymer matter on land can also be made by other natural and anthropogenic sources. The proposed hypothesis requires, of course, confirmation by observational data.

The distribution of nitrate between gas and aerosol phases affects significantly the course of chemical reactions in the troposphere and the formation of biogeochemical cycles of nitrogen compounds. It is important, in particular, that nitric acid (HNO_3) vapour can react with ammonium (NH_3) giving a fine aerosol of ammonium nitrate. The content of gas-phase nitric acid in the troposphere is affected by the present mineral aerosol and aerosol formed in biomass burning, since aerosol serves as a chemical sink for HNO_3. The removal of HNO_3 from the atmosphere can also be determined by sea salt aerosol, an interaction with which leads to the formation of coarse aerosol nitrate. An appearance of aerosol nitrate can be connected with the nocturnal heterogeneous reactions with participation of NO_3 and N_2O_5 on the wet surface of sulphate aerosol particles. Both natural and anthropogenic aerosols survive a long-range transport with the resulting detection of sulphate and nitrate aerosol in the remote regions of the Pacific Ocean, and over the tropical Indian Ocean where aerosol particles consisting of sulphates, nitrates, organic compounds, black carbon, ash, and other components were recorded.

Ma *et al.* [51] discussed the results of measurements of size distributions of ionic aerosols (NH_4^+, Na^-, K^+, Ca^{2+}, Mg^{2+}, SO_4^{2-}, Cl^-, NO_3^-, CO_3^{2-}, and formic, acetic, and oxalic acids), concentrations of particles of organic and black carbon, gas-phase HNO_3, and SO_2 at Waliguan Observatory located in the north-eastern Tibetan plateau (36°17′N, 100°54′E, 3,816 m a.s.l.) in October–November, 1997 and January, 1998. The main goal of processing of the observational results was to analyse the distribution of nitrate between gas and liquid phases. The analysis has shown that nitrate was present mainly in the aerosol phase with a typical ratio of concentrations of particle to total nitrate ($NO_3^-(p)/(HNO_3^-(p) + HNO_3(g))$ equal to about 0.9. The size distribution of various samples of nitrate aerosol turned out to be variable, the amount of fine-mode nitrate (diameter of particles $D_p < 2\,\mu m$) being, as a rule, greater or comparable with that corresponding to course-mode nitrate.

The results of the numerical simulation of the distribution of the aerosol nitrate between two modes using the model of chemical balance between gas and aerosol phases agreed well (within measurement errors) with observational data. An analysis has been made [51] of the possible chemical reactions of the formation of nitrate aerosol with due account of the measured dependence of the content of ionic aerosol on the size of particles. This analysis has led to the conclusion that the ionic composition of fine aerosols is determined by the reaction between gas-phase nitric acid and ammonium, whereas coarse-mode particles form through condensation of nitric acid vapour on the surface of mineral aerosols. The presence of black carbon and an accumulation of potassium and oxalate in fine aerosol particles are a signature of the contribution of biomass burning (and subsequent long-range transport).

5.2 PHOTOCHEMICAL PROCESSES WITH THE PARTICIPATION OF AEROSOL

Martin *et al.* [54] studied the sensitivity of the content in the troposphere of such oxidants as OH, O_3, and O_3 presursors to atmospheric aerosols – something not considered before in the global models of chemical processes in the atmosphere. In this connection, the following phenomena were considered:

(1) aerosol induced scattering and absorption of UV radiation; and
(2) uptake of MGCs in the course of chemical reactions with the participation of HO_2, NO_2, and NO_3.

Calculations of sensitivity were made using the global 3-D model of chemical processes in the troposphere (GEOS-CHEM) with prescribed 3-D global distributions of sulphate, black and organic carbon, sea salt, and dust aerosols calculated with the use of the global model GOCART – with the same prescribed global fields of meteorological parameters. Only the contribution of N_2O_5 hydrolysis in aerosol particles was left out of account, the importance of which had been studied earlier and was taken into account in present models of chemical processes in the troposphere.

A substantial effect of aerosol results in a decrease of photolysis frequency and attenuation of chemical assimilation of HO_2. Changes in photolysis frequencies are mainly connected with the effect of absorbing aerosol (the effect of scattering sulphate aerosol is negligible). Aerosols determine, in particular, a decrease of photolysis frequency $J(O(^1D))$ in the case of the transformation $O_3 \rightarrow O(^1D)$ of 5–15% near the surface over most of the northern hemisphere, which is mainly caused by dust (mineral) aerosols. This also occurs, to a greater extent (up to double), in the regions of biomass burning (in these cases due to black carbon). As a result of HO_2 uptake by aerosols, more than a 1% loss of total HO_x (=OH + peroxides) occurs in the atmospheric boundary layer (ABL) in most of the continental regions.

The impact of heterogeneous chemical reactions on the surface of aerosol particles shows most strongly in the case of high concentrations of fine aerosols capable of stimulating the gas-to-particle transfer. Under conditions of the polluted atmosphere of east Europe (mainly due to sulphate and organic carbon) the aerosol-induced uptake of HO_x reaches 50–70% of the total loss of HO_x and exceeds 70% in regions of biomass burning in the tropics (in this case we mean particles containing organic carbon). Aerosols cause a 5–35% decrease of OH concentration over most of the northern hemisphere (in August) and decrease by a factor of 4 in India (in March). The chemical uptake of NO_2 and NO_3 is relatively weak. Such processes lead to more than 10% of changes in HNO_3 concentration only in the tropical North Atlantic, Sahara, and over the southern hemisphere oceans.

In the continental regions only a small aerosol induced increase of NO_x concentration is observed as a result of decreasing OH concentration. The aerosol induced annual-mean global-mean OH concentration constitutes 9% (this is comparable with the contribution of hydrolysis due to N_2O_5) and 13% in the northern hemisphere. About 60% of the global decrease of OH concentration is connected with radiative forcings determined by dust aerosols. The annual-mean global-mean level of O_3 formation decreases by 6%, but the O_3 budget in the troposphere remains practically constant. The concentration of CO increases by 5–15 ppb over most of the northern hemisphere. A decrease of O_3 in the ABL constitutes 15–45 ppb in southern Asia during the season of biomass burning in March; 5–9 ppb in summer in northern Europe; and only 1–3 ppb in the USA. The surface air concentration of ozone in western Europe and other industrial regions grows if aerosols are emitted to the atmosphere without an accompanying decrease in the concentration of ozone precursors. The global budget of tropospheric ozone is characterized by the level of its chemically induced formation which is equal to 4,924, and loss of 4,377 g O_3 yr^{-1}.

Ma *et al.* [52] performed an analysis of the results of aircraft measurements of the composition of the plume of polluted air masses from China to the western Pacific Ocean on 24 February–10 April, 2001 within the field experiment TRACE-P. They came to the conclusion that concentration of K ions (K^+) in fine aerosols can serve as a unique tracer of biosmoke (products of biomass burning on the continent). From data of the flying laboratories DC-8 and P-3B from airports in Hong Kong, China, and from the USAF base in Yokota (Japan) the presence of K^+ in fine aerosols was recorded in ~20% of all observations. There was detected a correlation of K^+ with absorbing aerosol particles ($r^2 = 0.73$), ammonium ($r^2 = 0.77$), and CO ($r^2 = 0.61$). In the observational data considered, the correlation between concentrations of K^+ and CO turned out to be higher than between methyl chloride (CH_3Cl) and CO ($r^2 = 0.50$) which was often used as a biosmoke tracer. The absence of a correlation between K^+ and natrium or calcium shows that contributions of sea salt and mineral dust aerosols to the formation of K^+ concentration were insignificant.

An apparent presence of biomass burning products was observed in the latitude belt 15°–25°N at altitudes 2–4 km. A consideration of 'back' trajectories of air masses and satellite information suggested the conclusion that the biomass burning plumes originate in the regions of South-East Asia. In these cases there were increased concentrations (with a mutual correlation) of K^+, NH_4^+, NO_3^-, and absorbing aerosol particles. In the latitudinal belt >25°N the polluted ABL plume was characterized by a mixed composition including both biosmoke and products of fossil fuel burning (SO_4^{2-}). During TRACE-P, 5 plumes, of a total of 53, were observed north of 25°N. A similar situation was also observed a month later during the field observational experiment ACE-Asia. Analysis of 'back' trajectories showed that such plumes were connected with air masses subject to an advection in a rather narrow latitudinal belt (34°–40°N) along the eastern coast of China, where such megacities as Beijing and Tyan-zin are located.

In these cases, centres of biofuel burning were apparently sources of biosmoke. To characterize the relative contribution of biosmoke to the formation of plumes of a mixed composition, data were used on molar ratios of dK^+/dSO_4^{2-} and SO_2/CO for fine aerosols. According to data on dK^+/dSO_4^{2-}, there is a correlation between a relative contribution of biosmoke and concentrations of inorganic fine aerosols ($r^2 = -0.93$) as well as the volume of fine aerosols ($r^2 = 0.85$). The mixed plume, with maximum values of volume (mass) of fine aerosols, K^+, NH_4^+, NO_3^-, SO_4^{2-}, absorbing (soot) aerosol, and CO, was recorded in the Yellow Sea. About 60% of the plume's composition is from biosmoke. On average, the contribution of biosmoke constitutes about 35–40% with respect to the mass of inorganic fine aerosol in the mixed plumes observed north of 25°N.

Numerous studies have been recently accomplished of the impact of tropospheric aerosol on chemical reactions (the budget of various MGCs in the atmosphere) and processes of photolysis both at industrial locations and on regional scales. However, no attempts have been made so far to generalize the obtained results on global scales and over long periods of time (of the order of decades), which is necessary for the evaluation of global budgets of MGCs and their contributions to the formation of the climatic impact of the atmospheric greenhouse effect. This situation is partly explained by the absence of required reliable information about the properties of aerosols formalized as global aerosol models. The main source of such information can only be data of satellite remote sensing of aerosol.

Bian *et al.* [8] performed an analysis of the aerosol impact on the formation of global budgets of O_3, OH, and CH_4 through aerosol-induced changes in photolysis frequencies and direct effects of heterogeneous chemical reactions on the surface of aerosol particles. The impact on photolysis was estimated using the global model of MGCs transport in the troposphere, with due account of the respective chemical reactions with prescribed global models of aerosol over the ocean from data of satellite observations, as well as (over land and ocean) calculated models of aerosol. With global averaging, the impact of aerosol on photolysis manifests itself through an increase of O_3 in the troposphere by 0.63 Dobson units and an increase of CH_4 concentration by 130 ppb due to an 8% decrease of OH concentration. The increase of the content of the two greenhouse gases in the troposphere results in an indirect effect of aerosol (both natural and anthropogenic) on the radiative forcing constituting $+0.08\ \mathrm{W\,m^{-2}}$.

Though the estimate of CH_4 concentration is globally averaged, it should be borne in mind that changes in the content of OH and CH_4 are of regional nature, showing most in north-western Africa (in January) and India and south Africa (in July) – this effect being greater in July than in January and in the northern hemisphere compared to the southern hemisphere. The prevailing effect is determined by aerosol over land. The globally averaged contribution of aerosol over the World Ocean constitutes less than one-third.

Lefer *et al.* [46] undertook a comparison of measured frequencies of photolysis using the scanning spectro-radiometer SAFS to determine the actinic radiation flux with the results of numerical modelling under cloud-free atmospheric conditions using the box photochemical model CFM. On average, the ratio of the measured photolysis frequencies jNO_2 (from data of aircraft observations in the period of accomplishing the field observation experiment TRACE-P) to those calculated turned out to be equal to 0.943 ± 0.271. Since cloud-free conditions were only present in 40% of cases (during TRACE-P), the results of calculations should not be considered representative, even more so that the impact of clouds on photolysis frequencies show stronger (varying from −90% to 200%) than that of aerosol (not more than ±20%).

Calculated and measured values of jNO_2 and $j(O^1D)$ differed from data of aircraft sounding of vertical profiles by 9% and within 0–7%, respectively, for the cloud-free atmosphere and the atmosphere with a small aerosol optical thickness (AOT). It means that it is impossible to ensure the difference between measured and calculated values of photolysis frequencies not exceeding 10% without reliable input information on aerosol characteristics even with low AOT values. Under conditions of the atmospheric chemical composition observed in the period of TRACE-P, values of the content of OH, NO, and HO_2 were more sensitive to changes (or errors) in photolysis frequencies than the levels of concentrations of other compounds (e.g., $CH_3C(O)O_2$, CH_3OOH).

Various compounds including NO_2, PAN, and HCHO were characterized by a different sensitivity to changes in photolysis frequencies j below and above the ABL. There is a linear increase or decrease of the rates of formation and destruction of O_3 depending on changes (or errors) in photolysis frequencies which eventually manifests itself as an increasing trend of O_3 concentration. According to TRACE-P observational data, the net effect of clouds and aerosol on photochemical processes manifested itself through a strong decrease of photochemically induced O_3 formation in the ABL.

An increasing level of atmospheric pollution determines an urgency of air quality model (AQM) improvement. In this context, [61] carried out an analysis of global data on AQM sensitivity to pollutant emissions to the atmosphere using a numerical simulation model CACM of the mechanisms of chemical reactions in the atmosphere developed at the Californian Institute of Technology. The emphasis has been laid on quantitative estimates of uncertainties which characterize the concentration of secondary organic aerosol (SOA). The Monte Carlo method was used to estimate the sensitivity, to a prescribed scheme, of the totality of chemical reactions which determine the SOA concentration.

With three summertime situations as an example, an analysis has been made of the uncertainty of the rate of changes in the various quantities under consideration. The discussed situations cover rather broad ranges of initial values of concentrations of different chemically active organic gas-phase components and nitrogen oxides corresponding to the conditions of an increased ozone concentration in the polluted urban atmosphere. In addition to calculations of uncertainties due to the gas-phase SOA precursors concentration, similar estimates have been obtained with O_3, HCHO, H_2O_2, and PAN (peroxiacetyl nitrate) taken into account.

The results obtained [61] have shown that concentrations of gas-phase SOA precursors, precalculated with due regard to nominal values of the rate of changes in different parameters determined from CACM model, are close to values obtained using the Monte Carlo method. Uncertainty estimates of the gas-phase SOA precursors using the CACM model lead to relative errors from 30% for the concentration ratio VOC/NO_x equal to 8 : 1 to 49% if this ratio grows to 32 : 1. Though the results discussed demonstrate that reactions connected with photo-oxidation of aromatic compounds and direct transformation of aldehydes into semivolatile organic acids are not very substantial for SOA formation, the estimates of the effect of changing rates of reactions agree with those obtained earlier.

A new important conclusion is that the most significant factor of reliability of the precalculated SOA concentration is the reliability of information on NO_2 and HCHO photolysis as well as on reactions between aromatic compounds and OH with a VOC/NO_x ratio $< 17:1$. Concentrations of SOA precursors depend more strongly on OH losses as a result of its reaction with NO_2 with a low value of VOC/NO_x. However, in the case of large values of this ratio, the reaction of O_3 with NO plays the most significant role.

Within the programme of monitoring the characteristics of the urban atmosphere at the supersite in Atlanta (USA) performed in August, 1999, a 3-channel chemical composition monitor PCM was used to measure the mass concentration and chemical composition of fine aerosols PM-2.5 (particles with diameter less than 2.5 μm). This device enables one to obtain time averaged (over 10–25 hours) estimates of concentrations of gas-phase NH_3, HCl, HNO_3, HONO, and SO_2 as well as acetic, formic, and oxalic acids. The observational period was characterized by a stable (stagnant) atmosphere with high levels of temperature, relative humidity, and UV radiation, favouring an increase of photochemical process activity, and formation of surface–air ozone and fine aerosol PM-2.5. An analysis of observational data carried out by [4] revealed photochemical sources of HNO_3, HONO, acetic, and formic acids. It was shown, however, that specific features of the method of measurements determined an underestimation (overestimation) of the measured concentration of $[HNO_3]$ ([HONO]). Specific processing of measurements of the content of organic carbon (OC) in aerosol was connected with the introduction of an empirical correction for omitted organic components (OOC): concentration [OOC] = 0.6 [OC]. However, despite this correction, about $13 \pm 10\%$ of measurements of the organic aerosol mass concentration still included unconsidered components. Obtaining additional filter samples has made it possible, nevertheless, to achieve the complete closing of the balance of aerosol mass.

The photochemically rather 'old' and well mixed air masses are supposed to contain more oxidized organic and less volatile functional groups of compounds, whereas stagnant air masses are characterized by less oxidized and therefore more volatile aerosol organics. Sulphates and all organic compounds (including lightweight organic acids, organic carbon, and OOC contained in aerosol particles) contribute most to the formation of the mass of fine aerosol PM-2.5, constituting $33 \pm 8\%$ and $39 \pm 12\%$ of the mass, respectively. Estimates of acidity based on consideration of SO_4^{2-}, NO_3^-, and NH_4^+ as aerosol components have led to the conclusion about the presence of slightly acid conditions.

Since long ago (especially in the context of the problem of acid rain) studies of pH for water in the atmosphere have attracted special attention. Mechanisms of acid rain formation have now been adequately studied. It was shown, in particular, that the value of pH for droplets is one of the most significant parameters characterizing the dynamics of chemical reactions in the water phase in the atmosphere. In cloud droplets the rate of oxidation processes depends strongly on pH, which is mainly caused by decreasing solubility of MGCs with decreasing pH. Changes in pH also affect the solubility of trace metals and, hence, change the catalytic cycles in droplets, in which metals take part. The level of a droplets pH is determined by the balance between present acid and neutralizing components.

The main acid components, mostly of anthropogenic origin, are H_2SO_4 and HNO_2 whose precursors are, respectively, SO_2 and NO_x, whereas the neutralizing components are ammonium (NH_4^+), mostly in gas phase, and particles of mineral aerosol. While the process of gas dilution is satisfactorily described with due regard to Henry's law and processes of diffuse accommodation, the process of aerosol dilution remains unclear in many respects.

Desboeufs *et al.* [12] discussed the results of laboratory studies of pH, which made it possible to simulate the process of aerosol particle dilution in cloud droplets and to analyse how these particles can affect the pH of droplets. The effect of two types of soil and anthropogenic aerosols has also been studied. The tests performed with the use of a well-drained cell have shown that due to aerosol dilution, pH changes for up to 30 minutes, which corresponds to a characteristic lifetime of a cloud droplet. Changes in pH range from 3–5, its maximum variability taking place at high pH values.

A relationship has been obtained [13] between the neutralizing capability of aerosol (NCA) (i.e., an amount of non-compensated reducing components) and pH after neutralization. The NCA value depends on the aerosol composition: the presence of silicates determines a sufficiently remarkable NCA level, whereas graphite C is responsible for negative NCA values (i.e., it causes a protoxidation). The chemical composition of aerosol can vary in the processes of evapocondensation of clouds, especially with an uptake of sulphate or sulphuric acid on the surface of aerosol particles. A considerable change in the ABL is observed when the amount of dissolved acid exceeds the neutralizing capability of aerosol.

The results of laboratory experiments carried out by [37] have demonstrated that photo-oxidation of organic compounds results in the formation of a substantial aerosol fraction consisting of polymers – polymerization taking place due to reactions between carbonyls and their hydrates. Having performed an analysis of natural organic compounds emitted to the atmosphere by wet tropical forests of Amazonia, [10] found a considerable amount of polar organic compounds (polyols) not observed before. The low vapour pressure of such polyols favours their condensation on aerosol particles. Estimation has shown that due to isoprene photo-oxidation the global-scale amount of resulting polyols can reach 2 Tg, which constitutes a considerable share of the net biogenic production of secondary aerosol (8–40 Tg). The problem of the secondary aerosol formation due to gas-phase oxidation of organic compounds has been discussed in detail by [20, 21]. Laskin *et al.* [44] carried out laboratory studies of reactions on the surface of aerosol particles which are the source of formation of sulphates in sea salt aerosol particles.

5.3 TROPOSPHERIC OZONE VARIATIONS IN THE HIGHLY POLLUTED ATMOSPHERE

There are numerous measurement data on the surface air ozone concentration in highly polluted atmospheres [2, 32, 33, 41, 58, 60]. Practically in all large cities of the world systematic simultaneous observations of the composition and state of the atmosphere in the surface layer are carried out at several locations. The role of nitrogen oxides in the processes of photochemical formation of ozone has been studied in detail [40]. It was established that it is observed at NO_x mixing ratios of the order of 3×10^{-9} and its intensity is directly proportional to concentration of hydroxyl [OH] and solar radiance (SR) in the short-wave spectral region (280–400 nm).

The dependence between variations of ozone and other polluting components, both gas-phase (hydrocarbons, sulphur and chlorine-containing compounds) and aerosol (organic, sulphate, soil–mineral) is exibited less vividly. In particular, this is explained not only by meteorological factors but also by different chains of photochemical, especially photocatalytic, reactions taking place in daylight and in reactions occurring at night, below the clouds, and in fog particles. A combined averaged effect of all these factors is most apparent in large cities and megalopolies. Analysis of such averaged dependencies has been made using data on pollution for twenty cities and megalopolies at different latitudes for the period 1981–1991 [33]. Data on the level of atmospheric pollution with the following components were used: dust, ozone, sulphur dioxide, nitrogen dioxide, and carbon monoxide. For 15 of the 20 cities, a certain dependence was observed – an increase of atmospheric dust load was followed by a decrease of ozone concentration in the surface layer. In five cities (Beijing, Los Angeles, Saõ Paulo, Mexico, and Jakarta) this dependence was absent. For the atmosphere of Mexico this can probably be explained by the altitude of the city's location (2.0–2.5 km), where photochemical processes of ozone generation prevail over the processes of heterogeneous destruction. These cities, except Beijing, are also characterized by an atmosphere heavily polluted with NO_2 and CO, which testifies to the prevailing effect of motor vehicles on atmospheric pollution. However, in cities located at high latitudes, with the atmosphere sufficiently polluted with NO_2 and CO (Moscow and London) the concentration of ozone does not increase.

Also, there is no apparent dependence between concentrations of ozone and sulphur dioxide. For Seoul, Shanghai and Rio-de-Janeiro an increased concentration of sulphur dioxide corresponds to low concentrations of ozone, and for the atmospheres of Los Angeles, Karachi, and Jakarta with an increased concentration of ozone the concentration of sulphur dioxide is relatively low (i.e., the partial sink of ozone takes place due to sulphur dioxide oxidation (formation of sulphuric acid and sulphate aerosols)). This mechanism is more efficient at low latitudes. However, Beijing and, to some extent, Jakarta data do not confirm these sufficiently clear ideas about mechanisms of interactions between various pollutants in the lower atmosphere. For Jakarta this is probably connected with more intensive photochemical processes compared with higher latitudes, for Beijing the reason for unusual ratios of pollutant concentrations should be sought in the chemical composition of pollutants, aerosols in particular.

Detailed studies of the variations of surface layer ozone and other atmospheric characteristics at certain locations give additional information about the nature of their interaction. In the case of measurements in the heavily polluted air in cloudless weather conditions, clear maxima of $[O_3]$ are observed at noon. However, the mutual dependence of $[O_3]$ and aerosol number density N_a is manifested in a complicated way and interpreted ambiguously. For instance, with a low dust loading of the urban atmosphere it is difficult to find any dependence between concentrations of aerosols and ozone, and with moderate levels of pollution there is an indirect dependence (i.e., ozone is spent on the formation of aerosol matter). It should be noted that the depth of the morning minimum of ozone mixing ratio correlates with air pollution. At heavily polluted locations the values $[O_3]_{min} = 2$ ppb are observed, differing by an order of magnitude from a minimum value of $[O_3]$ in the clean air. On some days, the diurnal course exhibits quasiperiodic oscillations of $[O_3]$ with a period of $\tau = 180$ and $\tau = 200$ min as well as small-scale oscillations with periods from 3–7 min, growing in mid-day, which testifies, first of all, to the existence of non-linear photochemical mechanisms governing the process of ozone molecule formation in the polluted urban atmosphere.

With heavily polluted urban air, the dependence between aerosol characteristics and ozone is determined not only by the accompanying high content of aerosol particles but also by nitrogen oxides and various hydrocarbons (aldehydes, aromatic and unsaturated hydrocarbons). Most probable is a mechanism of formation of aerosol particles and hydroxyl OH from hydrocarbons. This assumption is favoured by the often observed increase of fine aerosol particle concentrations. Analysis of aerosol samples on filters, performed in parallel with measurements by photoelectric counters, confirms an increase of the content of organic matter in aerosol particles [16].

The inversely proportional dependence of ozone and aerosol concentrations reflects, apparently, ozone molecule destruction on aerosol particles, first of all, dust-containing aluminum and zinc oxides – with a sufficiently high efficiency of heterogeneous destruction of ozone molecules. The dependence between the contents of ozone and aerosol particles in the surface layer can be vague because of different effects of other atmospheric parameters on the concentration of ozone and aerosols. In particular, the diurnal courses of ozone and aerosols are not similar. The morning maximum of aerosol content is usually expressed more strongly compared with ozone. The ozone concentration in the surface layer increases practically from morning to afternoon. A maximum of concentration can be observed from 14:00 hours to about 17:00 hours and then the concentration of ozone decreases until the next morning. Sometimes the concentration grows until nocturnal hours. After the morning maximum, the concentration of aerosols in the surface layer is almost always at a minimum in the period approximately from 12:00 hours to 15:00 hours and then a second, more powerful maximum of aerosol number density follows which, depending on meteorological prehistory and the extent of heating of the surface and air, can shift from 16:00 to 21–22:00 hours.

The presence of overcast cloudiness leads to a disappearance of the ozone concentration diurnal course and deformation of the diurnal course of the aerosol particle concentrations, mostly determined by the diurnal course of the anthropogenic pollution intensity.

Under cloud-free conditions in the summer the diurnal course of aerosols and ozone concentrations is expressed most vividly. This especially shows at low latitudes. For instance, in Alma-Ata the amplitude of diurnal variations of ozone mixing ratio reached 100–120 ppb (i.e., from a practically total absence of ozone in the morning to maximum values in the afternoon ($[O_3]_{max} = 150$ ppb)). In St Petersburg, ozone does not disappear completely and its maximum concentration in relatively pure regions of the city does not exceed 70 ppb. Strong variations of the ozone concentration are characteristic of the central part of the city [58].

In relatively pure air and for almost cloudless sky conditions (the cloud amount varied from 1–3) the diurnal course of concentrations of both ozone and aerosol particles shows vividly.

With an increasing pollution and increasing cloud amount (from 6–9) the diurnal course of both ozone and aerosol is absent and marked oscillations of concentrations takes place caused, probably, by changing air masses with different levels of pollution. The ozone concentration does not exceed 26 ppb. Photochemical processes of ozone molecule formation in the polluted troposphere may be simulated via the following reactions of formation and destruction of ozone molecules:

$$CO + 2O_2 + h\nu(\lambda = 410\,\text{nm}) = CO_2 + O_3$$

$$CO + H_2O + 2O_3 + h\nu(\lambda = 320\,\text{nm}) = CO_2 + 2O_2 + 2OH$$

In the process of these photochemical reactions in the troposphere, the rate of the ozone concentration change is mainly determined by the behaviour of nitrogen oxide and dioxide, which depends on the intensity and spectral composition of solar radiation. The maximum rate of ozone formation is observed when the concentration of nitrogen oxides constitutes about 3 ppb. As a result, a certain diurnal course of ozone concentration is observed in the lower atmosphere [10, 22, 58]. In the cities the processes of formation and destruction of ozone molecules on aerosol particles are especially significant. The most efficient photocatalysts of the processes of destruction of ozone molecules are aluminosilicates. Heterogeneous reactions of the ozone molecules destruction strongly distorts the diurnal course of ozone concentration in the surface layer.

Observational studies of the behaviour of ozone in the lower tropical atmosphere, with different levels of dust load, were performed in [28]. Measurements of aerosol characteristics, ozone and sulphur dioxide concentrations, as well as the solar irradiance in the UV spectral region were carried out in Mexico city, Guadalahara, Colima, Mansaniyo, and in the Playon hollow between the two peaks of the Nevada del Colima and Fuego del Colima at a height of 3,600 m. The impact of different types of pollution on the diurnal course of the concentration of the surface layer ozone has been found: Mexico and Guadalahara – strong anthropogenic pollutions, first of all, due to the transport; Colima and Mansaniyo – soil dusting, emission of gases and aerosols from sea surface, pollution by power stations and plants (sulphur dioxide, metals oxides, soot); and the Playon hollow – emission of gases and aerosol by the Fuego del Colima volcano, soil dusting (secondary volcanic dust).

All measurements revealed a persistent diurnal course of the surface layer ozone concentration slightly differing in clear weather and in the presence of clouds. In the daytime maximum, the clear weather values of ozone concentration (March–May) reached 50–60 ppb. The lowest values in this maximum were observed in Mansaniyo and Guadalahara (from 20–30 ppb), and the highest values on the outskirts of Colima in the wet season in the periods after strong scavenging of atmospheric pollutants by cloud and rain (mid-August) up to 70 ppb.

The rate of ozone decrease at sunset varied from 0.01 ppb s^{-1} to 0.2 ppb s^{-1}. It was rather difficult to estimate the share of aerosol decrease from these observations. A comparison of measurements made at different locations suggests the conclusion that a considerable share of ozone was spent on reactions of sulphur dioxide oxidation, especially under cloud conditions or high atmospheric humidity.

Special emphasis should be laid on intensive destruction of ozone molecules in the surface layer at night. In some cases, the concentration dropped to 0.5 ppb, which was characteristic of measurements in cities with the highest air pollution: Guadalahara and Mexico. A comparison of the observational data obtained in Alma-Ata (1988–1999) reveals a marked similarity of the ozone concentration diurnal course.

The effect of the heavily polluted atmosphere on variations of ozone over water surfaces has been studied for the coastal region of the Norwegian Sea. The observed mixing ratio changed by an order of magnitude, from 4 ppb to 41.4 ppb. In general, it is much below the concentration of ozone observed over land [23]. It should be noted that since the chemical composition of oceanic aerosols differs substantially from that of continental aerosols, their role in the processes of ozone destruction and formation should not be identical. In particular, a high air humidity and high hygroscopicity of aerosol matter should lead to the formation, mainly, of droplets of a very saturated solution of salts.

The determination of dependences of $[O_3]$ on meteorological visibility S_m leads to the conclusion similar to that drawn from measurements on continents: maximum values of $[O_3]$ are observed at the lowest values of S_m (Figure 5.1) (i.e., when the air is heavily polluted, the processes of ozone formation take place). A minimum of ozone concentration is usually observed at $S_m \approx 8$–12 km, and at $S_m \approx 15$–20 km the concentration of ozone again grows slightly. This dependence is most vividly manifested in the morning and before mid-day. In the evening, dependences of ozone concentration on meteorological visibility (i.e., aerosol concentration (more precisely, specific surface of aerosol particles)) is not observed. The behaviour of $[O_3]$ at different times of the day and in heavily polluted air ($S_m \approx 0$–3 km) is different.

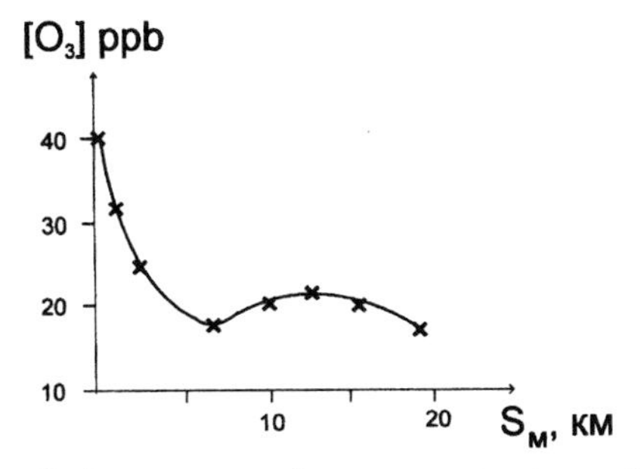

Figure 5.1. Dependence of the water–surface concentration of ozone on meteorological visibility (the North Sea, September–October 1987).

An analysis of multiple correlations with other meteorological characteristics (wind speed, sea roughness, air–water temperature difference, relative humidity, as well as direct solar and global radiation) has not revealed any significant correlations.

Also, no correlation has been observed with relative humidity. It should only be noted that maximum values of ozone concentration were observed at large relative humidities ($f > 95\%$) and for offshore wind directions.

The conclusion can be drawn that in cases of the heavily polluted air, there are processes favouring an increase of the ozone concentration over water. It is most probable that this effect is connected with the presence in the polluted air of organic compounds and nitrogen oxides [23]. The arrival of pure air masses from land does not lead to a marked increase of the water surface ozone concentration.

Of special interest are observations of the ozone behaviour over the central Atlantic during powerful dust outflows from the African continent [27]. Before a dust outflow there is a marked decrease of atmospheric pressure and relative humidity, with a strong decrease of ozone concentration during the dust outflow. The ozone concentration very slowly decreases to values observed before the formation in the atmosphere of a dust layer, which confirms the substantial role of ozone sinking on fine dust particles. Such a sink in the surface layer reaches $2 \times 10^7\,cm^3\,s^{-1}$ with the total surface area of aerosol particles $4 \times 10^{-7}\,cm^2\,cm^{-3}$. An approximate estimate of efficiency of active collisions of ozone molecules with the surface of aerosol particles gives $\gamma = 2.5 \times 10^{-6}$, which agrees well with the earlier values obtained for aluminosilicates [22].

5.4 MODELLING ATMOSPHERIC PROCESSES IN THE THE HIGHLY POLLUTED ATMOSPHERE

5.4.1 General principles

Field observations of the ozone behaviour in the atmosphere cannot give unique answers to the question about the role of various atmospheric processes in its spatial-temporal variability. Therefore, a numerical modelling of the processes governing this variability and chemical reactions, in particular, is needed. A complete modelling of chemical and photochemical processes determining the content of ozone in the atmosphere requires the consideration of a large number of heterogeneous chemical reactions whose characteristics are not always known. Besides this, knowledge of the physico-chemical processes of the transport and sinking of chemicals is necessary. This makes 3-D numerical modelling very complicated. The difficulty of such modelling may be overcome through decreasing the dimensionality of the problem and decreasing the number of reactions (i.e., using the so-called method of 'family preservation').

This method is based on the consideration of a large number of reversible reactions in atmospheric chemistry, the rates of which do not differ much at large values. If we group the components, for which such reactions take place, in opposite directions, into conditional families, then, first, the number of equations in the system decreases, and second, rapid reversible reactions will be excluded from the expression for the source, which leads to weakening the bonds between equations.

Distributions of concentrations within the family are calculated using formulas taking into account rapid reactions leading to the transfer of the family components from one into another. It is necessary that the lifetime of the components included in the family be much shorter than the transfer time constant, and the rates of reactions determining the distribution of concentration within the family be much greater than the rates of reactions affecting the total concentration for families. At an initial stage it is sufficient to analyse the behaviour of such a system for a 1-D case (the transfer takes place only either in horizontal or vertical directions).

The numerical and laboratory modelling of photochemical processes in the polluted atmosphere in the presence of various organic compounds [5, 60, 71] has led to the conclusion that various hydrocarbons participate in reactions of formation and destruction of ozone in different ways. So, high concentrations of ozone exceeding the background quantities by two orders of magnitude are observed near oil refineries. Here ozone, as a secondary atmospheric pollutant, is the product of photochemical and 'dark' reactions of oxidation of substances, including unsaturated aliphatic and aromatic hydrocarbons and nitrogen and sulphur oxides. In the processes of ozone formation, saturated hydrocarbons practically do not participate. Ozone participates in oxidation of several hydrocarbons, terpenes in particular. There is a clear correlation between concentration of unsaturated hydrocarbons and nitrogen oxides, which, first of all, are responsible for the formation of ozone molecules. Peroxiacetyl nitrates in the anthropogenically polluted lower atmosphere are reservoirs for nitrogen oxides NO_x. As a result, a high correlation is usually observed in the lower atmosphere between air polluted with organic compounds and concentrations of ozone and fine aerosols.

The ozone content in the polluted troposphere can change as a result of both the gas-phase reactions not necessarily leading to aerosol formation and of heterogeneous reactions. The process of generation of ozone molecules in the troposphere is mainly determined by the presence in it of nitrogen oxides:

$$NO_2 + h\nu = NO + O_2 \rightarrow O_3 \tag{5.2}$$

$$NO + O_3 = NO_2 + O_2 \tag{5.3}$$

The O_3 formation is regulated by the presence in the atmosphere of hydroxyl HO_2 and nitrogen oxide NO [51]:

$$HO_2 + NO = NO_2 \tag{5.4}$$

With the atmosphere polluted with products of incomplete combustion, carbon monoxide (CO) in particular, the latter will determine the intensity of OH and HO_2 formation, and eventually the intensity of ozone formation [51]:

$$CO + OH = CO_2 + H \tag{5.5}$$

$$H + O_2 + M = HO_2 + M \tag{5.6}$$

The net reaction is:

$$CO + 2O_2 + M = CO_2 + O_3 \tag{5.7}$$

These reactions can proceed at different rates depending on the NO concentration, since the reaction of ozone molecule destruction is the competitor of the reaction of O_3 formation:

$$HO_2 + O_3 = OH + 2O_2 \tag{5.8}$$

Only at $p = 1$ ppb, HO_2 reacts mainly with NO but not with O_3. The reaction of ozone destruction in polluted air will eventually be [51]:

$$CO + H_2O + 2O_3 + hv(\lambda = 320\,\text{nm}) = CO_2 + 2O_2 + 2OH$$

The sink for the main component responsible for O_3 generation results from reactions of nitric acid formation [11, 28, 34]:

$$NO_2 + OH = HNO_3$$

5.4.2 Interaction between tropospheric aerosols and ozone

Since the composition of tropospheric aerosols, especially of industrial origin, is rather variable both in time and space, studies of their interaction with ozone are very complicated, even more so since this interaction can lead both to O_3 molecule formation and destruction.

Consider the ozone sink on the surface of aerosol particles from the data of observations in the atmosphere [32, 54]. Since the heterogeneous catalytic destruction of ozone is the reaction of the first order, the change in ozone concentration due to its sink on the surface of aerosol per unit volume can be described with the expression [48]:

$$\frac{dm}{dt} = -\frac{mv}{4} S_{\gamma}\gamma \tag{5.9}$$

where m is the mass concentration of ozone in g cm^{-3}; v is the average rate of O_3 molecules, cm s^{-1}; S is the total area of the aerosol particles surface, cm^2 cm^{-3}; and γ is the coefficient of destruction of ozone molecules on the surface of aerosol particles.

Belan *et al.* [6] performed an analysis of the results obtained during the dust storm of the joint American–Soviet experiment in September, 1989 in Dushanbe. Using the flying laboratory of the Institute of Atmospheric Optics, data were obtained on the vertical profiles of meteorological parameters, ozone concentration, and optical and microphysical characteristics of aerosols in the 0–6-km altitude interval for the period 14–25 September. In a clear atmosphere, before the dust storm, variations of the ozone content at altitudes 2–6 km were negligible. A powerful dust storm took place on the nights of 19–20 September, which reduced the O_3 concentration in the lower atmosphere. Figure 5.2 presents the vertical profiles of the ozone mass concentration and coefficient of aerosol scattering α ($\lambda = 0.52\,\mu$m) for both a relatively pure atmosphere (18 September) and during

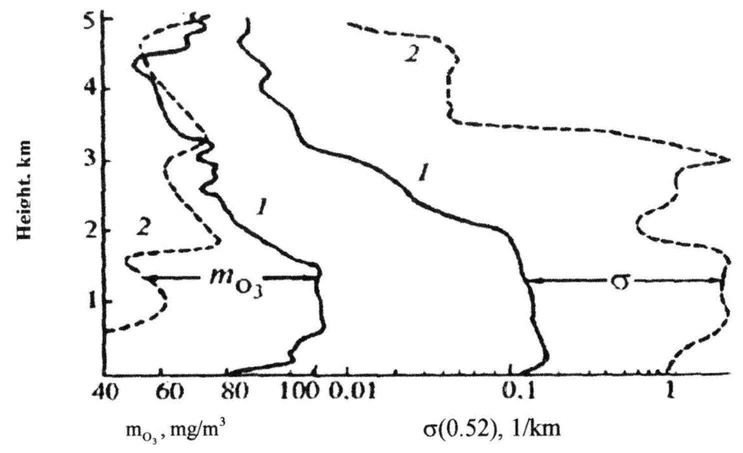

Figure 5.2. Vertical profiles of the ozone mass concentration and coefficients of aerosol scattering: 1 – for a relatively pure atmosphere; 2 – during a dust storm.

the dust storm itself (21 September). It is seen that in the subinversion layer of the atmosphere ($H < 3.4$ km) the concentration of ozone decreased strongly during the dust storm, when the optical aerosol thickness was 3. At a maximum of aerosol optical thickness (20 September) equal to 10 there were no flights because of conditions that were deemed unfit for flying at the airport in Dushanbe. Above the layer of the temperature inversion the ozone mass concentration practically did not change.

During the dust storm the total depletion of ozone in the lower atmosphere was due to an interaction with aerosol particles – a coefficient being estimated at $\gamma = 3.9 \times 10^{-11}\,\mathrm{g\,cm^{-2}\,s^{-1}}$.

Estimates of the contribution of aerosols to the destruction of O_3 molecules under usual atmospheric conditions show that it decreases substantially near the Earth's surface. In the surface layer this contribution cannot exceed several tens of % compared with other mechanisms of destruction, including, first of all, processes at the surface level. So, for instance, the summertime measurements in the regions of Tomsk showed that in this case the contribution of aerosols in the sink of ozone constituted from 2–20%.

Note that the ozone concentration can decrease not only because of its heterogeneous destruction, but also due to a recombination of atomic oxygen on the surface of aerosols. The equation for the rate of reactions was included into the Crutzen 1-D model [6].

The concentration of particles above 25 km is not equal to zero but it is much lower than at lower altitudes. Calculations were made of the dependence of ozone decrease on the efficiency of collisions γ for particles in the range $0.1 < r < 1.0\,\mu\mathrm{m}$ (this range is characteristic of stratospheric aerosols). At $r = 0.1\,\mu\mathrm{m}$ and $\gamma = 10^{-5}$ the ozone concentration in the lower stratosphere should decrease by 5%. Heterogeneous reactions between O_3 and aerosols should, of course, affect other minor gases in the atmosphere included those in the model.

Olszyna *et al.* [56] carried out laboratory studies of the efficiency of the ozone molecules destruction in 75% solutions of sulphuric acid at 500°C and in the presence of ions of various metals (Fe_3^+, Al_3^+, Cu_2^+, Ni_2^+, Cr_3^+, and Mn_2^+) as well as ammonium, and arrived at the conclusion that stratospheric acid aerosols cannot be a reason for the marked changes in the concentration of stratospheric ozone since γ does not exceed 10^{-9}. However, it should be noted that this result does not consider the possibility of ozone molecule destruction when H_2SO_4 particles are irradiated by sunlight. Besides this, the atmosphere contains not only particles of H_2SO_4 but also particles with oxides of metals.

Laboratory measurements of the coefficient of efficiency of ozone molecule destruction on some substances, simulating natural solid aerosol particles [15, 71], suggested the conclusion that some of these substances, Al_2O_3 in particular, can effectively promote the heterogeneous destruction of ozone molecules $((Al_2O_3) = 10^{-5})$. The estimates suggest the conclusion that with an emission to the atmosphere of aerosols containing aluminosilicates and ammonium oxide in concentrations corresponding to those observed, a considerable share of stratospheric ozone can be destroyed on them (up to 20–40%). From these data the heterogeneous coefficient of the rate of ozone sink in the troposphere is of the order of $1 \times 10^{-5}\,s^{-1}$.

In assessing the possibility of ozone removal from the stratosphere one should consider the mechanism of adsorptive absorption of ozone molecules by both liquid and solid aerosol particles. The intensity of the adsorption mechanism is determined to a considerable extent by the physico-chemical properties of the surface (specific surface, the presence of other adsorbents) for liquid particles of the solution as well as by differences in temperatures of the surface of a particle and its environment. The latter can be considerable at altitudes above 30 km as well as at the moment of a sharp change of the insolation regime, for instance, at solar eclipse. With a well-developed surface of particles ($S = 100\,m^2/g$) and mass concentration $m = 10^{-6}$–$10^{-5}\,g\,cm^{-3}$ the adsorption sink during several days can cause a substantial ($\approx$5%) change in the content of ozone in the air column.

Adsorption and partial desorption of ozone molecules determine a characteristic variability of the ozone content in the atmosphere. Probably, an adsorption of ozone molecules in water droplets also plays an important role in temporal variations of the stratospheric ozone concentration [65]. Practically, all simultaneous changes in the concentration of ozone and water vapour in the atmosphere demonstrate a clear anticorrelation. For instance, [53] published extensive data on simultaneous measurements of concentrations of O_3, CO, and H_2O at the observatory at Pic du Midi. Though these data are given as proof of the existence of photochemical processes regulating the content of ozone in the troposphere, the obtained dependences can be explained by the partial loss of ozone at adsorption and their subsequent destruction in water droplets [65]. In particular, the role of ions of hydroxyl HO_2 and hydrogen peroxide H_2O_2 in the processes of ozone molecule destruction in water is discussed. From the estimates obtained, the lifetime of ozone molecules in water droplets varies from 1 hour to 24 hours.

In the above-ocean surface atmosphere the specific character of heterogeneous processes are explained by a sufficiently high relative humidity of the lower tropospheric layers and by the presence of sea salt in them. However, the field observations testify to the prevalence in the submicron fraction of sulphate particles over chloride particles. If in seawater $[SO_4^{2-}]/[Na^+] = 0.25$, in the air this ratio exceeds unity. Measurements have shown that an excess content of $[SO_4^{2-}]$ agrees well with $[SO_2]$; the ratio $[SO_4^{2-}]/[Na^+] > 0.5$ is characteristic of particles with $D < 0.6\,\mu m$, and for gigantic particles it tends to 0.25. This fact can be considered evidence in favour of the photochemical gas-phase mechanism of SO_2 conversion with participation of ammonium, oxygen, and water. Note that nucleation of sulphate particles begins on positive ions. Apparently for this reason there are no strong variations of the ozone content over the ocean, except regions in the cases where the surface layer is heavily polluted with anthropogenic pollutants – where a clear connection is observed between meteorological visibility S_m and ozone concentration. Maximum changes in the ozone concentration in the marine troposphere were observed in the case of dust outbreaks, which took place during measurements in the central and northern Atlantic [27, 28].

In this connection, of interest are studies of the effect of spectral composition of radiation on the atmospheric ozone formation [3, 7, 16, 29–31, 57]. A high correlation is observed for the total radiation flux. This dependence is not only due to direct participation of solar radiation in photochemical processes, but also due to the temperature dependence of the rate of some reactions with participation of hydrocarbons as well as by the effect of solar radiation on the physico-chemical processes in the atmosphere (transfer processes – diffusion, advection, convection, phase conversions). Also note that the effect of radiation on the concentration of ozone in the whole atmosphere, and in the lower atmospheric layers, can be different. For instance, an increase of the total ozone content can lead to a decrease of the ozone content in the surface layer due to decreasing intensity of the ozone-forming UV radiation [57].

In the stratosphere, rapid variations of ozone are determined mainly by the mechanism of dissociation on aerosol particles containing nitrogen oxides and, possibly, by photocatalytic oxidation of sulphur dioxide. Ivlev *et al.* [29] considered the results of model calculations of H_2SO_4 aerosol accumulation and changes of the ozone content after a volcanic emission of sulphur dioxide to the stratosphere. The chemical scheme of the model used the photochemical reaction of SO_2 oxidation by ozone. The resulting calculated kinetics of H_2SO_4 aerosol accumulation in the lower stratosphere correlates well with the results of observations of the content of H_2SO_4 aerosols in the stratosphere.

5.5 OZONE LATITUDINAL VARIATIONS IN THE SURFACE LAYER OF THE ATMOSPHERE

Estimates of intensity of the stimulated sink of ozone and methane on the surface of aerosol particles suggests the possibility of detecting their effect on the spatial distribution of ozone not only in vertical (in the presence of powerful aerosol layers in the atmosphere) but also in horizontal directions [3, 30–33].

Since an important factor in the processes of ozone formation and destruction is the spectral composition of solar irradiance, the latitudinal dependence of the surface ozone concentration forms for a relatively homogeneous surface and similar air pollution levels [7]. Model calculations of the surface ozone concentration for some latitudes and time have been made repeatedly [1, 40, 64, 67]. There are observational estimates of the intensity of ozone sinking on the surface from gradient measurements of ozone concentration in the surface layer [1]. Deviations of model calculations of ozone concentration from that measured are probably explained by different intensities of ozone sinks on aerosol particles. It should be at a maximum in the morning and noon hours, when the intensity of photocatalytic processes is sufficiently high. The effect of various meteorological factors can completely annihilate the effect of the aerosol sink of ozone. Therefore, it is important to find out whether it is possible to introduce, into the climate models of the lower atmosphere, the parameter taking into account the aerosol sink of ozone and whether this parameter improves the climate model.

With this aim in view, an analysis was made of the data on the aerosol content in the atmosphere over the European territory of Russia in the summer–autumn periods of 1983–1989 [63].

From the data of actinometric network observations, isolines of aerosol optical thickness were calculated which satisfactorily reflected the total content of aerosols in the atmosphere (geometrical area of the aerosol particle cross section with $r > 0.03\,\mu m$). With the height of heterogeneous aerosol layer thickness H assumed to be 1.5 km, the specific area of the aerosol particle surface was estimated ($cm^2\,cm^{-3}$) in the surface layer ($\delta H = 200\,m$) actively participating in the formation of the aerosol sink for surface ozone.

The most heavily dust loaded atmosphere is observed in the south-western and southern parts of the European territory of Russia. These results agree well with data on dust load of the surface layer obtained from direct observations of the dust loaded atmosphere. Note should be taken that the electron microscopic analysis of aerosol samples testifies to a very complicated morphological structure of solid aerosol particles. Complex aggregates of fine particles constitute a considerable portion, therefore, their surface is much larger than the area of their cross section – reaching about $500\,m^2\,g^{-1}$.

To study the quantitative characteristics of the effect of heterogeneous processes on the surface ozone distribution, a 2-D zonal-mean photochemical model of the northern hemisphere troposphere was used, in which all photodissociation processes and chemical reactions, usually included in the tropospheric models, were taken into account. The 2-D equations of continuity were solved using a combined method (method of families preservation + A-stable methods of problems solution) [64]. The dark and photocatalytic aerosol sinks of ozone were estimated on the assumption of first-order reactions and a free-molecular regime [31]:

$$\frac{d[O_3]_{\text{dark}}}{dt} = \frac{1}{4}\bar{\upsilon}S_{sp}\eta\gamma[O_3]$$

$$\frac{d[O_3]_{\text{phot}}}{dt} = \frac{1}{16}\bar{\upsilon}S_{sp}\eta\Gamma[O_3]$$

where υ is the arithmetic mean rate of the thermal motion of an ozone molecule; S_{sp} is the specific area of the aerosol particle surface ($cm^2\,cm^{-3}$); η is the share of the surface of particles actively participating in molecule destruction (e.g., the share of Al_2O_3 particles); γ is the efficiency of interaction of ozone molecules with aerosol matter (probability of ozone molecule destruction); and Γ is the efficiency of the photostimulated destruction of an ozone molecule on the surface of an aerosol particle, parametrically calculated from laboratory data [31].

Uncertain values of S_{sp}, η, γ, and Γ makes calculations of the aerosol sink intensity unreliable for estimating absolute rate of aerosol sink, but a priori knowledge of aerosol particle size distribution enables one to specify these estimates and to make them more reliable. With assumed heterogeneous aerosol size distribution in the European territory of Russia for the period considered, the effect of aerosol on the surface ozone sink was assessed [63].

$$\chi = \frac{[O_3]_\Gamma - [O_3]_{a+\Gamma}}{[O_3]_\Gamma}$$

where $[O_3]_\Gamma$ is the concentration of ozone calculated with only gas-phase reactions taken into account, and $[O_3]_{a+\Gamma}$ is the concentration of ozone calculated with both gas-phase reactions and dark and photocatalytic aerosol sinks of atmospheric ozone considered. A comparison of the results obtained shows that maximum deviation of experimental values of ozone concentration from calculated ones corresponds to a maximum value of x. This convincingly testifies to a significant role of the aerosol sink of ozone in the formation of the spatial–temporal field of ozone concentration in the lower atmosphere.

Analysis of the data obtained from long-term observations of ozone variations, together with data on solar irradiance, the content of some trace gases (first of all, nitrogen oxides and hydrocarbons), as well as aerosols, suggests some conclusions about the nature of the diurnal and smaller scale variations of the ozone content in the atmosphere.

In particular, the significant role of heterogeneous reactions of ozone destruction on aerosol particles and a relatively weak effect of these reactions at the surface level are apparent (relatively small rates of the nocturnal sink of ozone onto soil). The diurnal variations of the surface ozone are substantially affected by cloud amount (i.e., the intensity and spectral composition of solar radiation reaching the lower atmosphere).

The ozone concentration varies mainly with periods about 180 minutes and 5.5–6.5 minutes and harmonics thereof. Variations with the period of about 180 minutes may be due to photochemical processes of ozone and aerosol-forming trace gases (nitrogen oxides, hydrocarbons, etc.). The photochemical nature of short-period variations of atmospheric ozone is confirmed by a strong dependence of ozone concentration on air pollution, in particular, observed over the sea surface during expeditions in the North Atlantic [57, 72].

5.6 BIOMASS BURNING AND ITS IMPACT ON TROPOSPHERIC OZONE AND OTHER MGCs

At present a maximum of total tropospheric ozone (TTO) in the annual course observed in the southern hemisphere tropical Atlantic Ocean is considered to be determined by spring period biomass burning in the regions of Africa and Brazil. During biomass burning the atmosphere receives a considerable amount of such tropospheric ozone precursors as CO, NO_x, and hydrocarbon compounds, which leads to an increase of TTO of 10–15 DU [18]. The interpretation of data for the South Atlantic is complicated by the fact that variability of a number of meteorological parameters is characterized by a structure with a zonal wavenumber equal to unity, similar to the spatial distribution of TTO. This makes it possible to assume the presence of a considerable impact of meteorological factors on the TTO field. The numerical simulation has demonstrated, for instance, that a maximum of TTO can also form in the absence of biomass burning, being determined by sedimentation in the troposphere and horizontal air transport.

On the other hand, in the region of Indonesia, TTO anomalies of the order of 10–20 DU were observed in the 1997–1998 El Niño period (at this time there were also large-scale fires in tropical rain forests of Indonesia), which reflects the necessity to consider both biogenic and dynamic factors The effect on the troposphere of the stratosphere with its characteristic impact of solar irradiance and quasi-biennial oscillations (QBO) can also play a significant role. Changes in the total stratospheric ozone content affect the UV radiation coming to the troposphere and thereby the course of photochemical processes in the troposphere.

In connection with these circumstances, an analysis has been made of the causes of TTO variability using data of satellite observations (Nimbus-7) over a period of 20 years (1979–1998) including an assessment of the role of biomass burning and large-scale transport. Data of observations in the tropics revealed three regions characterized by specific annual course and interannual variability of TTO: the eastern and western sectors of the Pacific Ocean and the Atlantic Ocean [40]. For the latter region, TTO maxima are reached approximately at the same time (September–October) both north and south of the equator, whereas from north to south of the equator the amplitude of the TTO annual course increases from 3–6 DU. On the other hand, the TTO annual course in both sectors of the Pacific Ocean was weak, with TTO maxima in March–April (northern hemisphere) and September–November (southern hemisphere).

The interannual variability of TTO in these three regions also substantially differs. In the region of the Atlantic Ocean the prevailing factor of variability is the QBO which is in counterphase with manifestations of the QBO in the stratosphere. Such regularity agrees with the assumption of the impact of UV radiation modulation in the troposphere on photochemical processes taking place in the troposphere on timescales typical of the QBO and determining the total ozone content variations in the stratosphere. However, TTO variations predicted by the photochemical model are much below those observed, which reflects the important role of dynamic factors.

Apparently, biomass burning has a considerable (but less significant than atmospheric dynamics) impact on TTO variability over the Atlantic Ocean. The prevailing factor of the interannual TTO variations over the Pacific Ocean is El Niño, when in the eastern Pacific the TTO value is at a minimum, and in the western sector at a maximum. Such a signature of variability reflects the combined effect of the convection induced transport and intensive biomass burning in the region of Indonesia. To draw maps of TTO with a high spatial resolution and to study long-term series of TTO data, the technique was proposed based on the low TTO variability near the Date Line change.

In the regions of the Amazon River and central Brazil during the dry season of the southern hemisphere (July–October), biomass burning occurs in the form of man-made fires of tropical forests ecosystems and serrado – bringing forth emissions of numerous MGCs (CO, CO_2, NO_x, hydrocarbon compounds, etc.) and aerosols. Also, secondary MGCs form due to reactions between the products of burning. Tropospheric ozone is one such secondary products. In this connection, in the period from 16 August to 10 September, 1995, in the region of Cuiaba (central Brazil, 16°S, 56°W) a field observational experiment SCAR-B was undertaken with the main objective to study the physical properties of smoke formed during biomass burning and its impacts on the radiation budget and climate [20].

The ozone balloon observations carried out during the period of the experiment revealed an increase in tropospheric ozone concentration (as compared with average dry season conditions) which strongly varied depending on the specific atmospheric circulation. Data of observations of the aerosol optical thickness and surface concentration of soot carbon enabled one to quantitatively characterize the conditions of atmospheric smoke loading. Analysis of air mass trajectories suggested the conclusion that an especially strong increase of ozone concentration took place on 26–29 August and was apparently connected not only with a direct effect of biomass burning products but also with emissions of various MGCs to the atmosphere in large cities located on the Atlantic coast of Brazil. This conclusion was confirmed by lidar data of cloud sounding on board the high-altitude aircraft ER-2.

Fires in savannas are one of the most important factors of ecodynamics in the regions of South Africa, whose impact on chemical processes in the atmosphere and the biogeochemical cycle can manifest itself not only on regional but also on global scales. Such fires occur mainly during the dry season (May–October) and are characterized by considerable interannual variations in location and intensity. It is supposed that fires in savannas do not cause substantial changes in the grass cover–atmosphere CO_2 exchange, since CO_2 emissions are practically compensated for by CO_2 consumption by the process of vegetation growth. However, emissions of other MGCs and aerosols are an important consequence of fires.

One of the most important characteristics of pyrogenic emissions into the atmosphere is the emission factor (EF) for various minor gas and aerosol components, which is the mass of emitted component calculated per unit dry mass of consumed fuel ($g\,kg^{-1}$). To estimate EF, [42] generated artificial fires of grass and forest during the period 5 June–6 August, 1996 at 13 sites (2 ha in size) located at a distance of 7.5 km south-east of Kaoma (Zambia, 14°52′S, 24°49′E, 1,170 m a.s.l.). Results of EF estimations obtained in the beginning of the dry season (early June–early August) have been considered [42] for carbon dioxide (CO_2), carbon monoxide (CO), methane (CH_4), non-methane hydrocarbon compounds (NMHC), and PM-2.5 aerosol particles with $D < 2.5\,\mu m$.

It follows from the obtained results that there is a correlation between EF for grass fires in savannas and the share of green vegetation used as an indicator of water content in the burned fuel – with EF values being greater in the case of products of incomplete burning in the beginning of the dry season compared with those at the end of the dry season. Models of EF for NMHC and PM-2.5, depending on the modified consumption efficiency (MCE), are statistically different for grass and forest ecosystems.

Korontzi *et al.* [42] performed a comparison of predicted estimates of EF using the observational data under discussion, as well as results obtained within the SAFARI-92 programme of studies of the impact of forest fires in South Africa in 1992 and SAFARI-2000 with the data for different ecosystems, in order to analyse the reasons for using the results which refer to regionally averaged information or to various specific ecosystems. The grounds for using the model SAFARI-92 to estimate the dependence of EF on MCE as applied to observational data for the onset of the dry season have also been studied. The comparison has shown that most significant differences appear at extremely low (0.907) or high (0.972) values of MCE. To calculate the EF for individual oxidized volatile organic compounds, whose concentration was measured during SAFARI-2000, models of MCE dependence on the share of green grass in the case of grass fires have been used. The results obtained reflect the importance of the annual course of emissions due to fires in savannas in the numerical modelling of emissions on scales from regional to continental.

5.7 CONCLUSION

An apparent conclusion from a brief review of heterogeneous chemical processes in the atmosphere is that the latest studies, especially studies of interaction of aerosols and MGCs in the atmosphere, are only in their initial stages. In particular, the climatic role of heterogeneous processes in the destruction of ozone molecules and in their transport from the stratosphere is very significant, and there is no doubt that modelling the heterogeneous processes of chemical transformations of atmospheric matter requires a further specification of the processes of generation of initial products, as well as numerical modelling of the rates of heterogeneous reactions.

The field observations of the ozone content in the atmosphere under different meteorological conditions and at different levels of pollution, as well as an overview of theoretical and experimental data, show that the problem of the impact of aerosols, and in particular dust outflows, on variations of the ozone content in the atmosphere is far from being unique. Even repeatedly observed marked decreases of ozone content in the layers with aerosol pollutants can be caused both by direct destruction of an ozone molecule on dust particles or by other processes. Therefore, further parallel synchronous observations of the content of ozone, aerosols, and other admixtures in the atmosphere together with measurements of several meteorological parameters are needed.

5.8 BIBLIOGRAPHY

1. Aldaz L. Flux measurements of atmospheric ozone over land and water. *J. Geophys. Res.*, 1969, **74**(28), 6943–6946.
2. Almogeya Kh. R., Kobrera K. A., and Svistev P. F. Some data on the Cuban surface ozone. *Reports at the Working Meeting on Study of Atmospheric Ozone. Tbilisi, 23–27 November 1981.* Metsniereba, Tbilisi, 1982, pp. 247–251 [in Russian].
3. Bashlykov V. M., Vakhtel A. V., Maslayeva N. I., and Ivanov A. A. Studies of the diurnal variations in concentrations of NO_2 and O_3 in the atmosphere of Moscow using the trace spectral–optical method. *Physics of the Atmos. and Ocean*, 1994, **30**(1), 53–58 [in Russian].
4. Baumann K., Ilt F., Zhao J. Z., and Chameides W. L. Discrete measurements of reactive gases and fine particle mass and composition during the 1999 Atlanta Supersite Experiment. *J. Geophys. Res.*, 2003, **108**(D7), SOS4/1–SOS4/20.
5. Bazhanov V. M. and Petrov V. N. The ozone content in the North Atlantic surface layer. *Meteorology and Hydrology*, 1986, **6**, 60–67 [in Russian].
6. Belan B. D. and Panchenko M. V. An estimation of the ozone sink on aerosol particles. *Optics of the Atmos. and Ocean*, 1992, **5**(6), 647–651 [in Russian].
7. Belan B. D. and Skliadneva T. K. Changes in the tropospheric ozone concentration as a function of solar radiance. *Optics of the Atmos. and Ocean*, 1999, **12**(8), 725–729 [in Russian].
8. Bian H. and Zender C. S. Mineral dust and global tropospheric chemistry: Relative roles of photolysis and heterogeneous uptake. *J. Geophys. Res.*, 2003, **108**(D21), ACH8/1–ACH8/14.
9. Bian H., Prather M. J., and Takemura T. Tropospheric aerosol impacts on trace gas budgets through photolysis. *J. Geophys. Res.*, 2003, **108**(D8), ACH4/1–ACH4/10.
10. Chameides W. L. and Walker J. C. G. A time-dependent photochemical model for ozone near the ground. *J. Geophys. Res.*, 1976, **81**(3), 413–420.
11. Claeys M., Graham B., Vas G., Wang W., Vermeylen R., Pashynska V., Cafmeyer J., Guyon P., Andreae M. O., Artaxo P., *et al.* Formation of secondary organic aerosols through photo-oxidation of isoprene. *Science*, 2004, **303**(5661), 1173–1176.
12. Baev A. A. and Svirezhev Yu. M. (eds). *Consequences of a Nuclear War. Physical and Atmospheric Effects.* Mir Publishing, Moscow, 1988, 391 pp. [in Russian].

13. Desboeufs K. V., Losno R., and Colin J. L. Relationship between droplet pH and aerosol dissolution kinetics: Effect of incorporated aerosol particles on droplet pH during cloud processing. *J. Atmos. Chem.*, 2003, **46**(2), 159–172.
14. *EUROTRAC-2 TOR-2. Tropospheric Ozone Research Annual Report 1998*. Inter. Sci. Secretariat, Muenchen, 1999, 163 pp.
15. Fishman J. and Crutzen P. J. A numerical study of tropospheric model. *J. Geophys. Res.*, 1977, **82**(37), 5897–5906.
16. Fishman J., Solomon S., and Crutzen P. J. Observational data in support of significant in situ photochemical source of tropospheric ozone. *Tellus*, 1979, **31**, 432–446.
17. Gelenczér A., Hoffer A., Krivácsy Z., Kiss G., Molnár A., and Mészáros E. On the possible origin of humic matter in fine continental aerosol. *J. Geophys. Res.*, 2002, **107**(D21), ICC2/1–ICC2/6.
18. Gleason J. F., Hsu N. C., and Torres O. Biomass burning smoke measured using backscattered ultraviolet radiation: SCAR-B and Brazilian smoke interannual variability. *J. Geophys. Res.*, 1998, **103**(D24), 31969–31978.
19. Gordov E. P., Rodimova O. B., and Fazliyev A. Z. *Atmospheric Optical Processes: Simple Non-linear Models*. Institute of Optics Atmosphere, Siberian Branch of RAS Publ., Tomsk, 2002, 251 pp. [in Russian].
20. Griffin R. J., Dabdub D., and Seinfeld J. H. Secondary organic aerosol. 1. Atmospheric chemical mechanism for production of molecular constituents. *J. Geophys. Res.*, 2002, **107**(D17), AAC3/1–AAC3/26.
21. Griffin R. J., Nguyen K., Dabdub D., and Seinfeld J. H. A coupled hydrophobic-hydrophilic model for predicting secondary aerosol formation. *J. Chemistry*, 2003, **44**, 171–190.
22. Guirgzhdene R. V., Shopauskas K. K., and Guirgzhdis A. I. On nocturnal ozone maxima in the atmospheric surface layer. *Reports at the Working Meeting on Study of Atmospheric Ozone, Tbilisi, 23–27 November 1981*. Metsniereba, Tbilisi, 1982, pp. 295–299 [in Russian].
23. Hynes R. G., Fernandez M. Z., and Cox R. A. Uptake of HNO_3 on water-ice and coadsorption of HNO_3 and HCl in the temperature range 210–235 K. *J. Geophys. Res.*, 2003, **108**(D24), AAC18/1–AAC18/11
24. Iacobellis S. F., Frouin R., and Somerville R. C. J. Direct climate forcing by biomass-burning aerosols: Impact of correlations between controlling variables. *J. Geophys. Res.*, 1999, **104**(D10), 12031–12045.
25. Ivlev L. S., Sirota V. G., Skoblikova A. L., and Khvorostovsky S. N. Spectrophotometric studies of the ozone-oxides interaction kinetics. *Proc. Central Aerological Observatory*, 1982, **149**, 85–90 [in Russian].
26. Ivlev L. S., Kondratyev K. Ya., Maksimenko O. V., Sirota V. G., and Shashkin A. V. Periodic variations of ozone and solar radiation in the atmospheric surface layer. *Atmospheric Optics*, 1988, **1**(11), 81–88 [in Russian].
27. Ivlev L. S. On the impact of dust outflows on the atmospheric ozone concentration. *Proc. of Leningrad State Univ. Press*, 1988, **101**, 63–79 [in Russian].
28. Ivlev L. S. and Chelibanov V. P. Measurements of variations of ozone concentration in the North Atlantic surface layer. *The All-Union Conference on Atmospheric Ozone, 2–6 October 1998*. Suzdal, Moscow Phys. Techn. Institute, Dolgoprudny, pp. 111–112 [in Russian].
29. Ivlev L. S., Sirota V. G., and Khvorostovsky S. N. The impact of oxidation of volcanic sulphur dioxide on the content of sulphuric-acid aerosols and ozone in the stratosphere. *Optics of the Atmos.*, 1990, **3**(6), 37–43 [in Russian].

30. Ivlev L. S. and Mikhailov E. F. Variations of the surface ozone concentration and their connection with atmospheric pollutions under urban conditions. In: *Complex analysis of atmospheric pollution* (ANZAG-87, Part 1). Gylym, Alma-Ata, 1990, pp. 119–125 [in Russian].
31. Ivlev L. S., Basov L. L., Sirota V. G., and Smyshlayev S. P. The photostimulated aerosol sink of atmospheric ozone and methane. *J. Ecol. Chemistry*, 1992, **1**, 77–86 [in Russian].
32. Ivlev L. S. and Chelibanov V. P. A short-period variability of the atmospheric ozone content, and the role of aerosols in this variability. In: L. S. Ivlev (ed.), *Natural and Anthropogenic Aerosols*. St Petersburg State Univ. Press, St Petersburg, 2003, pp. 383–407 [in Russian].
33. Ivlev L. S. Temporal variability of the ozone content in the lower atmosphere. In: V. I. Osechkin (ed.), *Atmospheric Ozone*. Leningrad Politechnic Institute Publ., Leningrad, 1992, pp. 82–106.
34. Jacobson M. Z. Analysis of aerosol interactions with numerical techniques for solving coagulation, nucleation, condensation, dissolution, and reversible chemistry among multiple size distributions. *J. Geophys. Res.*, 2002, **107**(D19), AAC2/1–AAC2/23.
35. Jordan C. E., Dibb J. E., Anderson B. E., and Fuelberg H. E. Uptake of nitrate and sulfate on dust aerosols during TRACE-P. *J. Geophys. Res.*, 2003, **108**(D21), GTE38/1–GTE38/10.
36. Jourdain B. and Legrand M. Year-round records of bulk and size-segregated aerosol composition and HCl and HNO_3 levels in the Dumont d'Urville (coastal Antarctica) atmosphere: Implications for sea-salt aerosol fractionation in the winter and summer. *J. Geophys. Res.*, 2002, **107**(D22), ACH 20/1–ACH 20/13.
37. Kalberer M., Paulsen D., Sax M., Steinbacher M., Dommen J., Prevot A. S. H., Fisseha R., Weingartner E., Frankevich V., Zenobi R., *et al.* Identification of polymers as major components of atmospheric organic aerosols. *Science*, 2004, **303**(5664), 1659–1662.
38. Kawamura K., Umemoto N., and Mochida M. Water-soluble dicarboxylic acids in the tropospheric aerosols collected over east Asia and western North Pacific by ACE-Asia C-130 aircraft. *J. Geophys. Res.*, 2003, **108**(D23), ACE7/1–ACE7/7.
39. Kondratyev K. Ya. *Climate Shocks: Natural and Anthropogenic*. Wiley. New York, 1988, 296 pp.
40. Kondratyev K. Ya. and Varotsos C. A. *Atmospheric Ozone Variability: Implications for Climate Change, Human Health, and Ecosystems*. Springer–Praxis, Chichester, UK, 2000, 617 pp.
41. Kondratyev K. Ya. The aerosol-induced radiative forcing. *Optics of the Atmos. and Ocean*, 2003, **21**(1), 5–16 [in Russian].
42. Korontzi S., Ward D. E., Susott R. A., Yokelson R. J., Justice C. O., Hobbs P. V., Smithwick E. A. H., and Hao W. M. Seasonal variation and ecosystem dependence of emission factors for selected trace gases and PM-2.5 for southern African Savanna fires. *J. Geophys. Res.*, 2003, **108**(D24), ACH7/1–ACH7/14.
43. Krueger B. J., Grassian V. H., Laskin A., and Cowin J. P. The transformation of soil atmospheric particles into liquid droplets through heterogeneous chemistry: Laboratory insights into the processing of calcium containing mineral dust aerosol in the troposphere. *Geophys. Res. Lett.*, 2003, **30**(8), 48/1–48/4.
44. Laskin A., Gasper D. J., Wang W., Hunt S. W., Cowin J. P., Colson S. D., and Finloyson-Pitts B. J. Reactions at interfaces as a source of sulfate formation in sea salt particles. *Science*, 2003, **301**, 340–344.

45. Lee Y.-N., Weber R., Ma Y., Orsini D., Maxwell-Meier K., Blake D., Meinardi S., Sachse G., Harward C., Chen T.-Y., *et al.* Airborne measurements of inorganic ionic components of fine aerosol particles using the particle-into-liquid sampler coupled to ion chromatography technique during ACE-Asia and TRACE-P. *J. Geophys. Res.*, 2003, **108**(D23), ACE14/1–ACE14/14.
46. Lefer B. L., Shetter R. E., Hall S. R., Crawford J. H., and Olson J. R. Impact of clouds and aerosols on photolysis frequencies and photochemistry during TRACE-P. 1. Analysis using radiative transfer and photochemical box models. *J. Geophys. Res.*, 2003, **108**(D21), GTE42/1–GTE42/12.
47. Liao H., Adams P. J., Chung S. H., Seinfeld J. H., Mickley L. J., and Jacob D. J. Interactions between tropospheric chemistry and aerosols in a united general circulation model. *J. Geophys. Res.* 2003, **108**(D1), 1/1–1/23.
48. Liu S.C., Kley D., McFarland M., Mahlman J. D., and Levy H., II. On the origin of tropospheric ozone. *J. Geophys. Res.*, 1980, **85**(C12), 7546–7552.
49. Liu W., Hopke P. K., and Van Curen R. A. Origins of fine aerosols mass in the western Unites States using positive matrix factorization. *J. Geophys. Res.*, 2003, **108**(D23), AAC1/1–AAC1/18.
50. Logan J. A. Troposphere-ozone: Seasonal behavior, trends, and anthropogenic influence. *J. Geophys. Res.*, 1985, **D90**(6), 10463–10482.
51. Ma J., Tang J., Li S.-M., and Jacobson M. Z. Size distributions of ionic aerosols measured at Waliguan Observatory. 1. Implication for nitrate gas-to-particle transfer processes in the free troposphere. *J. Geophys. Res.*, 2003, **108**(D17), ACH8/1–ACH8/12.
52. Ma Y., Weber R. J., Lee Y.-N., Orsini D. A., Maxwell-Meier K., Thornton D. C., Bandy A. R., Clarke D. R., Sachse G. W., Fuelberg H. E., *et al.* Characteristics and influence of biosmoke on the fine-particle ionic composition measured in Asian outflow during the Transport and Chemical Evolution over the Pacific (TRACE-P) experiment. *J. Geophys. Res.*, 2003, **108**(D21), GTE37/1–GTE37/16.
53. Marenco A. Variations of CO and O_3 in the troposphere evidence of O_3 photochemistry. *Atmospher. Environm.*, 1986, **5**, 911–918.
54. Martin R. V., Jacob D. J., Yantosca R. M., Chin M., and Ginoux P. Global and regional decreases in tropospheric oxidants from photochemical effects of aerosols. *J. Geophys. Res.*, 2003, **108**(D3), ACH6/1–ACH6/14.
55. Marenco A. Variations of CO and O_3 in the troposphere evidence of O_3 photochemistry. *Atmospher. Environm.*, 1986, **20**(5), 911–918.
56. Meszaros E. *Fundamentals of Atmospheric Aerosol Chemistry*. Akademiai Kiado, Budapest, 1999, 308 pp.
57. Olszyna K., Cadle R. D., and De Pena R. J. Stratospheric heterogeneous decomposition of ozone. *J. Geophys. Res.*, 1979, **84**(C4), 1771–1775.
58. Perov S. I. and Khrguian A. Kh. *Present Problems of Atmospheric Ozone*. Gidrometeoizdat, Leningrad, 1980, 280 pp. [in Russian].
59. Popov V. A. and Chernykh L. N. The photochemical type of the urban atmospheric pollution. In: K. Ya. Kondratyev (ed.), *Problems of the Environmental Pollution Control*. Gidrometeoizdat, Leningrad, 1981, pp. 54–59.
60. Poppe J., Koppmann R., and Rudolph J. Ozone formation in biomass burning plumes: Influence of atmospheric dilution. *Geophys. Res. Lett.*, 1998, **25**(20), 3823–3826.
61. Reshetov V. D. *Variability of Meteorological Elements in the Atmosphere*. Gidrometeoizdat, Leningrad, 1973, 249 pp. [in Russian].
62. Rodriguez M. A. and Dabdub D. Monte Carlo uncertainty and sensitivity analysis of the CACM chemical mechanism. *J. Geophys. Res.*, 2003, **108**(D15), 2/1–2/9.

63. Seinfeld J. H. and Pandis S. N. *Atmospheric Chemistry and Physics. From Air Pollution to Climate Change.* Wiley, New York, 1998, 1327 pp.
64. Sirota V. G. The impact of environmental pollution on the content of secondary aerosols, ozone and other minor gaseous components of the atmosphere. Doctoral thesis. St Petersburg State Univ., St Petersburg, 1992, Abstract, 30 pp. [in Russian].
65. Smyshlayev S. P. A Numerical simulation of the impact of anthropogenic factors on atmospheric ozone. Ph.D. thesis. St Petersburg State Univ., St Petersburg, 1993, Abstract, 19 pp. [in Russian].
66. Staehelin J. and Holgne J. Decomposition of ozone in water: Rate of initiation by hydroxide ions and hydrogen peroxide. *Environ. Sci. Techology*, 1982, **16**, 676–681.
67. Tegen I., Harrison S. P., Kohfeld K., Prentice I. C., Coe M., and Heimann M. Impact of vegetation and preferential source areas on global dust aerosol: Results from a model study. *J. Geophys. Res.*, 2003, **108**(D21), AAC14/1–AAC14/27.
68. Chelibanov V. P. (ed.) *The Joint American–Soviet Experiment on Study of Characteristics of Atmospheric Aerosol, Radiation and Ozone. Observation Materials and Experimental Data* (Section: Chemiluminescent ozone probe). Moscow Phys. Techn. Institute, Dolgoprudny, 1988, 117 pp. [in Russian].
69. Bolin B., Döös B. R., Yaeger J., and. Worrin P. (eds). *The Greenhouse Effect, Climate Changes and Ecosystems.* Gidrometeoizdat, Leningrad, 1989, 557 pp. [in Russian].
70. Trockine D., Iwasaka Y., Matsuki A., Yamada M., Kim Y.-S., Nagatani T., Zhang D., Shi G.-Y., and Shen Z. Mineral aerosol particles collected in Dunhuang, China, and their comparison with chemically modified particles collected over Japan. *J. Geophys. Res.*, 2003, **108**(D23), ACE10/1–ACE10/11.
71. Wang S., Ackermann R., Spicer C. W., Fast J. D., Schmeling M., and Stitz J. Atmospheric observations of enhanced NO_2–HONO conversion on mineral dust particles. *Geophys. Res. Lett.*, 2003, **30**(11), 49/1–49/4.
72. Wang Y., Jacob D. J., and Logan J. A. Global simulation of the tropospheric O_3—NO_2–hydrocarbon chemistry. 1. Model simulation. 2. Model evaluation and global ozone budget. *J. Geophys. Res.*, 1998, **103**(D9), 10713–10725, 10727–10755.
73. Zuev V. V. The behaviour of ozone in the surface layer: Possible explanations. *Optics of the Atmos. and Ocean*, 1998, **11**(12), 1356–1357 [in Russian].
74. Zvenigorodsky S. G. and Smyshlayev S. P. On possible change in ozone with an intensive disturbance of the aerosol component. *Proc. of Academy Sci. Soviet Union, Physics of the Atmos. and Ocean*, 1985, **21**(10), 1056–1063 [in Russian].

6

Interactions between aerosols and clouds

6.1 INTRODUCTION

Clouds are the most variable component of the climate system and play the key role in atmospheric energetics. Therefore, studies and assessment of the impact of clouds and atmospheric aerosols on the atmospheric radiation budget and radiative flux divergence are of primary interest from the viewpoint of analysis of the role of various factors in climate formation [39, 56, 73–75]. First, aerosols, their number density and physico-chemical properties, determine substantially the structure, evolution, and optical properties of clouds [3, 77, 129]. Second, cloud systems, in turn, affect the concentration and structure of atmospheric aerosol [8, 39, 41, 93]. Third, the presence of aerosol in the cloud system, both outside and inside cloud droplets, affects in a complicated way, the processes of solar radiation transfer in clouds [5, 23, 51, 111, 129]. And fourth, the processes of phase transition of water take place, which cause not only continuous changes of structural and radiative characteristics of the cloud system and its thermodynamic properties, but also dynamic processes both in the atmosphere and on the surface (upward and downward air fluxes, water transport, and changes in surface albedo and temperature) [3, 63, 72–81, 84, 124, 126].

The effect of atmospheric aerosols and cloudiness, and their interaction with solar radiation in numerical simulations of global changes in surface air temperature, are taken into account only as a remainder term to reach a coincidence with the observed quantities [39]. An analysis of the vast accumulated material convincingly proves that the process of solar radiation absorption in the dust-loaded and cloudy atmosphere is much more substantial than it has been suggested by model calculations [50–53, 74]. In many publications, experimentally established high values of solar radiation absorption are classified as an effect of abnormal (excessive) absorption of short-wave radiation in clouds [2, 119]. The term used testifies to the absence of a single explanation of the revealed contradiction between theoretical calculations and observational data. Therefore, of special interest is a correct interpretation of measurement results based on the theory of radiation transfer and light scattering in the atmosphere.

As far back as 1969, [78] stated that cloud microphysics should be studied together with macrophysics, especially with air motions. However, this statement was assumed as theoretical, since theoretically, the formation of condensation spectra of droplets has been successfully studied without consideration of the impact of condensation on cloud dynamics. At present, it is understood that not only these two processes are interconnected but that they are also connected with radiative processes of heating and cooling of cloud layers and adjacent cloud-free layers, as well as electric processes. This understanding of the problem is very important for solving the problem of climate prediction, and this sets a new task of a complex study of the problem of interaction between aerosol, cloudiness, and radiation.

6.2 THE IMPACT OF CLOUDINESS ON SOLAR RADIATION TRANSFER

The most urgent developments in the field of numerical climate modelling are: creation of high-resolution models for limited spatial scales useful for parameterization of cloud dynamics in climate models; assessment of contributions of the cloud radiation impact (including its diurnal change) and microphysical processes (with the role of aerosol taken into account) to the formation of the properties and structure of the cloud cover; and the improvement of algorithms to retrieve cloud characteristics from the data of remote sensing. The enumerated problems are solved using methods of calculations of solar radiation characteristics (irradiance, radiance, and absorption) and determination of the optical characteristics of clouds from the data of radiation measurements [30, 51, 131].

The process of solar radiation transfer in clouds is described with the well studied equation of transport for scattering media, but in contrast to the cloud-free atmosphere, multiple scattering plays the leading role [73, 74, 87]. The theory of radiation transfer in horizontally heterogeneous cloudiness has been considered [1, 23, 111, 124–126]. In a simplest case, a horizontally homogeneous cloudy atmosphere is considered [87, 89, 92] (i.e., a consideration of the model of an infinitely extended and horizontally homogeneous cloud layer). In nature, stratus clouds correspond best to such a model. Therefore, we will dwell upon the properties of stratiform cloudiness, which make it possible to use theoretical methods as applied to real clouds described in [87, 91, 92].

The low-level stratus clouds include stratus (St), stratocumulus (Sc), and nimbostratus (Ns) clouds; the mid-level stratus clouds include altocumulus (Ac) and altostratus (As); and the high-level stratus clouds include cirrostratus (Cs) clouds as well as frontal systems of clouds Ns–As, As–Cs, and Ns–As–Cs [8, 23, 79–81, 84].

Extended stratus clouds play an important role in the chain of feedbacks of the climatic system, affecting substantially the albedo and radiation budget of the surface atmosphere system as well as atmospheric general circulation [5, 73, 74, 79, 93]. Stratus clouds propagating over large territories can affect the Earth's radiation budget (ERB) not only on a regional but also a global scale [80].

The cloud albedo is higher than the albedo of the ocean or land without snow cover. Based on this fact and assuming that clouds prevent a heating of the surface and atmospheric subcloud layer in low and mid-latitudes, the conclusion has been usually drawn that cloudiness makes a negative contribution to the ERB. However, in high latitudes clouds do not intensify the reflection of light, since the snow surface albedo is also high, and in this case their role is supposed to prevail as a means of atmospheric heating [3, 74].

During the last decades it has become clear that the situation is more complicated: clouds themselves absorb some portion of the incident solar radiation [30, 51, 56, 87], favouring thereby atmospheric heating at all latitudes. Thus, in studies of stratus clouds, a first-priority problem is the cloud–radiation interaction [9, 33, 50, 73]. To construct numerical radiation models of clouds, adequate optical models are needed (i.e., corresponding to the nature of the considered phenomenon), therefore, it is necessary to assess the optical parameters of clouds including: volume coefficients of scattering and absorption. Atmospheric aerosols included in the processes of short-wave solar radiation and cloudiness interaction, play an ambiguous role in the formation of the thermal regime of the atmosphere and surface. There are direct and indirect impacts of atmospheric aerosols [3, 21, 28, 36, 41, 61, 103–106, 120]. The direct impact is determined by solar radiation absorption by soot and other atmospheric aerosols [6, 11, 15, 19, 21–24, 30, 34–36, 51]. The indirect impact is explained in that atmospheric aerosols, especially hygroscopic, are needed for water vapour condensation and formation of droplets [6, 15, 19–26, 34–36, 115–117, 121]. Therefore, a higher concentration of aerosols increases the amount of droplets, and hence, the optical thickness of cloud, which, in turn, enhances solar radiation reflection and decreases its absorption in the atmosphere and on the surface [129]. The results of aircraft radiation observations during the last decade [39] have shown that direct and indirect aerosol impacts on increase or decrease of solar radiation absorption is different in different regions and in different clouds.

A detailed analysis of the role of the greenhouse effect in global climate change [59] indicates that at the present stage it is impossible to reliably assess the radiative forcing of clouds and aerosols on global warming.

6.2.1 Repeatability of stratus cloudiness

From the results of aircraft and satellite observations and observations at the Rosgidromet network, data have been given [8, 80, 81] on the repeatability of stratus clouds and it has been shown that an average lifetime of overcast clouds over the territory of Europe is 13–15 hours in winter and about 5 hours in summer. The probability of preservation of overcast cloudiness for different time intervals over the European territory of Russia (ETR) is given in Table 6.1 [80].

Table 6.1. The probability (%) of the conservation of cloudiness, with the cloud amount equal to 1, above the European territory of the former USSR.

	Duration of cloudiness (h)				
Probability (%)	1	3	6	12	24
Winter	93	87	83	78	74
Summer	80	64	52	41	35

Table 6.2. The average altitudes of cloud z_b (bottom) and z_t (top) and geometrical thickness $\Delta H = z_t - z_b$ (km).

Type of cloudiness	z_b		z_t		ΔH	
	Winter	Summer	Winter	Summer	Winter	Summer
St	0.25	0.29	0.55	0.58	0.30	0.29
Sc	0.85	1.26	1.14	1.59	0.29	0.33
As	3.80	3.93	4.73	4.83	0.93	0.90

As has been justly mentioned in [81], in studies of the cloud characteristics from results of field experiments, it is necessary to understand that the obtained characteristics and parameters of clouds refer to a given realization and to a given time period. Nevertheless, there is a certain repeatability of individual parameters of stratus cloudiness, characteristic of various climatic zones. So, for instance, the prevailing altitude of stratus clouds in polar and mid-latitude zones is within 2 km, in the tropical zone 3 km. An averaging of the data of aircraft and balloon observations has led to the most typical values of stratus clouds bottom and top altitudes as well as their vertical thickness (Table 6.2).

Stratus clouds appearing in the atmosphere during the formation and passing of the atmospheric front, which are at a distance of not more than 200 km from the surface front line, are called frontal. From the data of satellite observations, the width of the frontal zone in central Europe can reach 1,000 km. The length of this zone constitutes 7,000 km. Cloud zones are divided into macrocells several hundreds of kilometres in size, which, in turn, consist of cloud bands or an overcast cloud field with cloud mass cells which are inhomogeneous in density and tens of kilometres in size [80, 81]. Information about repeatability of the width of the zone of frontal clouds of the upper and lower levels, from results of aircraft observations, is given in Table 6.3.

The main conclusion from the data of this table is that the upper level cloud fields <200 km in size are typical of the cold front, and 500–600 km of the warm front. Most often are observed the frontal clouds of the lower level with horizontal sizes within 50 km under conditions of the cold front and 75 km in the case of the

Table 6.3. Recurrence (%) of the extension of the frontal upper cloudiness above the European territory of the former USSR.

Width of zone (km)	Type of front		Width of zone (km)	Type of front	
	Cold	Warm		Cold	Warm
<50	2.4	–	501–600	–	25.0
51–100	12.2	–	601–700	–	20.8
101–200	26.8	–	701–800	–	11.7
201–300	29.3	6.5	801–900	–	6.2
301–400	22.0	10.4	901–1,000	–	2.6
401–500	7.3	12.9	1,001–1,500	–	2.6

Table 6.4. Recurrence (%) of cloud areas with different extensions of the frontal lower cloudiness.

Extension (km)	Type of front	
	Cold	Warm
<10	20	14
10–20	28	19
20–30	19	19
30–50	18	21
50–75	9	11
75–100	4	8
100–150	0	3
150–200	1	4
200–300	1	1

warm front (Table 6.4). It follows from these data that in the studies of stratus cloudiness it should be modelled by infinitely extended layers. Besides this, stratus cloudiness is rather stable in time, therefore, the method of determination of its optical parameters based on surface measurements of radiation flux for different solar zenith angles (i.e., in different time moments with intervals ~1–2 hours) can be applied.

6.2.2 Spectral radiation measurements under conditions of overcast cloudiness

The field experiments in the cloudy atmosphere were carried out within the scientific programmes CAENEX, GAREX, GARP, GATE, etc. Results from these experiments have been discussed in a number of monographs and papers [11, 30, 50–53, 73, 74, 97, 131]. Measurements of solar radiation were made with the use of spectrometers K-2 and K-3 developed in Leningrad University. For each of the field experiments in a cloudy atmosphere corresponded an experiment in the cloud-free atmosphere in the same region, at the same altitudes, and as close as possible to the same time of measurement. The working spectral range of the instrument covered 0.33–0.98 μm, the time of spectrum recording 10 s. The spectrometer had three overlapping working ranges: UV, visible, and near-IR, for which photomultipliers with a maximum spectral sensitivity in the ranges indicated were used as radiation sensors. In early measurements (the 1970s) the output signal of the spectrometer was recorded in analogue form with subsequent laboratory digitization. Then, starting in the 1980s, the measuring complex included an electronic digitization system, which made it possible to record the output signal in digital form, for further computer processing. In the 1980s an improved modification of the spectrometer appeared – the K-3 model – with a modern elemental basis, which enabled one to substantially decrease the size, weight, and time of spectrum recording, though from the viewpoint of specific character of measurement processing, models K-2 and K-3 were identical.

The spectral instrument function for the K-3 spectrometer was obtained in the laboratory from recorded lines of laser radiation (in the visible) and can be approximated by the triangular function with a half-width equal to 3 nm. The accuracy of this approximation was $\sim$1%. Note that in a considerable part of the spectral range of the K-3 spectrometer the recorded signal does not change sharply, and at a half-width $\Delta\lambda = 3$ nm, in further processing and interpretation of measurements, one can leave out of account the spectral instrument function as well as calibration errors. An exception is made for a narrow band of absorption of oxygen (the 760-nm band) as well as for some strong Fraunhofer lines in the solar UV spectral region.

The whole spectral complex, on the basis of K-2 and K-3 spectrometers, has been initially created for field measurements. This has been reached due to compactness, small weight, reliability, simple arrangement, and exploitation of all components of the complex. In the period 1970–1990 the complex was repeatedly used to measure various spectral characteristics of the atmosphere and surface radiation fields. At surface measurements of spectral atmospheric transparency and sky brightness, the instrument was mounted on a special support, which made it possible to direct it both for angular and azimuth viewing. Most important measurements performed with the use of the described complex were aircraft measurements of spectral irradiance (hemispherical fluxes) and radiance of scattered and reflected solar radiation, both to study the spectral radiance of natural surfaces and to obtain spectral radiation flux divergence for different atmospheric layers. For this purpose, downward and upward fluxes of solar radiation were measured at different levels in the atmosphere.

To measure short-wave radiance, the spectrometer was installed in an aircraft photo-hatch, on a special rotator, which made possible measurements at various viewing angles. The viewing azimuth was changed by respective changing of the flight route. The viewing angle of the instrument with respect to measurements of radiance was 2°. To measure downward and upward radiation fluxes, a special device was used – a lightguide at the ends of which were integrating spheres.

In the 1980s, the measurement complex included flexible lightguides which made it possible to use the complex for shipborne subwater measurements of irradiance and radiance of scattered short-wave radiation at different depths in the water.

To control the stability of the instrument sensitivity in the process of measurements a built in reference was used – an electric lamp with a stable feeding. To correct possible deviations of sensitivity in measurements, the reference spectrum was measured periodically and then compared with the average spectrum of the reference recorded during calibration. The software for primary processing with a personal computer, made it possible to perform all primary processing (including calibration) in the visual–interactive regime, with the predicted interferences of the operator into the processing process: from complete control of all operations – simulation of manual processing – to the automated regime of processing, with an immediate output of the measured spectrum. Further 'secondary' computer processing of the results of spectral measurements was made to obtain the more complicated characteristics of the atmosphere and surface, in particular, aerosol spectral optical thickness, coefficients of spectral brightness of the surface, and spectral radiative flux divergences in atmospheric layers.

In these experiments only the extended stratus cloudiness has been studied. An analysis has been made of the results of an experiment carried out in the period 1971–1985 over homogeneous surfaces (sea surface, snow cover, and desert). The surface albedo was calculated from observed values of reflected and incident radiation fluxes from low-altitude flights. Information about some experiments, the data of which have been used to assess the optical parameters of clouds, is given in Table 6.5. The table gives the cosine of solar zenith angle, geographical latitude, type and albedo of surface, and total radiative flux divergence (R) in the cloudy and cloud-free atmosphere. The table also gives the f_s value that characterizes the amount of solar radiation absorbed in the cloudy atmosphere–surface system compared with the cloud-free atmosphere–surface system.

The data of spectral radiation measurements carried out over the Lake Ladoga on 20 April, 1985 is exemplified in Figure 6.1 (experiment 7 in Table 6.5). Spectral values of radiative flux divergences in the cloudy (20 April, 1985) and cloud-free atmosphere (29 April, 1985) carried out in the same place over the Lake Ladoga are shown in Figure 6.2. These results demonstrate that solar radiation absorption in the cloudy atmosphere considerably exceeds the absorption of radiation in the same atmospheric layer under cloud-free conditions. For instance, a comparison of measurements made on 20 April, 1985 and 29 April, 1985 over Lake Ladoga point to considerably higher values of absorbed solar radiation in the cloudy atmosphere. Note also that the values of the flux incident onto the top of the cloud layer from measurements on 20 April, 1985 exceed those measured on 29 April, 1985 at the same altitude. This can be explained either by the presence of aerosol layers or by visually transparent cirrus clouds in the stratosphere and upper troposphere, which reduce the incident solar radiation above the level of measurements on 29 April, 1985.

Table 6.5. Results of the airborne radiative observation in a cloudy atmosphere.

No.	Experiment	μ_0	φ (°N)	Date	As	Other condition	f_s Surface water	f_s Cloud	R (Wm^{-2}) Clear	R (Wm^{-2})
1		0.966	16	12 July, 1974	0.1	Above the Atlantic	1.74	3.2	18.9	
2	Atlantic Ocean, cloud	0.966	17	4 August, 1974	0.06	after dust intrusion	1.45	2.9	26.1	
	Atlantic Ocean, clear	0.966	17	13 August, 1974	0.02	from the Sahara Desert				2.43
3	Atlantic Ocean, clear	0.819	44	10 April, 1971	0.06		1.11	1.2	2.86	
4	Azov Sea, cloud	0.616	47	5 October, 1972	0.06	Above sea surface	1.16	2.5	12.8	
	Azov Sea, clear	0.616	47	6 October, 1972	0.08	Industrial pollution				3.60
5	City Rustavy, cloud	0.438	42	5 December, 1972	0.18	Above the ground	1.07	1.3	15.0	
	City Rustavy, clear	0.438	42	4 December, 1972	0.22	Industrial pollution				2.35
6	Ladoga Lake, cloud	0.440	60	24 September, 1972	0.20	Above water surface	1.13	1.8	3.61	
	Ladoga Lake, clear	0.440	60	20 September, 1972	0.10	Above water surface				3.73
7	Ladoga Lake, cloud	0.647	60	20 April, 1985	0.64	Above ice with snow	1.10	1.5	4.5	
	Ladoga Lake, clear	0.669	60	24 April, 1985	0.55	Above ice with snow				0.40
8	Ladoga Lake, clear	0.276	75	1 October, 1972	0.40	Above water with ice	1.00	1.1	4.63	
	Kara Sea, clear	0.276	75	30 September, 1972	0.40	Industrial pollution				1.97
9	Kara Sea, cloud	0.483	75	29 May, 1976	0.40		0.90	0.95	7.25	
10	Kara Sea, cloud	0.483	75	30 May, 1976	0.40	Above water with ice	1.00	1.2	1.1	
	Kara Sea, clear	0.460	75	21 April, 1976	0.05	Above water surface				1.87

Notes: $\mu_0 = \cos\theta_0$; As = surface albedo.

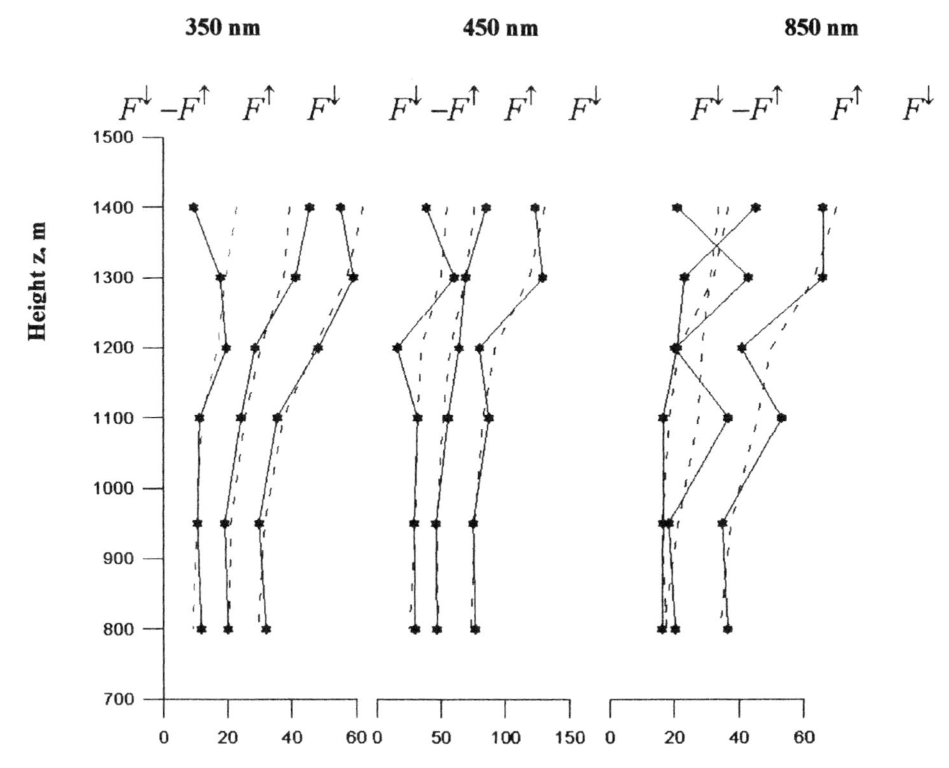

Figure 6.1. Vertical profile of net, downward, and upward fluxes of solar radiation in cloud for three wavelengths. *Solid lines* are the original measurements and *dashed lines* are the smoothed values. Observation made on 20 April, 1985, overcast stratus cloudiness. Cloud top 1,400 m, cloud bottom 900 m, solar incident zenith angle $\upsilon_o = 49^\circ$ ($\zeta_o = 0.647$), snow surface.

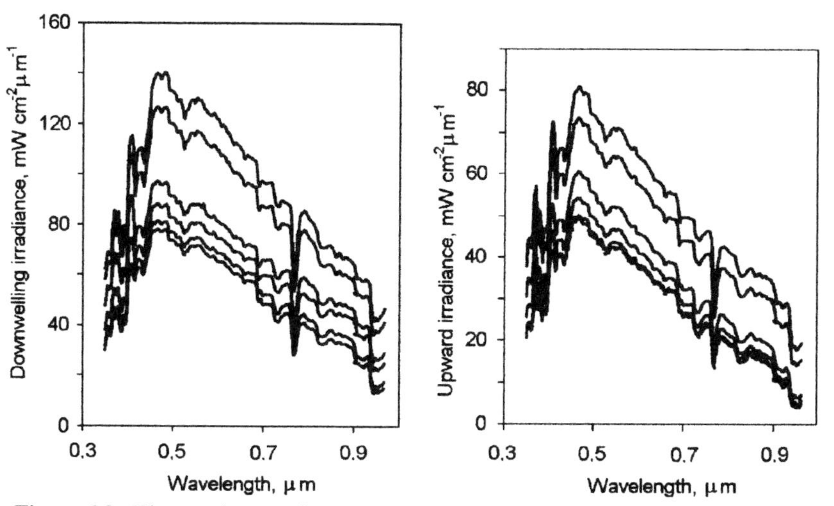

Figure 6.2. The resultants of airborne sounding under overcast sky conditions.

6.3 THE DISTRIBUTION OF AEROSOL IN THE CLOUDY ATMOSPHERE

Numerous aircraft measurements of the aerosol number density and size distribution, with the use of the Photoelectrical counter A3–5M performed by the scientists of the Institute of Atmospheric Optics of the RAS Siberian Branch in the cloudy atmosphere, have made it possible to reveal a number of specific features of the aerosol properties in clouds [8]. In the size distribution of cloud aerosols a minimum does not exceed $r = 0.1$–$1.5\,\mu m$, and their number density grows sharply for particles with $r < r_{\min}$. In the presence of clouds in the layer up to 2,000 m the aerosol concentration increases by a factor of 2–3 (Figure 6.3). This is a natural manifestation of condensation growth of particles with $r < 0.2\,\mu m$ to $r \geq 0.2\,\mu m$ and respective shifting of the maximum in the spectrum to its larger part under the influence of increasing relative humidity.

To obtain statistical and comparable results, a cloud was vertically divided into three parts: lower, approximately equal to one-third of the thickness of its layer H, denoted in the tables as $Z/H = 0.1$–0.3; central, denoted as $Z/H = 0.5$; and upper, also one-third of the layer, $Z/H = 0.7$–0.9. Also selected were layers of subcloud and above-cloud hazes, $Z/H = 0.0$–0.05 and $Z/H = 0.95$–1.00, respectively, as well as layers of the clear atmosphere near the cloud, under the cloud $Z/H < 0.0$, and above the cloud $Z/H > 1.00$. Hence, all profiles of number density with the appearance of clouds were divided into seven zones. The minimum thickness of clouds under study constituted 500 m.

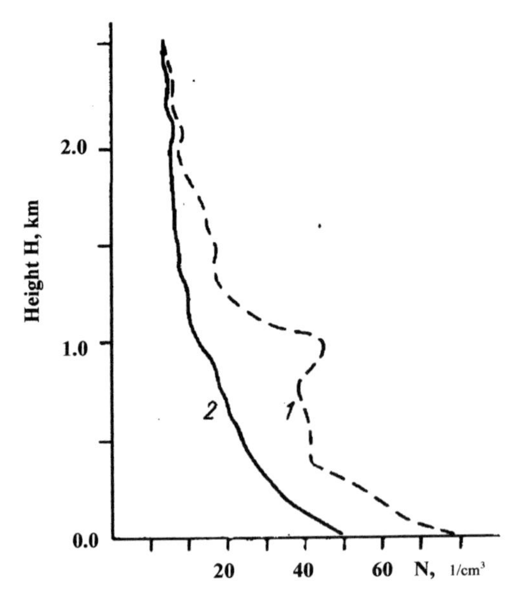

Figure 6.3. Aerosol number density profiles under cloudy (1) and cloud-free (2) conditions.

Table 6.6. Number density (cm^{-3}) of aerosol particles ($R \geq 0.2\,\mu m$) in different types of clouds.

Types of cloud	Level (Z/H)						
	<0.0	0.0–0.05	0.1–0.3	0.5	0.7–0.9	0.95–1.00	1.0
As and Ac	1.9 ± 1.9	5.4 ± 4.2	9.5 ± 12.8	14.4 ± 19.8	9.9 ± 10.4	4.7 ± 4.1	0.9 ± 0.6
$n = 12$	(0.2–6.0)	(1.0–15.0)	(0.7–50.0)	(1.2–75.0)	(1.4–40.0)	(0.7–20.0)	(0.3–3.0)
Cb and Ci	26.1 ± 21.0	39.5 ± 38.2	53.4 ± 51.2	41.2 ± 46.0	25.1 ± 28.9	10.3 ± 8.6	2.6 ± 1.8
$n = 14$	(0.02–70.0)	(0.4–200.0)	(2.0–180.0)	(3.0–150.0)	(0.9–110.0)	(0.7–50.0)	(0.6–6.0)
St and Sc	15.2 ± 14.0	32.5 ± 31.0	60.2 ± 69.1	73.9 ± 80.6	53.2 ± 70.0	23.8 ± 25.5	4.4 ± 3.8
$n = 46$	(0.5–90.0)	(1.0–210.0)	(2.6–270.0)	(3.0–300.0)	(0.8–300.0)	(0.2–190.0)	(0.1–15.0)
Ns and Frnb	31.0 ± 19.3	80.1 ± 68.9	135.9 ± 98.6	131.5 ± 88.8	90.1 ± 80.7	30.0 ± 23.2	7.3 ± 7.8
$n = 32$	(7.0–90.0)	(2.0–255.0)	(20.0–300.0)	(7.0–300.0)	(6.0–300.0)	(1.0–135.0)	(0.3–45)

Note: upper line – $N \pm \sigma_N$; lower line – N_{min} and N_{max} for all data.

In all types of clouds under consideration an increase of aerosol number density to a maximum took place, as a rule, in the middle part of the cloud ($Z/H = 0.5$). An exception were clouds of the Cu type where a maximum of concentration was observed in the lower one-third of the cloud ($Z/H = 0.1$–0.3). The mean square deviation of aerosol concentration follows the same pattern. There are differences in aerosol concentrations depending on the type of clouds: a maximum of concentration is observed in clouds of the type Ns and Frnb, and a minimum in clouds of types As and Ac.

In clouds of the As and Ac types a change of concentration is interconnected in all parts, including also the above-cloud space (Table 6.6). In clouds of the Cb and Cu types, on the contrary, concentrations correlate only in adjacent parts of the cloud, which reflects their convective nature. St and Sc clouds appearing with regular upward motions, reveal a well correlated change in aerosol concentration almost in all their parts, including the circum-cloud space. Finally, Frnb and Ns clouds exhibit a correlation of concentrations inside the cloud and complete independence of changes in aerosol concentration both in clouds and in the circum-cloud space.

The aerosol concentration N in clouds of the same type can change by an order of magnitude: $N = 282 \pm 32$ in a warm period (w.p.) over south of the ETR and $N = 28 \pm 7$ in a cold period (c.p.) over Kazakhstan. There is a distinct annual change in the content of the fraction 0.2–5 μm in clouds. Usually the ratio $N_{w.p.}/N_{c.p.}$ constitutes 2–3. Maximum aerosol concentrations are observed in clouds which appear over the south of the ETR, and minimum concentration over the regions of Kazakhstan. Of interest is the fact of an increased concentration of aerosol in the under-cloud space ($Z/H < 0$) in a warm period over western Siberia and south of the former USSR, which is determined by part of the measurements performed in rain, where the concentration is almost the same as in cloud.

At altitudes above 2,000m or in the above-cloud space ($Z/H > 1.0$) the values of aerosol concentration differ little among themselves and from values characteristic of the clear atmosphere. This corresponds to the known hypothesis [84] of the impact of clouds on the aerosol environment:

(a) Around the cloud the atmosphere is gradually getting rid of small particles giving way to large aerosols.
(b) The air gets to the cloud in its lower part and leaves it in its upper part. The transformation of aerosol particles in the cloud results in the fact that leaving air should contain less small particles and more large ones.
(c) The cloud droplets size distribution changes with time: the droplets grow and their number density decreases.

The first two statements of Mazin's hypothesis are confirmed by measurements: concentrations of aerosol under the cloud and above it differ by almost an order of magnitude. The air leaving the cloud should contain less small particles and more large ones – concentration of particles with $D = 4$–$10\,\mu m$ turns out to be even higher than in the cloud. The filtering ability of clouds was observed in all flights and in observations between cloud layers.

6.3.1 Vertical distribution of aerosol scattering under different synoptic conditions

Studies of the vertical distribution of aerosol scattering have been carried out by scientists of the Tomsk Institute of Atmospheric Optics using the aircraft nephelometer in the cloud-free atmosphere with the near-surface visibility 12–40 km during 1982–1985 [8]. The main statistical characteristics for different synoptic situations in the Siberian region, by the scattering coefficients, are shown in Table 6.7.

Similar studies of the vertical distribution of backscattering were carried out with lidar instrumentation developed in the Institute of Atmospheric Optics of the RAS Siberian Branch (Table 6.8). Measurements of the vertical profiles of directed light scattering coefficients have shown that most often the change of $\beta(H)$ and $\alpha(H)$ has a stratiform structure. The values of coefficients of total scattering and backscattering are at a maximum at $\lambda = 0.37\,\mu m$ and at a minimum at $\lambda = 0.7\,\mu m$ – this ratio is valid for all altitudes of measurements.

Table 6.7. Basic statistical characteristics of the scattering coefficient α.

Synoptic conditions	α (km^{-1})	$\pm\Delta\alpha$ (km^{-1})	$\sigma^2(\alpha)$ (km^{-2})	$\pm\Delta\sigma^2(\alpha)$ (km^{-2})
Anticyclone	0.32	0.24	0.058	−0.019 +0.021
Warm front	0.15	0.05	0.0025	−0.0012 +0.0013
Cold front	0.13	0.05	0.0025	−0.0013 +0.0014
Low-gradient field	0.21	0.12	0.014	−0.007 +0.008

Table 6.8. Basic statistical characteristics of the backscattering coefficient β_{176} (at an angle of 176°) for the four types of atmospheric circulation.

Synoptic conditions	$\beta_{176} \times 10^{-3}$ ($km^{-1}\,sr^{-1}$)	$\pm\Delta\beta_{176} \times 10^{-3}$ ($km^{-1}\,sr^{-1}$)	$\sigma^2(\beta_{176}) \times 10^{-6}$ ($km^{-2}\,sr^{-2}$)	$\pm\Delta\sigma^2(\beta_{176}) \times 10^{-6}$ ($km^{-2}\,sr^{-2}$)
Anticyclone	0.98	0.63	0.39	$-0.12 + 0.16$
Warm front	0.39	0.27	0.073	$-0.022 + 0.028$
Cold front	0.34	0.21	0.044	$-0.011 + 0.012$
Low-gradient field	0.51	0.37	0.14	$-0.04 + 0.06$

Note that the gradient of optical characteristics in the surface layer is almost the same for all wavelengths, which indicates a slight change in the spectrum of particle sizes at these altitudes. It follows from this that for this layer the vertical gradient of the coefficients of directed light scattering is mainly determined by a change in the aerosol particle number density. The layer of turbulent mixing is characterized, first of all, by weakly changing vertical profiles of average values of light scattering coefficients. Variations of the optical characteristics within this layer are negligible and most often are within 10–20%, maximum fluctuations are observed at the bottom and top boundaries of this atmospheric layer, where deviations from an average can reach 50% and higher.

The upper boundary of the turbulent layer (UBTL) borders on the zone of the free atmosphere, which determines an ambiguity of the optical characteristics of this tropospheric layer. Depending on external conditions, the UBTL is located at altitudes of 2–5 km. Usually, the height of the upper border of the UBTL correlates well with specific features of the vertical gradient of the average temperature of the atmosphere. Measurements of the vertical profiles of average values of the directed light scattering coefficients, and averaged temperature for the summer season, were performed in the cloud-free atmosphere at a surface air temperature (SAT) of ~(15–22)°C. The vertical temperature change is different for two layers of the atmosphere. The lower layer is characterized by the vertical distribution of average temperature when the temperature gradient exceeds the dry adiabatic temperature gradient for respective altitudes. Conditions are created in the atmosphere for development of convection and convective turbulence. This, in turn, favours a good mixing of air masses and leads to homogeneous optical properties of air masses both horizontally and vertically.

The upper layer mentioned can be characterized as a layer in which the temperature decreases with altitude more slowly than due to the the dry adiabatic temperature gradient. The temperature stratification of the atmosphere suppresses the convective motion, and the energy of convective turbulence is spent on overcoming the negative buoyancy force. The mechanism of aerosol turbulent mixing in this atmospheric layer is of low efficiency. Therefore, the optical properties of the atmosphere in this layer differ substantially from the properties of the underlying troposphere. The point of inflection on the vertical profile of temperature corresponds to an altitude at which the coefficients of directed light scattering change spasmodically under anti-cyclonic conditions in northern Kazakhstan and western Siberia. Above the layer of turbulent mixing over a large city, there is a distinct layer with an increased aerosol concentration (the urban 'cap'), for which a jump in the temperature profile is a detaining layer in which anthropogenic aerosol rapidly accumulates.

The optical characteristics of the atmosphere are strongly affected by the diurnal variability of aerosol which is at a maximum in the surface layer. This is connected with the impact of the temperature inversion causing a substantial change in the spectrum of aerosol particle sizes in the surface layer of the atmosphere. Stronger variations in average values of the directed light scattering coefficients in the lower troposphere are observed with changing synoptic situations.

The first type of situation most often taking place in the atmosphere corresponds to a small gradient change of the directed light scattering coefficients. A distinctive feature of this change in the scattering characteristics of the atmosphere is that the value of the scattering coefficients for the surface layer and high altitudes differs by not more than a factor of 2–3. As a rule, this change in atmospheric scattering properties is accompanied by the thermal regime of the atmosphere, when the temperature gradient is below the dry adiabatic lapse rate. The probability of observations of this type of profile in the atmosphere from the data of statistical processing constitutes 0.72 for the regions of northern Kazakhstan and 0.58 for western Siberia. A weak variability in the vertical profile of the directed light scattering coefficient enables one to approximate it by linear function:

$$\beta(\varphi, H) = \beta(\varphi, 500) - k(\varphi)H$$

Here $\beta(\varphi, H)$ is the value of the scattering coefficient at an angle φ and height H; $\beta(\varphi, 500)$ is the value of the directed light scattering coefficient at 500 m; and $k(\varphi)$ is the coefficient of proportionality. The k value averaged over the multitude of realizations for the period 1981–1985 is $0.35 \times 10^{-8}\,\mathrm{m^{-2}\,sr^{-1}}$.

Most often, observations in the atmosphere were made under anticyclonic conditions. The passing of cold fronts and cyclonic situations in the regions of measurements were followed by a deformation of the vertical profile of the scattering characteristics of the atmosphere. A linear approximation of the profiles of light scattering coefficients became unacceptable. An analysis of statistical characteristics of the backscattering coefficients vertical profiles has shown that the behaviour of the curves depends substantially on both the type of synoptic situation and on the surface visibility at which the experiment is carried out. An exception is the data for anticyclones (note that a maximum of data has been obtained for such a situation). In this synoptic situation the atmosphere is more heavily loaded with aerosols than under other conditions expressed through an increase of backscattering coefficients by several times for all high-altitude clouds.

Another characteristic feature of the data obtained under anticyclonic conditions is a relatively weak altitudinal variability of vertical profiles. The vertical profiles for the anticyclone refer to the most typical conditions which are characterized by small gradients in the optical characteristics. The values of near-surface visibility determine the optical properties of only the lowest layer (~500 m thick).

Above this layer, the scattering properties of the atmosphere are equalized and practically do not depend on characteristics of the underlying layers. Profiles obtained under the conditions of the small-gradient field and in the regions of increased pressure are characterized by the same features. The most important difference consists of a smaller magnitude of β for the surface layer. Apparently, it is connected with differences in the formation of the aerosol structure and anticyclonic conditions, under conditions of a small-gradient atmosphere.

There is much in common between the analysis of observational data obtained at the moment of, and after the passing of, cold and warm fronts. They are characterized by stepwise profiles with high gradient of scattering profiles of the atmosphere. As for other synoptic situations, they are characterized by a high gradient of optical characteristics in the lower atmospheric layer. However, above it there is an almost homogeneous atmospheric layer within which the main mechanism affecting the spatial structure of atmospheric aerosol is turbulent diffusion. Usually, in the summer–autumn period this layer stretches for 2.5–4 km. Above this layer the scattering properties of the atmosphere sharply decrease.

Note that the passing of both cold and warm fronts is followed, as a rule, by precipitation which cleans the air. For this reason, the atmosphere becomes more transparent, which is reflected in the β magnitude, which in this situation at all altitudes is several times less than under anticyclonic conditions. The structures of eigenvectors and eigen-numbers of correlation matrices under conditions enumerated above have much in common. Firstly, there is a high convergence of the obtained expansions in the system of empirical orthogonal components. The d_k parameters, characterizing a relative magnitude of accumulated mean square deviations (MSDs), already for the first three vectors exceeds 0.9 for all situations met during the experiment. The sum converges especially rapidly in the situation when measurements are made whilst passing the rear of the warm front and at a visibility of <20 km. These facts suggest the conclusion that with this type of approximation, random processes of scattering in the atmosphere can be described with two or three parametric presentations depending on the required accuracy.

The maximum share of the variability taken into account falls on the first eigenvector. Note that first eigenvectors behave differently, depending on respective synoptic situations. The most characteristic feature of first eigenvectors is their almost constant magnitude at altitudes above 1.5–2 km. These vectors are most variable in the lower layer up to 1 km. This fact testifies to the impact of the surface on the vertical distribution of aerosol in the atmosphere, which is most strong in the lower layers. The vertical profiles of second and third eigenvectors, corresponding to various meteorological conditions of the experiment, differ substantially. As a rule, these vectors in all cases can change sign. They cross zero 2–3 times, which testifies to the strong dynamics of the observed processes in the altitude range 0–5 km. An important characteristic of eigenvectors is their stability (i.e., the ability to preserve their structure in transition from one sample to another). To check the stability of eigenvectors, the initial data were divided into two samplings corresponding to 1983 and 1984–1985. An analysis of these samplings has shown that only the system of eigenvectors obtained for anticyclonic conditions, with a high surface visibility is relatively stable. All three eigenvectors for different samplings mainly preserve their structure.

Obviously, changes in the structure of the vertical distribution of scattering coefficients are determined by changes in concentrations of scattering particles and changes in spectral and chemical compositions of aerosol contained in the atmosphere.

6.4 THE EMPIRICAL DIAGNOSTICS OF AEROSOL–CLOUD INTERACTION

Although there are serious grounds for believing that sulphate aerosol affects substantially the formation of global climate, the respective quantitative estimates remain very uncertain. There are both a 'direct' impact manifested through aerosol-induced backscattering of short-wave solar radiation, with mean global values from -0.2 to $-0.9\ \mathrm{W\,m^{-2}}$, and 'indirect' impacts of aerosol (potentially more significant and much more uncertain) leading to a change in the optical properties and lifetime of clouds (the so-called first and second indirect impacts). As for the first indirect impact, its estimates vary from 0 to $-2\ \mathrm{W\,m^{-2}}$, and in case of the second indirect impact they remain too uncertain for discussion. Clouds take part in the processes of both formation and removal of sulphate aerosol (SA); sulphates form almost exclusively via oxidation of SO_2 in both gas and water phases (in the latter case in cloud droplets). The gas-to-particle formation of sulphates under clouds is suppressed by an oxidant such as the OH hydroxyl. The water-soluble sulphates are removed from the atmosphere mainly through washing out by precipitation (wet deposition) and, to a lesser extent, via dry deposition. So far, these processes have been poorly studied and, in particular, the following questions remain unanswered: (1) do the processes in clouds enhance the formation of sulphates or, on the contrary, do they favour the washing out of sulphate aerosol from the atmosphere?; and (2) do clouds strongly limit the process of gas-to-particle conversion of sulphate aerosol?

To analyse the ratio of contributions of the processes of SA formation in the liquid phase, washing out of SA by precipitation, and limitation of processes of gas-to-particle conversion of SA under clouds, [49] considered correlations between clouds and SA. A statistical analysis of the data of 2-year daily observations of characteristics of cloud cover and SA concentration near the surface in Europe and North America has demonstrated the presence of substantial negative correlation between clouds and SA. It means that the impact of clouds on the washing out of SA from the atmosphere by precipitation and/or limitation of the gas-to-particle conversion of SA is manifested stronger than an increase of SA concentration due to its formation in the water phase of clouds. A stable sulphate/cloud anticorrelation on timescales of 8–64 days reflects the effect of large-scale dynamic processes on clouds and sulphate aerosol formation.

On the other hand, results of numerical simulation with the use of the atmospheric general circulation model (GCM) developed at the Goddard Space Flight Centre (USA) testify to the existence of only a weak coherence between SA and cloud cover. However, there is a strong anticorrelation between calculated SA concentration with the gas-to-particle conversion taken into account. The sulphate/cloud anticorrelation determined by GCM data is enhanced if the gas-to-particle conversion of SA under clouds is excluded, as it should take place under the influence of photochemical formation of OH. The only way to ensure a strong anticorrelation between SA and clouds consists in a correlation of the process of SA formation in the water phase in the model mentioned above.

This model, like many other models of global tracers, simulates emissions of dissolved components (including sulphates) from clouds at each time step instead of their description as resulting from the evaporation of clouds. A correction introduced with this fact taken into account will intensify the washing out of sulphates and, as a result, the total amount of SA and its lifetime will decrease by 25% (0.54 TgS and 4.2 days, respectively). On the whole, the sulphur cycle model should guarantee a greater amount of SA in the clear atmosphere and a smaller amount under cloudy conditions, which should seriously tell upon indirect radiative forcing (RF). The introduced correction results in a slight decrease of indirect (anthropogenic) RF from -1.7 to $-1.5\,\mathrm{W\,m^{-2}}$. A change of direct anthropogenic RF is proportional to an attenuation of SA formation from -0.66 to $-0.47\,\mathrm{W\,m^{-2}}$. As for the second indirect impact, here a positive correlation is possible. On the whole, the results considered testify to an expediency of consideration of aerosol/cloud negative correlation when substantiating the global model of sulphate aerosol.

With the use of satellite remote sensing data obtained with AVHRR and POLDER instrumentation, [114] analysed a correlation between characteristics of aerosol and clouds to assess the indirect impact of aerosol on the formation of RF. A consideration of global statistical data has shown that there is a correlation between an efficient radius of aerosol particles r_e, optical thickness of low-level cloudiness τ_c, and total number of aerosol particles in the atmosphere, reflecting an indirect impact of aerosol on RF and, hence, on climate. There was also observed a correlation between cloud amount and aerosol number density, whereas attempts to reveal a correlation between temperature at cloud top level and total aerosol content in the atmosphere turned out to be fruitless.

Regional statistical data indicate that, as a rule, there is a positive correlation between τ_c, cloud amount, and total aerosol content in the atmosphere, which agrees with the global statistical data. However, in the case of r_e, trends identical to a global mean correlation are manifested only in coastal regions. The use of such correlations and a supposition of a 30% increase of total aerosol content (number of particles) in the atmosphere from the beginning of the industrial era, gives an estimate of global mean indirect impact of aerosol on the RF within -0.6 to $-1.2\,\mathrm{W\,m^{-2}}$, whereas the direct impact of aerosol on the RF over the ocean (from the data of satellite observations) constitutes $-0.4\,\mathrm{W\,m^{-2}}$.

Thus, the indirect impact of aerosol on RF exceeds substantially the direct impact, though one should remember, of course, a considerable uncertainty of the estimates of indirect impact. Although the sensitivity of the cloud top temperature to aerosol number density has not been observed, a distinct negative correlation was revealed between aerosol number density and temperature, at which the radius of cloud droplets increases to 14 μm. This specific dependence indicates that aerosols affect clouds in such a way that changes take place in the size of droplets near the cloud top, in τ_c, and in cloud amount, with the cloud top temperature remaining at a level which ensures the absence of considerable changes in RF due to long-wave radiation.

Based on the use of the data of aircraft and surface observations near the Canary Islands (the eastern sector of the North Atlantic), carried out in 1997 within the second field observational experiment ACE-2 to study aerosol characteristics, [31] obtained estimates of the indirect climatic impact of aerosol through aerosol induced changes in cloud properties. The criterion of the impact is the value of the cloud induced radiative forcing. With this aim in view, a comparison has been carried out of the results of aircraft and surface observations of condensation nuclei (CN), aerosol accumulation mode, and aerosol size distribution. Also, a comparison was made of measured and calculated spectra of aerosol sizes with statistical data on the vertical velocity inside and under the stratocumulus clouds.

On the whole, there is a satisfactory (within measurement errors) agreement between aircraft and surface data. However, substantial differences have been detected in comparisons of calculated and observed spectra of the sizes of 'wet' aerosols, which can be explained either by the presence of systematic errors in measurements (made with the use of the particles counter FSSP-300) or by the errors of estimation of the size of dry aerosol particles. An analysis of data on the vertical velocity indicates that there is no marked change in the first and third moments of distribution of the vertical rate frequency between subcloud and intra-cloud atmospheric layers. However, there is an increase of the values of the second moment when crossing the lower boundary of cloudiness. The results obtained reflect the representativeness of surface observational data on the physical and chemical properties of aerosols as well as aircraft data on the vertical rate for numerical simulation of the processes of aerosol activation.

Within the scientific programme of the field observational experiment CRYSTAL-FACE on studies of cirrus cloud dynamics in the south-western part of Florida (USA) in July, 2002, [130] performed aircraft measurements of cloud condensation nuclei (CCN). Two thoroughly calibrated CCN counters measured CCN number density at two levels of oversaturation S equal to 0.2 and 0.85%. An analysis of observational results has shown that the studied aerosol was mainly of marine origin. The CCN concentration averaged 233 cm^{-3} (at $S = 0.2\%$) and 371 cm^{-3} ($S = 0.85\%$).

The data for three of the flights turned out to be abnormal (compared with all others) in a sense that during two of the flights on 18 July aerosols were observed mainly of continental origin, and on 28 July there was an abnormally high and spatially variable concentration of CCN. Analysis of the data of simultaneous measurements of aerosol size distribution and CCN as well as comparisons with results of calculations of CCN concentration based on the Köhler theory (on the assumption that aerosol consisted of ammonium sulphate) have demonstrated a good agreement between observed (N_o) and calculated (N_p) values of CCN concentration. At $S = 0.2\%$ the ratio $N_p/N_o = 1.047$ (the correlation coefficient

$R^2 = 0.911$) and when $S = 0.85\%$, this ratio is equal to 1.201 ($R^2 = 0.835$). With abnormal data for 28 July excluded, agreement becomes more complete: at $S = 0.85\%$ the ratio $N_p/N_o = 1.085$ ($R^2 = 0.770$).

Analysis of the data of satellite observation has led to the conclusion that the presence of high levels of urban and industrial atmospheric pollution, concentration of dust aerosol from deserts, and smoke aerosol from biomass burning leads, as a rule, to a decrease of cloud droplet size, and hence, to an increase of cloud albedo and to a decrease of efficiency of precipitation formation. On the contrary, large salt particles of dust and salt aerosol of marine origin, observed in the atmosphere over land, favour the growth of cloud droplets. Besides this, the aerosol polluting the atmosphere affects substantially the processes of formation of clouds and precipitation. Results of numerical modelling have demonstrated, for instance, that black carbon aerosol observed in China affects strongly the formation of precipitation on a regional scale, connected with an intensified absorption of solar radiation due to such aerosols. A similar situation was observed in the region of the Mediterranean Sea.

In this context, [113] analysed the impacts on the microphysical characteristics of smoke aerosol emitted to the atmosphere during the 1991 oilfield fires in Kuwait. With this aim in view, a comparison was made of microphysical properties of clean and polluted clouds from data of satellite observations. This comparison has shown that: (1) near the upper boundary of the smoke plume, clouds formed due to atmospheric heating as a result of an enhanced absorption of solar radiation and intensified convection; (2) salt particles resulting from fires functioned as gigantic CCN stimulating a fusion of droplets in heavily polluted clouds; (3) far from the sources of smoke aerosol, a very high concentration of CCN of small and middle sizes affected strongly the development of cloudiness due to suppression of the processes of droplet fusion and growth; and (4) freezing of small droplets took place in clouds affected by smoke aerosol at a low temperature at a considerable altitude, which hindered the development of the processes of formation of liquid and solid precipitation.

The observational results discussed indicate that in the atmosphere over land there is no efficient washing out of smoke aerosol particles, determining the possibility of their long-range transport and, hence, broadening the spatial scales of the region affected by smoke aerosol. An enhanced absorption of solar radiation due to smoke aerosol favours an intensification of convection over the smoke aerosol plume, which leads to the formation of clouds above the plume. The impact of this process prevails compared with the impact of cloud evaporation as a result of the smoke-induced enhancement of solar radiation absorption.

A great amount of the Saharan dust (mineral) aerosol about 200 Mt yr^{-1} (which constitutes 20–30% of the total mass of aerosol) suffers long-range transport in the tropics of the Atlantic Ocean in the northern hemisphere. Therefore, according to the data of observations on Barbados, the concentration of dust aerosol (DA) turns out to be 10–100 times greater in summer (in the period of the most intense long-range transport) than in winter. However, on the other hand, the observed DA concentrations on Sal Island (the north-eastern sector of the Pacific Ocean) turned out to be higher in winter than in summer. This contrast reflects the specific feature of atmospheric general circulation due to shifting of the Azores anticyclone and the inter-tropical convergence zone (ITCZ). The most substantial aspect of this specificity consists in that the DA transport in winter takes place mainly at altitudes below 1 km (in the layer of the trade wind inversion), and in summer in the 3–5-km layer (within the Saharan aerosol layer located above the layer of the trade wind inversion).

The earlier estimates of DA impact on the formation of direct RF have led to the conclusion that this impact is manifested comparatively weakly compared with the contribution of anthropogenic sulphate aerosol, since the DA share in the aerosol optical thickness constitutes about 17%, and that of sulphate aerosol 20%. The situation can, however, change substantially if: (1) the albedo of the underlying reflecting surface (e.g., clouds) is high; or (2) dust aerosols move with clouds.

In this connection, [103] discussed the results of observations of the outgoing radiation in the visible and IR spectral regions from Meteosat data for six years, which revealed that in any season, the cloud albedo decreases more strongly in the regions of a high content of dust aerosol (and, respectively, the aerosol optical thickness (AOT) increases). In spring, during the period of powerful outbreaks of Saharan DA, the decrease of albedo can exceed 20%. In summer, the presence of DA can lead to a change in the reflectance of the surface–atmosphere–cloud system, and the wintertime decrease of cloud albedo can be partially explained by the impact of atmospheric drying due to air masses responsible for the long-range transport of the Saharan DA. Also, the DA–cloud mixing taking place in winter can play an important role. This mixing results in a change of cloud droplet size distribution affecting the cloud albedo (DA emissions always cause a decrease of cloud albedo in the tropics of the Atlantic Ocean in the northern hemisphere) and, hence, is a substantial climate-forming factor.

Organic films appearing on water-soluble aerosol particles and on cloud droplets determine a decrease of the water surface tension. This can suppress the water vapour/minor gas component (MGC) exchange between particles (droplets) and the atmosphere, which is important for the microphysical and heterogeneous chemical processes characterizing the aerosol–cloud interaction. Based on consideration of rheological properties of surface films, [18] studied the characteristics of solubility of surface-active substances (SASs) in samples of aerosols and clouds (or fogs). Changes in surface tension due to a rapid expansion or compression of films were found from the data of measurements with the use of a tensometer expressed through a capability of the SAS to suffer an exchange between the surface layer and the main mass of the solution, and, eventually, through SAS solubility. The results obtained in [18] agreeing with data on standard soluble SAS, can be interpreted in terms of the theory of formation of hydrophilic adsorbing layers. It follows from the results discussed that water-soluble organic compounds contribute most to the formation of films on the surface of cloud and fog droplets. Covering the surface with film-forming compounds is mainly determined by the water-soluble organic compound concentrations, independent of the presence of a surface available for film formation. The conclusion can be drawn that the observed decrease of surface tension also appears in the real atmosphere.

The earlier measurements have demonstrated an increase of cloud droplet number density with growing concentration of anthropogenic aerosol particles, but the related changes of the optical thickness (τ_c) and reflectivity of clouds remained unclear. Such changes are difficult to detect because of a strong variability of the cloud liquid water content (LWC) – the main factor in the formation of variations in τ_c and albedo of clouds, whose contribution masks the impact of anthropogenic aerosol. To assess the dependence of LWC on τ_c, [48] performed a surface remote sensing of the atmosphere to retrieve τ_c (from the data of narrow-band radiometric observations) and LWC (with the use of the microwave range, which limits the possibilities of retrieving for the case of 1-layer overcast cloudiness). Observations were made in the north of the central region of Oklahoma (USA) during 13 days of the year 2000 characterized by a high variability of LWC and τ_c on each of these days. Reliably approximated by linear dependence, the data of observations of τ_c (LWC) enable one to reliably reproduce 97% of τ_c variability on individual days and ~63% of the total variability observed. The slope of the τ_c (LWC) curve is inversely proportional to an efficient radius of cloud droplets r_e, which varied from 5.6 ± 0.1 to $12.3 \pm 0.6\,\mu m$. The negative correlation between r_c and aerosol coefficient of scattering near the surface was observed, but the correlation coefficient was low ($R^2 = 0.24$), which can partially be explained by the absence of a correlation between the aerosol content in different atmospheric layers as well as by specific meteorological conditions (the vertical wind shear). The cloud albedo and the RF with constant LWC are very sensitive to the efficient radius of droplets. At a solar zenith angle of 60° and with a typical LWC value $100\,g\,m^{-1}$, on different days r_e changes within 5.8–10.2 μm, and due to an r_e decrease, a substantial reduction of short-wave RF at the top of the atmosphere (the Tuomi effect) took place [108].

6.5 NUMERICAL SIMULATION OF THE AEROSOL–CLOUD INTERACTION

Based on the use of the parameterization technique of radiation transfer in the atmosphere, developed at the European Centre of the Medium-Range Weather Forecasts, [104] assessed the impact of the DA layer formed under Sc clouds due to DA long-range transport during the dust storms in the Sahara desert, on the albedo of the ocean–atmosphere–cloud system. Calculations have demonstrated a decrease of the system's albedo reaching, on average, 15%, which closely corresponds to data of summertime satellite observations. In winter, when dust aerosol is concentrated in the trade winds layer, the effect of superposition of clouds and DA residing in the same layer as clouds, cannot explain the observed change in the system's albedo. An agreement with the observational data is ensured only with due regard to the complicated interaction of DA with the cloud aerosol (cloud droplets) and some MGCs.

The intracloud 'processing' of DA results in a substantial change of its properties. The numerical simulation has shown that the presence of DA in clouds can lead to an initial decrease of CCN number density, an increase of an efficient radius of droplets, and, eventually, to a decrease of cloud albedo. Under certain conditions, this decrease can exceed 10%, but, as a rule, a decrease of albedo varies within 3–10%.

An analysis of numerical simulation results, obtained with the use of a mesoscale model, revealed the presence, at some altitude, of carbon (soot) particles appearing in winter during fires in the African savannahs. The DA transformation and the presence of absorbing carbon aerosol determine a decrease of albedo exceeding 10% and thus explain the results of satellite observations in winter in the southern sector of the region under study. The results obtained reveal an important role of dust aerosol as a climate-forming factor manifesting itself west of the African coastline. According to estimates, the DA-induced radiative forcing in the presence of clouds: (1) is always positive under global scales for the whole atmosphere; and (2) manifests itself near the cloud top (in winter) or at some altitude (in summer). In summer, the RF near the cloud surface is negative, but it is mixed with the contribution of positive RF at the level of the dust layer's location.

Yin *et al.* [136] undertook a numerical modelling of the impact of dust (mineral) aerosol particles functioning as CCN on the dynamics of cloud cover, precipitation, and a change of the optical properties of clouds. With this aim in view, a 2-D non-hydrostatic model of cloud dynamics was used with detailed consideration of micro-physical processes. An initial size distribution of aerosol prescribed in the model is a superposition of background CCN and dust aerosol.

The results obtained indicate that insoluble dust aerosol particles become efficient CCN after passing through a convective cloud. Their efficiency as CCN in this case grows, since on the surface of particles a layer of sulphate is formed when they are first captured by growing water droplets or ice crystals – then (eventually) sulphate is emitted to the atmosphere after hydrometeor evaporation. Penetrating a cloud, such particles favour the growth of concentration of activated droplets and an expansion of their size distribution.

Calculations have shown that in continental clouds the impact of DA 'processed' by clouds consists of an acceleration of the formation of precipitation particles, though the rain rate depends, first of all, on large and gigantic CCN. In marine clouds, an addition of aerosol and DA 'processed' by clouds weakly affects precipitation, since the formation of clouds begins in the presence of a large amount of large particles. An additional input of CCN to marine and continental clouds leads to the growth of their optical thickness even when the amount of precipitation had already grown.

As a result of intense burning of biomass in the Amazon basin, a heterogeneous layer of smoke aerosol is formed in the atmosphere covering, during the dry season (June–December), a territory of the order of several million square kilometres. A complicated interaction of water vapour, CCN, and RF due to short-wave radiation absorption by 'black' aerosol appearing under such conditions, affects strongly the processes of convection, cloud formation, and cloud properties. The earlier measurements in the dry season period revealed a stable spatial distribution of water vapour, but a varying distribution of aerosol concentration. The latter means that the optical properties of clouds in the region of the Amazon River should depend less on meteorological conditions than on distribution of CCN concentration which grows strongly under the influence of smoke aerosol.

To quantitatively estimate the impact of aerosols on the optical properties of clouds, [109] used a 1-D model of cloudiness, making it possible to analyse the dependence of conditions of CCN growth on the physical and chemical properties of aerosol. The CCN concentration was measured with oversaturation estimated at 0.15% and 1.5%. The measured values of CCN concentration during the period of the rainy season were small and resembled values corresponding to marine conditions. The presence of aerosol in the rainy season, formed due to biomass burning, has determined a radical growth of CCN concentration. Changes in such characteristics of clouds as an effective radius of droplets and a maximum of oversaturation turned out to be most substantial at a low concentration of CCN. It follows from this that the most substantial interannual variability of cloud properties can be expected for periods of wet but not dry seasons.

It follows from the observational data that the difference between spectra of CCN sizes in forested and deforested regions during the rainy season is moderate and causes only small changes in cloud properties compared with the difference due to contrasts between wet and dry periods. The results obtained suggest the conclusion that the impact of deforestation on the moisture cycle and convection activity during the rainy season is determined by changes in surface albedo due to deforestation but not clouds. On the other hand, during the dry season the cloud droplet concentration can increase 7 times, which leads to almost a doubling of the effective radius of droplets predicted by numerical modelling. This transformation can cause a maximum of indirect aerosol RF, reaching $-27\,\mathrm{W\,m^{-2}}$ in the case of non-absorbing clouds.

The presence of substances absorbing short-wave radiation in smoke aerosol determines the cloud darkening in the region of the Amazon River and thus, a decrease of the total RF. Approximate estimates show that radiation absorption by smoke aerosol can compensate for about 50% of the maximum indirect aerosol radiative forcing. However, further studies are needed, since even small changes in chemical and physical properties of aerosol can substantially affect the kinetics of aerosols and their activity as CCN. For the reliable assessment of the impact of smoke aerosol on climate, it is insufficient to take into account only data on the CCN size spectrum.

In the 1950s and later, various numerical models of individual clouds and cloud systems were developed with due regard to cloud dynamics depending on the size distribution of droplets and on the chemical reactions taking place in clouds. To save computer time, different assumptions were used in these models limiting the commonness of consideration. Jacobson [41, 42] has started a new stage of numerical simulation aimed at reproducing the cloud and precipitation dynamics with due regard to a multiple size distribution of aerosol and, on the other hand, a description of the impact of the cloud and precipitation dynamics or the washing out of aerosol from the atmosphere.

The new model makes it possible to take into account:

(1) the impact of simultaneous transformation of the liquid and ice phase of clouds on the variety of aerosol size distribution;
(2) diffused–phoretic, thermal–phoretic, gravitational, and other processes, as well as processes of coagulation taking place in the presence of cloud droplets, ice particles, hail, and aerosol;
(3) the process of 'contact freezing' (CF) of droplets due to intracloud aerosol;
(4) heterogeneous and homogeneous freezing;
(5) cloud droplet destruction;
(6) coagulation between cloud droplets and ice crystals (with aerosol included in them) as well as intracloud aerosol;
(7) coagulation of precipitation hydrometeors with intra and subcloud aerosol, which determines the aerosol washing out;
(8) the subcloud processes of evaporation/sublimation with subsequent formation of smaller particles of hydrometeors and aerosol nuclei;
(9) washing out of MGCs; and
(10) water-phase chemical reactions.

Results of numerical simulation have led to the following conclusions:

(1) processes of coagulation of hydrometeors play an important role as factors that control the aerosol particle number density on global scales;
(2) the subcloud processes of washing out (as a result of coagulation of hydrometeors and aerosol) can be a more important mechanism for decreasing the aerosol number density in the atmosphere than washing out under the influence of precipitation (an opposite situation takes place with respect to the aerosol mass);
(3) the existence of dual maxima of the observed cloud droplet size distribution can partly be determined by different characteristics of activation for aerosols with different size distributions;
(4) a cooling due to evaporation from the droplet surface under conditions of non-saturated air can be the cause of droplet freezing (this process can be called 'transpiration freezing'); and
(5) the impact of heterogeneous–homogeneous freezing of droplets can be manifested at higher altitudes in the troposphere than the CF, but none of these processes can substantially affect the warm cloud dynamics or washing out of aerosol.

Clouds observed in the upper troposphere at a temperature ranging between -20°C and -85°C are usually known as cirrus-like (cirrus, cirrostratus, and cirrocumulus). It has also been established that stratus, stratocumulus, altostratus, and cirrus clouds affect the climate most substantially due to their large spatial extent. Within the project 'Studies of clouds using surface and aircraft radar and lidar instrumentation' (CARL) financed by the European community, [82] developed a regional model RAMS for the numerical modelling of the regional-scale processes in the atmosphere, which was used to study the formation of ice crystals as well as formation and evolution of cold clouds observed during completion of the programme of field observations at Palaiseau (France) from 26 April to 16 May, 1999.

The Doppler radar (3.2 mm – 94.9 GHz) and lidar (532 nm) soundings of clouds were supplemented with *in situ* measurements of their characteristics from the flying laboratory 'Merlin'. Numerical experiments to analyse the sensitivity of the model to the parameter of the shape of a gamma-distribution of particle sizes and to the data for initialization have made it possible to reveal the most substantial parameters which determine the cirrus cloud dynamics as well as to check an adequacy of the model by comparison with observational data.

Calculation results have demonstrated the ability of the model to reliably reproduce the cirrus cloud characteristics such as their spatial–temporal variability and geometrical parameters. Detailed comparisons with the data of aircraft observations revealed considerable differences between calculated and observed values of cloud water content and number density of small particles (2–47 μm), which can be explained by the impact of inadequate 'microphysical' algorithms and measurement errors, but a good agreement in the case of particles of intermediate (25–800 μm) and large (200–6,400 μm) sizes.

The differences observed, sometimes in the case of large particles, can be explained by uncertainties of determination of cloud boundaries or by disturbances introduced by aircraft. The time spent on initialization of the model has a considerable impact on the results of the first several hours of numerical simulation. On the whole, the model RAMS can adequately reproduce most of the parameters which determine the formation of cold clouds with the prescribed standard meteorological information as well as observational data which determine initial and boundary conditions.

Kreidenweis *et al.* [57] performed a comparison of the models of processes of aerosol washing out from the atmosphere and liquid-phase SO_2 oxidation with the participation of H_2O_2 and O_3 taking place in the upward air fluxes that accompany clouds. The approximate models in which only cloud droplets of the same size are taken into account have been compared with the models ensuring a consideration of aerosol and cloud droplet size distribution. All models simulate the growth of cloud droplets with a prescribed log-normal size distribution of ammonium bisulphate particles as well as subsequent chemical reactions in the water phase during adiabatic air lifting.

According to the earlier results, the use of the models with an effect of cloud droplet size distribution on chemical reactions taken into account suggests a conclusion about the presence in this case of more intense (2–3 times) oxidation due to $SO_2 + O_3$ reaction, which is explained by the dependence of the cloud water pH on the size of cloud droplets. All models reveal the existence of an intense washing out of dry aerosol particles, but calculations of cloud droplet number densities give different values varying between 275–358 cm^{-3}.

Such differences are determined by different parameterizations of gas component assimilations, thermodynamics of solutions, and some other processes. Differences between calculated values of the number density of droplets can transform into substantial changes in the optical thickness (up to 9%) and albedo (up to 2%) of clouds. Numerical modelling has shown that changes in the aerosol size distribution are sensitive to the cloud droplet number density. Differences of the transformed size distribution of aerosol particles can cause changes in the coefficient of light extinction reaching 13%. The results obtained illustrate an important role of the adequacy of parameterization of these processes in climate models.

Feingold and Kreidenweis [22] performed a theoretical analysis of the processes of aerosol 'processing' by clouds with due regard to collisions between aerosol particles and cloud droplets as well as chemical reactions in the water phase. With this aim in view a numerical modelling has been carried out which has made it possible to reproduce the impact of large-scale vortices as well as: (1) aerosol size distribution; (2) microphysical processes depending on the size of particles; and (3) transformation of SO_2 into sulphate in the course of the chemical processes in the water phase.

The earlier estimates have shown that results of aerosol 'processing' in the marine atmospheric boundary layer (MABL) covered with stratocumulus clouds depends strongly on cloud water content, concentration of aerosol and MGCs, as well as the duration of the aerosol particle/cloud droplet interaction. The new model takes into account the variability of all these parameters on scales of large vortices (of the order of several hundreds of metres) and within timescales of several seconds.

To analyse the role of the various processes of the aerosol–cloud interactions, several scenarios have been considered:

(1) a relatively low aerosol number density at which the aerosol processing via chemical reactions in the water phase does not affect substantially the formation of drizzle;
(2) the intermediate level of relatively large particle concentrations when chemical reactions in the water phase suppress the development of drizzle; and
(3) intermediate values of particle concentrations with respect to small sizes and an enhanced formation of drizzle due to reactions in the water phase.

The numerical modelling results referring to the enumerated situations testify to the importance of chemical processes in the water phase, capable of affecting substantially the dynamics and microphysics of stratocumulus clouds and illustrating the complicated interaction of the processes considered. The latter reflects the difficulties in parameterization of the processes of aerosol processing by clouds, characterized by a large number of feedbacks.

For reliability of the numerical simulation of atmospheric aerosol dynamics and for the solution of the related problems of the determination of meteorological visibility, and studies of heterogeneous chemical reactions in the atmosphere and radiative forcing, of great importance is a prescribed aerosol size distribution. Within the model of 'small' stratus clouds observed during the second aerosol experiment, [137] studied the impact of parameterization of aerosol size distribution on the results of numerical simulation of aerosol–cloud interaction with the use of a 1-D 'climate–aerosol–chemical processes' model. The model (with predicted number density and mass of aerosol – Modal-NM or number density, surface area and mass of particles – Modal-NSM) and sectional approaches (with 12 or 36 gradations of particles sizes taken into account) have been considered making it possible to precalculate the total number density and mass of aerosol particles. However, an application of the modal approach (Modal-M), when the prognostic mass of aerosol is considered, but number density is prescribed, does not provide reliable results.

The reliability of the aerosol size distribution parameterization depends substantially on its detailed presentation. The values of normalized differences reach 12% and 37% with 36 and 12 gradations of particles sizes, respectively. As applied to modal presentations Modal-NSM, Modal-SM, and Modal-M, the respective differences are 30%, 39%, and 179%. All presentations (except Modal-M) enable one to correctly calculate the water content and optical thickness of clouds, as well as aerosol extinction. Further development of the approaches discussed supposes consideration of a more realistic model of aerosol with due regard to totality of various modes and varying chemical compositions of particles (the latter can strongly affect the number of activated aerosol particles which is important for reliability of indirect RF estimates). Processes not considered in the study discussed can also play a substantial role in the evolution of the aerosol size distribution. These are intracloud washing out of aerosol particles, coalescence of droplets, and wet deposition. Parameters of the process of aerosol activation need specification. The impact of aerosol–cloud interaction on aerosol and cloud droplets remains unstudied.

The origin of marine aerosol particles (including CCN) also remains inadequately studied. One of its most important components is salt particles getting to the atmosphere in sea spray 'bursts'. Their size varies within 0.1–300 μm, and the number density depends on wind speed. Besides, there are processes of gas-to-particle formation of non-sea salt (NSS) aerosol particles taking place as secondary and tertiary nucleation, heterogeneous nucleation, and condensation. In the MABL, in the regions of the Arctic and Antarctic, there was also the *in situ* formation of ultrafine nuclear aerosol particles with $D < 3$ nm. One of the possible sources of aerosol in these cases is the oxidation of dimethylsulphide (DMS) leading to the formation of sulphate particles ($D < S$ is emitted to the atmosphere by marine phytoplankton). Along the coastlines (at low tides) there was observed almost a daily formation, and in the boreal forests nearly 50 times per year, of ultrafine particles of biogenic origin.

Based on observational data and numerical modelling results with the use of the model AEROFOR-2 of gas-to-particle conversion chemical processes, [100] studied the formation and evolution of properties of new particles of nuclear aerosol in the MABL near the coastline. The coastal zones are known as an intense natural source of aerosol particles as well as biogenic gas-phase compounds, which can condense on aerosol particles, determining thereby their growth. A numerical modelling was made of instantaneous rate of nucleation with different prescribed levels of emissions of biogenic gas-phase compounds, whose results demonstrated the possibility of formation of nucleation mode aerosol particles with number density $\sim 10^5$–$10^6\ \mathrm{cm^{-3}}$, corresponding to that observed.

According to the estimates obtained, the instantaneous rate of nucleation should vary (for reproduction of the observed concentration values) within

3×10^5–$3 \times 10^6\,cm^{-3}\,s^{-1}$, and the power of the sources of condensed gas-phase compounds should reach $5 \times 10^7\,cm^{-3}\,s^{-1}$. A considerable part of the newly formed particles of nuclear aerosols can further transform into CCN in the process of subsequent evolution in the time period >3 days under clear sky conditions at the observed levels of oversaturation in the MABL clouds. Such CCN can substantially contribute to the formation of the aerosol indirect impact on RF. The level of CCN concentration depends mainly on the degree of aerosol coagulation and condensation of biogenic gas-phase compounds and, to a lesser extent, on the condensation of H_2SO_4 formed due to DMS oxidation.

In all the cases of numerical simulation considered, the CCN number density increased by more than 100% with an oversaturation exceeding 0.35%. In the period of the low-level tide the composition of particles of the Aitken and nucleation modes changes substantially leading to complete insolubility of particles, whereas changes in the accumulation mode particles are less substantial. Due to further enhancement of DMS emissions to the atmosphere, the share of the soluble fraction of particles of the Aitken and nucleation modes increases gradually and thus their potential as the source of CCN grows.

The available observational data show that atmospheric aerosol (like precipitation) contains, as a rule, bacteria. The 'cloud' bacteria can actively grow and be reproduced at a temperature below 0°C. Bauer *et al.* [6] studied the ability of bacteria existing in aerosol and cloud water, whose samples have been taken in a remote mountain region of Austria, to function as CCN. Average values of bacteria concentration constituted 8 colony forming units (CFU) per cubic metre in the case of aerosol samples and 79 $CFU\,ml^{-1}$ for cloud water samples. A totality of studied bacteria included both gram-positive and gram-negative species (but they cannot form ice particles).

Experiments with the use of the CCN counter have shown that at an oversaturation within 0.07–0.11% all types of bacteria can function as CCN. Since the size of bacteria was less than the Kelvin diameter at respective levels of oversaturation, the physico-chemical properties of their external cells should have intensified their activity as CCN. Bearing in mind that the bacteria number density in atmospheric aerosol is by orders of magnitude less than the usual concentration of CCN (100–$200\,cm^{-3}$), it becomes clear that the real contribution of bacteria as CCN to the formation of water clouds is, apparently, negligible. However, if bacteria can nucleate ice particles, their substantial role in the processes of cloud freezing is not excluded.

Iversen and Seland [40] performed a numerical climate modelling based on the use of the NCAR CCM3 model containing units interactively simulating the cycles of sulphate (SO_4) and black carbon (BC). A parameterization of the sulphur cycle takes into account emissions of DMS by the World Ocean, SO_2, as well as natural and anthropogenic sulphate emissions. The BC cycle is simulated as determined by biomass and fossil fuel burning. Chemical and physical processes responsible for the formation and transformation of aerosol are simulated with a priori given levels of concentration of oxidants and background aerosol of marine, continental, and polar origin. In the case of chemical processes in the water phase, the dependence on the rate of exchange in the cloudy and cloud-free atmosphere has been considered.

At the levels of global emissions of different components determined by the data of the Intergovernmental Panel on Climate Change (IPCC), the lifetimes (cycle) are (in days): 1.5 (SO_2), 3.5 (SO_4), and 4.7 (BC) in 2000 and (by the prognostic scenario of emissions (A2) 1.6 (SO_2), 4.0 (SO_4), and 4.7 (BC) in 2100. Calculated values of BC and SO_4 concentrations turned out to be underestimated by a factor of 10, compared with those measured in winter in the Arctic, which can be partly explained by a very strong variability of cloudiness. On the contrary, the calculated values of the surface concentration of BC and SO_4 in low latitudes are 10 times overestimated. Such substantial differences are explained by the fact that the model leaves out of account the processes of transport, and that the efficiency of washing out in convective clouds is underestimated. In this connection, special numerical experiments have been carried out to assess the sensitivity of the numerical modelling results to a consideration of the processes taking place in clouds. Analysis of these estimates has shown that the levels of SO_4 and BC concentrations are very sensitive to processes of the vertical transport and washing out in convective clouds. Thus methods of parameterization of processes in clouds and aerosol–cloud interactions should be improved.

Atmospheric aerosol consists of hundreds or even thousands of organic and inorganic compounds. An important subset of aerosol particles is CCN which play a critically important role in the formation of cloud droplets. Inadequate information about the organic fraction of aerosol and inadequate reliability of the description of interactions between organic and inorganic components of aerosol seriously complicate an understanding of the processes of CCN functioning. Laboratory studies of CCN were limited mainly by consideration of either polydisperse aerosol of the complicated chemical composition existing in the real atmosphere or individual components of aerosol created in the laboratory.

The earlier studies revealed a widely varying ability of different individual components to stimulate the formation of aerosol particles. While for some organic compounds this ability is weak, other compounds can be active as inorganic salts. The CCN usually consist of a mixture of many organic and inorganic compounds whose interaction in the process of cloud droplet formation is far from being studied.

In this connection, [106] undertook a study of the ability of CCN particles, with a multicomponent composition, to regulate the process of cloud droplet formation and its properties. Laboratory experiments were performed with the use of internally mixed multicomponent particles as well as particles consisting of a nucleus covered with hexadecane. Components of chemical composition of particles were potassium chloride; ammonium sulphate; pinonic, pinic, norpinic, and glutamic acids; and leucine and hexadecane. The combined use of the tandem differential mobility analyser (TDMA) and thermal diffusion CCN counter (CCNC) has made it possible to measure the diameter of particle activation. Experiments have been carried out with a 'dry' diameter of particles within 0.02–0.2 μm for oversaturation levels 0.3% and 1%.

The data of laboratory measurements of parameters of the process of cloud droplet formation are compared with results of theoretical calculations proceeding from the assumption of additive contributions of individual components of the CCN chemical composition. This supposition can be considered adequate in the case of calculations of the activative diameter of CCN with a mixed composition. The results obtained show that the covering of particles with hexadecane, the volume of which constitutes up to 96% of the total volume of particles, does not affect the activity of particle nuclei as CCN, which agrees with the results obtained earlier for particles whose nuclei are ammonium sulphate. An addition of a 1% solution of cooking salt strongly affects the diameter.

Kärcher and Lohmann [46] proposed a technique for parameterization of cirrus cloud formation based on consideration of homogeneous freezing of particles which takes into account the impact of aerosol functioning as CCN, on the process of freezing during an adiabatic lifting of air particles. This impact can be substantial when the time constant of the process of freezing is short compared with the timescale of the process of growth of initial ice particles. A new scheme of parameterization has been validated with the use of results of the numerical simulation of air particle dynamics. In this context, an analysis has been made of the connection between number densities of aerosol particles and ice particles in cirrus clouds formed via homogeneous freezing. This connection turned out to be weaker than for water clouds consisting of liquid droplets. It has been demonstrated [46] that even the participation of volcanic aerosol characterized by a high concentration, in the process of freezing, does not affect substantially the formation of cirrus clouds. Thus, in the case of crystal clouds, the Tuomi mechanism [129] of the impact of aerosol on cloud properties is not so strong as in the case of water clouds.

Gao *et al.* [25] performed *in situ* measurements of relative air humidity with respect to ice (RH_i) and concentration of nitric acid vapour (HNO_3) in natural cirrus clouds and in the 'inversion trail' of aircraft in the upper troposphere. According to these measurements, at a temperature $T < 202\,K$, in both cases considered, the RH_i sharply increases to average values $>130\%$. Such an increase of RH_i can be explained by a new class of ice particles (Δ-ice) containing HNO_3. It is assumed that HNO_3 molecules on the surface of particles hinder the reaching of equilibrium in the ice–water vapour system due to functioning of the mechanism similar to that observed at a lowering of the freezing point of proteins used as antifreeze. The presence of Δ-ice particles determines a new connection between global climate and natural and anthropogenic emissions of nitrogen oxides. Hence, a consideration of Δ-ice in numerical climate models should lead to changes in the prescribed properties of cirrus clouds and water vapour distribution in the upper troposphere.

Studying the processes in cirrus clouds, which form in anvil clouds' vertical development, [24] have come to the conclusion that, as a rule, ice crystals in cirrus clouds form on CCN–aerosol particles in the middle troposphere but not in the ABL, as was supposed earlier. It means that aerosol, as the product of long-range transport from distant sources of pollution, can strongly affect the formation of anvil clouds.

A perspective direction of development in the context of aerosol–cloud interaction appears in connection with the recently detected interaction between noctilucent clouds in the mesosphere and a layer of iron atoms located at the same altitude [37].

The prevalent features of the Arctic climate are temperature and humidity inversions, especially during the cold-half of the year. Temperature inversions are caused by radiative cooling, warm air advection, downward air motions, processes in clouds, thawing at the surface level, and topography. Humidity inversions are caused by precipitation in the form of ice crystals, which leads to a decrease of humidity in the lower atmospheric layers. Information about cloud amount in the Arctic is obtained mainly from the data of surface observations, according to which the annual change of total cloudiness is at a maximum in summer (up to 90%) and at a minimum in winter (40–60%). A comparison of satellite data on cloud amount with surface data has shown that satellite data give higher estimates (within 5–35%).

Although to simulate the dynamics of the climate in the Arctic, rather complicated models of atmospheric general circulation have been used, there is no doubt that a parameterization of cloudiness and radiation dynamics cannot be considered realistic. To study the real dynamics of mixed clouds and ice clouds as well as their effect on radiation transfer, [65] undertook a numerical modelling with the use of the mesoscale model GESIMA. The modelling results have been illustrated in concrete situations from the data of observations on 16 and 28–29 April, 1998 at the drifting station SHEBA within the Arctic experiment on cloud studies (ACE) as part of the programme FIRES of the first regional experiment on cloud cover satellite climatology.

A comparison of numerical modelling results with observational data has demonstrated the ability of the GESIMA model to adequately reproduce the location of cloud cover boundaries, ice and water content of clouds, and an efficient radius of cloud particles. Calculations for the clear and anthropogenic aerosol-loaded atmosphere have shown that the presence of aerosol changes substantially the microphysical parameters of clouds (it leads to the growth of small droplet number density) and thus affects precipitation. In contrast to the earlier results, it turned out that in a polluted cloud an accretion of snow crystals and cloud droplets intensifies due to a higher number density of droplets observed in the polluted cloud. Besides this, there is a decrease of the rate of autoconversion of cloud droplets and accretion of drizzle droplets by snow due to a cessation of the processes of collisions and coalescence in the polluted cloud.

The total result of two oppositely directed changes with the growth of aerosol concentration from 100 to 1,000 particles per cubic centimetre turns out to be a decrease of precipitation in the polluted atmosphere. Precipitation amount critically depends on the shape of crystals. With the assumed presence of crystal aggregates, a 10-fold increase of aerosol concentration leads to a 40% decrease of the amount of accumulated snow after a 7-hour numerical simulation, whereas in the case of flat crystals the snow amount decreases by 30%. This contrast is explained by a higher efficiency of snow crystal accretion with cloud droplets in the presence of aggregates.

In the context of the discussed problems, of great interest are polar stratospheric clouds (PSC) which play a dual role in the process of stratospheric ozone depletion in the polar regions: (1) on the surface of PSC particles an activation of halogens takes place; and (2) the sedimentation of very large PSC particles ($r > 5\mu m$) containing nitric acid favours denitrification, which weakens a deactivation of chlorine and enhances a catalytically induced destruction of ozone. Large particles containing HNO_3 and with a small number density ($n \approx 10^{-4}\,cm^{-3}$, $r < 10\,\mu m$) were observed in the synoptic-scale Arctic PSC. On the contrary, the PSC appearing in orographic waves are characterized by relatively small horizontal scales (of the order of several thousands of square kilometres) and appear under cold climate conditions due to adiabatic shifting of air masses.

Different types of particles were observed in orographic PSC. While the process of cooling is sufficiently slow at a temperature of 2–3 K above the ice point (T_{ice}), the stratospheric H_2SO_4 aerosol (H_2O/H_2SO_4) assimilates a considerable amount of nitric acid and water vapour, which determines the formation of droplets of the supercooled triple solution (STS). At a temperature ~3 K below T_{ice}, ice nucleation takes place in STS particles. In the direction of air flux from the Arctic ice clouds, solid particles were observed which, most likely, consist of nitric acid trihydrate ($NAT = NHO_3 \times 3H_2O$).

According to available observational data, NAT is included in PSC and plays an important role in sedimentation, denitrification and intensifying ozone depletion catalysed by chlorine. However, the mechanism of NAT formation is still unknown. Luo *et al.* [67] have shown that though a strong cooling observed in orographic waves limits the saturation level which characterizes the NAT ($S_{NAT} = P^{part}_{HNO_3}/P^{var}_{HNO_3} \gtrsim 30$, where P is the partial pressure due to particles (part) and vaporous (vap) NAT) in the STS of droplets, under gas-phase conditions, it is possible that $S_{NAT} \lesssim 500$. As laboratory experiments have shown, at high levels of oversaturation, a sedimentation and nucleation of NAT on the surface of ice or other solid particles takes place until they are covered with STS.

Based on observations in the Arctic during three winters together with the numerical model of microphysical and optical properties of particles, a simple technique has been proposed [67] to parameterize the process of deposition and nucleation of NAT, whose rate depends only on the level of NAT oversaturation and temperature. The results obtained testify to the possibility of explaining the formation of NAT particles with their number density of order $10\,cm^{-3}$, with due regard to deposition and nucleation of NAT on the surface of ice particles in the presence of gas-phase NAT under conditions of extremely high levels of oversaturation.

Luo *et al.* [67] discussed the possibilities of functioning of the meteorite substance as the factor ensuring the development of heterogeneous reactions and nucleation of NAT. Particles of micrometeorites >0.1 μm, for number density about $10^{-5}\,cm^{-3}$, cannot provide a high concentration of NAT particles observed in PSC in orographic waves. In the presence of meteor dust with particle radii about 5–10 μm and a high number density, the total surface of meteor particles is 2–3 orders of magnitude less than the maximum surface of ice particles. Since, however, the NAT over the surface of non-ice particles is 4 times greater and the rate of NAT nucleation on the surface of meteor particles is unknown, it is not excluded that NAT nucleation can in this case be substantial. The observational data confirming this supposition are still absent.

6.6 CONCLUSION

The presence of many unsolved problems of interaction between atmospheric aerosols and clouds testifies to the necessity of further studies, the urgency of which is determined, first of all, by the key importance of these problems for understanding and quantitative estimation of the aerosol–cloud impact on climate. The programme of further studies should include coordinated field observational experiments and numerical simulations. Such efforts can be exemplified by [36] who described related developments within the programmes FEBUKO – the field studies of the budgets and transformation of organic aerosol in the context of cloud cover dynamics in the troposphere and MODMEP – a numerical modelling of multiphase processes in the troposphere with due regard to the dynamics and chemical reactions.

6.7 BIBLIOGRAPHY

1. Ackerman S. A. and Cox S. K. Aircraft observations of the shortwave fractional absorptance of non-homogeneous clouds. *J. Appl. Meteor.*, 1981, **20**, 1510–1515.
2. Ackerman S. A. and Stephens G. L., The absorption of solar radiation by cloud droplets: An application of anomalous diffraction theory. *J. Atmos. Sci.*, 1987, **44**, 1574–1588.
3. Hobbs P. V. (eds). *Aerosol–Cloud–Climate Interactions.* Academic Press, New York, 1993, 233 pp.
4. Andreae M. O., Rosenfeld D., Artaxo P., Costa A. A., Frank G. P., Longo K. M., and Silva-Dias M. A. F. Smoking rain clouds over the Amazon. *Science*, 2004, **303**(N5662), 1337–1342.
5. Asano S. Cloud and Radiation Studies in Japan. Cloud Radiation Interactions and Their Parameterization in Climate Models. *WCRP-86 (WMO/TD No. 648).* WMO, Geneva, 1994, pp. 72–73.
6. Bauer H., Giebl H., Hitzenberger R., Kasper-Giebl A., Reische G., Zibuschka F., and Puxbaum H. Airborne bacteria as cloud condensation nuclei. *J. Geophys. Res.*, 2003, **108**(ND21), AAC2/1–AAC2/5.
7. Baumann K., Ilt F., Zhao J. Z. and Chameides W. L. Discrete measurements of reactive gases and fine particle mass and composition during the 1999 Atlanta Supersite Experiment. *J. Geophys. Res.*, 2003, **108**(ND7), SOS4/1–SOS4/20.
8. Belan B. D., Grishin A. I., Matvienko G. G., and Samokhvalov I. V. *The Spatial Variability of the Atmospheric Aerosol Characteristics.* Nauka Publ., Siberian Branch, Novosibirsk, 1989, 152 pp. [in Russian].
9. Bian H., Prather M. J., and Takemura T. Tropospheric aerosol impacts on trace gas budgets through photolysis. *J. Geophys. Res.*, 2003, **108**(ND8), ACH4/1–ACH4/10.
10. Bian H. and Zender C. S. Mineral dust and global tropospheric chemistry: Relative roles of photolysis and heterogeneous uptake. *J. Geophys. Res.*, 2003, **108**(ND21), ACH8/1–ACH8/14.

11. Binenko V. I. and Kondratyev K. Ya. The vertical profiles of typical cloud formations. *Proc. of the Main Geophysical Observatory (St Petersburg)*, 1975, issue 331, pp. 3–16 [in Russian].
12. Boers R., Jensen J. B., Krummel P. B., and Gerber H. Microphysical and short-wave radiative structure of wintertime stratocumulus clouds over the Southern Ocean. *Q. J. R. Meteorol. Soc.*, 1996, **122**, 1307–1339.
13. Bott A., Trautmann T., and Zdunkowski W. A numerical model of the cloud-topped planetary boundary layer: Radiation, turbulence and spectral microphysics in marine stratus. *Q. J. R. Meteorol. Soc.*, 1996, **122**, 635–667.
14. Bréon F.-M. and CNES Project Team, *POLDER Level-2 Product Data Format and User Manual* (PA.MA.O.1361.CEA Edition 2 – Revision 2). CNES Publ., Paris, 1998, 45 pp.
15. Brooks S. D., Wise M. E., Cushing M., and Talbart M. A. Deliquescence behavior of organic/ammonium sulfate aerosol. *Geophys. Res. Lett.*, 2002, **29**(N19), 23/5–23/4.
16. Claeys M., Graham B., Vas G., Wang W., Vermeylen R., Pashynska V., Cafmeyer J., Guyon P., Andreae M. O., Artaxo P., *et al.* Formation of secondary organic aerosols through photooxidation of isoprene. *Science*, 2004, **303**(N5661), 1173–1176.
17. Curry J. A., Hobbs P. V., King M. D., Randall D. A., Minnis P., Isaac G. A., Pinto J. O., Uttal T., Bucholtz A., Cripe D. G., *et al.* FIRE Arctic Clouds Experiment. *Bull. Amer. Meteorol. Soc.*, 2000, **81**(1), 5–29.
18. Decesari S., Facchini M. C., Mircea M., Cavalli F., and Fuzzi S. Solubility properties of surfactants in atmospheric aerosol and cloud/fog samples. *J. Geophys. Res.*, 2003, **108**(ND21), AAC6/1–AAC6/9.
19. Desboeufs K. V., Losno R., and Colin J. L. Relationship between droplet pH and aerosol dissolution kinetics: Effect of incorporated aerosol particles on droplet pH during cloud processing. *J. Atmos. Chem.*, 2003, **46**(N2), 159–172.
20. Deschamps P. Y., Bréon F. M., Leroy M., Podaire A., Bricaud A., Buriez J. C., and Sèze G. The POLDER Mission: Instrument Characteristics and Scientific Objectives. *IEEE Trans. Geosc. Rem. Sens.*, 1994, **32**, 598–615.
21. Feingold G. Modeling of the first indirect effect: Analysis of measurement requirements. *Geophys. Res. Lett.*, 2003, **30**(N19), ASC7/1–ASC7/4.
22. Feingold G. and Kreidenweis S. M. Cloud processing of aerosol as modulated by a large eddy simulation with coupled microphysics and aqueous chemistry. *J. Geophys. Res.*, 2002, **107**(ND23), AAC6/1–AAC6/15.
23. Polar aerosol, extended cloudiness, and radiation. In: K. Ya. Kondratyev and V. I. Binenko (eds), *First Global GARP Experiment* (Volume 2). Gidrometeoizdat, Leningrad, 1981, pp. 89–91 [in Russian].
24. Fridlind A. M., Ackerman A. S., Jensen E. S., Heymsfield A. J., Poellot M. R., Stevens D. E., Wang D., Miloshevich L. M., Baumgardner D., Lawson R. P., *et al.* Evidence for the predominance of mid-tropospheric aerosols as subtropical anvil cloud nuclei. *Science*, 2004, **304**(N5671), 718–722.
25. Gao R. S., Popp D. J., Fahey D. W., Marey T. P., Herman R. L., Weinstock E. M., Baumgardner D. G., Garrett T. J., Rosenlof K. H., Thompson T. L., *et al.* Evidence that nitric acid increases relative humidity in low-temperature cirrus clouds. *Science*, 2004, **303**(N5657), 516–520.
26. Gelenczér A., Hoffer A., Krivácsy Z., Kiss G., Molnár A., and Mészáros E. On the possible origin of humic matter in fine continental aerosol. *J. Geophys. Res.*, 2002, **107**(ND21), ICC2/1–ICC2/6.

27. Glantz P., Noone K. J., and Osborne S. R. Scavenging efficiencies of aerosol particles in marine stratocumulus and cumulus clouds. *Quart. J. Roy. Meteorol. Soc.*, 2003, **129**(Part A, N590), 1329–1350.
28. Graf H.-F. The complex interaction of aerosols and clouds. *Science*, 2004, **303**, 1309–1311.
29. Grassl H. Albedo reduction and radiative heating of clouds by absorption aerosol particles. *Beitr. Phys. Atmos.*, 1975, **48**, 199–209.
30. Grishechkin V. S. and Melnikova I. N. A study of radiation flux divergences in stratus clouds in the Arctic regions. In: K. Ya. Kondratyev (ed.), *Rational Use of Natural Resources*. Leningrad Polytechn. Inst., Leningrad, 1989, pp. 60–67 [in Russian].
31. Guibert S., Snider J. R., and Brenguier J.-L. Aerosol activation in marine stratocumulus clouds. 1. Measurement validation for a closure study. *J. Geophys. Res.*, 2003, **108**(ND15), CMP2/1–CMP2/17.
32. Hanson D. R. and Eisele F. L. Measurement of prenucleation molecular clusters in the NH_3, H_2SO_4, H_2O system. *J. Geophys. Res.*, 2002, **107**(ND12), AAC10/1–AAC10/18.
33. Harries J. E. The greenhouse Earth: A view from space. *Quart. J. Roy. Meteor. Soc.*, **122**(April 1996, Part B, No. 532), 799–818.
34. Hegg D. A., Majled R., Yuen P. F., Baker M. B., and Larson T. V. The impacts of SO_2 oxidation in cloud drops and in haze particles on aerosol light scattering and CCN activity. *Geophys. Res. Lett.*, 1996, **101**(N19), 2613–2616.
35. Hegg D. A. Impact of gas-phase HNO_3 and NH_3 on microphysical processes in atmospheric clouds. *Geophys. Res. Lett.*, 2000, **27**(N15), 2201–2204.
36. Herrmann H., Wolke R., and the FEBUKO/MODMEP-team. A coupled field and modelling study on aerosol-cloud interaction – Field Investigations of Budgets and Conversions of Particle Phase Organics in Tropospheric Cloud Processes (FEBUKO) and – Modelling of Tropospheric Multiphase Processes: Tools and Chemical Mechanisms (MODMEP). *AFO 2000 Newsletter*, 2004, **N6**, 3–10.
37. Hunten D. M. An iron deficiency in polar mesospheric clouds. *Science*, 2004, **304**(N5669), 395–396.
38. Hynes R. G., Fernandez M. Z., and Cox R. A. Uptake of HNO_3 on water–ice and coadsorption of HNO_3 and HCl in the temperature range 210–235 K. *J. Geophys. Res.*, 2003, **108**(ND24), AAC18/1–AAC18/11.
39. *IPCC Third Assessment Report. Vol. 1. Climate Change 2001. The Scientific Basis.* Cambridge University Press, Cambridge, UK, 2001, 881 pp.
40. Iversen T. and Seland Ø. A scheme for process-tagged SO_4 and BC aerosols in NCAR CCM 3: Validation and sensitivity to cloud processes. *J. Geophys. Res.*, 2002, **107**(ND24), AAC4/1–AAC4/30.
41. Jacobson M. Z. Development of mixed-phase clouds from multiple aerosol size distributions and the effect of the clouds on aerosol removal. *J. Geophys. Res.*, 2003, **108**(ND8), AAC4/1–AAC4/23.
42. Jacobson M. Z. Analysis of aerosol interactions with numerical techniques for solving coagulation, nucleation, condensation, dissolution, and reversible chemistry among multiple size distributions. *J. Geophys. Res.*, 2002, **107**(ND19), AAC2/1–AAC2/23.
43. Jordan C. E., Dibb J. E., Anderson B. E., and Fuelberg H. E. Uptake of nitrate and sulfate on dust aerosols during TRACE-P. *J. Geophys. Res.*, 2003, **108**(ND21), GTE38/1–GTE38/10.
44. Jourdain B., Legrand M. Year-round records of bulk and size-segregated aerosol composition and HCl and HNO_3 levels in the Dumont d'Urville (coastal Antarctica)

atmosphere: Implications for sea-salt aerosol fractionation in the winter and summer. *J. Geophys. Res.*, 2002, **107**(ND22), ACH20/1–ACH20/13.

45. Kalberer M., Paulsen D., Sax M., Steinbacher M., Dommen J., Prevot A. S. H., Fisseha R., Weingartner E., Frankevich V., Zenobi R., *et al.* Identification of polymers as major components of atmospheric organic aerosols. *Science*, 2004, **303**(N5664), 1659–1662.
46. Kärcher B. and Lohmann U. A parameterization of cirrus cloud formation: Homogeneous freezing including effects of aerosol size. *J. Geophys. Res.*, 2003, **108**(ND23), AAC9/1–AAC9/10.
47. Kerkweg A., Wurzler S., Reisin T., and Bott A. On the cloud processing of aerosol particles: An entraining air-parcel model with two-dimensional spectral cloud microphysics and a new formulation of the collection kernel. *Quart. J. Roy. Meteorol. Soc.*, 2003, **129**(3), 567–569.
48. Kim B.-G., Schwartz S. E., Miller M. A., and Min Q. Effective radius of cloud droplets by ground-based remote sensing: Relationship to aerosol. *J. Geophys. Res.*, 2003, **108**(ND23), AAC8/1–AAC8/20.
49. Koch D., Park J., and Del Genio A. Clouds and sulfate are anticorrelated: A new diagnostic for global sulfur models. *J. Geophys. Res.*, 2003, **108**(ND224), AAC10/1–AAC10/15.
50. Kondratyev K. Ya., Binenko V.I., and Melnikova I. N. *Absorption of solar radiation by clouds and aerosols in the visible wavelength region at different geographic zones.* CAS/WMO working group on numerical experimentation, WMO, 1996, 6 pp.
51. Kondratyev K. Ya., Binenko V. I., and Melnikova I. N. Solar radiation absorption by clouds in the visible spectral region. *Meteorology and Hydrology*, 1996, (2), 14–23 [in Russian].
52. Kondratyev K. Ya., Binenko V. I., and Melnikova I. N. Absorption of solar radiation by clouds and aerosols in the visible wavelength region. *Meteorology and Atmospheric Physics*, 1997, **68**(1–2), 1–10.
53. Kondratyev K. Ya. *Multidimensional Global Change.* Wiley–Praxis, Chichester, UK, 1998, 761 pp.
54. Kondratyev K. Ya., Grigoryev Al. A., and Varotsos C. A. *Environmental Disasters: Anthropogenic and Natural.* Springer–Praxis, Chichester, UK, 2002, 484 pp.
55. Kondratyev K. Ya. Global changes of climate: Observational data and numerical modelling results. *Study of the Earth from Space*, 2004, (2), 61–96 [in Russian].
56. Koren I., Kaufman Y. J., Remer L. A., and Martins J. V. Measurement of the effect of Amazon smoke on inhibition of cloud formation. *Science*, 2004, **303**(N5662), 1342–1345.
57. Kreidenweis S. M., Walcek C. J., Feingold G., Gong W., Jacobson M. Z., Kim C.-H., Liu X., and Penner J. E. Modification of aerosol mass and size distribution due to aqueous-phase SO_2 oxidation in clouds: Comparisons of several models. *J. Geophys. Res.*, 2003, **108**(ND7), AAC4/1–AAC4/12.
58. Krueger B. J., Grassian V. H., Laskin A., and Cowin J. P. The transformation of soil atmospheric particles into liquid droplets through heterogeneous chemistry: Laboratory insights into the processing of calcium containing mineral dust aerosol in the troposphere. *Geophys. Res. Lett.*, 2003, **30**(N8), 48/1–48/4.
59. Kurosu T., Rozanov V. V., and Burrows J. P. Parameterization schemes for terrestrial water clouds in the radiative transfer model GOMETRAN. *J. Geoph. Res.*, 1997, **102**(D18), 21809–21823.

60. Laskin A., Gasper D. J., Wang W., Hunt S. W., Cowin J. P., Colson S. D., and Finloyson-Pitts B. J. Reactions at interfaces as a source of sulfate formation in sea salt particles. *Science*, 2003, **301**, 340–344.
61. Leck C., Norman M., Bigg E. K., and Hillamo R. Chemical composition and sources of the high Arctic aerosol relevant to cloud formation. *J. Geophys. Res.*, 2002, **107**(ND12), AAC1/1–AAC1–17.
62. Lefer B. L., Shetter R. E., Hall S. R., Crawford J. H., and Olson J. R. Impact of clouds and aerosols on photolysis frequencies and photochemistry during TRACE-P. 1. Analysis using radiative transfer and photochemical box models. *J. Geophys. Res.*, 2003, **108**(ND21), GTE42/1–GTE42/12.
63. Levin Z., Teller A., Ganor E., Graham B., Andreae M. O., Maenhaut W., Falkovich A. H., and Rudich Y. Role of aerosol size and composition in nucleation scavenging within clouds in a shallow cold front. *J. Geophys. Res.*, 2003, **108**(D22), AAC5/1–AAC5/14.
64. Liao H., Adams P. J., Chung S. H., Seinfeld J. H., Mickley L. J., and Jacob D. J. Interactions between tropospheric chemistry and aerosols in a united general circulation model. *J. Geophys. Res.*, 2003, **108**(ND1), 1/1–1/23.
65. Lohmann U., Feichter J., Penner J., and Leatich R. Indirect effect of sulfate and carbonaceous aerosols: A mechanistic treatment. *J. Geophys. Res.*, 2000, **105**, 12193–12206.
66. Lohmann U., Zhang J., Pi J. Sensitivity studies of the effect of increased aerosol concentrations and snow crystal shape on the snow fall rate in the Arctic. *J. Geophys. Res.*, 2003, **108**(ND11), AAC6/1–AAC6/17.
67. Lubin D., Chen J-P., Pilewskie P., Ramanathan V., and Valero P. J. Microphysical examination of excess cloud absorption in the tropical atmosphere. *J. Geophys. Res.*, 1996, **101**(D12), 16961–16972.
68. Luo B. P., Voigt C., Fueglistaler S., and Peter T. Extreme NAT supersaturations in mountain wave ice PSCs: A clue to NAT formation. *J. Geophys. Res.*, 2003, **108**(ND15), 4/1–4/10.
69. Ma J., Tang J., Li S.-M., and Jacobson M. Z. Size distributions of ionic aerosols measured at Waliguan Observatory. 1. Implication for nitrate gas-to-particle transfer processes in the free troposphere. *J. Geophys. Res.*, 2003, **108**(ND17), ACH8/1–ACH8/12.
70. Ma Y., Weber R. J., Lee Y.-N., Orsini D. A., Maxwell-Meier K., Thornton D. C., Bandy A. R., Clarke D. R., Sachse G. W., Fuelberg H. E., *et al.* Characteristics and influence of biosmoke on the fine-particle ionic composition measured in Asian outflow during the Transport and Chemical Evolution over the Pacific (TRACE-P) experiment. *J. Geophys. Res.*, 2003, **108**(ND21), GTE37/1–GTE37/16.
71. Maenhaut W., Schwarz J., Cafmeyer J., and Chi X. Aerosol chemical mass closure during the EUROTRAC-2 AEROSOL Intercomparison 2000. *Nucl. Instum. and Math. Phys. Res. B.*, 2002, **189**, 2.
72. Manankova A. V. A stochastic description of the kinetics of new phase formation. In: K. Ya. Kondratyev (ed.), *Problems of Atmospheric Physics. Physics and Chemistry of Atmospheric Aerosols.* St Petersburg State Univ. Press, St Petersburg, 1997, **20**, pp. 25–53 [in Russian].
73. Marchuk G. I., Kondratyev K. Ya., Kozoderov V. V., and Khvorostyanov V. I. *Clouds and Climate.* Gidrometeoizdat, Leningrad, 1986, 512 pp. [in Russian].
74. Marchuk G. I., Kondratyev K. Ya., and Kozoderov V. V. *The Earth's Radiation Budget: Key aspects.* Nauka Publ., Moscow, 1988, 224 pp. [in Russian].

75. Marchuk G. I. and Kondratyev K. Ya. *Priorities of Global Ecology*. Nauka Publ., Moscow, 1992, 264 pp. [in Russian].
76. Marshak A., Davis A., Wiscombe W., and Cahalan R. Radiative smoothing in fractal clouds. *J. Geophys. Res.*, 1995, **100**(D18), 26247–26261.
77. Mason B. *Physics of Clouds*. Gidrometeoizdat, Leningrad, 1961, 542 pp. [in Russian].
78. Mason B. J. Some outstanding problems in cloud physics. *Quart. J. Roy. Meteor. Soc.*, 1969, **95**(405), 449–485.
79. Matveev L. T. *Fundamentals of General Meteorology: Physics of the Atmosphere*. Gidrometeoizdat, Leningrad, 1965, 876 pp. [in Russian].
80. Matveev L. T., Matveev Yu. L., and Soldatenko S. A. *The Global Field of Cloudiness*. Leningrad, Gidrometeoizdat, 1986, 287 pp. [in Russian].
81. Matveev Yu. L. A physico-statistical analysis of the global field of cloudiness. *Proc. USSR Acad. Sci., ser. FAO*, 1984, **20**(11), 1205–1218 [in Russian].
82. Mavromatidis E. and Kallos G. An investigation of cold cloud formation with a three-dimensional model with explicit microphysics. *J. Geophys. Res.*, 2003, **108**(ND14), AAC14/1–AAC14/21.
83. Mayer, B., Kylling A., Madronich S., and Seckmeyer G. Enchanced absorption of UV radiation due to multiple scattering in clouds: Experimental evidence and theoretical explanation. *J. of Geophys. Res.*, 1998, **103**(D23), 31241–31254.
84. Mazin I. P. and Shmeter S. M. *Clouds, Composition and Physics of Formation*. Gidrometeoizdat, Leningrad, 1983, 279 pp. [in Russian].
85. Mazin I. P. and Khrgian A. X. (eds). *Clouds and the Cloudy Atmosphere. Reference Book*. Gidrometeoizdat, Leningrad, 1989, 648 pp. [in Russian].
86. Meier A. and Hendricks J. Model studies on the sensitivity of upper tropospheric chemistry to heterogeneous uptake of HNO_3 on cirrus ice particles. *J. Geophys. Res.*, 2003, **108**(ND23), ACH9/1–ACH9/16.
87. Melnikova I. N. Light absorption in cloud layers. *Problems of Atmospheric Physics*, **19**, 1989, 18–25 [in Russian].
88. Melnikova I. N. and Mikhailov V. V. The optical parameters in strati on basis of aircraft spectral measurements. Thesis of International Radiation Symposium, Tallinn, August, 1992.
89. Melnikova I. N., Domnin P. I., Varotsos C., and Pivovarov S. S. Retrieval of optical properties of cloud layers from transmitted solar radiance data. *Proceeding of the International Society for Optical Engineering, v. 3237, 23rd European Meeting on Atmospheric Studies by Optical Methods. September 1996, Kiev, Ukraine*. 1997, pp. 77–80.
90. Melnikova I. N. and Mikhailov V. V. Vertical profile of spectral optical parameters of stratus clouds from airborne radiative measurements. *J. Geophys. Res.*, 2001, **106**(D21), 27465–27471.
91. Melnikova I. N. and Nakajima T. Single scattering albedo and optical thickness of strati obtained from measurements of reflected solar radiation by the POLDER instrument. *Studies of the Earth from Space*, 2000, (3), 1–6 [in Russian].
92. Melnikova I. N. and Nakajima T. The spatial distribution of the cloud optical parameters obtained from measurements of reflected solar radiation with the POLDER instrument from the satellite ADEOS. In: L. S. Ivlev (ed.), *Natural and Anthropogenic Aerosols*. St Petersburg State Univ. Press, St Petersburg, 2000, pp. 78–85 [in Russian].
93. Monin A. S. *Introduction to Climate Theory*. Leningrad, Gidrometeoizdar, 1982, 247 pp. [in Russian].

94. Nakajima T. Y. and Nakajima T. Wide-area determination of cloud microphysical properties from NOAA AVHRR Measurements for FIRE and ASTEX regions. *J. Atm. Sci.*, 1995, **52**, 4043–4059.
95. Nakajima, T., King M. D., Spinhirne J. D., and Radke L. F. Determination of the optical thickness and effective particle radius of clouds from reflected solar radiation measurements. II. Marine stratocumulus observations. *J. Atmos. Sci.*, 1991, **48**, 728–750.
96. O'Dowd C. D., Geever M., Hill M. K., Smith M. H., and Jennings S. G. New particle formation: Nucleation rates and spatial scales in the clean marine coastal environment. *Geophys. Res. Lett.*, 1998, **25**(N10), 1661–1664.
97. Petrov I. V. Study of the processes of formation and transformation of sulphuric acid aerosols. In: K. Ya. Kondratyev (ed.), *Problems of Atmospheric Physics. Physics and Chemistry of Atmospheric Aerosols*. St Petersburg State Univ. Press, St Petersburg, 1997, pp. 5–24 [in Russian].
98. Pfeilsticker, K., Erle F., Funk O., Marquard L., Wagner T., and Platt U. Optical path modification due to tropospheric clouds: Implications for zenith sky measurements of stratospheric gases. *J. Geoph. Res.*, 1998, **103**(D19), 25323–25335.
99. Pfeilsticker, K. First geometrical path length probability density function derivation of the skylight from high-resolution oxygen A-band spectroscopy. 2. Derivation of the Levy index for the skylight transmitted by midlatitude clouds. *J. Geoph. Res.*, 1999, **104**(D43), 4101–4116.
100. Pirjola L., O'Dowd C. D., and Külmälä M. A model prediction of the yield of cloud condensation nuclei from coastal nucleation events. *J. Geophys. Res.*, 2002, **107**(ND19), PAR3/1–PAR3/15.
101. Plane J. M. C., Murray B. J., Chu X., and Gardner C. S. Removal of meteoric iron on polar mesospheric clouds. *Science*, 2004, **304**(N5669), 426–428.
102. Poetzsch-Heffter C., Liu Q., Ruprecht E., and Simmer C. Effect of cloud types on the Earth radiation budget calculationwith the ISCCP C1 dataset: Methodology and initial results. *J. Climate*, 1995, **8**, 829–843. 899.
103. Pradelle F., Cautenet G., and Jankowiak I. Radiative and microphysical interactions between marine stratocumulus clouds and Saharan dust. 1. Remote sensing observations. *J. Geophys. Res.*, 2002, **107**(ND19), AAC15/1–AAC15/12.
104. Pradelle F. and Cautenet G. Radiative and microphysical interactions between marine stratocumulus clouds and Saharan dust. 2. Modeling. *J. Geophys. Res.*, 2002, **107**(ND19), AAC16/1–AAC16/15.
105. Feigelson E. M. (ed.). *Radiative Properties of Cirrus Clouds*. Nauka Publ., Moscow, 1989, 224 pp. [in Rusian].
106. Raymond T. M. and Pandis S. N. Formation of cloud drops by multicomponent organic particles. *J. Geophys. Res.*, 2003, **108**(ND15), AAC10/1–AAC10/8.
107. Reus M., Strom J., Curtius J., Pirjola L., Vignati E., Arnold F., Hansson H. C., Kulmula M., Leiieveld J., and Raes F. Aerosol production and growth in the upper free troposphere. *J. Geophys. Res.*, 2000, **105**(ND20), 24751–24762.
108. Rissman T. A., Nenes A., and Seinfeld J. H. Chemical amplification (or dampening) of the Twomey effect: Conditions derived from droplet activation theory. *J. Atmos. Sci.*, 2004, **61**(8), 919–930.
109. Roberts G. C., Neues A., Seinfeld J. H., and Andreae M. O. Impact of biomass burning on cloud properties in the Amazon Basin. *J. Geophys. Res.*, 2003, **108**(ND2), AAC9/1–AAC9/19.
110. Rodriguez M. A. and Dabdub D. Monte Carlo uncertainty and sensitivity analysis of the CACM chemical mechanism. *J. Geophys. Res.*, 2003, **108**(ND15), 2/1–2/9.

111. Romanova L. M. Spatial variations of radiative characteristics of the horizontally-heterogeneous clouds. *Proc. RAS, ser. FAO*, 1992, **28**(3), 268–276 [in Russian].
112. Rublev A. N., Trotsenko A. N., and Romanov P. Yu. The use of the data of satellite radiometers AVHRR to determine the optical thickness of clouds. *Proc. RAS, ser. FAO*, 1997, **33**(5), 670–675 [in Russian].
113. Rudich Y., Sagi A., and Rosenfeld D. Influence of the Kuwait oil fires plume (1991) on the microphysical development of clouds. *J. Geophys. Res.*, 2003, **108**(ND15), AAC14/1–AAC14/8.
114. Sekiguchi M., Nakajima T., Suzuki K., Kawamoto K., Higurashi A., Rosenfeld D., Sano I., and Mukai S. A study of the direct and indirect effects of aerosol using global satellite data sets of aerosol and cloud parameters. *J. Geophys. Res.*, 2003, **108**(ND22), AAC4/1–AAC4/15.
115. Shchekin A. K., Rusanov A. I., and Kuni F. M. Thermodynamics of condensation in formation of the film on a soluble nucleus. *Colloid. J.*, 1993, **55**(5), 185–193 [in Russian].
116. Shchekin A. K., Yakovenko T. M., and Kuni F. M. Determination of stationary rate of nucleation on soluble aerosols containing surface-active substances. In: L. S. Ivlev (ed.), *Natural and Anthropogenic Aerosols*. St Petersburg State Univ. Press, St Petersburg, 2000, pp. 20–28 [in Russian].
117. Shchekin A. K., Grinin A. P., and Kuni F. M. Condensation of nuclei of mixed composition containing soluble and insoluble component. In: L. S. Ivlev (ed.), *Natural and Anthropogenic Aerosols*. St Petersburg State Univ. Press, St Petersburg, 2003, pp. 217–218 [in Russian].
118. Smirnov V. I. Cloud and precipitation droplets size distribution. In: K. Ya. Kondratyev (ed.), *Progress in Sci. and Technol., Ser. Meteorology and Climatology* (Volume 15). VINITI Publ., Moscow, 1987, 195 pp [in Russian].
119. Stephens G. How much solar radiation do clouds absorb? Technical comments. *Science*, 1996, **271**, 1131–1133.
120. Suzuki K., Nakajima T., Numaguti A., Takemura T., Kawamoto K., and Higurashi A. A study of the aerosol effect on a cloud field with simultaneous use of GCM modeling and satellite observations. *J. Atmos. Sci.*, 2004, **61**(2), 179–194.
121. Tatyanenko D. V., Shchekin A. K., and Kuni F. M. On conditions for the spreading coefficient and the nucleus size in the theory of nucleation on wettable insoluble nuclei. *Colloid. J.*, 2000, **62**(4), 536–545 [in Russian].
122. Tegen I., Harrison S. P., Kohfeld K., Prentice I. C., Coe M., and Heimann M. Impact of vegetation and preferential source areas on global dust aerosol: Results from a model study. *J. Geophys. Res.*, 2003, **108**(ND21), AAC14/1–AAC14/27.
123. Thomas A., Borrmann S., Kiemle C., Cairo F., Volk M., Buermann J., Lepuchov B., Santacesaria V., Matthey R., Rudakov V., *et al.* In situ measurements of background aerosol and subvisible cirrus in the tropical tropopause region. *J. Geophys. Res.*, 2002, **107**(ND24), AAC8/1–AAC8/14.
124. Titov G. A. and Zhuravleva T. B. The spectral and total absorption of solar radiation in the broken cloudiness. *Optics of the Atmos. and Ocean*, 1995, **8**(10), 1419–1427 [in Russian].
125. Titov G. A. and Zhuravleva T. B. Absorption of solar radiation in broken clouds. *Proceedings of the Fifth ARM Science Team Meeting. San Diego, California, USA. March, 19–23, 1995*, pp. 397–340.
126. Titov G. A. Radiative effects of heterogeneous stratocumulus clouds. 1. The horizontal transport. II. Absorption. *Optics of the Atmos. and Ocean*, 1996, **9**(10), 1295–1307 [in Russian].

127. Trockine D., Iwasaka Y., Matsuki A., Yamada M., Kim Y.-S., Nagatani T., Zhang D., Shi G.-Y., and Shen Z. Mineral aerosol particles collected in Dunhuang, China, and their comparison with chemically modified particles collected over Japan. *J. Geophys. Res.*, 2003, **108**(ND23), ACE10/1–ACE10/11.
128. Twohy C. H., Durkee P. A., Huebert B. J., and Charlson R. J. Effects of aerosol particles on the microphysics of coastal stratiform clouds. *J. Climate*, 1995, **8**, 773–783.
129. Twomey S. Aerosols, clouds and radiation. *Atmos. Environ.*, 1991, **25A**, 2435–2442.
130. Van Reken T. M., Rissman T. A., Roberts G. C., Varutbangkul V., Jonsson H. H., Flagan R. C., and Seinfeld J. H. Toward aerosol/cloud condensation nuclei (CCN) closure during CRYSTAL-FACE. *J. Geophys. Res.*, 2003, **108**(ND20), AAC2/1–AAC 2/18.
131. Vasilyev A. V., Melnikova I. N., and Mikhailov V. V. The vertical profile of spectral scattered solar radiation fluxes in a stratus cloud from results of aircraft measurements. *Physics of the Atmos. and Ocean*, 1994, **30**(5), 661–665 [in Russian].
132. Vasilyev S. L. A method of changing convective clouds properties. *Int. App. J.*, N PCT/GB/02148, 1995.
133. Wagner T., Erle F., Marquard L., Otten C., Pfeilsticker K., Senne T., Stutz J., and Platt U. Cloudy sky optical paths as derived from differential optical absorption spectroscopy observations. *J. Geoph. Res.*, 1998, **103**(D19), 25307–25321.
134. Wang S., Ackermann R., Spicer C. W., Fast J. D., Schmeling M., and Stitz J. Atmospheric observations of enhanced NO_2 – HONO conversion on mineral dust particles. *Geophys. Res. Lett.*, 2003, **30**(N11), 49/1–49/4.
135. Wylie D. P. and Hudson J. G., Effects of long-range transport and clouds on cloud condensation nuclei in the springtime Arctic. *J. Geophys. Res.*, 2002, **107**(D16), AAC13/1–AAC13/11.
136. Yin Y., Wurzler S., Levin Z., and Reisin T. G. Interactions of mineral dust particles and clouds: Effects on precipitation and cloud optical properties. *J. Geophys. Res.*, 2002, **107**(ND23), AAC19/1–AAC 19/14.
137. Zhang Y., Easter R. C., Ghan S. J., and Abdul-Razzak H. Impact of aerosol size representation on modeling aerosol-cloud interaction. *J. Geophys. Res.* 2002. **107**(ND21), AAC4/1–AAC4/17.
138. Zuberi B., Bertvam A. K., Cassa C. A., Molina L. T., and Molina R. J. Heterogeneous nucleation of ice in $(NH_4)_2SO_4$—H_2O particles with mineral dust immersions. *Geophys. Res. Lett.*, 2002, **29**(N10), 142/1–142/4.

Part Three

Numerical modelling of the processes and properties of atmospheric aerosol

7

The optical properties of atmospheric aerosol and clouds

7.1 INTRODUCTION

Studies of the optical characteristics of atmospheric aerosols and clouds with the use of *in situ* measurements of radiation attenuation by the atmosphere or from radiation scattering by the atmosphere have provided rich material for statistical modelling [20, 21, 32, 37, 69, 80, 83, 95–97, 106, 116, 120, 124, 130, 145, 151, 153, etc.]. Obtaining data on the spatial and temporal distribution of atmospheric aerosol and clouds with the use of lidar soundings [10, 32, 90, 162–164] has been especially successful. At present, it is clear that the aerosol substance in most of particles is well mixed up, having almost no absorption bands in the visible spectral interval. With an increase of relative air humidity, particles rapidly grow, and the main contribution to the aerosol attenuation of radiation in the short-wave region is made by the so-called submicron particles ($r < 1.0\,\mu$m), which are approximately spherical in shape – their size distribution N can be approximated by an exponential function $dN/dr = A \times r^{-\nu}$. This information is sufficient to interpret the data with respect to simple measurements and to extrapolate their results onto other conditions in order to determine, in particular, the exponent ν and, with due regard to the spectral dependence of coefficients of attenuation and scattering, the dependence on relative humidity and meteorological visibility.

Such a scheme, however, should be corrected when the nature of particles changes: their chemical composition that affects the magnitude and the spectral change of the complex refraction index [59–61], as well as the shape of particles [66] and complicated structure of particles, in particular, for 2-layer particles [33, 63, 146–148]. Of importance is the appearance of particles with size distributions markedly differing from the Junge distribution [6–8]. This correction is introduced with due regard to the nature of air masses (synoptic characteristics, season, powerful sources of aerosol substances) [162]. This considerably broadens the range of the optical aerosol models. However, there is a problem of extrapolation of data onto the IR spectral region [62, 67].

Modelling the optical characteristics of aerosol proceeding from its microphysical characteristics makes it possible to get rid of the drawbacks of statistical empirical modelling, and to better understand the physical nature pattern of transformation of optical properties of aerosol in the atmosphere. This approach has first been realized by the scientists of the Leningrad University – supported by the scientists from Tomsk [6–9, 62–68, 79, 90, 162].

The combined use of the data of optical and microphysical studies with subsequent comparisons of results of calculations with the observational data has been accomplished in the process of the 'closed' modelling of the optical characteristics of atmospheric aerosols of different nature [60–68]. The results of fundamental studies on modelling the optical properties of aerosols of different nature has been published [5]. To solve the problems connected with the consideration of atmospheric aerosol characteristics, verified optical models of the atmosphere have been used [43, 44, 54, 73–77, 133, 157, 162].

Modelling the optical characteristics of cloud formations differs in that the main component in this case is liquid or solid water. The impact of substances dissolved in droplets on the complex refraction index of a droplet is usually negligible. Insoluble inclusions tell mainly on the droplet's absorbing properties. The macrostructure of cloud formations is more unstable due to dynamic processes followed by phase transformations of water. Therefore, the statistical modelling of the spatial distribution and optical properties of cloud formations faces serious difficulties [37, 50, 54, 105–111].

7.2 STATISTICAL MODELLING OF THE OPTICAL CHARACTERISTICS OF ATMOSPHERIC AEROSOL

7.2.1 Variability of the optical characteristics of atmospheric hazes in the surface layer of the atmosphere

Studies of the spectral dependence of the attenuation coefficient $\varepsilon(\lambda)$ and its variability in the surface layer of the atmosphere, especially in the short-wave spectral region, have been intensively carried out from the 1930s with the aim of determining visibility in the lower atmospheric layers [111]. Below are given the statistical characteristics of the variability of spectral dependences of the attenuation coefficients:

$$\varepsilon(\lambda) = \alpha(\lambda) + \sigma(\lambda)$$

where α is the absorption coefficient and σ is the scattering coefficient of atmospheric haze in the transparency windows 0.37–12 μm based on multiyear data of measurements in the region of Zvenigorod (Moscow region) along the horizontal route [95–97, 130].

Present ideas about aerosol make it possible to apply this information to most of the globe, at least to its forest–steppe zone, where the impact of the submicron

Table 7.1. The spectral characteristics of haze.

λ (μm)	$\varepsilon(\lambda)$ (km^{-1})	d_ε	π_1	π_2	λ (μm)	$\varepsilon(\lambda)$ (km^{-1})	d_ε	π_1	π_2
0.37	0.41	0.30	0.51	−0.29	8.85	0.17	0.07	0.10	0.20
0.42	0.37	0.28	0.48	−0.30	8.96	0.18	0.08	0.11	0.21
0.50	0.30	0.23	0.40	−0.24	9.12	0.15	0.07	0.10	0.20
0.63	0.24	0.15	0.26	−0.11	9.38	0.16	0.07	0.10	0.21
−1.03	0.17	0.11	0.18	−0.04	9.61	0.16	0.07	0.09	0.20
1.2	0.15	0.09	0.15	−0.01	10.2	0.13	0.68	0.09	0.20
1.6	0.12	0.07	0.11	0.05	10.53	0.15	0.73	0.10	0.19
2.2	0.10	0.06	0.08	0.03	10.9	0.18	0.80	0.11	0.20
3.16	0.20	0.09	0.14	0.15	11.1	0.17	0.85	0.12	0.21
3.7	0.09	0.06	0.08	0.11	11.6	0.16	0.88	0.12	0.22
3.9	0.10	0.05	0.06	0.12	12.0	0.18	0.90	0.13	0.23
8.33	0.22	0.08	0.11	0.23	12.7	0.22	1.00	0.14	0.25
8.63	0.20	0.07	0.10	0.20	13.05	0.28	1.09	0.15	0.28

aerosol fraction prevails. Outside this zone, as well as in industrial zones, the data mentioned below are supposed to be valid for the submicron fraction of aerosol, but they need an additional consideration of the impact of the coarse fraction of aerosol on its evolution and optical manifestations.

Table 7.1 contains average values of $\varepsilon(\lambda)$, mean square deviations d_ε, and two first eigenvectors $\pi_i(\lambda_k)$ of the correlation matrix $B_{\varepsilon\varepsilon}(\lambda_k, \lambda_i)$. Its eigen-numbers are $\mu_1 = 0.333$ and $\mu_2 = 0.041$, and the trail $SpB_{\varepsilon\varepsilon} = 0.385$:

$$\varepsilon(\lambda) = \langle\varepsilon(\lambda)\rangle + C_1\pi_1(\lambda) + C_2\pi_2(\lambda)$$

the mean standard deviation (MSD) for coefficients C_1 and C_2 are equal, respectively, to μ_1 and μ_2, and the MSD is estimated as:

$$1 - \frac{\mu_1 + \mu_2}{SpB_{e\varepsilon}} = 0.03$$

Although the correlation matrices are unstable, which testifies to the instability of the statistical process, their eigenvectors $\pi_i(\lambda)$, reflecting properties of the factors determining the state of atmospheric aerosols, are conservative.

Since statistical characteristics of $\varepsilon(\lambda)$ obtained in different years agree, they can be considered as objective parameters of the aerosol haze.

A high correlation between variations of $\varepsilon(\lambda)$, in different spectral regions, ensures the reliability of the determination of the total spectral dependence of $\varepsilon(\lambda)$ from the results of a single measurement for some $\lambda = \lambda_0$, namely:

$$\begin{aligned}\langle\varepsilon(\lambda)\rangle &= \Sigma(\lambda)\langle\varepsilon(\lambda_0)\rangle, \varepsilon(\lambda) - \langle\varepsilon(\lambda)\rangle \\ &= \eta(\lambda)[\varepsilon(\lambda) - \langle\varepsilon(\lambda)\rangle] \quad \text{or} \quad \varepsilon(\lambda) = \eta(\lambda)\varepsilon(\lambda_0) + E(\lambda)\end{aligned}$$

Table 7.2. Spectral functions Σ, η, and E.

λ (μm)	$\Sigma(\lambda)$	$\eta(\lambda)$	$E(\lambda)$	λ (μm)	$\Sigma(\lambda)$	$\eta(\lambda)$	$E(\lambda)$	λ (μm)	$\Sigma(\lambda)$	$\eta(\lambda)$	$E(\lambda)$
0.37	1.37	1.26	0.03	3.7	0.3	0.17	0.04	10.2	0.43	0.18	0.07
0.42	1.23	1.2	0.01	3.9	0.33	0.13	0.06	10.53	0.50	0.20	0.09
0.55	1.0	1.0	0.00	8.33	0.73	0.22	0.15	10.9	0.60	0.22	0.11
0.63	0.8	0.63	0.05	8.63	0.66	0.18	0.15	11.1	0.57	0.26	0.09
1.03	0.57	0.42	0.04	8.85	0.57	0.20	0.11	11.6	0.53	0.26	0.08
1.2	0.5	0.37	0.04	8.96	0.6	0.22	0.11	12.0	0.60	0.26	0.10
1.6	0.4	0.26	0.04	9.12	0.5	0.22	0.08	12.7	0.73	0.26	0.14
2.2	0.33	0.18	0.05	9.38	0.53	0.20	0.10	13.05	0.93	0.31	0.19
3.16	0.67	0.31	0.11	9.61	0.47	0.18	0.09				

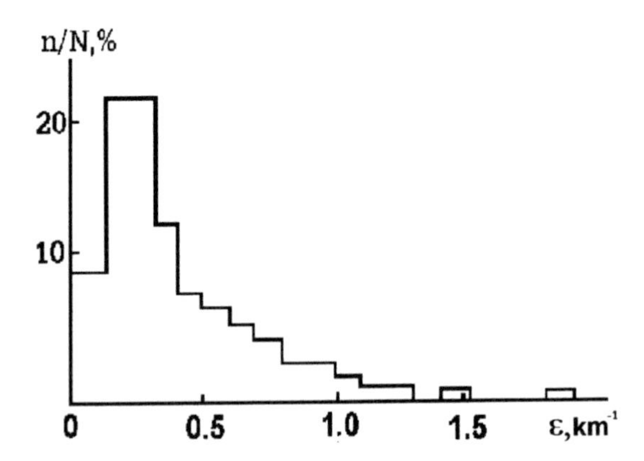

Figure 7.1. Relative frequency of occurrence of different values of ε ($\lambda = 0.5\,\mu$m).

Functions $\Sigma(\lambda)$:

$$\eta(\lambda) = [B_{e\varepsilon}(\lambda, \lambda_0)]/[B_{e\varepsilon}(\lambda_0, \lambda_0)], E(\lambda) = [\Sigma(\lambda) - \eta(\lambda)]\langle\varepsilon(\lambda)\rangle$$

are given in Table 7.2 for $\lambda_0 = 0.55\,\mu$m. The MSD for $\varepsilon(\lambda)$ does not exceed 30% of the MSD in the interval $\lambda \leq 2.2\,\mu$m.

Figure 7.1 is a histogramme of the probability of occurrence of different values of ε ($\lambda_0 = 0.55\,\mu$m) for conditions at Zvenigorod.

With the growth of relative air humidity, the attenuation coefficient in the short-wave spectral region ($\lambda \leq 2\,\mu$m) increases due to the condensation growth of particles. In the long-wave region ($\lambda \geq 2\,\mu$m) this phenomenon is practically not observed. A concrete type of dependence of $\varepsilon(r)$ is determined by the chemical nature of particles and therefore it is ambiguous [130]. However, under stable weather the effect is characterized by values tabulated in Table 7.3, where average, maximum and minimum values of ε are given for $\lambda = \lambda_0$ at different relative

Table 7.3. The impact of relative humidity on the attenuation coefficient ε.

RH (%)	ε (km^{-1})	ε_{max}	ε_{min}	f (%)	ε (km^{-1})	ε_{max}	ε_{min}
40–50	0.10	0.35	0.01	70–80	0.25	0.80	0.03
50–60	0.12	0.35	0.02	80–90	0.45	1.0	0.03
60–70	0.15	0.50	0.02	90–99	0.60	4 and more	0.03

Table 7.4. Seasonal variations in P and τ.

Multiyear		Winter		Spring		Summer		Autumn	
P	τ	P	τ	P	τ	P	τ	P	τ
0.533	0.629	0.813	0.207	0.563	0.629	0.718	0.331	0.617	0.482
0.815	0.204	0.856	1.55	0.780	0.248	0.799	0.224	0.821	0.197
0.883	0.124	0.879	0.123	0.859	0.151	0.871	0.138	0.883	0.182

humidities. For rough estimates we can assume:

$$\varepsilon(r) = 2r^2(1 + (RH)^2\varepsilon(RH)70\%)$$

where RH is relative humidity, with a MSD up to 100%.

For the whole atmosphere, Table 7.4 contains information about maximum, average, and minimum values of $\tau^*(\lambda = \lambda_0)$ and $P = e^{-\tau*}$ on cloud-free days from the data for the Alma-Ata region.

7.2.2 The spectral dependences of aerosol optical characteristics

An extensive observational database on the spectral change of radiation attenuation through the whole atmosphere and along the air routes in the surface layer obtained in different climatic regions and in different seasons has made it possible to construct many optical aerosol models both in the visible and in the IR spectral regions. In the latter case, it was done with the use of data from microphysical measurements [5, 6, 42, 43, 75–79, 89, 90, 113, 162]. Certain dependences have been established of the spectral change of coefficients on relative air humidity, time of day, synoptic processes, and air masses origin [32, 58, 113, 120, 124]. The main problem in the interpretation of these data is to separate the aerosol component of radiation absorption [8, 9, 75–79].

Using a 3-D macrostructure of the surface aerosol and a special technique of optical data conversion, a statistical analysis has been performed [124] of the seasonal and diurnal variability of the size distribution of surface haze for a large observational database. It was found that the spectrum of particle size distribution for the accumulation fraction reveals a direct boundary on the side of large r, and at the same time, the spectrum in the middle range of sizes ($r < 0.8\,\mu$m) is filled sporadically and exhibits dynamic changes different from its adjacent range ($r > 0.8\,\mu m$), most often coinciding with the activity of sources of thermal sublimation of aerosols (forest fires and peat bogs decay). The dynamics of changes in the size range $0.8 < r < 2.0\,\mu m$ are similar to those for coarse particles, which coincides with enhanced turbulent mixing. The correlated changes of the size spectrum for the surface haze over short time intervals testify to the important role of persistent (diurnal) variations of meteorological parameters in changes of the optical and microphysical characteristics of aerosol along the horizontal routes near the surface.

7.2.2.1 *Specific absorption in the visible spectral region*

The data on specific absorption $\beta = \alpha/\sigma$ of aerosol under natural conditions from *in situ* optical measurements are fragmentary, poorly systematized, and refer mainly to the visible spectral region. Using the data in Table 7.1 for haze, that is, in fact, for the submicron fraction of aerosol, and assuming $\beta \approx 0.1$–0.2 for $\lambda_0 = 0.55\,\mu m$, we obtain the mean coefficient of absorption $\langle\alpha\rangle = 5 \times 10^{-2}\,km^{-1}$.

7.2.2.2 *The spectral change of $\alpha(\lambda)$ in the short-wave spectral region*

For the extrapolation of β measurements onto the whole short-wave spectral region one can use measurement data for the spectral change of $\alpha(\lambda)$ for aerosol samples. Since measurements are made with the use of samples of deposited (i.e., transformed) aerosol, they provide reliable information only on the relative spectral change of $\alpha(\lambda)$, but not on absolute values of this parameter.

According to observational data obtained [38, 39], the light absorption due to a totality of all the fractions of aerosol in the interval 0.4–2.4 μm turned out to be mainly 'grey' – the spectral dependence of $\alpha(\lambda)$ is expressed rather weakly and is irregular. According to the data for the interval 0.4–1.0 μm, the absorption spectra of the submicron fraction particles deposited on filters in various geographical regions are of the same character. The mineral particles exhibit similar absorption spectra, but in a much weaker form. A totality of observational data suggests the conclusion that in this case, absorption is determined by the sticking to the surface of some amount of absorbing particles (of the submicron fraction) on non-absorbing large particles. In other words, the aerosol absorption in the short-wave spectral region can be attributed (at least, partly) to the submicron fraction, especially to its small-size subfraction, including soot and organic components.

Due to a neutral spectral change of $\alpha(\lambda)$, the spectral dependence of $\beta(\lambda)$ will be:

$$\beta(\lambda) = \beta(\lambda_0)\sigma(\lambda_0)/\sigma(\lambda)$$

and since in the short-wave region, according to Table 7.1, we have approximately $\sigma(\lambda_0)/\sigma(\lambda) = (\lambda/\lambda_0)^{\xi}$, where $\xi \approx 0.8$, then:

$$\beta(\lambda) \cong \beta(\lambda_0)(\lambda/\lambda_0)^{0.8}$$

Thus β rapidly grows with λ increasing at $\lambda \leq 1$–$1.5\,\mu m$ $\beta \ll 1$ (i.e., scattering prevails), whereas at $\lambda \geq 2.0$–$2.5\,\mu m$ $\beta \geq 1$ absorption prevails, which is manifested through a delayed decrease of ε with increasing λ (Table 7.2). According to the data

[38, 39], the value of α is practically independent of relative humidity, and as a result, an increase of σ with increasing humidity determines a decrease of β.

There is a high correlation between variations in $\varepsilon(\lambda_0)$ and hence, in $\sigma(\lambda_0)$, and variations in the concentration of submicron particles, however, there is practically no correlation between variations in $\varepsilon(\lambda_0)$ and $\sigma(\lambda_0)$, on the one hand, and the concentration of dust particles, on the other hand [130].

7.2.2.3 *The aerosol absorption transparency windows of the longwave range*

Figure 7.2 shows typical IR spectra of absorption for natural aerosols deposited on filters with the divisions of aerosols into fractions.

The coarse ($r \geq 3\,\mu m$) and submicron ($r \sim 0.2$–$0.4\,\mu m$) fractions have distinctly differing absorption bands which reflect the difference in their chemical nature. At present, the absorption spectra for individual fractions and their numerous components as well as the laws defining their variations remain almost unstudied. Therefore, only qualitative judgements are possible.

The absorption spectra for the coarse fraction practically coincide with those for mineral dust and are characterized by a broad absorption band centred at about $10\,\mu m$. The absorption spectra for the submicron fraction have a much more complicated structure and are the result of superposition of, at least, four components – soot, SO_4 groups (bands centered at 2.8 and $8.9\,\mu m$), NH_4 groups (bands centered at 6.2 and $9.6\,\mu m$), and organic compounds (bands centered at 3.2 and $7.1\,\mu m$ and a rapidly growing absorption with decreasing wavelength, starting from $6\,\mu m$) [95].

Vast territories of mid-latitudes are characterized by the relatively weakly developing dust fraction and, hence, its contribution to the aerosol absorption of radiation in the long-wave interval is of secondary importance. An exception

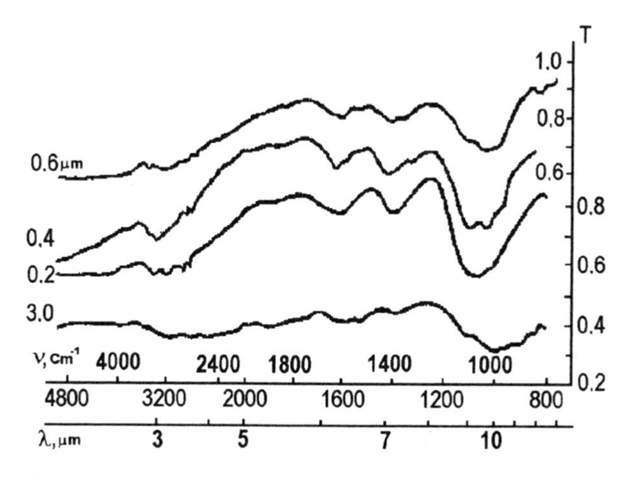

Figure 7.2. IR absorption spectra for different samples of natural aerosol (Lithuania, 1976). [130].

applies for fogs and clouds, where large droplets condense on dust particles, playing the determining role in the IR region.

An analysis of the extensive and diverse observational data has shown that in the long-wave transparency windows ($4 < \lambda < 13\,\mu m$) the light attenuation by aerosol takes place exclusively due to its absorption by submicron particles. Their scattering ability in this range is negligibly small ($\beta \gg 1$) (i.e., $\varepsilon \cong \alpha$). There is a high correlation between the concentration of these particles and the magnitude of absorption (as well as emissivity) of the air in the transparency windows 8–12 μm [95, 130].

Since in the range $\lambda = 4$–$13\,\mu m$, for particles of the submicron fraction the dimensionless radius $\rho = 2\pi a/\lambda < 1$, their absorption is described by the Rayleigh approximation:

$$\alpha \Big| \cong \frac{2n\chi\lambda^2\langle\rho\rangle^3}{(n^2+2)\pi} N = \frac{16\pi n\chi}{(n^2+2)\lambda} V$$

where V is the factor of filling. In the surface layer, under very clean natural conditions, $V = 10^{-11} - 10^{-12}$, but under very turbid conditions, $V \cong 10^{-9}$ – giving on average, $V \cong 10^{-11}$ [95].

In the transparency windows, even under very clean conditions without aerosol, the contribution of aerosol absorption is comparable with, or exceeds, the contribution of molecular absorption (by water vapour line wings). In other words, in the transparency windows 8–13 μm, the main factor regulating the light attenuation is its real absorption by aerosol particles of the submicron fraction and, primarily, its organic, sulphuric acid and nitrogen–hydrogen-containing components (Figure 7.2). Averages and MSDs of $\varepsilon(\lambda)$ in the 8–12-μm transparency window should be considered as an estimate of the upper limit for the average and MSD of the $\alpha(\lambda)$ parameter under the assumption of the absence of absorption by water vapour.

7.2.2.4 *Modelling structural aerosol characteristics*

In a first approximation such optical characteristics as $\log D_{11}$ and f_{ik} (D_{11} is the coefficient of directed light scattering, f_{ik} is the element of the matrix of light scattering by the atmosphere) depend linearly on $\log \varepsilon$, its value being the main parameter which determines the state of aerosol.

The conversion of totality of angular dependences of $\log D_{11}$ and f_{ik} corresponding to different values of $\log \varepsilon$ (i.e., different horizontal visibilities) has led to the determination of characteristic parameters of real aerosols with different ε values [120, 130].

In a first approximation, observed totalities of the optical parameters of atmospheric aerosols in the whole range of their change in the visible spectral region are satisfactorily described as a result of the coexistence of two fractions: (a) the submicron fraction with a small, practically imperceptible absorption ($\chi \leq 0.01$–0.02) whose parameters, as a function of ε, are shown in Figure 7.3; and (b) the strongly absorbing and practically non-scattering submicron fraction of very small particles with parameters $N = (1$–$5) \times 10^4\,cm^{-3}$, $\alpha_M = 0.03\,\mu m$, $\chi \cong 0.3$–0.7, and

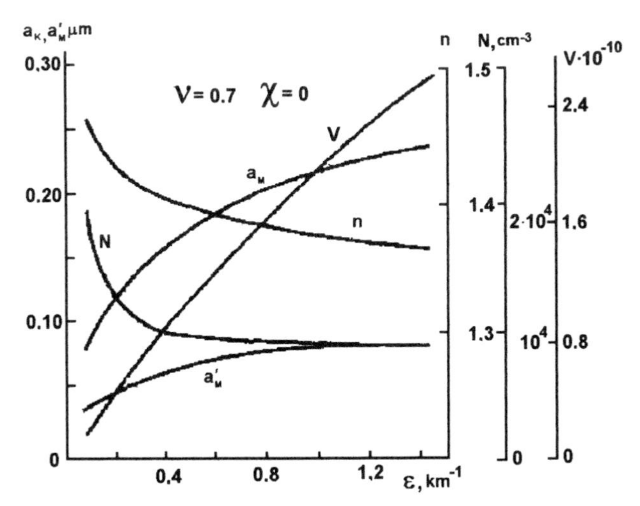

Figure 7.3. Dependences of parameters of the microphysical optical model of aerosol on the attenuation coefficient.

$\nu' \cong 0.5$–0.6 being almost completely responsible for light absorption by aerosol in the short-wave spectral region. From the data of optical measurements, the factors determining the behaviour of the absorbing fraction are unknown. Its connection with the weakly absorbing fraction is also unknown. Therefore, in modelling, the concentrations of both fractions are assumed to be proportional, which is possibly valid at a low relative humidities. Since this modelling was based on the data of observations in mid-latitudes, only the submicron fraction of aerosol was modelled.

7.2.3 The vertical stratification of atmospheric aerosols

Processes of generation of the aerosol submicron fraction take place at all altitudes, but their nature and intensity change with altitude. The optical sounding data have shown that the spatial distribution of aerosol has a distinct patchy or cloudy structure, rapidly changing under the influence of dynamic, thermal, radiative, and chemical factors. This variability results in unstable optical weather with characteristic altitudinal dependence. A complicated totality of factors determining the variability of the optical weather at different levels, cannot practically be taken into account sufficiently completely. Therefore, with the present knowledge, the statistical description of the spatial distribution of aerosol is inevitably connected with a neglect of detail and only gives a description of the general features of the vertical stratification of aerosol.

As an example of the typical statistical model based on optical measurements, we consider the model of the Institute of Atmospheric Physics (Moscow). The schematic representation of the vertical distribution of aerosol is based on the data of multiyear searchlight, twilight, and satellite soundings of the atmosphere, as well as data of searchlight and aircraft soundings [130]. These data agree well both with each other and with the data of other studies [58, 62, 66–68, 74–79, 89, 90, 162, 163]. The existence of several layers of increased air turbidity has been discovered, separated by comparatively stable zones of relative clean air. Such a multilevel stratification of aerosol results from the existence of several levels at which optimal conditions are created for generation and accumulation of aerosol, at each level the processes of generation being specific, determining the specific features of the optical properties of the resulting aerosol.

(1) Within the proposed schematic model, the properties of aerosol at strato-, meso-, and thermospheric levels are assumed to be constant. The properties of aerosol at the tropospheric level are described by two independent parameters $\varepsilon_a(\lambda_0, h = 0)$ and $\tau_a(\lambda_0, h = 0)$ which are unambiguously connected with the input parameters of the horizontal visibility S model:

$$S = \frac{3.9}{\varepsilon(\lambda_0, h = 0)}$$

and vertical transparency:

$$P = \exp\left[-\tau(\lambda_0, h = 0)\right]$$

(2) The model refers exclusively to transparency windows, where the aerosol haze plays a substantial role and does not consider the impact of light absorption by gas components.
(3) The model is constructed individually for each layer.

It is essential that the aerosol stratification and thereby the vertical optical thickness $\tau(\lambda)$ do not usually correlate with horizontal visibility (i.e., with the surface coefficient of scattering $\sigma(\lambda, h = 0)$).

Figure 7.4 demonstrates the average vertical change of the scattering coefficient for three wavelengths. The σ parameter changes at all levels by a factor of 3 with respect to the average value (the MSD of $d(\log\sigma)$) varies within 0.3–0.5).

$$\sigma_a^{\text{mpon}}(\lambda_0, h) \cong \sigma_a(\lambda_0, h = 0)\exp\left[\frac{\sigma_a(\lambda_0, h = 0)}{\tau_a(\lambda_0)} h\right]$$

where $\sigma_a(\lambda_0, h = 0)$ and $\tau_a(\lambda)$ are independent input parameters of the model. Since the nature of aerosol is assumed to be identical through the whole atmosphere, the relationship:

$$\frac{\sigma_a(\lambda_0, h = 0)}{\tau_a(\lambda)} = \frac{\sigma_a(\lambda_0, h = 0)}{\tau_a(\lambda_0)}$$

is supposed to be independent of λ and is attributed to λ_0.

From the surface to a level of about 10–12 km there a lack of effective presence of tropospheric aerosol – mainly the submicron fraction generated in the surface layer and scattered by turbulence. Its vertical distribution can be, on average, roughly approximated with the relationship shown below (at $\lambda_0 = 0.55\,\mu\text{m}$).

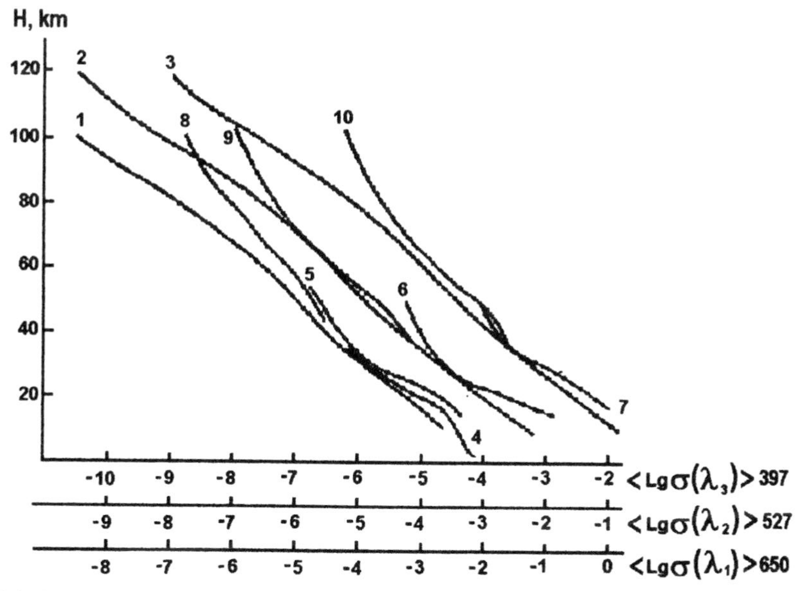

Figure 7.4. Vertical profiles of the scattering coefficient from the data of twilight measurements for $\lambda_1 = 650\,\mu\text{m}$ (8), $\lambda_2 = 527\,\mu\text{m}$ (9), and $\lambda_3 = 397\,\mu\text{m}$ (10), from data of Elterman for $\lambda_4 = 530\,\mu\text{m}$ (4), from the data of 'Soyuz-3' (5–7), and results of calculations of the coefficient of molecular scattering (1–3) for three wavelengths of light [130].

For the model of tropospheric aerosol ($h < 12\,\text{km}$) the following relationship is supposed to be valid:

$$\varepsilon(\lambda, h) = \eta(\lambda)\varepsilon(\lambda_0) + E(\lambda)\exp\left[\frac{\varepsilon(\lambda_0, h = 0)}{\tau(\lambda_0, h = 0)}h\right]$$

where the values of $\eta(\lambda)$ and $E(\lambda)$ are taken from Table 7.4:

$$\alpha(\lambda, h) \cong \begin{cases} 0.15(\lambda)\exp\left[-\dfrac{\varepsilon(\lambda_0 h = 0)}{\tau(\lambda_0 h = 0}h\right] \text{ at } \lambda \leq 4\,\mu\text{m} \\ \varepsilon(\lambda, h) \text{ at } \lambda = 8\text{–}12\,\mu\text{m} \end{cases}$$

$$\sigma(\lambda, h) = \varepsilon(\lambda, h) - \alpha(\lambda, h) \cong \left(\frac{\lambda_0}{\lambda}\right)\varepsilon(\lambda_0, h = 0)\exp\left[-\frac{\varepsilon(\lambda_0, h = 0)}{\tau(\lambda_0, h = 0)}\right]$$

$$\beta(\lambda) \approx \begin{cases} 0.15(\lambda/\lambda_0) \text{ at } \lambda \leq 4\,\mu\text{m} \\ \gg 1 \text{ at } \lambda = 8\text{–}12\,\mu\text{m} \end{cases}$$

The contribution of molecular scattering is negligibly small. To estimate variations due to changes in relative humidity, we can approximately assume:

$$\sigma(\lambda, h, \mathrm{RH}) = 2(\mathrm{RH}^2 + \mathrm{RH}^4)\sigma(\lambda, h, \mathrm{RH} = 70\%)$$

$$\sigma(\lambda, h, \mathrm{RH}) = \alpha(\lambda, h, \mathrm{RH} = 70\%)$$

$$D_{11}(\varphi, h) = D_{11}^{\mathrm{mol}}(\varphi, h) + \frac{\varepsilon(\lambda, h)}{4\pi} C(\varphi) + \left[\frac{\varepsilon(\lambda_0 h)}{\sigma_M(\lambda_0, h)} \sigma(\lambda_0, h = 0)\right]^{K(\varphi)^{-1}}$$

$$D_{ik}(\varphi, h) = D_{ik}^{\mathrm{mol}}(\varphi, h) + \frac{\varepsilon(\lambda, h)}{4\pi} C(\varphi) \times \left[\frac{\varepsilon(\lambda_0, h)}{\sigma_M(\lambda_0, h)} \sigma_M(\lambda_0, h = 0)\right]^{K(\varphi)^{-1}} \xi_{ik}(\varphi)$$

$$\xi_{ik}(\varphi) = \langle \xi_{ik}(\varphi) \rangle + \left[a_{ik} \frac{\varepsilon(\lambda_0, h)}{\sigma_M(\lambda_0, h)} \sigma_M(\lambda_0, h = 0) + b_{ik}\right] \pi_{ik}^1(\varphi)$$

$$a_{12} = a_{21} = -0.773 \qquad a_{33} = a_{44} = 1.683 \qquad a_{43} = a_{34} = 1.06$$

$$b_{12} = b_{21} = -0.296 \qquad b_{33} = b_{44} = 0.645 \qquad b_{43} = -b_{34} = 0.406$$

The functions $C(\varphi)$, $K(\varphi)$, and $\pi_{ik}^1(\varphi)$ have been tabulated in [130].

Aerosol of the following stratospheric level fills efficiently the altitude range ~4–40 km (with a maximum at about 19 km). From present knowledge, here the submicron fraction also prevails generated at altitudes about 20 km and distributed by turbulent fluxes both up and down. From the data of impactor samples analysis, it consists mainly of sulphuric acid, sulphate, and ammonium persulphate and has a log-normal size distribution of particles averaging $\langle a \rangle \sim 0.15\,\mu\mathrm{m}$ [130]. For the model of stratospheric aerosol ($4 < h < 40$ km) the following relationships are assumed:

$$\varepsilon(\lambda, h) = \sigma_M(\lambda_0, h)\left\{\left(\frac{\lambda_0, h}{\lambda}\right)^4 + 4\Sigma(\lambda) \exp\left[\frac{h - 19}{7,2}\right]\right\}$$

where $\Sigma(\lambda)$ is taken from Table 7.2:

$$\alpha(\lambda, h) \cong \begin{cases} 0.15[\varepsilon(\lambda_0, h) - \sigma_M(\lambda_0, h)] \text{ for } \lambda \leq 4\,\mu\mathrm{m} \\ \varepsilon(\lambda, h) \text{ for } \lambda = 8\text{–}12\,\mu\mathrm{m} \end{cases}$$

$$\sigma(\lambda, h) = \sigma_M(\lambda_0, h)\left\{\left(\frac{\lambda_0}{\lambda}\right) + \frac{4\lambda_0}{\lambda} \exp\left[-\frac{|h - 19|}{7,2}\right]\right\}$$

$$D_{11}(\varphi, h) = D_{11}^M(\varphi, h) + \frac{\sigma_M(\lambda_0, h)}{\pi}\left(\frac{\lambda_0}{\lambda}\right) \exp\left[\frac{|h - 19|}{7.2}\right] \times C(\varphi)\left\{\left[\frac{\varepsilon(\lambda_0, h)}{\sigma_M(\lambda_0, h)} - \right]\sigma_M(\lambda_0, h = 0)\right\}$$

$$D_{ik}(\varphi, h) = D_{ik}^M + [D_{11} - D_{11}^M]\xi_{ik}(\varphi)$$

$$\xi_{ik}(\varphi) = \langle \xi_{ik}(\varphi) \rangle$$

The altitudinal change of the scattering coefficients is well approximated, on average, for λ_0 by the relationship:

$$\sigma_a^{\text{strat}}(\lambda_0, h) = 4\sigma_{\text{mol}}(\lambda_0, h) \exp(-|h - 19|/7.2)$$

where σ_{mol} is the coefficient of molecular scattering and h is expressed in km. In view of the poorly studied statistical laws of the variability of stratospheric aerosol, only averaged estimates have been used in the modelling.

At an altitudinal range of 30–70 km, aerosols of the mesospheric layer are present (a maximum near 50 km), whose nature remains poorly studied. However, there are certain indications that it is generated mainly at a level of about 50 km. It consists mainly of sulphate and nitrogen compounds, and forms, apparently, not a very broad lognormal distribution with $\langle a_M \rangle = 0.07\,\mu\text{m}$. For the model of mesospheric aerosol ($30 < h < 70$ km) we have:

$$\sigma(\lambda, h) = \sigma_M(\lambda_0, h)\left(\frac{\lambda_0}{\lambda}\right)^4 \left\{1 + 4\exp\left[-\frac{|h-50|}{7}\right]\right\}$$

$$\alpha(\lambda, h) = 0.15[\sigma(\lambda_0, h) - \sigma_M(\lambda_0, h)] \text{ at } \lambda \leq 4\,\mu\text{m}$$

$$D_{ik}(\varphi, h) = D_{ik}^M \left(\frac{\lambda_0}{\lambda}\right)^4 \left\{1 + 4\exp\left[-\frac{|h-50|}{7}\right]\right\}$$

$$\xi_{ik}(\varphi) = \xi_{ik}^M(\varphi)$$

The vertical distribution of σ_a in this region can be schematically approximated, on average, by the following relationship:

$$\sigma_a^{\text{mes}}(\lambda_0, h) = 4\sigma_{\text{mol}}(\lambda_0, h)\exp(-|h - 50|/7)$$

Above the ~50-km level one can observe the presence of aerosol of meteoric origin, the role of which rapidly increases with increasing altitude and becomes prevalent in the thermosphere. Data on its chemical composition, size distribution, and optical properties are practically absent. The average vertical change of the scattering coefficient for aerosol at the thermospheric level can be approximately described by the relationship:

$$\sigma_a^{\text{therm}}(\lambda_0, h) = \sigma_{\text{mol}}(\lambda_0, h = 65\text{ km})\exp(-|h - 65|/17)$$

For the model of thermospheric aerosol ($50 < h < 150$ km):

$$\varepsilon(\lambda, h) = \sigma_M(\lambda_0, h)\left(\frac{\lambda_0}{\lambda}\right)^4 + \sigma(\lambda_0, h = 65\text{ km})\left(\frac{\lambda_0}{\lambda}\right)^4 \exp\left[\frac{|h-65|}{17}\right]$$

$$D_{11}(\lambda, \varphi, h) = \frac{\sigma(\lambda, h)}{4\pi} f_{11}^M(\psi)$$

The transitional zones of a stable relative cleaning of the air are located at altitudes of about 7 km (4–10 km), 33 km (27–40 km), and 60 km (55–65 km). Within these zones, aerosol is present both in underlying and overlying layers, but

Table 7.5. The vertical profile of $\sigma_{mol}(\lambda = \mathbf{0.55}\,\mu m)$.

H (km)	σ_{mol}	H (km)	σ_{mol}	H (km)	σ_{mol}	H (km)	σ_{mol}
0	1.16–2	30	1.75–4	60	2.88–6	90	2.98–8
5	6.99–3	35	8.03–5	65	1.57–6	95	9.51–9
10	3.92–3	40	3.79–5	70	8.24–7	100	4.68–9
15	1.85–3	45	1.87–5	75	4.08–7	105	1.70–9
20	8.44–4	50	9.74–6	80	1.93–7	110	9.25–10
25	3.80–4	55	5.88–6	85	7.50–8	115	2.93–10

outside the zones, only the aerosol of a given layer is substantial. In the transitional zones where the aerosol of adjacent (upper and lower) zones are superimposed:

$$\alpha(h,\lambda) = \alpha^B(h,\lambda) + \alpha^H(h,\lambda)$$

$$\sigma(h,\lambda) = \sigma^{B,A} + \sigma^{H,A} + \sigma_{mol}$$

$$\beta(h,\lambda) = [\alpha(h,\lambda)/\sigma(h,\lambda)]$$

$$D_{ik}(\varphi,h) = D_{ik}^{B,A} + D_{ik}^{H,A} + D_{ik}^{mol}$$

$$\xi(\varphi) = (\sigma^{B,A} f_{ik}^{B,A} + \sigma^{H,A} f_{ik}^{H,A} + \sigma_{mol} f_{ik}^{M})/\sigma(h)$$

The input parameters S and P are determined either by measurements in a given place at a given time, or estimated from the data of Figure 7.1 and Table 7.4.

The relationships given above, like the calculated tabulated data from [130] are valid for a narrow spectral interval – near $\lambda = 0.55\,\mu m$. (They can be used in the interval from approximately 0.4 to 1.0 μm.) An extrapolation onto long-wave and UV spectral regions should be based on the microphysical modelling results.

The average vertical dependence of the coefficient of molecular scattering of light by the air for $\lambda = \lambda_0 = 0.55\,\mu m$ is given in Table 7.5, with $\sigma(\lambda,h) = (\lambda_0/\lambda)^4 \sigma_{mol}(\lambda_0,h)$.

7.3 THE BASIC PARAMETERS OF THE OPTICAL AND MICROPHYSICAL MODELS OF ATMOSPHERIC AEROSOLS

Let us briefly consider now the results of some later studies on the optical and microphysical models of atmospheric aerosols [62, 67].

7.3.1 The shape and morphological structure of particles

Numerical experiments aimed at calculation of the optical characteristics of polydisperse aerosol systems with due regard to non-spherical particles (spheroids and cylinders) show that in most cases in the real atmosphere a consideration of particle non-sphericity provides corrections at the level of measurement errors both for the coefficients of attenuation, scattering, and absorption, and for phase functions. Therefore, the use of an approximation of particle sphericity is almost always justified.

However, in the higher layers of the atmosphere, conditions can be met under which non-spherical particles are strictly oriented – their non-sphericity cannot then be ignored. The data of optical measurements in the stratosphere indicate the existence of a considerable amount of non-spherical particles at altitudes above 15 km.

In the lower atmospheric layers, neglection of particle non-sphericity is impossible in the case of the formation of small ice particles at negative temperatures and microcrystals of anthropogenic origin (urban aerosols and mining).

Modelling the optical properties of non-spherical particles is a complicated problem and has been carried out for some approximations: spheroids of different oblateness and flatness, infinite cylinders, and point-contact spheres [34, 46, 66, 67]. Use of the results of these calculations is hindered by increasing volumes of required data (a consideration of a spheroid's orientation with respect to the source and receiver of radiation) and almost complete absence of reliable data on size distributions of spheroids and their oblateness.

The use of models of spheroidal particles makes it possible, in particular, to explain the observed effect of overestimated backscattering in the higher layers of the atmosphere.

7.3.2 The homogeneous chemical composition of aerosol and the complex refraction index for particle substances

Differences in the chemical composition of various fractions of atmospheric aerosols of different sizes can be substantial, especially for middle size and coarse fractions. Since the interaction of a particle substance of both fractions with water vapour is different, the dependences of complex refraction indices on the relative humidity value will be different, also [34, 46, 67, 146, 148].

Particles of a certain range of sizes can differ in their chemical composition. Numerical experiments suggest the conclusion that for atmospheric aerosols the substitution of representation of particles with different refraction indices for their presentation by an ensemble of particles with an averaged efficient refraction index is sufficiently representative: for the polydisperse system of particles with refraction indices differing by 0.2 an error does not exceed 5% for the coefficients of attenuation, scattering, and absorption, and also for scattering functions [148].

The observational data show that the spectral change of the complex refraction index for the atmospheric aerosol substance is characterized by specific features inherent to almost all types of aerosol. They are determined, first of all, by the presence in the substance of particles of sulphates, silicates, and water as well as by the fact that the absorption bands of other chemical compounds coincide mainly with the absorption bands of the substances mentioned. A maximum variability of the complex refraction index $\tilde{m}(\lambda, r) = n - i\chi$ is connected with the different water content in the substances comprising the particles. In this connection, the parameter of the dependence on relative humidity is introduced in the model approximation of $\tilde{m}$ [66]. Of particular importance for model approximations of $\tilde{m}(\lambda, r)$ is the magnitude of the imaginary part of the refraction indices in the visible spectral region and at the wavelengths of laser soundings. Results of most studies in the visible spectral region give values of χ close to 0.05 and less. Large values of $\chi(\lambda, r)$ are observed for anthropogenic sources, primarily, urban aerosols containing a substantial share of soot [8, 24, 44, 77, 79 90, 162]. Changes in the chemical composition and, hence, in the value of the complex refraction index for the substance of the dispersed phase, in particular, χ, take place over 24 hours, in the main, as a result of photochemical reactions and processes of condensation. In the case of lidar measurements the value of $\chi(r)$ determines the ratio of the coefficient of backscattering to the coefficient of scattering, and in the case of $\chi < 0.01$ depends strongly on aerosol particle size distribution. With increasing relative air humidity the spectral change of the coefficient of attenuation in the IR spectral region gets radically transformed.

The heterogeneous composition of the particles substance manifests itself with watering of insoluble particles, covering of particles with a soot layer, and in the presence of absorbing inclusions in transparent particles (e.g., Fe^{3+} in minerals and soot). In such cases the role of the radiation-absorbing component increases compared with its share in the total content of aerosols.

The heterogeneous distribution of substances affects the optical properties depending on how the different chemical compositions are distributed: uniformly, by layers, or by separate inclusions [27, 33, 97, 146, 148]. To calculate the optical characteristics, it is sufficient to use two models of distribution of substances in particles: homogeneous and two-layer distributions which satisfactorily describe the polydisperse ensembles forming due to a coagulation and a condensation growth of particles.

Using the data of optical and microphysical measurements and calculations of the optical characteristics for spheres, one can in some cases detect the formation of the surface film of a condensing substance [66]. An adsorption of organic molecules on anisotropic particles is reliably detected with the electro-optical method: the magnitude and shape of the signal change at the modulation of particles' turning with an electric current [58].

7.3.3 Number density and the function of particle size distribution

Studies of the impact of variations in the size distribution function in the intervals $2\pi r < \lambda$ and $2\pi r > \lambda$, along with the impact of the upper and lower boundaries of the aerosol particle size distribution on the optical characteristics of the polydisperse ensembles of particles, as well as the comparison of results of these studies and data on the accuracy of assessment of the distribution functions of atmospheric aerosols, have led to the conclusion that for most real cases, the particle size distribution is satisfactorily described with the use of two or three-modal distributions [62, 67]. The choice of the analytical function of distribution is not an important factor, but a consideration of the modal radius of particles and the steepness of the decrease of particle concentration with the size of particles exceeding the modal one is substantial. In most cases the modal radii for surface aerosols changes within narrow limits: $r_{01} \cong 0.02$–$0.08\,\mu m$, $r_{02} \cong 0.15$–$0.35\,\mu m$, and $r_{03} \cong 1.5$–$5.00\,\mu m$, where $i = 1$ – corresponds to aerosols of natural condensation origin, 2 – to photochemical aerosols mainly of anthropogenic origin, and 3 – to aerosols of dispersed origin. An increase in air humidity exceeding 70% can lead to a marked growth of modal radii beyond the upper limits of the ranges of their indicated size change [67]. Since the second-mode results from the evolution of the first-mode, there is a certain similarity between the chemical compositions of the substances of particles of the first two modes. The spatial–temporal variations of the aerosol content in the atmosphere are determined by the presence of powerful local sources of particles whose intensity changes during a season or a day, as well as by the impact of numerous physical and physico-chemical processes which control the distribution and removal of aerosol particles from the atmosphere.

The vertical aerosol distribution is characterized by a stratified structure which is explained mainly by the thermodynamic state of different atmospheric layers. For the horizontal distribution of aerosols on a global scale, the most important factors are the substantially different character of particle generation over the continents and sea surfaces and the latitudinal dependence of particle concentration more distinctly manifesting itself with an increasing altitude of observations [12, 13, 62, 67, 149]. Especially heterogeneous is the horizontal structure of aerosols in industrial regions: the stratified structure is observed as well as the dependence on temperature stratification, humidity, and advective air fluxes [12, 13, 32, 49, 66, 153]. The main mass of aerosols is contained in the lower atmospheric layers while the height of the homogeneous aerosol atmosphere varies mainly within $H_0 \cong 1.5$–$2\,km$.

One should note the independence of the vertical distributions of tropospheric and stratospheric aerosols. The latter are characterized by specific features of distribution connected with season and latitude. Some dependence is possible on the local relief in the regions with high mountains (Pamirs, the Himalayas). Drastic changes in the structure and optical properties of stratospheric aerosols occur after powerful volcanic eruptions. Variations of number density of small particles is followed by sharp changes in their size distribution. In some cases the presence of large particles is accompanied by a decrease of the small particle concentration (also, due to the coagulation trapping of these particles) [62, 67].

The aerosol particle number density is a relatively uninformative and rather uncertain characteristic. This also concerns the results of *in situ* microphysical measurements of aerosol characteristics: either the accuracy of the measured number density of particles, with sizes $r < 0.1\,\mu m$, drops sharply or measurements are impossible because of a low sensitivity of measuring complexes and stronger variations of particle concentrations, in particular, due to coagulation processes (for $N > 10^6\,cm^{-3}$, the lifetime of the system constitutes a few minutes). The number density of particles of the size $r > 10\,\mu m$ is also unstable and not representative. Most stable are values of the number density of particles within $1\,\mu m > r > 0.1\,\mu m$ – representing the state of aerosol loading of the atmosphere adequately enough.

The use of the value of mass concentration to describe the spatial–temporal structure of aerosols is worthwhile when dividing the aerosols into two fractions: (1) $r \leq 5\,\mu m$ and (2) $r > 5\,\mu m$.

7.4 STUDIES OF THE MORPHOLOGICAL STRUCTURE OF ATMOSPHERIC AEROSOLS

The impact of the sources and mechanisms of aerosol particle formation on their morphological structure is common knowledge. After first attempts to perform a morphological classification of aerosols from the data of electron microscopic analysis it was established that there is a certain dependence of the type of particles on the altitude of the atmospheric layer in which they are sampled. In the upper atmospheric layers ($H \geq 30$ km) one can mainly observe conglomerates of complicated shapes of spherical particles of submicron sizes; in the Junge stratospheric layer ($25 \geq H \geq 15$ km) are spherical particles with a large central nucleus and a high content of sulphates. Tropospheric particles have a denser structure and, in the main, a spherical or oval shape; particles at the altitudes of noctilucent clouds are often a nucleus representing a large central particle surrounded with a multitude of smaller particles. Subsequent studies have led to the conclusion that one of the most important factors determining the morphological structure of particles is the magnitude of the relative humidity level [66].

Using the optical microscope one can observe the watering of some particles with comparatively small values of relative humidity (RH = 50–60%). Under the electron microscope such particles are seen as spheres with a central nucleus or nuclei and loose substance on the periphery. Apparently, such particles are not of the condensation origin, as is usually understood, but of condensation–coagulation origin (i.e., loose conglomerates of small particles formed at coagulation of the latter which are filled either with water or with another liquid (via a capillary effect)). With the impactor capture of large particles and their fractionation, the fractions of particles not watered even at high relative humidities is 70%. This favours the initial coagulation origin of large conglomerates that are loosely packed. The filling of microcapillaries with water in such particles can take place at relative humidities greater than 20%. The surface tension of the water compresses the substance of the particle, which results in a denser solid substance on the particle–air border. When particles move to drier atmospheric layers, a 'dry' coagulation can be observed, and as a result, a more loose layer grows. The electron microscopic analysis of aerosol samples has made it possible to detect particles repeatedly moving from layers with low humidity to layers with high humidity and back again. Such particles have a multilayered structure and are observed mainly in the upper troposphere.

There is a direct dependence of the morphological structure and size of aerosol particles formed in the atmosphere as a result of the photochemical and catalytic reactions of sulphur and nitrogen-containing compounds with water vapour: at low contents of water vapour, smaller spherical particles form, and in the presence of metal-containing substances even crystal particles form, whereas at high contents of water vapour the formation takes place of large and relatively more dense liquid droplet particles. In the first case, small particles coagulate creating loose aggregates of a fractal structure. Similar transformations are observed at the condensation or sublimation of the products of burning and subsequent rapid coagulation of the resulting particles at $N \geq 10^8\,\text{cm}^{-3}$.

The fractal aggregates are of a specially organized structure in which each selected element is similar to the system as a whole. As applied to the problem of atmospheric aerosols, aggregates formed of a multitude (10^2–10^3) of particles with similar physical–chemical properties can be considered as fractals. The size of the particles is substantially less than the size of the system, and differ little from individual (initial) particles. The distribution within the aggregate is described by general statistical laws (i.e., a self-similarity and a scale invariance of the spatial structure within wide limits being observed). Concrete values of the parameters of the fractal structure depend on the mechanism of aggregate formation. For such formations the relationship $R = aN^{1/D}$ is valid, where R is the size of the cluster. The range of possible variations of D is limited and for the volume clusters $1 \leq D \leq 3$. The value of D for smoke aggregates averages 1.78 and reveals an increasing trend with high humidity values. The radius of initial particles depending on the source of smoke (temperature of burning, etc.) constitutes $a = 0.01$–$0.05\,\mu\text{m}$, and the size of aggregates varies within several tenths of a micron to $\sim 10\,\mu\text{m}$. Bearing in mind the loose structure of such particles, the process of gravitational sedimentation for them is of low efficiency. They are removed from the atmosphere mainly through washing out or being driven by downward fluxes [7, 27].

It is known that if the substance evaporates in the electric field and this process is accompanied by the formation of weakly ionized vapour, then in the process of aggregation, clusters of filamentary structures of chains and hollow tubes are formed. Such structures can form also with strong air fluxes in place of the formation of initial aerosol particles. The electron microscopic analysis of aerosol samples taken in the atmosphere at different altitudes and under various conditions of formation and evolution of aerosols, confirms these conclusions with respect to the morphological structure of the particles.

One of the morphological classifications of the types of atmospheric aerosols was motivated by the problems of aerosol optics [66–67]. Based on the analysis of observational data, eight morphological types of particles have been proposed: 1 – dense spheres; 2 – loose spheres; 3 – particles with an envelope of very small particles (with the 'fir-coat'); 4 – dense non-spherical (debris of rocks and other objects); 5 – loose non-spherical (agglomerates); 6 – chain structures, including fractals; 7 – loose with dense nuclei (e.g., sulphate particles of the Junge layer); and 8 – crystals and particles with dry envelopes.

The relative share of different types of particles changes widely in the atmosphere, depending on the altitude of sampling and on the range of the observed particles sizes. Tables 7.6 and 7.7 illustrate the data for samples taken over the Bering Sea in March, 1973 and in the surface layer of the equatorial Atlantic in July–August, 1974.

Table 7.6. The number of different shapes of particles (%).

Number of samples	H (m)	Types of particles 1	2	3	4	5	6	7	8
4	1,000	18	65	–	8.0	–	7.5	–	1.5
10	4,000	41	30	2.0	2.5	10	3.7	7.0	3.8
3	9,000	12	40	–	15	9.0	13	7.0	4.0

Table 7.7. The number of particles of different morphological structure in the surface layer of the equatorial Atlantic (GATE-74) (%).

	Type of particles (1,2,4,5)												
	20 July, 1974				30 July, 1974					10 August, 1974			
r (μm)	1	4	5	Σ	1	2	4	5	Σ	1	2	4	Σ
0.05–0.10	0.1	14.6	5.8	23.5	14.8	29.1	2.8	2.0	48.7	10.2	–	–	10.2
0.10–0.15	0.8	15.2	7.6	23.6	13.6	11.4	0.3	0.3	25.6	21.0	17.5	4.0	42.5
0.15–0.20	0.6	7.1	10.9	18.6	4.3	3.2	0.1	–	7.7	6.2	4.6	7.3	18.1
0.20–0.25	0.8	5.0	5.5	11.3	2.7	2.4	–	–	5.1	0.8	1.6	4.7	7.1
0.25–0.30	0.3	2.2	4.1	6.6	2.0	1.0	–	0.1	30	0.5	1.3	2.3	4.1
0.30–0.35	–	2.6	1.6	4.2	2.1	1.3	–	–	3.5	1.3	0.6	4.5	6.4
0.35–0.40	0.6	1.6	1.4	3.6	1.0	0.6	–	–	1.6	0.3	–	1.2	1.5
0.40–0.50	–	2.7	1.6	4.3	0.4	1.1	–	–	1.6	0.6	0.3	2.1	3.0
0.50–0.75	0.3	1.4	1.3	3.0	1.6	0.7	–	–	2.3	0.8	0.6	3.6	5.0
0.75–1.00	–	0.7	0.4	1.1	0.8	–	–	–	0.8	–	0.3	1.0	1.3
1.00–1.25	–	0.2	–	0.2	0.1	–	–	–	0.1	–	–	0.8	0.8
Total	*6.5*	*53.3*	*40.2*	*100*	*43.2*	*50.8*	*3.2*	*2.4*	*100*	*41.7*	*6.8*	*31.5*	*100*

The share of spherical particles increases with decreasing altitude (i.e., with the growth of relative humidity (at altitudes 1 km – 83%, 4 km – 71%, 9 km – 52%), with the maximum share of dense spheres (41%) at an altitude of 4 km, where the 'oldest' particles reside). The fractal structures and agglomerates are observed mainly at the 9-km altitude, and 'debris' in the lower layer and at the 9-km altitude. The presence of complicated structures at the 1-km altitude testifies to the relatively recent formation of these clusters.

In the surface layer of the equatorial Atlantic mainly four types of morphological structure of particles are observed. This is determined by specific features of the sources and conditions of formation of aerosols particles: dust from the Sahara desert and sea salt from the ocean surface. The fourth and fifth types of dust origin prevail over the particles of the first and second types (spherical particles possibly of condensation origin). The size of spherical particles is mainly $<0.2\,\mu m$, whereas dust non-spherical particles have relative maxima in distributions for different size ranges (by mass, in the case of gigantic particles). Most particles of the eighth type (crystals, tetrahedrons, plates, etc.) during modelling of the optical characteristics of aerosols can be changed for equivalent spheres, bearing in mind the chaotic orientation of such particles in a sufficiently turbulent medium. However, in industrial regions among particles of this type, are often observed particles of the 'asterisk' type formed of industrial gases in photochemical processes. Studies of the morphological structure of atmospheric aerosols in different climatic regions of the Earth confirmed the earlier conclusion about the impact of relative humidity on the morphology of particles as a factor determining the optical properties of aerosols.

7.5 STUDIES AND MODELLING OF THE CHEMICAL COMPOSITION OF ATMOSPHERIC AEROSOLS

For an adequate description of the physico-chemical properties of atmospheric aerosols, it is necessary either to know the aerosols characteristics, such as their concentration, size distribution, morphological structure, and chemical composition or to know the parameters by which these characteristics can be unambiguously described (i.e., to solve the inverse problem with use of the required a priori information). Numerous studies have been dedicated to this problem [42–44, 75–79, 111, 113, 116, 152].

It should be noted that certain promising results have recently been obtained in this field, for instance global and regional aerosol models with a small number of parameters have been substantiated [5–7, 32, 62–68, 74–78]. Here one usually proceeds from the following suppositions: aerosol particles are homogeneous in their chemical composition and have a spherical shape; there are two or three modes of the size distribution described by simple analytical expressions of the type of inverse λ-function: $N_i(r) = A_i e^{-b_i r} r^{-\nu_i}$, where A, b, and ν are prescribed parameters of the distribution function. The evolution of the modes is determined by the condensation process of the growth of particles with increasing relative humidity.

The condensation mechanism of particle growth is substantial at the initial stages after their instant generation (explosion, burning, etc.) and it does not practically affect the long (more than 24 hours) evolution of the modes. This approach, with the relative humidity of the medium and condensation activity of particles known, makes it possible to move on to values of mass concentrations of the dry basis (i.e., to estimates of the amount of aerosol substance or gases emitted to the atmosphere which are transformed into aerosol). However, the real pattern of the formation of the aerosol structure is more complicated and does not always permit the assumprions mentioned above. The coagulation mechanism of the growth of aerosol particles, especially in the presence of an electric field and electric discharges, when gas immediately transforms into solid state, leads to the formation of the fractal structures which are difficult to describe with spheres. Heterogeneous reactions on the surface of, and within, aerosol particles change their chemical composition. Besides this, particles of different chemical natures, especially fractal, coagulating with each other, do not allow one to unambiguously recognize their initial components.

7.5.1 The elemental composition

The composition of the Earth's crust can serve as a standard of the conservative elemental composition of soils. However, for some elements, the published estimates of their content in the Earth's crust differ strongly. Probably, this reflects the mobility of these elements in the Earth's crust. Table 7.8 demonstrates the data on the elemental composition of the Earth's crust taken from different publications, with an indicated scattering of the content of elements and supposed average values. For comparison purposes, the table also gives an averaged elemental composition of meteorites. Similar data for global soils are given in Table 7.9 [33]. Of interest are the very strong oscillations of the content in soil of such elements as

Table 7.8. The content of elements in the Earth's crust (% by mass).

Element	Earth's crust	Meteorite	Andesite	Element	Earth's crust	Meteorite	Andesite
O	47 ± 1.5	nm	nm	N	$(2.2 \pm 0.2) \times 10^{-3}$	nm	nm
Si	27.8 ± 0.2	nm	nm	Li	2.0×10^{-3}	nm	nm
Al	8.0 ± 0.2	1.3	8.8	Nb	$(1.8 \pm 0.3) \times 10^{-3}$	nm	nm
Fe	4.2 ± 0.8	25.0	5.8	Ga	$(1.6 \pm 0.3) \times 10^{-3}$	nm	nm
Ca	3.4 ± 0.8	nm	nm	Sc	$(1.5 \pm 0.7) \times 10^{-3}$	6×10^{-4}	2.5×10^{-4}
Na	2.5 ± 0.7	0.7	3.0	Pb	$(1.4 \pm 0.2) \times 10^{-4}$	2×10^{-5}	1.5×10^{-3}
K	2.5 ± 0.5	nm	nm	B	1.0×10^{-3}	nm	nm
Mg	2.4 ± 1.0	nm	nm	Th	8.0×10^{-4}	nm	nm
Ti	0.5 ± 0.06	5×10^{-2}	0.8	Sm	$(7.4 \pm 1.5) \times 10^{-4}$	nm	nm
P	0.1 ± 0.03	nm	nm	Cs	$(4.5 \pm 1.5) \times 10^{-4}$	nm	nm
Ni	0.1 ± 0.025	1.3	5.5×10^{-3}	Hf	$(3.7 \pm 1.7) \times 10^{-4}$	nm	nm
Mn	0.09 ± 0.01	0.2	0.12	Yb	$(3.5 \pm 0.5) \times 10^{-4}$	nm	nm
F	0.063 ± 0.033	nm	nm	Sn	$(2.2 \pm 0.3) \times 10^{-4}$	nm	nm
Ba	0.06 ± 0.02	nm	nm	As	$(1.7 \pm 0.3) \times 10^{-4}$	3×10^{-5}	2.4×10^{-4}
S	0.043 ± 0.017	nm	nm	Ge	$(1.7 \pm 0.2) \times 10^{-4}$	nm	nm
C	0.034 ± 0.014	nm	nm	Ea	$(1.7 \pm 0.5) \times 10^{-4}$	nm	nm
Sr	0.034 ± 0.007	nm	nm	W	1.6×10^{-4}	nm	nm
Cl	0.018 ± 0.006	7×10^{-3}	nm	Mo	1.5×10^{-4}	nm	nm
Zr	0.018 ± 0.09	nm	nm	Ru	$(7.5 \pm 2.5) \times 10^{-4}$	nm	nm
V	0.014 ± 0.006	7×10^{-3}	1×10^{-2}	Sb	$(3.5 \pm 1.5) \times 10^{-5}$	1×10^{-5}	2.0×10^{-5}
Rb	0.012 ± 0.004	nm	nm	I	$(3.2 \pm 1.8) \times 10^{-5}$	nm	nm
Cv	0.01 ± 0.01	0.25	5×10^{-3}	Br	$(1.8 \pm 1.5) \times 10^{-5}$	5×10^{-5}	4.5×10^{-4}
Zn	$(7.5 \pm 2.0) \times 10^{-3}$	5×10^{-3}	7.2×10^{-3}	Cd	$(1.6 \pm 0.5) \times 10^{-5}$	1×10^{-5}	3×10^{-5}
Ca	$(6.8 \pm 0.2) \times 10^{-3}$	nm	nm	In	$(7.5 \pm 2.6) \times 10^{-6}$	nm	nm
Cu	$(5.2 \pm 2.0) \times 10^{-3}$	1×10^{-2}	nm	Ag	$(7.5 \pm 0.5) \times 10^{-6}$	9×10^{-6}	7×10^{-6}
La	$(3.8 \pm 1.8) \times 10^{-3}$	nm	nm	Hg	$(6.5 \pm 1.5) \times 10^{-6}$	3×10^{-4}	7×10^{-6}
Nd	$(3.3 \pm 0.5) \times 10^{-3}$	nm	nm	Se	$(4.9 \pm 0.2) \times 10^{-6}$	1×10^{-3}	7×10^{-5}
Y	$(3.2 \pm 0.3) \times 10^{-3}$	nm	nm	Te	5×10^{-7}	nm	nm
Gd	$(3.2 \pm 0.4) \times 10^{-3}$	nm	nm	Au	$(2.8 \pm 1.7) \times 10^{-7}$	nm	nm
Co	$(2.3 \pm 0.7) \times 10^{-3}$	8×10^{-2}	1×10^{-3}				

Note: nm = not measured.

Table 7.9. The global mean elemental composition of soils (%).

Element	Average content	Max	Min	Element	Average content	Max	Min
O	50	5.1	46.4	Cr	0.007	0.010	0.001
Si	28	33	2.4	Sn	0.006	0.010	0.001
Al	7.0	8.2	4.2	Ce	0.005	0.008	0.001
Fe	3.94	4.72	3.0	La	0.004	0.005 6	0.000 65
Ca	2.7	30.2	1.37	Ni	0.004 5	0.008 0	0.002 0
Mg	2.3	4.7	0.56	Cu	0.003	0.005 0	–
Na	1.4	4.8	0.04	Co	0.002 1	0.002 8	0.000 8
K	1.57	2.66	0.27	Ga	0.002 3	0.003 0	0.001 3
H	1.2	1.3	0.88	Sc	0.001 5	0.002 1	3.4×10^{-4}
Ti	0.48	0.56	0.40	Pb	0.001 2	0.002 0	6×10^{-4}
Rb	0.23	0.45	0.014	Y	9×10^{-4}	1×10^{-3}	8×10^{-4}
Gd	0.20	0.20	–	B	4.5×10^{-4}	nm	nm
Li	0.20	0.23	–	Th	3.0×10^{-4}	1×10^{-3}	6×10^{-5}
C	0.15	11.4	0.10	Cs	3.0×10^{-4}	nm	nm
P	0.15	0.17	0.004	Sm	2.6×10^{-4}	nm	nm
S	0.10	0.17	0.024	Hf	7.0×10^{-4}	1×10^{-3}	4×10^{-4}
Cl	0.045	0.17	0.001	Yb	5.5×10^{-4}	nm	nm
N	0.050	0.070	0.001	Sb	7.5×10^{-4}	1×10^{-3}	5×10^{-4}
F	0.020	0.025	–	Lu	2.7×10^{-4}	5×10^{-4}	4×10^{-5}
Br	0.004	0.108	0.000 8	Mo	1×10^{-4}	nm	nm
Mn	0.086	0.120	0.036	Zu	8×10^{-5}	1.6×10^{-4}	6.0×10^{-6}
Ba	0.036	0.060	0.001	I	3.5×10^{-5}	nm	nm
Zn	0.036	0.152	0.002	Ag	2.1×10^{-5}	5×10^{-5}	1×10^{-6}
Sv	0.032	0.061	0.005	Se	1.0×10^{-5}	3.1×10^{-4}	8×10^{-6}
Zr	0.010	0.015	0.005	Cd	7.5×10^{-6}	1×10^{-5}	5×10^{-6}
V	0.010	0.016	0.002	In	1×10^{-6}	nm	nm

Note: nm = not measured.

Ca, C, Na, and K, reflecting the known fact of the existence of different types of soils: sand, clay, lime-carbonate, saline lands, etc. The content in soils of the elements O, Al, Fe, Ti, Li, Cd, Zr is stable. A normalized representation of such data is more obvious. Usually the content of elements in the Earth's crust is normalized (clarks). In the absence of data for all the elements, the normalization is made by one of the relatively stable elements: Al, Fe, Si, Ti, and Sc, to assess enrichment with respect to the Earth's crust. In a similar way, the normalization can be carried out for soil and seawater, as well as other possible sources of the material. Apparently, in the case of measurements of the elemental composition of aerosols in the surface layer of the atmosphere, the normalization by elemental composition of the surrounding surface is most logical.

In this case a high value of εF_{crust} for a certain element indicates the existence of a further source of this element. With the phase transformations of water in the atmosphere and precipitation the aerosol particles are captured, and an adsorption of gas products (NH_3, SO_2, O_3, etc.) takes place with subsequent reactions in cloud elements and precipitation. Therefore, the elemental composition of atmospheric water also contains information about the components polluting the atmosphere.

Since the lifetime of water in the atmosphere is about 6 days, and cloud droplets can both wash out and emit the aerosol substance to the atmosphere, a consideration of these processes is also needed to assess the existence of various elements in the atmosphere.

Table 7.10 gives the modelled contents of various elements, of basic types, in atmospheric aerosol: global, background, urban, industrial, continental, marine, dust (Saharan), and Antarctic. The table gives elements which are either very seldom identified or found not providing important information for identification of sources.

Of special interest are elements Be, Li, F, P, and I as well as elements having radioactive isotopes with sufficiently long half-lifes. However, their behaviour in the atmosphere is not discussed in this book. The review of the presented data reveals a wide range of observed concentrations. One should note that the tables do not contain extreme cases of both maximum and minimum values of concentrations of elements. Even for industrial aerosols a maximum value of the mass concentration of the solid residual of aerosols given in the Table 7.10 is 424.5 $\mu g\, m^{-3}$, which is close to the usual values of aerosol concentrations in heavily polluted cities, and may be overestimated by an order of magnitude or more. The same can be observed during dust storms. Note that the size distribution of aerosols and the ratio of elements vary widely.

A change in the size distribution and in the ratio of elements also takes place with changing altitudes of aerosol sampling. Most of the instruments for the elemental analysis of atmospheric aerosols are mounted at a height of 1.0–1.5 m above the surface. During a long sampling (diurnal) the impact of the height of the inlet on the elemental composition of aerosols is smoothed by the diurnal change in the content of aerosols in the surface layer.

The ratio of elements in aerosols is a more conservative property of aerosols than absolute concentrations of elements. Therefore, at normalization of the concentration of elements by their total concentration or by the concentration of the most conservative element, one can obtain more stable patterns of the state of aerosols in the atmosphere. The content of elements in the Earth's crust, soils, seawater, etc., can serve as the basis for normalization.

On the assumption that there is an element of substance which is completely transformed to the aerosol state, normalization by this element, considering the content of this element and identified elements in the original substance, makes it possible to estimate the enrichment coefficient from a given element with respect to the original.

$$EF_{\text{orig.}}(M/R) = \frac{[M]_a}{[R]_a} \Big/ \frac{[M]_{\text{orig.}}}{[R]_{\text{orig.}}}$$

where $[M]_{\text{orig.}}$ and $[M]_a$ are concentrations of the identified element-original in aerosol state and $[R]_{\text{orig.}}$ and $[R]_a$ – for the element-reference. The recalculation of

Table 7.10. The elemental composition of the basic types of atmospheric aerosols (ng m^{-3}).

Element	Global	Continental			Marine			Industrial			Antarctic			Dust (Sahara)		Back-ground
		Min.	Max.	Aver.	Min.	Max.	Aver.	Min.	Max.	Aver.	Min.	Max.	Aver.	Max.	Aver.	
Na	1,100–2,850	10	200	115	220	5,100	1,900	1,000	15,000	8,200	2.3	11	7	1,700	350	874
Si	1,800–5,000	500	10,000	2,000	100	4,800	1,500	4,400	8,400	6,400	0.1	1.0	0.3	120,000	24,000	500
S	2,300–5,400	300	13,500	6,750	220	7,500	3,750	4,800	28,000	14,000	1.5	10	3.0	6,000	1,200	1,200
Cl	2,200–5,400	7	7,000	1,000	200	2,600	1,200	1,100	20,000	14,000	0.8	3.2	1.6	80	16	1,075
Al	600–1,800	50	1,500	900	3	100	50	1,800	17,000	13,000	0.3	1.2	0.8	12,500	7,250	80
Fe	430–2,000	9	5,100	500	30	1,000	580	2,200	25,000	12,000	0.05	0.2	0.09	22,000	4,500	124
Ca	330–2,150	100	10,000	500	50	3,650	1,200	2,000	5,000	2,950	0.05	0.35	0.20	12,000	2,410	198
Mg	980–1,300	20	3,000	190	100	1,700	1,400	1,100	2,000	1,400	0.25	2.4	1.2	9,400	1,870	169
K	780–1,200	50	3,000	400	35	1,500	900	300	8,000	1,200	0.10	0.23	−0.15	8,000	1,580	116
Ti	10–160	0.85	180	60	0.08	2.7	240	190	2,200	1,800	–	0.10	0.01	3,500	690	60
Mn	15–90	0.66	730	15	0.05	14	8	60	1,500	960	0.004	0.02	0.01	480	96	3.3
V	4.9–45	0.20	30	3	0.024	14	10	52	1,500	1,100	0.000 6	0.002 4	0.001 5	60	12.3	0.75
Zn	170–450	2.0	70	60	0.036	56	1.4	50	6,500	4,600	0.020	0.060	0.040	120	23.4	11
Br	5–7.8	0.17	120	0.54	1.0	27	9.9	8.9	1,000	740	0.000 9	0.002	0.001 5	4	0.80	0.5
Cr	6.4–18	0.6	37	5	0.001	7	3.5	10	500	140	0.003	0.010	0.006	100	20.5	1.3
Cu	24–120	0.5	90	12	0.017	10	5	23	1,300	1,000	0.025	0.064	0.045	5.0	1.05	9.5
Ba	75	0.8	50	14	–	–	28	–	–	38	–	–	–	70	14	10
Pb	47–125	0.2	40	15	0.05	64	37	300	7,000	4,800	0.2	1.2	0.7	24	4.8	13
Ni	20–720	0.15	100	42	0.01	7.2	2.5	240	1,200	870	–	–	–	140	28.2	0.86
Ce	4	0.7	20	1.4	–	–	3.3	–	–	61	–	–	–	63	12.6	
Co	1.8	0.004	3	0.24	0.003	0.5	0.08	2.9	27	14	0.000 4	0.001 2	0.000 8	10	2.0	0.02
Cd	3.7–6.9	0.02	30	3.5	0.003	0.6	0.29	3.4	120	85	0.003	0.6	0.3	12	2.4	0.46
Sc	0.4	0.008	15	0.12	0.0008	0.18	0.10	–	12	4.3	0.000 6	0.002	0.001 8	5.5	1.1	0.10
Sb	7.2	0.18	3	1.8	0.001	1.6	0.75	14	200	160	0.000 5	0.003	0.001 8	1.0	0.2	1.0
As	12.5	0.02	15	3.6	0.01	0.8	0.30	5.2	100	81	0.007	0.04	0.024	6.8	1.36	0.05
Hg	0.83	0.02	32	0.22	0.02	5.0	1.3	1.2	23	17	–	–	–	2.0	0.4	0.05
Se	3.4	0.05	4	0.5	0.02	0.5	0.35	2.3	22	18	0.004	0.008	0.006	2.4	0.48	0.10
Ag	1.5	0.01	3.8	0.08	0.002	0.08	0.018	1.7	–	17	–	0.007	0.004	0.86	0.17	0.05
N	680–2,000	440	1,300	800	4.1	137	69	1,880	17,800	13,600	0.28	1.12	0.75	3,060	1,700	60
C	8,400–25,200	530	16,000	9,800	5.1	170	85	3,450	32,000	24,500	4.0	16	10.6	6,080	2,800	720
O	9,000–27,000	610	18,300	11,000	15.0	500	250	25,000	230,000	180,000	3.0	12	8.0	152,700	45,900	2,000
H	770–2,300	54	1,620	970	1.6	54	27	2,300	21,000	16,000	0.28	0.84	0.56	12,800	5,200	330
Σ	30,200–83,800	2,690	92,100	35,400	950	29,000	12,800	49,650	424,500	315,500	13.3	62.7	35.4	371,000	100,000	7,560

the data in Table 7.10 for the Earth's crust by the element Al suggests the conclusion that the elements, by their characteristics, may be divided into several groups: (a) elements similar to Al in their properties and with enrichment factors close to unity with respect to the Earth's crust; (b) elements in aerosol state, markedly enriched with respect to the Earth's crust, with an enrichment factor from ten to several hundreds, (c) elements with enrichment factors of 10^3–10^5, and (d) elements with very variable enrichment factors (Na, Si, S, Cl, Br).

With the interpretation of the behaviour of the factors of enrichment of elements one should take into account the imperfectness of the behaviour of Al as a normalization element due to: (a) the variability of its content in the substance-original; (b) the dispersion of substances containing Al, not corresponding to the dispersion pattern for the substance itself (an accumulation of Al in a certain fraction of particles, for instance, in the coarse fraction), and (c) the incomplete transformation of Al from its original state to an aerosol state. These factors can be assumed to be able to change the factors of enrichment for other elements by 2–3 times (mainly, toward an increase). Now one should understand the causes of high factors of enrichment for elements not determined by a strong difference in the elemental composition of the substance-original and the Earth's crust. Table 7.10 demonstrates a strong change of elemental composition of the substance-original for marine and industrial aerosols. In another case, the main causes of high factors of enrichment are either by very long lifetimes of aerosols containing these elements with a long-lived activity of the source or by the formation of aerosols directly in the atmosphere from vapour or gases containing these elements. In this case, vapours and gases may have other sources. The resulting aerosols will be mainly fine particles and, hence, long-living.

Complex studies of atmospheric aerosol confirm that the elements Se, Pb, Ag, Cd, As, Sb, Br, and Hg are contained mainly in fine particles, and the elements Cl, S, N, C, and Br are present in substances which form directly in chemical and photochemical reactions in the atmosphere (organic matter, acids, and salts). In particular, with a very high emission of these substances to the atmosphere, for instance, volcanic eruptions, the factors of enrichment for these elements forming aerosols *in situ* change in time [66].

The water content in the model substance does not exceed 1% by mass. In the real atmosphere, it can increase up to 90% or more, depending on relative air humidity (i.e., the total mass concentration of aerosol substance can increase by an order of magnitude and more). (The growth of concentration, in particular, is limited by deposition of fog droplets onto the surface.) The growth of total aerosol mass concentration can be calculated from known empirical formulas, for instance:

$$\tilde{m}_{aw} = \tilde{m}_{ao}(1 - \mathrm{RH})^{-\varepsilon}$$

where $\tilde{m}_{ao}$ is the mass concentration of aerosol solid residual; $\tilde{m}_{aw}$ is the aerosol mass concentration the observed relative air humidity; and the exponent ε is an empirically determined parameter approximately equal to 0.25 ± 0.03.

7.5.2 Modelling the chemical composition of aerosols

The chemical composition of particles can be different and is determined both by initial materials and by mechanisms of particle generation. In some cases these are composite particles consisting of several chemical substances. For instance, acid rains consist of a water solution of acid; radioactive particles formed in accidents with dispersion of nuclear materials include not only radioactive substances but also materials of soil, etc.

By chemical composition, there are marine, continental, urban, and stratospheric aerosols. Although chlorides are the main component of marine aerosols, the content of sulphates and organic matter can be in these cases rather substantial, also. The chemical composition of the mineral component of atmospheric aerosols over the continents (% by mass) has been studied in most detail: SiO_2 – 40 ± 50%, Al_2O_3 – 15%, Fe – 5.4 ± 6.0%, CaO – 2.4 ± 0.9%, Na_2O – 2 ± 1.5%, K_2O – 1.9 ± 1.1%, MgO – 1.5 ± 1.2%, TiO – 1.0 ± 0.2%, and MnO – 0.08 ± 0.05%. Apart from these components, the content of Cu, Ba, Ni, Sc, Cr, and Zn is relatively significant.

A large cycle of measurements of the elemental composition of continental aerosols has been carried out by the scientists of the St Petersburg University. Measurements were made in the summer–autumn periods in different climatic zones of Russia. The data obtained revealed a very strong temporal–spatial variability of the aerosol chemical composition, in particular, the diurnal change of the content of different chemical components determined by diurnal variations of the intensity of aerosol formation, as well as convective and advective processes in the atmosphere.

The specific behaviour of aerosols of different chemical nature is responsible for the different chemical composition of aerosols in the atmosphere at different altitudes. The vertical distribution of the chemical composition of aerosols in the troposphere is characterized, as a rule, by a weakly decreasing mass concentration of most elements contained in inorganic aerosols. For some elements (e.g., Fe), this decrease can be absent. The concentration of SiO_2 decreases most strongly with altitude, which is apparently connected with relatively large sizes of particles containing silicon.

The chemical composition of aerosols in the upper troposphere and stratosphere is characterized by a high content of anion SO_4^{-2} in the aerosol substance, a relatively high content of Fe and other elements characteristic of mineral substances. In the Junge layer at altitudes of 17–21 km, sulphuric acid and sulphates are the main component of the aerosol substance.

7.5.3 Modelling the complex refraction index of the atmospheric aerosol substance

In modelling the complex refraction index, two approaches have been used: statistical, based on the use of data of laboratory measurements of the spectral absorption of aerosol samples and the refraction index; and synthesis of the complex refraction index of a mixture of pure substances with known ratios of the constituting

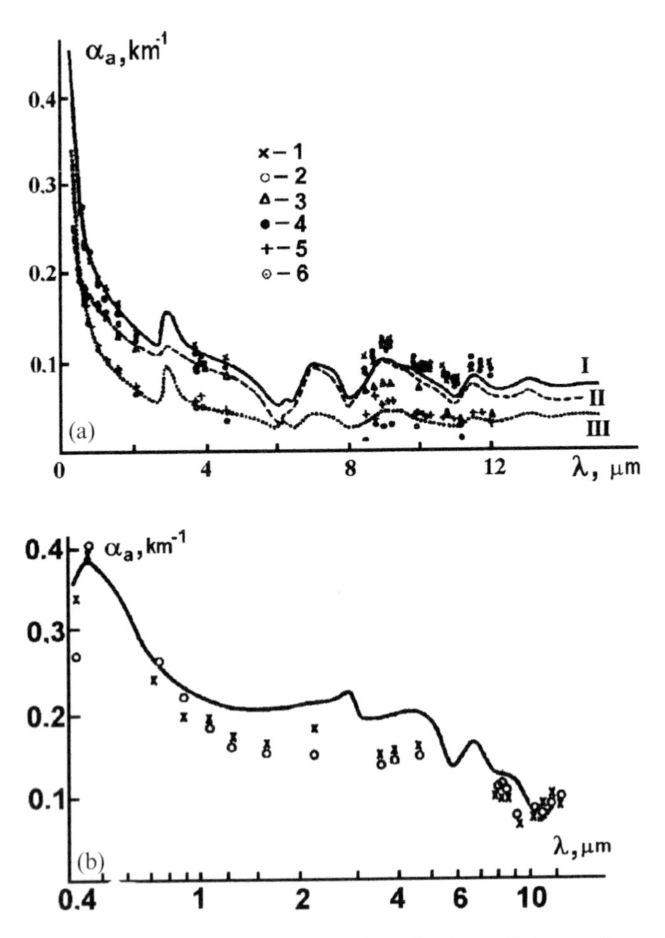

Figure 7.5. (a) Calculated (curves) and observed (points) coefficients of aerosol attenuation for different situations in the atmosphere. Observational data: 1, 2, and 5 – measurements made in Tomsk; 3, 4, and 6 measurements made in the settlement of Voeykovo. I–$S_m = 11\,\mu$m, RH = 44%; II–$S_m = 14\,\mu$m, RH = 45%; III–$S_m = 19\,\mu$m, RH = 81%. (b) Comparison of calculated (curves) and observed (points) estimates of the coefficients of aerosol attenuation for coastal regions in the Crimea.

components [59–61]. Both drawbacks and advantages of the two approaches are apparent. The first approach was successfully used for a certain type of aerosols (sulphates, of desert origin). The second approach is characterized by wider possibilities of application, making it possible to analyse the impact of changes in the chemical composition on variations of the magnitude and spectral change of the complex refraction index [61], first of all, the water content in the substance of particles (Figure 7.5) [60]. The specific feature of the spectral change of the complex refraction index for different components of the aerosol substance enables one to use certain simplifications in their modelling and apply them to substantiate the models and to solve inverse problems of aerosol optics.

From the numerical modelling of the aerosol optical properties carried out during recent years, first of all, light absorption by particles with different size distributions [21, 63, 74–79, 91, 146, 150] revealed a marked impact on the optical properties of aerosol of not only the complex refraction index and size of particles but also parameters characterizing the internal size distribution of a chemical substance, the shape of particles and their orientation with respect to the source of emission, and the state of the particle's surface. In particular, a possibility has been discovered of the photophoretic effect in the real atmosphere [91].

A model has been proposed [7] of specific aerosol in the middle atmosphere, where specific features of the size distribution of particles distinctly manifest themselves in their optical and aerodynamic properties.

7.5.4 Inverse problems of the atmospheric aerosols optics

Progress toward solution of inverse problems of aerosol optics, as determined by a priori information on the spatial–temporal variability of atmospheric aerosols, has become more complete and definite [6, 62, 66, 79, 90, 124, 162].

To avoid repetition (inverse problems have been discussed in detail in [110, 164]), note only that one should remember that in the real atmosphere, situations can be observed when the use of a priori representation can lead to a wrong solution.

The classical methods of inversion of the optical characteristics of aerosol into their size distributions can give results far from reality, especially with a multimodal structure of atmospheric aerosol [113, 116]. Therefore, it is necessary to combine the results of remote optical soundings of aerosol characteristics with *in situ* microphysical measurements of aerosol characteristics, which simplify the solution of inverse problems and enable one to obtain additional information about aerosol properties from discrepancies in data of the optical measurements and results of calculations of the optical characteristics from the data of *in situ* measurements [62].

7.5.5 The submicron fraction and mechanisms of its transformation

The molecular composition of the submicron fraction of aerosol is very complicated, variable, and poorly studied. In the main, these are sulphur, nitrogen, and chlorine-containing salts and acids, metal oxides, water containing compounds of nitrogen and sulphur, and hydrocarbons – mainly products of oxidation and polymerization of volatile oils and aromatic compounds occurring under natural and urban conditions [8, 66, 79, 90, 95–97]. The problem is that we have to deal with both mixed compounds and particles which are a superposition of condensates of different natures. The proportions, in which various radicals are contained, as well as concrete forms of their combinations, are rather variable. Therefore, in modelling the aerosol properties it makes sense to take only the group characteristics of compounds of various types.

The state of the submicron fraction practically determining the optical weather is the product of the combined impact of several factors: the process of generation with distinct diurnal seasonal and geographical changes, which are different for organic and inorganic components and responsible for an accumulation of submicron aerosol during periods of weeks; the process of condensation transformation regulated by local variation of relative air humidity; the slow process of aerosol 'ageing' as a result of coagulation growth and conglomeration of particles; and the process of particle removal from the atmosphere by washing out and sedimentation.

Note that in the process of formation of large-droplet fogs and clouds after reaching dew point, particles of the submicron fraction practically do not participate, being part of large-droplet formations as an independent fraction. Although the existence and the important role of these processes have been detected, their laws, remaining poorly studied, cannot be used in model construction. An exception is a distinctly expressed process of condensation transformation affecting many optical parameters of aerosol. The most important feature of the submicron fraction turned out to be a stability of particle size distribution (i.e., its low sensitivity to the origin and chemical individuality of particles). Such a distribution is well approximated by not an exponential law, as had been thought earlier, but by a lognormal law. From the viewpoint of substantiation of the mechanisms of generation, evolution, and sinking of aerosol in the atmosphere, most substantiated is an inverse gamma distribution $f(r) = Ar^{-\nu} \exp(br_0/r^s)$, where ν corresponds to the Junge index, $r_0 = a_m$ is the modal radius, and $A = N_0$, b, and s are parameters of distribution.

Under real conditions, in the surface layer of the atmosphere, the range of variability of the distribution parameters is determined, approximately, by the following quantities: $V = V_a/V_{\text{air}} = 10^{-9}$–$10^{-12}$; $N = (2$–$10)10^3\,\text{cm}^{-3}$; $a_M = 0.1$–$0.35\,\mu\text{m}$; and $\nu = 0.5$–0.7. At the 20-km level $V \cong 10^{-14}$–10^{-15} and $a_M \cong 0.07$–$0.15\,\mu\text{m}$.

Measurements show that the half-width of the distribution ν is comparatively stable, because all variations of conditions of the formation and existence of the submicron fraction manifest themselves mainly in variability of only two parameters of distribution, namely, the factor of filling V and location a_M of distribution maximum. The first of these parameters mainly affects the quantitative characteristics of aerosol of the optical medium (coefficients of scattering, absorption, and attenuation), whereas the second parameter affects mainly the characteristics such as phase function, polarization characteristics, and spectral dependences.

In fact, the submicron fraction is the product of a combination of several submicron fractions of different origins and different (sometimes mixed-up) chemical compositions with different parameters of distribution V and a_M and, respectively, different optical constants of substances – indices of refraction and absorption $n(\lambda)$ and $k(\lambda)$. In the process of condensation growth or drying of particles due to evaporation of the liquid component of an aerosol substance, the parameters a_m and V change with the particle number density N remaining preserved. Besides this, with a condensation change in particles, marked changes in $n(\lambda)$ take place, which affect the aerosol optical properties. This determines the difficulty of the microphysical modelling of the optical properties of aerosol, since it is connected with the necessity of flexible manoeuvering with the model's parameters depending on meteorological and geographical conditions.

Specific features of the structure and behaviour of the submicron fraction of real aerosols determine also the specific feature of variability of its optical parameters. First of all, these are the comparatively high stability of connections between various optical parameters and the possibility of a description of the optical weather using few parameters.

7.6 ABSORPTION OF SOLAR AND LONG-WAVE RADIATION BY ATMOSPHERIC AEROSOL

7.6.1 Specific features of spectral characteristics

A great number of parameters determining the optical properties of atmospheric aerosols complicates the numerical modelling of its optical characteristics. However, the use of a priori information obtained from the data of microphysical measurements together with results of spectral optical measurements suggest the following main conclusions on the specific spectral change of aerosol coefficients of attenuation, scattering, absorption, and backscattering:

(1) In the visible and near-IR regions, the spectral dependences of the coefficients are determined, first of all, by the type of particles size distribution [42–44].
(2) Anomalous spectral dependences of the aerosol coefficient of absorption are possible, being determined by the presence of iron oxides in the atmospheric aerosol particles [67].
(3) Anomalous spectral dependences of the coefficient of aerosol attenuation can be determined by a specific type of particles size distribution in the range $r < 1\,\mu m$ at low number densities (most often in mountains).
(4) The presence of water in aerosol particles defines the absorption bands in the spectra of aerosol attenuation in the wavelength intervals 2.7–3.0 μm and 5.8–6.2 μm [6–9].
(5) The presence in aerosol particles of sulphates and sulphuric acid determines the presence of absorption bands near $\lambda = 3.3\,\mu m$ and $\lambda \cong 9.6\,\mu m$ [9, 79].
(6) The presence of silicate particles manifests itself in the strong absorption band 9.0–9.6 μm [32, 61].
(7) The presence of organic substances in atmospheric aerosols manifests itself relatively weakly through the existence of specific absorption bands ($\lambda \cong 6\,\mu m$). More important is the manifestation of specific features of the size distribution of organic aerosol particles and their absorption band in the visible spectral region ($\lambda > 0.35\,\mu m$) [7, 8].
(8) The presence of soot as individual particles and as inclusions into other aerosol particles is very important for the absorbing properties of atmospheric aerosol in the visible and near-IR spectral regions [7, 24, 97].
(9) An existence of maximum aerosol attenuation near $\lambda \cong 9\,\mu m$ due to the presence of both sulphate and silicate particles results in the ratio of aerosol absorption to attenuation in this wavelength interval varying markedly (approximately from

0.5–0.02, depending on the region and altitude of the atmospheric layer where measurements are made).

Of great interest are the results of complex optical–microphysical studies which make it possible to obtain a complete pattern of formation and evolution of aerosol systems under different external conditions. Of importance are simultaneous measurements of the optical characteristics of aerosols in the visible and IR spectral regions. In the case of calculations, it is assumed that water is uniformly distributed in a particle (microcapillary penetration and swelling of particles), and the dependence of particle size on relative humidity is described with the Kasten empirical formula [76].

These results enable one to construct prognostic models of aerosol characteristics with the use of the following input parameters: meteorological visibility and relative humidity. The first parameter makes it possible to express the calculated concentration of aerosols via the aerosol attenuation in the visible spectral region, and the second parameter to determine modal radii of particles and the type of particle size distribution.

7.6.2 The role of submicron and dust fractions of aerosol in radiation absorption

The role of various fractions as absorbing substances in various spectral regions is determined by their molecular composition and mass concentration. In the short-wave spectral region the role of the coarse dust fraction is weak: as measurements show, the fraction of particles with $r = 0.05$–$0.50\,\mu m$ (i.e., the submicron fraction) is mainly responsible for absorption [77, 89, 90, 95–97]. There are *in situ* spectroscopic, chemical, and microscopic data indicating that one of the main absorbing substances in the visible and near-IR spectral regions is carbon – present everywhere in the form of fine soot particles. These soot particles can exist in the form of tiny branchy flakes (fractal particles) with an average radius $<0.1\,\mu m$ or be part (possibly as condensation nuclei) of particles of different natures belonging to the submicron fraction. However, here the carbon particles also appear as an independent fraction in a certain sense.

The absorbing ability of aerosol particles formed of organic compounds manifests itself most with $\chi \geq 0.2$ in light hazes and $\chi \leq 0.02$ in heavy hazes, which is quite realistic for many chemical components that form particles. Therefore, it might be expected that the organic component plays a marked role in the aerosol light absorption in the visible and near-IR spectral regions, especially under conditions of strong development of this component in smog episodes (natural or anthropogenic). The behaviour of scattering matrices indicates that the absorbing component is concentrated not in the whole submicron fraction but only in its finest subfraction with an average size of particles $r < 0.1\,\mu m$ [95]. Especially important is the role of the organic component in light absorption in the region of the strong absorption band near $3.2\,\mu m$ characteristic of volatile oils [8, 96]. One can expect its strong increase as it moves to the UV spectral region.

The role of sulphur and nitrogen-containing compounds in radiation absorption in the short-wave spectral region is weak. However, they start playing an important, and possibly, prevailing role in the long-wave region, where the main contribution to aerosol absorption is made by the submicron aerosol fraction [120]. Here the absorption by the dust-fraction particles becomes substantial (at least, in certain spectral intervals), and grows with increasing dust-loading of the atmosphere, especially under specific conditions of desert and loess-forming regions (including the region of the so-called 'Sea of Gloom' near the western coastline of equatorial Africa).

The true light absorption by aerosol becomes substantial against a background of other optical effects at $4\chi\rho \geq 1$ [120]. Therefore, the absorption of light by condensed water in the wavelength interval 0.3–13 μm is substantial only in clouds and large-droplet fogs. In hazes and foggy hazes where in the shortwave interval ($\lambda = 0.3$–$4\,\mu$m), $\rho \leq 15$, and in the longwave interval ($\lambda \geq 8\,\mu$m), $\rho \leq 1$, the absorption of light is substantial due to non-water components with strong bands of molecular absorption, namely, soot, organic, sulphuric acid, and hydronitric components. In dust formations the absorption is determined by the mineral fraction of aerosol. Therefore, the concrete type of the spectrum of light absorption by aerosol varies strongly both quantitatively and qualitatively, depending on the nature and parameters of aerosol – the influence of molecular composition of the latter being insufficient.

A complex comprehensive approach to modelling the optical characteristics of aerosol of the surface layer of the atmosphere has been demonstrated [6, 113]. Based on the data on aerosol size distribution and complex refraction index for its individual fractions, calculations have been made of spectral characteristics of these fractions considering the effect of humidity both on the spectrum of particle sizes and on optical constants. By comparing the calculated data with observations made with the use of normalized coefficients of aerosol attenuation, the ratio of concentrations of particles of different fractions was estimated and particle size distributions were specified proceeding from 3-modal parameterization which was compared with microphysical data. Then, a parameterization was made of the dependence of $N(r)$, $n(r)$, and $\chi(r)$ on S_m and relative humidity, and spectral characteristics $\varepsilon(\lambda)$, $\alpha(\lambda)$, and $\sigma(\lambda)$ as well as elements of the scattering matrix were calculated.

Figure 7.6 demonstrates for comparison purposes the results of *in situ* spectral measurements of $\varepsilon(\lambda)$ and model calculations from microphysical parameters with the use of the technique of synthetic complex refraction index [6]. An analysis of results of comparisons of calculated values of the optical characteristics of surface aerosol with observation data in a broad spectral region and under various atmospheric conditions has led to the conclusion that this model makes it possible to predict, with an accuracy of 30–50%, the spectral change and absolute values of the optical characteristics of atmospheric aerosols in the spectral interval 0.3–15 μm. For the spectral interval $\lambda < 1\,\mu$m the accuracy of prediction is much higher (10–15%).

7.6.3 The impact of tropospheric aerosols on surface atmosphere albedo and radiative flux divergence in the atmosphere

For tropospheric aerosols it is of interest to construct a kinetic model [120] based on the scheme of basic processes: synthesis of vapour of aerosol-forming compounds

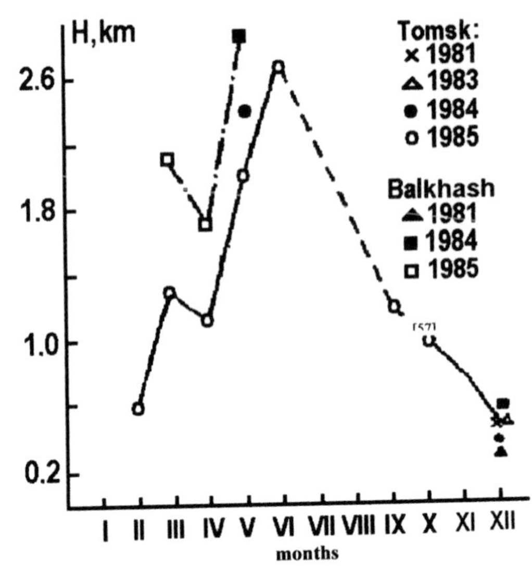

Figure 7.6. The dynamics of altitudes of the mixing layer for Tomsk and Balkhash.

and clusters, heterogeneous condensation, coagulation, and moisturizing and removal of fine particles. The basic input parameters of the model are the volume concentration of such vapour and the characteristic time of the existence of the submicron fraction of fine aerosol ($r \sim 1$–$100\,\mu$m). The submicron fraction of aerosol is of intra-atmospheric origin. Continuously synthesized from gas phase in the atmosphere, it is characterized by the concentration 10^5–$10^6\,\mathrm{cm}^{-3}$ with modal radius $0.04\,\mu$m and average mass concentration $\sim 1\,\mu\mathrm{g\,m}^{-3}$. The global mass averages ~ 50 Mt, at a global rate of formation of $\sim 5 \times 10^2\,\mathrm{Mt\,yr}^{-1}$.

The submicron fraction (10–800 nm) of aerosol most stable and long-lived is assumed to be a mode mainly responsible for the optical state of the atmosphere. Characteristic values of the number concentration of the aerosol submicron fraction constitute $(2$–$10) \times 10^2\,\mathrm{cm}^{-3}$ with a modal radius of 0.1–$0.35\,\mu$m. In this fraction, soot formed of hot vapour of hydrocarbons in the high-temperature processes of burning plays an important role.

Crystal carbon present in submicron aerosol in the form of various types of soot, plays an important role in the radiative regime of the atmosphere. These particles combine in flakes (several hundredths of μm) and deposit onto the surface of dust particles, droplets, etc., and then are removed from the atmosphere together with the carriers after several tens of hours (sometimes 1–2 weeks). The soot density in the air changes from $1\,\mu\mathrm{g\,m}^{-3}$ under especially pure conditions to 10–$30\,\mu\mathrm{g\,m}^{-3}$ in heavily polluted regions – its share approaches 10% of the total mass of submicron aerosol. The annual mean amount of anthropogenic aerosols emitted to the atmosphere constitutes about 30 Mt [66, 79, 95]. The soot component of aerosol is to a great extent responsible for its role as a radiative factor, at any rate, in the short-wave spectral region. The characteristic non-selective absorption by soot often prevails over the absorption of other components in the visible spectral region and in the near and mid-IR range, except for absorption bands of organic components in the UV spectral region.

The short-wave absorption due to the coarse aerosol fraction (dust and water) with sizes of particles >1 μm is also mainly determined by particles of the submicron fraction deposited on the surface of large particles. So, for instance, for aerosol samples taken in mountains (Ambastumani, Georgia) in the course of the Soviet–American background experiment, spectral dependences were measured of the background coefficient of aerosol absorption in the visible and IR spectral regions, which in the wavelength interval 0.25–0.8 μm change from 0.096–0.13 km^{-1}, which is several times less than in the presence of dust–sand outbreaks and by an order of magnitude less that the coefficient of aerosol absorption in industrial cities [95]. In the visible, the aerosol absorption due to the submicron fraction of aerosol is connected with the presence of crystal carbon, and in the UV region with the presence of organic constituents in particles. For comparison purposes, the average absorption coefficient of the urban aerosol constitutes $6.5 \pm 2.9\ km^{-1}$, with the light scattering coefficient $9 \pm 6\ km^{-1}$ (i.e., the single scattering albedo constitutes about 0.6), which favours the heating of the atmosphere and surface. The background value is 0.92, whereas 0.85 is the critical level separating the aerosol effects of radiative heating and cooling of the atmosphere. In mountains, under background natural conditions, the mass concentration of soot varies from 0.1–1 $\mu g\, m^{-3}$. The maximum contribution to total absorption is made by particles with radius 0.16–0.3 μm. For background conditions of Ambastumani, based on the complex optical, chemical, and mass spectrometric analysis of aerosol samples, the contribution was estimated, of the organic constituent of submicron particles to radiation absorption by aerosol, in the range of 2.5–14 μm. Maximum values of the absorption coefficient were observed near 3.4 μm (8–10 km^{-1}) and 5.8 μm (3.3–10 km^{-1}).

Calculations of the imaginary part of the refraction index for atmospheric aerosol for the regions of Ambastumani and Moscow have shown the prevailing role of the submicron fraction of aerosol. A considerable contribution to the absorption by the aerosol submicron fraction in the IR region (3 μm and 6 μm) is made by organic constituents along with soot (5.05 μm) and ammonium sulphate (7.07 μm and 9.17 μm) [95–96]. The anthropogenic aerosol contributes strongly to the radiation attenuation in the 8–14-μm region. An analysis of characteristics of total and aerosol attenuation (in the range 5–2.2 μm) in the surface layer of the atmosphere and their connection with meteorological parameters has shown that in the averaged spectrum of aerosol particle concentration, maxima are observed for particle sizes of 0.2 and 0.5 μm, and in the averaged spectrum of aerosol attenuation, an absorption band is distinctly seen near 9.6 μm at a low relative humidity (25%). With the growth of humidity, the averaged spectrum of aerosol attenuation changes substantially, which is connected with aerosol watering and reduced absorbing properties. A negative correlation has been detected between temperature and the submicron fraction of aerosol (0.2–0.5 μm), but positive between temperature and concentration of particles 0.5–2 μm, which points to different mechanisms of generation of these fractions of aerosol particles in the atmosphere [130].

It follows from the data of numerous observations that under background conditions (both over the sea and over land) the change of the aerosol optical thickness of the atmosphere ranges within 0.05–0.15, and for its average value one can assume a magnitude comparable with the value of optical thickness due to Rayleigh scattering (~0.125). The aerosol optical thickness in the case of dust–sand outbreaks over a city in the presence of anthropogenic aerosol is characterized by values from 0.2–0.45 with an average of 0.25. This value can serve as the reference point in assessment of the anthropogenic impact on the atmosphere. With growing altitude H (0.2–8.4 km) in clear weather, a combined impact of background aerosols over a weakly reflecting surface leads to an increase of the system's albedo over the surface of the Chuckchee and Okhotsk Sea – with a gradient of $\Delta A/\Delta H = 0.004$ and $0.006\,\mathrm{km}^{-1}$ at $h_{\oplus} > 35^{\circ}$, and concentration N of particles with $r > 0.3\,\mu\mathrm{m}$ equal to $\sim 2\,\mathrm{cm}^{-3}$; over the Azov and Black Seas with $\Delta A/\Delta H = 0.006\,\mathrm{km}^{-1}$ at $h_{\oplus} = 40^{\circ}$ and $N = 3\,\mathrm{cm}^{-3}$; over the Atlantic Ocean with $\Delta A/\Delta H = 0.09\,\mathrm{km}^{-1}$ at $h_{\oplus} = 80^{\circ}$ and $N = 8\,\mathrm{cm}^{-3}$; and at a dust outbreak ($N = 30\,\mathrm{cm}^{-3}$) – $\Delta A/\Delta H = 0.037\,\mathrm{km}^{-1}$.

The surface atmosphere system's albedo over a strongly reflecting surface (ice, snow, and clouds) decreases with the altitude, especially strongly in the boundary layer with a gradient of $\Delta A/\Delta H = 0.01\,\mathrm{km}^{-1}$. Based on the field experiments CAENEX, GATE, GAAREX, and FGGE and complex measurements in the free atmosphere of total and spectral radiative flux divergence as well as aerosol optical thickness of the atmosphere (in some cases it was obtained from actinometric observations) there was constructed a dependence of absorptivity b, at the wavelength $\lambda = 500\,\mathrm{nm}$, for semiarid agricultural regions of Texas and California. For background conditions at an aerosol optical thickness up to 0.125, the relative aerosol absorption b at $\lambda = 500\,\mathrm{nm}$ reached 0.04 with total absorption changing from 0.02–0.08 in the layer from 0.5–7–8 km. With the aerosol optical thickness growing to 0.4, the aerosol absorption constitutes 0.2 and can reach 80% of total absorptivity in the spectral interval 0.3–3.0 μm, with an average of 50%. For powerful aerosol layers, the heating due to short-wave radiation absorption reaches $0.4^{\circ}\mathrm{C\,hr}^{-1}$ and prevails over the long-wave cooling [113].

Lidar soundings at $\lambda = 0.69\,\mu\mathrm{m}$ and at other wavelengths gave values of atmospheric optical thickness, under background conditions, close to 0.12, and at dust–sand outbreaks of ~0.8. Surface spectral measurements of short-wave radiation under urban conditions have shown that the anthropogenic impact manifests itself at optical thicknesses $\tau > 0.10$ at $\lambda = 500\,\mathrm{nm}$ through a marked attenuation of direct solar and total radiation and an increasing share of scattered radiation [76]. The dependence of $b(\lambda)$ makes it possible from the data of filter actinometric measurements to assess the aerosol absorption of short-wave radiation without using an expensive means of atmospheric sounding from aircraft. Observational data on absorptivity and albedo of the surface atmosphere system agree well with results of calculations of the radiative impact of aerosol.

The effect of tropospheric aerosol in radiative flux divergence has been considered by [59] from the data on sensitivity of the absorptivity b and albedo A of the atmosphere to the real n and imaginary χ parts of the refraction index $\tilde{m}$ as well as on median radius r and half-width of particles size distribution with the use of the

Table 7.11. The measured and calculated values of $b = f(\lambda)$ for $A = 0.9$ and 0.6 at $\cos\theta = 0.2$.

Aerosol optical thickness	b_{meas}	b calculated at	
		$A_0 = 0.9$	$A_0 = 0.6$
0.1	0.01	0.024	0.05
0.2	0.07	0.041	0.11
0.3	0.13	0.058	0.16
0.4	0.19	0.080	0.21

Eddington delta approximation with two input parameters: total number of aerosol particles N_0 with respect to background number and the ratio of the number of absorbing particles N_n to the background one N_n/N_b. The aerosol particle size distribution is assumed to be log-normal. The optical characteristics of the elemental volume of the medium (the optical thickness, coefficients of scattering, attenuation, and phase function) are determined by parameters n, χ, N_b, and N_o. The respective log-normal distributions with average median radii 0.03, 0.3, and 6.3 µm are attributed to three fractions of aerosol: I – 0–0.1 µm, II – 0.1–1.0 µm, and III – 1–100 µm.

By solving the equation of radiation transfer, radiative characteristics were calculated as functions of aerosol parameters n, χ, N_b, and N in the real range of changes of the latter. An analysis of the dependence of A on n, χ, N_b, and N revealed an importance of the contribution of the submicron fraction and a possibility to obtain, at $\lambda = 0.55$ µm, values of $n = 1.5$ and $\chi = 0.65$, so that N remains a variable parameter. An introduction of absorption $\chi = 0$ reduced all the dependences enumerated. The soot aerosol or hydrocarbons of the general type (organics) should be taken into account as absorbing aerosols with respect to the background aerosol. Then the backscattering albedo for the aerosol atmosphere $A = 0.92$–0.98. The value of N_0 is assumed to be $10\,\text{cm}^{-3}$. These data enable one to calculate the main radiative characteristics A_0 and P depending on the total number of soot particles N, the degree of increasing number of soot particles N_s/N_o, and the optical thickness.

Table 7.11 gives the results of measurements of absorptivity b of the troposphere on the aerosol optical thickness compared with results of calculations of single scattering albedo equal to 0.9 and 0.6, which corresponds to most typical values for background and anthropogenic aerosols.

The observational data and calculations demonstrate the closeness of the values of total albedo A and A_0. But it is most important to take into account the aerosol absorption in the visible spectral region. In the IR region the water vapour absorption always prevails.

On average, total albedo $A_t = 0.8\,A_0$, and the total albedo of the system $A_s = KA_0P(m)$, where $K = 0.8$ for surface albedo $A_s = 0$ and $H = 1$ for $A_s = 0.3$. Here $P(m)$ is the water vapour transmission function for water content m. A

Table 7.12. Averaged latitudinal variations of albedo A_t for background aerosol. (φ is the latitude.)

φ	5	15	25	35	45	55	65	75	85
A_t	0.094	0.094	0.0090	0.0077	0.0074	0.0070	0.0075	0.0001	−0.0036

correction for water vapour is introduced at $m > 0.2\,\mathrm{g\,m^{-3}}$. A decrease of albedo takes place with growing aerosol turbidity, especially for strongly absorbing aerosol at $A_0 > 0.1$. The A_t value can decrease due to a large contribution of scattering on weakly absorbing aerosols (see Table 7.12). At large A_0 values this decrease is substantial at any value of A_0 in the presence of absorption.

The impact of background aerosol on the long-wave (thermal) radiation is mainly weak and often turns out to be negligible compared with its impact on solar radiation. Radiative effects of aerosol are characterized by regional specific features which manifest themselves due to a dependence not only on aerosol concentration, but also on surface albedo and solar zenith angle. Within a certain latitudinal zone, aerosol over the oceans with a low albedo can increase the planetary albedo, enhancing thereby the local cooling of the surface atmosphere system. Over a surface with comparatively high albedo it can reduce the albedo, favouring thereby the system's heating. This especially manifests itself with the equator-to-pole transition. Note that tropospheric aerosol is concentrated mainly in the atmospheric boundary layer, which favours a surface cooling and a boundary layer heating.

As seen, the impact of aerosol leads to a cooling in all latitudinal zones except latitudes >85°. The contingency of the impact of tropospheric aerosol on the surface atmosphere system's albedo and surface temperature should be also taken into account, especially in the transition to the polar regions with snow and ice.

Results of complex studies of basic radiative characteristics of the surface atmosphere system based on measurements of total and spectral radiation fluxes are comparable with reference calculations of radiation fluxes and flux divergences in the turbid atmosphere based on the use of the Monte Carlo method.

7.7 COMPLEX AEROSOL–RADIATION OBSERVATIONS IN THE FREE ATMOSPHERE

The amount of combined observations of aerosol and radiative characteristics performed in the free atmosphere is rather limited. As a rule, the aerosol and radiation measurements are made separately with the use of instrumentation carried by flying laboratories [12, 13, 17, 99, 110, 139, 149, 150–154]. A unique example of the complex aerosol–radiation sounding of the atmosphere is the Soviet–American experiments in Rylsk accomplished with balloon-carried instruments together with aircraft measurements of total and spectral hemispheric radiation fluxes, which have made it possible to obtain data both on radiative characteristics of the atmosphere and surface and on aerosol size distribution as well as optical characteristics of the atmosphere. The instrumentation and technique for aircraft observations in the free atmosphere were the same as in the Complete Radiation Experiment [153]. To compare the obtained data with radiative properties of the atmosphere over a large industrial city, aircraft measurements were made on 9 August, 1975 over the city of Donetsk.

Actinometric aircraft soundings were made up to an altitude of 7.2 km. The hemispherical short-wave radiation flux at this level was 16% less than the extra-atmospheric flux at solar zenith angle 48°. This attenuation is apparently connected with scattering by the overlying atmosphere in the presence of the aerosol layer above 10 km, as well as with the impact of stratospheric aerosol (droplets of sulphuric acid solution, meteorite dust, etc.) and, possibly, with the eruption of the Tolbachik volcano at Kamchatka on 26 July, 1975.

Aircraft measurements over Rylsk on 3 August, 1975 were carried out after disintegration of the cumulus cloud fields (Cu, cloud amount 8) in the presence of a strong haze in the surface layer. Despite the fact that the absorptance of the atmospheric layer (7.2–0.5 km) on this day was somewhat higher compared with the finest day on 5 August, 1975, the absorption was small and constituted only 2.3%. The use of surface and aircraft actinometric data has demonstrated an active role in radiation absorption of the atmospheric surface layer with a height of 0.5 km above ground level.

The absorptance of this layer turned out to be 3.8%, which was connected with the location of haze along the Seim River valley. The rate of radiative change of temperature in the surface layer was by an order of magnitude greater than in the overlying layer (7.2–0.5 km). According to the spectral analysis of aerosol samples, the content of Fe, Mg, and Al at altitude 0.5 km was 4.4, 0.47, and 0.1 $\mu g\, m^{-3}$, respectively. The concentration of soil dust constituted $\sim 50\, \mu g\, m^{-3}$. Filter measurements of aerosol number concentration at the 300-m level have shown that for particles with $r > 0.2\, \mu m$ the concentration constitutes $2.80\, cm^{-3}$ and decreases with altitude. The total transparency coefficient at that time was 0.85, the aerosol optical thickness for $\lambda = 500$ nm constituted 0.16 at meteorological visibility $S_m = 30$ km. (For comparison, data for St Louis (USA), $S_m = 8$ km, Los Angeles 16 km, Denver and London 6–25 km, and in the Arctic 100–150 km [79, 90, 120, 157].)

The surface-atmosphere system's albedo increased with altitude. The spectral albedo of the land (ploughed)–atmosphere system has a small maximum at about 500 nm which is explained by the coloration of agricultural crops in the fields and grass cover. The spectral albedo of the city–atmosphere system from altitude 7.2 km at the same sun elevation, in the visible spectral region, increased with wavelength, and in the near-IR region, except the bands of molecular absorption, exhibited a neutral character. The total albedo was about 20%. The spectral distribution of relative radiative flux divergence in the visible was of neutral character and connected mainly with molecular absorption.

Over the large industrial city of Donetsk (the Ukraine), the aerosol layer is located at a much higher altitude. Therefore, the absorptance of the 7.2–0.5-km layer turned out to be 14%, which is connected with the impact of emissions of industrial enterprises and with coal mining in this region. The spectral change of the relative radiative flux divergence was similar to the spectral distribution of absorptance over Zaporozhye during CAENEX-72. This spectral change in the short-wave spectral region is due to absorption on particles containing coal dust, soot, and iron oxides. The chemical composition of aerosol in the surface layer over the city was characterized by an iron content of up to $50\,\mu g\,m^{-3}$.

Within the Soviet–American cooperation in the summer of 1976 in the region of Laramie (USA), balloon observations of aerosol and radiation were undertaken, aimed at: (1) a comparison of the vertical profiles of atmospheric aerosol obtained in different ways (with the use of the filter trap and impactor (USSR) and optical particle counter (USA)); and (2) a consideration of the impact of aerosol on the long-wave radiation transfer from the data of actinometric sounding of the atmosphere together with the data of aerological measurements and information about the vertical distribution of ozone and other optically active gas components of the atmosphere. Each sounding was accompanied by measurements of aerological parameters of the atmosphere. All of them were made in the absence of cloudiness and with a light breeze. The balloon instrumentation included the impactor, which gives information about particles in the size range 0.2–2.5 μm. At the same time, filter samplings were made in three layers of the atmosphere (1.5–23, 5.5–10.5, 10.2–15 km). The vertical resolution was 300 m with a 20% relative error in aerosol concentration measurements. The optical dustsonde measured particles with the radius >0.15 and >0.25 μm with a vertical resolution of 250 m. During the rise of the balloon, the aerosol number density was measured every 15 minutes, the error of concentration measurement constituted about ±5%. Simultaneously, the condensation nuclei counter was used with an expansion chamber in which an oversaturation with respect to water constituted 200–300%. The actinometric radiosonde ARS-1 made it possible to measure long-wave radiation fluxes with a MSD = $0.005\,cal\,min^{-1}\,cm^{-2}$.

Impactor measurements and measurements with the photoelectric dustsonde for particles with $r > 0.15$ and 0.25 μm revealed a good agreement in the case of measurements in the troposphere and poor agreement in the lower stratosphere. The Junge layer located at altitudes from 17–22 km was recorded using both methods, but a number concentration from impactor data, especially for smaller particles was 3 times less than that measured with the optical counter of particles, which is explained by the following:

(1) particles with $r = 0.2$ μm are extremely small for the optical microscope used in analysis of impactor samples;
(2) several small particles ($r < 0.15$ μm) could be recorded with the optical counter as one particle; and
(3) together with solid particles, droplets can reside in the stratosphere, which evaporate from the impactor samples and, at best, leave traces that can be detected in the electron microscopic analysis.

The aerosol concentration in the lower stratosphere is characterized by the presence of a maximum, with the vertical change of concentration being of a fine stratified structure. The results of the joint Soviet–American balloon observations confirmed the presence of a constant and long-lived stratospheric layer of aerosol both in the eastern and western hemisphere. Simultaneous measurements of condensation nuclei (N_n) and aerosol (N_a) have shown that the concentration of condensation nuclei is at a maximum in the troposphere and by absolute value exceeds the aerosol concentration at the same level. In the stratosphere it is at a minimum and either comparable with or higher than the concentration of particles with $r > 0.15\,\mu m$. A change of relative humidity and the presence of condensation nuclei can lead to both condensation and coagulation formation of stratospheric aerosol.

The background content of stratospheric aerosol in the Junge layer constituted $\sim$103 particles cm^{-3} (for particles with $r > 0.15\,\mu m$) and had a maximum of concentration at a maximum tropospheric altitude. The content of aerosol in the stratosphere depends on the season and the latitude of location. It is determined by atmospheric circulation and volcanic activity and characterized by comparable concentrations of aerosol at the North and South Poles. The size distribution of stratospheric aerosol in the range 0.2–2.5 μm is approximately described by the Junge formula with exponent 3.5–4.0, and is relatively constant.

The filter sampling did not perform chemical analysis for sulphur-containing compounds. The performed elemental analysis has shown a compatibility of the obtained data with the previous measurements. So, for instance, the mass concentration of Ca constituted $\sim 0.3\,\mu g\,m^{-3}$ and Mn $0.001\,\mu g\,m^{-3}$. Differences in aerosol concentration from the data of the photoelectrical counter and impactor in the submicron range of sizes can be attributed to registration of sulphuric acid droplets by the optical sonde ($r \sim 0.2\,\mu m$). According to the data available, the stratospheric aerosol consisted mainly of droplets of a concentrated solution of sulphuric acid. The real part of the complex refraction index of aerosol can change from 1.34 (water) to 1.54 (sulphate particles), and for stratospheric aerosol from 1.4 to 1.5 with the upper limit of the imaginary part of the refraction index $\sim$0.01 and optical thickness of the stratosphere $\sim$0.02–0.1 (with an average of 0.045) [62, 67].

Improved construction of the balloon 2-cascade impactor and amyl acetate film as a substratum enabled one to obtain data on the altitudinal variability of the aerosol size distribution and on the concentration of some elements (Fe, Na, Hg, Ca, and R) and their stratified structure [66]. The cyclonic activity during the aerosol–radiation studies led to a decrease in tropopause altitude from 120–130 hPa to 160–180 hPa with its simultaneous stratification. This fact is especially important since the altitude of the tropopause affects substantially the vertical profile of aerosol concentration. The aerosol layers in the atmosphere under consideration are partly determined by exchange processes near the tropopause and, to a certain degree, by volcanic activity. Here a similarity of the profiles of the aerosol and ozone mixing ratio is observed, as well as a decrease of the temperature lapse rate. All this affects the long-wave radiation transfer in the atmosphere and the vertical distribution of the aerosol number density N (a decrease with altitude in the troposphere, the presence of a detaining layer near the tropopause, above which a minimum of concentration is observed, and then the presence of the aerosol layer with a maximum between 15 and 20 km).

Measurements of long-wave radiation fluxes were made at night two hours after sunset and two hours before sunrise, to exclude the impact of short-wave radiation on measurement data. Analysis of actinometric sounding results has shown that the troposphere is characterized by an increase of a net radiation flux with altitude and, respectively, a radiative cooling. In inversion layers, in the lower stratosphere (above the tropopause), where the aerosol layers were located, there was a radiative heating. For instance, at the 460–520-hPa level (i.e., below 7,000 m), there was an aerosol layer with a concentration of particles $\sim 3\,\mathrm{cm}^{-3}$. Here, near the bottom of the temperature inversion, there was a radiative heating of up to $0.43^{\circ}\mathrm{C\,hr}^{-1}$, and above it a strong radiative cooling. In the lower stratosphere (at the 70–80-hPa level) the concentration of aerosol particles reached $1.7\,\mathrm{cm}^{-3}$. Here, with the altitudinal growth of temperature and partial pressure of ozone, there is a radiative heating of up to $0.27^{\circ}\mathrm{C\,hr}^{-1}$ in the lower part of the aerosol layer and a cooling in its upper part (down to $0.06^{\circ}\mathrm{C\,hr}^{-1}$). For the stratospheric layer with the main mass of aerosol (80–45 hPa), the rate of radiative heating reached $0.09^{\circ}\mathrm{C\,hr}^{-1}$ (the heating took place in 60% of the cases).

Calculations of short-wave and long-wave radiation fluxes have shown that the presence of stratospheric aerosol favours the heating of the troposphere and stratosphere, and in the Jung layer the rate of radiative heating reaches $0.08^{\circ}\mathrm{C\,hr}^{-1}$. This agrees with observational data despite the difference between the initial data for calculations and observation conditions. One can suppose that optically active aerosol layers, like clouds, have radiatively active boundary layers, where the radiative flux divergence changes its sign. The latter fact may be an indirect indicator for detection of aerosol layers from the data of actinometric soundings. There was an attempt to construct the dependence of the rate of radiative change of temperature on aerosol concentration in the layer above 300 hPa, but the correlation between these parameters was very weak, which is determined by a multiparametric impact of the atmosphere on the long-wave radiation transfer. One can assume that such correlations should be more stable in the region of atmospheric transparency windows.

Based on complex measurements of radiative characteristics, data were systematized on the spectral albedo of basic natural formations and on the ratio between spectral and total albedo. Under background conditions with the prevailing scattering aerosol, the surface-atmosphere system's albedo increased with increasing altitude of sounding, and the presence of absorbing aerosol favoured a decrease in its absolute value. Tropospheric aerosols under cloud-free background conditions both over the sea and over land (rural location) are characterized by efficient values of the imaginary part of the refraction index within 0.004–0.01, and the urban aerosol (from large industrial centres) by values from 0.01–0.03, which is connected with the effect of iron oxides and soot. The optical thickness of aerosol changes from 0.05 in the Arctic to 0.4 over the city and in the case of dust–sand outbreaks. Over the Mexico City megalopolis the optical thickness of aerosol can reach 2.

The spectral changes of absorption in the atmosphere in the presence of aerosol layers enabled one to select the aerosol part of the absorbed radiation on the basis of both observations and calculations. The higher the technogenic constituent connected with the impact of the aerosol soot component (which totally manifests itself with an increasing optical thickness of aerosol) the stronger is the absorptance of the atmosphere over the city. The obtained dependence of atmospheric absorptance on the aerosol optical thickness makes it possible, on the basis of filter measurements of direct solar radiation at $\lambda = 500\,\text{nm}$, to estimate the aerosol contribution to radiation absorption by the atmosphere. The rate of temperature change due to short-wave and long-wave radiation is at a maximum in the atmospheric boundary layer and decreases with the altitude of sounding. The presence in the atmosphere of powerful aerosol layers (of the type of the Saharan dust layer or industrial layer) leads to a radiative heating up to $0.4^{\circ}\text{C}\,\text{hr}^{-1}$ and a decrease of cooling due to long-wave radiation. The presence of the Junge stratospheric aerosol and the thermal stratification leads to a radiative heating of $0.1^{\circ}\text{C}\,\text{hr}^{-1}$ in the 50–100-hPa layer.

7.8 THE DYNAMIC AND REGIONAL AEROSOL MODELS

The dynamic and regional aerosol models have been recently developed [1–4, 12, 13, 66, 92, 124]. Of great interest for substantiation of the dynamic aerosol models is a consideration of temporal variations of aerosol characteristics on annual and diurnal scales. The seasonal change of the vertical distribution of stratospheric aerosol concentrations has been studied by [1–4, 124] and the values of parameters obtained by different authors with the use of their models agree rather well [1–13, 62, 67, 73, 79, 90, 120, 123, 130, 162]. Differences in the values of parameters are observed, however, at altitudes above 25 km, where reliable data are scarce.

The statistical processing of observational data on aerosol size distribution in the surface layer on the assumption of the presence of two main modes of distribution, and in their description by inverse γ-functions [66], revealed a relatively narrow range of changes in modal radii for different climatic zones, and rather regular changes of modal radii in the surface layer during 24 hours. This suggests the conclusion that the diurnal change of particle size distribution is possible and should be taken into account in the model construction and calculations of the atmospheric aerosol characteristics.

A large volume of observational data on the size distribution and chemical composition of surface and tropospheric aerosols, as well as data on backscattering at $\lambda = 0.55\,\mu\text{m}$ for two geographical regions (western Siberia and Kazakhstan), has made it possible to substantiate the statistical microphysical and optical models of tropospheric aerosol for these regions [1–4, 12, 13, 90, 100, 124, 163] under different synoptic situations: with the Arctic, moderate, and subtropical air masses. Synoptic formations were divided into the following types: cold front and the rear of the cold front; warm front and the rear of the warm front; small-gradient field and high-pressure domains; and anticyclone periphery. The surface level visibility was divided into two ranges: $S_m < 20$ km and $S_m > 20$ km.

7.8.1 The mixing layer dynamics

Numerous measurements of the vertical profiles of the pollutants concentration show that their major mass is concentrated in the lower, 1.5–2-km layer of the atmosphere – the mixing layer. The altitude of the upper boundary of the layer decreases rapidly in January. Then it starts increasing. The aerosol concentration in the mixing layer reaches a maximum in May, and the altitude of the mixing layer reached a maximum in June. Beginning from May, the aerosol concentration in the whole mixing layer is reduced, and from June the altitude of the mixing layer itself decreases, also. In December, changes in aerosol concentration correlate only within the 0–0.4-km layer or above 1.0 km, which, for this season, can be considered as a free atmosphere. In May, correlations are stable up to 1.6 km. The altitude of the mixing ratio over the western Siberia is characterized by a distinctly expressed annual change with a maximum in summer and a minimum in winter, which agrees well with the dynamics of the processes of turbulent mixing determining this pattern. The interannual variability of the mixing layer altitude for western Siberia constitutes 0.2–0.3 km.

The height of the mixing layer in Kazakhstan, in general, follows the pattern for western Siberia, but differs in amplitude. While in December, the mixing layers over western Siberia and Kazakhstan have similar heights, in warm seasons over Kazakhstan the height of the mixing layer is much larger. This is determined by the fact that surface heating in a warm season in Kazakhstan is stronger than in western Siberia; aerosol is driven by convective fluxes to higher levels (Figure 7.7).

7.8.2 The synoptic factor

Substantial changes in the aerosol concentration are connected with the character of air masses, which at a certain time determines the weather in the region. Over western Siberia the Arctic air masses are most clear, and moderately cold air masses are most heavily polluted. The latter is explained by the fact that such air masses in this region are, as a rule, the 'old' Arctic air which gets enriched with aerosol when in this region.

Under conditions of a stable stratification, the layers are observed in which the turbulent mixing is preserved, judging by increased MSD values of air temperature. Also, the layers with increased aerosol concentrations coincide with the layers in which air temperature fluctuations grow. The non-uniform vertical distribution of turbulence intensity can lead to the formation of aerosol layers. Note that the layers with a more intense turbulence also affect the humidity field, though not as strongly as the aerosol field.

From microphysical data, aerosol correlation matrices were calculated for altitudes from ground level to 2.4 km for particle number densities (Table 7.13).

Of interest are statistically processed data on particle size distribution in the

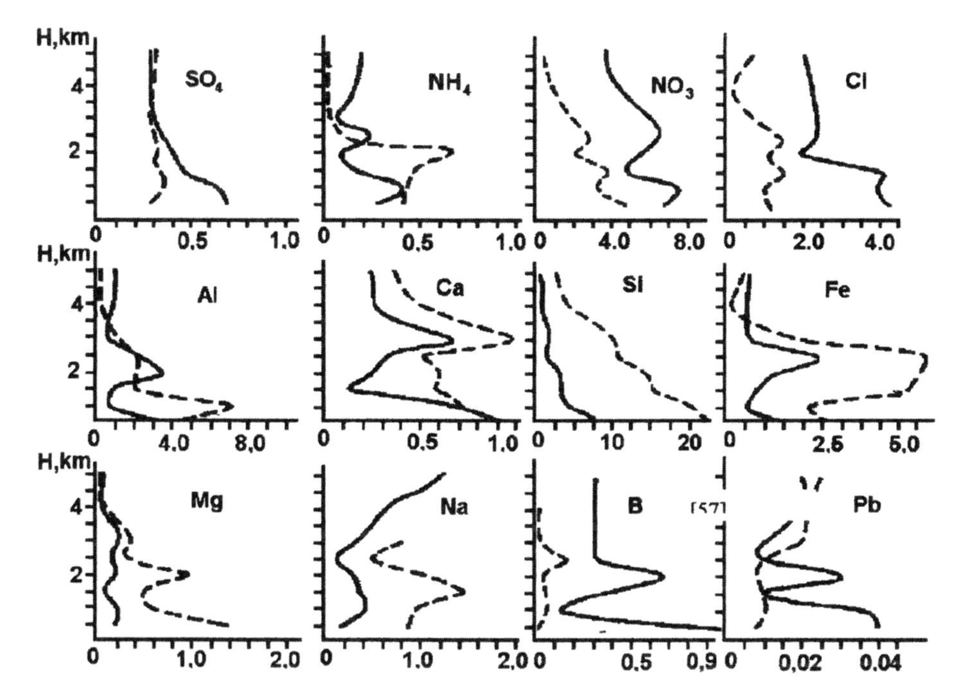

Figure 7.7. The vertical distribution of elements ($\mu g\,m^{-3}$) in aerosols over western Siberia (solid) and Kazakhstan (dashed).

range 0.4–10 μm up to an altitude of 4.5 km for western Europe and 3 km for Kazakhstan (Table 7.14).

The results of statistical processing of data on the chemical composition of aerosol particles are given in Tables 7.15 and 7.16.

Figure 7.7 demonstrates the mean statistical curves of the vertical distribution of some aerosol chemical components. The statistical processing of the lidar sounding data enabled one to construct correlation matrices, to calculate eigenvectors of the correlation matrix and mean statistical coefficients of backscattering to an altitude of 4.5 km. The profiles of the backscattering coefficients are characterized by a sharp decrease in their values in the lowest 500-m layer. Respectively, the optical characteristics MSD in this layer is most often at a maximum, also.

During recent years, an interesting complex of lidar studies of the atmospheric aerosol structure was carried out in central Asia at the lidar station Teploklіuchenka in central Tien Shan [92]. A large database has been obtained on the spatial–temporal structure of the backscattering coefficient for altitudes above 30 km, as well as its transformation after powerful emissions of aerosol substances and aerosol-forming gases of natural and anthropogenic origin (volcanic eruptions and the Asian aerosol brown cloud).

Measurements at three wavelengths during the passage of the Asian aerosol brown cloud over the sounding station made it possible to approximately estimate

Table 7.13. The aerosol correlation matrices.

H (km)	0.0	0.2	0.4	0.6	0.8	1.0	1.2	1.4	1.6	1.8	2.0	2.2	2.4
Western Siberia, winter													
0.0	1	0.70	0.52	0.13	0.34	−0.03	0.01	−0.01	−0.12	−0.15	−0.19	−0.20	−0.20
0.2		1	0.71	0.04	0.21	−0.03	0.03	0.07	−0.08	−0.09	−0.07	−0.12	−0.10
0.4			1	0.14	0.34	0.19	0.17	0.18	0.14	0.13	0.15	0.08	0.07
0.6				1	0.58	0.32	0.42	0.36	0.09	0.09	−0.03	0.03	0.06
0.8					1	0.58	0.52	0.46	0.12	0.11	0.11	0.10	0.10
1.0						1	0.69	0.72	0.92	0.85	0.54	0.49	0.50
1.2							1	0.96	0.78	0.89	0.67	0.70	0.62
1.4								1	0.89	0.96	0.70	0.79	0.75
1.6									1	0.93	0.75	0.65	0.63
1.8										1	0.82	0.80	0.83
2.0											1	0.95	0.86
2.2												1	0.90
2.4													1
Western Siberia, summer													
0.0	1	0.80	0.64	0.63	0.62	0.62	0.59	0.51	0.56	0.30	0.15	−0.06	−0.08
0.2		1	0.90	0.89	0.70	0.59	0.50	0.50	0.57	0.35	0.27	−0.05	−0.20
0.4			1	0.95	0.85	0.72	0.66	0.67	0.76	0.47	0.25	0.00	−0.14
0.6				1	0.92	0.69	0.80	0.57	0.68	0.45	0.27	0.04	−0.09
0.8					1	0.75	0.66	0.54	0.54	0.49	0.30	0.11	−0.02
1.0						1	0.80	0.79	0.71	0.65	0.41	0.26	0.05
1.2							1	0.88	0.75	0.63	0.23	0.43	0.27
1.4								1	0.86	0.74	0.40	0.44	0.21
1.6									1	0.72	0.41	0.40	0.18
1.8										1	0.75	0.62	0.39
2.0											1	0.49	0.28
2.2												1	0.91
2.4													1

the particles size distribution at different altitudes and to observe a shift of the modal radius from 0.5 μm in the lower cloud layer at altitudes 6–8 km to 0.05 μm in the upper 8–10-km layer.

7.9 THE PROBLEM OF 'EXCESSIVE' ABSORPTION OF SHORT-WAVE RADIATION IN CLOUDS

During the last decade of the last century many publications appeared concerning an increase of solar radiation absorption in the cloudy atmosphere compared with the cloud-free atmosphere based on the combined processing of data on total solar radiation measured from the surface and satellites (see [81, 82, 88]).

Table 7.14. The size distribution of aerosol number concentration (N (cm^{-3})) at different altitudes.

	Diameter of particles (μm)																					
	0.4–0.5		0.5–0.6		0.6–0.7		0.7–0.8		0.8–0.9		0.9–1.0		1.0–1.5		1.5–2.0		2.0–4.0		4.0–7.0		7.0–10.0		
H (km)	$\bar{N}$	σ_N	$\bar{N}$	σ_N	$\bar{N}$	σ_N	$\bar{N}$	σ_N	$\bar{N}$	σ_N	$\bar{N}$	σ_N	$\bar{N}$	σ_N	$\bar{N}$	σ_N	$\bar{N}$	σ_N	$\bar{N}$	σ_N	$\bar{N}$	σ_N	n
Western Siberia																							
0.0	13.094	19.75	5.385	10.14	2.508	5.374	1.315	3.590	1.426	4.528	0.631	1.886	0.624	1.676	0.223	0.587	0.175	0.482	0.041	0.181	0.006	0.031	106
0.3	6.893	12.26	2.575	6.698	0.758	2.027	0.338	0.794	0.347	0.745	0.281	0.564	0.350	0.693	0.150	0.247	0.088	0.160	0.016	0.038	0.003	0.010	215
0.6	8.476	14.03	3.195	7.984	1.169	2.241	0.267	0.680	0.412	0.584	0.309	0.406	0.402	0.454	0.354	1.525	0.317	0.582	0.101	0.282	0.015	0.043	222
0.9	4.770	8.633	1.190	2.213	0.457	1.158	0.149	0.202	0.157	0.254	0.118	0.170	0.162	0.187	0.083	0.131	0.080	0.185	0.019	0.070	0.004	0.022	224
1.2	4.986	9.366	1.081	1.965	0.381	0.538	0.213	0.297	0.193	0.236	0.154	0.183	0.239	0.306	0.146	0.200	0.153	0.332	0.050	0.132	0.007	0.021	143
1.5	5.121	8.917	1.135	2.181	0.451	1.056	0.214	0.612	0.202	0.449	0.150	0.320	0.214	0.289	0.141	0.336	0.140	0.793	0.023	0.058	0.003	0.009	366
1.8	3.235	7.952	0.778	2.912	0.193	0.501	0.093	0.214	0.092	0.230	0.064	0.138	0.103	0.314	0.052	0.122	0.052	0.171	0.010	0.046	0.001	0.007	378
2.1	2.839	6.303	0.597	1.698	0.174	0.429	0.076	0.163	0.065	0.152	0.050	0.107	0.075	0.140	0.044	0.113	0.045	0.147	0.011	0.049	0.001	0.008	467
2.5	1.533	4.251	0.329	0.795	0.129	0.303	0.058	0.128	0.063	0.188	0.044	0.086	0.071	0.149	0.039	0.099	0.034	0.100	0.007	0.029	0.004	0.068	632
3.0	1.330	5.052	0.252	0.663	0.253	2.491	0.053	0.170	0.058	0.222	0.054	0.221	0.104	0.508	0.070	0.492	0.054	0.419	0.038	0.519	0.005	0.073	298
3.5	0.205	0.318	0.073	0.135	0.039	0.082	0.028	0.084	0.025	0.048	0.021	0.046	0.037	0.078	0.016	0.038	0.019	0.061	0.003	0.008	0.002	0.006	94
4.0	0.652	2.163	0.153	0.351	0.067	0.124	0.034	0.052	0.038	0.057	0.033	0.049	0.058	0.098	0.035	0.076	0.018	0.030	0.006	0.033	0.001	0.006	97
4.5	0.277	0.583	0.124	0.331	0.077	0.075	0.038	0.049	0.032	0.065	0.029	0.039	0.046	0.045	0.018	0.018	0.019	0.035	0.002	0.035	0.000 5	0.001	24
Kazakhstan																							
0.5	8.222	8.681	1.700	1.758	0.666	1.333	0.600	1.309	0.400	1.226	0.377	1.161	0.383	1.530	0.377	1.239	0.937	1.358	0.200	0.840	0.033	0.120	69
1.0	7.667	1.269	2.122	1.862	0.611	1.428	0.644	1.380	0.373	1.458	0.311	1.191	0.483	1.249	0.377	0.935	0.997	1.224	0.238	0.880	0.044	0.200	119
1.5	1.949	1.044	0.693	0.575	0.395	0.407	0.243	0.274	0.346	0.335	0.304	0.383	0.636	0.867	0.394	0.583	0.319	0.483	0.016	0.030	0.004	0.007	81
2.0	1.953	5.726	0.623	0.869	0.380	0.582	0.254	0.411	0.326	0.543	0.337	0.584	0.626	0.988	0.398	0.796	0.281	0.629	0.008	0.018	0.000 8	0.008	82
2.5	0.414	0.205	0.206	0.301	0.115	0.162	0.064	0.104	0.078	0.145	0.072	0.141	0.121	0.176	0.058	0.062	0.020	0.036	0.004	0.010	0.006	0.000 6	53
3.0	0.360	0.447	0.166	0.155	0.076	0.040	0.047	0.044	0.053	0.077	0.046	0.058	0.079	0.158	0.040	0.082	0.022	0.036	0.005	0.010	0.000 2	0.000 09	37

Table 7.15. The chemical composition of aerosol over the western Siberia and Kazakhstan ($\mu g m^{-3}$).

	Western Siberia		Kazakhstan			Western Siberia		Kazakhstan	
Component	X	N	X	N	Component	X	N	X	N
SO_4^2	0.44	234	0.28	203	Mn	0.10	224	0.06	135
NH_4	0.22	233	0.32	204	Sn	0.05	230	0.002	120
NO_3	5.95	234	3.01	203	Ti	0.03	292	0.08	138
Cl^-	3.01	235	1.15	201	Na^{4-}	0.34	236	0.92	205
Al	2.04	286	3.29	138	K^{4-}	0.09	235	0.02	203
Ca	0.42	146	0.67	69	Cu	0.09	305	0.13	167
Si	3.57	221	17.64	126	Pb	0.02	222	0.01	146
Fe	1.30	225	2.81	135	Ag	0.10	153	0.19	71
Ni	0.03	232	0.05	88	B	0.77	101	0.58	125
Mg	0.18	224	0.73	86	Cr	0.03	49	0.08	98

Note: X is the average content of the ith element or ion; N is the number.

A study and assessment of the impact of clouds and atmospheric aerosols on the atmospheric radiation budget and radiative flux divergence is of primary interest from the viewpoint of analysis of the role of various factors in climate formation [80–87]. The difference in radiation budgets for the cloudy and cloud-free atmosphere $F_{\text{cloud}} - F_{\text{clear}}$ determines the short-wave radiative forcing due to cloudiness (CRF) $C_s(\text{S})$ at the surface level (S) and at the atmospheric top boundary C_s (top of the atmosphere – TOA). The global annual mean value of short-wave cloud-ratiative forcing $C_s(\text{TOA})$ changes from -45 to $-50\,\text{mW}\,\text{cm}^{-2}$. Clearly, the value $C_s(\text{TOA}) = C_s(\text{S}) + C_s(A)$ is determined by the sum of contributions of CRF of the surface $C_s(\text{S})$ and CRF of the atmosphere $C_s(A)$.

The data of surface observations of total and reflected solar short-wave radiation (SWR) enable one to assess the net solar radiation flux for the cloudy atmosphere and for clear sky F_{cloud} and F_{clear} (the SWR budget) at the surface level, and satellite measurements give a net radiation flux at the level of the atmospheric upper boundary [19]. If the SWR absorption by the surface–atmosphere system in the presence of clouds grows compared with absorption in clear sky, then $C_s(A) > 0$. A parameter was introduced [19] which characterizes the impact of cloudiness on SWR absorption in the atmosphere estimated from the relationship:

$$f_s = \frac{C_s(\text{S})}{C_s(\text{TOA})} = \frac{(F_{\text{cloud}} - F_{\text{clear}})_{LB}}{(F_{\text{cloud}} - F_{\text{clear}})_{UB}},$$

where LB is the lower boundary and UB is the upper boundary of the atmosphere. It was shown that the mean diurnal values of $Cs(\text{S})$ for Boulder (USA) constitute: $C_s(\text{S}) = -92.6\,\text{mW}\,\text{cm}^{-2}$ and $C_s(\text{TOA}) = -63.2\,\text{mW}\,\text{cm}^{-2}$, which gives $f_s = 1.46$ [19]. From model calculations, the f_s parameter is approximately equal to unity, from which it follows that calculations substantially underestimate the value of

Table 7.16. The chemical composition of aerosol in different air masses ($\mu g\,m^{-3}$).

Region	Air mass	SO_4^{2-}	NH_4	NO_3	Cl^-	Al	Ca	Si	Fe	Ni	Mg
Western Siberia	Arctic	$\frac{0.32}{80}$	$\frac{0.03}{80}$	$\frac{6.35}{79}$	$\frac{2.36}{81}$	$\frac{1.77}{136}$	$\frac{0.42}{82}$	$\frac{2.51}{103}$	$\frac{0.75}{106}$	$\frac{0.03}{101}$	$\frac{0.24}{106}$
	Mid-latitude	$\frac{0.65}{95}$	$\frac{0.18}{95}$	$\frac{7.22}{95}$	$\frac{4.60}{95}$	$\frac{2.42}{101}$	$\frac{0.19}{32}$	$\frac{4.77}{71}$	$\frac{2.08}{94}$	$\frac{0.02}{82}$	$\frac{0.13}{93}$
	Subtropical	$\frac{0.32}{26}$	$\frac{0.06}{25}$	$\frac{2.86}{26}$	$\frac{1.79}{26}$	$\frac{2.61}{23}$	$\frac{0.14}{8}$	$\frac{2.63}{24}$	$\frac{0.44}{23}$	$\frac{0.02}{23}$	$\frac{0.04}{23}$
Western Siberia (2,000 m)	Arctic	$\frac{0.33}{19}$	$\frac{0.02}{18}$	$\frac{7.81}{19}$	$\frac{1.25}{20}$	$\frac{2.67}{39}$	$\frac{0.33}{21}$	$\frac{2.98}{29}$	$\frac{1.51}{30}$	$\frac{0.04}{27}$	$\frac{0.31}{30}$
	Mid-latitude	$\frac{0.39}{18}$	$\frac{0.02}{18}$	$\frac{4.52}{18}$	$\frac{3.34}{18}$	$\frac{5.34}{17}$	$\frac{0.09}{9}$	$\frac{3.43}{17}$	$\frac{1.40}{17}$	$\frac{0.02}{16}$	$\frac{0.08}{17}$
	Subtropical	$\frac{0.44}{10}$	$\frac{0.06}{10}$	$\frac{0.82}{10}$	$\frac{1.46}{10}$	$\frac{3.85}{8}$	$\frac{0.15}{3}$	$\frac{2.33}{8}$	$\frac{0.20}{8}$	$\frac{0.04}{8}$	$\frac{0.02}{8}$
Kazakhstan	Arctic	$\frac{0.25}{8}$	$\frac{0.20}{8}$	$\frac{1.92}{8}$	$\frac{0.001}{8}$	$\frac{2.93}{20}$	$\frac{1.23}{8}$	$\frac{18.9}{20}$	$\frac{6.38}{20}$	$\frac{0.13}{12}$	$\frac{0.56}{13}$
	Mid-latitude	$\frac{0.29}{173}$	$\frac{0.34}{174}$	$\frac{3.05}{173}$	$\frac{1.07}{171}$	$\frac{3.64}{106}$	$\frac{0.68}{51}$	$\frac{19.9}{103}$	$\frac{2.43}{103}$	$\frac{0.04}{65}$	$\frac{0.90}{62}$
	Subtropical	$\frac{0.38}{12}$	$\frac{0.15}{12}$	$\frac{3.33}{12}$	$\frac{2.65}{12}$	$\frac{0.89}{11}$	$\frac{0.18}{9}$	$\frac{0.10}{11}$	$\frac{0.17}{11}$	$\frac{0.02}{11}$	$\frac{0.02}{11}$

Region	Air mass	Mn	Sn	Ti	Na^+	K^+	Cu	Pb	B	Cr	Σ
Western Siberia	Arctic	$\frac{0.09}{106}$	$\frac{0.05}{101}$	$\frac{0.03}{138}$	$\frac{0.29}{82}$	$\frac{0.08}{81}$	$\frac{0.09}{137}$	$\frac{0.14}{20}$	$\frac{1.30}{44}$	$\frac{0.03}{30}$	16.59
	Mid-latitude	$\frac{0.07}{93}$	$\frac{0.09}{80}$	$\frac{0.04}{105}$	$\frac{0.09}{95}$	$\frac{0.04}{95}$	$\frac{0.07}{118}$	$\frac{0.13}{135}$	$\frac{0.12}{14}$	$\frac{0.02}{16}$	22.80
	Subtropical	$\frac{0.19}{23}$	$\frac{0.01}{23}$	$\frac{0.01}{23}$	$\frac{0.40}{26}$	$\frac{0.46}{26}$	$\frac{0.03}{26}$	$\frac{0.05}{11}$	$\frac{0.08}{12}$	$\frac{0.02}{3}$	12.17
Western Siberia (2,000 m)	Arctic	$\frac{0.17}{30}$	$\frac{0.001}{27}$	$\frac{0.03}{39}$	$\frac{0.12}{20}$	$\frac{0.10}{19}$	$\frac{0.09}{39}$	$\frac{0.02}{26}$	$\frac{6.41}{7}$	$\frac{0.03}{7}$	23.50
	Mid-latitude	$\frac{0.04}{17}$	$\frac{0.001}{16}$	$\frac{0.02}{18}$	$\frac{0.15}{18}$	$\frac{0.01}{18}$	$\frac{0.02}{21}$	$\frac{0.008}{20}$	$\frac{0.08}{6}$	$\frac{**}{2}$	18.94
	Subtropical	$\frac{0.07}{8}$	$\frac{0.007}{8}$	$\frac{0.02}{8}$	$\frac{0.91}{10}$	$\frac{0.86}{10}$	$\frac{0.01}{8}$	$\frac{0.13}{8}$	$\frac{0.11}{6}$	–	11.48
Kazakhstan	Arctic	$\frac{0.05}{20}$	$\frac{0.003}{8}$	$\frac{0.22}{20}$	$\frac{0.001}{9}$	$\frac{0.005}{8}$	$\frac{0.14}{20}$	$\frac{0.002}{9}$	$\frac{0.004}{8}$	$\frac{0.03}{19}$	33.00
	Mid-latitude	$\frac{0.07}{103}$	$\frac{0.002}{100}$	$\frac{0.06}{106}$	$\frac{0.96}{174}$	$\frac{0.02}{173}$	$\frac{0.13}{135}$	$\frac{0.01}{125}$	$\frac{0.74}{97}$	$\frac{0.10}{69}$	34.73
	Subtropical	$\frac{0.03}{11}$	$\frac{0.001}{11}$	$\frac{0.02}{11}$	$\frac{0.80}{12}$	$\frac{0.03}{12}$	$\frac{0.05}{11}$	$\frac{0.005}{11}$	$\frac{0.10}{10}$	$\frac{0.02}{9}$	8.93

Note: numerator represents the average concentration of the ith element or ion; denominator represents the number of samples.

the SWR absorbed by the cloudy atmosphere by $C_s(\mathrm{TOA}) - C_s(\mathrm{S}) \approx 30\,\mathrm{mW\,cm^{-2}}$ [19]. The discrepancy detected has initiated discussion, since it drastically changes the idea of the role of clouds in the formation of weather and climate [18, 19, 22–25, 71, 121, 122, 125, 136–138, 141, 144].

7.9.1 The concepts of the effect of 'excessive' (anomalous) SWR absorption in the cloudy atmosphere

Explanations of the effect of 'excessive' absorption of the SWR proposed in numerous recent studies can be divided into six basic groups:

(1) An 'excessive' absorption is an artifact appearing due to measurement errors and inadequacy of techniques for observational data processing [40, 121–123, 137, 141, 156]. This conclusion has been confirmed by some results of measurements in the cloudy atmosphere not revealing any radiation absorption in the short-wave spectral region. Clouds are most variable in their optical and radiative characteristics depending on physical mechanisms of their formation and on different geographical regions, and in many cases they do not increase the total SWR absorption by the atmosphere and surface, but decrease it. This is explained by the fact that clouds reflect a considerable part of the incident radiation preventing from it being absorbed by underlying atmospheric layers and the surface. Therefore, in some radiation experiments, an excessive absorption by clouds has not been observed. It should be emphasized that in many cases, measurements did not permit obtaining a database sufficient to be reliably processed. Often, measurements in the cloudy atmosphere were not accompanied by respective measurements in the cloud-free atmosphere within a few days, the surface albedo was not always measured, and only reflected radiation was measured. All this hinders an adequate assessment of the radiative characteristics of the cloudy atmosphere.

(2) A greater absorption of the SWR in the cloudy atmosphere compared with the cloud-free conditions can be explained by the fact that in broken cloudiness, radiation comes out through the side boundaries of clouds and is not considered in measurements at the top and bottom boundaries. To this group of studies are referred both field and model experiments [14, 23, 51, 140–143]. A technique has been proposed [23] to estimate the amount of radiation coming out through the side boundaries of the cloud, a priori supposing the absence of the true absorption of SWR in the cloud.

(3) An excessive SWR absorption is an apparent effect caused by horizontal transport of radiation in the cloudy layer due to the horizontal heterogeneity of the layer (the stochastic structure of the layer). A detailed description of this theoretical approach has been given in [55, 101, 142]. It has been proposed to distinguish between the roughness of the upper boundary of the layer (case 1) and heterogeneous internal structure of the layer (variations of the attenuation coefficient, case 2). A numerical analysis [55, 142] has shown that the horizontal SWR transport is most strongly expressed in the case of the stochastic upper

boundary of the cloud layer compared with the case of variations of its internal parameters. The size of the horizontal extent at which an averaging should be made in measurements of reflected and transmitted fluxes for the correct assessment of radiation absorption in the layer constitutes, respectively, 30 km (case 1) and 60 km (case 2). The case of the stochastic upper boundary corresponds in nature to stratocumulus clouds, and variability of the attenuation coefficient to stratus clouds. Different combinations of the values of absorption and scattering in the layer, heterogeneous both horizontally and vertically, have been considered in [55], and it has been found that 'the most important factors are the value of absorption and the absorbing substance–scattering particles or the medium in which they are dispersed'.

(4) An anomalous absorption in the cloud can be explained, along with other factors, by the absorption of the SWR by water vapour (which is not taken into account) in the near-IR spectral region [29, 35, 50, 118, 127, 132]. However, a thorough consideration of the calculated molecular absorption in the near-IR spectral region does not lead to the observed value of absorption due to cloudiness. Besides this, the results of spectral measurements [141] have shown that the anomalous absorption manifests itself most strongly in the visible spectral region, where the radiation absorption by water molecules is very weak.

(5) To explain an excessive absorption of the SWR, many scientists [52, 151, 159] proposed consideration of specific features of the microphysical structure of cloudiness. The presence of very large droplets in a cloud increases the radiation absorption, but rather too weakly and insufficiently to explain an anomalous effect. Detailed calculations have been carried out [52] of the optical parameters and respective radiative characteristics for the systems of 2-layer particles with an absorbing nucleus. These studies have not revealed any excessive absorption of radiation by clouds. In all the models considered, a marked absorption of solar radiation by clouds in the visible spectral region can be obtained only in the presence of a considerable amount of atmospheric aerosols [16, 148, 159].

(6) A combined impact has been considered [71, 101, 127] of the possible causes mentioned above, and on some supposition, calculated and observed values of radiation absorption are close. But all series of observations could not be adequately explained. Thus so far, the problem remains unsolved – a fact which has been emphasized by various scientists [16, 28, 82–88, 99, 126, 159].

7.9.2 A comparison of results of measurements of the SWR absorption from the data of different aircraft experiments

In studies of radiation absorption by clouds (either confirming or denying the presence of the effect of excessive absorption), data of observations from satellites and from surface meteorological networks were mainly used. These observations were made with the use of different instruments during a long time period – requiring further complicated statistical data processing. As a result, an averaged pattern was obtained which included various types of cloudiness. However, the absence of a single method of data selection and processing has often led to contradictory conclusions in a number of studies mentioned above.

Let us analyse the results of aircraft observations in terms of the parameter f_s introduced above. From the data of aircraft measurements of the SWR fluxes we calculate the radiative flux divergence $R = (F^{\downarrow} - F^{\uparrow})_{\rm top} - (F^{\downarrow} - F^{\uparrow})_{\rm base}$ in the atmospheric layer in the presence, and in the absence, of stratified cloudiness. The results of aircraft measurements have made it possible in some cases to record the effect of a strong anomalous SWR absorption ($f_s > 1$), in other cases the absence of any impact of clouds on radiation absorption ($f_s = 1$), and sometimes a reflection of radiation by clouds which prevented its absorption by underlying atmospheric layers and the surface, and reduced the total SWR absorption in the atmosphere ($f_s < 1$).

7.9.3 The dependence of solar radiation absorption on the optical thickness of the cloud layer

From the results of field experiments in the cloud-free atmosphere, polluted atmosphere, and under overcast cloudiness conditions, a relative amount of the SWR flux divergence $b(\zeta, \tau) = R/\pi S\zeta$ was calculated [81, 82, 88, 145] as a function of the optical thickness of the layer.

An interpolation of experimental points revealed linear dependences of $b(\tau)$, which is confirmed by an analytical expression for SWR absorption in the scattering layer given in [111]:

$$b(\zeta, \tau) = (1 - \omega_0)(2\tau u(\zeta) + v(\zeta)) \tag{7.1}$$

where the functions $u(\zeta)$ and $v(\zeta)$ depend on the cosine of solar zenith angle ζ.

The parameter $(1 - \omega_0)$ is connected with the probability of quantum survival ω_0 and is determined by the magnitude of true radiation absorption compared with scattering in the atmospheric layer. In the cloud-free atmosphere this quantity is approximately equal to 0.5 (direct and single scattered radiation prevails), in the cloudy atmosphere -0.005 (the radiation field being formed mainly by multiple scattering). The optical thickness constitutes in the cloud-free atmosphere $\tau_0 \sim 0.5$, and in the cloud layer $\tau_0 \sim 25$. The functions $v(\zeta)$ and $u(\zeta)$ differ little in the cloud-free and cloudy atmosphere. The second component in equation (7.1) is below zero because $v(\zeta) < 0$, and the product $v(\zeta)(1 - \omega_0)$ constitutes approximately 0.3–0.4 for the cloud-free atmosphere and 0.005 for cloudy conditions, which contributes differently to the magnitude of the parameter $b(\zeta, \tau)$ in these cases. Thus the impact of atmospheric absorbing aerosols which determines the magnitude of $1 - \omega_0$, turns out to be more significant under overcast cloudiness conditions.

Note that the parameter f_s is close to unity when the optical thickness of the cloud layer is small ($\tau \leq 7$). In the cases (experiments 1, 2, and 4), when the content of soil or soot aerosols in the atmosphere was great, the values of $f_s \geq 2$ pointed to a considerable solar radiation absorption in the atmosphere and at the surface level. This confirms the conclusion that the effect of excessive, or anomalous, absorption manifests itself most under conditions of an increased content of atmospheric aerosols and cloudiness with a high optical thickness (>15) at low solar zenith angles, and do not show at all in clear clouds of a small optical thickness at high solar zenith angles.

7.9.4 The dependence of solar radiation absorption on geographical latitude and solar zenith angle

The values of the parameter f_s and radiative flux divergence R in an atmospheric layer decrease from tropical latitudes to polar ones, with its values being especially high in the tropical zone, which agrees with the results of analysis carried out in [81, 82, 88, 145]. This dependence is broken in industrial regions characterized by the heavily polluted atmosphere as well as in experiment 6 accomplished in the presence of 2-layer cloudiness.

A thorough analysis has been performed [161] of a series of monthly mean values of total SWR fluxes obtained from satellite and surface observations during 46 months (from March, 1985 until December, 1988). They include a latitudinal dependence of the magnitude of f_s shown in Figure 7.8(a). In this figure, results are shown obtained from analysis of the data of aircraft observations. Magnitudes of f_s for total radiation practically coinciding with the curve obtained in [161] are marked with asterisks, and circles indicate its magnitudes in the short-wave spectral interval for $\lambda = 0.5\,\mu m$, which substantially exceed the averaged values in the whole spectrum but qualitatively preserve the same dependence on geographical latitude.

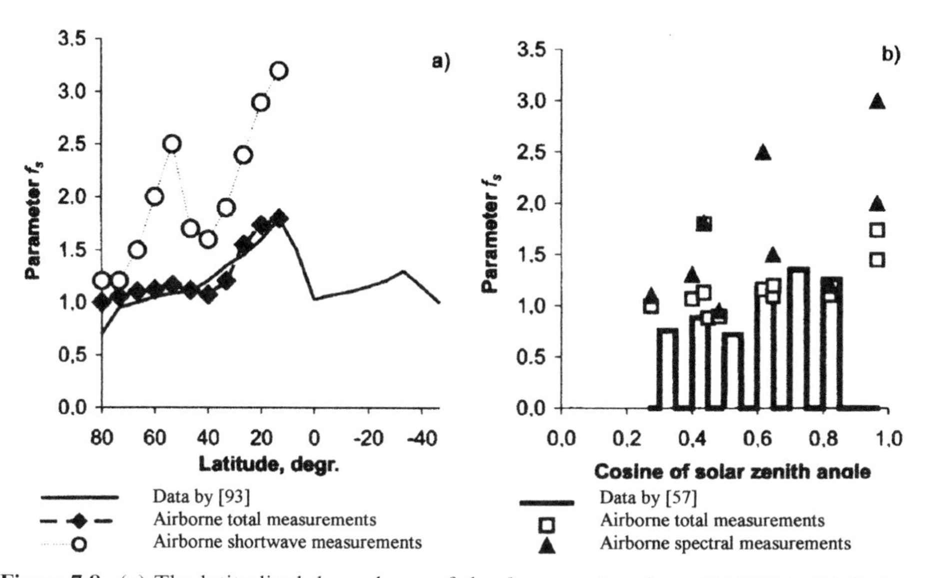

Figure 7.8. (a) The latitudinal dependence of the f_s parameters from [57,112] (solid line) and from aircraft measurements (dashed line). Asterisks mark the values of f_s obtained from aircraft measurements over the whole spectrum, circles in the short-wave interval. (b) Dependence of the f_s parameters on the solar zenith angle from [88] (nomograph) and from aircraft measurements. Asterisks mark the total data, triangles are for $\lambda = 0.5\,\mu m$.

The dashed curve is drawn by values of f_s for $\lambda = 0.5\,\mu\text{m}$ averaged over each latitudinal zone. As has been mentioned above, the values of f_s exceeding 2 correspond to a high content in the atmosphere of radiation absorbing aerosol particles together with the optically thick cloudiness.

A study has been carried out [57, 112] of the dependence of the effect of anomalous absorption by clouds, characterized by the parameter f_s, on the solar zenith angle. Figure 7.8(b) demonstrates the results obtained in [57] in the form of a nomogramme. The results of aircraft radiation experiments are shown in the same figure. Asterisks indicate the data obtained from total measurements, triangles for $\lambda = 0.5\,\mu\text{m}$. It is seen that the values of f_s for total measurements practically coincide, despite different observational techniques. Note that the results obtained in [57, 112] and the data analysed were obtained with the use of quite different techniques of measurements and data processing as well as in different time periods and in different geographical regions.

The data for the short-wave spectral interval demonstrate a much stronger effect of excessive absorption. If we consider the values of the solar zenith angle cosine and geographical latitude (bearing in mind that radiation experiments are usually carried out at noon), common features of these dependences become apparent. But it is difficult to draw a final conclusion from these results, what is the initial cause of the effect – the Sun elevation or the optical properties inherent in the stratified cloudiness in different latitudinal zones. It has been found out [112] that the CRF near the surface in the Arctic decreases linearly with the growing cosine of solar zenith angle. At low Sun elevations (the cosine of zenith angle $\varsigma < 0.15$) the presence of stratus clouds leads to a surface heating, and at higher Sun elevations to a surface cooling. The presence of atmospheric soot aerosol leads to an increase of the CRF at the upper boundary of the atmosphere with the growing cosine of the solar zenith angle [94]. On the other hand, sulphate aerosol does not affect the angular dependence of the CRF. The conclusion can be drawn that it is the impact of soot aerosol in clouds that favours an anomalous absorption.

According to the results of numerous calculations, with the oblique incidence of sun rays onto the cloud layer, the share of reflected radiation increases, and the flux of radiation entering the cloud layer as well as its absorption in this layer decreases. However, together with a high surface albedo, the cloud layer reflects radiation that has already reached the surface and been reflected upward. Therefore, clouds play the role of the heating factor at large solar zenith angles and a high surface albedo. Besides, it is clear that the effect of cloudiness on the SWR absorption in the atmosphere manifests itself in the presence of a sufficient amount of clouds. In this context, it is of interest to compare data given in Figure 7.9 describing the zonal field of clouds, and data shown in Figure 7.8(a). As seen, the latitudinal dependences of the cloud amount and the parameter f_s, characterizing the CRF, qualitatively coincide.

Studies carried out within the programmes CAENEX, GAAREX, FGGE, and GATE have distinctly demonstrated, on the basis of analysis of the data of aircraft observations, the presence in most cases of substantial SWR absorption by clouds. An analysis of the results of studies performed in recent decades, suggests the

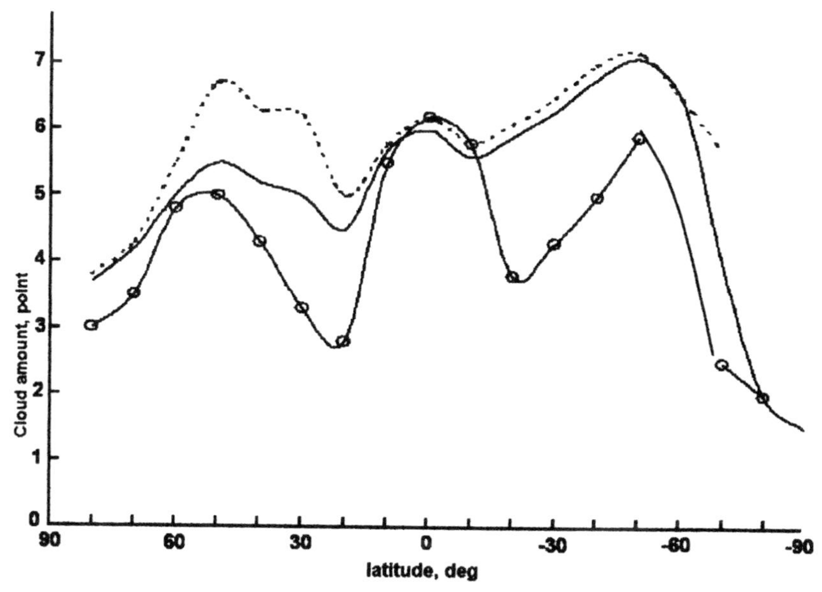

Figure 7.9. The global mean cloud amount averaged over the latitudinal belt (1), over the water surface (2), and land (3) for 1971–1990.
[57, 112].

following conclusions:

(1) An excessive SWR absorption is determined by the optical properties of clouds but not by the errors of measurements and methods of statistical processing of results, as some authors try to prove.
(2) In the tropical regions, an excessive SWR absorption is rather substantial ($f_s \sim 3$), whereas in the polar region, this effect is almost unobserved ($f_s \leq 1$). It has been proved experimentally that the absorbing and scattering properties of stratus clouds depend not only on geographical latitude but also on the type of the absorbing atmospheric aerosol present in the clouds.
(3) A consideration of the properties of clouds in terms of the optical parameter – single-scattering albedo – shows also that the higher values of radiation absorption (lower values of single-scattering albedo) are characteristic of vast cloud layers in middle and low latitudes. The results of satellite data processing confirm the conclusions drawn from analysis of the data of aircraft measurements.
(4) An analysis of radiative characteristics (radiation fluxes and radiative flux divergences) in cloud layers and in the cloud-free atmosphere revealed a considerable increase of solar radiation absorption in the cloudy atmosphere, which especially manifests itself in powerful cloud layers together with a high content of dust or soot aerosols in the atmosphere. A linear dependence of radiation absorption on the optical thickness has been observed.

(5) An effect of excessive (anomalous) SWR absorption in the cloudy atmosphere is observed over the whole wavelength spectrum, but it is especially strong in the short-wave interval. The proven existence of the effect of an excessive absorption changes considerably the existing ideas of the heat balance of the atmosphere. In this connection, it is necessary to take into account the atmospheric heating due to this effect in climatic calculations and in predicting a climate change.

7.10 THE OPTICAL CHARACTERISTICS OF STRATIFIED CLOUDINESS

7.10.1 The single-scattering albedo and the volume absorption coefficient

It is seen from Figure 7.10, which illustrates the spectral dependence of the value of $1 - \omega_0$, that the bands of molecular absorption for atmospheric gases are strongly pronounced but have different intensity in different cloud layers. The band of radiation absorption by atmospheric aerosols is detected from the results of aircraft data processing at $\lambda = 0.42\,\mu m$. This absorption band can be identified as radiation absorption by atmospheric sand aerosols as a result of sand storms in the deserts of the Kara Kum and Sahara shortly before radiation measurements were made. Such particles contain haematite [152–154]. Weak aerosol absorption bands at $\lambda \approx 0.5\,\mu m$ and $0.8\,\mu m$ are seen on the curves obtained from the data of aircraft experiments carried out over the sea surface, where the content of sea salt in atmospheric aerosols is high. In the Arctic regions the atmosphere is clearer, and scattering is almost conservative within a broad wavelength interval ($\omega_0 = 1$, Figure 7.10).

The results of retrieval of the spectral values of the $1 - \omega_0$ parameter from the data of aircraft observations, surface, and satellite measurements show a monotonous increase of absorption with wavelength characteristic of solar radiation absorption due to products of organic fuel burning [76, 79]. The spectral dependence of the absorption coefficient (single-scattering albedo) obtained from the results of processing aircraft observational data and most of the pixels of satellite images demonstrated a neutral change typical of the soot and dust aerosols.

The results of calculations of the volume absorption coefficients κ of individual layers in a cloud demonstrate a strong vertical heterogeneity of the cloud. Besides this, on the curves of spectral dependences of the volume absorption coefficient κ in the upper parts of cloud layers, apart from the absorption bands for oxygen and water vapour, the Chappuis ozone band ($0.65\,\mu m$) manifests itself, whereas in the lower parts of clouds this band is not seen. This is explained by a higher content of ozone at higher levels in the troposphere.

In the figures obtained from processing satellite data, several areas (containing up to several tens of pixels) are seen, where the value of $1 - \omega_0$ reaches 0.05. Hence, this value is the result of an increased error along the edges of the image. In other images considered here, pixels with an increased absorption are observed over industrial regions and can indicate the presence in clouds of soot aerosol or other

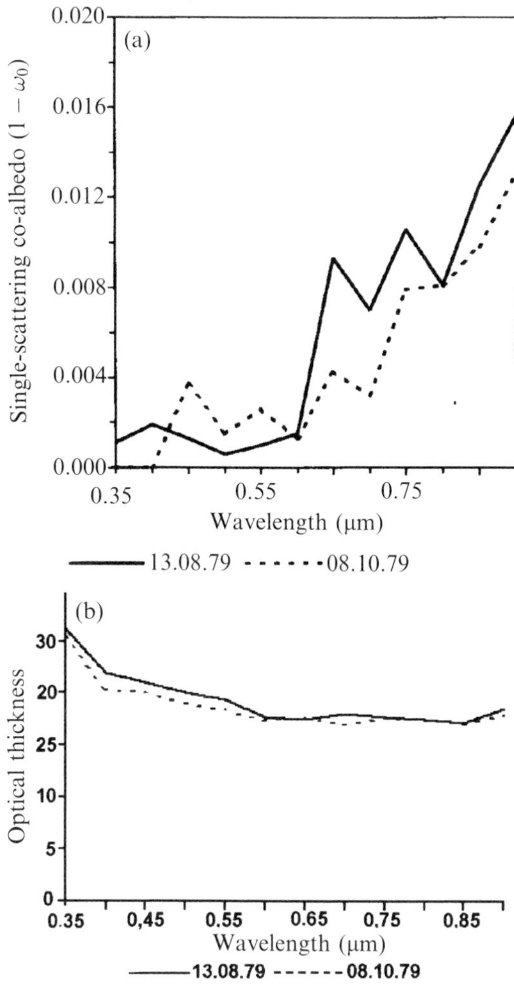

Figure 7.10. The spectral dependence of (a) the single-scattering albedo and (b) the optical thickness τ_0 from the results of measurements in the Arctic: 11–13 August, 1979 and 12–18 October, 1979.

aerosols strongly absorbing radiation. Only sparse individual pixels in the images demonstrate a conservative scattering.

7.10.2 The optical thickness τ_0 and the volume scattering coefficient

The values of the volume scattering coefficient differ strongly under different conditions. The spectral dependences of the volume scattering coefficient in Figure 7.11 demonstrate a marked vertical heterogeneity of stratus cloud. In different layers both the value of the single-scattering albedo and its spectral dependence are different. This results from the heterogeneous microphysical structure of the stratus cloud. The volume scattering coefficient determined for a cloud from measurements of fluxes at

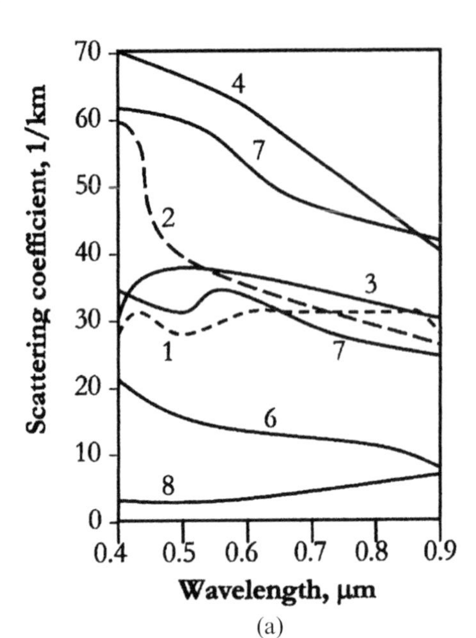

(a)

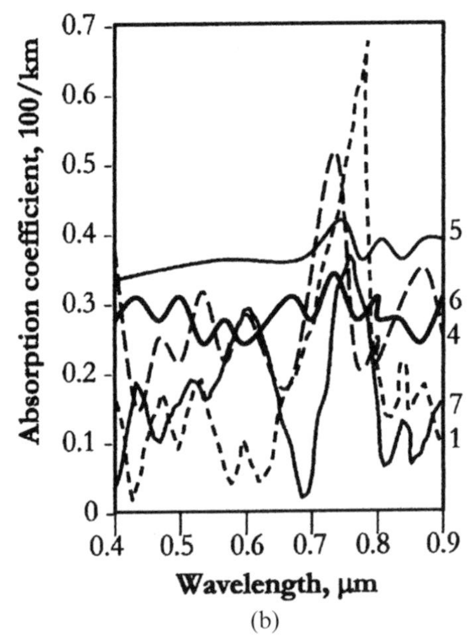

(b)

Figure 7.11. The volume coefficients of (a) scattering and (b) absorption obtained from observational data. Numbers on curves correspond to the following regions: 1 and 2 – Atlantic Ocean (cloud); 3 – Atlantic Ocean (clear); 4 – Azov Sea; 5 – Rustavy City; 6 – Lake Ladoga (cloud); 7 – Lake Ladoga (clear); 9 – Kara Sea.

the boundaries of cloud layers coincides, within measurement errors, with the values obtained for individual layers and averaged over the whole cloud layer. The scattering coefficient is at a maximum in the middle cloud layers and is shifted towards its upper boundary. The obtained vertical profile of the scattering coefficient is similar to the results of aircraft observations [14] obtained for stratocumulus clouds formed over the ocean in the southern hemisphere, and results of the Arctic cloud experiment FIRE [30]. Since the stratus clouds differ strongly in their properties, the results obtained can be considered as coinciding with the data of *in situ* measurements [102].

For most of the pixels of satellite images the optical thickness is within 10–25, but in some sectors containing several pixels, the value of the optical thickness reaches 70–80 and even exceeds 100. The spatial changes of the optical thickness are rather monotonic, which is an indirect confirmation of sufficiently small errors both in measurements and in the method of their processing.

Note that here the results are given of an application of the method of retrieval of the extended cloudiness optical parameters to the data of radiation measurements [151], which is more advantageous than the approaches used earlier to solve the problem considered [114, 115, 129, 131]. This is determined by the fact that: (1) two parameters are evaluated simultaneously – single-scattering albedo and optical thickness – for each wavelength, for each pixel independently; and (2) there are no additional limitations on the value or spectral dependence of the retrieved parameters. The impact of the upper boundary roughness of cloudiness was approximately taken into account in solving an inverse problem in the case of processing satellite measurements. It turned out that the introduced parameter of shadowing also takes into account both the effect of the overlying atmospheric layers and the errors of approximation of the scattering function with the use of the Henye–Greenstein formula [151].

Note that a more accurate presentation of the scattering function can change the results [147]. Therefore, it is necessary to develop methods to reliably determine the scattering function for the cloud or at least the asymmetry parameter from radiation measurements to use observed values but not modelled ones.

The optical properties of crystal clouds can be determined by the orientation of crystals toward the source of radiation. Usually, it is assumed that the spatial orientation of crystal cloud particles is chaotic. Studies of such clouds, with the use of lidar measurements of the backscattering matrix (BSM), have shown that the state of the crystal cloud in which the symmetry with respect to vertical rotation is broken, is more probable than symmetry preservation. This means that there is an almost constantly functioning (at least, for some share of cloud particles) mechanism of orientation in some prevalent azimuth directions (i.e., across prevalent wind directions at the altitudes of crystal clouds) [70].

7.10.3 The optical properties of aerosol in a cloudy medium – polluted clouds

Both observational and theoretical studies of the radiation regime of the cloudy atmosphere have shown that optically active anthropogenic aerosols affect substantially the optical characteristics of clouds. Measurements of the optical properties of stratified cloudiness give values of single-scattering albedo of the cloud in the visible spectral region, which are much lower than it follows from calculations using Mie theory [146–148, 162, 163]. In connection with the recent attempts to give a non-standard explanation of anomalous absorption (from considering it to be the result of incorrect interpretation of observational data to a reconsideration of mechanisms of interaction of electromagnetic waves with cloud particles), the following question is pertinent: are possibilities of explanation of anomalous absorption within the classic theory of scattering completely exhausted?

Since the absorption due to pure water in the visible spectral region is negligibly small, it can be assumed that the cloud contains an absorbing aerosol either independently mixed with water droplets or interacting with them. Numerical estimates show that the only aerosol with sufficient realistic concentration values are soot particles [63, 65, 146–148].

The following morphological structures existing in polluted clouds are: (1) fine soot particles modelled as spheres; (2) cloud droplets covered with the finest soot particles; (3) water droplets with a central soot nucleus, which can be approximately modelled as spheres; (4) soot particles of the type of fractals; and (5) soot fractals either compressed by the high air humidity with several large globules or transformed into a particle, which can be modelled as a stretched spheroid.

Following [150], we consider versions 1–3. Case 1 is a standard model of independently mixed water and soot particles considered as homogeneous spheres. Case 2 is modelled as an approximation of a 2-layer sphere with a shell whose complex refraction index is calculated as weighted-mean by mass of soot and water in the surface layer of a droplet (determined by the radius of soot particles). Here it should be explained that laboratory studies show that at the initial stage, soot adsorption on a droplet takes place not uniformly but in patches. Hence, the 2-layer spheres with the finest soot layers are not likely to exist. Finally, case 3 is a standard model of a 2-layer spherical particle with a soot nucleus and water shell. An electron microscopic analysis of aerosol particles suggests the conclusion that such structures are sufficiently often observed [66].

To calculate the optical characteristics of 2-layer particles (models 2 and 3), algorithms have been used described in [146, 148]. To minimize the number of parameters to be calculated, the radii of all particles were considered to be similar, which made it possible to reduce the number of varied parameters to three: the ratio of soot mass to water mass per unit volume Q; the radius of water droplets; and the radius of soot particles. It is obvious that the transition to particle ensembles will not lead to a principal change of the results with respect to estimates of Q at which the absorption in the cloud is comparable with an anomalous absorption measured in experiments.

As shown by the results of calculations, model 2 gives a minimum of single-scattering albedo, model 3 gives a maximum, and model 1 is intermediate. From the estimates available in the literature [37, 88, 102], the content of soot in the atmosphere, with different degrees of pollution, can vary from $10^{-8}\,\mathrm{g\,m^{-3}}$ to $10^{-5}\,\mathrm{g\,m^{-3}}$, and the characteristic values of stratus cloud water content are of the order of $10^{-1}\,\mathrm{g\,m^{-3}}$. Hence, possible variations of the Q values range between 10^{-7} and

10^{-4}. According to calculations, at Q close to 10^{-5} to 10^{-4}, values of single-scattering albedo of 0.999 and even 0.99 are quite possible, which, by estimates in [150, 151, 160], are sufficient to confirm theoretical and observational values of absorption in stratus clouds. It is of interest that the 'needed' values of single-scattering albedo can be obtained not only with a 'special' model 2 (a water particle with a soot shell) but also classical model 1 of independently mixed homogeneous water and soot particles.

Thus, with a sufficiently large concentration of soot but possible heavily polluted clouds, model 2 and even 1 can give the values of single-scattering albedo coinciding with those observed. Hence, an anomalous absorption in heavily polluted clouds can be explained by the classical theory. On the other hand, questions remain open: how often can such clouds be observed?; how far can they move from the source of pollution?; and is the process of soot accumulation inside stratus clouds possible?

7.10.4 The coloration of clouds with aerosol

As a rule, clouds and fogs are of a 3-phase system:

- a sprayed gas medium;
- water droplets or crystals resulting from a non-equilibrium condensation of moistrure on dust fractions; and
- atmospheric haze.

Characterized by a marked absorption in the short-wave spectral region, intensified in the cloud by multiple scattering, this fraction causes a marked coloration of clouds in the transparency windows of water and affects substantially the cloud albedo and, hence, the radiative heat exchange of the planet with space [120, 129].

The fluxes of upward radiation near the cloud top $F\uparrow$ and downward radiation near its bottom $F\downarrow$ are equal to:

$$F\downarrow = F_0 \cos\xi \times T \qquad F\uparrow = F_0 \cos\xi \times R$$

where R is the albedo; T is the cloud transmittance; S is the solar zenith angle; and F_0 is the spectral density of solar illumination of the platform orthogonal to the Sun's rays, and according to [120, 129]:

$$T = g(\cos\xi)\frac{shy}{sh(x+y) - Ashx}$$

$$R = e^{-y}[1 - e^{-x}(1 - Ae^{\nu})T]$$

$$x = \tau\gamma, y = l\lambda, \gamma = 2\sqrt{\beta/l},$$

$$l = 4/3(1 - \bar{\eta}) \cong 10 \cdot g(\cos\xi) \cong 0.26 + 1.1\cos\xi$$

where A is the surface albedo and η is the weighted mean cosine of the scattering angle for the cloud scattering function.

In the transparency windows of all modifications of water, the absorption of cloud droplets is negligibly small, like light scattering by smoke particles introduced into the cloud; that is:

$$\alpha = \alpha_{\text{smoke}} \quad \sigma = \sigma_{\text{cloud}} \quad \beta = \alpha_{\text{smoke}}/\sigma_{\text{cloud}} \ll 1$$

Under conditions of a small β, y and $x = \tau y/l$ the following approximation is correct:

$$\frac{shy}{sh(x+y) - Ashx} \cong \frac{1}{1 + (1-A)\tau/l} \times \left\{1 - \frac{2}{3}\frac{\tau}{l}\beta\left[1 + \frac{1-(1-A)\tau/l}{1+(1-A)\tau/l}\right]\right\}$$

These formulae make it possible to assess the impact of haze in the cloud on its albedo R and transmission T as a function of surface albedo A. Thus the impact of atmospheric pollution (anthropogenic including) on the regional albedo of the planet and its heat exchange with space is determined. The impact of the statistical structure of cloudiness can be taken into account with due regard to [140–142].

Variations of brightness or albedo of clouds observed from space make it possible to reveal, based on the equations mentioned above, regions and centres of increased atmospheric pollution due to a submicron fraction of aerosol and to obtain regional estimates of the degree of this fractions development. As a result of small sizes of haze particles, for an approximate estimation one can use the Rayleigh approximation [129–130], from which it follows:

$$\beta = \frac{16\pi n}{(n^2+2)}\frac{æ}{\lambda}\frac{V}{\sigma_{\text{cloud}}} \qquad y \cong \frac{8}{(n^2+2)}\left(\frac{næVl}{\pi\sigma_{\text{cloud}}}\right)^{1/2} \qquad æ = \tau\frac{8}{(n^2+2)}\left(\frac{næVl}{\pi\sigma_{\text{cloud}}}\right)^{1/2}$$

which permits us to relate R and T to χ, V, and σ_{cloud}.

7.10.5 Empirical formulas to estimate the volume coefficients of scattering and absorption under conditions of multiple scattering of radiation in cloud layers

The results of spectral radiation measurements performed from aircraft, the ground surface, and satellites in the presence of stratified cloudiness have been discussed above. Specific features of the results obtained consist in the low values of the single-scattering albedo and spectral dependence of the optical thickness, contradicting the results expected according to the Mie theory of scattering. It follows from Mie theory that the volume scattering coefficient (and the related τ_0 value for particles >5 μm) should not depend on wavelength in the short-wave spectral region, and the values of the volume absorption coefficient in a cloud are within $10^{-5} - 10^{-8}$ (the single-scattering albedo is about 0.999 99–1.0). Let us consider a possible mechanism explaining the spectral dependence of the scattering coefficient of the cloud layer and high values of the absorption coefficient. This mechanism is connected with the multiple scattering of radiation in the cloud.

In calculations of the radiation field in the cloud and descriptions of the process of multiple scattering, a cloud layer is considered to consist either of droplets or to be additively superimposed on the molecular atmosphere. Molecular scattering for the case of a cloud layer is taken into account by summing up the scattering coefficients. Since the molecular (Rayleigh) scattering coefficient is, by about 2 orders of magnitude, less than the scattering coefficient on droplets, its contribution turns out to be negligibly small.

As is known, due to multiple scattering in an optically thick cloud layer, an average number of collisions of photons penetrating the cloud layer at a conservative scattering is proportional to τ_0^2 [111] (for reflected photons the average number of the acts of scattering is proportional to τ_0). Thus, the path of the photon in the cloud strongly increases compared with the cloud-free atmosphere, and therefore the number of collisions of photons with air molecules increases as does the contribution of molecular scattering. The radiation absorption extracts some of the photons from the process of photon transport and, thus, partially reduces the effect of the increasing role of molecular scattering. Therefore, it should be borne in mind that the cloud layer is not simply 'superimposed' onto the molecular atmosphere, but affects the process of scattering on molecules, enhancing it. Note that an increase of molecular absorption in the oxygen absorption band $\lambda = 0.76\,\mu\text{m}$, caused by an extension of the path of photons in the cloud because of multiple scattering, has been considered in a number of studies [101, 111, 151]. The same is valid for scattering and absorption of radiation by atmospheric aerosols, whose particles are among the cloud droplets.

Of course, the theory of multiple scattering of radiation and the equation of radiation transfer take into account all processes of scattering and absorption. But this is valid only if they are correctly taken into account in the model of the scattering and absorbing medium. Usually, into the equation of radiation transfer, average values are introduced of the input parameters for the elemental volume of the scattering medium, and then the equation is solved using some method of the theory of radiation transfer. However, physically, it is not correct to average the parameters for the elemental volume of the medium at an initial stage of solving the physical problem. This inaccuracy grows when the scales of the elemental volume selected in calculations for each component of the medium differ substantially [151]. Strictly speaking, for the correct description of the scattering phenomena in the cloud, Maxwell's equation should be solved for the whole air–droplet–aerosol medium. However, it is not our goal to discuss the mathematical problem of description of multiple scattering of radiation in a multicomponent medium, and therefore an empirical method proposed in [103–109] will be described below.

The coefficient of scattering (or absorption) for the multicomponent medium is usually the sum of the coefficients of scattering (or absorption) for the respective components. Let us denote the optical parameters for the molecular component by M, aerosol component by A, and droplet component by D. Then, for the short-wave range, without account of the extended path of the photon in the cloud due to multiple scattering we can write:

$$\alpha = \alpha_M + \alpha_A + \alpha_D$$

$$k = k_M + k_A$$

Bearing in mind the multiple impact of scattering and absorption by different components, let us introduce the following empirical relationships:

$$\left.\begin{aligned} \alpha &= (\alpha'_R + \alpha'_A) C \tau_D^p \omega_0^q + \alpha_D \\ k &= (k'_M + k'_A) C \tau_D^p \omega_0^q \end{aligned}\right\} \tag{7.2}$$

where ω_0 is the single-scattering albedo; C is the coefficient of proportionality; τ_D and α_D are the optical thickness of cloud and the volume coefficient of scattering due to only scattering on droplets; α'_R, α'_A, k'_M, and k'_A are the coefficients of scattering and absorption of molecules and aerosol particles outside the cloud; and p and q are the empirical coefficients estimated earlier [151]. In these relationships the coefficient of scattering on droplets α_D is presented without a multiplier since the equation of radiation transfer and respective asymptotic formulas were written for one component, droplets, and the impact of multiple scattering on the α_D value is considered in the equation. The component $k'_M \tau_D^p \omega_0^q$ in the second line of equation (7.2) is not zero in the region of the molecular absorption bands and α_R is the Rayleight scattering coefficient at respective wavelength and attitude in the atmosphere. Note that consideration is made only under conditions of a large optical thickness of cloud $\tau_0 \gg 1$. From the results of these estimations [151], it turned out that $p = 2$ and $q = \tau_0^2$, which satisfactorily agrees with the fact mentioned above: the average number of photon collisions in a cloud for transmitted radiation is $\sim \tau_0^2$ [111, 151], and the constant C is equal to unity. Note that values of the p and q exponents were obtained on the basis of analysis of the values of the volume coefficients of scattering and absorption determined from the data of two experiments at two wavelengths. The growth of the volume coefficient of scattering with increasing wavelength agrees well with the results of observations which demonstrate that the cloud optical thickness in the UV spectral region reaches values of several hundreds [151].

Using relationships (7.2) one can transform the parameters $[\alpha(\lambda) - \alpha(0.8)]$ and $\kappa(\lambda)$ into parameters calculated by the Mie theory of scattering and usually ascribed to the elemental volume [111]. Figure 7.12(a,b) demonstrate the spectral dependences of the volume coefficient of absorption $\kappa(\lambda)$ and scattering $[\alpha(\lambda) - \alpha(0.8)]$ obtained from observational data and transformed with equations (7.2). The results shown in Figure 7.12 practically coincide with the values calculated by the Mie theory. The values of the volume coefficient of absorption correspond to the value of single-scattering albedo $\omega_0 = 0.999\,98$ usually ascribed to the cloud medium. Transformed with equation (7.2), the remainder $[\alpha(\lambda) - \alpha(0.8)]$ differs little from the coefficient of Rayleigh scattering in the cloud-free atmosphere.

Note that the arguments given above refer to the so-called 'external mixture' (i.e., to the case when the aerosol particles are between droplets). If they are inside droplets (the 'internal mixture') then the aerosol absorption is correctly considered in calculations with the formulas for a 1-component medium. Based on the results obtained, the conclusion can be drawn that the observed anomalous absorption points to an external mixture of atmospheric aerosol and cloud droplets. The radiation absorbing aerosol in the cloud in the cases considered is outside of the

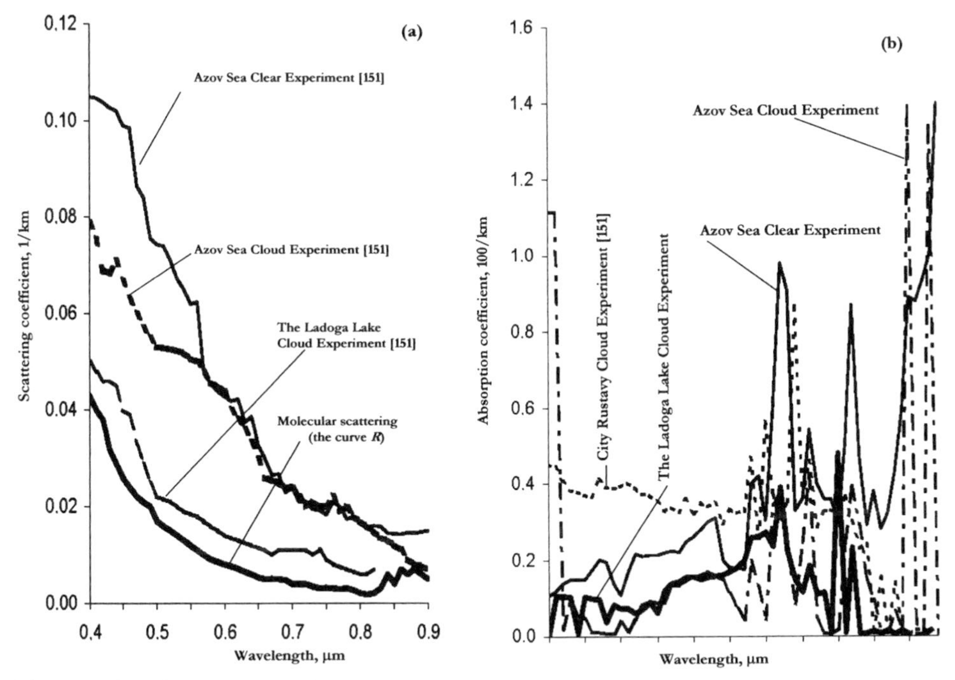

Figure 7.12. The volume coefficients of (a) scattering and (b) absorption transformed with the use of relationships (7.2). The curve R characterizes the molecular scattering at an altitude of 1 km.

droplets in the 'dry' form. Otherwise, the coefficient of absorption obtained from radiation measurements would coincide with conventional model values.

The transformation of the absorption coefficient k obtained from radiation observations, into the coefficient of aerosol absorption k_{aer}, calculated by the theory of scattering, can be made with due regard to multiple scattering using the formula [107, 109]:

$$k_{\text{aer}} = \frac{k}{\tau^2 q} - k_{\text{mol}}$$

where τ is the optical thickness of cloud; q is the coefficient depending on the true radiation absorption and on geometrical thickness of cloud; and k_{mol} is the coefficient of molecular absorption, which differs from zero in the absorption bands for atmospheric gases.

The result of this transformation leads to values given in Table 7.17, which also shows the results, for comparison, of *in situ* measurements and the results of radiation absorption measurements [151]. It follows from Table 7.17 that the absorption coefficient retrieved from radiation measurements practically coincides with results of *in situ* measurements. Hence, the role of atmospheric aerosols absorbing solar radiation located outside the cloud droplets has been underestimated in studies of radiation–cloud interaction due to not taking account of the multiple scattering of radiation by cloud droplets.

Table 7.17. Coefficients of absorption by atmospheric aerosol in a cloud layer.

Optical parameters	Type I Clear cloud	Type II Cloud above continent	Type III Polluted cloud
k_{eff} (km^{-1})	<0.02	0.08	0.25
τ	16	9.5	8.7
ω_0	0.9993	0.997	0.956
k_{aer} (km^{-1}) obtained from radiation measurements	$<1 \times 10^{-4}$	1.2×10^{-3}	$\sim 1 \times 10^{-1}$
k_{aer} (km^{-1}) *in situ* measurements	$(1\text{–}10) \times 10^{-5}$ $(1\text{–}12) \times 10^{-5}$	7×10^{-4}	$(2.3\text{–}7.7) \times 10^{-2}$

The data of satellite, aircraft, and surface observations have demonstrated a, greater than supposed, content of carbon and mineral constituent in atmospheric aerosols and their great contribution to the formation of the radiation regime of the cloudy atmosphere which had been previously underestimated [103, 104]. Of special interest are watered atmospheric aerosols from industrial emissions and dust storms over the deserts. Apparently, these sources are insufficient for the effect of anomalous absorption to manifest itself on global scales, though the outflows of mineral and carbon aerosols propagate as far as 3,000 km, preserving their activity in the optical range [64]. The prevalence of ice clouds in circumpolar regions leads to a more intense washing out of aerosols, which leads, in turn, to a smaller content of dust aerosol in the cloud and a weaker absorption of radiation. Thus, the proposed mechanism explains the basic features of formation and geographical propagation of the effect of anomalous absorption by cloudiness.

The aerosol particles presented by hydrophobic particles (sand, soot, etc.) can most probably be in the cloud among the droplets. Such atmospheric aerosols, in contrast to hydrophilic particles (in many cases consisting of sulphur-containing substances) increase the short-wave radiation absorption in the cloud, not promoting an increase of the number of droplets in the cloud and, thereby, its optical thickness. An approximate estimation of a combined contribution to the solar radiation absorption in the cloud considering the mechanism proposed above for the spectral interval 0.4–1.0 μm, with average values of the coefficient of aerosol absorption 0.08 km^{-1}, the coefficient of scattering in the cloud 30 km^{-1}, and the cloud thickness $\Delta z = 1$ km, shows that in the cloud (only due to aerosol) absorption grows by approximately 15%. An increase of absorption in the ozone band for the cloud constitutes ~6–10%, which corresponds to the results of [76]. In the oxygen band $\lambda = 0.76$ μm the absorption grows by about 10–12%, which agrees with the results of [76]. For more powerful clouds, the effect of the growth of absorption due to multiple scattering will be stronger and can explain an anomalous absorption of short-wave radiation by clouds.

In conclusion it should be emphasized that in construction of an optical model of clouds it is necessary to take into account the volume coefficients of absorption and scattering of not only droplets but also other components of the cloud layer (molecules and aerosol particles) [106].

7.11 BIBLIOGRAPHY

1. Panchenko M. V. (ed.). Aerosols of Siberia (thematic issue). *Optics of the Atmos. and Ocean*, 1996, **9**(6), 701–892.
2. Zuev V. E. and Panchenko M. V. (eds). Aerosols of Siberia (thematic issue). *Optics of the Atmos. and Ocean*, 2000, **13**(6–7), 547–706.
3. Panchenko M. V. (ed.). Aerosols of Siberia (thematic issue). *Optics of the Atmos. and Ocean*, 2003, **16**(5–6), 375–534.
4. Panchenko M. V. (ed.). Aerosols of Siberia (thematic issue). *Optics of the Atmos. and Ocean*, 2004, **17**(5–6), 375–534.
5. Almeida G. A., Koepke P., and Shettle E. Atmospheric aerosols: global climatology and radiative characteristics. A. Deepak Publishing, Hampton, VA, 1991, 549 pp.
6. Andreev S. D. and Ivlev L. S. Modelling the optical characteristics of aerosols of the surface layer of the atmosphere in the spectral region 0.3–15 μm. Part I. Principles of model construction. Part II. The choice of the model parameters. Part III. Modelling results. *Optics of the Atmos. and Ocean*, 1995, **8**(5), 788–795; **8**(8), 1227–1235 and 1236–1243.
7. Andreev S. D. and Ivlev L. S. Modelling the optical characteristics of some specific forms of aerosols of the middle atmosphere. *Natural and Anthropogenic Aerosols. Proc. of the 2nd Int. Conf., St Petersburg, 27 September–1 October 1999*. SPbU Institute of Chemistry, St Petersburg, 2000, pp. 110–112 [in Russian].
8. Andreev S. D. and Ivlev L. S. The IR radiation absorption by different fractions of atmospheric aerosols. *Proc. Russian Acad. Sci., Physics of the Atmos. and Ocean*, 1980, **16**(9), 907–916 [in Russian].
9. Andreev S. D., Ivlev L. S., and Poberovsky A. V. The aerosol attenuation of radiation in the 8–13 μm transparency window. *Proc. Russian Acad. Sci., Physics of the Atmos. and Ocean*, 1974, **10**(10), 1104–1107 [in Russian].
10. Ansmann A., Wagner F., Müller D., Althausen D., Herber A., von Hoyninggen-Huene W., and Wandinger U. European pollution outbreakes during ACE-2: Optical particle properties inferred from multiwavelength lidar and star-Sunphotometry. *J. Geophys. Res.*, 2002, **107**(ND15), AAC8/1–AAC8/14.
11. Arking A. Absorption of solar energy in the atmosphere: Discrepancy between a model and observations. *Science*, 1996, **273**, 779–782.
12. Belan B. D., Grishin A. I., Matvienko G. G., and Samokhvalov I. V. *The Spatial Variability of the Atmospheric Aerosol Characteristics*. Novosibirsk, Nauka Publ., 1989, 152 pp. [in Russian].
13. Belan B. D., Sakerin S. M., Skliadneva T. K., and Kabanov D. M. The urban impact on the aerosol, radiative and meteorological characteristics. *Proc. 3rd Int. Conf. Natural and Anthropogenic Aerosols, St Petersburg, 25–27 September 2001*. St Petersburg State Univ. Publ., St Petersburg, 2001, pp. 35–37 [in Russian].
14. Boers, R., Jensen J. B., Krummel P. B., and Gerber H. Microphysical and short-wave radiative structure of wintertime stratocumulus clouds over the Southern Ocean. *Q. J. R. Meteorol. Soc.*, 1996, **122**, 1307–1339.

15. Bokoye A. I., Royer A., O'Neill N. T., and McArthur L. J. B. A North American Arctic aerosol climatology using ground-based sunphotometry. *Arctic*, 2002, **55**(3), 215–228.
16. Bott A. A numerical model of cloud-topped planetary boundary-layer: Impact of aerosol particles on the radiative forcing of stratiform clouds. *Q. J. R. Meteorol. Soc.*, 1997, **123**, 631–656.
17. Cattrell C., Carder K. L., and Gordon H. R. Columnar aerosol single-scattering albedo and phase function retrieved from sky radiance over the ocean: Measurements of Saharan dust. *J. Geophys. Res.*, 2003, **108**(D9), AAC10/1–AAC10/11.
18. Cess R. D. and Zhang M. H. How much solar radiation do clouds absorb? Response. *Science*, **vol. 271**, 1133–1134.
19. Cess, R. D., Zhang M. H., Minnis P., Corsetti L., Dutton E. G., Forgan B. W., Garber D. P., Gates W. L., Morcrette J. J., Potter G., *et al.* Absorption of solar radiation by clouds: Observation versus models. *Science*, 1995, **267**, 496–499.
20. Chapursky L. I. *The Reflection Properties of Natural Objects in the 400–2500-nm Range Part I.* USSR Ministry of Defense Publ., 1986, 160 pp. [in Russian].
21. Chapursky L. I., Chernenko A. P., and Andreeva N. I. The spectral radiative characteristics of the atmosphere under dust storm conditions. *Proc. of the Main Geophysical Observatory (St Petersburg)*, 1975, **366**, 77–84 [in Russian].
22. Charlock T. P., Alberta T. L., and Whitlock C. H. GEWEX data sets for assesing the budget for the absorption of solar energy by the atmosphere. *GEWEX News, WCRP*, 1995, **5**(4), 9–11.
23. Chou M.-D., Arking A., Otterman J., and Ridgway W. L. The effect of clouds on atmosphetric absorption of solar radiation. *Geoph. Res. Lett.*, 1995, **22**, 1885–1888.
24. Clarke, A. D. Aerosol light absorption by soot in remote environments. *Aerosol Sci. Technol.*, 1989, **10**, 161–171.
25. Clarke, A. D. Integrating sandwich: A new method of measurement of the light absorption coefficient for atmospheric aerosols. *Appl. Opt.*, 1982, **21**, 3011–3020.
26. Colarco P. R., Toon O. B., Torres O., and Rasch P. J. Determining the UV imaginary index of refraction of Saharon dust particles from Total Ozone Mapping Spectrometer data using a three–dimensional model of dust transport. *J. Geophys. Res.*, 2002, **107**(D16), AAC4/1–AAC4/18.
27. Colbeck J., Appleby L., Hardman E. J., and Harrison R. M. The optical properties and morphology of cloud-processed carbonaceous smoke. *J. Aeros. Science*, 1990, **21**, 527–538.
28. Collins W. A global signature of enhanced shortwave absorption by clouds. *J. Geophys. Res.*, 1998, **103**(24), 31669–31679.
29. Crisp D. and Zuffada C. Enhanced water vapor absorption within tropospheric clouds: A partial explanation for anomalous absorption. *IRS '96. Current problems in Atmospheric Radiation. Proc. of the Int. Radiation Symposium, August, 1996, Fairbanks, Alaska, USA.* A. Deepak Publishing, VA, 1997, pp. 121–124.
30. Curry J. A., Hobbs P. V., King M. D., Randall D. A., Minnis P., Isaac G. A., Pinto J. O., Uttal T., Bucholtz A., Cripe D. G., *et al.* FIRE Arctic Clouds Experiment. *Bull. Amer. Meteorol. Soc.*, 2000, **81**(1), 5–29.
31. Deschamps P. Y., Bréon F. M., Leroy M., Podaire A., Bricaud A., Buriez J. C., and Sèze G. The POLDER Mission: Instrument characteristics and scientific objectives. *IEEE Trans. Geosc. Rem. Sens.*, 1994, **32**, 598–615.
32. Diabin Yu. P., Ivanov L. S., Tantashev M. V., and Filippov V. L. The optical properties of tropospheric aerosols. *Aerosol and Climate*, 1981, **1**, 99–112 [in Russian].

33. Donchenko V. K. and Ivlev L. S. On the identification of aerosols of different origin. *Natural and Anthropogenic Aerosols, Proc. 3rd Int. Conf., St Petersburg, 24–27 September 2001*, pp. 41–51 [in Russian].
34. Dubovik O., Holben B. N., Lapyonok T., Sinyuk A., Mishcherko M. I., Yang P., and Slutsker I. Non-spherical aerosol retrieval method employing light scattering by spheroids. *Geophys. Res. Lett.*, 2002, **29**(10), 54/1–54/4.
35. Evans W. F. J. and Puckrin E. Near-infrared spectral measurements of liquid water absorption by clouds. *Geophys. Res. Lett*, 1996, **23**, 1941–1944.
36. Farafonov V. G. and Farafonov Viach. G. The variable separation method in the problem of the electromagnetic radiation scattering by ellipsoids. *Proc. 3rd Int. Conf. Natural and Anthropogenic Aerosols, St Petersburg, 24–27 September 2001*. St Petersburg State Univ. Publ., St Petersburg, 2001, pp. 104–109 [in Russian].
37. Feigelson E. M. (ed.). *Radiation in the Cloudy Atmosphere*. Gidrometeoizdat, Leningrad, 1981, 280 pp. [in Russian].
38. Fisher K. Measurement of absorption of visible radiation by aerosol particles. *Beitrage Phys. Atmos.*, 1970, **43**(4), 244–254.
39. Fisher K. Mass absorption coefficient of natural aerosol particles in the 0.4–2.4 mm wavelength. *Beitrage Phys. Atmos.*, 1973, **46**(2), 89–97.
40. Francis P. N., Taylor J. P., Hignett P., and Slingo A. Measurements from the U.K. Meteorological office C-130 aircraft relating to the question of enhanced absorption of solar radiation by clouds. *IRS '96. Current problems in Atmospheric Radiation. Proc. of the Int. Radiation Symp., August, 1996, Fairbanks, Alaska, USA*. A. Deepak Publishing, Fairbanks, USA, 1997, pp. 117–120.
41. Franke K., Ansmann A., Müller D., Althausen D., Ventkataraman C., Reddy M. S., Wagner F., and Scheele R. Optical properties of the Indo-Asian haze layer over the tropical Indian Ocean. *J. Geophys. Res.*, 2003, **108**(D2), AAC6/1–AAC6/17.
42. Gavrilova L. A. and Ivlev L. S. The aerosol models to calculate the radiative characteristics of the atmosphere. *Proc. Acad. Sci., Physics of the Atmos. and Ocean*, 1995, **31**(3), 667–678 [in Russian].
43. Gavrilova L. A. and Ivlev L. S. The parameterization of microphysical characteristics of aerosol in the radiation models of the atmosphere. *Proc. Acad. Sci., Physics of the Atmos. and Ocean*, 1996, **32**(2), 172–182 [in Russian].
44. Gavrilova L. A. and Ivlev L. S. The radiative models of atmospheric aerosols. In: K. Ya. Kondratyev (ed.), *Problems of Atmospheric Physics and Chemistry of Atmospheric Aerosols*. St Petersburg State Univ. Publ., St Petersburg, 1997, **20**, pp. 3–178 [in Russian].
45. Gillette D. and Nagamoto C. Size distribution and single particle composition for two dust storms in Soviet Central Asia in September 1989. In: K. Ya. Kondratyev (ed.), *Joint Soviet–American Experiment on Arid Aerosol*. Hydrometeoizdat, St Petersburg, 1993, **16**(5–6), 340–498 [in Russian].
46. Ginoux P. Effects of nonsphericity on mineral dust modeling. *J. Geophys. Res.*, 2003, **108**(D2), AAC3/1–AAC3/10.
47. Gomes L. Chemical composition by size of dust collected in dust storms in SW Tadzhikistan, September, 1989. In: K. Ya. Kondratyev (ed.), *Joint Soviet–American Experiment on Arid Aerosol*. Hydrometeoizdat, St Petersburg, 1993, **17**(5–6), 375–534 [in Russian].
48. Gorodetsky V. V., Maleshin M. N., Petrov S. Ya., Sokolova E. A., Pchelkin V. I., and Solovyev S. P. Small-size multi-channel optical spectrometers. *Optical J.*, 1995, **18**(7), 3–9 [in Russian].

49. Grigoryev Al. A. and Kondratyev K. Ya. *Ecodynamics and Geopolicy* (Volume 2). *Ecological Disasters*. St Petersburg Sci. Centre Publ., St Petersburg, 2001, 684 pp. [in Russian].
50. Harshvardhan M., Ridgway W., Ramaswamy V., Freidenreich S. M., and Batey M. J. Spectral characteristics of solar near–infrared absorption in cloudy atmospheres. *J. Geophys. Res.*, 1998, **103**(D22), 28793–28799.
51. Hayasaka T., Kikuchi N., and Tanaka M. Absorption of solar radiation by strato-cumulus clouds: Aircraft measurements and theoretical calculations. *J. Appl. Meteor.*, 1994, 1047–1055.
52. Hegg D. Comments on 'The effect of very large drops on cloud absorption. Part I: Parcel models.' *J. Atmos. Sci.*, 1986, **43**(4), 399–400.
53. Heintzenberg J., Okada K., and Luo B. P. Distribution of optical properties among atmospheric submicrometer particles of given electrical mobilities. *J. Geophys. Res.*, 2002, **107**(D11), AAC2/1–AAC2/10.
54. Hess M., Koepke P., and Schult I. Optical properties of aerosols and clouds: The software package OPAC. *Bull. Amer. Meteorol. Soc.*, 1998, **79**(5), 831–844.
55. Hignett P. and Taylor J. P. The radiative properties of inhomogeneous boundary layer cloud: Observations and modelling. *Q. J. R. Meteorol. Soc.*, 1996, **122**, 1341–1364.
56. Holben B. N., Tanré D., Smirnov A., Eck T. F., Slutsker I., Abuhassan N., Newcomb W. W., Schafer J. S., Chatenet B., Lavenu F., *et al.* An emerging ground-based aerosol climatology: Aerosol optical depth from AERONET. *J. Geophys. Res.*, 2001, **106**(D11), 12067–12097.
57. Imre, D. G., Abramson, E. N., and Daum, P. H. Quantifying cloud-induced short-wave absorption: An examination of uncertainties and recent arguments for large excess absorption. *J. Appl. Met.*, 1996, **35**, 1191–2010.
58. Ivanov V. P. *Applied Atmospheric Optics in Thermal Imaging*. Kazan′ New Knowledge Publ., Kazan′, 2000, 356 pp. [in Russian].
59. Ivlev L. S. and Popova S. I. The complex refraction index of the dispersed phase of atmospheric aerosol. *Proc. USSR Acad. Sci., Physics of the Atmos. and Optics*, 1973, **9**(10), 1034–1043 [in Russian].
60. Ivlev L. S. and Popova S. I. The impact of humidity of the values of the optical constants of the atmospheric aerosol substance. *Proc. of the USSR High School, ser. Physics*, 1974, **5**, 11–15 [in Russian].
61. Ivlev L. S. and Popova S. I. The optical constants of the atmospheric aerosol substant. *Proc. of the USSR High School, ser. Physics*, 1975, **5**, 91–97 [in Russian].
62. Ivlev L. S. and Andreev S. D. *The Optical Properties of Atmospheric Aerosols*. Leningrad State Univ. Publ., Leningrad, 1986, 359 pp. [in Russian].
63. Ivlev L. S. and Korostina O. M. Calculations of the optical characteristics of strato-spheric aerosol particles of a 2-layer structure. *Proc. USSR Acad. Sci., Physics of the Atmos. and Optics*, 1994, **30**(6), 802–806 [in Russian].
64. Ivlev L. S., Zhukov V. M., Korostina O. M., Leyva Contrares A., Mulia Velaskes A., and Bravo-Cabrera J. L. Specified optical characteristics of aerosols in the surface layer over Mexico City. *Optics of the Atmos. and Ocean*, 1994, **7**(9), 1202–1206.
65. Ivlev L. S., Melnikova I. N., Korostina O. M., and Schultz A. I. An assessment of the microphysical parameters of a stratus cloud based on aircraft radiation measurements. *Optics of the Atmos. and Ocean*, 1996, **9**(10), 1379–1385.
66. Ivlev L. S. and Dovgaliuk Yu. A. *Physics of the Atmospheric Aerosol Systems*. St Petersburg State Univ. Press, St Petersburg, 1999, 258 pp. [in Russian].

67. Ivlev L. S. Modelling the optical characteristics of atmospheric aerosol. *Optical J.*, 2001, **68**(4), 9–15 [in Russian].
68. Ivlev L. S., Vasilyev A. V., Belan B. D., Panchenko M. V., and Terpugova S. A. The optical-microphysical models of the urban aerosols. *Proc. 3rd Int. Conf., Natural and Anthropogenic Arerosols, Sankt-Petersburg, 24–27 September 2001*. St Petersburg State Univ. Press, St Petersburg, pp. 161–170 [in Russian].
69. Jarzembski M. A., Norman M. L., Fuller K. A., Srivastava V., and Cutten D. R. Complex refractive index of ammonium nitrate in the 2–20 μm spectral range. *Appl. Opt.*, 2003, **42**(6), 922–930.
70. Kaul B. V., Volkov S. N., and Samokhvalov I. V. The results of studies of crystal clouds via lidar soundings of light backscattering matrices. *Optics of the Atmos. and Ocean*, 2003, **16**(4) 354–361.
71. Kiehl J. T. *et al.* Sensitivity of a GCM climate to enhanced shortwave cloud absorption. *J. Clim.*, 1995, **8**, 2200–2212.
72. King M. D., Si-Chee Tsay, and Platnick S. In situ observations of the indirect effects of aerosols on clouds. In: R. J. Charlson and J. Heitzenberg (eds), *Aerosol Forcing of Climate*. Wiley, Wshington, 1995, pp. 227–248.
73. Kneizis F. X., Abreu L. W., Anderson G. P., Chetwynd G. H., Shettle E. P., Berk A., Bernstein L. S., Robertson D. S., Acharya P., Rothman L. S., *et al. The Modtran 2/3. Report and Lowtran 7 Model*. Phillips Laboratory, Hanscon, Massachusetts, 1996, 230 pp.
74. Kondratyev K. Ya., Ivlev L. S., and Nikolsky G. A. Complex studies of stratospheric aerosol. *Meteorology and Hydrology*, 1974, **9**, 16–26 [in Russian].
75. Kondratyev K. Ya., Moskalenko N. I., and Terzi V. F. A closed modelling of the optical characteristics of atmospheric aerosol. *Annals USSR Acad. Sci.*, 1980, **253**(6), 1354–1356 [in Russian].
76. Kondratyev K. Ya., Moskalenko N. I., Terzi V. F., and Skvortsova S. Ya. Modelling the optical properties of industrial aerosol. *Annals USSR Acad. Sci.*, 1981, **259**(6), 1354–1356 [in Russian].
77. Kondratyev K. Ya., Moskalenko N. I., Terzi V. F., and Skvortsova S. Ya. Modelling the optical characteristics of atmospheric aerosol. In: K. Ya. Kondratyev (ed.), *Aerosol and Climate*. Gidrometeoizdat, Leningrad, 1981, pp. 130–153 [in Russian].
78. Kondratyev K. Ya., Moskalenko N. I., and Terzi V. F. Modelling the optical characteristics of stratospheric aerosol. *Annals USSR Acad. Sci.*, 1982, **262**(5), 1092–1095 [in Russian].
79. Kondratyev K. Ya., Moskalenko N. I., and Pozdniakov D. V. *Atmospheric Aerosol*. Gidrometeoizdat, Leningrad, 1983, 224 pp. [in Russian].
80. Kondratyev K. Ya. *Climate Shocks: Natural and Anthropogenic*. Wiley, New York, 1988, 296 pp.
81. Kondratyev K. Ya., Binenko V. I., and Melnikova I. N. Solar radiation absorption by clouds in the visible spectral region. *Meteorology and Hydrology*, 1996, **2**, 14–23 [in Russian].
82. Kondratyev K. Ya., Binenko V. I., and Melnikova I. N. Absorption of solar radiation by clouds and aerosols in the visible wavelength region. *Meteorology and Atmospheric Physics*, 1997, No. 0/319, 1–10.
83. Kondratyev K. Ya. and Galindo I. *Volcanic Activity and Climate*. A. Deepak Publ., Hampton, U.S.A., 1997, 382 pp.
84. Kondratyev K. Ya. *Multidimensional Global Change*. Wiley–Praxis, Chichester, U.K., 1998, 761 pp.

85. Kondratyev K. Ya. Global climate change and the Kyoto Protocol. *Időjárás*, **106**(2), 1–37.
86. Kondratyev K. Ya. High-latitude environmental dynamics in the context of global change. *Időjárás*, 2003, **107**(1), 1–29.
87. Kondratyev K. Ya., Krapivin V. F., and Varotsos C. A. *Global Carbon Cycle and Climate Change*. Springer–Praxis, Chichester, U.K., 2003, 385 pp.
88. Kondratyev K. Ya., Binenko V. I., and Melnikova I. N. *Absorption of Solar Radiation by Clouds and Aerosols in the Visible Wavelength Region at Different Geographic Zones*. CAS/WMO working group on numerical experimentation, WMO, Geneva, 6 pp.
89. Kondratyev K. Ya. *Properties, Formation Processes and Consequences of the Impact Atmospheric Aerosol*. St Petersburg Centre of Ecological Safety, St Petersburg, 2005, 450 pp. [in Russian].
90. Krekov G. M. and Rakhimov R. F. *The Optical-Radar Model of the Continental Aerosol*. Nauka Publ., Novosibirsk, 1982, 198 pp. [in Russian].
91. Latfulova L. B., Starikov A. V., and Beresnev S. A. The absorbing properties of atmospheric aerosol: Analysis of microphysical optical characteristics. *Optics of the Atmos. and Ocean*, 2001, **14**(1), 69–75 [in Russian].
92. Lelevkin V. M., Orozobakov T. O., and Chen B. B. Studies of the stratospheric aerosol layer over Central Asia using lidar sounding. *Dastan* (Bishkek), 2001, **31**, 8–21 [in Russian].
93. Li F. and Ramanathan V. Winter to summer monsoon variation of aerosol optical depth over the tropical Indian Ocean. *J. Geophys. Res.*, 2002, **107**(D16), AAC2/1–AAC2/13.
94. Liao H. and Seinfield J. H. Effect of clouds on direct aerosol radiative forcing of climate. *J. Geophys. Res.*, 1998, **103**(D4), 3781–3788.
95. Liubovtseva Yu. S., Gabelko L. B., and Yaskovich L. G. The aerosol absorption in the 0.25–25-μm spectral region. In: E. M. Feggelson (ed.), *Optics of the Atmosphere and Aerosol*. Nauka Publ., Moscow, 1986, pp. 174–204 [in Russian].
96. Liubovtseva Yu. S., Yudin N. P., and Yaskovich L. G. The composition and optical properties of the submicron fraction of atmospheric aerosol. In: E. M. Feggelson (ed.), *Optics of the Atmosphere and Aerosol*. Nauka Publ., Moscow, 1986, pp. 65–81 [in Russian].
97. Liubovtseva Yu. S., Yudin N. I., and Yaskovich L. G. *Proceedings of Second All-Union Conference on the Atmospheric Optics, Tomsk, 2–5 April 1983*. Siberian Branch of Russian Academy of Sciences Publ. Tomsk, 1983, pp. 67–69 [in Russian].
98. Livingston J. M., Russel P. B., Reid J. S., Rodemann J., Schmid B., Allen D. A., Torres O., Levy R. C., Remer L. A., Holben B. N., *et al.* Airborne Sun photometer measurements of aerosol optical depth and columnar water vapor during the Puerto Rico Dust Experiment and comparison with land, aircraft, and satellite measurements. *J. Geophys. Res.*, 2003, **108**(D19), PRD4/1–PRD4/23.
99. Lubin D., Chen J.-P., Pilewskie P., Ramanathan V., and Valero P. J. Microphysical examination of excess cloud absorption in the tropical atmosphere. *J. Geophys. Res.*, 1996, **101**(D12), 16961–16972.
100. Makienko E. V., Kabanov D. M., Rakhimov R. F., and Sakerin S. M. The microphysical features of the aerosol component in different regions of the Atlantic. *Optics of the Atmos. and Ocean*, 2004, **17**(5–6), 437–443 [in Russian].
101. Marshak A., Davis A., Wiscombe W., and Cahalan R. Radiative smoothing in fractal clouds. *J. Geophys. Res.*, 1995, **100**(D18), 26247–26261.

102. Mazin I. P., Monakhova N. A., and Shugaev V. F. The vertical distribution of water content and optical characteristics in continental stratus clouds. *Meteorology and Hydrology*, 1996, **9**, 14–34 [in Russian].
103. Melnikova I. N. The spectral optical parameters of cloud layers. Theory. Part I. *Optics of the Atmos. and Ocean*, 1992, **5**(2), 178–185 [in Russian].
104. Melnikova I. N. and Mikhailov V. V. An assessment of the optical characteristics of cloud layers. *Annals of RAS*, 1993, **328**(3), 319–321 [in Russian].
105. Melnikova I. N. A study of the impact of multiple scattering on the true light absorption in clouds with the use of calculations by the Monte Carlo method. *Natural and Anthropogenic Aerosols, Proc. 1st Int. Conf., Sankt-Petersburg, 29 September–4 October 1997*. St Petersburg State Univ. Publ., St Petersburg, 1998, pp. 298–307 [in Russian].
106. Melnikova I. N. and Fedorova E. Yu. The vertical profile of the optical parameters in the cloud layer. In: LSU Coll. *Problems of Atmospheric Physics* (Issue 20), 1997, pp. 261–272 [in Russian].
107. Melnikova I. N. and Domnin P. I. An assessment of the optical parameters of a homogeneous optically thick cloud layer. *Optics of the Atmos. and Ocean*, 1997, **10**(7), 734–740 [in Russian].
108. Melnikova I. N. The vertical profile of the spectral coefficients of scattering and absorption of the stratified cloudiness. *Optics of the Atmos. and Ocean*, 1998, **11**(1), 5–11 [in Russian].
109. Melnikova I. N., Domnin P.I., and Rodionov V. F. An assessment of the optical parameters of the cloud layer from measurements of reflected and transmitted solar radiation. *Izv. Russian Acad. Sci., ser. Physics of the Atmos. and Ocean*, 1998, **34**(5), 669–676 [in Russian].
110. Melnikova I. N. and Vasilyev A. V. *Short-Wave Solar Radiation in the Earth's Atmosphere*. Springer, Berlin, 2005, 303 pp.
111. Minin I. N. *Theory of Radiation Transfer in Planetary Atmospheres*. Nauka Publ., Moscow, 1988, 264 pp. [in Russian].
112. Minnet P. The influence of solar zenith angle and cloud type on cloud radiative forcing at the surface in the Arctic. *J. Clim.*, 1999, **12**, 147–158.
113. Moskalenko N. I., Terzi V. D., and Skvortsova S. Ya. The optical characteristics of aerosol formations. In: K. Ya. Kondratyev (ed.), *Aerosol and Climate*. Gidrometeoizdat, Leningrad, 1991, pp. 154–165 [in Russian].
114. Nakajima T. Y. and Nakajima T. Wide-area determination of cloud microphysical properties from NOAA AVHRR Measurements for FIRE and ASTEX regions. *J. Atmos. Sci.*, 1995, **52**, 4043–4059.
115. Nakajima T., King M. D., Spinhirne J. D., and Radke L. F. Determination of the optical thickness and effective particle radius of clouds from reflected solar radiation measurements. II. Marine stratocumulus observations. *J. Atmos. Sci.*, 1991, **48**, 728–750.
116. Nikitinskaya N. I., Barteneva O. D., and Veselova L. K. On the variability of the spectral optical aerosol thickness of the atmosphere under high transparency conditions. *Proc. USSR Acad. Sci., Physics of the Atmos. and Ocean*, 1973, **10**(4), 437–442 [in Russian].
117. Noone K. J., Targino A., Olivares G., Glantz P., and Jansson J. Aerosols and their role in the earths energy balance. *Global Change Newsletter*, 2004, No. 59, 7–10.
118. O'Hirok, Gautier C. Modelling enhanced atmospheric absorption by clouds. IRS '96. *Current problems in Atmospheric Radiation. Proc. Int. Radiation Symp., August, 1996, Fairbanks, Alaska, USA*. A. Deepak Publishing, VA, 1997, pp. 132–134.

119. Ostrom E. and Noone K. J. Vertical profiles of aerosol scattering and absorption measured in situ during the North Atlantic Aerosol Characterization Experiment (ACE-2). *Tellus Ser. B – Chemical and Phys. Meteorology*, **52**(2), 526–545.
120. Panchenko M. V., Sviridenkov M. A., Terpugova S. A., and Kozlov V. S. Active spectro-nephelometry in studies of microphysical characteristic of submicron aerosol. *Optics of the Atmos. and Ocean*, 2004, **17**(5–6), 428–436.
121. Pilewskie, P. and Valero F. P. J. How much solar radiation do clouds absorb? Response. *Science*, 1996, **271**, 1134–1136.
122. Pilewskie P. and Valero F. P. J. Direct observations of excess solar absorption by clouds. *Science*, 1995, **267**, 1626–1629.
123. Poetzsch-Heffter C., Liu Q., Ruprecht E., and Simmer C. Effect of cloud types on the Earth radiation budget calculation with the ISCCP C1 dataset: Methodology and initial results. *J. Climate*, 1995, **8**, 829–843.
124. Rakhimov R. F., Uzhegov V. N., Makienko E. V., and Pkhalagov Yu. A. The micro-physical interpretation of the seasonal and diurnal variability of the spectral dependence of the coefficient of aerosol attenuation along the surface routes. *Optics of the Atmos. and Ocean*, 2004, **17**(5–6), 386–404 [in Russian].
125. Ramanathan V., Subasilar B., Zhang G. J., Conant W., Cess R. D., Kiehl J. T., Grassl G., and Shi L. Warm pool heat budget and shortwave cloud forcing: A missing physics? *Science*, 1995, **267**, 500–503.
126. Ramanathan V. and Vogelman A. M. Greenhouse effect, atmospheric solar absorption and the Earth's radiation budget: From the Arrhenius-Langley era to the 1990s. *Ambio*, 1997, **26**(1), 38–46.
127. Ramaswamy V. and Freidenreich S. M. A high-spectral resolution study of the near-infrared solar flux disposition in clear and overcast atmospheres. *J. Geophys. Res.*, 1998, **103**(D18), 23255–23273.
128. Reid J. S. and Maring H. B. Foreword to special section on the Puerto Rico Dust Experiment (PRIDE). *J. Geophys. Res.*, 2003, **108**(D19), PRD1/1–PRD1/2.
129. Rosenberg G. V., Malkevich M. S., Malkova V. S., and Siachenov V. I. An assessment of the optical characteristics of clouds from measurements of the reflected solar radiation by Kosmos-320 satellite. *Proc. USSR Acad. Sci., Physics of the Atmos. and Ocean*, 1974, **10**, 14–24 [in Russian].
130. Rosenberg G. V., Gorchakov G. I., Georgievsky Yu. S., and Liubovtseva Yu. S. *The Optical Parameters of Atmospheric Aerosol*. Nauka Publ., Moscow, 1980, 261 pp.
131. Rublev A. N., Trotsenko A. N., and Romanov P. Yu. Use of the AVHRR satellite radiometer data to assess the optical thickness of cloudiness. *Izv. Russian Acad. Sci.*, ser. *Physics of the Atmosphere and Ocean*, 1997, **33**(5), 670–675 [in Russian].
132. Savijärvi H., Arola A., and Risnen P. Short-wave optical properties of precipitating water clouds. *Q. J. R. Meteorol. Soc.*, 1997, **123**, 883–899.
133. Shettle E. P. The data were tabulated by E. P. Shettle of the Naval Research Laboratory and were used to generate the aerosol models which are incorporated into the LOWTRAN, MODTRAN and FASCODE computer codes [data from HITRAN-96 CD ROM media, 1996].
134. Shimota A., Kobayashi H., and Wada K. Retrieval for physical parameters of aerosols in an urban area by ground-based FTIR measurement. *J. Geophys. Res.*, 2002, **107**(D14), AAC6/1–AAC6/10.
135. Skuratov S. N., Vinnichenko N. K., and Krasnova T. M. Measurements of upward and downward shortwave radiation fluxes with the use of stratospheric aircraft 'Geofizika' in the tropics (Seychelles, February–March, 1999). In: The CIS Int. Symp. '*Atmospheric*

Radiation' (ISAR-99). St Petersburg State Univ. Press, St Petersburg, 1999, pp. 58–59 [in Russian].

136. Stephens G. and Tsay S. C. On the cloud absorption anomaly. *Quart. J. Roy. Meteorol. Soc.*, 1990, **116**, 671–704.
137. Stephens G. Anomalous shortwave absorption in clouds. *GEWEX News. WCRP*, 1995, **5**(4), 5–6.
138. Stephens G. How much solar radiation do clouds absorb? Technical comments. *Science*, 1996, **271**, 1131–1133.
139. Taylor J. P., Edwards J. M., Glew M. D., Hignett P., and Slingo A. Studies with a flexible new radiation code. II. Comparison with aircraft short-wave observations. *Q. J. R. Meteorol. Soc.*, 1996, **122**, 839–861.
140. Titov G. A. Numerical modelling of the radiative characteristics of a broken cloudiness. *Optics of the Atmos. and Ocean*, 1988, **1**(4), 3–18.
141. Titov G. A. and Zhuravleva T. B. Absorption of solar radiation in broken clouds. *Proc. of the Fifth ARM Science Team Meeting San Diego, California, USA, 19–23 March, 1995*, pp. 397–340.
142. Titov G. A. and Kasyanov E. I. Radiative properties of heterogeneous stratocumulus clouds with stochastic geometry of the upper boundary. *Optics of the Atmos. and Ocean*, 1997, **10**(8), 843–855.
143. Twohy C. H., Clarke A. D., Warren S. G., Radke L. F., and Charlson R. J. Light-absorbing material extracted from cloud droplets and its effect on cloud albedo. *J. Geophys. Res.*, 1989, **94**(D6), 8623–8631.
144. Valero F. P. J., Cess R. D., Zhang M., Pope S. K., Bucholtz A., Bush B., and Vitko J., Jr. Absorption of solar radiation by the cloudy atmosphere: Interpretations of collocated aircraft measurements. *J. Geophys. Res.*, 1997, **102**(D25), 29917–29927.
145. Vasilyev A. V., Melnikova I. N., and Mikhailov V. V. The vertical profile of spectral fluxes of scattered solar radiation in a stratus cloud from results of aircraft measurements. *Izv. Russian Acad Sci., ser. Physics of the Atmos. and Ocean*, 1994, **30**(5), 661–665 [in Russian].
146. Vasilyev A. V. and Ivlev L. S. A universal algorithm to calculate the optical characteristics of 2-layer spherical particles with the homogeneous nucleus and shell. *Optics of the Atmos. and Ocean*, 1996, **9**(12), 1552–1561.
147. Vasilyev A. V. and Ivlev L. S. Numerical modelling of the spectral aerosol light scattering indicatrix. *Optics of the Atmos. and Ocean*, 1996, **9**(1), 129–133.
148. Vasilyev A. V. and Ivlev L. S. Empirical models and optical characteristics of aerosol ensembles of 2-layer spherical particles. *Optics of the Atmos. and Ocean*, 1997, **10**(8), 856–868.
149. Vasilyev A. V. and Ivlev L. S. An optical statistical aerosol model of the atmosphere for the region of the Ladoga Lake. *Optics of the Atmos. and Ocean*, 2000, **13**(2), 198–203.
150. Vasilyev A. V. and Ivlev L. S. On the optical properties of polluted clouds. *Optics of the Atmos. and Ocean*, 2002, **15**(2), 157–159.
151. Vasilyev A. V. and Melnikova I. N. *The Shortwave Solar Radiation in the Earth's Atmosphere. Calculations. Measurements. Interpretation.* St Petersburg State Univ. Press, St Petersburg, 2002, 388 pp. [in Russian].
152. Vasilyev O. B., Grishechkin V. S., Kashin F. V., and Kondratyev, K. Ya. Studies of the atmospheric spectral transparency, spectral scattering indicatrices, and determination of aerosol parameters. In: I. N. Minin (ed.), *Problems of Atmospheric Physics* (Issue 17). LSU Publ., Leningrad, 1982, pp. 230–246 [in Russian].

153. Vasilyev O. B. Spectral shortwave radiation fluxes in the atmosphere and some applied problems. Doctoral thesis. Tomsk State Univ., Tomsk, 1986, 344 pp [in Russian].
154. Vasilyev O. B., Grishechkin V. S., Kovalenko A. P. The spectral information-measurement system to study the shortwave radiation field in the atmosphere from the surface and aircraft. In: K. Ya. Kondratyev (ed.), *The Complex Remote Sensing of Lakes*. Nauka Publ., Leningrad, 1987, pp. 225–228 [in Russian].
155. Vasilyev O. B., Ivlev L. S., Muhlia Velazquez A., Leyva Contreras A., and Peralta y Fabi R. Influence of aerosol on radiative transfer in the polluted atmosphere. *IRS – 92: Current problems in atmospheric radiation. Proc. of the Int. Radiation Symp., Tallinn, Estonia, 3–8 August 1992*. A. Deepak Publ., Hampton, USA, 1993, pp. 195–198.
156. Yamanouchi T. and Charlock T.P. Comparison of radiation budget at the TOA and surface in the Antarctic from ERBE and ground surface measurements. *J. Climate*, 1995, **8**, 3109–3120.
157. Waggoner, R. E., Weiss A. P., Ahlquist N. C., Covert D. S., and Charlson R. J. Optical characteristics of atmospheric aerosols. *Atmos. Environ.*, 1981, **15**, 1891–1909.
158. Wiscombe W. J., Welch R. M., and Hall W. D. The effect of very large drops on cloud absorption. Part I: Parcel models. *J. Atmos. Sci.*, 1984, **41**, 1336–1355.
159. Wiscombe W. J. An absorbing mystery. *Nature* (GB), 1995, **376**, 466–467.
160. Zhang M. H., Lin W. Y., and Kiehl J. T. Bias of atmospheric shortwave absorption in the NCAR Community Climate Models 2 and 3: Comparison with monthly ERBE/GEBA measurements. *J. Geophys. Res.* 1998, **103**, 8919–8925.
161. Zhanging Li, Barker H. W., and Moreau L. The variable effect of clouds on atmospheric absorption of solar radiation. *Nature*, 1995, **376**, 486–490.
162. Zuev V. E. and Krekov G. M. *Optical Models of the Atmosphere*. Gidrometeoizdat, Leningrad, 1986, 256 pp. [in Russian].
163. Zuev V. E. and Kabanov M. V. *Optics of Atmospheric Aerosol*. Gidrometeoizdat, Leningrad, 1987, 264 pp. [in Russian].

8

Aerosol long-range transport and deposition

8.1 INTRODUCTION

A comparatively long (up to 2–3 weeks) lifetime of aerosol particles in the atmosphere determines the possibility of their long-range transport [1–128]. The best known and studied situations of this kind are emissions to the atmosphere of dust aerosol (DA) during dust storms in northern Africa and subsequent trans-Atlantic transport of particles, with occasional meridional DA fluxes to western Europe. Another known situation is the long-range DA transport to the north-western sector of the Pacific Ocean during dust storms in north-western China and Mongolia (a brief review of the dust storm problems can be found in [38, 52]).

Understanding the key role of aerosol in climate formation has stimulated further studies of the long-range transport of aerosol. For example, a new programme ITCT-Lagrangian-2k4 to study the intercontinental transport and chemical transformations of aerosol has been undertaken.

Parrish and Law [90] have briefly characterized the content of the ITCT programme whose main objective is to study the long-range (intercontinental) transport and chemical transformations of aerosol and oxidants as well as their precursors. To obtain observational data for this purpose, it would be worthwhile to use instruments installed on the Lagrangian platform moving together with air masses under study. However, only 'pseudo-Lagrangian' observations are possible with the use of one or more flying laboratories performing a multiple sounding of a certain air volume. This is the basic aim of the ITCT programme, the first stage of which consisted of aircraft observations in the summer of 2004 in the region of the North Atlantic concentrated on studies of emissions of aerosol precursors and tropospheric ozone in North America, including an analysis of their long-range transport and chemical transformation over the North Atlantic basin as well as their subsequent impact on the atmosphere in western Europe.

The key objectives of the ITCT programme include: (1) determination of the potentials of photochemical oxidants and aerosol formation in polluted air masses formed in North America and moving across the Atlantic Ocean to the region of western Europe; (2) analysis of atmospheric dynamics responsible for the long-range transport of pollutants from the planetary boundary layer to North America; and (3) determination of the quantitative characteristics of the transport of North American pollutants to the background atmosphere, their subsequent evolution, and impact on climate. More than ten flying laboratories from different countries are planned to take part in the ITCT programme.

To study the laws of the intercontinental transport of atmospheric pollutants, [109, 110] performed a numerical simulation for a period of one year (conditions for the year 2000 were considered) for six passive tracers emitted on different continents (to characterize the levels of emissions, data of inventory of emissions of carbon monoxide CO were taken). Calculations have shown that emissions from the Asian continent are characterized by a most rapid vertical propagation, whereas a trend to remain within the lower troposphere is typical of the European emissions. The European emissions move mainly to the Arctic where they contribute most to the formation of Arctic haze. The tracers come from the continent where emissions take place to another continent in the upper troposphere, as a rule, in about four days. After this, tracers can also appear transported in the lower troposphere.

With characteristic lifetimes taken to be two days, it turns out that local tracers are a dominating component in the atmosphere over all continents except Australia, where the share of 'foreign' tracers constitutes about 20% with respect to the whole mass of tracers. On the assumption that the tracers' lifetime constitutes 20 days, even on the continents with a high level of 'home' emissions the share of tracers from other continents exceeds 50%. Since three regions, where the tracers are transported to and where the tracers 'dissolve' slowly, require special attention, further studies should be concentrated in three areas:

(1) wintertime accumulation of tracers from Asia over Indonesia and the Indian Ocean;
(2) concentration of the Asian tracers over the Middle East in summer, which is at a maximum; and
(3) distribution over the Mediterranean Sea of tracers coming in summer from North America.

Simulations of the spatial distribution of dust (mineral) aerosol is a difficult problem because of the episodic nature of DA sources and its long-range transport. Based on the data of the observational time series for 22 years at different locations, but mainly from data of satellite remote sensing, [68] performed comparisons between the observed spatial–temporal variability of the aerosol distribution and results of numerical simulation with the combined use of the MATCH model of aerosol transport in the atmosphere, taking into account chemical reactions determining the transformation of its properties as well as the DEAD model simulating the processes of formation and transformation of DA.

The results of comparisons revealed a good agreement, though with some differences. To analyse the reasons for these differences, a consideration has been undertaken of the dependence of the numerical simulation results on the variability of various input meteorological parameters and on the choice of the scheme of parameterization of the process of particle 'mobilization' (their input to the atmosphere as a result of the saltation process). The discussed analysis of sensitivity suggested the conclusion that not far from Australia, differences between the calculated spatial distribution of aerosol optical thickness and the observed one are explained by an inadequacy of data on the wind field near the surface and on the sources of aerosol. According to the obtained estimates, the total emissions of DA to the atmosphere as a result of dust storms constitute 1,654 Tg yr^{-1}. The most powerful sources of DA emissions are the African deserts whose contribution to the total content of DA in the atmosphere is 73%. Eastern Asia contributes most to the formation of the field of aerosol content over the Pacific Ocean in the northern hemisphere, whereas in the southern hemisphere, Australia is the main source of dust aerosol. The characteristic DA lifetime in the atmosphere constitutes about six days.

Important results characterizing the effect of aerosol on the atmosphere in the process of long-range transport have been obtained by [79]. In May–June, 1999, on Lampeduza island (the Mediterranean Sea: 35.5°N, 12.6°E) the second phase of observations was accomplished within the PAUR II field experiment to study the photochemical processes in the atmosphere and the UV solar radiation. The surface complex of instruments included the Brewer spectrophotometer, the aerosol lidar at the 532-nm wavelength, and the multichannel sun photometer with shadow screen (MFRSR) to measure the net and scattered radiation fluxes as well as the aerosol optical thickness (AOT) at wavelengths 415, 500, 615, 671, 866, and 937 nm. Simultaneously, aircraft measurements were made (at altitudes up to 4.5 km) of actinic flux of short-wave radiation (AFR) in the bands of NO_2 and O_3 photodissociation, which made it possible to estimate the level of O('D) formation.

Meloni *et al.* [79] discussed the results of observations obtained during three days: 18 May, 1999, when an intrusion of a plume of the Saharan dust aerosol took place at altitudes up to 7 km, and the AOT reached 0.51 at $\lambda = 415$ nm , as well as 25 and 27 May, when the aerosol content of the continental and marine origin in the atmosphere (at altitudes up to 3 km) was moderate (the AOT at 415 nm constituted about 0.2). A comparison of the data of AFR measurements with results of calculations has made it possible to analyse the dependence of AFR on the vertical profiles of aerosols and ozone concentration (in all cases considered there was a good agreement between observational data and numerical simulation results).

Calculations for fixed solar zenith angles (SZA) under conditions of the Saharan dust intrusion revealed an AFR decrease in the lower and middle troposphere reaching 24% (as compared with conditions of a moderate content of aerosol in the atmosphere). When the SZA increased, this decrease grew reaching a maximum with the highest content of aerosol. The values of AFR fluxes at a surface level were practically independent of the specific character of the vertical profile of aerosol concentration (only the total content of aerosol in the atmosphere is important). With the mean climatic (instead of the observed) profile of aerosol concentration used to calculate the AFR, this causes changes in the AFR not exceeding ±3% under conditions from the low-to-moderate total content of aerosol.

However, with an intrusion of the Saharan dust aerosol, the calculated values of AFR turned out to be overestimated, as a rule, by 19%. An analysis of the calculated (with different prescribed aerosol profiles) variability of AFR in the spectral interval responsible for the process of ozone photodissociation ($O_3 + h\nu \rightarrow O_2 + O(^1D)$) has led to the conclusion that the specific character of the vertical profile of ozone concentration affects substantially the AFR in the spectral interval of photodissociation, especially at low SZA values and near the surface. The aerosol-induced enhancement of scattered radiation results in AFR values calculated with the prescribed mean climatic profile of aerosol turning out to be overestimated by 7.7% compared with respective real conditions with SZAs of $\sim$37°. This difference decreases with altitude and with increasing SZA.

8.2 THE AFRICAN AEROSOL

Reid *et al.* [95–98] analysed the results of observations of the properties of the Saharan dust aerosol from data of aircraft and surface studies of the elemental composition of total DA and its individual particles. To sample aerosols, polycarbone filters were used, and the chemical analysis was made with the use of electron microscopic and X-ray (EDAX) instruments. The chemical composition of individual DA particles was determined with their size and morphology taken into account. At the main surface observation point located on Cabras Island (18.21°N, 65.60°W) samples were obtained with the cascade impactor (DRUM) providing measurements for eight intervals of sizes and diameters ranging between 0.1 and 12 μm (0.09. 0.24, 0.34, 0.56, 0.74, 1.1, 2.5, and 5 μm) with a time resolution of 4 hours. To analyse the elemental composition of DA samples (from Al to Zn), the X-ray fluorescence (XRF) method was used.

This gave results in agreement with the DRUM data and results of analysis of the aircraft DA samples. The cluster analysis of data for individual particles revealed the presence of 63 statistically substantial clusters. Several clusters have been reliably identified as belonging to different mother mineralogical components of soil composition. However, since clusters are, as a rule, aggregates, most of them are characterized by a complicated mineralogical composition. With 65.5 thousand DA particles on the airborne filters, it is possible to obtain statistically substantial results for a great number of particles. The modal radius of particles was estimated at $\sim$5 μm, with an average aspect ratio 1.9 (thus the non-spherical shape of particles was evident).

Data on DA particle morphology reflects changes in the sources of DA emitted to the atmosphere and in the content of the non-alumo-silicate components, whereas information on the ratio of the content of aluminium and silicates does not indicate such changes, which suggests the conclusion about the process of aerosol homogenization in the course of its long-range transport. During the whole period of continuous observations (3–24 July, 2000) the DRUM data revealed four episodes of a considerable increase in the level of atmospheric dust load (5, 10, 15–16, 21

July), which correlated with results of AOT observations. 'Flares' of the aerosol surface concentration, not reflected in AOT changes, were also noticed on 13 and 24 July. The mineral composition of DA was characterized by prevailing illite.

Reid and Maring [98] characterized the main goals and most substantial results obtained during the PRIDE field observational experiment in the region of Puerto Rico during intensive observations on 28 June–24 July, 2001 aimed at complex studies of the long-range transport of atmospheric dust aerosol. The basic surface point of observations was located on Cabras Island at a distance of several hundred meters from the main island of Puerto Rico. Here a mobile aerosol laboratory of the Miami University and a complex of instruments from the Goddard Space Flight Centre were located. Simultaneously, two flying laboratories were functioning (Piper Navajo and Cessna-172; data from C-130 were also used) as well as the research ship *Chapman* and two observational sites of the AERONET network. Of importance were also data of the satellite remote sensing of aerosol with the use of instruments such as MODIS, MISR, CERES, GOES, and AVHRR.

The main goals of PRIDE were to obtain complex information about the microphysical and optical characteristics as well as the chemical composition of DA getting to the atmosphere in northern Africa and then transported great distances across the Atlantic Ocean. Questions to be answered by analysis of PRIDE data were as follows:

(1) how adequate are the existing numerical models of the long-range transport of DA;
(2) to what degree is the available information about DA sources reliable;
(3) how reliable are data on DA properties;
(4) what new (from the viewpoint of understanding DA properties) information does the analysis of the complex observational data give;
(5) what is the ratio between the *in situ* observational data and remote sensing results;
(6) what role does the consideration of DA particle non-sphericity play; and
(7) how important are the obtained results for the solution of various applied problems?

As Reid *et al.* [95–97] have shown, during the PRIDE field observational experiment accomplished in the region of Puerto Rico, six episodes were observed of the input of Saharan dust suffering long-range transport across the Atlantic Ocean: on 28–29 June, 4–6 July, 9–10 July, 15–16 July, 21 July, and 23 July, 2000. In these periods the AOT values in the middle part of the visible spectral region varied from those for the clear marine atmosphere (0.07) to the heavily dust-loaded atmosphere (AOT $>$ 0.5). The AOT value averaged 0.24. At this time, near the African coast, the AOT values reached, on average, $\sim$0.45 with a maximum up to 0.8. The AOT values obtained in the region of Puerto Rico turned out to be somewhat below those observed during the previous three years, which was explained by the effect of a stronger than usual, washing out of DA from the atmosphere by precipitation. The vertical distribution of DA concentration was characterized by a strong temporal variability.

Along with the formation of the well-known Saharan aerosol layer (SAL), there existed also conditions of a more weakly expressed long-range transport of DA. The dust aerosol often reached an altitude of about 5 km, and the presence of dust in the marine atmospheric boundary layer (MABL) did not correlate with any 'typical' profile of the vertical distribution of DA concentration. In particular, there was no correlation with the intensity of trade wind inversions observed in the region considered. Since the formation of trade wind inversions is mainly determined by the impact of local conditions, this can explain the absence of a correlation between SAL and inversions. The transition from the regime of DA transport in the MABL to the regime characteristic of the lower troposphere took place gradually, but in the presence of sporadic fluctuations. However, in some cases, distinct dust layers were observed over the trade wind inversion. Such dust layers could be observed at distances up to hundreds of kilometres both in the direction of the prevailing transport and perpendicular to it.

As a rule, there were not observed any substantial altitudinal changes in the size of particles (vertical gradients), except the uppermost part of the dust-loaded troposphere about 200 m thick. This conclusion is also valid for conditions when aerosols are strongly stratified. One exception was for data of 4–6 July, 2000, when there was a strong vertical gradient of the sizes of DA particles in the range of altitudes from the layer of the trade wind inversion to the maximum altitude of aerosol propagation (about 5 km). An analysis of the factors of formation of the vertical profile of DA concentration has led to the conclusion that the impact of the particles' gravitational deposition in the process of their long-range transport was not substantial, as a rule. This, in turn, suggests a supposition that the long-range transport across the Atlantic Ocean has not caused any substantial changes in the vertical profile of DA concentration formed over the African continent. However, the combined impact of the processes of large-scale deposition and the convection-induced vertical mixing (together with a differential advection at different levels) results in chaotic changes sometimes observed in the DA vertical profile.

Numerous studies have been dedicated to the analysis of sensitivity of the aerosol optical properties to the variability of size distribution, chemical composition, and shape of aerosol particles. As for the latter factor, it is not that essential for such particles such as hydrated salt particles or particles resulting from biomass burning, but is very important in the case of dust aerosol particles, when a consideration of the shape of particles strongly tells upon their optical properties and, hence, on the estimates of their climatic impact. In this context, [97] performed a comparison of data on DA size distribution obtained with the use of various measurement techniques during the PRIDE field experiment undertaken at Puerto Rico to study the size distribution of Saharan dust received from Africa between 28 June and 24 July, 2000.

This comparison revealed the presence of rather substantial differences, especially with the use of optical counters of particles and aerodynamic techniques. Analysis of the results obtained with different techniques has shown that the method of electronic microscopy gives substantial systematic errors when estimating the number density of larger particles (4–10 μm), which leads to a very uncertain size distribution of particles. The deficiency of the cascade impactors (MOUDI and DRUM) resulted in the loss of particles at input reaching 40%. In the case of the DRUM impactor, large particles are also destroyed at input. Reliable measurements of the size of particles are provided with the APS–3300 aerodynamic size counter, which, however, is also deficient since it cannot distinguish between particles of dust and sea salt aerosols. The aircraft optical counter FSSP–10 almost doubles the size of DA particles (this is also typical of other counters of large-size aerosol particles).

Parameters of size distribution retrieved from the AERONET data on sky brightness and attenuation of direct solar radiation with the prescribed spheroidal shape of particles are sufficiently reliable, on the whole, but always require a critical analysis from the viewpoint of the possible effect of invisible cirrus clouds and insufficient reliability of the normalization in solving an inverse problem. Neither of the techniques considered guarantees adequate values (as compared with those observed) of mass coefficients of extinction and efficiency of scattering by dust aerosols with the prescribed sphericity of particles. Therefore, aerosol scattering turns out to be underestimated by about 30%. Ideas have been expressed about the possible ways of improving techniques of determination of aerosol particle size distribution.

Livingston *et al.* [67] discussed the results of aircraft observations (21 flights) of aerosol optical thickness of the atmosphere and total content of water vapour (TCWV) in the atmosphere using the 6-channel (380.1, 450.9, 525.7, 864.5, 941.9, and 1021.3 nm) sun photometer (AATS-6) equipped with a system of automated pointing toward the sun. Observations were made over the period 28 June–24 July, 2000 during the PRIDE field experiment to study the properties of dust aerosols reaching the atmosphere in northern Africa and suffering long-range transport across the Atlantic Ocean. As a rule, there was good agreement between the results of retrieved total number density of aerosol from the measured vertical profile of extinction at altitudes up to 5 km using the AATS-6 with synchronous aircraft *in situ* observations of total number density of aerosol and with data from surface lidar soundings.

The AOT spectral distribution for different atmospheric layers corresponds to the results of calculations considering an attenuation of direct solar radiation due to large DA particles (or DA with an addition of sea salt aerosol) for Angström exponent values of about 0.20. The difference between the results of TCWP retrieval (from data for $\lambda = 941.9$ nm) and data of direct measurements did not exceed 4% (0.13 g cm^{-2}). Data of AOT surface observations (AERONET) using the AATS-6 on Cabras Island and from low-flying aircraft agree within differences not exceeding 0.004–0.030. However, the TCWP values retrieved from AERONET data, systematically exceed the values obtained from aircraft data by about 21%. A comparison of AOT values obtained from AATS-6 data (low-flying aircraft over the ocean) with AOT over the ocean retrieved from satellite data on the outgoing short-wave radiation (MODIS, TOMS, and the GOES-8 scanner) has led to contradictory conclusions – determining urgent further comparisons.

On the Turkish coast of the Mediterranean Sea (Erdemli: 36°33′N, 34°15′E) within the programme of the global observational network AERONET, with the use of sun photometers to retrieve the AOT of the atmosphere, [62] performed observations over the period from January, 2000 to June, 2001. The main goal of observations was to study the optical properties of dust (mineral) aerosol emitted to the atmosphere during dust storms in the Sahara desert (in the central part of the desert in spring and in the eastern part in summer) with subsequent long-range transport to the north-eastern sector of the Mediterranean Sea.

The DA transport in summer and autumn took place in the troposphere above the 700-hPa level. On the other hand, at lower levels (<850 hPa) the industrial and urban aerosol reached regions of observations from the north (Balkan Mountains, the Ukraine) and from Turkey. Apart from a high level of AOT, sometimes reaching 1.8, episodes of dust loading of the atmosphere were characterized by the following special features: (1) a sharp decrease (almost to zero) of the Angström coefficient; (2) the presence of strong scattering with the single-scattering albedo (SSA) $>0.95 \pm 0.03$ and refraction index 1.51 ± 0.07 at $\lambda = 0.44\,\mu m$ (in both cases there was a small increase of these parameters with increasing wavelength); (3) the imaginary part of the complex refraction index 0.0012 ± 0.0007 at $\lambda = 0.44\,\mu m$ and 0.0075 at $\lambda = 0.87\,\mu m$; and (4) almost neutral dependence of these optical parameters on wavelength.

The DA size distribution turned out to be bimodal with characteristic average values of the radius of particles of large and small-size aerosol equal to 2.2 μm and 0.08 μm. The dust aerosol from the Sahara and Arabian peninsula was characterized by different values of absorption index of 0.0015 and 0.0005, respectively, which was, apparently, determined by the different mineral composition of DA particles.

Based on the use of GOES-8 data for 30 days (the repetition of which was 30 minutes), [122] retrieved the fields of AOT of the atmosphere in the region of the PRIDE field experiment (14°N–26°N, 73°W–63°W) in order to study the Saharan dust aerosol and its long-range transport across the Atlantic Ocean. The AOT values were retrieved from data of satellite observations of the outgoing short-wave radiation (OSWR) in the visible spectral region through comparison of measured OSR values with results of calculations made with the method of discrete ordinates with prescribed aerosol size distribution and complex refraction index $1.53–0.0015i$ at $\lambda = 0.55\,\mu m$.

A comparison of retrieved AOT values with data of surface observations with the use of the sun photometer revealed a good agreement having correlation coefficients (for two sites of observations) 0.91 and 0.80. The retrieved monthly mean AOT values at $\lambda = 0.67\,\mu m$ constituted (for these two sites of observations) 0.19 ± 0.13 and 0.22 ± 0.12, whereas from the data of surface sun photometers they are 0.23 ± 0.13 and 0.22 ± 0.10, respectively.

Estimates of the errors caused by neglecting the non-sphericity of particles have led to the conclusion that the real sources of AOT errors ($\Delta\tau$) are from an uncertainty of the complex refraction index ($\Delta\tau \pm 0.05$) and inadequately prescribed surface albedo ($\Delta\tau = \pm(0.2–0.4)$). The results obtained suggest the conclusion that the AOT values over the ocean are reliably retrieved from satellite data (even with low AOT values) to monitor the spatial–temporal variability of the AOT fields and, hence, DA content in the atmosphere.

Ginoux [36] studied the effect of non-spherical particles on the gravitational deposition of mineral dust particles parameterized for the case of stretched ellipsoids and Reynolds numbers <2. The obtained dependence (a decrease of deposition rate due to non-sphericity) was used in the GOCART model of the impact of aerosol on chemical processes of formation of the ozone concentration global field in order to simulate the aerosol size distribution in April, 2001.

The use of two schemes of parameterization has shown that in the case of particles with $D > 5\,\mu m$ the calculated size distribution of aerosol is sensitive to the choice of the parameterization scheme. In the common case, the modelling results can depend on the substitution of the spherical form of the particle by an ellipsoid form. The ratio λ of its linear sizes is an important parameter. When $\lambda < 5$ the modelling results do not depend on this substitution. However, in the case of strongly stretched particles ($\lambda > 5$) such differences are substantial. But since, according to data of *in situ* measurements, most often $\lambda = 1.5$, the effect of particle shape on results of numerical simulation on global scales turns out to be negligible.

During the period 11–16 October, 2001 from the data of ten stations (EARLINET lidar aerosol soundings) in Europe there was a considerable increase of atmospheric dust load in the south-western, western, and central regions of western Europe determined by the long-range transport of dust (mineral) aerosol due to dust storms in the Sahara. For the first time, on such scales, an input of an optically dense DA plume was recorded with high-resolution data on the DA content in the vertical.

Ansmann *et al.* [4] have shown that the main DA layer was located over the ABL (above 1 km) at altitudes up to 3–5 km, but DA particles sometimes reached the 7–8-km levels. Typical values of AOT at $\lambda = 532\,nm$ for the atmospheric layer above 1 km varied within 0.1–0.5, reaching a maximum of about 0.8 over northern Germany. A comparison of the AOT lidar data with AOT retrieved from the data of spaceborne TOMS instruments revealed a qualitative agreement. An analysis of the results of calculations of backward trajectories of air masses for a period of 10 days has demonstrated that the Sahara was the DA source with aerosols coming to Europe 1–2 days after their emission. Values of the ratio of depolarization, Angström exponent, and the extinction/backscattering ratio obtained from data of lidar soundings constituted, respectively, 15–20%; from −0.5 to 0.5 and 40–80 sr. The results of numerical simulation of DA distribution in the atmosphere over Europe, on the whole, agreed well with the observational data considered.

Colarco *et al.* [24, 25] undertook a numerical modelling of emissions, the long-range transport of the Saharan dust aerosol across the Atlantic Ocean and DA deposition during the period of undertaking the PRIDE field experiment (June–July, 2000) in order to study dust aerosols in the region of Puerto Rico. Calculations were made with the use of a 3-D model of the aerosol long-range transport with the prescribed parameters characterizing the meteorological conditions. Analysis of the numerical simulation results has shown that the time taken to reach the high level of AOT values observed at Puerto Rico does not depend on the time of DA emissions in the Sahara region. It means that the atmosphere in North Africa always contains a huge reservoir of suspended dust particles, and the intensity of the DA long-range transport depends mainly on the specific character of the atmospheric circulation.

According to the model of DA emissions (including the size distribution parameters) proposed by [36], the emission of DA particles, with $D < 10$ nm in Africa and Arabia in July, 2000, constituted 214 Tg – 19% of the aerosol mass being transported westward from Africa (over the North Atlantic) with the share of particles reaching Puerto Rico constituting 20%. The calculated AOT values depend strongly on the prescribed aerosol size distribution and an adequacy of parameterization of the wet deposition process.

Until recently, the factors determining the long-range transport of the Saharan dust aerosol to the Caribbean region taking place in the tropics of the North Atlantic seemed to be well studied. Cold and wet air masses from the north-east to north Africa across the Mediterranean Sea are strongly heated over the African continent due to the turbulent heat exchange with the surface. Then follows a westward advection of the heated, dust-loaded, and well mixed air at altitudes up to 6 km. From the very beginning of the process of advection, the lower boundary of the dust-loaded layer rapidly rises due to the impact of north-eastern trade winds with their relatively clear and wet air. The layer of warm, dry, and dust-loaded air formed in the lower troposphere was called the Saharan aerosol layer (SAL). Its lower boundary is located at the level of about 850 hPa (1.5 km), and the upper boundary near 500 hPa (6 km). With the westward motion of SAL a gradual lifting of its lower boundary takes place to about 750 hPa (3 km) and lowering of its upper boundary to 550 hPa (5 km). Thus in the Caribbean region the SAL is about 2 km thick.

Studies of the nature of the SAL formation carried out by [24, 25] have led to the conclusion that its nature is more complicated than was initially supposed. It turns out, in particular, that the wintertime transport of DA takes place at lower altitudes than suggested by the SAL classic model. The Saharan aerosol near the coast of South America was detected even in the ABL. The data of satellite, aircraft, and surface observations performed during the period 28 June–24 July, 2000 in the region of Puerto Rico and including five episodes of an abnormally strong dust loading of the atmosphere, help to better understand the SAL's nature. Analysis of these data has shown that the 'classic' SAL was observed only during the last five days of observations. Before that, the vertical distribution of aerosol was more complicated and variable.

A comparison of results of numerical simulation and observations has shown that calculations sufficiently reliably reproduce the formation of the DA plume and its transport, though, according to calculations, the DA is transported in the western sector of the North Atlantic farther southward than it follows from analysis of satellite images. The numerical model reliably reproduces the spatial distributions of the uniformly mixed and stratified aerosols observed in the period of the PRIDE field experiment. Calculations suggested the conclusion that the evolution of the vertical profile of DA concentration taking place in the process of long-range transport was mainly determined by the effect of sedimentation of particles and downward atmospheric motions. The process of wet deposition here plays the key role. If we assume that the content of iron in the DA composition constitutes 3.5%, the total deposition of iron onto the ocean surface in July, 2000 was to reach 0.71–0.88 Tg, which agrees with data of surface observations of the DA mass concentration. With the prescribed annual deposition, which exceeded that observed in July, 2000 by 5 times, the numerical simulation gave a deposition of 1 m on the Florida peninsula over one million years.

Observations of the Saharan aerosol reaching Barbados due to long-range transport across the Atlantic Ocean have been made during the last 30 years and are a unique series of related long-term observations. The earlier processing of observational data has shown that during 1960–1980 the surface concentration of atmospheric aerosol changed 4 times.

Mahowald *et al.* [69] have tested the validity of the hypothesis according to which the main sources of dust aerosol are not distributed sources (like cultivated surface areas or new regions of the desert) as was supposed earlier, but dried-up topographic low ground. It follows from the obtained estimates that a consideration of the contribution of distributed areas is important but does not permit the explanation of the observed DA dynamics completely. However, the deficit of observational data prevents obtaining reliable quantitative estimates.

From data of surface observations of the cloud-free sky brightness in the solar vertical at 14 wavelengths in the visible spectral region, [17] retrieved the mean values of (SSA) for the whole atmosphere and scattering function for the Saharan mineral aerosol. Observations were made in July and August, 1998 at Fort Jefferson, on the island of Dry Tortugas (24°37.7′N, 82°52.5′W) located about 100 km west of Key West (Florida, USA) using the 512-channel spectrometer. An algorithm of retrieval consisted of an iterative adjustment of calculated values of brightness to those observed, which excluded a necessity to take into account the shape of aerosol particles. The error of SSA retrieval was estimated at 0.02. The retrieved values of SSA reflect the spectral features characteristic of minerals containing iron, and turned out to be much greater than those assumed in climate models. This shows that the real Saharan dust aerosol causes a more intensive climate cooling than is usually assumed.

The Saharan dust storms are not the single powerful source of aerosol emissions to the atmosphere from the African continent – another substantial source is biomass burning.

Fires in the African savannas are responsible for more than two-thirds of the global biomass burning of savannas. These fires result in emissions to the atmosphere of various minor gas components (MGCs) (including CO_2, CO, NO_x, SO_2, hydrocarbons, halocarbon compounds, oxidized organic compounds, and aerosol). In South Africa, fires in savannas occur mainly in the period of the dry season (April–October), when meteorological conditions are characterized by the formation of persistent air masses, south-eastern trade winds, and the subtropical belt of high pressure. The presence in the atmosphere of stable layers near 700 and 500-hPa levels limits the vertical motion of smoke aerosol and MGCs. Pollutants appearing in the atmosphere over South Africa are transported over long distances causing an increase of the tropospheric ozone content over the southern region of the Atlantic Ocean and reaching the region of the Indian Ocean.

In August–September, 2000, within the programme of the South African regional field Experiment (SAFARI-2000), aircraft (Convair-580) observations were carried out, one of the goals of which was the study of the consequences of biomass burning in South Africa from the viewpoint of the related pollution of the atmosphere, especially formation of aerosol haze layers. The observational programme described by [103] included the determination of the vertical profiles (in the lower troposphere) of air temperature and relative humidity, concentration of sulphur dioxide (SO_2), ozone (O_3), condensation nuclei (CN), and carbon monoxide (CO), as well as horizontal distribution of 20 MGCs and aerosol in five regions of South Africa during the period of biomass burning. The regions considered were semiarid territories of savannas of southern Zambia and the deserts of western Namibia.

Maximum averaged concentrations of CO_2, CO, CH_4, particles of black carbon, sulphate aerosol, and total carbon were recorded in Botswana and Zambia (respectively, 388 and 392 ppm; 369 and 463 ppb; 1753 and 1758 ppb; 79 and 88 ppb; 2.6 and 5.5 $\mu g\,m^{-3}$; 13.2 and 14.3 $\mu g\,m^{-3}$). This is explained by the impact of intense biomass burning in Zambia and the surrounding regions. In the sector of South Africa, maxima of average concentrations of SO_2, sulphate aerosol, and CN (respectively, 5.1 ppb; 8.3 $\mu g\,m^{-3}$, and 6400 cm^{-3}) were determined by contributions of both biomass burning and emissions from electric power stations and mining. Over arid Namibia, layers of pollutants in the troposphere were observed with average mixing ratios for SO_2, O_3, and CO (1.2 ppb; 76 ppb, and 310 ppb, respectively), which were close to those recorded in other, more polluted regions, determined by the contribution of pollutant transport (due to biomass burning) from the regions of biomass burning in southern Angola. Average concentrations of MGCs and aerosol over the whole region considered constituted: 386 ± 8 ppm (CO_2), 261 ± 81 ppb (CO), 2.5 ± 1.6 ppb (SO_2), 64 ± 13 ppb (O_3), $2.3 \pm 1.9\,\mu g\,m^{-3}$ (aerosol black carbon), $6.2 \pm 5.2\,\mu g\,m^{-3}$ (aerosol organic carbon), $26.0 \pm 4.7\,\mu g\,m^{-3}$ (total aerosol concentration), and $0.4 \pm 0.1\,\mu g\,m^{-3}$ (potassium particles). These values are comparable with those corresponding to the urban polluted atmosphere. Since these results were obtained, as a rule, at locations removed from industrial sources of pollution, biomass burning can be assumed to be the dominating factor for the high levels of MGCs. Sometimes, at an altitude of about 3 km, there appeared relatively thin ($\sim$0.5 km) layers of very clear air masses between the layers of heavily polluted atmosphere. Near the coast, cirrus clouds formed at an altitude of about 1 km.

Within the CARIBIC programme, with the use of civil aircraft for regular monitoring of the atmosphere, [89] performed measurements of the elemental composition of atmospheric aerosol at altitudes 9–11 km (near the tropopause layer) in the northern hemisphere, and discussed the results obtained during two years during intercontinental flights between Germany and Sri Lanka/Maldives (the Indian Ocean). Aerosol was sampled with impactors and analysed for the content of 18 elements with the use of the PIXE X-ray technique. At the same time, measurements were made of particle number densities, concentrations of ozone and carbon oxide, and for analysis of data of aerosol observations the backward trajectories of air masses were calculated.

Analysis of measurement results has shown that the prevalent component of the aerosol chemical composition was sulphur, whose concentration in the upper troposphere doubled, on average, in moving from the tropics to high latitudes and doubled in the lowest part of the stratosphere compared with the troposphere (in mid-latitudes). A consideration of the trajectory data has led to the conclusion that mountain rocks, burning of fossil fuels, and biomass burning contributed considerably to the formation of the aerosol composition, whereas meteor ablation was of secondary importance. The trajectory data reflect the presence of aerosol transport across the tropopause from the troposphere to the stratosphere along isentropic surfaces. The long residence times of aerosol particles and their gas-to-particle conversions in the lower stratosphere due to sulphur dioxide oxidation result in formation of a high-altitude source of aerosol, whose existence (with the lowering of stratospheric air) makes it possible to explain the vertical gradient of the aerosol sulphur mass concentration near the extratropical tropopause.

On the whole, one can state that aerosol sulphur appears due to the arrival both from below (the surface) and from the stratosphere. Concentrations of sulphur, potassium, and iron are characterized by different annual variations, most strongly expressed in the case of potassium with a very low concentration in the period July–January and with a maximum of concentration in spring. In contrast to this, a weak annual variation of sulphur concentration with low values is typical of July–September in the tropics and subtropics, whereas in the tropics a minimum of concentration is observed in July–January. The annual variation of iron concentration is weak.

8.3 THE ASIAN AEROSOL

At present, the Asian population constitutes about 3.4 billion (~60% of the global population) with about half of the 136 million children born after 2000 being from Asia. The growth of the population has resulted in a considerably enhanced pollution of the environment, including the atmosphere, where a comparatively rapid long-range transport of pollutants takes place. Since the meteorological conditions typical of the northern hemisphere spring favour the long-range eastward transport of air masses, this determines the arrival of polluted air to Asia from western Europe and even from North America. On the other hand, an intensive convection taking place over the Asian continent and the season dependent west–east transport favour the transport of air masses across the Pacific Ocean. The presence of the jet stream, which is especially intense near Japan (reaching the rate $>70\,m\,s^{-1}$), determines the time of air mass transport between Asia and North America, sometimes taking several days.

Martin *et al.* [71] discussed the results of studies of the aerosol long-range transport over the Pacific Ocean during the period of late winter in the northern hemisphere (25 February–19 April, 1999) accomplished under the auspices of NASA within the PEM West-B field observational experiment to monitor the transport of aerosol from the Asian continent to the Pacific Ocean (these observations were part of a more general programme of the Global Tropospheric Experiment – GTE). The PEM West-B experiment was undertaken as an addition to the PEM West-A mission carried out earlier and in another season (14 August–6 October, 1996). Data on the content in the atmosphere of such MGCs as non-methane hydrocarbon and hydrocarbon compounds as well as carbon monoxide were obtained with instruments carried by the NASA flying laboratories DC-8 and P-3B flying over the Pacific Ocean. To calculate the backward trajectories of air masses along the flight routes, meteorological information available in the European Centre for Middle-Range Weather Forecast was used.

Calculations have shown that MGCs recorded over the Pacific Ocean, which crossed the ocean during their long-range transport, came from Asia and even from regions farther west, depositing in the regions of the subtropical Pacific anticyclone with subsequent south-western transport driven by the trade winds in the lower troposphere. It took 20–25 days for particles to reach the western sector of the Pacific Ocean near the New Guinea coast. Apparently, similar ('mirrored') processes of the long-range transport also take place in the southern hemisphere, especially with the transport of biomass burning products in South Africa and South America.

The field observational experiments accomplished in remote regions of the globe during the last two decades revealed the presence of a considerable amount of anthropogenic pollutants in the atmosphere of these regions in an environment which was earlier considered to be free of pollutants. This refers, in particular, to the Arctic and the Antarctic, as well as to remote regions of the Atlantic and Pacific Oceans. Analysis of data from aircraft observations revealed large-scale plumes of aerosols and aerosol layers in the free atmosphere formed due to industrial pollution and biomass burning with subsequent long-range transport of atmospheric pollutants.

The aerosol plumes of continental origin are often characterized by an increased concentration of condensation nuclei of different compositions (sulphates, nitrates, dust, carbon, and soot aerosol) with an enlarged contribution of the accumulation mode (particle diameters of 0.1–1.0 μm) to the aerosol size distribution. Large-scale PEM-Tropics (PEMT) programmes of aircraft observations of atmospheric composition over the tropical Pacific Ocean were accomplished in 1996 (PEMT-A) and 1999 (PEMT-B).

An analysis of observational data performed by [82] revealed the existence of powerful 'rivers' of polluted air masses from the continents to remote regions of the ocean. Under these conditions, the aerosol characteristics differed strongly from those of the background. In the PEMT-A period the plumes of pollutants were observed mainly in the southern hemisphere, whereas during PEMT-B an opposite situation took place. Calculations of backward trajectories have demonstrated the presence of such sources of pollution as urban and industrial emissions and biomass burning, as well as (in the case of 'rivers' of Asian origin) dust storms. The aerosol size distribution strongly varied from plume to plume. Aerosol particles of industrial origin were an internal mixture of soot-like volatile components.

Studies of the dependence of volatility on the particles size suggested a conclusion about the presence of ammonium in aerosol particles. An estimation of SSA $\tilde{\omega}$ gave values from 0.68 (products of biomass burning in the absence of the large-size fraction) to 0.94 (particles of pollution and dust) in the free troposphere and 0.98 (particles of pollution and sea salts) in the MABL. Analysis of the vertical profiles of aerosol revealed the most dense plumes in dry air (at relative humidities <40%) with smaller values of $\tilde{\omega}$ than in the MABL.

The Pacific Ocean research mission PEM-Tropics was planned mainly to study the clear (background) atmosphere in remote tropical regions of the World Ocean, but early on, vast layers of polluted atmosphere were detected in the marine troposphere, reflecting a substantial pollution of the background atmosphere over the ocean in the southern hemisphere, especially over the tropical Atlantic Ocean. In particular, in the southern hemisphere subtropics there was observed an extended (>15,000 km) plume of CO meandering round the globe gradually propagating along the ~20°S latitudinal belt. Such a plume was especially clearly seen in the period of biomass burning in spring in the subtropics of both hemispheres.

Based on the use of the MM5 mesoscale model, which enables one to simulate (with a high spatial resolution) the global dynamics of the atmosphere, [20] undertook a numerical modelling for two situations including data of PEMT-A and B for the periods September–October, 1996 and March–April, 1999, respectively. In the case of PEMT-A, parts of the plume close to the source were clearly seen from data of the TOMS spectrometer (to map the total content of ozone) on the distribution of absorbing aerosols.

A detailed analysis of the data of aircraft observations enabled one to retrieve the general pattern of the plume's evolution. The plume in the southern hemisphere formed due to accumulation of pollutants near the surface in Africa and South America. In thunderstorm situations the plume moves upward, but intense emissions of pollutants can also take place due to synoptic-scale motions. Then the plume moves eastward under the influence of the subtropical jet stream until its fragmentation occurs caused by storms in the Pacific Ocean. A similar situation is observed in the northern hemisphere subtropics in the PEM-Tropics B period.

According to available observational data, events of the the long-range transport of dust aerosol and MGCs took place off the western coast of North America from the deserts of China and Mongolia, from industrial regions of eastern Asia as well as from Siberia, where large-scale forest fires occur. Holzer *et al.* [42] analysed the laws of long-range transport of dust aerosol and MGCs by estimating the probability density function (p.d.f.) of the time of air mass transport (G) in order to select (filter out) the transport against the background of other factors such as variations of the power of the emission sources and the processes of chemical transformation of DA and MGCs.

In contrast to the usually applied technique of calculating the backward trajectories of the long-range transport in order to forecast the emission sources, the solution to this problem, in terms of p.d.f. makes it possible to more completely analyse the totality of possible means of transport taking into account both the spatially resolved advection and the subgrid processes. The statistical information about diurnal mean values of G, was obtained with the use of the MATCH model of the processes of long-range transport with chemical transformation taken into account and with a priori prescribed input from meteorological information.

The calculated values of G make it possible to estimate the climatic share of air masses coming from the region of the source for the time interval of transport (or the age of air masses). Over the western coast of North America, this share reaches a maximum with the time of transport being ~8 days in the upper troposphere and 6 days later near the surface. Such estimates were obtained when considering the DA and pollutant transport – in the case of the long-range transport of forest fire products in Siberia the respective values reached 12–14 days. Analysis of G variability at a fixed time of transport enabled one to identify the DA and MGCs sources and to assess the probability of such a transport. So, for instance, one of the events refers to the Pacific north-west (PNW) region with coordinates (43.8°–53.3°N; 115.3°–124.7°W). The degree of correlation between G values and the wind field averaged over the PNW region makes it possible to reveal the structures of large-scale anomalies which correspond to favourable conditions of transport to the PNW region.

An analysis of the data of *in situ* aircraft measurements, lidar soundings, and calculations of reverse trajectories of air masses performed by [72] enabled one to monitor the process of the long-range transport and evolution of the properties of dust aerosol coming to the atmosphere in the east Asian region mainly during dust storms. Aircraft measurements performed over central Japan in 2000–2001 demonstrated the presence of a small-scale but persistent long-range transport of DA at altitudes of 2–6 km in spring, including days when distinct DA emissions from the Asian continent were absent. The dust aerosol which can be considered as the background one, was also observed in summer at altitudes above 4 km under conditions of prevalent west–east transport. A consideration of a long series of lidar soundings at Nagoya revealed a substantial change of aerosol characteristics in the free troposphere from spring to summer. The most probable source of the observed background DA was the desert of Taklamakan.

Therefore, during the last years the problem of the long-range transport of DA emitted to the atmosphere in deserts during dust storms has attracted great attention. The impact of such a transport is manifested not only on regional but also on global scales, being highlighted in the DA-induced changes in the processes of formation and properties of clouds, 'fertilization' of the ocean (due to the input of iron compounds), visibility deterioration, transport of pathogens, and stimulation of respiratory deceases. Direct and indirect impacts of dust aerosols on the formation of radiative forcing determines a substantial (and poorly studied, so far) impact of DA on climate.

The Sahara and deserts of central and eastern Asia are the most powerful sources of dust aerosol. An important contribution to the natural DA formation is made by the Taklamakan (western China) and Gobi (north-eastern China and Mongolia) deserts.

Abnormally strong dust storms took place on the Asian continent in 6–9 April, 2001. On 8 April the DA front reached the Korean peninsula, and on 9 April Japan. The atmosphere over the vast Pacific basin was loaded with dust aerosol. By 12–13 April, part of the gigantic dust plume reached the USA, and later on (19–20 April) it was observed in the atmosphere over the Atlantic Ocean. Based on the use of an interactive (for the atmosphere-ocean system) 46-layer (with the horizontal resolution 27 km) model COAMPS (mesoscale model of atmospheric circulation) developed by US Air Force specialists, [66] undertook a numerical simulation of the dynamics of the Asian dust storms for the period 5–15 April, 2001. In this case the main DA sources were the Taklamakan and Gobi deserts. An adequacy of the obtained results was verified with the use of results of satellite observations demonstrating the spatial structure of the DA long-range transport, data of surface observations of PM-10 particle number densities (with $D < 10\,\mu m$), and lidar soundings at different locations (Langzhou, Beijing, Heifei, Tsukuba, and Nagasaki).

This model made it possible to adequately simulate the evolution of dust storms and long-range DA transport. It turned out, in particular, that in Mongolia a cyclone (6–7 April) and cold front (8–9 April) which was followed by a secondary cyclone, determined an appearance in the atmosphere of a forcing impact causing the formation of a DA field in the troposphere. Both cyclones were responsible for dust transport to altitudes up to 8–9 km, at the cyclone top the DA transport being anticyclonic and directed to the north-east. Turbulent mixing and convection played the main role in the formation of dust aerosol and its transport to the upper boundary of the ABL. In the zone of the cyclone-determined prevalent transport, the vertical advection mainly determined the DA transport to high altitudes and the west–east transport controlled the process of the long-range transport. Calculations of the mass budget for the whole period considered have shown that about 75% of the total DA input (643 million tons of particles with $D < 36\,\mu m$) to the atmosphere returned to the surface in both desert and (20%) non-desert regions due to processes of wet and dry deposition. About 1.6% of deposited aerosol falls on the share of the east China and Japanese Sea basins, and 3.6% of the aerosol crosses the eastern and north-eastern boundaries of the model.

Intense studies of atmospheric aerosol within the ACE-Asia field observational experiment were carried out during the period 31 March–4 May, 2001 and concentrated in the regions of the Yellow and Japanese Seas. Based on the GOCART global model developed at Georgia Technical University and the Goddard Space Flight Centre, which simulates the processes of transformation of the tropospheric ozone and aerosol concentration field, as well as meteorological parameters (the GEOS DAS technique developed at the Goddard Centre), [21] predicted the 'aerosol situation' for the ACE-Asia period. The adequacy of this model has been tested by comparison with the data of observations made with the measuring complex carried by the flying laboratory C-130.

A perfect forecast of the fields of wind, relative humidity, and temperature was obtained with the GEOS DAS method. The GOCART model adequately simulated the 3-D field of DA concentration including emissions to the Pacific basin but failed to simulate the high DA concentration in the ABL, which can be explained by neglected recently developed sources of dust aerosol in the Inner Mongolia province (China). With these sources taken into account the calculated field of an enhanced concentration of dust aerosol in the ABL over the Yellow Sea turned out to be quite realistic. The GOCART model enables one to reliably simulate not only the large-scale intercontinental DA transport but also the small-scale spatial–temporal variability of the aerosol content, which is especially important in planning the field experiments.

From the data of analysis of aerosol samples obtained on Adak Island (Alaska), the southernmost island in the chain of the Aleutian Islands, as well as at Poker Flat (30 km north-east of Fairbanks, Alaska), [16] obtained information about the aerosol size distribution and chemical composition (chemical analysis provided an estimation of the content of 42 elements from natrium to lead using the technique of synchrotron X-ray fluorescence). Analysis of the elemental composition of aerosol together with the results of calculations of backward trajectories showed that the aerosol was the product of the long-range transport from the Asian continent. In the case of observations on Adak Island, the input of aerosol from the regions of northern Europe and Russia made a certain contribution. Maximum values of the concentration of large particles of soil (the aerodynamic diameter is between 1.15 and 2.5 μm) at Poker Flat constituted only 15% with respect to those observed on Adak Island. If we assume that in both cases the aerosol was sampled in similar air masses, then it means that during its long-range transport across the Pacific Ocean the aerosol deposited onto the ocean surface. However, in fact, the difference between the observed values of aerosol concentration could also be explained by different trajectories of the transport and conditions of the vertical mixing in the regions of the two observational stations considered. To assess the relative role of these factors, further studies are needed.

Uno *et al.* [118, 119] described the CFORS model of the long-range transport of atmospheric MGCs and aerosols taking into account their chemical transformation in the process of transport, designed for planning the field observational experiments and subsequent interpretation of observational data. This model ensures a simulation of 3-D fields of MGCs and aerosol concentrations with a high (~90 s) temporal resolution. The model was used to interpret the data of surface observations obtained within the VMAP programme of studies of the marine aerosol properties variability.

To analyse the laws of the transport and transformation of MGCs and aerosol taking into account both anthropogenic pollutants (sulphate, black carbon, and CO) and natural components (including radon and dust aerosol), the temporal evolution of the fields of MGCs and aerosol concentration was considered. The analysis revealed an important role of weather changes on synoptic scales as a factor of pollutant transport on continental scales in spring in eastern Asia. A complicated spatial–temporal structure of the fields of pollutants in the periods of emissions from the Asian continent to the Pacific Ocean basin has been discussed in detail [119]. The numerical modelling results agree well with data from aircraft observations.

At the surface network located at four Japanese islands (Rishiri, Sado, Hachijo, and Chichijima) in the western sector of the Pacific Ocean along the parallel 140°W (in the latitudinal band 25°–45°N), [74, 75] performed regular observations of different characteristics of atmospheric aerosol, including the content of elemental carbon (EC). Results of observations of the EC content were discussed [74] for the period from March to May, 2001, coordinated with observations within the ACE-Asia field experiment. Average values of EC mass concentration varied within 0.18–0.60 $\mu g\, m^{-3}$, changes in concentration on the northern islands (35°–45°N) Rishiri and Sado, turned out to be synchronous, like data for the southern islands (25°–35°N) Hachijo and Chichijima, but between the observational results at two pairs of islands there was no correlation, which reflects the difference between the trajectories of the long-range transport of aerosol to these two pairs of islands. As a rule, air masses from the Asian continent become polluted over the southern but not northern regions.

Uno *et al.* [119] performed a numerical simulation of the aerosol long-range transport and transformation of aerosol properties with the use of the CFORS regional model in order to retrieve the data on black carbon concentration within the VAMAP and APEX programmes on five remote islands of the Japanese archipelago during the period of undertaking the ACE-Asia field observational experiment in the spring of 2001. The black carbon (BC) aerosol is assumed to be completely fine-sized, which enables one to avoid the necessity to take into account gravitational deposition, wet deposition, and chemical ageing of carbon aerosol particles. The level of dry deposition onto the surface of the ocean (land) is taken to be 10^{-4} (10^{-3}) $m\, s^{-1}$.

Results of two numerical experiments were discussed in [119]:

(1) a control experiment in which all emissions of BC were taken into account; and
(2) a comparative experiment, in which emissions due to biomass burning were neglected, to evaluate (by comparing with the data of the first experiment) the contribution of biomass burning to the formation of the carbon aerosol concentration field in the considered western sector of the Pacific Ocean (the centre of this region has the coordinates 25°N, 115°E; calculations were made over the grid with an 80-km step for an atmospheric layer 0–23 km, with 23 non-uniformly distributed layers taken into account).

The comparison with observational data demonstrated that the CFORS model simulates satisfactorily many observed features of the spatial–temporal variability of BC concentration with different regimes of atmospheric circulation. The main features of the synoptic situation during observations (Rishiri and Sado) refer mainly to altitudes below 2 km, and the contribution of biomass burning to the BC formation turned out to be <20%. In the case of stations farther south (including Chichijima) this contribution reaches 23% near the surface and 52% in the free atmosphere. Thus, both the main sources and the means of transport of carbon aerosol to northern and southern stations are different.

Takemura *et al.* [114] used a combined model SPRINTARS of the aerosol long-range transport and radiation transfer to simulate a large-scale trans-Atlantic transport of dust storm products in Asia to North America as applied to conditions in the period of the spring seasons of 2001 and 2002. Calculations showed that about

10–20% of the Asian dust aerosol coming to the region of Japan reaches North America. The composition of aerosol is characterized by the presence of not only dust aerosol but also anthropogenic carbon and sulphate aerosol whose AOT is approximately equal to that for dust aerosol.

During completion of the ACE-Asia field observational experiment in the western sector of the Pacific Ocean from the flying laboratory C-130, samples were taken of submicron aerosol on teflon filters. An analysis of these samples was made for the content of aerosol organic carbon (OC) with the use of the IR Fourier spectrometer (FTIR) to record the transmission spectra. The spectral analysis revealed the presence of such components as silicates, carbonates, alkene, aromatic alcohols, carbonyl, amine, and organic silicate groups. To analyse the elemental composition of aerosol, the X-ray fluorescence method was used. Measurements of the OC mass concentration gave values within 0.4–14.2 $\mu g\,m^{-3}$, and organic compounds from 0.6–19.6 $\mu g\,m^{-3}$ which, on average, is equivalent to 36% of the whole mass of submicron aerosol.

A consideration of relationships between values of CO concentration and OC (i.e., an inclination of the straight line corresponding to the CO/OC ratio) performed by [70] made it possible to determine 10 categories of air masses characterized by specific inventory data for emissions from the Asian continent. The data on the CO/OC ratio were used to recognize the sources of emissions and their impact on the composition of the aerosol organic components. A total of 52% of aerosol samples obtained during the ACE-Asia experiment was characterized (from CO/OC data) by the presence of biomass burning products, with distinct regional specific chemical composition of large-scale emissions of aerosol produced on the Asian continent to the Pacific Ocean basin. This specificity is explained by an increased DA concentration in air masses coming from the Taklamakan desert, the growth of the content of nitrates and ammonium sulphate when air masses move from the region of Shanghai, and sulphates from the Hokkaido Island.

During the ACE-Asia field observational experiment (March–May, 2001) an extensive programme was carried out of surface observations of the chemical composition, size distribution, and transport of aerosol on Rishiri Island located near the northern point of Hokkaido Island (45.07°N, 141.12°E). According to observational data, average values of mass concentrations of non-sea salt (NSS) SO_4^{2-}, NO_3^-, NH_4^+, and NSS Ca^{2+} in aerosol particles constituted, respectively, 2.48, 0.64, 0.72, and 0.17 $\mu g\,m^{-3}$. Average values of the concentration of elemental and organic carbon in particles of the small-sized fraction of aerosol ($D < 2.5\,\mu m$) turned out to be 0.25 and 0.80 $\mu g\,m^{-3}$, respectively.

Under the influence of polluted air masses moving from the Asian continent to the Pacific Ocean basin the aerosol concentration substantially increased, but under background conditions it approached levels characteristic of the remote regions of the ocean. The results obtained by [73] indicate the coexistence of NSS SO_4^{2-} and NO_4^+ in small-sized aerosols, and NO_3^- and NSS Ca^{2+} in the large-sized fraction of aerosol.

Ions NO_3^- appearing on the continent are transported for long-distances as part of the large-sized fraction of aerosol. The long-range transport of anthropogenic pollutants and dust aerosol can be realized both in combination and separately. Anthropogenic particles of the small-sized fraction containing a considerable amount of NSS SO_4^{2-} often appear first, followed by large particles of dust (mineral) aerosol absorbing NO_3^- . In the periods of the arrival of air masses from the continent, often appear short-term intrusions of air masses containing a considerable amount of aerosol carbon compounds, whose source is, probably, biomass burning. Regimes of the long-range transport of sulphates, nitrates, and carbon compounds are characterized by specific features, though all of them are determined by the effect of the processes of internal combustion (motor vehicles).

Osada *et al.* [88] performed observations of the aerosol size distribution and ozone content during January, 1999–November, 2002 at Murododaira station (36.6°N, 137.6°E, 2,450 m a.s.l.) located on the western slope of Mt. Tateyama in the central region of Japan. Data of nocturnal observations (24:05 local time) were considered, which indicated a weak variability of the tropospheric ozone concentration (mean standard deviation (MSD) constituted 4 ppb) with an average concentration of O_3 in winter (October–February) of 40 ppb, strong variability (MSD 8 ppb) with a higher concentration (51 ppb) in spring (March–May), and strong variability (14 ppb) with a lower concentration (32 ppb) in summer (June–September). Most substantial monthly mean volume concentration of the accumulation mode particles ($0.3 < D < 1.0\,\mu m$) was observed in June ($2.7\,\mu m^3\,cm^{-3}$) whereas in winter (October–February) the concentration averaged only $0.7\,\mu m^3\,cm^{-3}$.

A statistical analysis of the results of calculations of the backward trajectories of air masses showed that the prevalent summertime conditions of the stagnant atmosphere in the coastal regions of the Yellow Sea and near the coasts of Japan favoured an increase in the accumulation mode particle number density. An important factor of such an increase was also SO_2 emission by the Miyakejima volcano starting from August, 2000. A maximum of monthly mean volume concentration ($11.2\,\mu m^3\,cm^{-3}$) of large-sized aerosol ($D > 1.0\,\mu m$) was recorded in April. Variability of the diurnal variation at night was also at a maximum in spring (MSD constituted $13.6\,\mu m^3\,cm^{-3}$) and low (about $2\,\mu m^3\,cm^{-3}$) during other seasons. The high level of volume concentration and its springtime strong variability was due to frequent inflows of yellow dust particles from the Asian continent. During the periods of such episodes called *cosa* a 30-fold increase of the volume concentration of the large-sized aerosol during 3 hours was often observed. The year 2001 was a period of especially intense activity of the *cosa* event – beginning in January through to early July.

Schmid *et al.* [101] performed a comparison (called the 'closed' analysis) of the total aerosol-induced attenuation of short-wave radiation evaluated from the data of observations at $\lambda = 550$ nm characterized by the values of extinction coefficient σ_{ep} with results of observations from the flying laboratory *Twin Otter* using the 14-channel sun photometer AATS-14 and data of *in situ* measurements of aerosol characteristics as well as ship lidar soundings. From 31 March–1 May, 2001, within the ACE-Asia programme, 19 flights were made over Japan at altitudes up to 3.8 km from the region of Hiroshima (34.15°N, 132.23°E).

A comparison of the observed and calculated (with the use of four different techniques) vertical profiles of σ_{ep} revealed a satisfactory agreement in the presence of differences between absolute σ_{ep} values which depended on altitude. The σ_{ep} values calculated from the measured size distribution and chemical composition of aerosol agree well with AATS-16 data in the MABL but they are considerably underestimated in the layers of dust aerosol when aerosol particles are assumed to be of spherical shape. A comparison of 14 vertical profiles of σ_{ep} determined from *in situ* measured characteristics of scattering and absorption of radiation by aerosol with those measured using the sun photometer has shown that the first are underestimated by ~13% with respect to the latter.

A combined analysis of the results of lidar and photometric observations revealed the presence of DA layers at altitudes up to 10 km, the optical thickness of which (at altitudes 3.5–10 km) varied within 0.1–0.2 (at $\lambda - 500$ nm), and the ratio of the coefficients of extinction and backscattering (at $\lambda = 523$ nm) varied within 59–71 sr. According to data of aircraft observations, the atmosphere in the period of observations was relatively dry with the total content of water vapour <1.5 cm and absolute humidity $\rho_w < 12\,g\,m^{-3}$. A comparison of *in situ* measured ρ_w values with those retrieved from AARS-16 data revealed a high correlation ($r^2 = 0.96$), but systematical underestimation of retrieved values by about 7%.

Based on the combined use of NOAA-14 AVHRR data and results of objective analysis performed by NCET–NCAR (USA), [94] studied the laws of the spatial distribution and specific character of the long-range transport of aerosol over the regions of the Arabian Sea, the Gulf of Bengal, and the Indian Ocean in the periods of the dry season (November–April) and the south-western monsoon during a wet season (June–September). Initial information was data on aerosol optical thickness retrieved from satellite observations for AVHRR channel 1 ($\lambda = 630 \pm 50$ nm) within the area 25°S–25°N, 40°E–100°E.

The results obtained illustrate distinct differences of the AOT fields in the periods of dry and wet seasons determined by specific air masses and differences of the types of aerosol in these periods. Surprising though it may be, during the whole year the AOT fields (and thus the aerosol content) are strongly affected by the Indian subcontinent despite a considerable migration of the inter-tropical convergence zone (ITCZ) during a year and the remoteness from the subcontinent of the seas located in the southern hemisphere.

Both the long-range transport of aerosol to the ocean during the dry season in the northern hemisphere and (to a greater extent) the transport of dust aerosol from the Arabian desert to the region of the Arabian Sea reveal itself during the wet season. In the region of the Gulf of Bengal and in the tropical Indian Ocean the effect of aerosol transported from the Indian subcontinent and from south-eastern Asia is very strong. The long-range transport of aerosol from the continent is confined, however, by the Arabian Sea, the Gulf of Bengal, and the Indian Ocean in the northern hemisphere. During the whole year, a strong AOT gradient is observed over the Arabian Sea and the Gulf of Bengal, especially in the meridional direction in the latitudinal band 5°S–10°N. A huge plume of aerosol (AOT > 1.0) in the tropical Indian Ocean in September–November, 1997 was connected with forest fires in Indonesia that occurred during an El Niño event.

Highly uncertain estimates of direct and indirect climatic impact of aerosol have stimulated a great deal of interest in the analysis of processes determining the effect of aerosol and clouds on climate formation. These uncertainties are mainly connected with complicated aerosol–cloud interactions, which depend on the cloud droplet size distribution, chemical composition, and type of clouds. The problem becomes even more complicated because of changes in properties of aerosol particles due to their interaction with cloud droplets in clouds. Interactions between aerosol and clouds manifest themselves in two ways: through functioning of aerosol particles as cloud condensation nuclei (CCN) and through the inverse impact of clouds on aerosol leading to changes in aerosol number density, size distribution, and chemical composition. Changes in the properties of aerosol particles are caused by the process of coagulation of cloud droplets during their growth due to an addition of substance as a result of oxidation of MGCs such as SO_2 taking place in the liquid phase (droplets) as well as by changes in the aerosol concentration after the washing out of particles from the atmosphere and by phoretic processes.

To understand the mechanisms of the aerosol climatic impact, information is needed about its concentration, properties and lifetime under different atmospheric conditions. In this context, [65] performed observations of the cloud droplet-induced nucleation washing out of aerosol particles when they pass a shallow cold front at a mountain station located in the northern part of Israel, at the top of Mt. Meron (35.41°E, 32.99°N, 1200 m above sea level) on 8–10 December, 2000. Analysis of air trajectories (long-range transport) has shown that before passing the cold front, air masses prevailed which originated in the north and brought anthropogenic aerosol from eastern Europe. After passing the front, the prevailing air came from the east bringing along with it mineral dust aerosol. Analysis of the aerosol chemical composition revealed prevalent aerosol components such as sulphate, nitrate, and ammonium. About 65% of particles containing sulphate took part in the process of cloud droplet nucleation, with the nucleation washing out of aerosol correlating with the size of aerosol particles. While particles $<0.14\,\mu m$ were not affected by this process, in the case of larger particles their number density decreased in proportion to cloud droplet concentration. Under conditions of overcast cloudiness, a nucleation washing out of 80% of the particles took place in the range of sizes 0.3–1 μm. After cloud dissipation the number density of particles with $D < 1\,\mu m$ returned to the initial level.

Using data of aircraft observations obtained during the regional experiment on studies of the Arctic clouds (FIRE/ACE) accomplished within the International Satellite Cloud Climatology Project (ISCCP), [123] studied the variability of the vertical profile of CCN concentration in the region of the Arctic Ocean. Results of calculations of backward trajectories of air masses and satellite images of cloud cover served as additional information. Analysis of all data has shown that an extended stratified cloudiness in the arctic ABL causes a decrease of CCN concentration in the ABL due to their washing out.

A decreased level of concentration was observed in all cases, when the air penetrated clouds, except cases of cirrus clouds whose effect on CCN varies widely. Washing out of CN and CCN by clouds reduces their concentration at any level of oversaturation. The earlier observed strong increase of CCN concentration in the atmosphere over the Atlantic and Pacific Oceans with the inflow of air masses from the continents was not observed over the Arctic Ocean. Maximum values of CCN concentration were recorded when the air was over the Arctic Ocean for not less than 6 days in the absence of the air–cloud interaction. The vertical profiles of CCN concentration in the Arctic differ strongly from those observed in other marine regions. Maximum values of concentration were observed at the level of the aircraft's sounding ceiling (6.5 km, near the level of the tropopause) (i.e., in contrast to the usual situation) the CCN concentration increased with altitude).

8.4 PROCESSES OF AEROSOL DEPOSITION

The closing stage of the aerosol cycle is its deposition onto a surface. With the use of the NARCM model of the long-range transport of dust aerosol taking into account the transformation of its properties determined by wet and dry deposition in the process of transport, [127] simulated the evolution of a DA plume appearing in the deserts of the Asian continent to the Pacific Ocean basin in the northern hemisphere reaching the western coast of America. It follows from the results of the numerical simulation that the process of dry deposition plays the prevalent role in DA removal from the atmosphere near the sources of emissions, whereas during the transport across the Pacific Ocean the wet deposition is more important, its contribution being about 10 times greater than that of dry deposition. Deposition of the Asian dust aerosol on the Pacific basin not only correlated with rain rate but also depended on the spatial structure of the long-range DA transport.

Monthly DA emissions from the Asian continent and their long-range transport were concentrated near 38°N (March), 42°N (April), and 41°N (May). In this connection, [127] described in detail the specific features of the spatial structure of the springtime DA transport across the Pacific Ocean. The location of the main axis of the transport varied within latitudes 30°–40°N in March, and in April the zonal transport of dust aerosol was concentrated along 40°N. May was characterized by prevailing trajectories of the long-range transport split in two: zonal (with eastward transport) and meridional – from the region of the DA source to the north-eastern region of Asia. Calculations of the DA budget observed in the spring of 2001 led to the conclusion that the main sources of the Asian dust aerosol were located in the regions of the deserts of China and Mongolia with a level of emissions of about $215\,\mathrm{t\,km^{-2}}$ (their export constituted, however, only $8.42\,\mathrm{t\,km^{-2}}$). The main sink regions for the Asian dust aerosol were the loess plateau and the Pacific Ocean basin.

Gallagher *et al.* [33] performed *in situ* measurements of fluxes of small ($D = 0.1$–$0.2\,\mu\mathrm{m}$) aerosol particles over heather and grass in order to analyse the impact of surface roughness on the rate of dry aerosol deposition. In the case of grass cover, observations were made before and after its mowing. Analysis of the data obtained revealed their agreement with similar results to the earlier development, especially with the similarity of conditions of atmospheric stability taken into account. An approximated method was proposed [33] of parameterization of the deposition rate V_{ds} as a function of the roughness parameter z_0:

$$V_{ds}/u^* = k_1 + k_2(-300\,z/L)^{2/3}$$

where $k_1 = 0.001\,222\,1\,g(z_0) + 0.003\,906$; and $k_2 = 0.009$. Here z is the Obukhov parameter of stability and u^* is the dynamic rate. Gallagher *et al.* [33] emphasized a necessity of further observations for different types of surface (especially with roughness ranging between 0.1 and 1.0 m), as well as a need to generalize the parameterization technique for particle sizes ranging between 0.5 and 1.0 μm.

Along with aerosols of soil origin, the sea salt aerosol emitted to the atmosphere by the World Ocean contributes most to the formation of the global mass of atmospheric aerosol (the dry mass of marine aerosol emissions was estimated at 10^3–10^4 Tg yr^{-1}). Hoppel *et al.* [43] performed a numerical simulation whose results showed that in the presence of a horizontally homogeneous source of particles as in the case of sea salt aerosol, the rate of aerosol dry deposition differed cardinally from that of deposition of particles driven by advection or by vertical mixing.

A technique was proposed [43] which made it possible to calculate the rate of deposition in these two cases. Use of the equilibrium method to estimate the function of the SSA source from data on aerosol concentration at a certain altitude, as well as the rate of deposition, has demonstrated the uselessness of such a technique for particles with $r < 5$–10 μm. This result is due to the following two reasons: (1) the time for establishing an equilibrium between the sources and losses due to dry deposition is much longer than the characteristic lifetime of small particles determined by washing out of particles due to precipitation; and (2) it is very difficult to take into account the effect of the synoptic-scale vertical transport as well as mixing between the MABL and the free atmosphere.

Therefore, a modified technique has been developed to assess the source function for the sea salt aerosol. This technique is based on the introduction of the notion of 'involvement rate' in order to take into account the effect of processes such as: (1) deposition or ascent determined by a large-scale divergence or convergence within the MABL; (2) vertical motions due to the lowering or broadening of the MABL; and (3) an exchange with the free troposphere across the upper MABL boundary.

A consideration in the models of the large-scale aerosol transport and their deposition, as shown by [128], is an important constituent of their accuracy. For the process of deposition, a linear approximation was used in the Euler-type model LRTAP (Long-Range Transport of Air Pollution): $H_s = (k_{s1} + k_{s2})c_s$, where c_s is the concentration of aerosol of the sth type, and k_{s1} and k_{s2} are coefficients of the rate of dry and wet deposition, respectively, calculated from the formulas: $k_{s1} = g_{s1}/H_{\text{mix}}$, $k_{s2} = g_s I/H_{\text{mix}}$. Here g_{s1} is the rate of dry deposition of aerosol of the sth type, H_{mix} is the height of atmospheric mixing, I is the rain rate, and g_s is the dimensionless coefficient of washing out of aerosol of the sth type.

8.5 NUMERICAL SIMULATION OF AEROSOL LONG-RANGE TRANSPORT

8.5.1 Relationships between the scales of atmospheric mixing processes and the choice of models

Experience gained from many studies on modelling the atmospheric processes of pollutant transport dictates a necessity to classify these processes in accordance with interrelationships between spatial and temporal scales [2, 27]. The need for this classification has been substantiated in many international programmes, such as 'Global Changes', 'Global Atmospheric Chemistry' (IGAC), 'Modelling of Biogeochemical Global Cycles' (MBGC), and others. The need and even necessity of classification of physical processes of atmospheric mixing is dictated by parameters of the systems for measurements (monitoring) of atmospheric characteristics, requirements of simplicity for the models of the aerosol and gas transport in the atmosphere, as well as by limitations of available databases. The interaction of these causes leads to a range of spatial scales, which provides an efficient parameterization of the processes of propagation of atmospheric pollution and agrees with international standards. At present, the GEODAS standard is the most widespread, it has nine levels of spatial resolution from one degree to half-seconds in longitude and latitude. Seven scales of cartographic presentation of data are provided (Table 8.1).

Data on these scales can be obtained by synthesis of satellite data and data of national monitoring systems. The latter are important because they specify the data domain and select priorities characteristic of a given region. For instance, for developed countries, of importance are operational assessments of air quality in the zones of megalopolises and large industrial enterprises. For developing countries, the control of transboundary transport of pollutants is of principal importance as well as an assessment of possible changes of atmospheric air in connection with the building of industrial plants. The ratio of scales given in Table 8.1 corresponds to the majority of situations of atmospheric monitoring.

Along with the problem of choosing the scales of spatial resolution, there is a problem of their timescale agreement. This problem is important for the formation of the structure of numerical models which describe the pollutant dynamics in the atmosphere. According to preliminary estimates of the International

Table 8.1. Scales of cartographic information presentation characteristic of the developed monitoring systems.

Spatial resolution (km)	Scale	Spatial resolution	Scale
0.5×0.5	1:1,250	40×40	1:50,000
1×1	1:2,500	250×250	1:250,000
5×5	1:10,000	500×500	1:625,000
10×20	1:25,000		

Geosphere–Biosphere Programme, there is a scale of transitions in temporal and spatial measurements between complexity and depth of the hierarchic structure of connections considered in the model. So far, the developed models practically ignored this fact, and therefore often it was impossible to apply them to the natural object under study. An agreed discretization scale for natural phenomena to be used in models, proposed by [84, 85], enables one to classify natural phenomena with due account of their hierarchic subordination on temporal and spatial scales. This classification is based on the fundamental understanding of the hierarchy in the general theory of systems. According to this theory, the behaviour of any complicated system is determined by the triad of frequencies of its variability, which provides an agreement between coherence and stability of the system. This makes it possible to exclude from consideration any unnecessary details in the model's structure according to the prescribed timescale, or to establish a minimum time step from data on spatial scales. For instance, if in the model the time step is chosen to be equal to one year, there is no sense in taking into account such processes as atmospheric turbulence. In other words, in this case the atmosphere can be described with a point model, and all attempts to develop complicated constructions to describe the processes of atmospheric motion cannot improve the reliability of the model but will increase its complexity.

A more strict theoretical substantiation of this approach is a combined description of dynamic processes with different characteristic timescales which results in parameters in the mathematical descriptions referring to different processes and strongly (by orders of magnitude) differing from one another. This enables one to divide these processes into three groups:

- processes referring to a chosen timescale;
- processes which, with respect to timescale, can be considered as being in dynamic equilibrium (quasistationary) and for which some parameterization can be introduced (fast processes); and
- processes which, with respect to the chosen timescale, can be considered constant (i.e., static (slow processes)).

As an example, we take a numerical model of the biogeochemical cycle of carbon which includes both the geological timescale processes and fast processes (photosynthesis, respiration, etc.). If the aim of the model is to study the dynamics of CO_2 content in the atmosphere over several decades, the geological processes with characteristic timescales of millions of years should be excluded from consideration and the processes should be parameterized with characteristic timescales of days, for instance, processes of living biomass formation with photosynthesis [59]. In general, the problems mentioned have not yet been solved, and there is no constructive mechanism for matching the spatial and temporal scales. Each scientist follows a principle of his own in choosing the model elements for their subsequent realization. Unfortunately, at this stage there appear unavoidable deviations of the model from reality.

Characteristic time intervals of variability for most natural processes are well known. Here are some of them:

- processes of deposition – minutes, hours;
- plant transpiration – hours, days;
- formation of plant biomass – days, months;
- changes in communities of plants and animals – months, years;
- soil formation – years, centuries; and
- geomorphological processes – centuries, millennia.

Therefore, a systematization of timescales for the processes considered should follow a synthesis of the system of models. As a result, priorities for models and their units as well as software structure can be determined.

Because of the difficulties of parameterizing atmospheric processes, there are many models of atmospheric dynamics. The type of the model correlates strongly with the spatial scale. A consideration in the models of the processes of the physical mixing of chemical processes taking place in the atmosphere is determined by accuracy level. The developed models, depending on the spatial scale, take into account processes of physical transformation of pollutants from microprocesses in clouds to large-scale atmospheric motions. Depending on this, models are divided into dispersive, Gaussian, Eulerian, and Lagrangian. Within this system there is a hierarchy of models taking into account or neglecting the vertical structure of the atmosphere, atmosphere–surface (land, water) interaction, exchange processes between clouds, and vertical air fluxes depending on relationships between the synoptic and the physical parameters of the atmosphere. One of the examples of such a study is a series of versions of the ICLIPS model with a spatial resolution 500×500 km and a time step of one year (Integrierte Abschatzung von Klimaschutz-strategien) [54, 117]. A more accurate model ECMWF (European Centre for Medium Range Weather Forecast) has a spatial resolution 150×150 km and a time step of six hours [37].

It is impossible to establish an unambiguous connection between the scale of the model and its internal infrastructure without taking into account various characteristics of the model. Therefore, the estimates given in Table 8.2 should be considered as recommendations, directed at the modeler's aim giving him a possibility to assess an expedience of including certain components into the model.

8.5.2 Interrelationship between the types of models and aerosol characteristics

The formation of the fields of atmospheric pollution from natural and anthropogenic sources depends strongly on the physical characteristics of pollutants. Clearly, for complete understanding of the processes of formation and growth of clouds and precipitation, it is necessary to consider all dynamic microphysical interactive processes. These processes are determined by a combination of physico–chemical parameters of the atmosphere itself and pollutant components, which are characterized by strongly variable characteristics such as weight, concentration, size, shape, phase state, and electric charge. For instance, the classification of atmospheric

Table 8.2. A fragment of the scale for development of the structure of the atmospheric pollution dynamics model.

Spatial resolution of the model	Processes recommended for consideration in the model
Industrial region, landscape, megapolis, city (up to 50 km)	Use of the Gauss-type models. Burning of wastes, deforestation and reconstruction of surface covers, contamination of drinking water and water basins, industrial emissions of aerosol, soil contamination, washing out of pollutants with rain, production processes, medico–biological assessment of the territory, and the division of the atmosphere into many levels.
Large region, district, country (up to 1,000 km)	Use of Lagrangian and Eulerian models. Large-scale atmospheric circulation with selection of the upper and lower atmosphere, irrigation and other aquatic systems, integral areal sources of biospheric pollution, biogeochemical cycles, erosion, large-scale fires, desertification and swamping, succession of surface cover, river runoff, and interactions on shelves.
Continent, globe (>1,000 km)	Use of block models. Averaged characteristics of the atmosphere and climate, oceanic circulation, interactions in the system that is 'atmosphere–land–ocean', biogeochemical cycles, and succession of large tracts of forests.

aerosol pollution adopted by the US National Oceanic and Atmospheric Administration (NOAA) includes three basic classes and eight subclasses. This classification is sufficient to be used in models which do not take into account the size of aerosol particles and do not include the ion level processes. The existing classification of particles by size covers particle diameters from 0.0001 μm to 1 cm. Within this range, solid particles with $D = 0.0001$–1 μm are considered as smoke components, while liquid particles are elements of fog. Particles exceeding 1 μm in diameter are interpreted as dust or spray. Depending on the size, the role of particles in the dynamic processes of atmospheric pollution changes also. Particles with a diameter less than 1 μm form smog, tens of μm – clouds and fog, hundreds of μm – haze and drizzle, thousands of μm – rain. This classification simplifies the choice of the model's structure, if the nature of processes of atmospheric pollution is known. In a more complicated situation, when the size spectrum of aerosol particles is sufficiently broad, the division of pollutants by their physical and chemical characteristics enables one to synthesize the complicated model as a totality of hierarchically submitted partial models and to simplify thereby the procedure of calculation of dynamic characteristics of the polluted atmosphere [1].

The physical characteristics of atmospheric pollution includes also the rate of gravitational deposition, residence in the atmosphere, and phase state. Some of the atmospheric gas components such as N_2, O_2, He, Ne, Ar, Kr, Xe, and H_2 have a very long lifetime. The lifetime of CO_2, O_3, N_2O, and CH_4 is from several years to decades. Such gases as H_2O, NO_2, NO, NH_3, SO_2, H_2S, CO, HCl, and I_2 live in the atmosphere for only several days or weeks. Depending only on this characteristic to describe the dynamic of various gases in the atmosphere, one can choose an adequate model with minimum requirements for a database [13, 84, 85].

For the choice of the type of polluted atmosphere dynamics model, of importance is the size of an aerosol particle. The mentioned intervals only partly cover a possible classification of aerosols. Additional information is needed about the source of pollutant, which further specifies the parametric space of the model. Knowledge of the cause of pollution simplifies the choice of model type. Of course, the classification and typification of aerosols and gases can be more detailed. There are tens of types of smoke, for instance. The size of smoke particles can be 1–0.01 μm for resin smoke, 0.15–0.01 μm for tobacco smoke, etc. Here, in the model, it is necessary to consider microprocesses connected with the motion of these particles. For instance, the run of a particle of carbonaceous smoke during t seconds averages 0.000 68 t/D cm. Filling the base of knowledge of the monitoring system with such dependences is one of the first-priority problems of ecoinformatics [60].

In modelling the scattering of gases and particles in the atmosphere it is important to know the difference between polluted and clear atmosphere properties. Also, one should always bear in mind the vertical heterogeneity of the atmosphere. On a global scale, the formation of air quality depends on processes at all levels of the atmosphere: the troposphere, stratosphere, hemosphere, and ionosphere. For instance, in the problem of the impact of aviation on the atmosphere it is necessary to take into account the interaction between the troposphere and stratosphere. In studying the fluxes of pollutants from surface sources, the motion of the lower atmosphere is considered first of all. Of course, here of importance is the spatial scale and, hence, the time interval of pollutant residence in the atmosphere. Data known for the clear atmosphere should be used in the control simulation experiments.

8.5.3 Passive and active aerosol transport in the atmosphere

Pollutants emitted to the atmosphere are subject to gravitational sedimentation, turbulent mixing, wind-driven transport, and washing out by rain. A totality of these impacts determines the character of behaviour of the polluting cloud, shape and type of the flux of pollutants, as well as the spatial structure of the distribution of aerosol number density over a given territory.

Smoke and other atmospheric aerosols are gravitationally influenced and interact with solar radiation, gases, and ions. In the surface layer, this interaction is supplemented with the different effects of the surface (vegetation cover, soil, land surface roughness, and sea surface roughness).

The role of sedimentation is more substantial in the case of large particles – larger in diameter than submicron particles. Small particles sediment much slower compared with their transport by the moving atmosphere, and therefore in many models this vertical constituent is neglected. Note that for the process of sedimentation the diameter of particles is less important than their density. For instance, soot structures with a low efficient density and large aerodynamic cross section are easily wind-driven and sediment much slower than the compact particles of the same mass. The rate of sedimentation of particles with $D = 0.1$–$1\,\mu m$ averages $0.001\,m\,s^{-1}$, which is negligibly small compared with the rate of atmospheric transport.

In the case of heavily polluted formations one can observe the process of photophtorose consisting of particle lifting due to non-uniform solar heating. However, the possibility of this phenomenon and its characteristics have been poorly studied, therefore in a first approximation, many experts leave it out of account, especially because during time intervals longer than a day, due to Brownian motion, an irregular heating of particles decreases. Finally, such physical process should be pointed out as particle coagulation, consisting of the capture of one particle by another due to different rates of motion. In this case, particles can either stick together or be repelled, so that diverse situations of interactions arise which determine the shape of the cloud of pollutants and can prolong their lifetime in the atmosphere.

Washing out is a very important process of removal of pollutants from the atmosphere. Here, two situations are possible. One is connected with a simple capture of particles by rain droplets, the other – with the so-called nucleation process. This process is connected with oversaturated water vapour condensation on the surface of aerosol particles, which leads to the formation of water droplets or ice crystals with subsequent deposition onto the land surface. Therefore, one of the ways to specify the model of aerosol dynamics in the atmosphere is to include within them a unit of parameterization of the water cycle in its different phase states.

The final estimation of the residence time for a given pollutant in the atmosphere is realized with the use of respective models. This estimation has been given in detail in [92]. Here the meteorological features of pollutant propagation in the atmosphere are described; scales of transport and scattering of pollutants are analysed; models are constructed which predict the concentration of pollutants; algorithms of parameterization of the processes of clouds and pollutant jet formations are simulated; and ratios are given to describe the vertical structure of the atmosphere. The Earth's radiation budget components are analysed and the simplest characteristics of relationships between pressure, wind, temperature, and humidity are given. The state of the atmosphere is classified as neutral, unstable, and stable by the scale of the vertical temperature lapse rate, which considerably simplifies the process of parameterization of the vertical gradients and rates. The scale of atmospheric phenomena is estimated over the time interval from one second to one month, and with a spatial scale varying from 20 km to 1,000 km. Within this scale, processes of transport and scattering of atmospheric pollutants are analysed from point sources, as well as moving and covering the final territory.

In general, a change of the concentration of any pollutant C is described with the following equation:

$$\frac{\partial C(t, \varphi, \lambda, h)}{\partial t} + \nabla \times \vec{V}C = \nabla D \times \nabla C + R \tag{8.1}$$

where $\vec{V}(V_\varphi, V_\lambda, V_h)$ is the wind speed; φ is the latitude; λ is the longitude; h is the height; t is the time; D is the coefficient of molecular diffusion; and R is the change due to atmospheric turbidity, emission, and mixing.

Detailed description of the terms of equation (8.1) requires analysis of specific processes of atmospheric propagation of pollutants and construction of respective units of a general model (dynamic, correlative, probabilistic, system, evolutionary, etc.). As examples of such units, we shall consider parameterizations used successfully in the models ICLIPS, ECMWF, and others.

Problems of chemical interaction of atmospheric pollutants are also important, and their consideration in modelling further complicates the study. Therefore, most of the models of pollutant propagation in the atmosphere assume a priori that all components are mutually neutral. However, in some cases a parameterization of the processes of chemical conversion of pollutants is possible due to the use of statistical characteristics of chemical reactions or by describing the laws of phase transitions. In particular, a simple model of SO_2 conversion into H_2SO_4 turned out to be sufficiently efficient [57]:

$$d[\mathrm{H_2SO_4}]/dt - d[\mathrm{SO_4}]/dt = W[\mathrm{SO_2}]$$

where $W = 0.1\%\ \mathrm{day}^{-1}$ in the daytime and $0.01\%\ \mathrm{day}^{-1}$ at night.

Numerous models have been created to simulate the process of sedimentation of pollutants. So, for instance, [104] proposed several parameterizations for the coefficient of aerosol washing out from the atmosphere:

$$r = 10^{-4} I^{1/2} \quad r = \theta \times I^{a} \quad r = -C^{-1}\frac{dC}{dt} \quad r = 3.3 \times 10^{-4} I^{0.9}$$

where $I = RR/(24N)$ is the rain rate ($\mathrm{mm\,hr^{-1}}$); RR is the precipitation amount per month (mm); N is the number of days precipitation; and θ and a are parameters.

The following diffusion equation is widely used:

$$\frac{\partial C}{\partial t} + V_\varphi \frac{\partial C}{\partial \varphi} + V_\lambda \frac{\partial C}{\partial \lambda} + V_h \frac{\partial C}{\partial h} = -\frac{\partial}{\partial \varphi}(\overline{V_\varphi C}) - \frac{\partial}{\partial \lambda}(\overline{V_\lambda C}) - \frac{\partial}{\partial h}(\overline{V_h C}) \tag{8.2}$$

If we suppose that in equation (8.2) an advection prevails over diffusion in the direction h (i.e., $\partial(\overline{V_h C})/\partial h \ll V_h \partial C/\partial h$), then equation (8.2) with respect to λ gives:

$$\frac{\partial C_\varphi}{\partial t} + V_h \frac{\partial C_\varphi}{\partial h} + V_\lambda \frac{\partial C_\varphi}{\partial \lambda} = -\frac{\partial}{\partial \lambda}\int_{-\infty}^{\infty} \overline{V_\lambda C}\, d\varphi$$

where

$$C_\varphi = \int_{-\infty}^{\infty} C\, d\varphi.$$

As a result of this transformation, the problem becomes two-dimensional.

Chobadian *et al.* [22] proposed two formulas to estimate the depth of the mixed atmospheric layer, which is important in determination of the model's vertical structure:

$$H = 8.8 x_1 U_a^{-1} \Delta\theta \quad H = \frac{U_*}{U_a}\left\{\frac{x_1|\theta_1 - \theta_2|}{|\Delta\theta/\Delta x_3|}\right\}^{1/2}$$

where x_1 is the rate of shifting with respect to land surface (m s^{-1}); $\Delta\theta$ is the vertical gradient of potential temperature in the inverse layer (°C); U_* is the rate of friction over the leeward surface; θ_1 is the lower level of potential temperature over the source of the pollution (K); and $|\Delta\theta/\Delta x_3|$ is the absolute value of the rate of motion over the source.

The desire to simplify the parameterization of individual subprocesses of atmospheric pollution dynamics led in many cases to the development of sufficiently simple and efficient models requiring a small database. Numerous studies have been dedicated to the classification of situations taking place in the real atmosphere with pollutants emitted from point sources (stacks, etc.). Sufficiently complete classification can be found in [102], where 10 types are given of the behaviour of the pollutant jets (smoke) in the vicinity of an isolated point source (stack): folded, spiral, flagwise; sedimenting, broadening, expanding, breaking, bifurcational, fumigating, and rising. In each of these cases a Gauss-type model can be used with a minimum of input information. The monitoring system should only be able to distinguish between these situations based on measurements of meteorological parameters. For instance, the folded shape of the jet can form due to a rough surface (high buildings, hilly locality) in clear sunny weather. Smoke particles are wind-driven and zig-zag inside the broadening and gradually sedimenting jet. The breaking jet is typical of the second-half of the day in summer, when the atmosphere is well heated, and convective motion of air masses prevail.

One of the units of the atmospheric pollution model describes the process of emission of aerosol to the atmosphere. In most cases the Gauss-type model is used:

$$C(r,h) = f(h)\frac{M}{2\pi\sigma_r^2}\exp(-r^2/(2\sigma_r^2)) \tag{8.3}$$

where r is the distance to the centre of the pollutant jet; σ_r is the horizontal size of the jet; $f(h)$ is the vertical distribution of the mass of pollutants; $M = Q\Delta t$ is the mass of pollutants in the jet; Q is the rate of emission from the source; and Δt is the time step. The f function can be approximated by the following dependence: $f(h) = 1/H$ for $0.5H < h < 1.5H$ and 0 in the opposite case (H is the effective height of the jet). For instance, to calculate H, the following formula has been proposed [26]:

$$H = \begin{cases} Z + 0.29V^{-1}Q_h^{1/2} & \text{normal conditions of stability} \\ Z + 0.49V^{-1}Q_h^{0.29} & \text{weak conditions of stability} \end{cases}$$

where Q_h is the rate of the heat flux from the source (stack; cal s^{-1}); and Z is the height of the source (m).

To simplify the process of simulation of the pollutant field in the zone of the source, the flux can be quantized into individual formations (small clouds, dust masses), each being considered separately as a homogeneous cloud. In particular, for quantization of the flux into n parts over the time Δt, the following formula can be used: $n = 2VDt/H$. For instance, at wind speed $V = 5$ m/s and $Z = 100$ m we obtain that some clusters can be emitted from the stack every 10 s. As a result, the process of simulating the dynamics of the whole emitted jet is much simplified. So, for instance, in modelling the propagation of sulphur one can additionally distinguish between sulphur-forming components.

8.5.4 Types of aerosol models and their information base

Models of the atmospheric pollution dynamics are divided into Gaussian, Eulerian, and Lagrangian methods of parameterization which describe processes of pollutant scattering. Within each type there are statistical, box, correlative, and determinate submodels. The most important classification of models is by unification of their methods applied to paramterize the atmospheric processes. In this case the model classes are distictive as a result of their databases.

The Gauss-type models need information about the height of the source of pollutants and dispersion characteristics inside the pollutant cloud. Different approximations of the Gauss jet in the vicinity of a point source depend on parameters of atmospheric surface layer stability. The model's configuration is affected by the geophysical characteristics. With the source present in the zone of transition from water surface to land, clusters can be selected with stable and unstable behaviour of the pollutant jet. With respect to the horizontal coordinate x on land, with wind direction along this coordinate, three zones can be selected with characteristic features: the zone of undisturbed dispersion, the zone of deposition, and the zone of delay.

In the zone of undisturbed dispersion, the pattern of distribution of the pollutant flux is formed under the influence of a homogeneous and stable layer:

$$C(\varphi, \lambda, h) = \frac{Q}{2\pi V \sigma_\lambda \sigma_h} \exp\left\{-\frac{1}{2}\left(\frac{\lambda}{\sigma_\lambda}\right)^2\right\} \times \left\{\exp\left[-\frac{1}{2}\left(\frac{h-Z}{\sigma_h}\right)^2\right] + \exp\left[-\frac{1}{2}\left(\frac{h+Z}{\sigma_h}\right)\right]\right\}$$

where φ is the wind direction; λ is the coordinate in the horizontal Cartesian system; h is the vertical coordinate; and σ_φ and σ_p are the horizontal and vertical dispersions (scattering).

In the second zone the behaviour of the cloud of pollutant forms under conditions of instability and therefore the calculations of the pollutant concentraton requires a broadened information base.

Denote: φ' – the leeward distance at the intersection of the cloud and the upper part of the boundary layer; σ' – average value of scattering at the boundary between the stable and unstable zones; and $p' = [L(\varphi) - Z]/\sigma_h$, $L(\varphi)$ – the height of the boundary layer in the leeward direction φ. Then:

$$C(\varphi, \lambda, h) = \frac{Q}{(2\pi)^{3/2}} \int_0^\varphi \frac{1}{\sigma' \sigma_h(V, \varphi, \varphi')V} \exp\left\{-\frac{(p')^2}{2} - \frac{1}{2}\left(\frac{\lambda}{\sigma'}\right)^2\right\} \times \left\{\exp\left[\frac{-(h-Z)^2}{2\sigma_h^2(V, \varphi, \varphi')}\right] + \exp\left[-\frac{(h+Z)^2}{2\sigma_h^2(V, \varphi, \varphi')}\right]\right\} \frac{\partial p'}{\partial \varphi'} d\varphi'$$

The third zone is quite specific due to conditions of cloud formation. This cloud is completely located within the ABL, and the concentration of pollutants in it can be calculated with the formula:

$$C(\varphi, \lambda, h) = \frac{Q}{2\pi V \sigma_\lambda \sigma_h} \exp\left[-\frac{1}{2}\left(\frac{\lambda}{\sigma_h}\right)^2\right] \times \left\{\exp\left[-\frac{1}{2}\left(\frac{h-Z}{\sigma_h}\right)^2\right] + \exp\left[-\frac{1}{2}\left(\frac{h+Z}{\sigma_h}\right)^2\right]\right\}$$

for $\sigma_h < 0.47L(\varphi)$, and:

$$C(\varphi, \lambda, h) = \frac{Q}{\pi V \sigma_\lambda L(\varphi)} \exp\left[-\frac{1}{2}\left(\frac{\lambda}{\sigma_\lambda}\right)^2\right]$$

for $\sigma_h > 0.8L(\varphi)$. Within the interval $0.47 \leq L(\varphi) \leq 0.8L(\varphi)$ the concentration C is calculated with the interpolation formula between the points $\sigma_h = 0.47L(\varphi)$ and $\sigma_h = 0.8L(\varphi)$.

This model describes well the fields of pollutants over territories on scales up to 100 km. For larger territories, other types of models are used.

The information base of the Euler and Lagrange-type models is formed according to the Cauchy problem for equation (8.2). Depending on a variety of real situations the database composition, ensuring a realization of the model of pollutant transport, becomes hierarchical. This hierarchy determines a relationship between spatial and temporal scales, it determines the accuracy of the model and its other characteristics. Numerous global databases consist, as a rule, of information levels with an object orientation.

8.5.5 Modelling the wind field

Knowledge of the wind field components is necessary for modelling the processes of atmospheric pollutant propagation over a territory of the region independent of a spatial scale. In the presence of many sources of atmospheric pollution in the region, information is needed about the 3-D wind field, and this is only possible with combined measurements and modelling. Usually, for large regions or in studies of conditions of pollutant propagation at the land–water boundary, the grid Euler-type model is used:

$$\partial u/\partial x + \partial v/\partial y + \partial w/\partial z = 0$$

$$\partial u/\partial t + u\partial u/\partial x + v\partial u/\partial y + w\partial u/\partial z = f(v - v_g) + \partial(K_M \partial u/\partial z)/\partial z$$

$$\partial v/\partial t + u\partial v/\partial x + v\partial v/\partial y + w\partial v/\partial z = f(u_g - u) + \partial(K_M \partial v/\partial z)/\partial z$$

$$\partial T/\partial t + u\partial T/\partial x + v\partial T/\partial y + w\partial u/\partial z = \partial(K_H(\Gamma + \partial T/\partial z)/\partial z$$

where f is the Coriolis parameter; Γ is the rate of dry adiabatic deposition; T is the air temperature; K_M and K_H are coefficients of diffusion; and u_g and v_g are the thermal wind components:

$$u_g = -\frac{gT}{f}\int_0^z \frac{1}{T^2}\left(\frac{\partial T(t,x,y,\xi)}{\partial y}\right)d\xi$$

$$v_g = \frac{gT}{f}\int_0^z \frac{1}{T^2}\left(\frac{\partial T(t,x,y,\eta)}{\partial x}\right)d\eta$$

For practical application, the following should be taken into account: deviation and rising of the flux over the hills, thermal friction, the tunnelling effect in valleys, and the thermal impacts of islands, lakes, and mountains. Additions to the given system of equations are made proceeding from an actual configuration and topology of the region.

8.5.6 The Gauss-type models

The Gauss-type models are used, as a rule, to parameterize the processes of pollutant propagation near high sources. The best studied modifications of the PGT dispersion model (Pasquill–Gifford–Turner) adequately describes the surface fluxes of pollutants and less accurately the fluxes formed from elevated sources. Raaschou-Nielsen *et al.* [93] have modified the PGT model providing it with more universal functions for an arbitrary point source. The modification is based on the base Gauss-type model to describe the concentration of pollutants in the surface layer of the atmosphere:

$$C(\varphi,\lambda,0) = \frac{Q}{\pi u \sigma_z \sigma_\lambda}\exp\left[-0.5\left(\frac{h_{ef}}{\sigma_z}\right)^2\right]\exp\left[-0.5\left(\frac{\lambda}{\sigma_\lambda}\right)^2\right] + \text{reflection terms} \quad (8.4)$$

where Q is the source power; h_{ef} is the average height of the pollutant cloud; and $\bar{u}$ is the effective rate of transport.

Important parameters of equation (8.4) are of the pollutant jet scattering. They are complex functions of the meteorological situations. Theoretical and observational studies of many authors have shown that the turbulence and diffusion in the convective boundary layer are controlled by two important parameters: the height of mixing z_i and the scale of convection rate w_*:

$$w_* = \left(\frac{g}{T\rho c_p} Y z_i\right)^{1/3} \quad (8.5)$$

where Y is the surface sensible heat flux; g is the acceleration due to the Earth's gravity; T is the air temperature; and c_p is the air specific heat at a constant pressure.

The vertical σ_w and horizontal σ_v constituents of the rate of deviation of the turbulent flux from the centre of the convective cloud are proportional to w_* and, hence, depend on z_i. The total energy of turbulence is a composition of two energies generated by mechanical (σ_{wm}) and convective (σ_{wc}) forces:

$$\sigma_w^2 = \sigma_{wm}^2 + \sigma_{wc}^2 \tag{8.6}$$

It is assumed here that mechanical and convective forces do not correlate. Then, by analogy, we can write:

$$\sigma_z^2 = \sigma_{zm}^2 + \sigma_{zc}^2 \tag{8.7}$$

The σ_{zc}^2 value is calculated on the assumption that:

$$\frac{d}{dt}\sigma_{zc} = \sigma_{wc}(z') \quad \text{where } t = \varphi/\bar{u}. \tag{8.8}$$

Here z' is the effective height at which the vertical turbulence is calculated. Dependences on the height for σ_{wc} are parameterized with relationships:

$$\sigma_{wc}^2 = \begin{cases} 1.54w_*^2(z/z_i)^{2/3} & \text{for} \quad z < 0.1z_i \\ 0.33w_*^2 & \text{for} \quad z \geq 0.1z_i \end{cases} \tag{8.9}$$

where the level $z = 0.1z_i$ corresponds to similar values of the vertical constituent of turbulence.

Let h_s be the height of the pollutant jet and t have the scale $x/\bar{u}$. Then using equations (8.8) and (8.9) we obtain the relationship between h_s and z_i. For $h_s \geq 0.1z_i$ the following approximation is valid:

$$\sigma_{zc}^2 = 0.33w_*^2t^2 \tag{8.10}$$

With $h_s < 0.1z_i$ we have:

$$\sigma_{zc}^2 = \begin{cases} 1.54w_*^2(h_s/z_i)^{2/3}t^2 & \text{for} \quad \sigma_{zc} < h_s \\ (0.83w_*z_i^{-1/3}t + 0.33h_s^{2/3})^3 & \text{for} \quad h_s \leq \sigma_{zc} < 0.1z_i \\ (0.581w_*t + 0.231h_s^{2/3}z_i^{1/3} - 0.05z_i)^2 & \text{for} \quad \sigma_{zc} \geq 0.1z_i \end{cases}$$

The mechanical constituent in equation (8.7) is calculated on the assumption that variations of the vertical gradient due to mechanical mixing are constant in the boundary layer and determined with the relationship:

$$\sigma_{wm}^2 = 1.2u_*^2 \tag{8.11}$$

where u_* is the friction rate.

For the parameter of mechanical scattering σ_{zm} the ratio to the rate of mechanical turbulence σ_{wm} is not so simple as in the case of convective turbulence. This is connected with the fact that the size of the scale characterizing the mechanical scattering, in contrast to the case of convective, changes with altitude, and therefore it is less than the convection scale. Therefore, for the unstable state the following dependences are used:

$$\sigma_{zm}^2 = \begin{cases} \sigma_{zmu}^2 = 1.2u_*^2t^2\exp(-0.6tu_*/h_s) & \text{for} \quad tu_*/h_s < 1 \\ \sigma_{zmu}^2 = 1.2u_*^2t^2\exp(-0.6) & \text{for} \quad tu_*/h_s \geq 1 \end{cases}$$

For stable conditions the following approximation is valid:

$$\sigma_{zm}^2 = \sigma_{zmu}^2/(1 + 1.11tu_*/L)$$

where L is the Monin-Obukhov length.

Thus both terms on the right in equation (8.7) are estimated. Now determine the σ_λ parameter in equation (8.4). The characteristic of horizontal scattering of the pollutant is well approximated with the formula:

$$\sigma_\lambda = \left(\frac{0.25w_*^2}{1 + 0.9\varphi w_*/uz_i} + u_*^2\right)^{1/2} \frac{\varphi}{\bar{u}} \tag{8.12}$$

In this expression the first and second terms in brackets represent the contribution of convective and mechanical turbulence, respectively. For nocturnal conditions, the first term is equal to zero. For the stable atmosphere and at weak winds the ratio $u_*/\bar{u}$ decreases rapidly with growing stability. Nevertheless, numerous observations have shown that the horizontal scattering with an hourly averaging can exceed the values for the unstable atmosphere. The horizontal rate fluctuates with constant amplitude about $0.5\,\mathrm{m\,s^{-1}}$. As the horizontal scattering remains proportional to fluctuations of horizontal wind, the u^* parameter in equation (8.12) for the case of stable stratification can be changed to $0.5\,\mathrm{m\,s^{-1}}$. It means that σ_λ will never be less than $0.5\varphi/\bar{u}$.

Estimate the parameter of horizontal scattering:

$$\sigma_{\lambda f} = \left(\Delta d/\sqrt{2\pi}\right)\varphi$$

where Δd is the change of wind direction in radians.

According to equation (8.4), the accuracy of calculation of the aerosol concentration at a point with coordinates (φ, λ) depends substantially on wind speed. An estimate of the aerosol density in a given space volume depends on the accuracy of its estimation at a given height. For instance, in the case of unstable stratification the wind speed is approximated with its value at a height of the source, and under stable conditions the wind speed is calculated by the vertical averaging of its values. According to [81], we have:

$$u(z) = \frac{u_*}{k}\left[\ln\left(\frac{z + z_0}{z_0}\right) - \psi_m\left(\frac{z}{L}\right) + \psi_m\left(\frac{z_0}{L}\right)\right] \tag{8.13}$$

where z_0 is the linear size of surface roughness, ψ_m are universal functions of similarity calculated, for instance, in [91]. Based on these calculations, the wind speed is estimated using equation (8.13) at $z \leq z_B$ and is assumed to be equal to $u(z) = u(z_B)$ for all heights $z > z_B$, where $z_B = \max\{0/1z_i, |L|$.

Practical application of the Gauss-type models has many aspects appearing depending on the multitude of the factors of the natural–anthropogenic environment. This includes an effect of buoyancy of aerosol particles in the atmosphere, intermittent atmospheric layers, and many others. Raaschou-Nielsen *et al.* [93]

proposed to calculate the h_{ef} parameters using the following formulas:

$$H_{ef} = h_s + \Delta h$$

where

$$\Delta h = \min\{\Delta h_{in}, \Delta h_f\} \quad \Delta h_{in} = 1.6\left(\frac{F}{u}\right)^{1/3} \varphi^{2/3} \quad F = \frac{V}{\pi}\frac{g}{T_e}(T_e - T_a)$$

Here F is the buoyancy flux; h_s is the stack height; V is the flux volume; T_e is the temperature of the flux emitted from the stack; and T_a is the air temperature.

The height of the aerosol jet due to the effect of its buoyancy is calculated with the formula:

$$\Delta h_f = \begin{cases} 1.3\dfrac{F}{uu_*^2}\left(1 + \dfrac{h_s}{\Delta h_f}\right) & \text{neutral conditions} \\ 4.3\left(\dfrac{F}{u}\right)^{3/5}\left(\dfrac{Yg}{T_a \rho c_p}\right)^{-2/5} & \text{convection} \\ 2.6\left(\dfrac{F}{us}\right)^{1/3} & \text{stability} \end{cases}$$

where $s = (g/T_a)\partial\theta_a/\partial z$. Here $\partial\theta_a/\partial z$ is the potential air temperature gradient at the stack's level.

A multitude of possible situations with the use of the Gauss-type model is determined by locality topology, the presence of the boundaries of the sea–land type, mountains, and others. A consideration of a concrete situation enables one to simplify the formulas and raise their reliability. In particular, [121] considered such situations at a lake–land boundary. At the water–land boundary the air flux is continuously adjusting itself to the wind field configuration forming into jets corresponding to mechanical and temperature conditions. Under conditions of a persistent coastal heat layer the gradient of aerosol concentration can be stable, which depends on the distance of the source from the shore. Here the following formula is valid:

$$\frac{3X}{\beta'^2 Usz_b^3} = \frac{3}{2}\left[\frac{z_{eq}}{z_b}\right]^2\left[\frac{z_{eq}}{z_b} - 1\right] + \frac{2}{9} \tag{8.14}$$

where z_{eq} is the stable height of the jet over the stack; z_b is the height of the bottom of the stable atmospheric layer over the stack; U is the wind speed at the height of the stack; β' is the effective coefficient of aerosol capture ($\beta' \cong 0.4$); s is the stability parameter; and X is the jet buoyancy parameter. The s parameter depends on the potential temperature gradient θ and air temperature T_a: $s = (g/T_a)\partial\theta/\partial z$ (g is the gravity coefficient). The F parameter can be calculated as the function of velocity v of the jet emitted from the stack, internal radius R of the stack, and temperature T of the emitted jet:

$$F = gvR_2(T - T_a)/T$$

With the z_d value known from equation (8.14) one can easily find z_{eq}, and the effective height Y of the jet is found as the sum: $Y = z_{eq} + h_s$, where h_s is the height of the stack. The whole zone of the aerosol cloud propagation with wind s blowing from the water toward land is divided into zones of stability, instability, and aerosol cloud lowering. Such a digitization of space simplifies the model reducing the required database and increases its reliability. In each zone, the use of the Gauss-type model leads to a simplified parameterization of the process of aerosol scattering and facilitates an interpretation of the modelling results.

In the undisturbed dispersion zone the distribution of the aerosol jet in the homogeneous stable atmospheric layer is described with the base Gaussian equation of scattering:

$$C(\varphi, \lambda, z) = \frac{Q}{2\pi U \sigma_\lambda \sigma_z} \exp\left[-\frac{1}{2}\left(\frac{\lambda}{\sigma_\lambda}\right)^2\right] \times \left\{\exp\left[-\frac{1}{2}\left(\frac{z-Y}{\sigma_z}\right)^2\right] + \exp\left[-\frac{1}{2}\left(\frac{z+Y}{\sigma_z}\right)^2\right]\right\}$$

where U is the wind speed; Y is the effective height of the jet; and σ_λ and σ_z are parameters of the horizontal and vertical dispersion.

In the zone of fumigation, forces become activated which cause a distortion of the jet due to instability of heat fluxes in the surface layer. In this zone the following approximation is valid:

$$C(\varphi, \lambda, z) = \frac{Q}{(2\pi)^{3/2}} \int_0^{\varphi} \frac{1}{\sigma' \sigma_z(u, \varphi, \varphi') U} \exp\left[-\frac{1}{2}\left(p'^2 + \left(\frac{\lambda}{\sigma'}\right)^2\right)\right] \times \left\{\exp\left[-\frac{(z-Y)^2}{2\sigma_z^2(u, \varphi, \varphi')}\right] + \exp\left[-\frac{(z+Y)^2}{2\sigma_z^2(u, \varphi, \varphi')}\right]\right\} \frac{dp'}{d\varphi'} d\varphi'$$

where φ' is the leeward distance of intersection of the aerosol jet with the upper boundary of the internal thermal boundary layer (ITBL); $p' = [L(\varphi) - Y]/\sigma_z(s, \varphi)$; $L(\varphi)$ is the height of ITBL at the distance φ; $\sigma_\lambda(s, \varphi)$ is the σ_λ value for the stable atmospheric layer at the distance φ; $\sigma_z(s, \varphi)$ is the σ_z value for the unstable atmospheric layer at the distance φ; $\sigma_\lambda(u, \varphi, \varphi')$ is the σ_λ value for the unstable atmospheric layer φ after passing the leeward distance φ'; and $\sigma_z(u, \varphi, \varphi')$ is the σ_z value for the unstable atmospheric layer at the distance φ after passing the leeward distance φ':

$$\sigma' = \sqrt{\sigma_\lambda^2(s, \varphi) + \sigma_\lambda^2(u, \varphi, \varphi')}$$

In the zone where the aerosol cloud top is close to $L(\varphi)$ and it is totally within the ITBL, the following approximations are valid:

$$C(\varphi, \lambda, z) = \begin{cases} xEG/\sigma_z(u, \varphi) & \text{for} \quad \sigma_z < 0.47L(\varphi) \\ xE/L(\varphi) & \text{for} \quad \sigma_z > 0.8L(\varphi) \end{cases} \tag{8.15}$$

where

$$x = Q/[\pi U \sigma_\lambda(u,\varphi)] \quad E = \exp\left[-\frac{1}{2}\left(\frac{\lambda}{\sigma_\lambda(u,\varphi)}\right)^2\right]$$

$$G = \exp\left[-\frac{1}{2}\left(\frac{z-Y}{\sigma_z(u,\varphi)}\right)^2\right] + \exp\left[-\frac{1}{2}\left(\frac{z+Y}{\sigma_z(u,\varphi)}\right)^2\right]$$

For the leeward distance φ, where $0.47L(\varphi) \leq \sigma_z(\varphi)\varphi 0.8L(\varphi)$, the concentration of aerosol is calculated by interpolation between its estimates in the boundary zones according to equation (8.15).

8.5.7 Modelling the planetary boundary layer

Modelling the processes of aerosol propagation in the surface layer is connected with the simulation of dependences on wind speed, air temperature, and humidity as 3-D functions of time and space coordinates. One such model of the planetary boundary layer (PBL) has been described in [12]. The model contains two units. The first unit analytically simulates the PBL processes. The field of aerosol is averaged and presented by a 1-D structure. Another transitional layer of the atmosphere is described by hydrodynamic and thermodynamic equations solved numerically with the use of the finite difference schemes. Equations for the atmospheric transition layer are derived on the assumption that the atmosphere is incompressible and hydrostatic, water vapour does not change its properties, the potential temperature is calculated with the formula $\theta = T + \Gamma z$, where T is the air temperature, Γ is the rate of dry adiabatic cooling, and z is the vertical coordinate. Equations for the atmospheric transition layer are as follows:

$$\frac{\partial u}{\partial t} + \frac{\partial(uu)}{\partial x} + \frac{\partial(vu)}{\partial y} + \frac{\partial(wu)}{\partial z} = -\frac{1}{\rho_a}\frac{\partial p_M}{\partial x} + f(v - v_g) + \frac{\partial}{\partial z}\left(K_m \frac{\partial u}{\partial z}\right) + K_H\left(\frac{\partial^2 u}{\partial x^2} + \frac{\partial^2 u}{\partial y^2}\right)$$

$$\frac{\partial v}{\partial t} + \frac{\partial(uv)}{\partial x} + \frac{\partial(vv)}{\partial y} + \frac{\partial(wv)}{\partial z} = -\frac{1}{\rho_a}\frac{\partial p_M}{\partial y} + f(u - u_g) + \frac{\partial}{\partial z}\left(K_m \frac{\partial v}{\partial z}\right) + K_H\left(\frac{\partial^2 v}{\partial x^2} + \frac{\partial^2 v}{\partial y^2}\right)$$

$$\frac{\partial u}{\partial x} + \frac{\partial v}{\partial y} + \frac{\partial w}{\partial z} = 0$$

$$g\frac{(T_M + T_n)}{\theta_a} - \frac{1}{\rho_a}\frac{\partial p_M}{\partial z} = 0$$

$$\frac{\partial\theta}{\partial t}+\frac{\partial(u\theta)}{\partial x}+\frac{\partial(v\theta)}{\partial y}+\frac{\partial(w\theta)}{\partial z}=\frac{\partial}{\partial z}\left(K_h\frac{\partial\theta}{\partial z}\right)$$
$$+K_H\left(\frac{\partial^2\theta}{\partial x^2}+\frac{\partial^2\theta}{\partial y^2}\right)+\frac{1}{\rho_a c_p}\frac{\partial Q_N}{\partial z}$$

$$\frac{\partial q}{\partial t}+\frac{\partial(uq)}{\partial x}+\frac{\partial(vq)}{\partial y}+\frac{\partial(wq)}{\partial z}=\frac{\partial}{\partial z}\left(K_q\frac{\partial q}{\partial z}\right)+K_H\left(\frac{\partial^2 q}{\partial x^2}+\frac{\partial^2 q}{\partial y^2}\right)$$

where

$$u_g=-\frac{1}{f\rho_a}\frac{\partial p_n}{\partial y}\qquad v_g=\frac{1}{f\rho_a}\frac{\partial p_n}{\partial x}$$

(u, v, w) are projections of wind speed onto rectangular coordinates x, y and z, respectively; K_H is the coefficient of horizontal turbulent mixing; ρ_a is the air density; and q is the specific humidity. A strict solution of these equations is impossible in general, and therefore it is necessary to choose a suitable discrete grid on which these equations are substituted with the finite difference equations. The step of time quantization and the size of discrete space grid should meet the condition:

$$\Delta t \le \min\{\Delta x/u, \Delta y/v, \Delta z/w\}$$

In practice, instead of this sufficiently cumbersome scheme, simulation schemes are often used based on the acceptable quantization of space. In this case the territory of a region or the globe is divided, as a rule, into homogeneous sites with linear sizes $\Delta\varphi$ in latitude and $\Delta\lambda$ in longitude: $\Omega=\{\Omega_{ij}; i=1,\dots,n; j=1,\dots,m\}$. Setting the vertical structure of the atmosphere and taking into account physical and chemical processes of aerosol transformation, relations recurrent with respect to the time step Δt are derived to recalculate their concentrations. With available control measurements at several points in space $\Xi=\{(\varphi,\lambda,z),(\varphi,\lambda)\in\Omega, 0\le z\le H\}$ the steps of quantization of the geographic grid and time are optimized on the basis of one or several criteria of accuracy. As a rule, such an approach provides the required stability of the modelling results [59].

8.5.8 The Euler-type models

Denote as $C_s(t,\varphi,\lambda,h)$ the concentration of aerosol of the sth type at the height h over the point with coordinates (φ,λ) at a time moment t. For the Euler-type model, general equations of aerosol transport in the environment according to equation (8.1) have the following form:

$$\frac{\partial C_s}{\partial t}+V_\varphi\frac{\partial C_s}{\partial\varphi}+V_\lambda\frac{\partial C_s}{\partial\lambda}+V_h\frac{\partial C_s}{\partial h}=\frac{\partial}{\partial\varphi}\left(K_\varphi\frac{\partial C_s}{\partial\varphi}\right)+\frac{\partial}{\partial\lambda}\left(K_\lambda\frac{\partial C_s}{\partial\lambda}\right)$$
$$+\frac{\partial}{\partial h}\left(K_h\frac{\partial C_s}{\partial h}\right)+E_s(t,\varphi,\lambda,h)$$
$$+P_s(C_1,\Lambda,C_q)-\nu_{1s}-\nu_{2s}$$

where K_φ, K_λ, and K_h are coefficients of the turbulent diffusion, E_s is the characteristic function of the sources of emission of aerosols of the s-th type, P_s is the operator describing physical and chemical transformations of aerosol, ν_{1s} is the rate of aerosol washing out with precipitation, ν_{2s} is the rate of dry deposition, $V = \{V_\varphi, V_\lambda, V_h\}$ are the wind speed components.

The model of this type is used to calculate the aerosol concentration in the atmosphere, as a rule, for scales of territories exceeding 50 km. Simplified schemes of calculation of the C function are drawn by dividing the space into units $\Delta\varphi \times \Delta\lambda \times \Delta h$, and at each height h_k a step-by-step calculation of concentration $C_s(t, \varphi_i, \lambda_j, h_k)$ is made. The calculation scheme can be further simplified by dividing the procedure into two stages. First, for each level of the vertical digitization of space the distribution of $C_s(t, \varphi_i, \lambda_j, h_k)$ and then the processes of the vertical transition of aerosols is taken into account. This scheme enables one to easily move on to vertically averaged levels depending on available information about the parameters of the atmospheric vertical stratification. Convergence of such a procedure depends on relationships between the parameters Δt, $\Delta\varphi$, $\Delta\lambda$, and Δh.

The Euler-type model contains many degrees of freedom including the consideration of various scenarios. The base of knowledge of the simulation system (the description of ESPhAP is given below) contains sets of parameterizations of partial processes of aerosol transformation, and a concrete choice is made by the user. Information in data and knowledge bases is structured according to a multitude of spatial and object identifiers – matrix structures $\{A_m\}$. In particular, in the 'default mode' regime the following parameterizations are used.

8.5.8.1 *Aerosol emission*

In accordance with the structure of the identifier describing the sources of pollution, in each compartment $\Delta\varphi \times \Delta\lambda$ the passport information about the rates of emission of the sth pollutant is input: minimum and maximum rates $E_{s,\min}$ and $E_{s,\max}$, respectively. In the absence of additional information about some source, the E_s value is calculated according to the procedure of the uniform distribution over the interval $[E_{s,\min}, E_{s,\max}]$ or another law of distribution is assumed (e.g., the Gauss law).

8.5.8.2 *Dry deposition*

The functional description for ν_{1s} and ν_{2s} is important, and the model's adequacy depends on it. Therefore, many studies have been dedicated to this problem. It has been proven that the following linear approximation of the functions ν_{1s} and ν_{2s} is most acceptable:

$$V_{is} = K_{is} C_s \quad i = 1, 2$$

The coefficient of proportionality K_{is} can be the function of time and space coordinates and depend on the physical state of the surface. With the use of the identifier $\|\mu_{ij}\|$ this dependence can be broadened. For instance, the version $\mu_{ij} = 0$ can mean that the dry deposition in the model is neglected completely. For $\mu_{ij} = 1$

the scheme proposed in [128] is taken into account:

$$K_{2s} = K_{2s}^* v_{ds} d / H_{\text{mix}}(t, \varphi, \lambda)$$

where K_{2s}^* is the coefficient of proportionality; v_{ds} is the rate of deposition of the sth type of aerosol (e.g., $v_{ds} = 0.008\ \text{m s}^{-1}$ for SO_2, $0.002\ \text{m s}^{-1}$ for SO_4, and $0.001\ \text{m s}^{-1}$ for NO_2 and HNO_3); H_{mix} is the height of the atmospheric mixed layer (m); and d is the parameter considering the physical state of the surface ($d = 1$ for dry surface, $d = 2$ for wet surface, etc.).

With the use of the μ_{ij} parameter the system of modelling can govern the choice of various dependences of the rate of dry deposition in a wide range of synoptic parameters and other environmental characteristics contained in the base of knowledge. For instance, for $\mu_{ij} = 2$ the following model is chosen:

$$v_{2s} = \begin{cases} 0.002U_* & \text{for} \quad L \geq 0 \\ 0.002U_* \left[1 + (-300/L)^{2/3}\right] & \text{for} \quad L < 0 \end{cases}$$

where L is the Monin-Obukhov length and U_* is the rate of friction.

8.5.8.3 Washing out

By analogy with the previous case, in modelling the rate of wet deposition of aerosols from the atmosphere it is expedient to consider the problem from different positions, since in general, the v_{1s} value is the function of many factors, some of them not always being estimated reliably. For the convenience of computer modelling, the identifier $\|v_{ij}\|$ is introduced and, for instance, for $v_{ij} = 0$ we assume automatically $v_{1s} = 0$ (no washing out). For $v_{ij} = 1$ there is a version when:

$$v_{1s} = K_{2s}^* C_s r$$

where r is the rain rate.

With $v_{ij} = 2$ we have the model of washing out:

$$\partial C_s / \partial t = -\phi r(t, \varphi, \lambda) C_s(t, \varphi, \lambda, h)$$

At $\phi = \text{const}$, we obtain a simple explicit scheme of C_s recalculation in each cell:

$$C_s(t + \Delta t, \varphi_i, \lambda_j, h_k) = C_s(t, \varphi_i, \lambda_j, h_k) \exp\{-\phi r(t, \varphi_i, \lambda_j,)\Delta t\}$$

8.6 AN EXPERT SYSTEM FOR THE PHYSICS OF ATMOSPHERIC POLLUTION

8.6.1 The expert system's structure

Figure 8.1 gives a list of units of the emission system for the physics of atmospheric pollution (ESPhAP). The software composition for its functioning is reproduced in Table 8.3. The input information in the ESPhAP is assumed to concentrate in thematic files which can be formed on the basis of various sources, such as remote

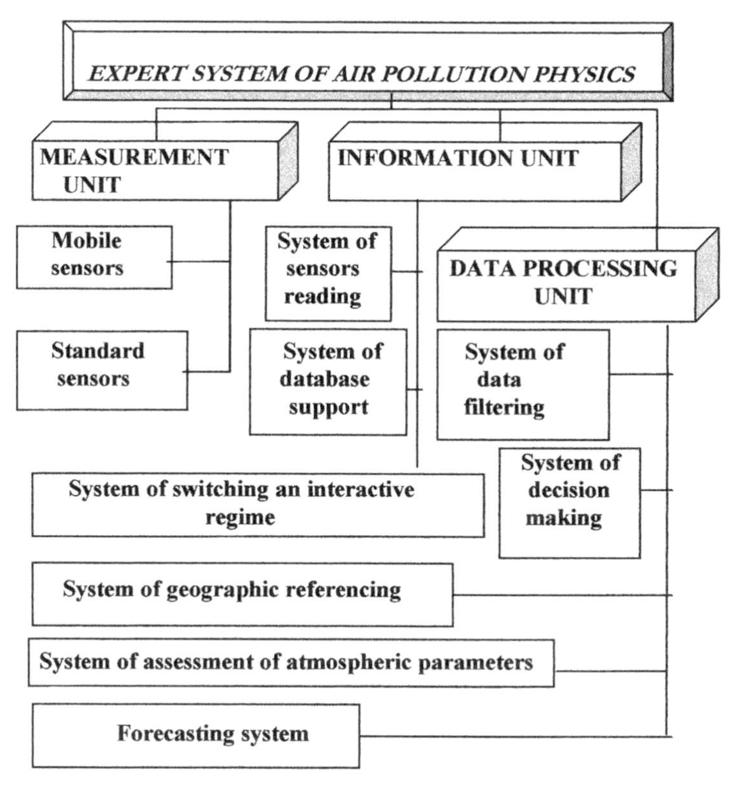

Figure 8.1. The ESPhAP structure with functions of complex assessment of air pollution over a given territory.

sensors, lines of information communication, transported standard and non-standard accumulators of observational data, and direct input of data from the keyboard or another compatible input unit. The base version of ESPhaP is calculated for the latter case. Switching of ESPhAP to other regimes requires additional software elements adjusting the medium and providing its transformation into the file unit. The interaction between all elements of the ESPhAP is schematically shown in Figure 8.2.

8.6.2 Formation of database components

Algorithms of formation and processing of spatial databases have been reliably verified in the geographical information system (GIS). To refer the ESPhAP to a concrete object, we choose the matrix hierarchical topologic structure (spatial identifier) which reflects the spatial and object structure of the territory Ω so that:

$$\Omega = \bigcup_{(i,j)} \Omega_{ij} \quad \Omega_{ij} = \bigcup_{(s,l)} \Omega_{i_s j_l}$$

Table 8.3. The ESPhAP software.

Identifier of software product	Characteristic of software product
CSII	Calibration and scaling of input information
IDF	Input data filtration
STA	Spatial–temporal agreement of various types of data
SIOD	Spatial interpolation of observational data and formation of spatial images
DR	Data restoration
GTM	Gauss-type model
ETM	Euler-type model
CSDB	Choice of the spatial digitization grid and transfer to block modelling
RFDB	Removal from database of information needed for modelling
ASI	Modelling the processes of pollutant transport due to atmosphere–sea interaction
APS	Modelling the processes of pollutant transport with due regard to the functioning of the 'atmosphere–plant–soil' system
MSF	Model of sulphur fluxes in the environment
CPMR	Cartographic presentation of modelling results
AMD	Analysis of modelling data and their visualization with MCL levels taken into account
UI	User interface
MNF	Model of nitrogen fluxes in the environment
MOB	Model of oxygen balance in the natural–anthropogenic medium
MCC	Model of CO_2 cycle in the environment
CI	Control and interaction of the ESPhAP units

Note: MCL = maximum (permissible) concentration level.

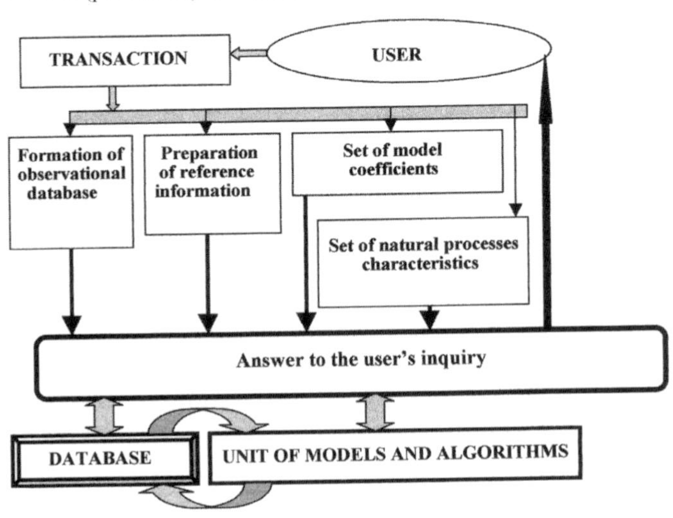

Figure 8.2. Approximate scheme of ESPhAP functioning in the regime of dialogue with the user.

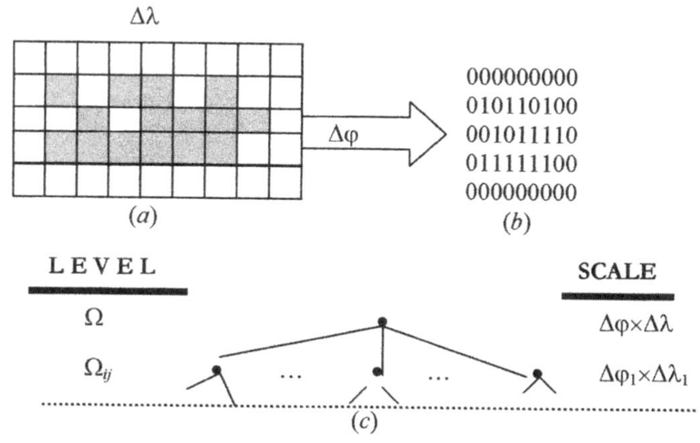

Figure 8.3. Procedure of topological referencing of the territory to the ESPhAP database components: (a) – contours of the territory Ω, (b) – the form of the A_1 identifier in the database, and (c) – hierarchy of database levels corresponding to the spatial and thematic structure of Ω.

Schematically this is shown in Figure 8.3. According to this algorithm, the territory Ω in the ESPhAP database is presented by a set of matrices:

$$A_k = \|a_{ij}^k\|, \quad k = \overline{1, N}$$

where

$$a_{ij}^1 = \begin{cases} 0 & \text{for} \quad (\varphi, \lambda) \notin \Omega \\ a_1 & \text{for} \quad (\varphi, \lambda) \in \Omega \end{cases}$$

$$a_{ij}^2 = \begin{cases} A & - \quad \text{forest} \\ B & - \quad \text{grass} \\ C & - \quad \text{agriculture (crops)} \\ D & - \quad \text{plough land} \\ E & - \quad \text{highway} \\ F & - \quad \text{boggy locality} \\ G & - \quad \text{urban territory} \\ H & - \quad \text{industrial zone} \\ J & - \quad \text{fruit trees} \\ I & - \quad \text{agricultural vegetation (others)} \end{cases}$$

$$a_{ij}^3 = \begin{cases} A & - \quad \text{no pollution sources} \\ B & - \quad \text{pollution source of the type `B'} \\ C & - \quad \text{pollution source of the type `C'} \\ & \quad\;\; \text{other type of pollution sources} \end{cases}$$

Table 8.4. Classification of the atmospheric pollution sources.

Category of the pollution source	Examples of enterprises of a given category	Atmospheric pollutants
Chemical plants	Oil refining, woodworking, superphosphate excavation, and cement production	Sulphur oxides, sulphides, fluorine compounds, organic vapour, particles, and aromatic substances
Spraying and dispersing in agriculture	Control of pesticides and weeds	Organic phosphates, chlorinated hydrocarbons, arsenic, and lead
Crushing, grinding, and sifting	Road-building plants	Minerals and organic substances
Destruction	Reconstruction of cities	Minerals and organic particles
Burning of fields	Burning of stubble and clearings	Smoke, ash, and soot
Fuel burning	Power stations	Sulphur and nitrogen oxides, carbon monoxide, smoke, and aromatic substances
Fuel production	Gas scattering	Fluorine compounds
Printer's ink production	Photography and printing	Fluorine compounds
Metallurgical works	Metal melting, and steel and aluminum production	Lead, arsenic, zinc, chlorine compounds, and sulphur oxides
Mills and car engines	Elevators and means of transport	Sulphur and nitrogen oxides, aromatic substances, carbon monoxide, and smoke
Use of nuclear fuel	Nuclear reactors	Argon-41
Extraction of ore	Grinding, crushing, and sifting	Uranium and beryllium dust
Burning of garbage and waste	Dust heaps, garbage-processing plants	Sulphur oxides, smoke, ash, organic vapour, and aromatic substances
Waste processing	Stores of scrap metal, car heaps, and processing of second-hand materials	Smoke, organic vapour, aromatic substances, and soot.

Matrices A_4 and A_8 contain geographic information about the territory levels with respect to the sea level and data on rain rate, respectively. The A_5 identifier gives the distribution of population density in the region. The A_6 and A_7 structures determine mean statistical data on wind speed and direction over the territory Ω. All information is prescribed for a certain time period and, hence, the ESPhAP database stores the sets of matrices $\{A_{ij}\}$ marked with the respective time interval. Each element of these matrix structures is hierarchically decoded by levels but not necessarily in the same way. The A_3 matrix takes into account the types of pollution sources enumerated in Table 8.4. For a better familiarization of ESPhAP, the database foresees possibilities to identify the atmospheric pollution sources with a set of pollutants.

The structure of the elements of the A_9 identifier identifies each spatial cell Ω_{ij} with the type of the polluting enterprise.

8.6.3 A subsystem for statistical solutions

A totality of units SIGNAL and IBMMEN performs the function of decision making about the signal output for the user in correspondence with the a priori prescribed elemental structure of the maximum concentration level (MCL). To preserve generality, it is assumed that the MCL can be prescribed in each cell Ω_{ij}, or in some, their totality by an independent set of elements. This information contains the A_{10} identifier.

In the process of measurements or calculations, for each pollutant of the type ξ the function of concentration distribution $P(\varphi, \lambda, h, t)$ is formed. Depending on time, at each point (φ, λ, h) a series of data $P_{\xi 1}, P_{\xi}^2, \Lambda, P_{\xi}^n, \Lambda$ is formed. Because of the measurement and calculation errors due to incorrect data, the choice of situations $P_\xi < \mathrm{MCL}_\xi$ or $P_\xi \geq \mathrm{MCL}_\xi$ is made at each time moment t on the basis of the respective statistical assessment in distinguishing between two versions: (1) the decision is made on the basis of a totality of indicators; and (2) the decision is made on the basis of each indicator separately. In the latter case the inequality $P_\xi \geq \mathrm{MCL}_\xi$ is tested for stability during a prescribed time interval Δt. In the case of meeting the condition $P_\xi \geq \mathrm{MCL}_\xi$ during this time interval the display gives a signal with the respective information. The time Δt is determined at expert level in the process of the system's exploitation to minimize the amount of false alarms or, from a condition of maximum probability of not missing a situation with a violated MCL. The actual choice of Δt depends on the multitude of local anthropogenic and other factors and refers to the user's competence. To make it easier to obtain an expert estimate or to make a decision in the real time regime, especially in the case of the use of an integral criterion, the user receives estimates of all indicators on a homogeneous normalized scale.

Let $q(l, i, j, k)$ be the indicator of the atmospheric quality in the compartment $\Omega_{i,j,k}$ by the lth parameter. To reduce all indicators to a single scale, we normalize them:

$$Q_l = Q(l, i, j, k) = q(l, i, j, k) / \max_{i,j,k} q(l, i, j, k)$$

Then the results of normalization are smoothed by height (k), latitude (i), longitude (j), and the whole territory Ω. The smoothed results are used to assess the air quality of territories of the user's choice. The final decision is made on the basis of comparison of Q_l estimates with a set of threshold values included in the database. The complex assessment of the state of the atmosphere over a given part of the territory Ω is made by the vector parameter $Q = (Q_1, \ldots, Q_n)$. This n-dimensional vector belongs to some indicator space X, each point in which corresponds to a concrete combination of MCL_ξ (prescribed by the user). The whole space X is divided into two parts: X_0 – pollution does not exceed acceptable

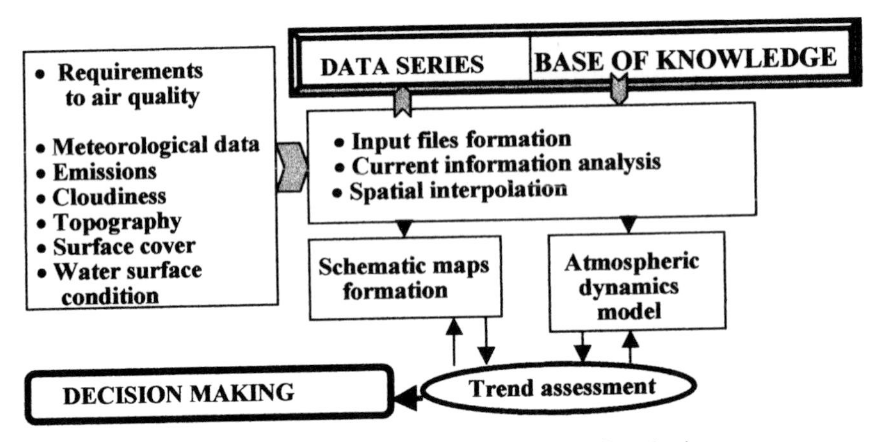

Figure 8.4. Basic components of ESPhAP functioning.

Table 8.5. Components of the ESPhAP database.

	Rate of dry deposition ($cm\,s^{-1}$)	
Element of A_2 identifier	SO_2	NO_2
Water surface	0.2–1.5	0.01–0.04
Lime surface	0.3–1.0	0.03–0.1
Acid dry soil	0.1–0.5	0.01–0.05
Acid wet soil	0.1–0.8	0.01–0.08
Plants <10 cm	0.1–0.8	0.02–0.25
Plants from 10 cm to 1 m	0.2–1.5	0.4–0.7
Forest	0.2–2.0	0.1–0.8

levels; X_1 – pollution reaches a dangerous level. A concrete realization of Q is estimated by belonging to one of these sub-spaces.

Thus the main stages of ESPhAP functioning are described by the scheme presented in Figure 8.4. Of course, the quality of the decisions made depends on the adequacy of information in the database. Table 8.5 shows a fragment of the database. These database fragments and various modifications of parametric descriptions of partial processes of aerosol transformation are controlled by a totality of identifiers which the user disposes of through such schemes of interaction with the system as scrolling windows and hierarchical menus. Therefore, a multitude of identifiers is used at the level of the system's manager who sees to the compatibility of its functions.

8.6.4 A subsystem for control and visualization

The ESPhAP provides a wide hierarchical dialogue for the user with the computer version of the simulation system of aerosol propagation in the atmosphere. An

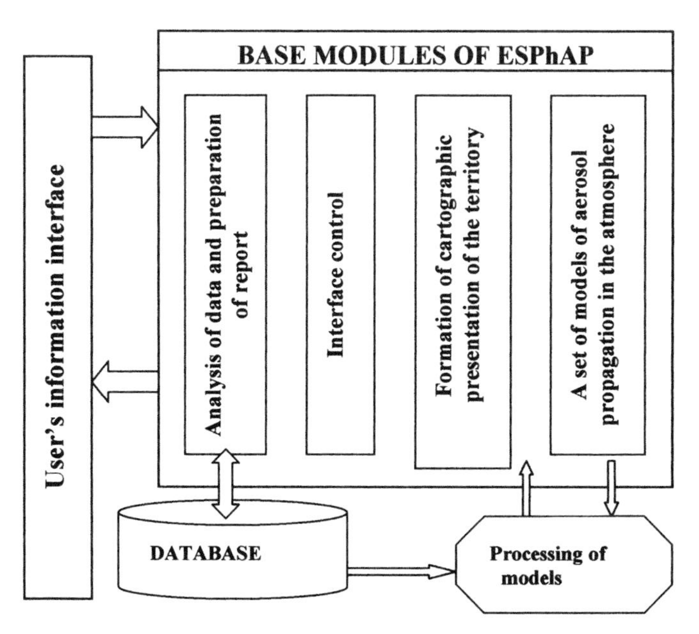

Figure 8.5. An approximate scheme of the operator–ESPhAP interaction in the regime of its adaptation to specific conditions of simulation experiments on the assessment of aerosol propagation over a given territory.

adaptation to the territory in the regime of dialogue is accomplished at the expense of identifiers which ensure the formation of database fragments responsible for the structure of surface cover, meteorological situation, and environmental parameters including information about the sources of aerosols and their characteristics. The operator's interaction is shown schematically in Figure 8.5.

The ESPhAP operates with schematic maps containing conditional images of the Earth's surface or distributions of atmospheric parameters. The schematic presentation of the modelling results is connected with their transformation into one of the following forms of distribution of the parameter to be estimated:

- schematic representation (1-D, 2-D, 3-D);
- symbolic representation of a 2-D image;
- tabulated representation by a set of formats;
- false-colour image reflecting the representation of modelling results by a fixed digital scale; and
- cartographic representation by isolines drawn on the basis of GIS technology.

8.7 GLOBAL SULPHUR CYCLE AND ITS SIMULATION MODELLING

8.7.1 Basic characteristics of the global sulphur cycle

One of the key problems of present global ecodynamics is the study of conditions for acid rain formation and prediction. For the first time, this problem was widely

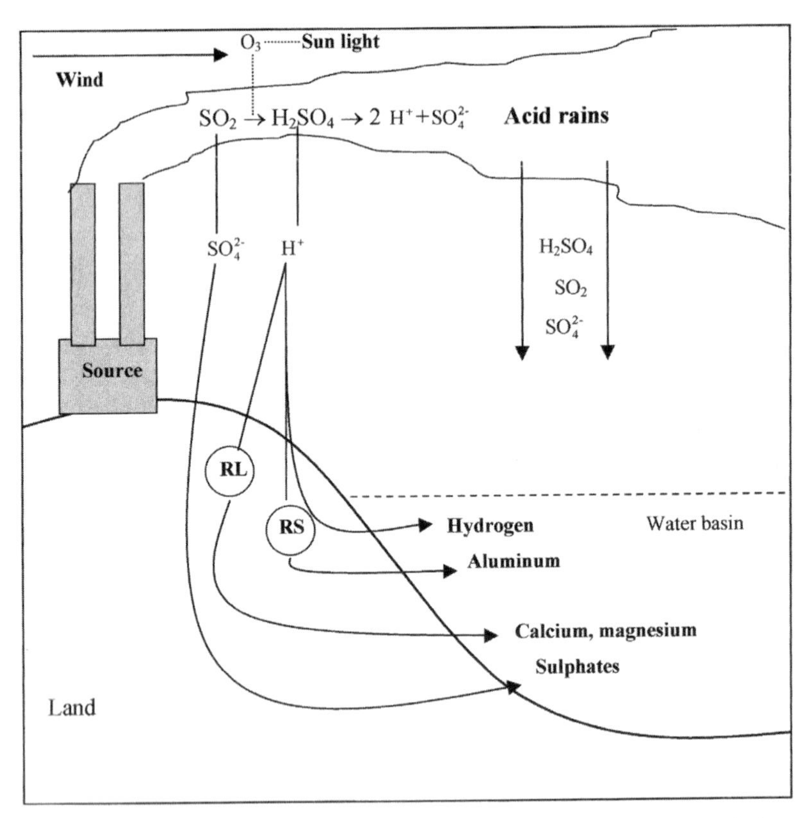

Figure 8.6. The conceptual scheme of the impact of anthropogenic sulphur emissions on the aquatic medium quality. Notations: RL – reactions with limestone minerals; RS – reactions with aluminum-containing silicate minerals.

discussed at the 28th General Assembly of the International Unit on Theoretical and Applied Chemistry held in Madrid in September, 1975. Subsequent conferences and various international programmes made it possible to accumulate data and knowledge in this sphere. It follows from these that sulphur compounds emitted to the atmosphere from natural and anthropogenic sources are an important precursor of acid rains strongly damaging the environment. Sulphur resides in the atmosphere mainly in the form of gas-phase SO_2 and H_2S as well as sulphate ion SO_4^{2-}.

Sulphur dioxide is a basic precursor of acid rain. Its concentration at the surface level is estimated at $1\,\mu g\,m^{-3}$. The participation of SO_2 in acid rain formation is realized in two ways: through dry deposition onto a wet surface and formation of H_2SO_4 directly in the atmosphere with subsequent deposition either onto the surface or in water basins [23]. Sulphur dioxide reacts with water and gives sulphuric acid: $SO_2 + H_2O + 1/2O_2 \rightarrow H_2SO_4$. Schematically these processes are shown in Figure 8.6. The SO_2 residence in the atmosphere depends strongly on means of its removal. Due to dry deposition, SO_2 is removed from the atmosphere over 7.6 days, and due to its transformation into SO_4^{2-} – over 13 days. The rate of dry deposition depends on the type of surface and many other environmental parameters. This rate averages $2\,cm\,s^{-1}$ over land and $0.9\,cm\,s^{-1}$ over the ocean. Due to the combination of the processes of SO_2 removal from the atmosphere its residence can shorten to 4.8 days. Berndt *et al.* [8], with the help of laboratory analysis, detected the earlier unknown mechanism for the transformation of SO_2 into H_2SO_4 in the real atmosphere due to the participation of O_3 in oxidation reactions, which shows the presence of new processes of acid rain formation whose study will make it possible to specify the rates mentioned above. It means that in modelling the sulphur cycle it is necessary to thoroughly describure the environmental parameters.

A simplified formula of acid rain is: *acid rain* $= H_2O + SO_2 + NO_2$. Natural and anthropogenic emissions of SO_2 to the atmosphere are responsible for 60–70% of acid rain over the globe. Deposition from the atmosphere of excessive sulphate is estimated at $360\,Tg\,yr^{-1}$ at an average rate of $110\,Tg\,yr^{-1}$, with 31% of excessive sulphate in case of rain water being of anthropogenic origin. On the whole, the anthropogenic sources emit to the atmosphere >90% S. Such sources are:

- coal burning (coal contains 2–3% S, and its burning gives SO_2);
- oil burning and its refining (the power of sulphur sources is 4–5 times lower than in the case of coal burning);
- ore melting to obtain metals, such as copper, nickel, and zink;
- volcanic eruptions [35];
- organic decomposition;
- weathering of sulphur-containing rocks ($\sim 15\,Tg\,yr^{-1}$);
- input of sulphur to the atmosphere via sea spray ($\sim 45\,Tg\,yr^{-1}$);
- fertilization of soil with sulphates and their subsequent input to the atmosphere with dust ($\sim 10\,Tg\,yr^{-1}$); and
- aviation and car engines [105].

Natural sources of sulphur compounds in the form of hydrocarbons, dimethylsulphide (DMS), carbonyl sulphide, and methyl mercaptan include soils, marshes, forests, volcanoes, the hydrosphere, and agricultural soils. It is known that from the World Ocean surface DMS gets to the atmosphere and is rapidly oxidized to give sulphates residing in the atmosphere over not more than five days. From available estimates, volcanoes emit annually to the atmosphere from 4–16 million tons of sulphides (recalculated for SO_2). The sulphur-containing compounds also form as a result of geothermal activity and living organism activity on land and in the water. Rivers bring sulphur to seas and oceans in volumes of $\sim 100\,Tg\,yr^{-1}$. Natural sources of sulphur are rather small. For instance, in the USA and Canada emissions of sulphur products from natural sources constitute, respectively, not more than 4 and 18% of total emissions of sulphur.

The spectrum of anthropogenic sources of sulphur compounds is diverse and varies with an addition of new names. Sulphur is known to be present in many useful minerals, such as coal, oil, iron, copper, and other ores. Their use by man leads to emissions of sulphur to the atmosphere despite the use of purification devices. The main by-product of industrial processes and fossil fuel burning is sulphur dioxide.

For instance, in the USA, SO_2 emissions are distributed by source types as follows: electric power stations – 67%, fossil fuel burning – 3%, industrial enterprises – 15%, transport – 7%, and other sources – 8%. On a global scale, these indicators vary strongly both in space and in time. In particular, in Canada, in contrast to the USA, electric power stations emit 20% SO_2, whereas non-ferrous metallurgy gives 43% SO_2. On the whole, in Canada the industrial enterprises are the main source of SO_2 (74%). It should be noted that the sources of sulphur in the USA are responsible for more than 50% of the acid rain in Canada, and territories bordering the Quebec province get up to 75% of their acid rain, due to SO_2 emissions, from the USA. From different estimates, the transboundary SO_2 transport from the USA to Canada varies within 3.5–4.2 million tons per year.

In most countries, with the intense use of petroleum, the problem of acid rain is closely connected with economic problems. Current types of petroleum contain sulphur from 150–600 ppm. The economic losses due to acid rain force developed countries to work on technologies to obtain petroleum with a low content of sulphur. It is expected that by the year 2005 a level of 30–50 ppm will be reached.

The structure of sulphur dioxide emissions is similar in various countries. In Germany the share of heat power stations and boiler houses constitutes about 90% of the total emissions of sulphur dioxide, and the share of industrial enterprises and transport reaches only 7.5% and 2.5%, respectively. Emissions of sulphur compounds to the atmosphere in high latitudes are of clearly seasonal nature. Estimates of these indicators are given in Table 8.6 and 8.7.

Getting to the atmosphere from different sources in a given territory, sulphur compounds can be transported by air masses for long distances and deposit themselves in other territories [34]. Figure 8.7 exemplifies the contribution of different countries to the sulphur balance for Norway and Sweden. It is clear that knowledge of the spatial distribution of the concentration of sulphur compounds, with meteorological information taken into account, will make it possible to predict acid rain [77]. The expert system ESPhAP which takes into account a totality of the models of atmospheric transport of pollutants, and the model of the sulphur cycle in the environment included into ESPhAP as an independent unit, make such predictions

Table 8.6. SO_2 emissions (mln t yr^{-1}) from anthropogenic sources in the USA and Canada [126].

Source	USA	Canada
Power generation	15.9	0.7
Transport	0.7	–
Industrial boiler houses	3.3	–
Household boiler rooms	–	–
Plant facilities	3.1	3.2
Other sources	1.5	0.7
Total	*24.5*	*4.6*

Table 8.7. Main sources of formation of primary sulphates in the USA (mln t SO_2 yr^{-1}) [126].

Source	Volume of emissions	Percent of total emissions
Thermal stations burning coal	0.34	39
Industrial boiler houses burning coal	0.06	7
Thermal stations burning oil	0.05	4
Industrial boiler houses burning oil	0.09	11
Transport	0.04	4
Industrial enterprises	0.09	15
Other sources	0.14	21
Total	*0.81*	*100*

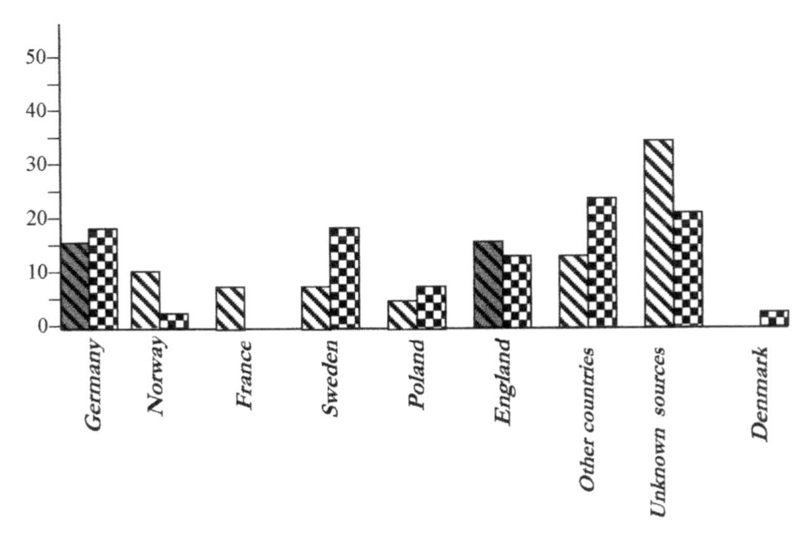

Figure 8.7. Sources of the transboundary transport of sulphur compounds in the territory of Norway and Sweden.

possible. Of course, certain limitations appear here connected both with the limited global database and with the absence of some functional descriptions. This difficulty can be overcome with the use of the global simulation model (GSM) within which the sulphur cycle is parameterized with due regard to the role of many subsystems of the biosphere and anthropogenic processes [53–60]. Moreover, an inclusion of the sulphur unit to the GSM broadens its functions, since it is dictated by the dependence of biotic processes on the content of sulphur in the biospheric compartments. Available data on supplies and fluxes of sulphur compounds in the atmosphere, soils, vegetation cover, and hydrosphere enable one to formulate mathematical relationships to simulate the global sulphur cycle.

On a global scale, the sulphur cycle is a mosaic structure of local fluxes of its compounds with other elements produced due to water migration and atmospheric processes. The conceptual schemes of global and regional sulphur cycles have been described in detail by many authors [84, 85]. However, the available models have been designed for autonomous functioning and application, which does not permit to include them in the GSM without substantial changes in their parametric spaces. Here is one of the solutions to this problem.

8.7.2 The role of sulphur in the environment

Sulphur as a non-metal is widely spread in nature and is one of the components of the global biogeochemical cycles. From the viewpoint of humans, sulphur belongs to elements which can negatively affect life systems. The harmful impact of sulphur on the environment is manifested mainly through acid rain. Water basins acidity due to acid rain and subsequent transport of sulphur compounds with runoff from adjacent lands are one such manifestation. The process of acidification is shown schematically in Figure 8.6.

Processes causing the acidification of water basins are mainly connected with anthropogenic sources of sulphur and, of course, other chemical elements such as nitrogen. In the preindustrial period the acidity of inland water basins had not exceeded $pH = 8$, as a rule. With the growing anthropogenic impact on the environment, the water basins acidity increased – overcoming for many of them $pH = 5.7$ in the middle of the last century. At present, in the zones of high industrialization, natural water basin acidity reaches $pH < 5$. The increasing trend of acidity causes serious problems for fish reserves. Many lakes and rivers in North America and Europe have been excessively acidified, and therefore the aquatic biota in them has suffered irreversible changes. For instance, in the USA [108] about 4.2% of lakes and 2.7% of river systems are in a state where their capability to neutralize a high acidity in a natural way is equal to zero. The chronic excess acidification of water basins ($pH < 5.2$) leads to irreversible changes in ecosystems.

As seen in Figure 8.6, many factors affect the composition of water, among these of importance being vegetation cover and type of soils in the aquatic system's basin. Therefore, to solve the problem of their acidity regulation all factors should be considered as a complex, which is only possible with the use of numerical models. Experimental technologies cannot be used for this purpose due to the unique character of natural systems. The curves in Figure 8.8 demonstrate the role of some factors in the formation of the pH scale.

Acid rains lead forests to destruction. Most damage is caused to coniferous forests. As a rule, forests grow in regions with sufficiently intensive precipitation and, hence, they can get large doses of harmful elements from acid rain. Acid rain damages the leaves and pine-needles as well as changes the soil properties. The vegetation productivity reduces drastically at $pH \in [2; 2.6]$. When the $pH \leq 2$ young shoots wither. On the whole, danger from acid rain appears at $pH \in [0.5]$. The level of danger depends on the climatic zone and the type of soil–plant formation present. At $pH \geq 5.6$, precipitation does not threaten the environment.

The soils of most forests have $pH \in [3.2, 5.5]$. This interval is provided by the stability of ion-exchange processes in soil, which preserves the living conditions for

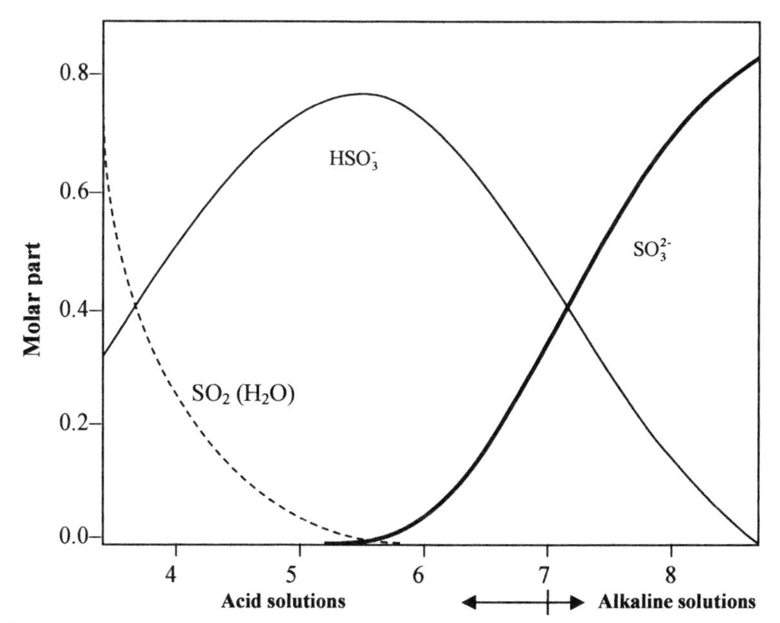

Figure 8.8. Relationship between molar parts of sulphite, bisulphate, and dissolved sulphur dioxide in the formation of pH level.

root systems of trees. An excess in soil of cations H^+, affecting the leaching of nutrients as well as the conversion of insoluble aluminum compounds to soluble ones, leads to a substitution of ion-forming centres and, hence, external conditions for the root system become strongly violated.

On the whole, for the global assessment of the role of acid rain, a database is needed which would characterize the regional level of pH and give the structure of pH-forming processes. Fragments of such a database are formed in many developed countries, but it is not enough for parameterization of the global pattern of acid rain formation. As shown in [100], such data are collected, for instance, for the territory of India. Using data of observations of the rain composition in India for the period 1984–2002, [100] found that during rainy seasons the content of SO_4 and NO_3 in rain water steadily grows over many regions of India, but the pH level remains within the interval of alkalinity. Such data and knowledge of the values of transboundary fluxes of sulphur make it possible to calculate the pH levels with due regard to the growth of industrial production and development of transport. Another example of database accumulation for acid rain control is an analysis of trends in changes of SO_2 and SO_4^{2-} concentrations in the atmosphere over the urban territories of the western and mid-Atlantic regions of the USA for the period 1990–1999. Here of importance is an assessment of the spatial variability of these concentrations (30–42%), making it possible to more reliably calculate the parameters of the respective equations in biogeochemical units in the models of the atmospheric aerosol transport.

8.7.3 Global sulphur cycle model

The model of the global sulphur cycle (MGSC) proposed here is a unit of a GSM with inputs and outputs compatible with other units of the global model. In contrast to hydrogen, sulphur compounds cannot be attributed to long-lived elements of the biosphere. Therefore, in the unit of sulphur the spatial digitization of its natural and anthropogenic reservoirs should be planned to reflect the local distributions of sulphur in the vicinity of its sources and to enable one to estimate the intensities of the inter-regional fluxes of sulphur compounds. The version of the sulphur unit proposed here, in contrast to the known hydrodynamic models of long-distance transport, takes into account the fluxes of sulfur compounds between the hydrosphere, atmosphere, soil, and biota. The model does not consider the vertical stratification of the atmosphere. The characteristics of sulphur fluxes over land and oceans averaged vertically are calculated. The spatial digitization of the biosphere and the World Ocean corresponds to the criterion inherent to the GSM. The block-scheme of the model of the biogeochemical cycle of sulphur is shown in Figure 8.9, a description of the fluxes of sulphur compounds is given in Table 8.8. This scheme is realized in every cell Ω_{ij} of the Earth's surface and in every compartment Ω_{ijk} of the

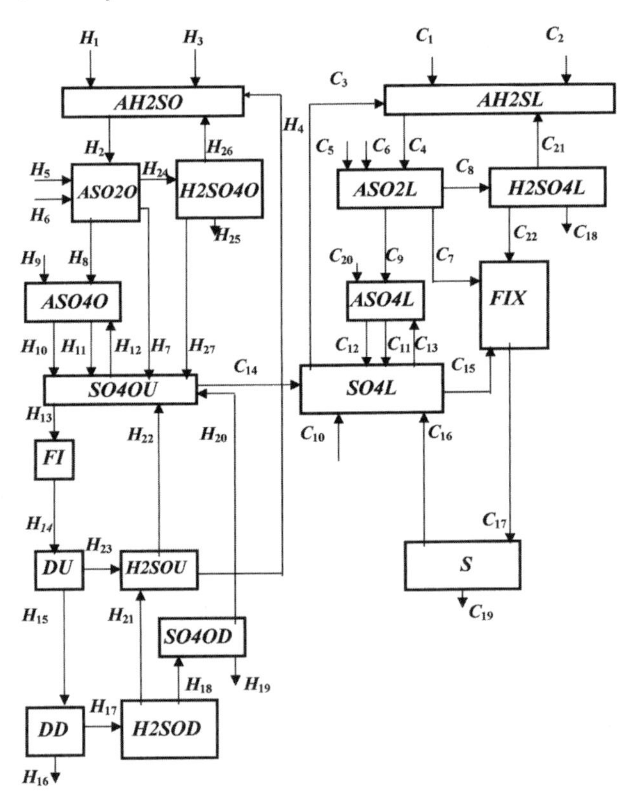

Figure 8.9. The scheme of sulphur fluxes in the environment considered in the MGSC. Notations are given in Table 8.8.

Table 8.8. Characteristics of the land and hydrospheric fluxes of sulphur shown in Figure 8.9. Assessments of fluxes ($mg\,m^{-3}\,day^{-1}$) obtained by averaging over the respective territories [58].

Sulphur flux	Land		Hydrosphere	
	Identifier	Estimate	Identifier	Estimate
Volcanic eruptions				
H_2S	C_1	0.018	H_3	0.0068
SO_2	C_5	0.036	H_5	0.0073
SO_4^2	C_{20}	0.035	H_9	0.0074
Anthropogenic emissions				
H_2S	C_2	0.072	H_1	0.00076
SO_2	C_6	0.92	H_6	0.038
SO_4^2	C_{10}	0.47		
Oxidation of H_2S to SO_2	C_4	1.13	H_2	0.3
Oxidation of SO_2 to SO_4^2	C_9	1.35	H_8	0.16
Dry sedimentation of SO_4^2	C_{12}	0.37	H_{11}	0.11
Fall-out of SO_4^2 with rain	C_{11}	1.26	H_{10}	0.38
Biological decomposition and emission of H_2S into the atmosphere	C_3	1.03	H_4	0.31
Assimilation of SO_4^2 by biota	C_{15}	0.41	H_{13}	1.09
Biological decomposition and formation of SO_4^2	C_{16}	1.13	H_{17}	0.43
			H_{23}	0.12
Sedimentation and deposits	C_{18}	0.22	H_{15}	0.98
	C_{19}	0.11	H_{16}	0.55
			H_{19}	0.0076
			H_{25}	0.036
Wind-driven return to the atmosphere	C_{13}	0.25	H_{12}	0.33
Replenishing sulphur supplies due to dead biomass	C_{17}	0.86	H_{14}	1.1
Assimilation of atmospheric SO_2	C_7	0.46	H_7	0.18
Washing out of SO_2 from the atmosphere	C_8	0.27	H_{24}	0.061
River runoff of SO_4^2 to the ocean	C_{14}	1.17		
Transformation of gas-phase H_2SO_4 to H_2S	C_{21}	0.018	H_{26}	0.0076
Assimilation of the washed out part of atmospheric SO_2 by biota	C_{22}	0.036	H_{27}	0.015
Oxidation of H_2S to SO_2 in the water medium			H_{18}	0.045
			H_{22}	0.19
Advection of SO_2			H_{20}	0.38
Advection of H_2S			H_{21}	0.37

World Ocean. The interaction between the cells and the compartments is organized through the climate unit of the GSM. Therefore, the equations of the sulphur unit lack the terms reflecting the dynamic pattern of the spatial transformation of the sulfur reservoirs. With due regard to notations assumed in Figure 8.9 and in

Tables 8.8 and 8.9, the equations describing the balance relationships between the reservoirs of sulphur compounds will be written in the form:

$$\frac{dAH2SL}{dt} = C_1 + C_2 + C_3 - C_4 + C_{21} \tag{8.16}$$

$$\frac{dASO2L}{dt} = C_4 + C_5 + C_6 - C_7 - C_8 - C_9 \tag{8.17}$$

$$\frac{dASO4L}{dt} = C_9 + C_3 + C_{20} - C_{11} - C_{12} \tag{8.18}$$

$$\frac{dS}{dt} = C_{17} - C_{16} - C_{19} \tag{8.19}$$

$$\frac{dSO4L}{dt} = C_{10} + C_{11} + C_{12} + C_{16} - C_3 - C_{13} - C_{14} \tag{8.20}$$

$$\frac{dFIX}{dt} = C_7 + C_{15} - C_{17} + C_{22} \tag{8.21}$$

$$\frac{dH2SO4L}{dt} = C_8 - C_{18} - C_{21} - C_{22} \tag{8.22}$$

$$\frac{dAH2SO}{dt} = H_1 + H_3 + H_4 + H_{26} - H_2 \tag{8.23}$$

$$\frac{dASO2O}{dt} = H_2 + H_5 + H_6 - H_7 - H_8 - H_{24} \tag{8.24}$$

$$\frac{dASO4O}{dt} = H_8 + H_9 + H_{12} - H_{10} - H_{11} \tag{8.25}$$

$$\frac{\partial SO4OU}{\partial t} + v_z \frac{\partial SO4OU}{\partial z} + k_z \frac{\partial^2 SO4OU}{\partial z^2} = \mathrm{H}_7 + H_{10} + H_{11} + H_{20} + H_{22} + H_{27} + C_{14} - H_{12} - H_{13} \tag{8.26}$$

$$\frac{\partial H2SOU}{\partial t} + v_z \frac{\partial H2SOU}{\partial z} + k_z \frac{\partial^2 H2SOU}{\partial z^2} = H_{21} + H_{23} - H_4 - H_{22} \tag{8.27}$$

$$\frac{\partial H2SOD}{\partial t} + v_z \frac{\partial H2SOD}{\partial z} + k_z \frac{\partial^2 H2SOD}{\partial z^2} = H_{17} - H_{18} - H_{21} \tag{8.28}$$

$$\frac{\partial SO4OD}{\partial t} + v_z \frac{\partial SO4OD}{\partial z} + k_z \frac{\partial^2 SO4OD}{\partial z^2} = H_{18} - H_{19} - H_{20} \tag{8.29}$$

$$\frac{\partial DU}{\partial t} + v_z \frac{\partial DU}{\partial z} + k_z \frac{\partial^2 DU}{\partial z^2} = H_{14} - H_{15} - H_{23} \tag{8.30}$$

$$\frac{\partial DD}{\partial t} + v_z \frac{\partial DD}{\partial z} + k_z \frac{\partial^2 DD}{\partial z^2} = H_{15} - H_{16} - H_{17} \tag{8.31}$$

$$\frac{\partial FI}{\partial t} + v_z \frac{\partial FI}{\partial z} + k_z \frac{\partial^2 FI}{\partial z^2} = H_{13} - H_{14} \tag{8.32}$$

$$\frac{\partial BOT}{\partial t} = H_{16} + H_{19} \tag{8.33}$$

where v_z is the velocity of the vertical water motion in the ocean, m day^{-1}; and k_z is the coefficient of the turbulent mixing, m^2 day^{-1}.

Table 8.9. Some estimates of the sulphur reservoirs that can be used as initial data.

Reservoir	Identifier in equations (8.16)–(8.33)	Quantitative estimate of the sulphur reservoir ($mg\,m^{-2}$)
The atmosphere over the oceans		
H_2S	AH2SO	10
SO_2	ASO2O	5.3
SO_4^2	ASO4O	2
The atmosphere over land		
H_2S	AH2SL	36.9
SO_2	ASO2L	17.9
SO_4^2	ASO4L	12.9
Land		
SO_4^2	SO4L	11.2
biomass	FIX	600
soil	S	5,000
Photic layer of the World Ocean		
H_2S	H2SOU	1.9
SO_4^2	SO4OU	19×10^7
biomass	FI	66.5
MOB	DU	730
Deep layers of the World Ocean		
H_2S	H2SOD	2×10^6
SO_4^2	SO4OD	3.4×10^9
MOB	DD	13,120

Equations (8.16) through (8.33) in each cell of the spatial division of the ocean surface are supplemented with initial conditions (Table 8.9). The boundary conditions for equations (8.26) to (8.33) are zero. The calculation procedure to estimate the sulphur concentration consists of two stages. At each time moment *ti*, first for all cells Ω_{ij}, equations (8.16)–(8.33) are solved by the method of quasilinearization, and all reservoirs of sulphur are estimated for $t_{i+1} = t_i + \Delta t$, where a time step Δt is chosen from the condition of convergence of the calculation procedure. Then at the moment t_{i+1}, with the use of the climate unit of the global model, these estimates are specified with account of the atmospheric transport and ocean currents over the time Δt.

The sulphur supplies in the reservoirs are measured in $mgS\,m^{-3}$, and the sulphur fluxes have the dimensionality $mgS\,m^{-3}\,day^{-1}$. The sulphur supplies in the water medium are calculated with the volumes of the compartments Ω_{ijk} taken into account. To estimate the sulphur supplies in the atmosphere, it is assumed that an effective thickness of the atmosphere h is an input parameter either introduced into the model by the user or prescribed as the constants from Table 8.9, or obtained from the climate unit of the global model.

The quantitative estimates of the fluxes in the right-hand parts of equations (8.16)–(8.33) are obtained in different units of the global model. The anthropogenic fluxes of sulphur H_1, H_6, C_2, C_6, and C_{10} are simulated in the unit of scenarios. The fluxes H_3, H_5, H_9, C_1, C_5, and C_{20} are prescribed either by the climate unit or formed in the unit of scenarios. The GSM user controls these processes in correspondence to the CON unit directions. The numerical experiments mentioned here used estimates of these fluxes that were published in the literature. The accuracy of different functional presentations of the fluxes in the equations (8.16)–(8.33) correspond to the accuracy of similar fluxes of the biogeochemical cycles of hydrogen, phosphorus, and nitrogen.

The rate of emission of H_2S into the atmosphere due to humus decomposition is described by the linear function $C_3 = \mu_1(\mathrm{pH}) \times \mathrm{SO}_4L \times T_L$, where μ_1 is the proportion coefficient depending on soil acidity, $\mathrm{day}^{-1} \times \mathrm{K}^{-1}$; and T_L is the soil temperature, °K. The initial value of SO_4L in equation (8.20) is estimated proceeding from the humus supply, considering that the content of sulphur in humus is prescribed by the parameter a_g, %. According to the available observations of the input of H_2S into the atmosphere from the ocean, the flux H_4 varies widely from low values to high values at the transition from stagnant waters to zones of upwelling. The flux H_4 is assumed to be the function of the ratio of the rates of H_2S oxidation in the photic layer to the rate of the vertical uplifting of water. Therefore, to describe the H_4 flux, the parameter t_{H2SU} is used, which reflects the lifetime of H_2S in the water medium: $H_4 = \mathrm{H_2SU}/t_{H2SU}$. We determine the value of t_{H2SU} as a function of the rate of the vertical advection v_z and concentration of oxygen O_2 in the upper layer Z_{H2S} thick: $t_{H2SU} = \mathrm{H_2SOU} \times v_z(\theta_2 + \mathrm{O}_2)/[\mathrm{O}_2(\theta_1 + v_z)]$, where the constants θ_1 and θ_2 are determined empirically, the value of O_2 is calculated by the oxygen unit of the global model.

The reaction of oxidation of H_2S to SO_2 in the atmosphere over land and over the water surface is characterized by a rapid process of the reaction of hydrogen sulphide with atomic and molecular oxygen. The lifetime of H_2S in this process constitutes 32 days. At the same time, the reaction of H_2S with O_3 in the gas phase is slow. It is impossible to simulate, within the global model, the diversity of the situations appearing here. However, an inclusion of the fluxes H_2 and C_4 into the unit of sulphur enabled one to take into account the correlation between the cycles of sulphur and oxygen. These fluxes are parameterized with the use of the indicator t_{H2SA} of the lifetime of H_2S in the atmosphere: $C_4 = A\mathrm{H_2S}L/t_{H2SA}$, $H_2 = A\mathrm{H_2SO}/t_{H2SA}$. The mechanism to remove SO_2 from the atmosphere is described by the fluxes H_7, H_8, H_{27}, C_7, C_8, and C_9. Schematically this mechanism consists of a set of interconnected reactions of SO_2 with atomic oxygen under the influence of various catalysts. A study of the succession of the reaction enables one to estimate the lifetime of SO_2 for oxidation over land t_{SO2L} and the water surface t_{SO2A1}. This makes it possible to assume the following parameterizations of the fluxes H_8 and C_9: $H_8 = A\mathrm{SO_2O}/t_{SO2A1}$ and $C_9 = \mathrm{ASO}_2L/t_{SO2L}$.

Sulphur dioxide is assimilated from the atmosphere by rocks, vegetation, and other Earth cover. Over the water surface this assimilation is connected with the intensity of turbulent gas fluxes and surface roughness. We describe a dry deposition of SO_2 over vegetation by the model $C_7 = q_2RX$, where $q_2 = q_2' \times A\mathrm{SO}_2L/(r_{tl} + r_s)$; r_{tl} is the atmospheric resistance to SO_2 transport over vegetation of l type, $\mathrm{day\,m}^{-1}$; r_s is the surface resistance of s type to SO_2 transport, day m^{-1}; RX is the product of vegetation of X type, $mg\,m^{-2}\,day^{-1}$ (calculated by the biogeocenotic unit of the global model for 30 types of vegetation); and q_2' is the proportion coefficient. The parameters r_{tl} and r_s are functions of the types of soil–vegetation formations estimated, respectively, at 0.05 and 4.5 for forests, 0.9 and 3 for grass cover, 0.5 for bushes, 0.8 and 1 for bare soils, 1.9 and 0 for water surface, and 2 and 10 for snow cover.

The process of washing out of SO_2 from the atmosphere with changing phase to H_2SO_4 and a subsequent neutralization on the surface of l type is described by the function: $C_8 = q_{1l}W \times ASO_2L$ with the Langmuire coefficient q_{1l} and precipitation intensity $W(\varphi, \lambda, t)$.

An interaction of acid rain with Earth surface elements is reflected in Figure 8.9 by the fluxes C_{18}, C_{21}, and C_{22} for land and H_{25}, H_{26}, and H_{27} for water surface. To parameterize these fluxes, we assume the hypothesis that the reservoirs of H_2SO_4L and H_2SO_4O are spent in proportion to the outfluxes, and the coefficients of this proportion are the controlling parameters of the numerical experiments: $C_{18} = h_1 \times H_2SO_4L$, $C_{22} = h_2 \times RX \times H_2SO_4L$, $C_{21} = h_3T_a \times H_2SO_4L$, $H_{25} = h_6 \times H_2SO_4O$, $H_{26} = h_4T_a \times H_2SO_4O$, $H_{27} = h_5 \times RFI \times H_2SO_4O$, $h_1 + h_2 \times RX + h_3T_a = 1$, $h_4T_a + h_5 \times RFI + h_6 = 1$, where $T_a(\varphi, \lambda, t)$ is the air surface temperature.

We parameterize the fluxes H_7 and H_{24} by the relationships: $H_7 = ASO_2O/t_{SO2A2}$ and $H_{24} = q_{1l}W \times ASO_2O$ where t_{SO2A2} is the lifetime of SO_2 over the water surface, in days.

Sulphates interacting with the ecosystems and establishing an interaction of the sulphur cycle with other biogeochemical processes are one of the most important elements in the global cycle of sulphur. Numerous complicated transformations of sulphates in the environment are described by a set of fluxes H_7, H_8, H_{10}, H_{11}, H_{12}, C_9, C_{11}, C_{12}, and C_{13} for the atmospheric reservoir and the fluxes H_{13}, H_{18}, H_{19}, H_{20}, H_{22}, C_3, C_{14}, C_{15}, and C_{16} for land and the World Ocean.

Physical mechanisms for the transport of sulphates from the atmosphere to the soil and water medium are connected with dry and wet sedimentation. An efficient model of the wet removal of particles and gases from the atmosphere was proposed by [63]: a substitution of the mechanism of the aerosols and gases by a simplified binary model enables one to match it with other units of the global model: $H_{10} = \mu W \times ASO_4O$, $H_{11} = \rho v_O \times ASO_4O$, $C_{11} = b_3W \times ASO_4L$, $C_{12} = d_1v_a \times ASO_4L$, where v_O and v_a are the rates of aerosol dry deposition over the water surface and land, respectively; and μ, b_3, ρ, and d_1 are constants.

The return of sulphates from the soil and water medium to the atmosphere is connected with rock weathering and spray above the rough water surface: $C_{13} = d_2 \times \mathrm{RATE} \times SO_4L$ and $H_{12} = \theta \times \mathrm{RATE} \times SO_4U$, where $\mathrm{RATE}(\varphi, \lambda, t)$ is the wind speed over the surface, $m\,s^{-1}$; and d_2 and θ are the empirical coefficients.

The flux C_{14} relates the surface and water reservoirs of sulphur. Let σ be the share of the river system area on land and d_3 – the proportion coefficient, then $C_{14} = d_3W \times SO_4L + (C_{11} + C_{12})\sigma$.

The surface part of the sulphur cycle is connected with the functioning of the atmosphere–vegetation–soil system. Plants adsorb sulphur from the atmosphere in the form of SO_2 (fluxes C_7 and C_{22}) and assimilate sulphur from the soil in the form of SO_4^{2-} (flux C_{15}). In the hierarchy of the soil processes two levels can be selected defining the sulphur reservoirs as 'dead organics' and 'SO_4^{2-} in soil'. The transitions between them are described by the flux $C_{16} = b_2 ST_L$, where the coefficient $b_2 = b_{2,1} b_{2,2}$ describes the rate $b_{2,1}$ of transition of sulphur contained in dead organics into the form assimilated by vegetation. The coefficient $b_{2,2}$ indicates the content of sulphur in dead plants.

The fluxes of sulphur in the water medium according to studies by [11] depend on the biological processes in the water bodies and constitute an isolated part of the global cycle of sulphur that contains only the fluxes that connect it with the atmospheric and surface cycles. Rough estimates show that the rates of the sulphur cycle in the seas and oceans do not play a substantial role for the remaining parts of its global cycle. Although this specific feature does exist, for the purity of the numerical experiment, in the proposed model the internal hydrospheric fluxes of sulphur compounds are separated in space and parameterized with the same details as other fluxes of sulphur in the atmosphere and on land. This excessiveness is important for other units of the GSM as well. In particular, it is important for the parameterization of photosynthesis whose rate RFI affects the closure of other biogeochemical cycles. Finally, assume: $H_{13} = \gamma \times RFI$, $H_{14} = b \times MFI$, $H_{15} = f \times DU$, $H_{16} = p \times DD$, $H_{17} = q \times DD$, $H_{18} = \mathrm{H_2}SOD/t_{H2SOD}$, $H_{19} = u \times \mathrm{SO_4}D$, $H_{20} = a_1 v_D \times \mathrm{SO_4}D$, $H_{21} = b_1 v_D \times H_2\mathrm{SO}D$, $H_{22} = H_2SOU/t_{H2SOU}$, $H_{23} = g \times DU$, where MFI is the mass of dead phytoplankton; t_{H2SOU} and t_{H2SOD} are the times of complete oxidation of H_2S in the seawater in the photic and deep layers, respectively; and $\gamma, b, f, p, q, u, a_1, b_1$, and g are constants.

The anthropogenic input to the sulphur unit is taken into account through the prescribed fluxes C_2, C_6, C_{10}, C_1, and C_6 in the form of the functions of spatial coordinates and time.

8.7.4 Modelling results

The GSM sulphur unit, like all units of biogeochemical cycles, has a similar level of accuracy of parameterization to other GSM units, and therefore there is no deregulation of the global model, and the stability of results of simulation experiments is ensured. To check this stability, we have undertaken some numerical experiments, taking the parameters of the sulphur unit from Table 8.10 and assuming $\Delta\varphi = 4°$, $\Delta\lambda = 5°$, $\Delta z_1 = 10\,\mathrm{m}$, and $\Delta z_2 = 100\,\mathrm{m}$. Results of modelling shown in Figures 8.3 and 8.4 enable one to conclude that sufficiently sharp changes in the initial conditions (±70%) and variations in parameters (±25%) affect weakly the system's dynamics. The distribution pattern in Figure 8.10 establishes itself, on average, over 1.5–2 years. As follows from Figure 8.11, an enhancement of initial sulphur supplies affects the system's dynamics during the first two years, but a decrease of initial sulphur supplies extends this interval up to five years.

The curves in Figure 8.12 characterize the dependence of acid rain on the level of anthropogenic activity. Calculations have shown that the pH value of precipitation stabilizes, on average, within 30 days from the moment of change in the regime of

Table 8.10. Qualitative estimates of the GSM sulphur unit parameters used in numerical experiments.

Parameter	Estimate	Parameter	Estimate
C_1	$6\,mg \times m^{-2}\,day^{-1}$	d_1	$73.379\,m^{-1}$
t_{SO2L}	13 days	d_2	$0.02\,m^{-1}$
h	3 km	f	$1.34 \times 10^{-3}\,day^{-1}$
v	$0.789 \times 10^{-11}\,m^{-1}$	g	$0.22 \times 10^{-4}\,day^{-1}$
p	$4.22 \times 10^{-5}\,day^{-1}$	C_6	$300\,mg \times m^{-2}\,day^{-1}$
v_D	$0.036\,m\,day^{-1}$	v_z	$1\,m\,day^{-1}$
H_3	$5\,mg \times m^{-2}\,day^{-1}$	$\mu_1(pH < 6.3)$	$1.02 \times 10^{-4}\,day^{-1}\,K^{-1}$
q_{1l}	0.9	t_{H2SA}	32 days
μ	$40\,m^{-1}$	t_{SO2A2}	27.9 day
γ	$0.483 \times 10^{-3}\,m^{-1}$	ρ	$1.655 \times 10^{-4}\,mm^{-1}$
u	$2.23 \times 10^{-12}\,day^{-1}$	b	$0.008\,8\,day^{-1}$
H_6	$35\,mg \times m^{-2}\,day^{-1}$	a_1	$3.1 \times 10^{-10}\,m^{-1}$
C_5	$8\,mg \times m^{-2}\,day^{-1}$	C_2	$28\,mg \times m^{-2}\,day^{-1}$
$\mu_1(pH \geq 6.3)$	$1.02 \times 10^{-3}\,day^{-1}\,K^{-1}$	t_{SO2A1}	5.6 day
a_g	0.35%	t_{H2SU}	4 days
r_s	$0.001\,2\,day\,m^{-1}$	d_3	$4.6\,m^{-1}$
b_3	$88.767\,m\,m^{-1}$	q	$3.24 \times 10^{-5}\,day^{-1}$
b_{22}	0.13%	t_{H2SOU}	10 days
Z	$2.28 \times 10^{-8}\,day^{-1}$	H_5	$4\,mg \times m^{-2}\,day^{-1}$
H_1	$0.7\,mg \times m^{-2}\,day^{-1}$	r_{tl}	0.58×10^{-4}–$0.23 \times 10^{-2}\,day\,m^{-1}$
C_{10}	$0.18\,mg \times m^{-2}\,day^{-1}$		

anthropogenic emissions of sulphur. The spatial distribution of pH of rain with an even increase of the fluxes of C_2, C_6, C_{10}, H_1 and H_6 by 0.2% yr^{-1} does not markedly change over 3 years. The ratio of acid rain rates for latitudinal bands 70°–90°N and 70°–90°S remains, on average, at a level of 2.5.

Of interest is an estimate of the contribution of various regions to Arctic pollution. With the stable state of the mean annual concentration of gas-phase H_2SO_4 in the Arctic atmosphere assumed to be 100%, the shares of participation in the formation of this level are as follows: the USA – 17%, Canada – 21%, Europe – 37%, and the eastern territory of Russia – 25%. These estimates correlate with the structure of anthropogenic sulphur fluxes (Table 8.9) prescribed by the scenario of anthropogenic activity and determined by the relationship of all parameters of the global sulphur cycle. Table 8.11 gives some estimates which suggest a conclusion about the stabilizing role of the World Ocean in the formation of the global distribution of sulphur. As seen from Figure 8.13, the intensity of the biological decomposition in water is distributed non-uniformly, and this means that hydrocarbon production in the oceans is a function of the vertical structure of their ecosystems. For instance, in the Indian Ocean there are two distinct maxima of H_2S production. A weak second maximum of hydrocarbon appears at depths ~1.5 km in the Atlantic Ocean. In other oceans, one maximum of the vertical distribution of H_2S is observed.

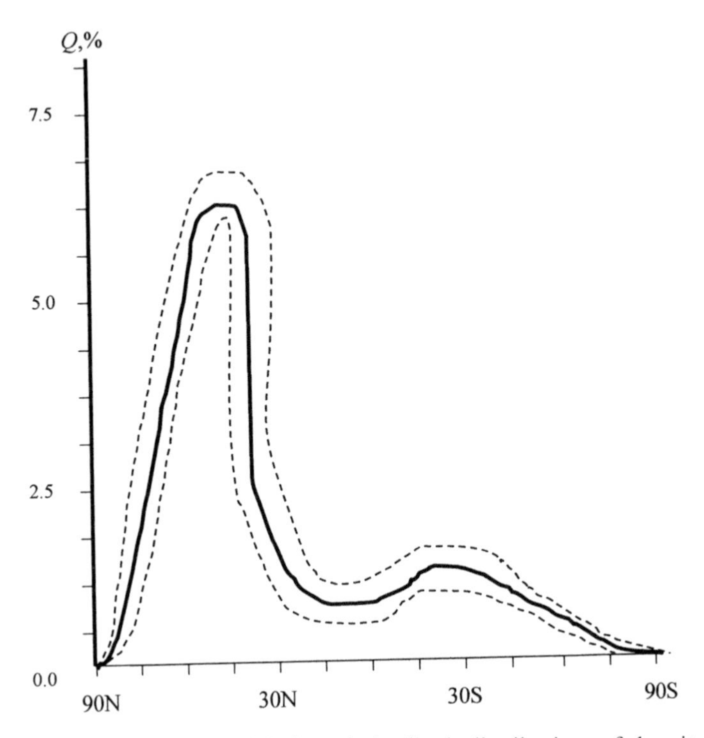

Figure 8.10. Variations in the equilibrium latitudinal distribution of longitude-summed sulphur supplies Q in the environment (%) to total sulphur reserve in the biosphere with initial conditions changing up to ±70% and parameters up to ±25%. Dashed lines denote the boundaries of variations. Solid line corresponds to parameters in Table 8.8 and initial data from Table 8.9.

Of course, calculations similar to those given in Figures 8.10–8.13 are only preliminary. Nevertheless, they enable one to assess the trends in changes of sulphur concentration in the atmosphere depending on real and anthropogenic processes in a given territory. As an example, we consider the territory of Vietnam, where the recent rates of industrial growth have increased, and the problem of atmospheric pollution has reached a national scale. From the data of the Vietnam National Centre of the Environment, the level of atmospheric pollution over its territory is determined by the background and local emissions of chemicals in the zones of large cities. For instance, in the cities of Ho Chi Minh and Wung Tau (Baria Province) the concentration of SO_2 reaches levels of 1.84 and 0.99 $\mu g\, m^{-3}$, respectively [15]. A considerable share of atmospheric pollutants constitute dust particles whose sources are factories producing building materials, the zones of construction themselves, and industrial enterprises. In the region of Thi Wai (Ding Nai Province) are concentrated: factories producing building materials as well as chemical, fuel–power, and food industries – creating a concentrated load on the environment. With emphasis only on sulphur compounds, we use the MGSC and ESPhAP for a complex assessment of the sulphur cycle over the territory of Vietnam.

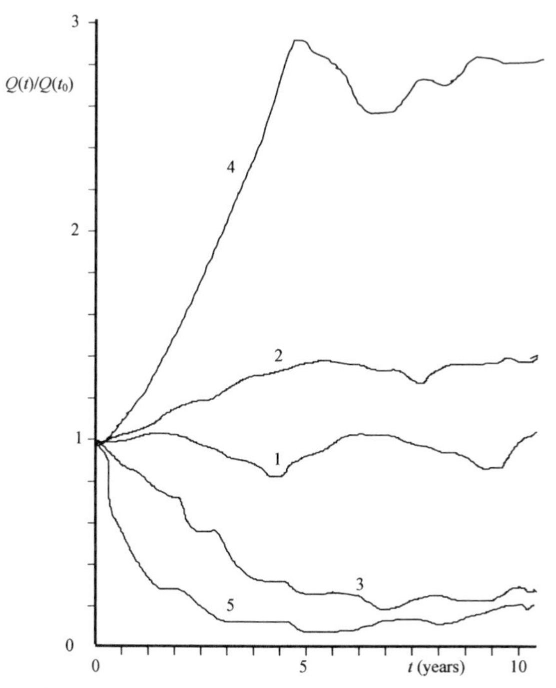

Figure 8.11. Dependence of dynamics of sulphur concentrations $Q(t)/Q(t_0)$, normalized to initial conditions averaged over Ω on initial conditions: 1-initial conditions correspond to data in Table 8.9; 2-reduced by 50%; 3-increased by 50%; 4- reduced by 70%; 5-increased by 70%. Parameters given in Table 8.8 were not changed.

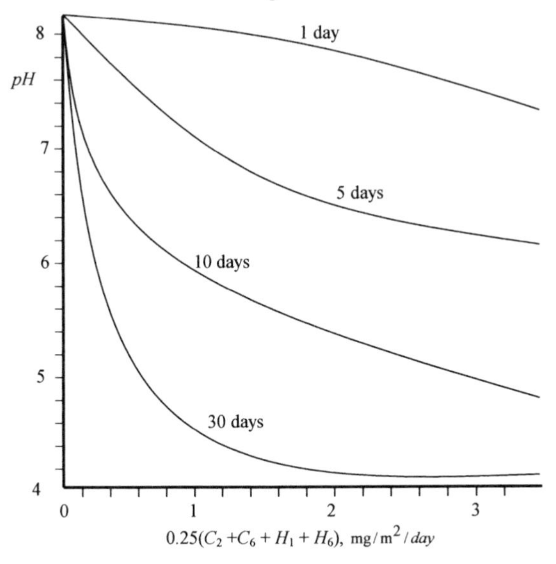

Figure 8.12. Dependence of average acidity of rain on anthropogenic sulphur fluxes. Change of anthropogenic activity is assumed to be homogeneous in all territories. The curves are marked with the time after the beginning of the experiment. pH is calculated with the formula pH $= \lg H^+$.

Table 8.11. Results of numerical experiments. Calculations of some characteristics of the global biogeochemical cycles within the GSM.

Characteristic	Result of numerical experiment	
	Northern hemisphere	Southern hemisphere
Total rate of sulphur deposition onto the World Ocean bottom ($mg\,m^{-2}\,day^{-1}$)	31.2	24.5
Sulphur transport from the oceans to land ($mg\,m^{-2}\,day^{-1}$)	74	21
Sulphur transport from land to the oceans ($mg\,m^{-2}\,day^{-1}$)	185	41
Deposition of sulphur onto a surface ($mg\,m^{-2}\,day^{-1}$)		
land (anthropogenic share, %)	3.5 (31)	2.7 (26)
ocean (anthropogenic share, %)	1.4 (18)	0.9 (14)
Sulphur flux absorbed by the oceans from the atmosphere ($mg\,m^{-2}\,day^{-1}$)	0.2	0.2
Time to half the SO_2 content in the atmosphere (days)	0.04	0.04
Share of anthropogenic sulphur in river runoff (%)	48	27
Rate of sulphur accumulation in soil–plant system formations ($mg\,m^{-2}\,day^{-1}$) for:		
forest	34.6	41.3
bush	2.7	3.4
grass	4.1	5.6
Sulphur sink from the continents to the World Ocean ($mg\,m^{-2}\,day^{-1}$) for:		
North America	21.9	–
South America	2.3	15.4
Europe and Asia	45.5	4.1
Africa	14.2	5.4
Australia		2.8

The anthropogenic constituent is determined from data in Table 8.12. Assume that the synoptic situation is of binary character: dry and rainy seasons. It follows from the modelling results that at the end of the rainy season the content of sulphur in the atmosphere decreases – with time it becomes more stable with regards to the distribution of sulphur in the atmosphere. There are no effects of sulphur accumulation. Figure 8.14 shows the curve of SO_2 concentration stabilization obtained with boundary effects left out of account. The scheme in Figure 8.15 shows the effect of the types of surface cover on the SO_2 content in the atmosphere.

The formation of the pH of rain over the territory of Vietnam is strongly affected by sulphur fluxes from the territories of adjacent countries (China, Laos, Kampuchea, etc.). Figure 8.16 demonstrates the $SO_2 + NO_2$ field formed over the territory of Vietnam on the assumption that outside this territory the level of atmospheric aerosol concentration over land is constant (0.8–$1.0\,\mu g\,m^{-3}$), and over the

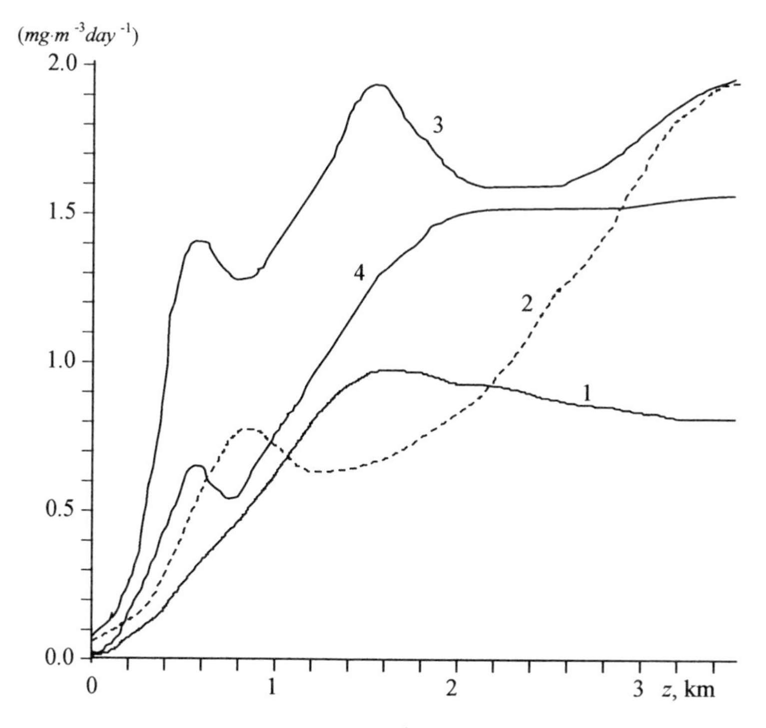

Figure 8.13. Average production of H_2S (mg m^{-3} day^{-1}) in the oceans: 1 – the Arctic Ocean; 2 – the Pacific Ocean; 3 – the Indian Ocean; 4 – the Atlantic Ocean.

Table 8.12. Estimates of intensities of atmospheric pollutant emissions in the territory of Vietnam.

	Intensity of pollutant emission (t km^{-2} yr^{-1})		
Source (city)	SO_2	NO_2	Dust
Hanoi	0.34	0.31	0.42
Ho Chi Minh	0.27	0.23	0.47
Wungtau	0.05	0.04	0.13
Danang	0.06	0.05	0.12
Haiphong	0.08	0.07	0.11

territory of Vietnam in January both SO_2 and NO_2 were absent. It is also assumed that the rainy season occurs between February–March, that until October the rainy season is moderate, and that the remaining time of the year there is no precipitation. Meteorological conditions are prescribed from the data of the National Meteorological Service with averaging over the last three years. As seen from the schematic map (Figure 8.16), the level of atmospheric pollution in the centre of northern Vietnam and along the whole coast below 16°N reaches 0.2–0.4 μg m^{-3} – this is

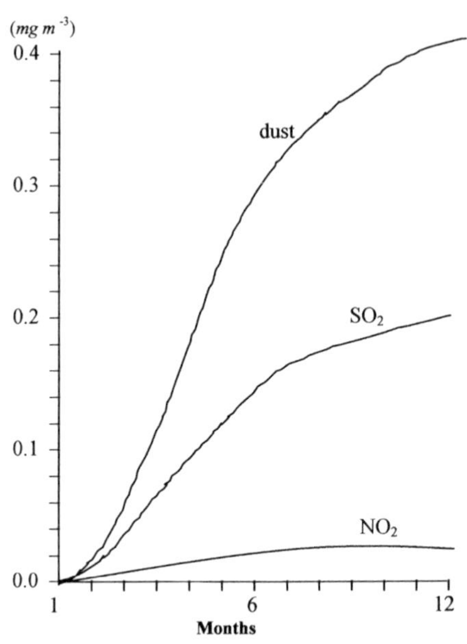

Figure 8.14. Dynamics of the average concentrations of SO_2, NO_2, and dust ($mg\,m^{-3}$) in the atmosphere over the territory of Vietnam. Initial data for January are assumed to be zero.

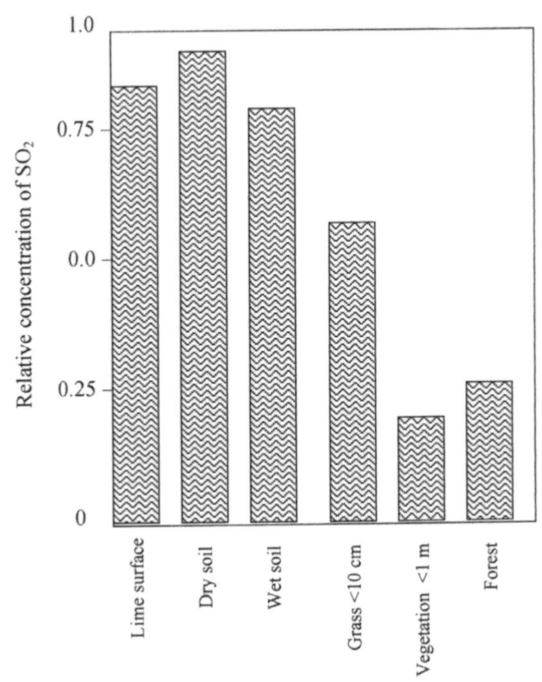

Figure 8.15. Calculated dependence of the ratio of the current SO_2 concentration to its maximum value on the type of surface cover for the territory of South Vietnam (Baria Province).

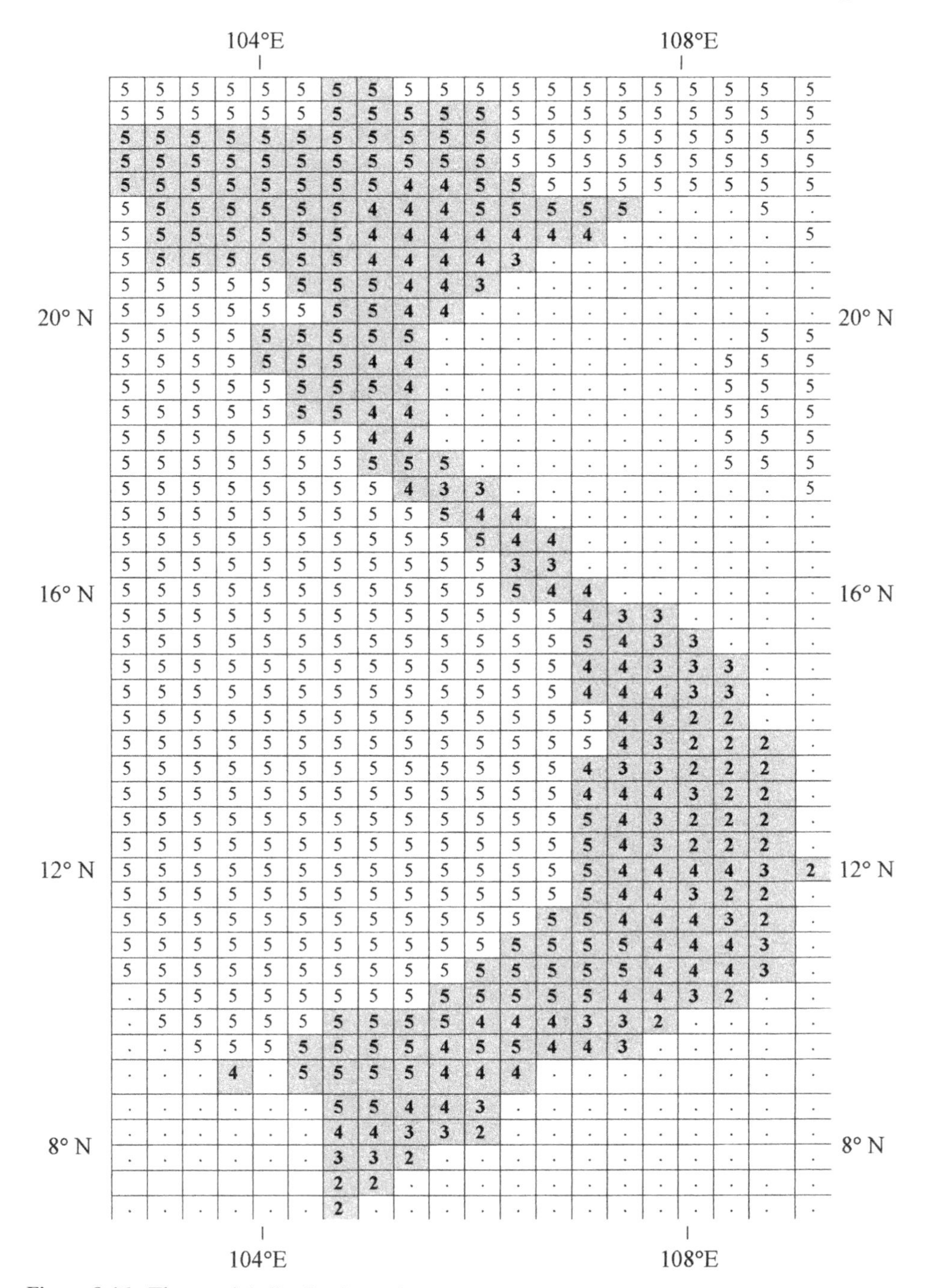

Figure 8.16. The spatial distribution of the impact of transboundary transports of sulphur over the territory of Vietnam. Scale (mg m^{-3}): 0 – <0.02; 1 – 0.02–0.2; 2 – 0.2–0.4; 3 – 0.4–0.6; 4 – 0.6–0.8; 5 – 0.8–10, etc.

due to transboundary transport only. In other parts of the territory this level increases reaching a maximum of pollution threshold. Similar calculations can be made for any global territory.

The problem of atmospheric pollution with sulphur compounds is vital for Europe. Knowledge of spatial distributions of sulphur over the European countries is also important for adjacent territories and especially for the northern territories of the Arctic basin. One of the examples of the use of models to solve the problem of assessment of the atmosphere over Europe is the calculation of significant components of atmospheric pollution described in [128] using the LRTAP (Long-Range Transport of Air Pollution) model of the Euler type. This model simulates 2-D distributions of atmospheric aerosols over the territory (including Europe) with the adjacent regions of the Atlantic Ocean, Asia, and Africa. Calculation is made over the 150 × 150-km geographical grid with a monthly time step. The model considers advection, diffusion, both types of deposition, emissions, and chemical reactions. The coefficients of diffusion are assumed to be constant and equal to $3 \times 10^4\,m^2\,s^{-1}$. The process of dry deposition of aerosols is parameterized by the annual dependence reflecting its day–night binary variability. For SO_2, the rate of dry deposition in the daytime is $0.4\,cm\,s^{-1}$ in January, $0.6\,cm\,s^{-1}$ in February and March, and $0.8\,cm\,s^{-1}$ at other times of the year. The rate of sulphur emission is estimated by the law of cosine, with a minimum of $0.65\,E_{mean}$ and a maximum of $1.35\,E_{mean}$, where E_{mean} is the annual mean estimate of sulphur flux from all sources located in a concrete cell of the territory digitization. It is assumed that 80% of the sulphur flux consists of SO_2 and 5% – SO_4^{2-}. The remaining 15% S comes from outside due to transboundary transport. The calculation of the seasonal variability of SO_2 and SO_4^{2-} concentration is exemplified in Figure 8.17. The SO_2 concentration ratios over the European countries is given in Figure 8.18.

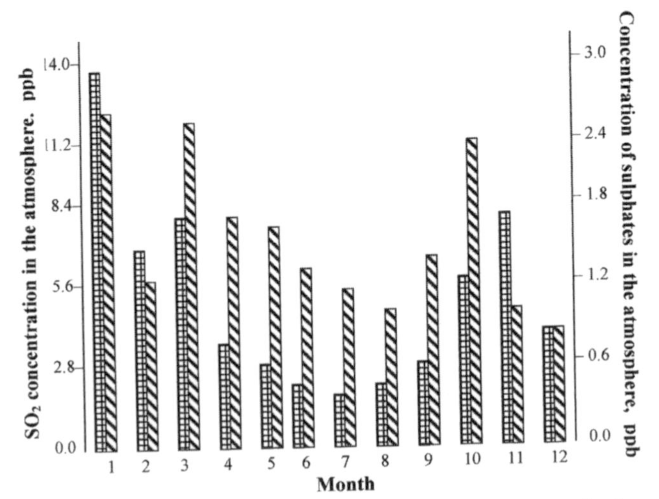

Figure 8.17. Seasonal variability of the area-averaged sulphur content in the atmosphere over the territory of western Europe calculated with the use of the LRTAP model as an ESPhAP unit.

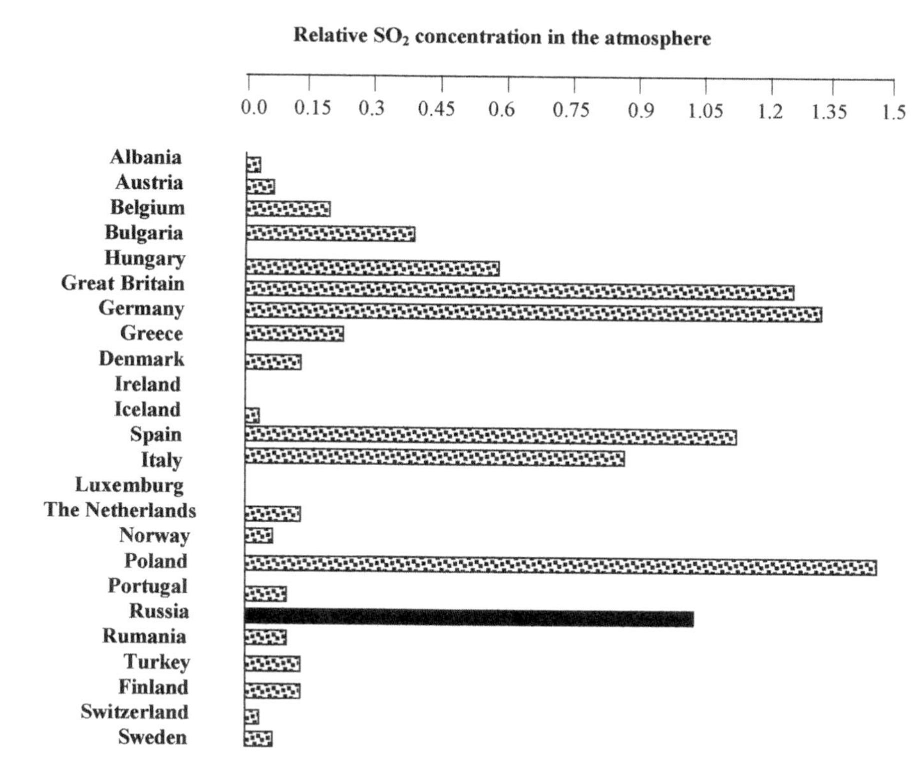

Figure 8.18. Distribution of SO_2 concentration in the atmosphere between some countries of western Europe calculated with respect to Russia.

8.8 CONCLUSION

The long-range transport of aerosol (especially transoceanic) determines various impacts of aerosol on the environment and climate on regional and global scales. In this context, of importance is high-latitude atmospheric pollution due to long-range transport of anthropogenic aerosol and MGCs, which has not been considered here to avoid repetition [10, 53]. The same explanation is true for the briefness of the discussion of the problem of long-range transport of biomass burning products and anthropogenic emissions to the Indian Ocean basin discussed earlier [50, 51].

Summarizing, one should emphasize an inadequacy of available information about the long-range transport of aerosol, which necessitates further development of both the observational systems (with special attention to lidar and satellite remote sensing) and numerical simulation of the global dynamics of atmospheric aerosol. It is clear that a reliable assessment of the means of aerosol propagation is possible with the combined use of the respective models and data from satellite monitoring [46, 51].

The above numerical modelling has shown that a reliable assessment of the spatial distributions of pH levels depends on many factors, among which the accuracy of parameters in equations of the MGSC unit and the form of its referencing to GSM are important and decisive. The very complicated means of parameterization of sulphur fluxes in space introduces great uncertainties into numerical simulations. As has been shown in [78], the use for this purpose of the Gauss-type models with different modifications makes it possible to obtain a highly accurate description of the spatial structure of sulphur fluxes with a minimum set of input parameters. Taking the four power stations located in the western part of Texas (USA) as an example, it was shown that the use of the modified Gauss-type model made it possible to solve the problem of assessment of individual contributions of point sources to the formation of pH levels with an accuracy of up to 60–80% depending on the distance and synoptic parameters.

To increase the reliability of acid rain forecasts, it is necessary to further improve the MGSC unit by filling it with new biogeochemical cycles of other chemicals, such as carbon bisulphide and sulphurous anhydride [9]. Consideration of only sulphur dioxide, sulphates, and hydrocarbons in the models of the global sulphur cycle limits the accuracy of these models. Unfortunately, most of the international and national programmes on the studies of the sulphur cycle are confined to these elements [51].

An increased role of the global model of the sulphur cycle in the assessment of the pH of rain and the calculation of water basin acidity, as well as the climatic impacts of aerosols, requires a development of perspective scenarios of anthropogenic activity such as the scenarios/strategies described in [59]. According to the Canadian Programme on acid rain, for instance, in the Ontario province, by the year 2015, the SO_2 emissions are planned to reduce by 50% (from 885 to 442.5 $\mathrm{kt\,yr^{-1}}$). Knowledge and consideration in the MGSC unit of such data will make it possible to calculate optimal strategies for all countries and give recommendations for planning sulphur emissions.

8.9 BIBLIOGRAPHY

1. Alastuey A., Querol X., Rodroguez S., Plana F., Lopez-Soler A., Ruiz C., and Mantilla E. Monitoring of atmospheric particulate matter around sources of secondary inorganic aerosol. *Atmos. Environment*, 2004, **38**(30), 4979–4992.
2. Aloyan A. E. Numerical modeling of minor gas constituents and aerosols in the atmosphere. *Ecological Modelling*, 2004, **179**(2), 163–175.
3. *2001 Annual Progress Report on the Canada-Wide Acid Rain Strategy for Post-2000.* CCME Publ., Quebec, 2002, 16 pp.
4. Ansmann A., Bösenberg J., Chaikovsky A., Camerón A., Eckhardt S., Eixmann R., Freudenthaler V., Ginoux P., Komguem L., Limné H., *et al.* Long-range transport of Saharan dust to northern Europe: The 11–16 October 2001 outbreak observed with EARLINET. *J. Geophys. Res.* , 2003, **108**(D24), AAC12/1–AAC12/15.
5. Augustine J. A., Cernwall C. R., Hodger G. B., Long C. N., Medina C. I., and De Luisi J. J. An automated method of MFRSR calibration for aerosol optical depth analysis with application to an Asian dust outbreak over the United States. *J. Appl. Meteorol.*, 2004, **42**(2), 266–278.

6. Ayras M., Niskavaara H., Bogatyrev I., Chekushin V., Pavlov V., Caritat P., Halleraker J. H., Finne T. E., Kashulina G., and Reimann C. Regional patterns of heavy metals (Co, Cr, Cu, Fe, Ni, Pb, V and Zn) and sulphur in terrestrial mossed samples as indication of airborne pollution in a 188,000 km^2 area in northern Finland, Norway and Russia. *J. of Geochem. Explor.*, 1997, **58**(2–3), 269–281.
7. Bates T. S., Lamb B. K., Guenther A., Dignon J., and Stoiber R. E. Sulfur emissions to the atmosphere from natural sources. *J. Atmos. Chem.*, 1992, **14**(1–4), 315–337.
8. Berndt T., Boge O., and Stratmann F. Atmospheric particle formation from the ozonolysis of alkenes in the presence of SO_2. *Atmos. Environment*, 2004, **38**(14), 2145–2153.
9. Binenko V. I., Khramov G. N., and Yakovlev V. V. *Emergency Situations in the Present World and Problems of Life Safety*. 'Integratsiya' Publ., St Petersburg, 2004, 398 pp. [in Russian].
10. Bobylev L. P., Kondratyev K. Ya., and Johanessen O. M. *Arctic Environment Variability in the Context of Global Change*. Springer–Praxis, Chichester, UK, 2003, 471 pp.
11. Bodenberder J., Wassmann R., Papen H., and Rennenberg H. Temporal and spatial variation of sulfur-gas-transfer between coastal marine sediments and the atmosphere. *Atmospheric Environment*, 1999, **33**(21), 3487–3502.
12. Bornstein, R. *Urban Induced Convergence Zones and Air Pollution Episodes*. Preprint Vol., Inter. Conf. cum Workshop on Air Quality Management, Brunei Darussalam,1999, 3 pp.
13. Bratus′ A. S., Meshcherin A. S., and Novozhilov A. S. Numerical models of interaction of pollution with the environment. *Herald of the Moscow State Univ.*, 2001, **15**(1), 23–28 [in Russian].
14. Brenninkmeijer C. A. M. Representative medium term observations of trace gases and aerosols in the upper troposphere and lower troposphere (CARIBIC, a Flying laboratory). *AFO Newsletter,* 2003, No. 4, 11–14.
15. Bui Ta Long. Application of GISM-technology to the numerical experiment problem of atmospheric pollution physics. *Problems of the Environment and Natural Resources*, 1998, **1**, 2–11 [in Russian].
16. Cahill C. F. Asian aerosol transport to Alaska during ACE-Asia. *J. Geophys. Res.*, 2003, **108**(D23), ACE32/1–ACE32/8.
17. Cattrall C., Carder K. L., and Gordon H. R. Columnar aerosol single-scattering albedo and phase function retrieved from sky radiance over the ocean: Measurements of Saharan dust. *J. Geophys. Res.*, 2003, **108**(D9), AAC10/1–AAC10/11.
18. Chan Lyi Kyong. A simulation system for the numerical experiment in atmosphsric pollution physics. PhD thesis. Institute of Radio Engineering and Electronics of the Russian Acad. Sci., 1994, 115 pp. [in Russian].
19. Chang L.-S., Park S.-U. Direct radiative forcing due to anthropogenic aerosols in East Asia during April 2001. *Atmos. Environment*, 2004, **38**(27), 4467–4482.
20. Chatfield R. B., Guo Z., Sachse G. W., Blake D. R., and Blake N. J. The subtropical global plume in the Pacific Exploratory Mission-Tropics A (PEM-Tropics A), PEM-Tropics B, and the Global Atmospheric Sampling Program (GASP): How tropical emissions affect the remote Pacific. *J. Geophys. Res.*, 2003, **108**(D16), ACH1/1–ACH1/20.
21. Chin M., Ginoux P., Lucchesi R., Huebert B., Weber R., Anderson T., Masonis S., Blomqvist B., Bandy A., Thornton D. A global aerosol model forecast for the ACE-Asia field experiment. *J. Geophys. Res.*, 2003, **108**(D23), ACE22/1–ACE22/17.

22. Chobadian A., Goddard A. J. H., and Gosman A. D. Numerical simulation of coastal internal boundary layer developments and a comparison with simple models. In: J. P. Lehman (ed.), *Air Pollution Modeling and Its Application* (Volume IV). Plenum Press, New York, 1985, pp. 343–358.
23. Civerolo K. L., Brankov E., Rao S. T., and Zurbenko I. G. Assessing the impact of the acid deposition control program. *Atmos. Environment*, 2001, **35**(24), 4135–4148.
24. Colarco P. R., Toon O. B., and Holben B. N. Saharan dust transport to the Caribbean during PRIDE. 1. Influence of dust sources and removal mechanisms on the timing and magnitude of downwind aerosol optical depth event from simulations of in situ and remote sensing observations. *J. Geophys. Res.*, 2003, **108**(D19), PRD5/1–PRD5/20.
25. Colarco P. R., Toon O. B., Reid J. S., Livingston J. M., Russel P. B., Redemann J., Schmid B., Maring H. B., Savoie D., Welton E. J., *et al.* Saharan dust transport to the Caribbean during PRIDE. 2. Transport, vertical profiles, and deposition in simulations of in situ and remote sensing observations. *J. Geophys. Res.*, 2003, **108**(D19), PRD6/1–PRD6/16.
26. Despres A., Rancillac F., and Bouville A. First result of the data processing of the VIth European campaign on remote-sensing of air pollution. In: Wispelaere, F. A. Schiermeier, and N. V. Gillani (eds.), *Air Pollution Modeling and Its Application* (Volume V). Plenum Press, New York, 1986, pp. 371–382.
27. Dimov I., Georgiev K., Ostromsky Tz., and Zlatev Z. Computational challenges in the numerical treatment of large air pollution models. *Ecological Modelling*, 2004, **179**(2), 187–203.
28. Dodenbender J., Wassmann R., Papen H., and Rennenberg H. Temporal and spatial variation of sulfur-gas-transfer between coastal marine sediments and the atmosphere. *Atmos. Environment*, 1999, **33**(21), 3487–3502.
29. Eckhardt S., Stohl A., Forster C., and James P. Climatology of ascending airstreams and their relation to the long-range transport of trace substances in the atmosphere (CARLOTTA). *AFO Newsletter*, 2003, No. 4, 3–6.
30. Fan Wan Bik. Some results of monitoring and assessment of the SRV atmosphere quality. In: K. Ya. Kondratyev (ed.), *Complex Global Monitoring of the State of the Biosphere*. Gidrometeoizdat, Leningrad, 1986, **2**, pp. 267–269 [in Russian].
31. Fast J. D., Zaveri R. A., Bian X., Chapman E. G., and Easter R. C. Effect of regional-scale transport on oxidants in the vicinity of Philadelphia during the 1999 NE-OPS field campaign *J. Geophys. Res.,* 2003, **108**(D16). AAC13/1–AAC13/11.
32. Fine J., Vuilleumier L., Reynolds S., Roth P., and Brown N. Evaluating uncertainties in regional photochemical air quality modeling. *Annual Review Environment Resources*, 2003, No. 28, 59–106.
33. Gallagher M. W., Nemitz E., Dorsey J. R., Fowler D., Sutton M. A., Flynn M., and Duyzer J. Measurements and parameterizations of small aerosol deposition velocities to grassland, arable crops, and forest: Influence of surface roughness length on deposition. *J. Geophys. Res.*, 2003, **108**(D12), AAC8/1–AAC8/10.
34. Gallardo L., Olivares G., Langner J., and Aarthus B. Coastal lows and sulfur air pollution in Central Chile. *Atmos. Environment*, 2002, **36**(23), 3829–3841.
35. Galle B., Oppenheimer C., Geyer A., McGonigle A. J. S., Edmonds M., and Horrocks L. A miniaturized ultraviolet spectrometer for remote sensing of SO_2 fluxes: A new tool for volcano surveillance. *Journal of Volcanology and Geothermal Research*, 2003, **119**(1–4), 241–254.
36. Ginoux P. Effects of nonsphericity on mineral dust modeling. *J. Geophys. Res.*, 2003, **108**(D2), AAC3/1–AAC3/10.

37. Gregory D., Morcrette J.-J., Jakob C., Beljaars A. C. M., and Stockdale T. Revision of convection, radiation and cloud schemes in the ECMWF Integrated Forecasting System. *Quart. J. Roy. Met. Soc.*, 2000, **126**, 1685–1710.
38. Grigoryev Al. A. and Kondratyev K. Ya. *Ecodynamics and Geopolicy* (Volume 2). *Ecological Disasters*. St Petersburg Sci. Centre Publ., St Petersburg, 2001, 684 pp. [in Russian].
39. Hadley A. and Toumi R. A simple model which predicts some non-linear features between atmospheric sulphur and sulphur emissions. *Environmental Pollution*, 2002, **119**(3), 365–374.
40. Hewitt C. N. The atmospheric chemistry of sulphur and nitrogen in power station plumes. *Atmos. Environment*, 2001, **35**(7), 1155–1170.
41. Holland D. M., Caragea P., and Smith R. L. Regional trends in rural sulfur concentrations. *Atmos. Environment*, 2004, **38**(11), 1673–1684.
42. Holzer M., McKendry I. G., and Jaffe D. A. Springtime trans-Pacific atmospheric transport from east Asia: A transit time probability density function approach. *J. Geophys. Res.*, 2003, **108**(D22), ACL11/1–ACL11/17.
43. Hoppel W. A., Frick G. M., and Fitzgerald J. W. Surface source function for sea-salt aerosol and aerosol dry deposition to the ocean surface. *J. Geophys. Res.*, 2002, **107**(D19), AAC7/1–AAC7/17.
44. Howarth R. W., Stewart J. W. B., and Ivanov M. V. (eds.) *Sulphur Cycling on the Continents Wetlands, Terrestrial Ecosystems, and Associated Water Bodies*. Wiley, Chichester, UK, 372 pp.
45. Hunton D. E., Ballenthin J. O., Borghetti J. F., Federico G. S., Miller T. M., Thorn W. F., and Viggiano A. A. Chemical ionization mass spectrometric measurements of SO_2 emissions from jet engines in flight and test chamber operations. *J. Geophys. Res.*, 2000, **105**(D22), 26841–26855.
46. Hutchison K. D., Smith S., and Faruqui S. The use of MODIS data and aerosol products for air quality prediction. *Atmos. Environment*, 2004, **38**(30), 5057–5070.
47. Hutjes R. W. A., Dolman A. J., Nabuurs G. J., Schelhaas M. J., Ter Maat H. W., Kabat P., Moors E., and Huygen J. Land use, climate and biogeochemical cycles: Feedbacks and options for emission reduction. *Dutch National Research Programme on Global Air Pollution and Climate Change* (Report no. 410 200 107). Rijksinstituut voor Volkgezondheid en Milieu, The Netherlands, 2001, 220 pp.
48. Ivanov A. P. and Chaikovsky A. P. *The Laser Ray Investigates the Atmosphere*. Institute of Physics of National Acad. Sci., Belarus', Minsk, 2002, 12 pp. [in Russian].
49. Kim K.-H. and Kim M.-Y. Comparison of an open path differential optical absorption spectroscopy system and a conventional in situ monitoring system on the basis of long-term measurements of SO_2, NO_2 and O_3. *Atmos. Environment*, 2001, **35**(24), 4059–4072.
50. Kondratyev K. Ya. From nano- to global scales: Properties, processes of formation and consequences of there atmospheric aerosol impact. 1. Field observational experiments. Africa and Asia. *Optics of the Atmos. and Ocean*, 2005, **18**, 3–18.
51. Kondratyev K. Ya. From nano- to global scales: Properties, processes of formation and consequences of atmospheric aerosol impact. 2. Field observational experiments. America, Western Europe and high latitudes. *Optics of the Atmos. and Ocean*, 2005, **18**, 19–31.
52. Kondratyev K. Ya., Grigoryev Al. A., and Varotsos C. A. *Environmental Disasters: Anthropogenic and Natural*. Springer–Praxis, Chichester, UK, 2002, 484 pp.

53. Kondratyev K. Ya., Krapivin V. F., and Phillips G. W. *Problems of the High-Latitude Environmental Pollution*. St Petersburg Sci. Centre of the Russian Acad. Sci. Publ., St Petersburg, 2002, 280 pp. [in Russian].
54. Kondratyev K. Ya., Krapivin V. F., and Phillips G. W. *Global Environmental Change: Modelling and Monitoring*. Springer-Verlag, Berlin, 2003, 319 pp.
55. Kondratyev K. Ya., Krapivin V. F., and Savinykh V. P. *Perspectives of Civilization Development: Multidimensional Analysis*. Logos Publ., Moscow, 2003, 573 pp. [in Russian].
56. Kondratyev K. Ya., Krapivin V. F., and Varotsos C. A. *Global Carbon Cycle and Climate Change*. Springer–Praxis, Chichester, UK, 2003, 568 pp.
57. Kondratyev K. Ya., Krapivin V. F., Savinykh V. P. and Varotsos C. A. *Global Ecodynamics: A Multidimensional Analysis*. Springer–Praxis, Chichester, UK, 2004, 686 pp.
58. Krapivin V. F. and Nazarian N. A. A numerical model to study the global sulphur cycle. *Numerical Modelling*, 1997, **9**(8), 36–50 [in Russian].
59. Krapivin V. F. and Kondratyev K. Ya. *Global Environmental Changes: Ecoinformatics*. Federal Ecological Fund, St Petersburg, 2002, 724 pp [in Russian].
60. Krapivin V. F. and Potapov I. I. *Methods of Ecoinformatics*. All-Russian Institute for Scientific and Technical Information Publ., Moscow, 2002, 496 pp. [in Russian].
61. Krapivin V. F., Mkrtchian F. A., Potapov I. I., and Nghuen Hong Son. Measurements of hydrophysical and geophysical parameters. *Problems of the Environment and Natural Resources*, 2003, **1**, 26–29 [in Russian].
62. Kubilay N., Cokacar T., and Oguz T. Optical properties of mineral dust outbreakes over the northeastern Mediterranean. *J. Geophys. Res.*, 2003, **108**(D21), AAC4/1–AAC4/10.
63. Langmann B. Numerical modelling of regional scale transport and photochemistry directly together with meteorological processes. *Atmos. Environment*, 2000, **34**(21), 3585–3598.
64. Lefohn A. S., Husar J. D., and Husar R.B. Estimating historical anthropogenic global sulfur emission pattern for the period 1850–1990. *Atmos. Environment*, 1999, **33**(21), 3435–3444.
65. Levin Z., Teller A., Ganor E., Graham B., Andreae M. O., Maenhaut W., Falkovich A. H., and Rudich Y. Role of aerosol size and composition in nucleation scavenging within clouds in a shallow cold front. *J. Geophys. Res.*, 2003, **108**(D22), AAC5/1–AAC5/14.
66. Liu M., Westphal D. L., Wang S., Shimizu A., Sugimoto N., Zhou J., Chen Y. A high-resolution numerical study of the Asian dust storms of April 2001. *J. Geophys. Res.*, 2003, **108**(D23), ACE21/1–ACE21/21.
67. Livingston J. M., Russel P. B., Reid J. S., Rodemann J., Schmid B., Allen D. A., Torres O., Levy R. C., Remer L. A., Holben B. N., *et al.* Airborne Sun photometer measurements of aeorsol optical depth and columnar water vapor during the Puerto Rico Dust Experiment and comparison with land, aircraft, and satellite measurements. *J. Geophys. Res.*, 2003, **108**(D19), PRD4/1–PRD4/23.
68. Luo C., Mahowald N. M., and del Corral J. Sensitivity study of meteorological parameters on mineral aerosol mobilization, transport, and distribution. *J. Geophys. Res.*, 2003, **108**(D15), AAC5/11–AAC5/21.
69. Mahowald N. M., Zender C. S., Luo C., Savoie D., Torres O., and del Corral J. Understanding the 30-year Barbados desert dust record. *J. Geophys. Res.*, 2003, **107**(D21), AAC7/11–AAC7/16.
70. Maria S. F., Russel L. M., Turpin B. J., Poreja R. J., Campos T. L., Weber R. J., and Huebert B. J. Source signatures of carbon monoxide and organic functional groups in

Asian Pacific Regional Aerosol Characterization Experiment (ACE-Asia) submicron aerosol types. *J. Geophys. Res.*, 2003, **108**(D23), ACE5/1–ACE5/14.

71. Martin B. D., Fuelberg H. E., Blake N. J., Crawford J. H., Logan J. A., Blake D. R., and Sachse G. W. Long-range transport of Asian outflow to the equatorial Pacific. *J. Geophys. Res.*, 2003, **108**(D2), PEM5/1–PEM5/18.
72. Matsuki A., Iwasaka Y., Osada K., Matsunaga K., Kido M., Inomata Y., Trochkine D., Nishita C., Nezuka T., Sakai T., Zhang D., and Kwon S.-A. Seasonal dependence of the long-range transport and vertical distribution of free tropospheric aerosols over East Asia: On the basis of aircraft and lidar measurements and isentropic trajectory analysis. *J. Geophys. Res.* 2003, **108**(D23), ACE31/1–ACE31/12.
73. Matsumoto K., Nagao I., Tanaka H., Miyaji H., Iida T., and Ikebe Y. Seasonal characteristics of organic and inorganic species and their size distributions in atmospheric aerosols over the Northwest Pacific Ocean. *Atmos. Environment*, **32**(11), 1931–1946.
74. Matsumoto K., Uematsu M., Hayano T., Yoshioka K., Tanimoto H., and Iida T. Simultaneous measurements of particulate elemental carbon on the ground observation network over the western North Pacific during the ACE-Asia campaign. *J. Geophys. Res.*, 2003, **108**(D23), ACE3/1–ACE3/6.
75. Matsumoto K., Uyama Y., Hayano T., Tanimoto H., Uno I., and Uematsu M. Chemical properties and outflow patterns of anthropogenic and dust particles on Rishiri Island during the Asian Pacific Regional Aerosol Characterization Experiment (ACE-Asia). *J. Geophys. Res.*, 2003, **108**(D23), ACE34/1–ACE34/15.
76. McGonigle A. J. S., Oppenheimer C. A., Hayes A. R., Galle B., Edmonds M., Caltabiano T., Salerno G., Burton M., and Mather T. A. Sulphur dioxide fluxes from Mt. Etna, Vulcano and Stromboli measured with an automated scanning ultraviolet spectrometer. *J. Geophys. Res.*, 2003, **108**(B9), 2455–2456.
77. McGonigle A. J. S., Thomson C. L., Tsanev V. I., and Oppenheimer C. A. simple technique for measuring power station SO_2 and NO_2 emissions. *Atmos. Environment*, 2004, **38**(1), 21–25.
78. Mehdizadeh F. and Rifal H. Modeling point source plumes at high altitudes using a modified Gaussian model. *Atmos. Environment*, 2004, **38**(6), 821–831.
79. Meloni D., diSarra A., Fiocco G., and Junkermann W. Tropospheric aerosols in the Mediterranean. 3. Measurements and modelling of actinie radiation profiles. *J. Geophys. Res.*, 2003, **108**(D10), AAC6/1–AAC6/12.
80. Meshkidze N., Chameides W. L., Nenes A., and Chen G. Iron mobilization in mineral dust: Can anthropogenic SO_2 emissions affect ocean productivity?. *Geophys. Res. Lett.*, 2003, **30**(21), ASC2/1–ASC2/5.
81. Monin A. S. and Obukhov A. M. Spatial characteristics of turbulence in the surface layer of the atmosphere. *Doklady of the USSR Acad. Sci.*, 1954, **93**, 223–226.
82. Moore K. G., II, Clarke A. D., Kapustra V. N., and Howell S. G. Long-range transport of continental plumes over the Pacific Basin: Aerosol physiochemistry and optical properties during PEM-Tropics A and B. *J. Geophys. Res.*, 2003, **108**(D2), PEM8/1–PEM8/27.
83. Nishizawa T., Asano S., Uchiyama A., and Yamazaki A. Seasonal variation of aerosol direct radiative forcing and optical properties estimated from ground-based solar radiation measurements. *J. Atmos. San.*, 2004, **61**(1), 57–72.
84. Nitu C., Krapivin V. F., and Pruteanu E. *Intelligent Systems in Ecology*. Onesti, Bucharest, 2004, 410 pp.
85. Nitu C., Krapivin V. F., and Bruno A. *System Modelling in Ecology*. Printech, Bucharest, 2000, 260 pp.

86. O'Dwyer M., Padgett M. J., McGonigle A. J. S., Oppenheimer C., and Inguaggiato S. Real-time measurement of volcanic H_2S and SO_2 concentrations by UV spectroscopy. *Geophys. Res. Lett.*, 2003, **30**(12), 1652–1657.
87. Olivares G., Gallardo L., Langner J., and Aarthus B. Regional dispersion of oxidized sulfur in Central Chile. *Atmos. Environment*, 2002, **36**(23), 3819–3828.
88. Osada K., Kido M., Iida H., Matsunaga K., Iwasaka Y., Nagatani M., and Nakada H. Seasonal variation of free tropospheric aerosol particles at Mt. Tateyama, central Japan. *J. Geophys. Res.*, 2003, **108**(D23), ACE35/1–ACE35/9.
89. Papaspiropoulos G., Martinsson B. G., Zahn A., Brenninkmeijer C. A. M., Hermann M., Heintzenberg J., Fischer H., and van Velthoven P. F. J. Aerosol elemental concentrations in the tropopause region from intercontinental flights with the Civil Aircraft for Regular Investigation of the Atmosphere Based on an Instrument Container (CARIBIC) platform. *J. Geophys. Res.*, 2002, **107**(D23), AAC3/1–AAC3/14.
90. Parrish D. and Law K. Intercontinental Transport and Chemical Transformation (ITCT-Lagrangian – 2k4). *J. Geophys. Res.*, 2003, **108**(D15), 8–13.
91. Paulsen C. A. The mathematical representation of wind and temperature profiles in the unstable atmospheric surphace layer. *J. Appl. Meteor.*, 1970, No. 9, 857–861.
92. Picket E. E. (ed.). *Atmospheric Pollution.* Springer-Verlag, 1987, Berlin, 286 pp.
93. Raaschou-Nielsen O., Hertel O., Vignati E., Berkowitcz R., Jensen S. S., Larsen V. B., and Lohse C. Evaluation of an air pollution model with respect to use in epidemiologic studies; comparison with measured levels of nitrogen dioxide and benzene. *J. Exp. Anal. and Env. Epi.*, 2000, No. 10, 4–14.
94. Rajeev K., Nair S. K., Parameswaran K., and Raju C. S. Satellite observations of the regional aerosol distribution and transport over the Arabian Sea, Bay of Bengal and Indian Ocean. *Indian. J. Mar. Sci.*, 2004, **33**(1), 11–29.
95. Reid E. A., Reid J. S., Meier M. M., Dunlap M. R., Cliff S. S., Broumas A., Perry K., and Maring H. Characterization of African dust transported to Puerto Rico by individual particle and size segregated bulk analysis. *J. Geophys. Res.*, 2003, **108**(D19), PRD7/1–PRD7/22.
96. Reid J. S., Jonsson H. H., Maring H. B., Smirnov A., Savoie D. L., Cliff S. S., Reid E. A., Livingston J. M., Meier M. H., Dubovik O., *et al.* Comparison of size and morphological measurements of coarse mode dust particles from Africa. *J. Geophys. Res.*, 2003, **108**(D19), 9/1–9/22.
97. Reid J. S., Kinney J. E., Westphal D. L., Holben B. N., Welton E. J., Tsay S.-C., Eleuterio O. P., Campbell J. R., Christopher S. A., Colarco P. R., *et al.* Analysis of measurements of Saharan dust by airborne and ground-based remote sensing methods during the Puerto Rico Dust Experiment (PRIDE). *J. Geophys. Res.*, 2003, **108**(D19), PRD2/1–PRD2/27.
98. Reid J. S. and Maring H. B. Foreword to special section on the Puerto Rico Dust Experiment (PRIDE). *J. Geophys. Res.*, 2003, **108**(D19), PRD1/1–PRD1/2.
99. Rissman T. A., Nenes A., and Seinfeld J. H. Chemical amplification (or dampening) of the Twomey effect: Conditions derived from droplet activation theory. *J. Atmos. Sci.*, 2004, **61**(8), 919–930.
100. Safai P.D., Rao P. S. P., Momin G. A., All K., Chate D. M., and Praveen P. S. Chemical composition of precipitation during 1984–2002 at Pune, India. *Atmos. Environment*, 2004, **38**(12), 1705–1714.
101. Schmid B., Hegg D. A., Wang J., Bates D., Redemann J., Russel P. B., Livingston J. M., Jonsson H. H., Welton E. J., Seinfeld J. H., *et al.* Column closure studies of lower tropospheric aerosol and water vapor during ACE-Asia using airborne Sun photometer

and airborne in situ and ship-based lidar measurements. *J. Geophys. Res.*, 2003, **108**(D23), ACE24/1–ACE24/22.

102. Scorer R. S. *Dynamics of Meteorology and Climate*. Wiley, New York, 1997, 350 pp.
103. Sinha P., Hobbs P. V., Yokelson R. J., Blake D. R., Gao S., and Kirchstetter T. W. Distributions of trace gases and aerosols during the dry biomass burning season in southern Africa. *J. Geophys. Res.*, 2003, **108**(D17), ACH4/1–ACH4/23.
104. Sinik N., Loncar E., and Vidic S. (1985) The use of field data in average wet deposition modeling. In: J. P. Lehman (ed.), *Air Pollution Modeling and Its Application* (Volume IV). Plenum Press, New York, 1985, pp. 155–161.
105. Sorokin A., Katragkou E., Arnold F., Busen R., and Schumann U. Gaseous SO_3 and H_2SO_4 in the exhaust of an aircraft gas turbine engine: Measurements by CIMS and implications for fuel sulfur conversion to sulfur (VI) and conversion of SO_3 to H_2SO_4. *Atmos. Environment*, 2004, **38**(3), 449–456.
106. Stein A. F. and Lamb D. The sensitivity of sulfur wet deposition to atmospheric oxidants. *Atmos. Environment*, 2000, **34**(11), 1681–1690.
107. Sting H. F. Hydrogen sulfide basins and stagnant period in the Baltic Sea. *J. Geophys. Res.,* 1963, **68**(13), 4009–4017.
108. Stoddard J. L., Kahl J. S., Deviney F. A., DeWalle D. R., Driscoll C. T., Herlihy A. T., Kellog J. H., Murdoch P. S., Webb J. R., and Webster K. E. *Response of Surface Water Chemistry to the Clean Air Act Amendments of 1990* (EPA/6201R–02/04). US Environmental Protection Agency, New York, 2003, 92 pp.
109. Stohl A., Eckhardt S., Forster C., James P., and Spichtinger N. On the pathways and timescales of intercontinental air pollution transport. *J. Geophys. Res.*, 2003, **108**(D23), ACH6/1–ACH6/17.
110. Stohl A., Huntrieser H., Richter A., Beirle S., Cooper O. R., Eckhardt S., Forster C., James P., Spichtinger N., Wenig M., *et al.* Rapid intercontinental air pollution transport associated with a meteorological bomb. *Atmos. Chem. Phys.*, 2003, **3**, 969–985.
111. Streets D. G., Guttikunda S. K., and Carmichael G. R. The growing contribution of sulfur emissions from ships in Asian waters, 1988–1995. *Atmos. Environment*, 2000, **34**(26), 4425–4439.
112. Streets D. G., Tsai N. Y., Akimoto H., and Oka K. Sulfur dioxide emissions in Asia in the period 1985–1997. *Atmos. Environment*, 2000, **34**(26), 4413–4424.
113. Suzuki K., Nakajima T., Numaguti A., Takemura T., Kawamoto K., and Higurashi A. A study of the aerosol effect on a cloud field with simultaneous use of GCM modeling and satellite observations. *J. Atmos. Sci.*, 2004, **61**(2), 179–194.
114. Takemura T., Uno I., Nakajima T., and Sano L. Modeling study of long-range transport of Asian dust and anthropogenic aerosols from East Asia. *Geophys. Res. Lett.*, 2002, **29**(24), 11/1–11/4.
115. Takeuchi M., Okochi H., and Igawa M. Characteristics of water-soluble components of atmospheric aerosols in Yokohama and Mt. Oyama, Japan, from 1990 to 2001. *Atmos. Environment*, 2004, **38**(28), 4701–4708.
116. Terada H., Ueda H., and Wang Z. Trend of acid rain and neutralization by yellow sand in East Asia: A numerical study. *Atmos. Environment*, 2002, **36**(3), 503–509.
117. Toth F., Bruckner T., Füssel H.-M., Helm C., Hooss G., Leimbach M., van Minnen J., Petschel-Held G., Schellnhuber H.-J., and Tothne-Hizsnyik E. *ICLIPS – Integrierte Abschätzung von Klimaschutzstrategien: Methodisch-naturwissenschaftliche Aspekte* (Research Report 01 LK 9605/0). Federal Ministry for Science and Research, Bonn, Germany, 2000, 115 pp. [in German].

118. Uno I., Carmichel G. R., Streets D., Satake S., Takemura T., Woo J.-H., Uematsu M., and Ohta S. Analysis of surface black carbon distributions during ACE-Asia using a regional-scale aerosol model. *J. Geophys. Res.*, 2003, **108**(D23), ACE4/1–ACE4/11.
119. Uno I., Carmichel G. R., Streets D. G., Tang Y., Yienger J. J., Satake S., Wang Z., Woo J.-H., Guttikunda S., Uematsu M., *et al.* Regional chemical weather forecasting system CFORS: Model descriptions and analysis of surface observations at Japanise island station during the ACE-Asia experiment. *J. Geophys. Res.*, 2003, **108**(D23), ACE36/1–ACE36/17.
120. Venkatram A., Isakov V., Yuan J., and Pankratz D. Modeling dispersion at distances of meters from urban sources. *Atmos. Environment*, 2004, **38**(28), 4633–4641.
121. Wang H. and Christiansen J. H. A re-entry plume fumigation model. In: C. De Wispelaere, F. A. Schiermeier, and N. V. Gillani (eds), *Air Pollution Modeling and Its Application* (Volume V). Plenum Press, New York, 1986, pp. 565–579.
122. Wang J., Christopher S. A., Reid J. S., Maring H., Savoie D., Holben B. N., Livinston J. M., Russel P. B., and Yang S.-K. GOES-8 retrieval of dust aerosol optical thickness over the Atlantic Ocean during PRIDE. *J. Geophys. Res.*, 2003, **108**(D19), 11/1–11/15.
123. Wylie D. P. and Hudson J. G. Effects of long-range transport and clouds on cloud condensation nuclei in the springtime Arctic. *J. Geophys. Res.*, 2002, **107**(D16), AAC13/1–AAC13/11.
124. Xie S., Qi L., and Zhou D. Investigation of the effects of acid rain on the deterioration of cement concrete using accelerated tests established in laboratory. *Atmospheric Environment*, 2004, **38**(27), 4457–4466.
125. Xu Y. and Carmichael G. R. An assessment of sulfur deposition pathways in Asia. *Atmos. Environment*, 1999, **33**(21), 3473–3486.
126. Zaikov G. E., Maslov S. A., and Rubailo V. L. *Acid Rains and the Environment*. 'Chemistry' Publ., Moscow, 1991, 139 pp [in Russian].
127. Zhao T. L., Gong S. L., Zhang X. Y., and McKendry I. G. Modeled size-segregated wet and dry deposition budgets of soil dust aerosol during ACE-Asia 2001: Implications for trans-Pacific transport. *J. Geophys. Res.*, 2003, **108**(D23), ACE33/1–ACE33/9.
128. Zlatev Z., Chrisensen J., and Hov O. A Eulerian air pollution model for Europe with nonlinear chemistry. *J. Atmos. Chem.*, 1992, **15**(1), 1–37.

9

Aerosol radiative forcing and climate

9.1 INTRODUCTION

The unprecedented growth of interest in the problems of the climate observed during the last decades has stimulated both scientific and applied developments which have considerably advanced our understanding of the causes of present climate changes and the laws of the paleoclimate. They have also advanced the development of scenarios for possible future changes of climate, though it must be stressed that scenarios are not predictions and that their potential for developing predictions must be assessed as doubtful. Unfortunately the growing interest in climate is, in part, explained by the important role played by various speculative exaggerations and apocalyptic 'predictions' (e.g., a complete melting of the Arctic sea ice in the first-half of this century). The problems of climate change, formulated as anthropogenic global warming, have therefore become a focus of geopolitics and global environmental policy. Presidents and Prime Ministers of several countries now discuss whether the Kyoto Protocol (KP) should be considered as a scientifically justified document. Confusion is caused, in particular, by the lack of sufficiently clear and agreed terminology. Ignoring the very complicated notion of the climate itself (which needs a separate discussion), one should remember that climate change was defined as being anthropogenically induced. One of the main unresolved problems is the absence of convincing quantitative estimates of the contribution of anthropogenic factors to the formation of global climate, though there can be no doubt that anthropogenic forcings of the climate do exist.

Some international documents containing analyses of the current ideas of climate refer to a consensus with respect to scientific conclusions contained in these documents. This wrongly assumes that the development of science is determined not by different views and relevant discussions, but by a general agreement and even voting. Apart from the question of definitions, the issue of uncertain conceptual estimates concerning various aspects of the climate problem remains of importance. In particular, this refers to the main conclusion of the summary of the International Panel on Climate Change, IPCC 2001, Report [73] which claims that: '... An increasing body of observations gives a collective picture of a warming world and most of the observed warming over the last fifty years is likely to have been due to human activities'.

It is to be regretted that the former Chairman of the IPCC Working Group-1 (WG-1), Professor J. Houghton, in a recent article (2003) in the British newspaper *The Guardian*, compared the threat of anthropogenic climate changes to weapons of mass destruction and admonished the USA for their refusal to support the concept of dangerous, anthropogenic global warming and thus the Kyoto Protocol. No matter how paradoxical it may seem, such claims are in fact being made against the background of an increasing understanding of the imperfections of the current global climate models and their still inadequate verification. This makes predictions on the basis of numerical modelling no more than conditional scenarios [78, 84–99, 159]. As for the USA, one should welcome the huge efforts of this country to support climate studies, manifested through both special attention to an improvement of observational systems and to developments in the field of climate problems in general [161, 166]. In 2004, the USA spent $4.5 billion on these problems.

The statement of the Intergovernmental Group (G-8) published on 2 July, 2003 [154] has justly emphasized that in the years to come efforts will be concentrated on three directions: (1) coordination of global observations strategies; (2) provision of a pure, more stable and efficient application of energy; and (3) provision of sustainable agricultural production and the preservation of biodiversity.

The Earth's climate system has indeed changed markedly since the industrial revolution started, with some changes being of anthropogenic origin. The consequences of climate change do present a serious challenge to the policy-makers responsible for the environmental policy and this alone makes the acquisition of objective information on climate change, of its impact and possible responses, most urgent. With this aim in mind, the World Meteorological Organization (WMO) and the UN Environmental Programme in 1988 organized the IPCC, divided into three working groups (WG) with spheres of responsibility for the: (1) scientific aspects of climate and its change (WG-I); (2) effects on and adaptation to climate (WG-II); and (3) analysis of possibilities to limit (mitigate) climate changes (WG-III).

The IPCC has so far prepared three detailed reports (1990, 1996, 2001) as well as several special reports and technical papers. Griggs and Noguer [60] have briefly reviewed the first volume of the Third IPCC Report (TIR) prepared by WG-I for the period June, 1998–January, 2001 with the participation of 122 leading authors and 515 experts. Four hundred and twenty experts reviewed the first volume and twenty-three experts edited it. Several hundred reviewers and representatives of many governments made additional remarks. With the participation of delegates from 99 countries and 50 scientists recommended by the leading authors, the final discussion of the TIR was held in Shanghai on 17–20 January, 2001. A 'Summary for decision-makers' was approved after a detailed discussion by 59 specialists.

Analysis of the observational data as contained in the TIR led to the conclusion that global climate change is taking place. The Report [73] gives a detailed review of the observational data of the spatial–temporal variability of the concentrations of various greenhouse gases (GHGs) and aerosol in the atmosphere. The adequacy of numerical models was discussed from the viewpoint of the climate-forming factors and the usefulness of models to predict climate change in the future. The main conclusion about anthropogenic impacts on climate was that 'there is new and stronger evidence that most of the warming observed during the last 50 years has been determined by human activity'. According to all prognostic estimates considered in the TIR, both surface air temperature (SAT) increase and sea level rise should take place during the 21st century.

When characterizing the IPCC data for the empirical diagnostics of climate, [52] drew attention to the uncertainty of the definitions of some basic concepts. According to IPCC terminology, climate changes are statistically substantial variations of an average state or its variability, whose stability is preserved for long time periods (for decades and longer). Climate changes can be natural in origin (connected both with internal processes and external impacts) and/or may be determined by anthropogenic factors, such as changes in the atmospheric composition or land use. This definition differs from that suggested in the Framework Climate Change Convention (FCCC) where climate changes are only of anthropogenic origin in contrast to natural climate change. In accordance with the IPCC terminology, climatic variability means variations of the average state and other statistical characteristics (mean standard deviation (MSD), repeatability of extreme events, etc.) of climate on every temporal and spatial scale, beyond individual weather phenomena. Hence climate variability can be both of natural (due to internal processes and external forcings) and anthropogenic origin: possess both internal and external variability. As [52] noted, seven key questions are most important for the diagnostics of observed changes and climate variability:

(1) How significant is climate warming?
(2) Is currently observed warming significant?
(3) How rapidly has the climate changed in the distant past?
(4) Have precipitation and atmospheric water content changed?
(5) Do changes in the general circulation of the atmosphere and ocean take place?
(6) Have climate variability and climate extremes changed?
(7) Are observed trends internally coordinated?

In order to answer the above questions, the reliability of observational data is fundamental. Without such observational data adequate empirical diagnostics of the climate remains impossible. Yet the information concerning numerous meteorological parameters, so very important for documentation, detection, and attribution of climate change, remains inadequate for the drawing of reliable conclusions. This is especially true for the global trends of those parameters (e.g., precipitation), which are characterized by a great regional variability.

Folland *et al.* [52] have answered some of the questions above. A comparison of the secular change of global average annual sea surface temperature (SST), land surface air temperature (LSAT), and nocturnal air temperature over the ocean

(NAT) for the period 1861–2000 on the whole revealed some similarity, though the warming in the 1980s from LSAT data turned out to be stronger, and the NAT data showed a moderate cooling at the end of the 19th century not demonstrated by SST data. The global temperature trend can be interpreted cautiously as equivalent linear warming over 140 years constituting 0.61°C at a 95% confidence level with an uncertainty range of ±0.16°C. In 1901 a warming of 0.57°C took place with an uncertainty range of ±0.17°C. These estimates suggest that beginning from the end of the 19th century, an average global warming of 0.6°C took place with the interval of estimates corresponding to a 95% confidence level equal to 0.4–0.8°C.

The spatial structure of the temperature field in the 20th century was characterized by a comparatively uniform warming in the tropics and by a considerable variability in the extratropical latitudes. The warming between 1910 and 1945 was initially concentrated in the northern Atlantic and adjacent regions. The northern hemisphere was characterized by cooling between 1946 and 1975, while in the southern hemisphere some warming was observed during this period. The temperature rise observed during the last decades (1970–2000) turns out, on the whole, to have been globally synchronous and clearly manifested across northern hemisphere continents in winter and spring. In some southern hemisphere regions and in the Atlantic, however, a small all-year-round cooling was observed. A temperature decrease in the northern Atlantic between 1960 and 1985 was later followed by an opposite trend. On the whole, the climate warming over the period of measurements was more uniform in the southern hemisphere than in the northern hemisphere. In many continental regions between 1950 and 1993, the temperature increased more rapidly at night than during daytime (this does not refer, however, to coastal regions). The rate of temperature increase varied from 0.1–0.2°C/10 years.

According to the data of aerological observations, the air temperature in the lower and middle troposphere was increasing after 1958 at a rate of 0.1°C/10 years, but in the upper troposphere (after 1960) it remained more or less constant. A combined analysis of the aerological and satellite information has shown that during the period 1979–2000 the temperature trend in the lower troposphere was weak, whereas near the land surface it turned out to be statistically significant and reached 0.16 ± 0.06°C/10 years. The statistically substantial trend of the difference between the Earth's surface and the lower troposphere constituted 0.13 ± 0.06°C/10 years, which differs from the data for the period 1958–1978, when the average global temperature in the lower troposphere increased more rapidly (by 0.03°C/10 years) than near the surface. The considerable differences between the temperature trends in the lower troposphere and near the surface are most likely to be real. So far, these differences cannot be convincingly explained. The climate warming in the northern hemisphere observed in the 20th century was according to [52] the most substantial over tha last 1,000 years.

Special attention has been paid in the IPCC 2001 Report to the possibility of predicting future climatic changes. The chaotic character of atmospheric dynamics limits the long-term weather forecasts to one or two weeks and prevents the prediction of a detailed climate change (e.g., it is impossible to predict precipitation in Great Britain for the winter of 2050). However, it is possible to consider climate projections (i.e., to develop scenarios of probable climate changes due to the continuing growth of GHG concentrations in the atmosphere). Such scenarios, if credible, may be useful for decision-makers in the field of ecological policy. The basic method to make such scenarios tangible involves the use of numerical climate models that simulate interactive processes in the climatic system that is the 'atmosphere–ocean–land surface–cryosphere–biosphere'. As Collins and Senior [34] noted, because there are so many such models, serious difficulty arises as to which is the best model to choose. As this problem of choice is insoluble, there remains the possibility of comparing the climate scenarios obtained by using various models.

According to the IPCC recommendations, four levels of projection reliability are considered:

(1) from reliable to very probable (in this case there is an agreement between the results for most of the models);
(2) very probably (an agreement of new projections obtained with the latest models);
(3) probable (new projections with an agreement for a small number of models); and
(4) restrictedly probable (model results are not certain but changes are physically possible).

A principal difficulty in giving substance to the projections is the impossibility of determining agreed predictions on how GHG concentrations will evolve in the future, which makes it necessary to take into account a totality of various scenarios. The huge thermal inertia of the World Ocean dictates a possibility of delayed climatic impacts of GHG concentrations, having already increased.

Calculations of annual average global SAT using the energy balance climate model with various scenarios of the temporal variations of CO_2 concentrations have led to SAT intervals in 2020, 2050, and 2100 to be 0.3–0.9°C, 07–2.6°C, and 1.4–5.8°C, respectively. Due to the ocean thermal inertia, a delayed warming should manifest itself within 0.1–0.2°C/10 years (such a delay can take place over several decades).

The following conclusions can be attributed to the category of projections with the highest reliability [34]:

(1) surface air warming should be accompanied by a tropospheric warming and stratospheric cooling (the latter is due to a decrease of the upward long-wave radiation flux from the troposphere);
(2) faster warming on land compared with oceanic regions (as a result of the great thermal inertia of the ocean); a faster warming in the high-mountain regions (due to albedo feedbacks);
(3) aerosol-induced atmospheric cooling holds a SAT increase (new estimates suggest the conclusion about a weaker manifestation of the aerosol impact);
(4) presence of the warming minima in the North Atlantic and in the circumpolar regions of the oceans in the southern hemisphere due to mixing in the oceanic thickness;
(5) decrease of the snow and sea ice cover extent in the northern hemisphere;
(6) increase of the average global content of water vapour in the atmosphere,

enhancement of precipitation and evaporation, as well as intensification of the global water cycle;

(7) intensification (on average) of precipitation in the tropical and high latitudes, but its attenuation in the subtropical latitudes;

(8) increase of precipitation intensity (more substantial than expected as a result of precipitation enhancement, on average);

(9) summertime decrease of soil moisture in the middle regions of the continents due to intensified evaporation;

(10) intensification of the El Niño regime in the tropical Pacific with a stronger warming in the eastern regions than in the western ones, accompanied by an eastward shift of the precipitation zones;

(11) intensification of the interannual variability of the summer monsoon in the northern hemisphere;

(12) more frequent appearance of high temperature extrema but infrequent occurrence of temperature minima (with an increasing amplitude of the diurnal temperature course in many regions and with a greater enhancement of nocturnal temperature minima compared with the daytime maxima);

(13) higher reliability of conclusions about temperature changes compared with those about precipitation;

(14) attenuation of the thermohaline circulation (THC) that causes a decrease of the warming in the North Atlantic (the effect of the THC dynamics cannot however compensate for the warming in west Europe due to the growing concentration of GHGs); and

(15) that a most intensive penetration of warming into the ocean depths occurs in high latitudes where the vertical mixing is most intensive.

As for the estimates characterized by a lower level of reliability, the conclusion (at level 4) about the lack of an agreed view on the changing frequency of storms in middle latitudes, is of special interest here, as is a similar lack of agreement about the changing frequency of occurrence of tropical cyclones under global warming. An important future task is to improve climate models aimed at eventually reaching a level of reliability that would enable the prediction of climatic change.

Allen [3] has discussed the basic conclusions contained in the 'Summary for policy-makers' (SPM) of the TIR and especially of its main conclusion that 'There is new and stronger evidence that most of the warming observed during the last 50 years should be attributed to human activity'. This conclusion supplements the statement according to which 'as follows from the present climate models, it is very unlikely that the warming taking place during the last 100 years was determined only by the internal variability' ('very unlikely' means that there is less than one chance in ten for an opposite statement to be well-founded).

Clearly, the reality of such a statement depends on an adequate modelling of the observed climatic variability. Analysis of the results of the relevant calculations using six different models has shown that three of six models reproduce climate variability on timescales from 10–50 years, which agrees with the observational data.

Another conclusion in the SPM (TIR) is that 'reconstruction of data on climate for the last 1,000 years shows that the present warming is unusual and it is unlikely that it can be of only natural origin' ('unlikely' means that there is less than one chance of three for an opposite conclusion).

This conclusion is supplemented with the following: 'Numerical modelling of the response to only natural disturbing forces ... does not explain the warming that took place in the second-half of the 20th century.' This view is based on the analysis of the results from the numerical modelling of changes in the average global SAT during the last 50 years. It follows from this that a consideration of natural forcings (solar activity, volcanic eruptions) has demonstrated a climatic cooling (mainly due to large-scale eruptions in 1982 and 1991) which has allowed the conclusion that the impact of only natural climatic factors is unlikely. However, there is only one chance in three that it was so: such carefulness is due to insufficient reliability based on indirect information concerning natural forcings in the past.

Results of numerical modelling cannot explain the pre-1940 climate warming with only anthropogenic factors taken into account, but are quite adequate considering both natural and anthropogenic impacts (GHGs and sulphate aerosol). As was mentioned in the SPM of the TIR, 'these results ... do not exclude possibilities of contributions of other forcings'. It is possible therefore that good agreement of the calculated and observed secular trends of SAT may in part be determined by a random mutual compensation of uncertainties. Another important illustration of the inadequacy of the numerical modelling results is their difference with observations concerning temperature changes near the Earth's surface and in the free troposphere. If, as according to models, the tropospheric temperature increases more rapidly than near the surface, then the analysis of observational data between 1979 and 2000 reveals that the temperature increase in the free troposphere is slower and probably completely absent.

When assessing the content of the IPCC 2001 Report, [60] argued that this report: (1) contains a most complete description of the current ideas about the known and unknown aspects of the climate system and the associated factors; (2) is based on the knowledge of an international group of experts; (3) is prepared based on open and professional reviewing; and (4) is based on scientific publications.

Sadly, none of these statements can be convincingly substantiated. The IPCC 2001 Report has therefore been strongly criticized in the scientific literature (see, in particular [2, 8, 13, 17, 44, 83–99, 102, 112, 120, 159, 173]), the most important items of which we shall now discuss.

9.2 EMPIRICAL DIAGNOSTICS OF THE GLOBAL CLIMATE

The main cause of contradictions in studies of the present climate and its changes is the inadequacy of the available observational databases. They remain incomplete and of poor quality. In this connection, [126] have carried out a thorough analysis of the evolution of the global observing system. As is well known, climate is characterized by many parameters: air temperature and humidity near the Earth's surface and in the free atmosphere; precipitation (liquid or solid); amount of cloud cover, the height of its lower and upper boundaries, and the microphysical and optical properties of clouds; radiation budget and its components; microphysical and optical parameters of atmospheric aerosols; atmospheric chemical composition; and others. However, the empirical analysis of climatic data is usually limited by the results of SAT observations, with data series available for no more than 100–150 years. Even these data series are heterogeneous, especially with regard to the global database; the main source of information for providing evidence for a global warming. Also, it should be borne in mind that the globally averaged secular trend of SAT values is based, to a large extent, on the use of imperfect observed data of SST.

The most important (and controversial) conclusion [73] concerning the anthropogenic nature of present-day climate change is based on analysis of the SAT and SST combined data (i.e., on the secular trend of mean annual global surface temperature (GST)). In this connection, two questions arise: first, about the information content of the notion of GST (this problem was formulated by [45]); second, about the reliability of GST values determined, in particular, by fragmentary data for the southern hemisphere, as well as the still unresolved problem of urban 'heat islands' [112].

Studies on the reliability of the SAT observations are continuing from the perspective of observational techniques. For more than 100 years SAT was measured with glass thermometers, but now arrangements to protect the thermometers from direct solar radiation and wind have been repeatedly changed. This dictates a necessity for filtering out SAT data to provide a homogeneous data series. In the period from April to August, 2000, at the station of the Nebraska State University, USA (40°83′N; 96°67′W), [70] carried out comparative SAT observations over smooth grass cover with the use of various means to protect the thermometers. At the same time, direct solar radiation and wind speed were measured. Analysis of observations has shown that differences of observed data can reach several tenths of a degree. Therefore, a technique has been proposed to increase the homogeneity of observational series. However, it does not permit the exclusion of the effect of calibration errors and drift of the temperature sensor's sensitivity.

For diagnostics of the observational data, emphasis should be placed on the analysis of climate variability in which a consideration, not of averages but moments, of higher orders is important. Unfortunately, there have been no attempts to use this approach. The same approach refers to estimates of the internal correlation of observational series. McKitrick [120], having analysed the secular trend of SAT, showed that by filtering out the contributions to temperature variations during the last several decades, at the expense of internal correlations (i.e., determined by the climatic system's inertia), it turns out that the temperature remains practically unchanged. There is a paradox: an increase of the global average SAT during the last 20–30 years is the principal basis for the conclusion concerning the anthropogenic contribution to the present-day climate change.

9.2.1 Air temperature

According to SAT observations discussed in [73], during the period from 1860 to the present-day, its annual and global averages increased by 0.6° ± 0.2°C. This is approximately 0.15°C higher than the value given in the IPCC 1996 Report, which was explained by a high SAT level between 1995 and 2000. The observed data revealed a strong spatial–temporal variability of the mean annual SAT over the globe. This manifested itself, for instance, with climate warming in the 20th century taking place during two time periods: 1919–1945 and from 1976–present-day. It follows from global climate diagnostics that the warming in the northern hemisphere in the 20th century was, apparently, the strongest during the last 1,000 years, the 1990s being the warmest decade, and the year 1998, the warmest year. An important feature of the climate dynamics was that, on average, the rate of increase of nocturnal (minimum) SAT values on land was almost twice as high compared with that of diurnal (maximum) SAT values, starting from 1950 (0.2°C against 0.1°C per ten years). This favoured an increase in the duration of frost-free periods in many regions of mid and high latitudes.

The IPCC 2001 Report [73] does not mention a previously assumed enhanced increase of climate warming in the northern hemisphere high latitudes as a characteristic indicator of anthropogenic global warming. However, an analysis of direct SAT measurements at the 'North Pole' stations during a period of 30 years [2], as well as of dendroclimatic indirect data for the last 2–3 centuries, shows that there had been no homogeneous enhancement of such warming. Climatic changes during the last century and the last decade were characterized by a strong spatial–temporal heterogeneity: in the Arctic, regions of both warming and cooling of climate were formed simultaneously (see also [127]).

From satellite observations (beginning from 1979), the trend of global average temperature for the lower troposphere (0–8 km) was +0.07°C per ten years [27a]. According to the data of aerological soundings there was an increase in the global average temperature of the lower troposphere by about 0.03°C per ten years, being much below the SAT increase (~0.15°C per ten years) [178]. This difference in warming manifests itself mainly in the oceanic regions in the tropics and subtropics – it is not clear why this is so [27a]. The results of the numerical climate modelling show that global warming should be stronger in the free troposphere than near the surface.

The difference of temperature trends near the surface and in the troposphere has caused heated discussion in the scientific literature [27a, 178]. Since the reliability of satellite remote sensing data raises no doubts, and their spatial representativeness (on global scales) is more reliable than that of the data of surface measurements, this difference should be interpreted as necessitating further analysis of the SAT and SST data adequacy.

Data on changes in the height of the tropopause have recently attracted attention [69, 149, 153]. As Santer *et al.* [153] noted, starting from 1979, the height of the tropopause increased by several hundred metres, agreeing with the results of numerical climate modelling, taking into account the growth of GHG concentrations, whose contribution prevails, again, in 'enigmatic' agreement of the observed and calculated data.

Studies of the dynamics of the tropical tropopause layer are of great interest for quantitative estimates of climate change and an understanding of the mechanisms of troposphere–stratosphere interaction. These circumstances have stimulated recent serious attention to studies of the climatic structure and variability of the tropical tropopause as well as mechanisms responsible for the formation of this structure. Serious attention has also been paid to the analysis of data on the content of water vapour in the stratosphere and mechanisms for the formation of thin cirrus clouds in the tropics. Randel *et al.* [149] undertook studies of the structure and variability of the temperature field in the upper troposphere and lower stratosphere of the tropics (at altitudes of about 10–30 km) from the data of radio-occultation observations for the period April, 1995–February, 1997 using a satellite system designed for geodetic measurements (GPS). A comparison with a large number (several hundreds) of synchronous aerological soundings has shown that a retrieval of the vertical temperature profiles from Global Positioning System/Meteorology (GPS/MET) data provides reliable enough information.

Analysis of the obtained results suggested that the spatial structure and variability of the tropopause altitude determined by a 'cool point' (minimum temperature) of the vertical temperature profile are governed mainly by wave-like fluctuations – like Kelvin waves. A strong correlation was observed between temperature from GPS/MET data and outgoing long-wave radiation, which can serve as an indirect indicator of penetrating convection in the tropics. This correlation confirms the reality of temperature fluctuation revealed from GPS/MET data and opens the possibilities of quantitative assessments of the response of large-scale temperature fields in the tropics to time-varying conditions of convection revealing coherent wave-like variations at altitudes between 12 and 18 km.

9.2.2 Snow and ice cover

Since the end of the 1960s, a 10% decrease in snow cover extent has been observed as well as a two-week reduction of the annual duration of lake and river ice cover in northern hemisphere middle and high latitudes, while in the non-polar regions, mountain glaciers are retreating. In 2002, northern hemisphere snow cover extent constituted 25.4 mln km^2, on average 0.2 mln km^2 less than during the preceding 30 years. The annual trend changes from 2.7 (August) to 46.9 (January) mln km^2 [178]. Also, starting in the 1950s, the extent of northern hemisphere ice cover in spring and summer has been decreasing by 10–15%. During the last decades (in the periods 'late summer–early fall') the Arctic sea ice cover thickness has probably decreased by about 40%, with the winter decrease being less substantial. From regular satellite observations (starting from the 1970s) no marked trend in the extent of ice cover in the Antarctic has been observed.

Numerical modelling using global climate models has shown (from considering the growing concentration of GHGs and aerosols) that climate warming should increase in the Arctic because of a feedback determined by the melting of the sea ice and snow cover causing a decrease in surface albedo. On the other hand, from the observed data, SAT has increased during the last decades over most of the Arctic. One of the regions where a warming has taken place is northern Alaska (especially in winter and in spring). In this regard, [160] have analysed the data on climatic changes in the north of Alaska to reveal their impact on the annual trend of the snow cover extent (SCE) and the impact of SCE changes on the surface radiation budget (SRB) and SAT.

9.2.3 Sea surface level and the ocean upper layer heat content

During the 20th century the World Ocean surface level rose by 0.1–0.2 m. Allegedly this was caused by the thermal expansion of seawater and ice melting on land due to global warming. The rate of the World Ocean level rise in the 20th century apparently exceeded that observed during the last 3,000 years by a factor of ten. Beginning from the end of the 1950s (when SST changes became large-scale), the heat content of the ocean upper layer has also been increasing.

Levitus *et al.* [105] analysed data on the warming of some components of the climatic system during the second-half of the 20th century. These data were derived from the growth of the heat content of the atmosphere and ocean as well as from estimates of the heat losses due to melting of some components of the cryosphere. These findings have led to the conclusion that the heat content of the atmosphere and ocean is rising. The growth of the heat content in the 3-km ocean layer between 1950 and 1990 exceeded, at least by an order of magnitude, the increase of the heat content in other components of the climate system. While the observed increase of the ocean heat content between 1955 and 1996 reached 18.2×10^{22} Joule, for the atmosphere it constituted only 6.6×10^{21} Joule. As for the values of latent heat due to water phase transformations, they were: 8.1×10^{21} Joule (a decrease of the mass of glaciers on land); 3.2×10^{21} Joule (a decrease of the sea ice cover extent in the Antarctic); 1.1×10^{21} Joule (melting of mountain glaciers); 4.6×10^{19} Joule (a decrease of the snow cover extent in the northern hemisphere); and 2.4×10^{19} Joule (melting of permafrost in the Arctic).

The observed data were compared [105] with results of numerical modelling using the GFDL interactive model of the 'atmosphere–ocean' system taking into account the radiative forcing (RF) due to: (1) the observed growth of GHG concentrations, changes of the sulphate aerosol content in the atmosphere, and extra-atmospheric insolation; (2) volcanic aerosol; as well as GHGs and sulphate aerosol only. The resulting comparisons have led to the conclusion that the observed changes of the ocean heat content can be explained, mainly, by the growth of GHG concentrations in the atmosphere, though one should bear in mind a substantial uncertainty concerning the estimate of RF due to sulphate aerosol and volcanic eruptions. The latter reduced the reliability of [105] with respect to the recognition of the anthropogenic warming.

Cai *et al.* [23] drew attention to the fact that the ocean dynamics can considerably affect future global-scale precipitation. Developments regarding these difficult problems are based on the use of both observed data and results of numerical modelling, and have led to quite different conclusions. The climatic warming of the last decades was characterized by a spatial structure similar to that of the El Niño–Southern Oscillation (ENSO) event. However, since there are no data on such a structure for the entire century, the observed structure of warming is assumed to be a manifestation of the multidecadal natural variability of climate, not the result of greenhouse forcing.

Moritz *et al.* [127] revealed a substantial inadequacy of climate models as applied to the Arctic conditions. In most cases the calculated Arctic oscillation (AO) trends turned out to be weaker compared with those observed. The calculated climate warming is greater in autumn over the Arctic Ocean, while the observed warming is at a maximum in winter and over the continents in spring.

9.2.4 Other climatic parameters

Data on soil temperature (GST) are important for climate diagnostics. As [115] have noted, an analysis of the GST data obtained in different regions of Canada by measuring the ground temperature in boreholes revealed considerable spatial differentiation both in the GST increase observed in the 20th century, and in the onset of the warming. For instance, from measurements in 21 boreholes, covering a period of the last 1,000 years, warming was detected (within 1–3°C) during the last 200 years. The warming was preceded by a long cooling trend in the region 80°–96°W, 46°–50°N, which continued until the beginning of the 19th century. According to data for ten boreholes in central Canada, the temperature had reached a minimum at about the year 1820 with a subsequent warming by about 1.5°C. In western Canada, during the last 100 years the warming reached 2°C.

An analysis has been made [115] of more adequate information on GST from data of measurements in 141 boreholes at a depth of several hundred metres. The holes were drilled in 1970–1990. The results obtained revealed an intensive warming that started in the 18–19th centuries following a long period of cooling (especially during the Little Ice Age) continuing for the rest of the millennium. The time of the onset of the present warming differed between regions. An analysis of the spatial distribution of the GST changes over the territory of Canada revealed a substantial delay in the onset of the present warming in the east-to-west direction, with a higher level of GST increase in the 20th century in western Canada. This conclusion is confirmed by data of SAT observations. It should be noted that the GST increase in eastern Canada had begun about 100 years before the industrial era.

Characteristics of the atmospheric general circulation are important components of climate diagnostics. Wallace and Thompson [175] pointed out that the west–east zonal wind component, averaged over the 55°N latitudinal belt, can be a representative indicator of the primary mode of surface air pressure anomalies – the annual mode of the northern hemisphere (NAM). Both NAM and a similar index SAM for the southern hemisphere are typical signatures of symbiotic relationships between the meridional profiles of the west–east transport and wave disturbances superimposed on this transport. Their index determined (using a respective normalization) as a coefficient for the first term of the NAM expansion in empirical orthogonal functions can serve as the quantitative characteristic of the modes. The presence of a positive NAM (or SAM) index denotes the existence of a relatively strong west–east transport.

In recent years it has been recognized that dynamic factors contribute much to observed temperature trends. For instance, in 1995 a marked similarity was observed between the spatial distributions of the SAT field and NAM fluctuations for the last 30 years, with a clear increase of the NAM index. The increasing trend of the index was accompanied by mild winters, changes of the spatial distribution of precipitation in Europe, and ozone layer depletion in the latitudinal belt >40°N. Similar data are available for the southern hemisphere. The main conclusion is that along with the ENSO event, both NAM and SAM are the leading factors in global atmospheric variability. In this regard, attention should be focused on the problem of the 30-year increasing trend in the NAM index that decreased after 1995. It is still not clear whether this trend is a part of a long-term oscillation.

The observational data show that during the 20th century an increase of precipitation constituted 0.5–1% per ten years over most of the land surface in the middle and high latitudes of the northern hemisphere, but a decrease (by about 0.3% per ten years) took place over most of the land surface in subtropical latitudes, which has recently weakened, however. As for the World Ocean, the lack of adequate observational data has not permitted the identification of any reliable trends in precipitation. In recent decades, intensive and extreme precipitation in the middle and high latitudes of the northern hemisphere has probably become more frequent. Beginning in the mid-1970s ENSO events have been frequent, stable, and intensive. This ENSO dynamic was reflected in specific regional variations of precipitation and SAT in most of the zones of the tropics and subtropics. Data on the intensity and frequency of occurrence of the tropical and extratropical cyclones, as well as local storms, still remain fragmentary and inadequate and do not permit conclusions to be drawn on any trends.

Changes in the biosphere are also important indicators of climate. One of them is the bleaching of corals. It is important to recognize that enhanced atmospheric forcings on coral reefs lead not to their disappearance but to their transformation into more resistant species [72]. Changes of seawater properties are one more indicator [20].

9.2.5 Concentrations of greenhouse gases and anthropogenic aerosol in the atmosphere

From 1750 to the present-day the CO_2 concentration in the atmosphere has increased by about one-third, reaching the highest level for the last 420,000 years (and, probably, during the last 20 million years), which is illustrated by the data from ice cores [73]. The growth of CO_2 concentration by about two-thirds during the last 20 years is explained by emissions to the atmosphere from fossil fuel burning (contributions of deforestation and, to a lesser extent, the cement industry constitute one-third). It is of interest that by the end of 1999, CO_2 emissions in the USA exceeded their 1990 level by 12%, and by 2008 their further increase should raise this value by another 10% [173]. Meanwhile, according to the Kyoto Protocol, emissions should be reduced by 7% by the year 2008 with respect to the level of 1990, which requires their total reduction by about 25% (which is, of course, utterly infeasible).

According to available observational data, both the World Ocean and land are currently global sinks for CO_2. In the ocean, both chemical and biological processes are responsible; on land we observe an enhanced 'fertilization' of vegetation due to increased concentrations of CO_2 and nitrogen, as well as with changes in land use. Yet much remains unclear about the global carbon cycle [91, 96, 98]. In particular, contradictions in the estimates of the role of the biosphere and ocean in the global carbon cycle remain to be resolved [98].

There is no doubt that fossil fuel burning will remain the main factor in the growth of CO_2 concentration in the 21st century. The role of the biosphere (both the ocean and land) as a barrier to the growth of CO_2 concentration will be reduced in time. According to the IPCC 2001 Report, the probable interval of CO_2 concentration values by the end of the century will constitute 540–970 ppm (pre-industrial and present values are, respectively, 280 ppm and 367 ppm). Changes of land use [73] are also an important factor of the global carbon cycle, but all carbon emitted to the atmosphere due to land use will be assimilated by the land biosphere. This could only lead to a decrease of CO_2 concentration of 40–70 ppm. As for prognostic estimates of other GHG concentrations by the year 2100, they vary widely. For instance, according to some estimates the role of CO as a GHG may become equal to the contribution of methane.

The concentration of methane in the atmosphere increased by a factor of 2.5, compared with that observed in 1750, and continues to grow. The annual rate of CH_4 increase was reduced, however, and became more variable in the 1990s compared with the 1980s. Beginning in 1750, nitrous oxide concentrations have increased by 16%. With the implementation of the Montreal Protocol, concentrations of several halocarbon compounds functioning as GHGs and ozone-destroying gases either have increased more slowly or started decreasing. However, concentrations of their substitutes and some other synthetic compounds started to grow rapidly (e.g., perfluorocarbons, PFC, and sulphur hexafluoride, SF_6).

As for the properties of atmospheric aerosol and its climatic impact, respective current information has been reviewed in detail in [6, 88, 95, 122, 171]. In this connection, it is pointed out again that the supposed anthropogenic nature of the present global climate warming was explained by the warming caused by the growth in GHG concentrations (primarily CO_2 and CH_4) as well as cooling due to anthropogenic aerosols. However, if the estimates of the 'greenhouse' warming can be considered as sufficiently reliable, then the respective calculations of RF due to aerosol are very uncertain. Of no less importance is the fact that while the global distribution of the 'greenhouse' RF is comparatively uniform, in the case of the 'aerosol' RF it is characterized by a strong spatial–temporal variability (including changes of the sign of RF).

9.2.6 Paleoclimatic information

Paleoclimatic information is an important source of data for the comparative analysis of the present and past climates. Analysis of the data of paleoclimatic observations reveals large-scale abrupt climate changes in the past when the climate system had exceeded certain threshold levels. Though some mechanisms for such changes have been identified, and the existing methods of numerical climate modelling are being gradually improved, the existing models still do not permit a reliable reconstruction of past climatic changes. With emphasis on the climatic implications of the growth of GHG concentrations in the atmosphere, less effort has been made to study the possible sudden climate changes that may be of natural origin though possibly intensified by anthropogenic forcing.

Since such changes lie beyond the problems addressed in the UN FCCC, [4] undertook a conceptual evaluation of the problem of large-scale abrupt climate changes. Though the available long-term stabilizing feedbacks have determined the existence on the Earth of comparatively persistent global climate for about 4 billion years, with characteristic timescales from one year to one million years, feedbacks prevailing in the climate system had favoured an enhancement of forcings on climate. So, for instance, changes of global average SAT within 5–6°C during the glaciation cycles apparently resulted from very weak forcings due to variations of orbital parameters.

Still more surprising is that during several decades, and in the absence of external forcings, regional changes have taken place reaching 30–50% of those that had taken place in the epochs of glaciations. Data from the period of instrumental observations have revealed abrupt climatic changes, quite often accompanied by serious socioeconomic consequences. So, for instance, the warming in many northern regions in the 20th century took place in two rapid 'steps', which enables one to suppose that in this case there was a superposition of the anthropogenic trend on interannual natural variability. Special attention was paid to the role of the ENSO event. The latter also refers to a sharp change of the climate system in the Pacific region in 1976–1977.

Considerable abrupt changes of regional climate in the period of Paleocene were detected from paleoclimatic reconstructions. They had been manifested as changes of the frequency of occurrence of hurricanes, floods, and especially droughts. Regional SAT changes reaching 8–16°C had happened in periods of 10 years or less. Dansgaard–Oeschger (DO) oscillations can serve as an example of large-scale sudden changes.

The climatic system involves numerous factors that intensify climatic changes with minimum forcings. The withering or death of plants, for example, may cause a decrease of evapotranspiration and hence lead to precipitation attenuation, which may further increase drought conditions. In cold-climate regions, the snow cover formation is accompanied by a strong increase of albedo, which favours further cooling (the so-called 'albedo effect'). Substantial climatic feedbacks are associated with the dynamics of the thermohaline circulation.

While the factors of enhancement of either changes or stability of climate are comparatively well known, understanding is very much weaker of the factors in the spatial distribution of anomalies over large regions, including global. In this connection, further studies of various modes of the general circulation of the atmosphere and the ocean (ENSO, DO oscillations, etc.) are important, as is the respective improvement of general circulation models. Most important here are the potential effects of abrupt climatic changes on both ecology and economies, as current estimates are generally based on the assumption of slow and gradual change.

Abrupt climate changes were especially substantial in periods of the transition of one climatic state to another. Therefore, if anthropogenic forcings of climate can favour the drifting of the climate system toward a threshold level, the possibility of raising the probability of abrupt climate changes also increases. Of great importance are not only the amount but also the rate of anthropogenic forcings on the climate system. So, for instance, a faster climate warming should favour a stronger attenuation of the thermohaline circulation as this may promote an acceleration of the shift to the threshold of climatic changes (it is important that under these conditions the thermohaline circulation dynamics becomes less predictable). To accept adequate solutions in the field of ecological policy, a deeper understanding of the whole spectrum of possible sudden climate changes is extremely important. Difficulties in the identification and quantitative estimation of all possible causes of sudden climate change and low predictability near threshold levels testify to the fact that the problem of abrupt climate changes will always be aggravated by more serious uncertainties than the problem of slow change. Under these conditions the development of ways to provide the stability and high adaptability of economics and ecosystems is of great importance.

9.3 RADIATIVE FORCING

Estimates of RF changes contained in IPCC 2001 Report [73], which characterize an enhancement of the atmospheric greenhouse effect and are determined by the growth of concentrations of minor gas components (MGCs), well mixed in the atmosphere, are $2.42\,W\,m^{-2}$, with the following contributions of various MGCs: CO_2 ($1.46\,W\,m^{-2}$), CH_4 ($0.48\,W\,m^{-2}$), halocarbon compounds ($0.33\,W\,m^{-2}$), and N_2O ($0.15\,W\,m^{-2}$). The ozone depletion observed during the last two decades could lead to a negative RF constituting $0.15\,W\,m^{-2}$, which can be reduced to zero in this century by using measures to protect the ozone layer. The growth of the tropospheric ozone content beginning in 1750 (by about one-third) could produce a positive RF of about $0.33\,W\,m^{-2}$.

From the time of the IPCC 1996 Report, the RF estimates have substantially changed, being determined not only by purely scattering sulphate aerosol considered above, but also by other types of aerosol, especially carbon (soot) characterized by considerable absorption of solar radiation as well as organic sea salt and mineral aerosol. The strong spatial–temporal variability of the aerosol content in the atmosphere, and its properties, seriously complicates an assessment of the climatic impact of aerosol [88, 122]. New results of numerical climate modelling have radically changed the understanding of the role of various factors of RF formation. According to [91], there is an approximate mutual compensation of climate warming due to the growth of CO_2 concentration and cooling caused by anthropogenic sulphate aerosol. Under these conditions, anthropogenic emissions of methane (mainly due to rice fields) and carbon (absorbing) aerosol should play a more important role.

Estimates of RF obtained with due regard to GHGs and aerosol are of importance in giving substance to conclusions concerning the contribution of anthropogenic factors to climate formation. The correctness of these conclusions is restricted, however, by three factors. One of them is that the interactivity of these factors seriously limits (if not excludes) the possibility of adequate estimates of contributions from individual factors. The second, not less important, factor is that the above calculated estimates refer to average global values and therefore are the results of a smoothing of the RF values characterized by a strong spatial–temporal variability. Finally, the most complicated problem is the impossibility of reliable assessment of the aerosol RF considering its direct and indirect components. According to estimates in [143], the value of direct RF at the surface level can increase to $50\,W\,m^{-2}$, and [27] obtained values exceeding $-100\,W\,m^{-2}$ during the period of forest fires in Indonesia. Vogelmann *et al.* [174] estimated the RF due to radiative heat exchange, from which it follows that during daytime near the surface the RF value is usually equal to several $W\,m^{-2}$. From the data of [136], the total RF at the surface level in the Antarctic varies within 0.4–$50\,W\,m^{-2}$. Yabe *et al.* [184] obtained the average value $85.4\,W\,m^{-2}$, and from [106], the RF in the USA constitutes 7–$8\,W\,m^{-2}$.

Weaver [180] has analysed the possible role of changes of the cloud RF (CRF) at the atmospheric top level, especially in extratropical latitudes, as a climate-forming factor whose role consists in its regulating the poleward meridional heat transport. The cloud dynamics in extratropical latitudes and related changes in CRF depend on the formation, in the atmosphere, of vortices responsible for the evolution of storm tracks. It is vortices determining the formation of storm tracks that contribute most to the meridional heat transport.

It has been shown [180] that the average annual radiative cooling of clouds in high latitudes has the same order of magnitude as a convergence of the vortices-induced meridional heat flux, but has an opposite sign. Since there is a close correlation between CRF and storm track dynamics, one can suppose two means of impact of storm track dynamics on poleward heat transport: (1) directly – via the vortices-induced heat transport in the atmosphere; and (2) indirectly – via CRF changes. The efficiency of heat transport by vortices is reduced by radiative cloud cooling. Changes of efficiency can be a substantial climate-forming factor. Various levels of efficiency can determine the possibility of the existence of different climatic conditions.

In the context of the problem of CRF formation due to long-wave radiation, [176] considered specific features of the spatial distribution of cloud cover in the period of an unusually intensive El Niño event in 1997–1998 from the data of observations from the SAGE-II satellite. Data on the cloud cover frequency of occurrence in this period and CRF are unique information for verification and specification of schemes of interaction parameterization in the 'clouds–radiation–climate' system used in the models of atmospheric general circulation.

Based on the use of the occultation technique of remote sensing, the SAGE-II data provide vertical resolution above 1 km and a quasiglobal survey (70°N–70°S). Analysis of the results under discussion revealed:

(1) An occurrence of the upper level opaque clouds exceeding the normal level in the eastern sector of the tropical Pacific and an opposite situation in the regions of the 'warm basin' of the Pacific; a combined distribution of anomalies of an opaque cloudiness located at altitudes above 3 km can be explained by the impact of the spatial structure of anomalies of SST fields and precipitation observed in the tropics.
(2) The same laws are characteristic of cloudiness near the tropical tropopause recorded at detection thresholds.
(3) The zonal mean distribution is characterized by a decrease of the amount of opaque clouds in low latitudes (except the southern hemisphere tropics at altitudes below 10 km) and an enhancement of clouds in high latitudes as well as by an increase (decrease) of cloud amount (at detection threshold) in the southern hemisphere tropics (in the upper troposphere of the horthern hemisphere subtropics).
(4) The geographical distribution of calculated CRF anomalies which agrees well with the data of satellite observations of the Earth's radiation budget. New estimates of direct and indirect RF have been obtained by [56].

Markowicz *et al.* [119] have undertaken a study to estimate the aerosol RF due to long-wave radiation (radiative heat exchange).

Rossow [152] has justly warned that attempts to isolate and describe a greater number of climatic feedbacks and to quantitatively estimate them using methods proposed earlier, have become confusing and disorienting, since an application of a simple linear theory consisting of many subsystems is completely unacceptable.

Changes in extra-atmospheric solar radiation is a climate-forming factor that should be taken into account. The contribution of these changes to RF starting from 1750 could have reached ~20% compared with the contribution of CO_2, which is mainly determined by an enhancement of extra-atmospheric insolation in the second-half of the 20th century (of importance is a consideration of the 11-year cycle of insolation). However, possible mechanisms of enhancement of the climatic impact of solar activity are still far from being understood [61, 86].

Shamir and Veizer [157] found, for instance, a high correlation between intensity of galactic cosmic rays and temperature for the last 500 million years. On this basis, it was concluded that 75% of the temperature variability in that period had been determined by the contribution of this factor (this problem has been also considered earlier in [86]).

9.4 RESULTS OF NUMERICAL CLIMATE MODELLING AND THEIR RELIABILITY

The problem of numerical climate modelling has been thoroughly analysed in numerous publications. Soon *et al.* [159] have recently published a critical review. Therefore, we will confine ourselves to brief comments.

Considerable progress has been made in the developments of more adequate numerical climate models taking the main components of the 'atmosphere-hydrosphere–lithosphere–cryosphere–biosphere' climate system into interactive account. The baffling complexity of climate models and of the schemes of empirical parameterization of various (especially subgrid) processes used in them, hinders the analysis of the models' adequacy, especially for their application to predict future climate. Attempts made so far to compare the results of numerical climate modelling with the observational data have therefore been rather schematic, controversial, and unconvincing. The problem of model verification remains an extremely urgent one.

For instance, the conclusions with respect to the secular trend of annual average global average SAT for the last one and a half centuries remain unconvincing. As claimed in IPCC 1996 Report, there is good agreement between the observed and calculated trends of SAT (with the growth of CO_2 concentration taken into account), but following [64], a consideration of methane and carbon aerosols should be given more importance. In both these cases the conclusions are based on arbitrary opinions, and agreement with observations is in fact no more than an adjustment. In addition, it is clear that a serious comparison of theory with observations should include the consideration of regional climatic changes (not only SAT), and not only of the average values of climatic parameters but also of their variability characterized by moments of a higher order.

According to [25], 'anthropogenic aerosols affect strongly the cloud albedo, with the values of global average RF being of the same order of magnitude (but opposite in the sign) as those determined by greenhouse gases ... the present developments indicate that the value of aerosol RF can be even higher than those assumed'.

The 'Achilles' heel' of the climate models is the parameterization of biospheric dynamics [97, 99, 185]. Numerous numerical experiments have been undertaken to assess the effect of deforestation in the Amazon basin. These have led to the conclusion that for complete deforestation of this region (a change of tropical rain forests for grass cover), both evaporation from the surface and precipitation should decrease, and surface temperatures would rise. The resulting increase of SAT would vary between 0.3° and 3°C, such changes being determined mainly by an increase of surface albedo and a decrease of soil moisture. The associated decrease of energy and water vapour fluxes to the atmosphere, reduction of moist convection, and the release of latent heat would result in a reduction of atmospheric heating, which in turn would produce two changes of atmospheric circulation: changes of the upward and downward air fluxes in the tropics and subtropics (Hadley circulation cells); and changes in the conditions of planetary wave generation (Rossby waves) propagating from the tropics to the middle latitudes.

Conclusions with respect to the observed and, more importantly, of the potential future climatic changes are very uncertain, both with respect to the data of diagnostics of present climate dynamics and to the results of numerical modelling. According to the IPCC 2001 Report [73], developments in the following six fields should be given top priority:

- To stop further degradation of the network of conventional meteorological observations.
- To continue studies in the field of global climate diagnostics in order to obtain long-term series of observational data with a higher spatial–temporal resolution.
- To seek a more adequate understanding of the interaction between the ocean climate system components (including its deep layers) and the atmosphere.
- To more realistically understand the laws of long-term variability of the climate.
- To broaden the application of an ensemble approach to climate modelling in the context of assessments of probabilities.
- To develop a totality ('hierarchy') of global and regional models with emphasis on the numerical modelling of regional impacts and extreme changes.

It should be added that in order to understand the physical laws governing present and future climates, studies of the paleoclimate are also important, especially of sudden short-term changes. An intensive development of spaceborne remote sensing has not provided adequate global information about the diagnostics of the climate system because the functioning of the existing system of spaceborne and conventional observation remains far from optimal.

9.5 SOME ASPECTS OF AEROSOL RADIATIVE FORCING RETRIEVAL TECHNIQUES

The reliability of aerosol radiative forcing (ARF) estimates depends on many factors, one of which being the reliability of information about the aerosol optical thickness (AOT).

As Chýlek *et al.* [32] noted, a maximum permissible error in the outgoing radiation flux determination from satellite data $\Delta F = 0.5\,\mathrm{W\,m^{-2}}$ determines the necessity to retrieve the AOT τ, with an error of not more than $\Delta\tau = 0.015$ on land and 0.010 over the oceans. However, this level of error cannot be achieved, so far, with the use of the AVHRR data, the MSD of τ values varies within 0.06–0.15, whereas in the case of MODIS data over land, $\Delta\tau = 0.05$–0.2τ, which corresponds to the interval of $\Delta\tau$ values from 0.07 to 0.21, with τ varying from 0.1 to 0.8. The use of the extranadir data of the multispectral thermal video-radiometer MTI for intermediate angles of scattering provides the level of error $\Delta\tau = 0.03$.

According to the numerical modelling results, the main obstacle to an increase of the τ retrieval accuracy is the unreliability of data on the scattering function (determined by the absence of reliable information about the aerosol size distribution, shape, and optical properties of particles). Such uncertainties affect more strongly the reliability of τ retrieval at large scattering angles (as a rule, close to nadir) than in the case of extranadir angles (this corresponds to moderate values of the scattering angle). From the experience of τ retrieval from MTI data, it was shown [32] that to provide the required accuracy of τ retrieval from the data of satellite observations, one should use a single or dual direction of viewing at extranadir scattering angles in the interval 50°–100°.

Myhre *et al.* [131] performed comparisons of retrieval algorithms for AOT over the ocean using the data of satellite observations during 8 months (November, 1996–June, 1997) made with the AVHRR, OCTS, POLDER, and TOMS instrumentation. Comparisons revealed the presence of considerable differences (by a factor of 2 or more) between retrieved AOT values. Most substantial were differences in the southern hemisphere, and their main source was, apparently, insufficient reliability of the cloud impact filtering.

According to the IPCC Report published in 1996, an indirect (connected with the impact of aerosol on the optical properties of clouds) globally averaged impact of aerosol on climate, characterized by the values of indirect RF, varies from 0 to $-1.5\,\mathrm{W\,m^{-2}}$. Six years later (in IPCC 2001), the range of uncertainties was broadened (0 to $-4.8\,\mathrm{W\,m^{-2}}$).

As Brenguier [18] noted, one of the factors of uncertainties is drizzle in clouds which form in the atmospheric boundary layer (ABL). In particular, this circumstance illustrates the importance of an adequate retrieval of the cloud cover dynamics in the ABL. Another problem is connected with the consideration (parameterization) of small-scale processes in the ABL and their non-linearity. So, for instance, an activation of aerosols as cloud concentration nuclei (CCN) is determined by upward motions at the level of the cloud bottom which should be reproduced to a spatial resolution (in the horizontal) of the order of 100 m. The present parameterization schemes are still far from meeting these requirements.

The necessity to take into account the interaction between various processes determining the cloud cover dynamics and its effect on the microphysical and optical properties of clouds brings forth serious difficulties. In this connection, the emphasis has been laid first on the aerosol-induced growth of cloud droplet number densities resulting in the respective changes of cloud albedo and indirect radiative forcing. This impact was called the first indirect impact of clouds on climate. However, it was also necessary to take into account the second indirect impact manifesting itself through a change of the precipitation formation intensity. The retrieval of this impact needs a description of the interaction between the microphysical characteristics of the clouds and ABL dynamics. Also, an analysis has been made of the significance of the 'semidirect' effect due to the short-wave radiation absorption by aerosol, which reduces the cloud cover development [79].

An important problem is to be able to provide an interactive consideration of the three types of aerosol impact on the ABL, clouds, and the indirect RF mentioned above. The solution to this problem has become one of the main objectives of the second field observational experiment (ACE-2) on studies of aerosol characteristics accomplished in 1997 in the region of the Canary Islands. Part of the programme of this experiment (CLOUDCOLUMN, CC) has been especially dedicated to a study of the indirect impact of anthropogenic aerosol on climate. In 1999, the European

Commission supported further studies in this direction within the PACE project on development of parameterization schemes for the impact of aerosol on climate.

As Mitra [125] pointed out, the Indian oceanic experiment (INDOEX) was the first complex problem-oriented observational international programme aimed mainly at the study of aerosol-induced radiative and climatic forcing of the regional and global climate with respective feedbacks taken into account. The preliminary stage of the accomplishment of INDOEX began in 1996–1997, and the basic part of the complex observations was accomplished in 1998–1999 with the participation of specialists from many different countries (India, USA, western Europe, Mauritius, and the Maldives). The obtained results were based on the use of surface, ship, aircraft, and satellite observational means.

The observational programme included obtaining information about the content and properties of atmospheric aerosol and the most substantial optically active MGCs (O_3, CO, NO_x, SO_2, etc.). It concentrated on the studies of direct and indirect aerosol radiative forcing (ARF). The most interesting, and in many respects unexpected, results of realization of this programme are the detection of a thick aerosol layer in the troposphere and the discovery of long-range transport of both aerosol and MGCs.

The available complex information about aerosol has opened the possibilities of analysing its impact on the climate, human health, and agriculture, and also making available various data on MGCs leading to prospects for developments on the 'chemical weather' problem. An important component of the INDOEX programme was observations from on board the Indian ship *Sagar Kanya* in January–March 1999 (prior to the American ship *Ronald H. Brown*) accomplishing an east–western voyage along the 20°S parallel in the region of the clear atmosphere south of the ITCZ, as well as in the region of the Arabian Sea (along 15°N, toward India). A comparison of the ARF data in the regions of the unpolluted atmosphere and in the presence of a thick aerosol layer has shown that in the second case the ARF was 6–10 times greater reaching $-35\,W\,m^{-2}$ (the coastal zone) and $-18.6\,W\,m^{-2}$ (the dust-loaded atmosphere over the ocean).

From the data of surface observations performed at a rural location in the region of the Great Plains, Oklahoma (USA) within the atmospheric radiation measurements (ARM) programme, [49, 50] tested the hypothesis of an indirect impact of atmospheric aerosol on climate proposed by Twomey. According to this hypothesis, aerosol particles, entering clouds and functioning as CCN, favour an increase of fine droplet number density and cause thereby an increase of cloud albedo, which causes a climate cooling (an important aspect of the hypothesis is that the cloud water content is assumed to be constant).

Solution of the problem of aerosol–cloud interaction is seriously complicated by numerous feedbacks appearing in the interacting microphysical, dynamic, and chemical processes. With the equivalent water content of clouds assumed to be constant, an analysis has been made [50] of the response of non-precipitating liquid-water clouds to changes in aerosol content. This response was characterized as a relative change of the effective radius of cloud droplets with a relative change of aerosol-induced extinction. The effective radius of droplets was retrieved from the data of radiation and microwave sounding (the spatial–temporal resolution of the observational results constituted, respectively, ~100 m and 20 s). The Raman lidar data served to retrieve the aerosol-induced extinction in the subcloud layer. An analysis of observational results has demonstrated that aerosol contained in marine air masses or in air masses from the north affects clouds more strongly than aerosol coming with air masses from the north-west. There is a sufficiently high correlation (0.67) between the response of cloudiness to aerosol and the intensity of turbulent mixing in clouds.

The interaction processes in the 'aerosol–cloud–radiation' system determining the indirect impact of aerosol on climate remain poorly studied, though they are an important factor of RF formation (the respective estimates vary between 0 and $-4.8\,\mathrm{W\,m^{-2}}$). Of importance is the contribution to indirect climatic impact of aerosol due to lower level stratus clouds, since: (1) their albedo is more sensitive to changes in the microphysical characteristics of clouds than the upper level high-albedo clouds (this is the first indirect impact of aerosol on clouds and climate); (2) a moderate geometrical thickness of clouds is often sufficient for droplets to reach the size of precipitating droplets, and therefore even a small increase of cloud droplet number density, N, can prevent precipitation, which affects the water content and albedo of clouds (this is the second indirect impact of aerosol on climate).

To analyse the formation and variability of the indirect climatic impact of aerosol within the second field experiment on the studies of aerosol (ACE-2) and the PACE programme to substantiate the parameterization of this impact, [124] undertook a comparison of six 1-D numerical models of the processes in the 'aerosol–cloud–radiation' system determining the climatic impact of aerosol under conditions of clear and polluted ABL. This development is divided into three stages. The first stage is aimed at an analysis of the adequacy of numerical modelling of aerosol activation as condensation nuclei, radiation transfer, and precipitation formation with high vertical resolution, as in the case of *in situ* observations in the ACE-2 period. During the second stage, similar tests of adequacy were made with a rougher vertical resolution. The objective of the third stage was numerical modelling for a 24–48-hour period to assess the possibilities to precalculate clouds in the ABL with a prescribed initialization of large-scale fields of meteorological parameters. To forecast the cloud droplets number density N, several schemes of parameterization have been used.

The results obtained in [124] revealed substantial differences in the cases of the use of physically substantiated prognostic schemes of aerosol activation taking into account the vertical velocity and empirical schemes based on the diagnostic data on the vertical velocity at the level of the cloud bottom. Prognostic schemes are characterized by a stronger variability of results compared with diagnostic ones because of differences in the consideration of the interaction between processes of aerosol activation and washing out of drizzle in the calculation of N.

In the case of initialization of 1-D models with a high vertical resolution, taking into account the observed vertical profiles of the cloud water content, a comparison was made of the results of numerical modelling with the observational data, which revealed a satisfactory agreement, deteriorating, however, if we confine ourselves to consideration of low-vertical resolution. Predicted precipitation turned out to be strongly underestimated, but this difference is reduced if we take into account the subgrid variability of the water content. It follows from calculations for the 24–48-hour period that, as a rule, estimates of cloud morphology turn out to be inadequate. Eventually, numerical modelling leads to considerable errors in assessing the optical properties of clouds. The precalculation of cloud morphology becomes more reliable with the use of schemes of the parameterization of the process of formation of cloud thickness as well as with a consideration of an external large-scale RF of the processes of cloud formation.

The accomplishment of the complex observational experiment LACE-98 has made it possible to obtain extensive information about atmospheric aerosol (aircraft measurements of the size distribution and number density of fine aerosols, coefficients of aerosol absorption, backscattering and depolarization, chemical composition of aerosol, as well as surface observations of the spectral optical thickness of the atmosphere and coefficients of extinction. Fiebig *et al.* [51] compared the obtained observational data on the optical parameters with results of numerical modelling for total H_2SO_4 aerosol near the tropopause as well as for ammonium sulphate/soot mixture in the remaining part of the air column (see also [135]).

This comparison has provided a closure (according to calculation results and observational data) with an error not exceeding 25% in the case of the AOT (with due regard to aerosol formed in biomass burning in North America and long-range transported with a 35% share of the soot component). With an assumed spherical shape of particles (with an average non-sphericity ratio of 1.3), the calculated depolarization of such aerosols agrees with the data of lidar soundings, whereas a comparison of calculated and observed values of the backscattering coefficient has shown that soot aerosol should be an external mixture component with non-absorbing particles. Based on the use of a 2-stream approximation of the theory of radiation transfer, the ARF at the level of the tropopause in the cloud-free atmosphere was estimated at $-5.8\,\mathrm{W\,m^{-2}}$ for an AOT $= 0.09$ (at $\lambda = 710\,\mathrm{nm}$) and solar zenith angle of 56°. The value of the ARF due to aerosol formed in biomass burning is equally sensitive to the state of the particle mixture (external or internal) and to surface albedo.

In the context of the problem of indirect climatic impact of atmospheric aerosol through the aerosol-induced changes of microphysical and optical characteristics of clouds, [137] discussed the results of the field studies of the impact of anthropogenic aerosol on the cloud droplet size distribution carried out in Canada. A comparison of calculated values of cloud albedo with observations has shown that the best agreement is observed when the calculated values are corrected taking into account the parameters of scaling β, which depend on relative MSD of the sizes of droplets ε that characterize an average radius of droplets and radius MSD. Here $\beta = (1 + 2\varepsilon^2)^{2/3}/(1 + \varepsilon^2)^{1/3}$. There is a positive correlation between the β parameter and droplet number density. This linear correlation has been used in numerical climate modelling by the ECHAM-4 model. Calculations have shown that the globally averaged value of the aerosol-induced indirect RF at the atmospheric top level decreased by $0.2\,\mathrm{W\,m^{-2}}$ as a result of correction with the β parameter being taken into account.

The indirect climatic impact of aerosol manifesting itself in the ABL is determined by numerous interactions between aerosol and the dynamics, microphysical properties, and optical properties of clouds. The input to the atmosphere of anthropogenic aerosol particles functioning as CCN, favours an increase of cloud droplet number density. As has been mentioned above, the related increase of the optical thickness and albedo of clouds, with a constant water content, was called the first indirect effect characterizing the climatic impact of aerosol.

On the other hand, also of importance is a change of cloud droplet size distribution, which affects its dynamics (first of all, via the process of precipitation formation leading to changes of cloud lifetime and spatial extent, on which the cloud albedo depends). Such a microphysical feedback affecting the cloud cover dynamics was called the second indirect effect, which determines the climatic impact of aerosol. Although a cloud albedo increase is comparatively small, being connected with manifestations of indirect effects, it can be substantial on global scales as a factor of attenuation of warming due to the atmospheric greenhouse effect. Therefore, studies of the indirect climatic impact of aerosol and its monitoring by satellites are extremely urgent.

Brenguier *et al.* [19] discussed the CLOUDCOLUMN (CC) project, one of the five projects accomplished within the ACE-2 programme of the second field experiment on the studies of aerosol characteristics with the aim of studying the indirect climatic impact of aerosol for marine stratocumulus clouds and to substantiate the strategy of 'closed' aerosol–cloud–radiation experiments. Observations within the CC project were made in June–July, 1997 in the region of the Canary Islands with the use of instruments carried by three flying laboratories and as part of a surface network.

The results have been discussed [19] of the eight series of aircraft measurements of microphysical characteristics of marine stratocumulus clouds under a broad range of observation conditions (different physico-chemical properties of aerosol, values of number density in the interval $50\text{–}25\,\mathrm{cm^{-3}}$, etc.). The obtained unique complex of synchronous observations of microphysical and radiative characteristics of cloud cover can be used to assess the indirect impact of aerosol on clouds and climate based on analysis of the ratio between the cloud optical thickness and effective radius of cloud droplets. Correlation between these values is usually negative, but in the heavily polluted atmosphere it can be positive. From the observational data obtained during ACE-2, polluted systems of clouds turned out to be somewhat drier and therefore thinner, which determined the formation of this positive correlation between indirect impact of aerosol on climate and an effective radius of droplets.

The product of incomplete burning of various fuels (primarily fossil fuels and biomass burning) is the so-called 'black carbon' (BC) – soot and smoke aerosol absorbing short-wave radiation. Estimates of direct RF due to BC and organic matter (OM) have led to values in the interval $+0.16$ to $+0.42\,\mathrm{W\,m^{-2}}$, and total absorbed radiation within $0.56\text{–}2\,\mathrm{W\,m^{-2}}$ (the parameter BC + OM is a soot component appearing due to fossil fuel burning). About 10% (by mass) of

Table 9.1. Annual average values of the total content of different types of aerosol in the atmosphere in the northern and southern hemispheres and globally.

Type of aerosol	NH	SH	Global
Anthropogenic sulphates SO_4^2	0.87	0.22	1.09
Natural sulphates SO_4^2	0.45	0.42	0.86
OC due to fossil fuels	0.39	0.03	0.41
BC due to fossil fuels	0.08	0.01	0.09
OC due to biomass burning	1.28	1.24	2.52
BC due to biomass burning	0.13	0.13	0.26
OC emissions at surface level	0.49	0.52	1.02
BC emissions at surface level	0.05	0.06	0.11
Natural OC	0.13	0.10	0.23
Dust ($r < 1\,\mu m$)	11.11	3.57	14.68
Sea salt aerosol ($r < 1\,\mu m$)	1.82	2.85	4.68

BC constitutes aerosol formed in biomass burning, for which RF values were -0.16 to $-0.74\,W\,m^{-2}$, whereas the radiation absorbed by aerosol varies within 0.75–$2\,W\,m^{-2}$.

Components of the products of biomass and fossil fuel burning responsible for radiation scattering (along with water soluble organic and inorganic components emitted to the atmosphere as smoke and soot compounds) can function also as CCN. This means that emissions of BC and OM participate in the formation of indirect RF due to the impact of CCN on the formation of clouds and their properties. It follows from available estimates that such a contribution can exceed 80% with respect to total indirect RF. The BC impact can also manifest itself through a local warming of the atmosphere and a decrease of cloud amount and cloud water content, leading to an albedo decrease. In connection with the earlier estimates of the related 'semidirect' RF, the conclusion was drawn about possible additional climate warming.

In connection with this, [139] evaluated the effect of soot and smoke aerosols on the climate with the use of the global climate model (GCM) with due regard to the effect of BC in cloud droplets on cloud albedo. The data of Table 9.1 characterize the annual average values of the total content of various types of aerosol used in this model. The numerical modelling of direct, semidirect, and indirect RF, with due regard to RF due to both short-wave and long-wave radiation, have led to the conclusion that the effect of the latter determines a decrease or even change of the sign of semidirect RF, but not an enhancement of warming.

The total RF substantially depends on the altitude of aerosol emissions since emissions at high altitudes cause an enhancement of negative long-wave RF. Besides this, emissions of absorbing aerosols at higher altitudes can enhance clouds at lower altitudes, where, as a rule, a temperature decrease can take place. According to estimates of the direct global average short-wave RF at the atmospheric top level

(TOA), it constitutes $0.17\,W\,m^{-2}$. With the use of the concept of 'quasi-RF' making it possible to partially take into account the impact of climatic feedbacks, the RF values are $0.28 \pm 0.32\,W\,m^{-2}$.

If, according to the earlier results, the radiation absorption by aerosol leads to a decrease of cloud amount and compensates for an aerosol-induced cooling, it follows from the numerical modelling results (based on the concept of quasi-RF) that the total RF (determined both by short-wave and long-wave radiation) turns out to be less (more negative) than before. Thus, the impact of smoke and soot aerosol manifests itself as a cooling effect taking into account changes of cloudiness and temperature taking place in the process of RF formation.

Calculations of the short-wave RF, considering all the anthropogenic aerosol in the regions where the content of BC exceeds $2\,mg\,m^{-2}$, gave values of -3.0 and $-3.1\,W\,m^{-2}$, for aerosol emissions due to biomass burning near the surface and in the middle troposphere, respectively. If the soot aerosol emissions take place near the surface then they determine, on average, a tropospheric warming at all levels and a decrease of cloud amount near the surface, where the emissions have occurred. The total 'quasi-RF' with due regard to both short-wave and long-wave radiation in case of soot aerosol is close to zero.

Recent estimates testify to a very strong impact of the phase state of atmospheric ammonium sulphate aerosol (with relative humidity $\sim 80\%$) on the level of the aerosol-induced RF. Also, of great interest is the impact of the aerosol phase state on the course of heterogeneous chemical reactions. Although the processes of deliquescence (water assimilation) and efflorescence (water loss) of pure ammonium sulphate have been well studied, the problem consists in the complexity of the chemical composition of aerosol particles, including up to 50% or more (by mass) of organic compounds.

Analysis of aerosol samples obtained at several locations across western Europe has shown that about 60% of the content of organic carbon in tropospheric aerosol is within the share of water-soluble organic compounds. According to observational data, at a rural location in Austria, mono and dicarboxylic acids constitute about 11% (with respect to the total content of organic carbon in cloud water). While insoluble organic compounds hamper an assimilation of water by aerosol, soluble organic matter, as a rule, favours water assimilation.

In view of insufficient information about the role of the phase transformations of aerosol, [21] carried out laboratory measurements in a running water vat with the use of the IR Fourier spectrometer to distinguish between particles' phase states as well as to study the processes of deliquescence and efflorescence in the case of ammonium sulphate, maleic acid, as well as internally mixed aerosol particles of maleic acid/ammonium sulphate mixtures. The obtained results indicate that in the case of aerosol particles of maleic acid, an assimilation of water begins at a low relative humidity of about 20% and continues until a maximum (89%) is reached, at which an assimilation of water is still possible. For particles of mixed composition (maleic acid/ammonium sulphate), an assimilation of water begins at a lower relative humidity than for particles consisting of just one of the components. Studies of efflorescence have led to the conclusion that crystallization of maleic acid particles takes place at relative humidities of < 30%. On the whole, the obtained results reflect the fact that the presence, in aerosol, of water-soluble organic matter internally mixed up with ammonium sulphate, broadens the range of conditions under which the aerosol remains in a liquid state.

Zender *et al.* [185] developed a numerical model which makes it possible to precalculate the number density and size distribution of atmospheric dust aerosol, which is to be used as a component of numerical models of climate and chemical processes in the atmosphere. The discussed Dust Entrainment and Deposition (DEAD) model has been used to simulate the global distribution of dust aerosol (DA) taking into account the processes of transformation of aerosol properties determined by involvement, dry and wet deposition, and chemical reactions with the participation of DA. Calculations have been made on the assumption that the soil texture is globally homogeneous and contains a sufficient amount of components which favour the process of saltation. The soil erodibility is parameterized with the use of a new physically substantiated geomorphic index, which is proportional to the river catchment area located up river from the region of each DA source. The processes of dry deposition are described with due regard to sedimentation and turbulent mixing. Processes of nucleation and size dependent washing out of DA particles in stratus and convective clouds are taken into account.

A comparison of the numerical modelling results with surface and satellite observational data revealed, on the whole, a satisfactory agreement. With the contribution of anthropogenic aerosol neglected, the DEAD model adequately simulates, for instance, the annual change of migration of the trans-Atlantic dust plume formed due to dust storms in Africa as well as a maximum of DA outflow from the Asian continent to the region of the Pacific Ocean. According to the results of numerical modelling, the global characteristics of DA and its variability in 1990 (particles with $D < 10\,\mu m$ were considered) are as follows: $1{,}490 \pm 160\,Tg\,year^{-1}$ (total DA emissions); $17 \pm 2\,Tg\,year^{-1}$ (the amplitude of the interannual variability of emissions); and 0.030 ± 0.004 (the optical thickness at $\lambda = 0.63\,\mu m$). The following data characterize the contribution of various continents ($Tg\,year^{-1}$): 980 (Africa), 415 (Asia), 37 (Australia), 35 (South America), and 8 (North America). All these estimates are substantially less than the values obtained earlier. The discussed results are characterized by underestimated transport and deposition of DA from eastern Asia and Australia to some regions of the Pacific Ocean, which is partially determined by the underestimated contribution of the long-range transport of particles $>3\,\mu m$. The results under discussion reflect the existence of 'hot spots' (positive anomalies of DA emissions) in the regions where easily saltating alluvial deposits are accumulated.

Strong anthropogenic emissions of great amounts of MGCs and aerosols in large cities attract attention to this problem in the context of both the possible impacts on the environment and humankind (in particular, on RF formation). One of the relevant examples is Mexico City located at an altitude of 2,240 m a.s.l., and its air basin (~18°–20°N, 98°–100°W) surrounded with mountains which serve as a barrier to atmospheric circulation. Processes of urban heat island formation are hindered in Mexico City by complicated relief. In the morning, when cold air masses have 'flow down' from the mountains to the urban territory, a situation of air stagnation occurs and, respectively, pollutant concentration. After sunrise, a warming of the south-western slopes of the mountains favours the input of wet air masses from the Gulf of Mexico. The high altitude of Mexico City causes a reduction of oxygen concentration and promotes an increase of the surface ozone concentration.

Based on the use of the NARCM regional model of the climate and the formation of the field of concentration and size distribution of aerosol, [128] calculated the transport, diffusion, and deposition of sulphate aerosol with the use of an approximate model of the processes of sulphur oxidation, not taking into account the chemical processes in urban air. However, the 3-D evolution of microphysical and optical characteristics of aerosol has been discussed in detail. The results of numerical modelling were compared with the data of observations near the surface and in the free troposphere carried out on 2, 4, and 14 March, 1997. Analysis of the time series of observations at Mexico City airport revealed low values of visibility in the morning due to the small thickness of the ABL, and a subsequent improvement in visibility with an increase of the ABL thickness. Estimates of visibility revealed its strong dependence on wind direction and aerosol size distribution. Calculations have shown that an increased detail of the size distribution presentation promotes a more reliable simulation of the coagulation processes and a more realistic size distribution characterized by the presence of the accumulation mode of aerosol with the size of particles $\sim 0.3\,\mu m$. In this case, the results of visibility calculations also become more reliable.

In connection with the important role of sulphate aerosol (SA) in the formation of the direct and indirect impact of anthropogenic SA on RF, which affects climate, great attention has been recently paid to ARF numerical modelling. The calculated values of global average direct and indirect ARF change, respectively, from -0.2 to $-0.8\,W\,m^{-2}$ and from 0 to $-1.5\,W\,m^{-2}$, that is, they are comparable with positive RF due to the growth of GHG concentrations. A wide range of uncertainties in calculated ARF are mainly determined by the approximate results of numerical modelling of global spatial–temporal variability of the aerosol content in the atmosphere as well as by the difficulties of adequate consideration of aerosol–cloud interaction.

Gong and Barrie [57] undertook a numerical modelling of the impact of aerosol on climate with the use of an aerosol module developed in Canada, describing the global variability of sea salt and sulphate (both natural and anthropogenic) aerosol (considering its size distribution) and the latest (third generation) model of the global climate. The size distribution of both types of aerosol is parameterized by dividing the spectrum of particle sizes into 12 intervals. The aerosol is assumed to be internally mixed. A comparison of calculated spatial–temporal variability of aerosol concentration with the observed one revealed satisfactory agreement. This refers, in particular, to the marine ABL (MABL). It has been shown in [57] that sea salt aerosol particles favour a suppression of the process of nucleation, an increase of surface area for condensation, and a change of the properties of clouds in the MABL, which determines a redistribution of mass concentration and number density of sulphate aerosol. The results of separate numerical modelling of the dynamics of sea salt and sulphate aerosol suggested the conclusion that the presence of sea salt aerosol almost doubles the diameter of sulphate aerosol particles in the MABL at high concentrations of sea salt aerosols but causes a decrease (on global scales) of the mass of sulphate aerosol in the surface layer of the MABL by 5–75% (depending on the distribution of sea salt aerosol particles).

Most substantial impacts of this kind take place in the mid-latitudes of the northern and southern hemispheres, and minimum of impacts near the equator. The anthropogenic pollution of the atmosphere in the regions of the Pacific and Atlantic Oceans in the northern hemisphere results in decreasing concentrations of sulphate aerosol within 10–30% due to sea salt aerosol. A maximum decrease (down to 50–75%) takes place in the southern hemisphere 'roaring forties' in the spring and autumn. The average global estimates of the impact of sea salt aerosol on decreasing mass and number of sulphate particles are, respectively, 9.13% and 0.76%.

The impact of sea salt aerosol determines a decrease in droplet number concentrations in marine clouds (CDNC) by 20–60%, with a maximum decrease in the northern hemisphere 'roaring fourties' (40–70%) and mid-latitudes (20–40%), which are characterized by a high concentration of sea salt aerosol. Some increase of CDNC due to sea salt aerosol was also observed in the equatorial band. The impact of changes of the aerosol content and cloud cover on global climate will be estimated in the future.

The first indirect climatic effect of aerosol (the Twomey effect) is based on the assumption that with a constant equivalent liquid water content (LWC) of clouds an increase of atmospheric aerosol number density (and, hence, concentration of CCN) should lead to an increase of cloud droplet concentrations and cloud albedo. In this connection, [49] carried out a numerical modelling to analyse the possibility of using the extinction coefficient α, retrieved from the data of surface remote sensing, for the subcloud atmosphere as an indirect indicator of the impact of aerosol on the size distribution of cloud droplets. An adiabatic model of cloud droplets limited by a consideration of a thin layer of non-precipitating stratocumulus clouds (it is in this case that the supposition with respect to adiabatic nature can be considered acceptable) makes it possible to reproduce the hygroscopic growth of CCN and to take into account the water vapour condensation on droplets (neglecting the growth of droplets due to coalescence). The model considered was used to calculate the cloud droplets size distribution at a given size distribution of ammonium sulphate aerosol when the masses move upward at a velocity of 20–300 cm s^{-1}. A one-modal size distribution of aerosol N is approximated by a log-normal size distribution of particles at $20\,\text{cm}^{-3} \leq N_a \leq 3{,}000\,\text{cm}^{-3}$, median radius of particles $0.03\,\mu\text{m} \leq r_g \leq 0.1\,\mu\text{m}$, and the distribution width $1.3 \leq \sigma \leq 2.2$. Different values of the mass share of ammonium sulphate and cloud water content are prescribed.

Estimates of sensitivity of the effective radius of droplets r_e to various input parameters have shown that r_e changes especially strongly depending on LWC (the impact of variations of these parameters has earlier been neglected when estimating the indirect effect of aerosol on clouds and climate). The relative role of other parameters changes depending on observation conditions, but the importance of N_a remains unchanged. The impact of vertical motions manifests itself most at a high concentration of aerosol. Analysis of all the results obtained has led to the conclusion that the use of the extinction coefficient as an indirect indicator of the impact of aerosol on cloud droplets size distribution can lead to underestimation of the importance of the first indirect impact. The levels of possible systematic errors of the extinction coefficient remain uncertain in view of their dependence on varying characteristics of aerosol (e.g., the aerosol number density N_a cannot be retrieved from the data of remote sensing), vertical motions, and LWC.

Fortmann [53] discussed in detail the microphysical and optical characteristics of atmospheric aerosol which determines the formation of aerosol radiative forcing (ARF), that is, the climatic impact of aerosol in the Arctic. In particular, the following problems have been considered: (1) physical and optical parameters of tropospheric aerosol and its specific impact on climate and on the formation of the Arctic haze; (2) general problems of measurements and numerical modelling of aerosol properties; (3) a regional climate model HIRHAM-4 with an aerosol unit for the Arctic (latitudes >65°N); (4) a 1-D model of radiation transfer and a consideration of the impact of aerosol and clouds on radiation transfer; (5) the use of the HIRHAM-4 climate model to assess the climatic impact of the Arctic haze (AH); and (6) an assessment of ARF from observational data.

Analysis of the numerical modelling results has shown that a consideration of AH determines an additional positive ARF within 1–8 $W\,m^{-2}$ at the atmospheric top level over the snow–ice surface of the Arctic Ocean, but over the open water of the Atlantic and Pacific Oceans the ARF values become negative, constituting about 2 $W\,m^{-2}$. The impact of AH on ARF at surface level manifests itself as an additional negative ARF reaching a maximum ($-7.5\,W\,m^{-2}$) over the open water of the ocean. An averaging of the short-wave ARF over the whole region of the Arctic gave a value of about $-2.4\,W\,m^{-2}$. Thus the total ARF at the surface level is $-1.7\,W\,m^{-2}$.

The estimates of aerosol-induced changes of SAT obtained with the use of the climate model have shown that these changes vary from -1 K (over the east Siberian Sea and Canada archipelago) to $+1$ K (in the region of Spitzbergen – Barents Sea, Kara Sea, and Taimyr peninsula). The possibility has been revealed of considerable interannual variations of monthly mean SAT reaching 2 K. It has been shown that both direct and indirect (through affecting clouds) impacts of aerosol on climate are equally important. Use of the data of observations of aerosol characteristics near Spitzbergen (within the ASTAR programme on studies of tropospheric aerosol and radiation) has demonstrated a strong regional variability of the impact of aerosol on SAT from a cooling of -2 K in the Baffin Gulf and in the Laptev Sea to a warming of about 3 K in the Beaufort Sea.

Feczkó *et al.* [46] estimated the radiative forcing due to aerosols and GHGs for Hungary in the early 1980s. The obtained results revealed considerable changes in RF due to ammonium sulphate and CO_2 for the last two decades. While the contribution to climate warming of the greenhouse effect as a result of increasing CO_2 concentration, increased by 60%, the anthropogenic contribution to climate cooling due to sulphate aerosol decreased by 45% (i.e., the impact of both these atmospheric components on climate manifested themselves through a warming).

Various kinds of vegetation emit to the atmosphere a great amount of non-methane hydrocarbons (NMHC). The total level of emissions in the period of vegetation of such NMHC as isoprene (C_5H_8), monoterpene ($C_{10}H_{16}$), sesquiterpenes ($C_{15}H_{28}$), and oxygen-containing compounds ($C_nH_{2n-2}O$) constitutes 825–1,150 TgC year^{-1}, exceeding the level of respective anthropogenic emissions (about 100 TgC year^{-1}).

As Barr *et al.* [9] have pointed out, the importance of such emissions is determined mainly by their impact on the three processes taking place in the atmosphere. The first process consists of the carboxylation of isoprene in plants, which influences the biospheric carbon cycle. The second process is connected with the fact that NMHC exhibits a high chemical activity with respect to such main oxidants as hydroxyl radicals (OH), ozone (O_3), and nitrate radicals (NO_3). Reactions with the participation of such components result in the formation of radicals of alkylperoxides (RO_2), which favour an efficient transformation of nitrogen monoxide (NO) into nitrogen dioxide (NO_2), which favours an increase of ozone concentration in the ABL. Finally, NMHC oxidation leads to the formation of such carbonyl compounds as formaldehyde (HCHO), which stimulates the processes of O_3 formation. An oxidation of monoterpenes and sesquiterpenes results in the intensive formation of fine carbon aerosol with diameters of $<0.4\,\mu m$ (the share of such aerosol production reaches 10–30%), substantially affecting various processes in the ABL.

Barr *et al.* [9] performed an analysis of the impact of phytogenic aerosol (PhA) which is defined as forming mainly due to monoterpene oxidation (first of all, α and β-pinenes) over the forests in the eastern part of Canada. In the forest ecosystem the level of emissions to the atmosphere of biogenic hydrocarbons was moderate, with the concentration of α and β-pinenes constituting about 1.6 ppb. The NMHC oxidation resulted in the formation of PhA at a number density of particles of about $5 \times 10^8\,cm^{-3}$. For a given concentration and size distribution of aerosol, its impact on the short-wave radiation transfer in the ABL has been assessed.

Under clear sky conditions in July and with aerosol number densities within $(2–5) \times 10^3\,cm^{-3}$, the diurnal mean attenuation of global radiation constituted $0.04\,W\,m^{-2}$ with the contribution of scattered radiation equal to $0.01\,W\,m^{-2}$. A maximum level of global radiation attenuation due to PhA reached $0.2\,W\,m^{-2}$. It follows from the obtained results that the PhA-induced reduction of the input of short-wave radiation can markedly compensate (on a regional scale) for the contribution of warming due to the enhanced atmospheric greenhouse effect. The PhA is also important as a factor of the impact on the optical properties of clouds functioning as CCN.

The global contribution of vegetation cover to emissions of volatile organic compounds (VOC) to the atmosphere constitutes about 90%. Biogenic emissions (BVOC) include isoprene (C_5H_8), monoterpenes ($C_{10}H_{16}$), and other chemically active carbon compounds. The total content of carbon in global emissions of BVOC can exceed 1 Pg year^{-1}. Many BVOCs react with surface ozone and other oxidants in the atmosphere, substantially affecting the chemical processes in the atmosphere on local, regional, and global scales. The presence of BVOC determines, for instance, an increase of the lifetime of methane in the troposphere by about 15%. Due to BVOC transformation, 13–24 Tg year^{-1} of secondary organic aerosol are formed in the global atmosphere, which is comparable with the level of carbon aerosol formation due to fossil fuel and biomass burning. As [104] have noted, biogenic aerosol seriously affects the formation of the radiation budget and CCN. This means that BVOC emissions strongly affect global climate formation through their effect on chemical processes, atmospheric aerosol concentration, and the global carbon cycle. Therefore, the IPCC has recommended VOC emissions be taken into account in scenarios for emissions used in numerical climate modelling. According to respective estimates, the level of anthropogenic VOC emissions in 1999 averaged about 140 TgC year^{-1}, and estimates for the year 2100 give possible limits from <100 to >550 TgC year^{-1}. BVOC emissions depend on the type of vegetation and environmental conditions (air temperature, solar radiation, water supply of plants, O_3 and CO_2 concentrations).

Levis *et al.* [104] described an algorithm based on the data of field and laboratory measurements, which enables one to calculate BVOC emissions being used as a component of the interactive climate model CCSM (version 2.0) for the 'atmosphere–ocean–land–sea ice cover' system developed by NCAR (USA) scientists. This development represents a first step in constructing a global climate model, taking into account the dynamics of biogeochemical cycles. To analyse the functioning of the CCSM model, two numerical experiments have been carried out, taking into account: (1) only land surface processes with the prescribed state of the atmosphere and satellite data on vegetation cover; (2) completely interactive processes in the climate system (including the precalculated dynamics of vegetation cover).

Calculation results have shown that in both cases an enhancement of BVOC emissions takes place in warm and forested regions (compared with other regions), agreeing with observed data. With the prescribed distribution of vegetation cover, global emissions of isoprene on land constitute 507 TgC year^{-1}, which closely corresponds to the results of interactive numerical modelling. In the case of the interactive model, BVOC emissions depend on climate change and vegetation cover, varying from year to year. The interannual variability of calculated anthropogenic emissions can exceed 10% with respect to annual anthropogenic emissions corresponding to IPCC scenarios. This necessitates a consideration of BVOC emissions in the interactive global climate model.

The validity of this conclusion is confirmed by results of studies of RF due to organic aerosols obtained by [117]. A consideration of carbonaceous components in dust aerosol and aerosol products of combustion detected in Africa, Asia, and North America, has shown that the rate of chemical reactions, with the participation of MGCs, is 3 times lower than characteristic values usually used in numerical climate modelling. Slower rates of transformation of volatile or hydrophobic organic compounds into condensed and hygroscopic ones results in an increasing amount of carbonaceous particles taken into account in climate models of up to 70%. As a result, the climatic RF due to secondary organic aerosol increases to $-0.8\,W\,m^{-2}$

(cooling due to scattering) and $+0.3\,\mathrm{W\,m^{-2}}$ (warming due to absorption). As a result, an absolute RF difference constitutes $1.1\,\mathrm{W\,m^{-2}}$ (i.e., reaches a half of the 'greenhouse' RF ($2.2\,\mathrm{W\,m^{-2}}$)).

9.6 THE IGAC PROGRAMME, ACE-ASIA, AND INDOEX PROJECTS

The international IGAC programme on the study of chemical processes in the global atmosphere included the ACE field observational experiments studying the atmospheric aerosol properties whilst bearing in mind estimates of the ARF of the climate. As [71] have noted, the fourth of seven related experiments (ACE-Asia) was aimed at:

(1) Complex observations of spatial–temporal variability of concentration and properties of aerosol (including the determination of ARF), as well as processes controlling the formation and evolution of aerosol properties.
(2) Long-term observations of aerosol characteristics for the period 2000–2003.

The system of observational means for ACE-Asia consisted of three flying laboratories, two ships, and a network of lidars and surface stations for aerosol sampling. Besides this, various data of satellite observations have been used. Accomplishment of the observational programme was combined with the numerical modelling of the processes of formation, transport, and transformation of aerosol using the CTM models which simulate the long-range transport of aerosol with its chemical transformation taken into account. The numerical modelling results were important for an adequate planning of observations, especially in the period of intensive observations when there was an anomalously great content of dust aerosol in the atmosphere. Observational data were used to test the reliability of the existing models of atmospheric aerosols used in calculations of radiation fluxes, application of CTM, and assessment of the impact of aerosol on global climate change.

The main result of the ACE-Asia field experiment was obtaining information about aerosol, which illustrates its widely varying physical and optical properties due to a complicated chemical composition of aerosol as a mixture of natural DA and anthropogenic components such as black carbon, sulphates, nitrates, and organic compounds. The observational data testify to the key role of aerosol in the formation of RF. So, for instance, under clear sky conditions the direct ARF in April reached $-30\,\mathrm{W\,m^{-2}}$. The ACE-Asia results have brought forth new questions to be answered, which necessitates a continuation of similar development.

In China, the Yangtze River delta is the main agricultural region. It is characterized by a strong aerosol pollution of the atmosphere. Xu *et al.* [183] discussed the results of measurements performed with the multichannel (wavelengths 380, 440, 500, 675, and 875 nm) sun photometer MICROTOPS-II. Measurements were made of AOT τ_λ, scattering coefficient σ_{sp}, absorption coefficient σ_{ap}, and downward flux of photosynthetically active radiation (PhAR, spectral interval 400–700 nm), during the period from 28 October to 1 December, 1999 in Linan (background station for atmospheric monitoring: 30°17′N, 119°45′E) to obtain quantitative data on optical properties and direct ARF. With the use of two schemes of radiation transfer and observational data, the ARF values for PhAR and for total short-wave radiation (TSWR (0.2–4 μm)) were estimated at surface and atmospheric top levels (SL, ATL).

The calculated average ARF value for the cloud-free atmosphere (for solar zenith angles <70°) for PhAR at the surface level constituted $-73\,\mathrm{W\,m^{-2}}$ ($0.2 < \tau_{500} < 1$), which agrees with the respective value ($-74.4\,\mathrm{W\,m^{-2}}$) found from measured PhAR and AOT at 500 nm (τ_{500}). According to observational data, the average value of $\tau_{500} = 0.6$, the diurnal average value of ARF at the surface level is $-11.2\,\mathrm{W\,m^{-2}}$, and the optical thickness and amount of clouds constitute 5.0 and 50%, respectively. This means that a PhAR decrease due to direct ARF constitutes −16%.

Calculations of direct ARF at the TOA level for TSWR averaged over 24 hours gave $-30.4\,\mathrm{W\,m^{-2}}$. Considering the presence of clouds, this estimate decreases to about $-12.1\,\mathrm{W\,m^{-2}}$. This value exceeds the global average ARF values approximately by an order of magnitude in the range −0.3 to $-1.0\,\mathrm{W\,m^{-2}}$ assumed in the IPCC 1996 Report. These results demonstrate a very strong impact of aerosol on the radiative regime of the troposphere, determined by such parameters as PhAR and TSWR at the surface level (SL) as well as the SWR budget at ATL.

Based on the use of the data of satellite and aircraft observations, as well as numerical modelling results, [132] have estimated the ARF at SL and TOA in the region of the east China Sea for April 2001. An analysis of observational results has shown that they change by up to 40% depending on the applied data processing technique (from prescribed input parameters, especially single-scattering albedo).

Monthly mean RF values under real cloud conditions vary from −5 to $-8\,\mathrm{W\,m^{-2}}$ at TOA and from −10 to $-23\,\mathrm{W\,m^{-2}}$ at SL at Gosan (33°28′N, 127°17′E) and at Amami-Oshima (28°15′N, 129°30′E). The respective average RF values are $-5.6 \pm 0.9\,\mathrm{W\,m^{-2}}$ (Gosan) and $-7.5 \pm 1.5\,\mathrm{W\,m^{-2}}$ (Amami-Oshima) at TOA and $-15.8 \pm 6.6\,\mathrm{W\,m^{-2}}$ and $-18.2 \pm 5.9\,\mathrm{W\,m^{-2}}$ at SL.

There is a high regional variability of RF due to variations of AOT and single-scattering albedo. The cloud RF (CRF) turned out to be equal to −20 to $-40\,\mathrm{W\,m^{-2}}$, from which it follows that direct aerosol RF (ARF) is comparable with the CRF at SL. Estimates of indirect ARF with the use of data of satellite observations and the SPRINTARS model of the long-range transport and evolution of properties of aerosol have led to values ranging between −1 and $-3\,\mathrm{W\,m^{-2}}$ both at TOA and SL. Since these values exceed the global average value ($\sim -1\,\mathrm{W\,m^{-2}}$), this means that the latter results from averaging the negative RF over the ocean and positive RF over land.

The 3-D global model SPRINTARS makes it possible to simulate the spatial–temporal variability of aerosol concentration and its optical properties based on a consideration of the totality of processes responsible for the formation of atmospheric general circulation and radiation transfer. Takemura *et al.* [163, 164] undertook a numerical modelling whose results were compared with the data of complex surface, aircraft, and satellite observations carried out in the spring of 2001 within the field experiment ACE-Asia in eastern Asia.

Analysis of the obtained results has shown that calculated values of aerosol concentration and optical thickness, as well as the Angström exponent, agree well with the observational data. A consideration of the data on the Angström exponent suggests the conclusion that the contribution of anthropogenic aerosol, determined by its sulphate and carbon components, to the formation of atmospheric AOT is of the same order of magnitude as the contribution of DA even in the case of dust storms.

Estimates of direct ARF gave negative values exceeding $-10\,\mathrm{W\,m^{-2}}$ at the tropopause level in the periods of powerful dust storms (in this case the contributions of anthropogenic and dust aerosols to the AOT formation are practically equal). Direct RF depends mainly on cloud water content and on the vertical distributions of aerosol and clouds. The numerical modelling has demonstrated that not only sulphate and sea salt aerosol, but also black carbon and soil–dust aerosol, which determine the absorption of both SWR and LWR, can cause a direct negative RF at SL which in the east Asian region can exceed the positive RF due to anthropogenic GHGs.

In the period of accomplishment of the ACE-Asia field observational experiment in April 2001 in the north-western region of the Pacific Ocean, [29] performed measurements of the content of elemental carbon (EC) in aerosol particles, which have shown that EC is concentrated mainly in large particles, which initially arrives in the atmosphere during dust storms in deserts or via aeolian origins, a coagulation of fine particles taking place in the process of long-range transport also play an important role. Calculations of diurnal averages of the mass coefficients of EC absorption gave, respectively, 12.6 ± 2.6 and $14.8 \pm 2.3\,\mathrm{m^2\,g^{-1}}$ for PM-10 (particles with $D < 10\,\mu\mathrm{m}$) and PM-1 ($D < 1\,\mu\mathrm{m}$). From the data for a limited number of days, the absorption coefficient (AC) for only large particles (PM-10 – PM-1) constituted $5.7 \pm 1.6\,\mathrm{m^2\,g^{-1}}$ in the dust-loaded atmosphere and $2.0 \pm 1.0\,\mathrm{m^2\,g^{-1}}$ in the clear atmosphere. It follows from the observational data that fine aerosol particles have been internally mixed. There was an inverse proportionality between AC and EC mass concentrations (apparently, this result cannot be explained by only the impact of air mass 'ageing').

If the obtained results are extrapolated onto global scales, it can be assumed that a reduction of EC emissions to the atmosphere would be a factor in reducing the efficiency of measures to limit undesirable climate changes. Numerical modelling of the radiative properties of DA particles with the prescribed idealized non-spherical shape has shown that taking into account the non-sphericity is important whilst considering the scattering properties of particles, but practically unimportant for determination of absorption. According to theoretical estimates, a coagulation of EC and DA can result in changes of short-wave radiation absorption by DA layers ranging between -42% and $+58\%$, depending on suppositions with respect to the initial mixing of EC and the optical properties of DA. It follows from observations, however, that such changes vary from −10 to −40%, with −10% being most probable.

During the ACE-Asia field experiment in Gosan (South Korea; 33.29°N, 126.16°E) during the period 25 March–4 May, 2001, surface actinometric observations were made to study the formation of direct ARF, especially under conditions of the input of DA to the atmosphere as a result of heavy dust storms on the Asian continent. The obtained observational data include information about total fluxes of global, direct solar, and scattered radiation as well as about respective radiation fluxes in the visible and near-IR spectral regions. To assess the AOT, τ_{500}, at $\lambda = 500$ nm, data from observations with the scanning radiometer were used.

Bush and Valero [22] obtained the following values of the RF/τ_{500} ratio averaged over the whole period of observations for the whole short-wave range, near-IR, and visible spectral regions, respectively: -73.0 ± 9.6, -35.8 ± 5.5, -42.2 ± 4.8 W m^{-2}/τ_{500}. Also, a new parameter has been estimated – a relative efficiency of RF defined as the ratio of RF to total SWR flux at the TOA. These values are: -18.0 ± 2.3, -16.2 ± 2.4, -26.7 ± 3.3%/τ_{500}. As seen, the relative efficiency of RF is at a maximum in the visible spectral region.

Conant *et al.* [35, 36] substantiated the model of the vertical profiles of size distribution, chemical composition, and hygroscopic properties of aerosol from the data of observations with the use of instruments carried by the flying laboratory *Twin Otter* of the centre for interdisciplinary developments with the use of aircraft, CIRPAS, and on board the ship *Ronald H. Brown*. The results of observations have been obtained in the Asia–Pacific region for the period of the ACE-Asia field observational experiment. The model takes into account such aerosol components as sulphates, black and organic carbon, sea salts, and mineral dust and the dependence of the content of these components on relative humidity and the state of mixing (internal or external mixing of particles, covering of DA particles with a polluting film).

The aerosol model together with the Monte Carlo method of calculation of radiation fluxes has made it possible to calculate the direct ARF for the short-wave spectral region. The calculated average regional values of ARF efficiency at the surface level (changes of short-wave radiation flux under clear sky conditions per unit AOT at $\lambda = 500$ nm) constituted -65 W m^{-2} during the whole period of intense observations within the ACE-Asia programme (only DA) and -60 W m^{-2} (aerosol due to pollutions). Based on the combined use of the CFORS model of aerosol transport with due regard to its chemical transformation and scheme of radiation transfer, the ARF was calculated for conditions of clear and cloudy skies during the period of ACE-Asia.

During the 5–15 April, 2001 dust storm, the direct short-wave ARF at TOA constituted -3 W m^{-2}, and at surface level -17 W m^{-2} (these estimates refer to real cloud conditions in the region 20°–50°N, 100°–150°E). The results of calculations of radiation fluxes and RF agree well with the data of ship observations. It was shown that the effect of clouds (compared with clear sky conditions) leads to a decrease of ARF at the TOA by a factor of 3. Under these conditions, clouds do not affect substantially the SWR absorption.

The estimate of the ARF ($-3\,W\,m^{-2}$) is close to that of RF due to GHGs but of an opposite sign. The main impact of aerosol on SWR transfer manifests itself as a redistribution of solar warming between the surface and the atmosphere (in favour of the atmosphere). Compared with the data of the INDOEX field experiment, according to which the averaged efficiency of RF at SL constitutes $-72\,W\,m^{-2} \times \tau_a$ (τ_a is the AOT), in the case of ACE-Asia, a lesser value was obtained, $-60\,W\,m^{-2} \times \tau_a$. But the ratio at TOA turned out to be -25 and $27.5\,W\,m^{-2}$, respectively.

The main difference between RF due to DA plumes in eastern Asia and southern Asia (INDOEX) manifested itself through the case of ACE-Asia: (1) the impact of DA turned out to be more substantial (in this case the anthropogenic contribution remains unknown, however); (2) the impact of the clouds of the upper and middle levels is more substantial (in contrast to prevalent low-level clouds typical of INDOEX); (3) a more complicated vertical distribution of aerosol was observed due to the effect of mid-latitude frontal systems. An interesting result of observations in the ACE-Asia period was that DA particles together with the secondary aerosols such as those consisting of sulphates and organic carbon, scatter short-wave radiation weaker than aerosol consisting of an external mixture of DA and secondary aerosol. However, if DA particles are covered with black carbon, the absorption does not change markedly.

Markowicz *et al.* [118] discussed the results of observations of chemical composition and optical properties of aerosol carried out on board the ship *Ronald H. Brown* during the period of the ACE-Asia field experiment (March–April 2001) in the western sector of the Pacific Ocean (mainly in the Japan Sea) in order to assess the ARF. The obtained average value of AOT at $\lambda = 500\,nm$ constituted 0.43 ± 0.25 over the Japan Sea, and the single-scattering albedo was 0.95 ± 0.03. There was observed a high correlation between single-scattering albedo and RH ($r^2 = 0.69$).

The presence of aerosol in the atmosphere determined a decrease of global radiation by $26\,W\,m^{-2}$, but an increase of absorbed atmospheric SWR and outgoing (reflected) SWR, respectively, by $13.4\,W\,m^{-2}$ and $12.7\,W\,m^{-2}$. An averaged efficiency of ARF (RF calculated per unit AOT) constituted $-60\,W\,m^{-2}$. A decrease of RH to 55% led to an increase of ARF efficiency reaching $6–10\,W\,m^{-2}$. The dependence on RH necessitates its consideration in comparing the data on ARF efficiency for different observational conditions.

Based on the use of satellite information (AVHRR and POLDER instruments), [158] obtained quasiglobal estimates of direct RF over the oceans ($-0.4\,W\,m^{-2}$) and indirect RF (from -0.6 to $-1.2\,W\,m^{-2}$), from which it follows that indirect RF turned out to be stronger than direct RF. According to estimates of [103], the global average ARF under clear sky conditions constitutes $-2.2\,W\,m^{-2}$.

Using the NJUADMS model of the long-range transport of sulphate aerosol and the regional climate model RegCM2, [177] obtained a value of direct RF due to sulphate aerosol averaged over the territory of China, equal to $-0.85\,W\,m^{-2}$, whereas the RF values in the regions of central and eastern China can reach $-7\,W\,m^{-2}$. The RF variability is characterized by a strong annual change.

Menon *et al.* [123] drew attention to the fact that under conditions of a climate warming over most of the globe during the last decades there was an increasing trend of floods in the south of China, an enhancement of droughts in the north of China, and a moderate climate cooling in China and India. With the use of the global climate model, estimates have been obtained from which it follows that a consideration of the climatic impact of aerosol consisting mostly of BC absorbing short-wave radiation, makes it possible to adequately reproduce the observed change of precipitation and temperature. The presence of absorbing aerosols determines an air warming. It changes the regional stability of the atmosphere and vertical motions, affect the large-scale atmospheric circulation and water cycle, causing considerable changes in regional climates.

Summarizing the results of the ACE-Asia project, [155] emphasized that the available estimates of the climatic impact of huge emissions of dust and anthropogenic aerosol from the territories of China and Mongolia to the Pacific basin (as has been mentioned above, sometimes products of these emissions reach the territory of North America) are still uncertain, though their considerable impact on radiation transfer, climate, chemical processes in the atmosphere, and cloud cover dynamics raises no doubts. The interaction between DA and atmospheric pollutants can increase the solubility of Fe and other nutrients, which can substantially affect the biodynamics of the ocean (e.g., almost a doubling of biomass was recorded in the mixed layer of the ocean due to the input of dust storm products, which can be explained by an induced 'fertilization' of the ocean).

The data of Table 9.2 illustrate the estimates of the contribution of various factors to the ARF formation at the SL and the TOA as well as short-wave radiation absorbed by the atmosphere (ASWR).

Detailed estimates of the errors in ARF calculations caused by insufficiently reliable data on sulphate aerosol have been carried out by [130].

The main objective of the INDOEX field observational experiment was to study anthropogenic aerosols transported to the region of the Indian Ocean from the south

Table 9.2. Estimates of the contribution of various factors to the ARF formation and ASWR ($W\,m^{-2}$) during the period 5–15 April, 2001 in eastern Asia (20°–50°N, 100°–150°E).

Factor	SL	ASWR	ATL
Dust aerosol	−9.3	3.8	−5.5
Sulphates	−3.6	0.3	−3.3
Organic carbon	−3.9	1.7	−2.2
Black carbon	−4.1	4.5	0.4
Sea salt	−0.4	0.0	−0.4
Internal mixture	−2.2	3.5	1.3
Long-wave RF	3.0	−2.3	0.7
Total RF (clear sky)	−20.5	11.5	−9.0
Total RF (real cloud conditions)	−14.0	11.0	−3.0

Asian continent and to assess the impact of these aerosols on the atmospheric radiative regime. Surface observations have been carried out at the climatic observatory Kaashidhoo (4.95°N, 73.5°E), in Minicoy (8.5°N, 73.0°E), and in Trivandrum (8.5°N, 77.0°E). The satellite instrumentation CERES was a source of information about the Earth's radiation budget (ERB) components.

An analysis of the results of long-term observations (for more than 15 years) in Trivandrum carried out by [153a] revealed a gradual increase of the aerosol optical thickness (AOT) in the visible spectral region from about 0.2 in 1986 to ~0.4 in 1999. The contribution of anthropogenic sources to AOT exceeded 70%. To estimate the annual change of aerosol characteristics, the observational data have been considered [153a] during, 'before' and 'after' monsoon periods. The respective variability turned out to be substantial. The available surface and satellite data on radiation fluxes have made it possible to study the impact of aerosol on SWR fluxes.

Analysis of information considered has led to the conclusion that to calculate the scattered radiation fluxes, it was not necessary to take into account the so-called 'excessive absorption'. The aerosol-induced SWR absorption by lower atmospheric layers reached $\sim 20\,W\,m^{-2}$, which corresponded to a radiative change of temperature by about 0.5 K per day. The effect of the mixed aerosol particles (internal or external) on the ARF value turned out to be negligibly small. A comparison of radiative effects of aerosol on land and over the ocean has demonstrated that an enhancement of radiative warming of the lower atmosphere by about 40% is accompanied by simultaneously increased cooling at the surface level constituting ~33% with respect to that observed over the ocean. In the case of the ocean, the ARF constitutes $-10\,W\,m^{-2}$ (TOA) and $-29\,W\,m^{-2}$ (SL), and radiation absorbed by the atmosphere is $19\,W\,m^{-2}$. On land, the respective values are $+7\,W\,m^{-2}$ (TOA) and $-21\,W\,m^{-2}$, whereas the absorbed radiation is $28\,W\,m^{-2}$. Wind intensification near the ocean surface results in an increase of cooling due to the growth of sea salt aerosol concentration which partially compensates for the warming of the lower atmospheric layers determined by the SWR absorption by aerosol.

Yabe *et al.* [184] obtained ARF values from the data of surface observations in Kyoto very close to those obtained during the INDOEX period. The average ARF value calculated per unit optical thickness constituted $-85.4\,W\,m^{-2}$. Calculations of ARF at TOA gave values 3 times less than those at the SL (this suggests the presence of strong SWR absorption by the atmosphere).

In September 2000, over the Atlantic Ocean near the western coast of Africa, a field observational experiment SHADE was accomplished in order to study the ARF from the data of aircraft measurements of characteristics of the Saharan dust aerosol in the atmosphere and radiation fields in the short and long-wave regions. An important part of the scientific programme was the simulation of observational data by numerical modelling, based on the use of the CTM-2 model of aerosol long-range transport with due regard to its chemical transformation with the input of meteorological information prescribed from the data of ECMRWF for September 2000.

Myhre *et al.* [129] calculated ARF for the short and long-wave regions. The results of numerical modelling of the DA long-range transport and ARF values agree well with observational data. The calculated maximum negative ARF value for the short-wave region constituted $-110\,W\,m^{-2}$ (this value refers to local noon on 26 August, 2000). The diurnal mean ARF values during the period of SHADE varied from -5 to $-6\,W\,m^{-2}$, and with a conditional global averaging the ARF was $-0.4\,W\,m^{-2}$. These results illustrate an urgency in taking into account the climatic impact of aerosol. Of course, the effect of the Saharan DA should also manifest itself over the globe. It is apparent that in view of the prevalent short-wave RF this effect manifests itself through a radiative cooling.

9.7 SOME DATA OF OBSERVATIONS IN WESTERN EUROPE

In connection with the preserved controversy of estimates of the so-called 'excessive absorption' (overestimated values of net radiation as compared with those observed), [181] undertook calculations of total and spectral SWR fluxes at the surface level with prescribed vertical profiles of aerosol size distribution and meteorological parameters (from the data of aircraft, radiosonde, and lidar observations).

Calculated values of fluxes were compared with results of pyranometric and pyrheliometric observations of total fluxes as well as with the data on spectral fluxes measured with the 512-channel spectroradiometer for the wavelength interval 500–920 nm. These observations were carried out in the spring of 1998 in the region of Berlin within the LACE programme on the study of atmospheric aerosol at the Observatory in Lindenberg. With the use of the results of observations performed under conditions of low and high aerosol content in the atmosphere, the sensitivity has been analysed of calculated SWR fluxes to changes in the chemical composition of aerosol particles (the complex refraction index) as well as aerosol size distribution and air humidity.

With the prescribed spectral dependence of the refraction index corresponding to ammonium sulphate, calculated global radiation turned out to be $10–20\,W\,m^{-2}$ (2–3%) greater than measured and direct solar radiation by $17–18\,W\,m^{-2}$ greater than those measured, whereas scattered radiation was $6–7\,W\,m^{-2}$ (4–10%) less than measured values. Calculated and observed spectral distributions of global radiation agree well in the central interval of the visible spectral region ($\lambda = 500–600$ nm), but with the growth of wavelength, calculated values of spectral fluxes become more and more overestimated compared with those observed. Thus the discussed results did not permit the inadequacy of numerical modelling of SWR transport in the atmosphere to be completely removed.

To study the impact of aerosol on the SWR transfer in the atmosphere, in the summer of 1998 in the region of the Lindenberg Observatory, [182] undertook a field observational programme LACE aimed mainly at completing complex ('closed') measurements of chemical, physical, optical, and radiative properties of aerosol in the clear and heavily clouded atmosphere. Using various instruments carried by three flying laboratories, simultaneous observations were carried out of the vertical profiles (in the altitudinal interval 0.2–11 km) of total and spectral (upward and downward) SWR fluxes as well as the microphysical and optical characteristics of aerosol. The observational results were compared with the data of numerical modelling. According to the obtained results, the averaged ARF at the TOA manifests itself through a radiative cooling of the surface–atmosphere system and varies from -4.7 to $-12.7\,\mathrm{W\,m^{-2}}$ in the cloud-free atmosphere. Over the northern hemisphere tropical Atlantic Ocean, in the summer, there is a long plume of aerosol between the Saharan coastline in the east and the Caribbean Islands in the west. Garrett *et al.* [55] studied the evolution of microphysical characteristics and radiative properties of atmospheric aerosol in this plume resulting from dust storms in the northern Sahara. The mass of the aerosol plume and its radiative properties refer to two layers located at different altitudes and covering most of the Atlantic basin. The prevalent component of the upper layer is dust aerosol initially entering the atmosphere in the Sahara whereas the lower layer located in the MABL, consists mainly of sea salt aerosol particles formed near the rough ocean surface. The contribution of a secondary but substantial value is made by particles of carbon, sulphate, and nitrate (CSN) aerosol.

The total contribution of CSN and sea salt aerosol to the formation of the ARF under clear sky conditions constitutes more than 50% with respect to total RF. The data of satellite observations and numerical modelling indicate that a decrease of the ARF due to an aerosol plume between the African coastline and the Caribbean region constitutes about 20% and is determined mainly by DA deposition. The considered RF decrease is held by three factors:

(1) Trade winds in the MABL providing a stable input of sea salt aerosol to the atmosphere.
(2) DA particles which initially reached altitudes of 2.5–5.5 km and therefore had a longer lifetime in the troposphere.
(3) Fine CSN and DA particles in the free troposphere. The observational data have shown that the CSN aerosol is the product of a polluted atmosphere in western Europe. Further studies are needed to obtain reliable data on the contribution of this aerosol to RF formation.

The complexity of the processes of direct and indirect (through affecting clouds) aerosol impact on climate is the main cause of a considerable uncertainty of the existing estimates of the contribution of aerosol-induced climate formation.

The key aspect of this complexity is a strong spatial–temporal variability of aerosol properties. Power and Goyal [146] carried out calculations to analyse the laws of spatial–temporal variability (including the annual and interannual variations) of atmospheric transparency (total aerosol content in the atmosphere) from the data of observations at 8 stations in South Africa and 13 stations in Germany.

From the observations in Germany, the Angström coefficient of turbidity β (the equivalent AOT of the atmosphere at $\lambda = 1\,\mu\mathrm{m}$) is characterized by a strong annual change of atmospheric turbidity with a maximum in the summer and a minimum in winter. In South Africa the annual change of β is much weaker with a maximum on average, in the spring and a minimum in the autumn and winter. During the last decades in Germany was observed a considerable reduction of atmospheric turbidity (within −5.62 to −16.56% for 10 years) and the effect of eruptions of El Chichon and Pinatubo volcanoes was clearly manifested (via β maxima). From the data of observations at four stations in South Africa, a long-term trend of turbidity was detected, but there was no universal trend. In Germany, changes of β values are more significantly determined by the impact of non-local sources of aerosol than in South Africa, where aerosol is mainly of local origin. The long-term monthly mean values of the Angström coefficient of turbidity in Germany vary within 0.019–0.143, and annual average values vary from 0.064 to 0.116. A much lower level of atmospheric turbidity is found in South Africa, where monthly mean (annual mean) values of β are within 0.004–0.071 (6.013–0.047).

With averaging of the data for all stations, the atmospheric turbidity in Germany is higher by a factor of 2.6–5.8 (depending on season) than in South Africa. The climate of South Africa is, of course, more humid than in Germany, but in both cases there is a strong annual change of water content in the atmosphere. During the last several decades the climate of South Africa has become even damper. The decreasing trend of atmospheric turbidity in Germany is connected, apparently, either with reduced emissions of aerosol to the atmosphere or with the impact of various other factors (apart from humidity). From the data of various observations (mainly in western Europe) for the last 50 years, the global radiation has decreased by 0.51 $W m^{-2}$ (2.7% for 10 years), which is determined, probably, by the impact of aerosol.

9.8 NUMERICAL MODELLING OF THE 3-D DISTRIBUTION OF AEROSOL AND CLIMATE

Developments concerning the 3-D field of concentration and properties of aerosol in the context of substantiation of air quality models have contributed much to numerical modelling of the role of aerosol in the formation of the climate.

The aerosol component of the CMAQ model described by [11] for multiscale assessments of air quality, is aimed at an efficient and time-saving simulation of the atmospheric aerosol dynamics. The aerosol size distribution is represented as a superposition of three log-normal modes of aerosol which include the small sized mode PM-2.5 ($D < 2.5\,\mu m$) consisting of two submodes – Aitken nuclei ($D < 0.1\,\mu m$) and accumulation submode ($D = 0.1$–$2.5\,\mu m$), as well as large sized mode PM-10 ($D = 2.5$–$10\,\mu m$). The process of aerosol property evolution is described by coagulation, growth, and formation of new particles.

Whilst considering the aerosol components, the PM-2.5 and PM-10 modes of primary emissions of elemental and organic carbon, as well as dust and other particles, have been taken into account. Secondary components of aerosol were sulphate, nitrate, ammonium, water, and secondary organic compounds, of natural and anthropogenic origin. A parameterization of aerosol light extinction in the visible spectral region has been made with the use of Mie formulae and empirical relationships verified by observational data. Algorithms have been described [11] which simulate interactions between aerosol and cloudiness. The obtained results have been illustrated using calculated data from box and 3-D models.

Mebust *et al.* [121] performed a preliminary analysis of the adequacy of Model-3 developed by a group of scientists for a multiscale assessment of air quality (CMAQ) by comparing calculated and observed values of visibility indices and concentration of various components of aerosol. Comparisons with the data of meteorological visibility observations at 130 airports of the USA for the period 11–15 July, 1995 have shown that calculated values of CMAQ parameters, on the whole, reliably reflect the main laws of the spatial–temporal variability of visibility, including the spatial gradients and extreme levels of visibility.

However, an application of both calculation techniques used in CMAQ to calculate light extinction (with Mie formulae and empirical methods) has led to an underestimation of visibility reduction (i.e., to its overestimation). In the case of calculations with Mie formulae, the normalized mean difference (NMD) and normalized mean error (NME) are −21.7% and 25.41%, respectively, and in the case of the empirical method −35.5% and 36.2%. In most cases the accuracy of the calculated values agree with those observed by coefficient 2, though the correlation coefficient is 0.25 and 0.24.

A special comparison has been made in [121] with the use of observed concentrations of sulphate, nitrate, PM-2.5, PM-10, and organic carbon at 18 stations in June 1995. In this case, comparisons have led to the conclusion that except for good results for sulphate (the mean systematic difference of the mass concentration constitutes $0.15\,\mu g\,m^{-3}$, and NMD = 3.1%), the model gave a systematic underestimation of concentration in aerosol of nitrate ($-0.10\,\mu g\,m^{-3}$, −33.1%), PM-2.5 ($-3.9\,\mu g\,m^{-3}$, −30.1%), PM-10 ($-5.66\,\mu g\,m^{-3}$, −29.2%), and organic carbon ($-0.78\,\mu g\,m^{-3}$, −33.7%). The adequacy of the simulation of concentrations of various components is as follows: sulphate ($r^2 = 0.63$, average error $1.75\,\mu g\,m^{-3}$, NME = 36.2%); PM-2.5 (0.55, $5.00\,\mu g\,m^{-3}$, 38.5%); organic carbon (0.25, $0.94\,\mu g\,m^{-3}$, 40.6%); PM-10 (0.13, $9.85\,\mu g\,m^{-3}$, 50.8%), and nitrate (0.01, $0.33\,\mu g\,m^{-3}$, 104.3%). Except for nitrate, in 75–80% of cases the calculated values of concentration agree with those observed within coefficient 2.

Solar radiation absorption in the atmosphere by BC-containing aerosol can cause a change of the radiative warming of the atmosphere and surface, which, in turn, affects the dynamics and hydrological processes responsible for the cloud cover formation. In this regard, [35] considered a new microphysical mechanism of the BC impact on climate consisting of the fact that the radiative warming due to solar radiation absorption in the presence of BC-containing CCN slows down or even prevents CCN from functioning as centres of cloud droplet formation. The temperature of BC-containing droplets is increased owing to their warming due to solar radiation absorption, which leads to an increase of water vapour pressure on droplets and limits an activation caused by most strongly absorbing CCN. In connection with this, it has been proposed to generalize the Köhler theory which determines the dependence of the equilibrium water vapour pressure on the level of water assimilation, with the size of CCN and the share of BC taken into account. This dependence manifests itself most clearly in the case of CCN with the volume of

BC greater than a sphere with D = 500 nm. In the presence of aerosol with a relative content of BC by mass of less than 10% (per each particle), a 10% decrease of CCN number density can occur due to a solar warming, with a 0.01% level of critical oversaturation. On the other hand, the effect of warming due to absorption by BC on activation with a −0.1% level of critical oversaturation is negligibly small.

Analysis of observational data has led to the conclusion about a global stratospheric cooling during the last decades, though the rate of cooling depends on the duration of the observational series and is specific for various regions of the globe. While in high latitudes of the southern hemisphere the cooling trend has been observed beginning in 1980, in the northern hemisphere most substantial changes took place after 1990. The common opinion about the nature of stratospheric cooling in the southern hemisphere is based on the fact that it is connected mainly with chemically induced ozone depletion. Here a mechanism of radiative–chemical feedback forms, since a cooling leads to ozone depletion. A quite different situation is observed in the northern hemisphere, where the internally induced climate system's dynamics should play an important role.

Manzini *et al.* [116] undertook a numerical modelling of the atmospheric circulation sensitivity to ozone depletion and to the growth of GHG concentrations, with chemical reactions in the atmosphere taken into account. With this aim in view, three series of numerical modelling have been carried out under fixed boundary conditions for the recent past (1960) and present conditions (1990 and 2000), considering changes of GHG concentrations, total organic chlorine, and average SST. Changes of the ozone content were calculated with the simulation of interactivity of the processes considered.

The numerical modelling results indicate a decrease in the ozone content in the stratosphere. Under Antarctic conditions, in 1990 and 2000 an 'ozone hole' formed, which was absent, however in 1960, as observations showed. In the stratosphere and mesosphere, the temperature decreased during the whole period under consideration, which was most clearly manifested at the level of the stratopause and in the lower stratosphere in the region of the South Pole, which agrees with observational data. In the lower stratosphere of the Arctic a cooling in March (compared with 1960 conditions) was detected only from calculations for 2000.

The activity of long waves propagating in winter from the troposphere in 1960 and 2000 turned out to be comparable. This suggests the conclusion about the contribution of ozone depletion and an increase of GHG concentrations to the cooling taking place in the lower stratosphere of the Arctic in 1960 and 2000. The obtained results suggest the conclusion that an extremely low temperature observed in March during the last decade could result from radiative and chemical processes, though the possible impact of other factors cannot be excluded.

A comparison of numerical modelling results for 1960 and 2000 revealed an enhancement of downward air fluxes in the mesosphere during the period of cooling of the lower stratosphere (in March in the Arctic and in October in the Antarctic). An enhancement of downward flux (downwelling) in the mesosphere can be explained as a response of gravity waves to wind intensification connected with a cooling of the lower stratosphere. The downward shift of the enhanced downwelling with a time shift of about one month can be partially explained by the impact of planetary waves. An enhancement of the dynamically induced warming connected with an enhanced downwelling favours a limitation of stratospheric cooling and intensification of the circumpolar vortex in the lower stratosphere and, thus, favours an ozone layer reconstruction due to such a feedback. Both in the Arctic and in the Antarctic a cooling due to ozone depletion covers a region where polar stratospheric clouds (PSC) form in spring, whose extent has increased between 1960 and 2000. The increase of PSC amount could lead to ozone depletion in 2000 as revealed by calculations.

Based on the use of the NCAR CCM-3 (version 3) model, [101] obtained new estimates of the indirect climatic impact of sulphate and BC aerosol due to the aerosol impact on cloud cover dynamics. Two versions of the aerosol impact on clouds have been considered. One of them (the first indirect effect or the effect of radius) is connected with the appearance of additional aerosol particles as CCN leading to a decrease of the size of cloud droplets. The second indirect effect manifests itself through a suppressed coalescence of droplets due to a decrease of the size of droplets and hence, an increase of cloud lifetimes (the effect of lifetime). Both these effects raise the cloud albedo.

The global fields of aerosol concentration have been simulated in [101] with the use of submodels of aerosol formation built in the climate model, with the respective 'life cycles' taken into account. Besides, characteristics of background aerosol and the dynamics of the background aerosol size distribution have been described. The droplets number density in liquid water clouds was calculated with the prescribed levels of oversaturation. The obtained size distribution of cloud droplets and the outgoing SWR fluxes agree well with the results of satellite observations.

With the use of the data on aerosol properties contained in the IPCC 2001 Report, it has been shown that in the case of global averaging a 5.3% decrease of the cloud droplets radius (by 0.58 μm for the average radius of droplets 10.31 μm) and a 4.9% increase of cloud water content due to the impact of anthropogenic aerosol take place. Maximum changes of these two parameters take place (in order of their significance) in the regions of south-eastern Asia (here the content of sulphate aerosol is at a maximum and the solar zenith angle is at a minimum), the northern Atlantic, Europe, Siberia, and the eastern USA. Similar changes are also observed in the values of indirect RF, whose global average value is $-1.8\,\mathrm{W\,m^{-2}}$, with contributions of changes in droplet radius and lifetime being, respectively, -1.3 and $-0.46\,\mathrm{W\,m^{-2}}$.

A repetition of the numerical experiment with the use of calculated data for 2100, according to the IPCC A2 scenario, has not changed the global average estimate of RF, but revealed a shift of the maximum indirect effect of aerosol to the tropical zone. Experiments on sensitivity have led to the conclusion that the impact of changes in cloud droplet radius is about 3 times stronger than that of changes in cloud lifetimes, as well as the fact that the contribution of carbon aerosol is small compared with the total indirect climatic impact of aerosol. An approximate character of the obtained results is determined, first of all, by neglecting some types of aerosol – for instance, organic carbon aerosol (this is connected with the absence of required information about aerosol) as well as an exclusion from consideration of ice clouds and the impact of the processes discussed on the LWR transfer. The climate model is planned to be further developed by taking into account the ocean–atmosphere interaction with a stronger emphasis on regional effects.

Cook and Highwood [37] were the first to undertake a numerical modelling of the climatic impact of aerosol absorbing solar radiation within the 'interim' model of atmospheric general circulation, IGCM, developed at Reading University (UK). The 22-layer global model IGCM of the 'atmosphere–2-m mixed layer of the ocean' system has been realized over the grid 5° lat. × 5° long. and ensures a simulation of equilibrium climate after integration for a period of five years (in fact, calculations have been made for 30 years). The results obtained in [37] indicate that the sign of direct ARF cannot be considered as a representative characteristic of the sign of real changes of global average annual SAT. The related important fact is that the climatic sensitivity to radiation absorbing aerosol is much stronger than the climatic sensitivity to the 'greenhouse' warming due to the growth of CO_2 concentration. The situation is determined by a change of cloud amount in the process of climate formation (SAT change).

When the aerosol single-scattering albedo is <0.95, the low-level cloud amount decreases, which determines the functioning of the respective positive feedback. On the other hand, changes of the upper level cloud amount manifest themselves as a process of negative feedback formation. The total impact of clouds is determined by the balance between the two effects of opposite sign and depends strongly on the choice of the parameterization scheme for the cloud cover dynamics. These results of numerical modelling revealed a distinct 'semidirect' impact of aerosol on climate. In this connection, a supposition has been made in [37] that in all GCM models, the aerosol–cloud feedback due to absorbing aerosols is implicitly present. This situation suggests the conclusion about the critical importance of an adequate parameterization of cloud cover dynamics (in particular, in the context of hypothesis of 'semidirect' ARF proposed by[63]). According to the IGCM model, clouds play an extremely important role in the formation of the climatic system's response to the atmospheric aerosol dynamics.

Takemura *et al.* [164] applied a coupled model of the long-range transport and atmospheric general circulation to estimate changes of cloud cover characteristics, precipitation, and temperature due to direct and indirect ARF. The obtained estimates of the efficient radius of cloud droplets, cloud-induced RF (CRF), are in satisfactory agreement with the data of satellite observations. The numerical modelling revealed a global-scale decrease of an efficient radius of droplets due to indirect anthropogenic ARF. On the other hand, changes of the cloud water and precipitation are determined by a variation of the dynamical hydrological cycle with a temperature change by aerosol direct and indirect effects rather than the second indirect effect itself. The global mean direct and indirect anthropogenic ARFs at the tropopause by anthropogenic aerosols are calculated to be -0.1 and $-0.9\,\mathrm{W\,m^{-2}}$, respectively. It has been suggested therefore that aerosol particles reduce the increase in SAT due to GHGs approximately by 40%, on the global mean. In accordance with the recent assessment by Lan *et al.* [102a] the total atmosphere–land system warms by 8 Wm^{-2} (due to aerosol), while the total atmosphere–ocean system cools by 78 Wm^{-2}.

Quaas *et al.* [148] have used three ensembles of simulations with the LMD general circulation model developed at the Laboratoire de Mètèorologie Dynamique to assess the radiative impacts of five GHGs (CO_2, CH_4, N_2O, CFC-11, and CFC-12) as well as sulphate aerosols for the time period 1930–1989 (sea surface temperature was prescribed). It has been shown that the GHG concentrations increase results in a reduction of clouds at all atmospheric levels, thus decreasing the total greenhouse effect in the long-wave spectrum and increasing absorption of solar radiation by reduction of cloud albedo. Since different changes in high (decrease) and low-level (increase) cloudiness have been obtained, the final result was a slightly positive net effect due to cloudiness changes. The total aerosol effect including the aerosol direct and indirect effects remains strongly negative.

Lohmann [114] has studied the global impact of aerosols on the riming rate of snow crystals of different shapes with cloud droplets in stratiform clouds using the AGCM to answer the question: can anthropogenic aerosols decrease the snowfall rate? The final answer was that in accordance with the present-day simulation there had been a slight increase of snowfall rate over preindustrial times.

The opinion expressed by [74], that control of emissions of BC aerosol and organic matter to the atmosphere due to fossil fuel burning can become an efficient method of slowing global warming, has caused hot discussion. Feichter *et al.* [48] noted in this connection that according to estimates obtained by Jacobson, emissions to the atmosphere of carbon aerosol (BC and organic carbon, OC) due to fossil fuel burning can cause an increase in global average SAT by 0.35 K. However, there is doubt with respect to the grounds for conclusions in this study connected with the fact that the emissions under discussion also contain sulphur dioxide (which determines the formation of sulphate aerosol). Besides this, emissions affect both absorption and scattering of short-wave radiation (i.e., they can cause both warming and cooling). Still more important is the fact that the climate model GATOR–GCMM used by Jacobson has not been tested for adequacy by comparing with other models and with observational data, which is necessary, especially since calculations for six years cannot ensure the simulation of an equilibrium climate. Penner [138] has made some comments on this.

Chock *et al.* [26] have made remarks on the results of Jacobson [74] concerning the adequacy of the applied climate model, reliability of input information, and substantiation of the required ecological policy. As for the input information, it was assumed that annual emissions of BC and OC constituted 6.4 and 10.1 $TgC\,yr^{-1}$. The assumed ratio OM : BC = 3.1 : 1 is equivalent to global emissions of OM and OM submicron aerosol reaching 19.9 and 15.9 $TgC\,yr^{-1}$, respectively (i.e., much higher than the estimates mentioned above). Besides this, drawbacks have been noted such as the incomplete consideration of climatic indirect ARF, which leads to a climate cooling, and an unacceptable simplification of the climate model. In his reply to these and other comments, Jacobson [75–77] rejected them as incorrect and groundless.

With respect to the discussion mentioned above, it should be emphasized that the most important thing, as follows from the above review, is that all earlier obtained estimates of the impact of aerosol on global climate are highly uncertain.

9.9 CONCLUSION

In connection with the Symposium 'Our "hazy" atmosphere: aerosol and climate' held on 14 February 2004 at California University (San Diego) (http://scrippsnews.ucsd.edu/article_num=620) it has justly been emphasized: 'Most probably, in several decades the scientists will look back at events of the early 21st century and interpret them as fundamental manifestations of the impact of aerosol on the Earth's climate'. This estimate can, of course, be considered only as a metaphor. The main thing raises no doubt, however: the climatic impact of natural and anthropogenic aerosol is, of course, substantial, but the existing quantitative estimates should be considered as preliminary. It is also apparent that here we face an exclusively complicated problem, which requires a study of physical, chemical, and biological processes for unprecedented broad spatial (from nano to global) and temporal scales. Assessing the significance of the problems, two conclusions can be drawn:

- The problem 'aerosol–clouds–climate' should be of key importance in the course of preparation of the fourth IPCC Report.
- Based on analysis of results obtained in the course of this preparation, a new interdisciplinary programme should be proposed and substantiated, which would provide an adequate understanding of the climate system's dynamics in the context of the 'aerosol–clouds–climate' problem.

Of course, developments should be continued to obtain more adequate and reliable observational data as well as further development of numerical modelling. Of critical importance is to take into account atmospheric and surface feedbacks on radiative forcing [179].

9.10 BIBLIOGRAPHY

1. Ackerman T. P., Braverman A. J., Diner D. J., Anderson T. L., Kahn R. A., Martonchik J. V., Penner J. E., Rasch P. J., Wielicki B. A., and Yu B. Integrating and interpreting aerosol observations and models within the PARAGON framework. *Bull. Amer. Meteorol. Soc.*, 2004, **85**(10), 1523–1533.
2. Adamenko V. N. and Kondratyev K. Ya. Global climate changes and their empirical diagnostics. In: Yu. A. Izrael, G. V. Kalabin, and V. V. Nikonov (eds), *Anthropogenic Impact on the Nature of the North and its Ecological Implications*. Kola Scientific Centre, Russian Academy of Sciences, Apatity, 1990, pp. 17–34 [in Russian].
3. Allen M. Climate of the twentieth century: Detection of change and attribution of causes. *Weather*, 2002, **57**(8), 296–303.
4. Alley R. B., Marotzke J., Nordhaus W. D., Overpeck J. T., Peteet D. M., Pielke R. A., Pierrehumbert R. T., Rhines R. B., Stocker T. F., Taley L. D., and Wallace J. M. Abrupt climate change. *Science*, 2002, **299**(5615), 1005–2010.

5. Alverson K. D., Bradley R. S., and Pedersen T. F. (eds). *Paleoclimate, Global Change, and the Future*. Springer, Heidelberg, 2003, 221 pp.
6. Anderson T. L., Charlson R. J., Schwartz S. E., Knutti R., Boucher O., Rodhe H., and Heintzenberg J. Climate forcing by aerosols – a hazy picture. *Science*, 2003, **300**, 1102–1104.
7. Anderson T. L., Charlson R. J., Winker D. M., Ogren J. A., and Holmèn K. Mesoscale variations of tropospheric aerosols. *J. Atmos. Sci.*, 2003, **60**(1), 119–136.
8. Babu S. S., Moorthy K. K., and Satheesh S. K. Aerosol black carbon over Arabian Sea during intermonsoon and summer monsoon seasons. *Geophys. Res. Lett.*, 2004, **31**(6), LOG104/1–LOG104/5.
9. Barr J. G., Fuentes J. D., and Bottenheim J. W. Radiative forcing of phytogenic aerosols. *J. Geophys. Res.*, 2003, **108**(D15), ACH15/1–ACH15/3.
10. Bauer S. E., Balkanski Y., Schulz M., Hauglustaine D. A., and Dentener F. Global modeling of heterogeneous chemistry on mineral aerosol surfaces: Influence on tropospheric ozone chemistry and comparison to observations. *J. Geophys. Res.*, 2004, **109**(2), DO2304/1–DO2304/17.
11. Binkowski F. S. and Roselle S. J. Models-3 Community Multiscale Air Quality (CMAQ) model aerosol component. 1. Model description. *J. Geophys. Res.*, 2003, **108**(D6), AAC3/1–AAC3/18.
12. Block A., Keuler K., and Schaller E. Impacts of anthropogenic heat on regional climate patterns. *Geophys. Res. Lett.*, 2004, **31**(12), L12211/1–L12211/4.
13. Boehmer-Christiansen S. A. Who and how determines the climate change concerning policy? *Proc. Russian Geogr. Soc.*, 2000, **132**(3), 6–22 [in Russian].
14. Boehmer-Christiansen A. and Kellow A. *International Environmental Policy: Interests and the Failure of the Kyoto Process*. Edward Elgar Publ. Co. Ltd., Cheltenham, UK, 2002, 214 pp.
15. Boer G. J., Yu B., Kim S.-J., and Flato G. M. Is there observational support for an El Niño-like pattern of future global warming? *Geophys. Res. Lett.*, 2004, **31**(6), L06201/1–L06201/4.
16. Bolin B. The WCRP and IPCC: Research Inputs to IPCC Assessments and Needs. *WCRP/WMO*, 1998, **904**, 27–36.
17. Borisenkov E. P. The greenhouse effect. Problems, myths, reality. *Astrakhan Herald of Ecological Education*, 2003, **1**(5), 5–12 [in Russian].
18. Brenguier J.-L. Introduction to special section: An experimental study of the aerosol indirect effect for validation of climate model parameterizations. *J. Geophys. Res.*, 2003, **108**(D15), CMP1/1–CMP1/3.
19. Brenguier J.-L., Pawlowska H., and Schüller L. Cloud microphysical and radiative properties for parameterization and satellite monitoring of the indirect effect of aerosol on climate. *J. Geophys. Res.*, 2003, **108**(D15), CMP6/1–CMP6/14.
20. Broecker W. S. Does the trigger for abrupt climate change reside in the ocean or in the atmosphere? *Science*, 2003, **300**(5625), 1519–1522.
21. Brooks S. D., Garland R. M., Wise M. E., Prenni A. J., Cushing M., Hewitt E., and Tolbert M. A. Phase changes in internally mixed maleic acid/ammonium sulphate. *J. Geophys. Res.*, 2003, **108**(D15), ACH23/1–ACH23/10.
22. Bush B. C. and Valero F. P. J. Surface aerosol radiative forcing at Gosan during the ACE–Asia campaign. *J. Geophys. Res.*, 2003, **108**(D23), ACE28/1–ACE 28/8.
23. Cai W. and Whetton P. H. Evidence for a time-varying pattern of greenhouse warming in the Pacific Ocean. *Geophys. Res. Lett.*, 2002, **27**(16), 2577–2580.
24. Chameides W. L. and Bergin M. Soot takes center stage. *Science*, 2002, **297**, 2214–2250.

25. Charlson R. J., Seinfeld S. H., Nenes A., Külmälä M., Laaksonen A., and Facchini M. C. Reshaping the theory of cloud formation. *Science*, 2001, **292**, 2025–2026.
26. Chock D. P., Song Q., Hass H., Schell B., and Ackerman I. Comment on 'Control of fossil-fuel particulate black carbon and organic matter, possibly the most effective method of slowing global warming' by M. Z. Jaconson. *J. Geophys. Res.*, 2003, **108**(D24), ACH12/1–ACH12/3; ACH13/1–ACH13/4.
27. Chou M.-D., Chan P. K., and Wang M. Aerososl radiative forcing derived from Sea-WiFS-retrieved aerosol optical properties. *J. Atmos. Sci.*, 2002, **59**(3), 748–757.

27a. Christy J. R. and Spencer R. W. Reliability of satellite data sets. *Science*, 2003, **301**(5636), 1046–1047.

28. Chuang C. C., Penner J. E., Prospero J. M., Grant K. E., Rau G. H., and Kawamoto K. Cloud succeptibility and the first aerosol indirect forcing: Sensitivity to black carbon and aerosol concentrations. *J. Geophys. Res.*, 2002, **107**(D21), AAC10/1–AAC10/23.
29. Chuang P. Y., Duvall R. M., Bae M. S., Jefferson A., Schauer J. J., Yang H., Yu J. Z., and Kim J. Observations of elemental carbon and absorption during ACE-Asia and implications for aerosol radiative processes and climate forcing. *J. Geophys. Res.*, 2003, **108**(D23), ACE2/1–ACE2/12.
30. Chung C. E. and Ramanathan V. Aerosol loading over the Indian Ocean and its possible impact on regional climate. *Indian J. Mar. Sci.*, 2004, **33**(1), 40–55.
31. Church T. M. and Jickels T. D. Atmospheric chemistry in the coastal ocean: A synopsis of processing, scavenging and inputs. *Indian J. Mar. Sci.*, 2004, **33**(1), 71–76.
32. Chýlek P., Henderson B., and Mushchenko M. Aerosol radiative forcing and the accuracy of satellite aerosol optical depth retrieval. *J. Geophys. Res.*, 2003, **108**(D24), AAC4/1–AAC4/8.
33. Coakley J. A., Jr., Tajuk W. R., Sayaraman A., Quinn P. C., Devauz C., and Tanrè D. Aerosol optical depths and direct radiative forcing for INDOEX derived from AVHRR: Theory. *J. Geophys. Res.*, 2002, **107**(D19), INX28/1–INX28/18.
34. Collins M. and Senior C. A. Projections of future climate change. *Weather*, 2002, **57**(8), 283–287.
35. Conant W. C., Nenes A., and Seinfeld J. H. Black carbon radiative heating effects on cloud microphysics and implications for the aerosol indirect effect. 1. Extended Köhler theory. *J. Geophys. Res.*, 2002, **107**(D21), AAC23/1–AAC23/9.
36. Conant W. C., Seinfeld J. H., Wang J., Carmichael G. R., Tang Y., Uno I., Flatau P. J., Markowicz K. M., and Quinn P. K. A model for the radiative forcing during ACE-Asia derived from CIRPAS Twin Otter and R/V Ronald H. Brown data and comparison with observations. *J. Geophys. Res.*, 2003, **108**(D23), ACE29/1–ACE29/16.
37. Cook J. and Highwood E. J. Climate response to tropospheric absorbing aerosols in an intermediate general–circulation model. *Quart. J. Roy. Meteorol. Soc.*, 2004, **130**A(296), 175–191.
38. Cziczo D. J., De Mott P. J., Brooks S. D., Prenni A. J., Thomson D. S., Baumgardner D., Wilson J. C., Kreidenweis S. M., and Murphy D. M. Observations of organic species and atmospheric ice formation. *Geophys. Res. Lett.*, 2004, **31**(12), L12116/1–L12116/4.
39. Dias M. A. F. S., Artaxo P., and Andreae M. O. Aerosols impact cloud in the Amazon basin. *GEWEX News*, 2004, **14**(4), 4–5, 19.
40. Diner D. J., Menzies R. T., Kahn R. A., Andersen T. L., Bösenberg J., Charlson R. J., Holben B. N., Hostetler C. A., Miller M. A., Ogren J. A., *et al.* Using the PARAGON

framework to establish an accurate, consistent, and cohesive long-term aerosol record. *Bull. Amer. Meteorol. Soc.*, 2004, **85**(10), 1535–1548.

41. Ebel A., Memmesheimer M., Friese S., Jacobs H. J., Feldmann H., Kessler C., and Piekorz G. Analysis of seasonal changes of atmospheric aerosols on different scales in Europe using sequentially nested simulations. *27th NATO/CCMS International Technical Meeting on Air Pollution and Its Application. Raniff Centre, Canada, 25–29 October 2004*, 8 pp.
42. Egorova T., Rozanov E., Manzini E., Haberreiter M., Schmutz W., Zubov V., and Peter T. Chemical and dynamical response to the 11-year variability of the solar irradiance simulated with a chemistry-climate model. *Geophys. Res. Lett.*, 2004, **31**(6), LO6119/1–LO6119/4.
43. Ekman A. M. L. Small-scale patterns of sulfate aerosol climate forcing simulated with a high resolution regional climate model. *Tellus*, 2002, **54D**, 143–162.
44. Ellsaesser H. W. *The current status of global warming*. The paper prepared at the request of the Marshall Institute. Washington, D.C., 2001, May, 5 pp.
45. Essex C. and McKitrick R. *Taken by Storm. The Troubled Science, Policy and Politics of Global Warming*. Key Porter Books, Toronto, 2002, 320 pp.
46. Feczkó T., Molnár Á, Mészáros E., and Major G. Regional climate forcing of aerosol estimated by a box model for a rural site in Central Europe during summer. *Atmos. Environment*, 2002, **36**, 4125–4131.
47. Feczkó T., Mészáros E., and Molnár Á. Radiative forcing tendency due to anthropogenic aerosol particles and greenhouse gases in Hungary. *Időjárás*, 2004, **108**(1), 1–10.
48. Feichter J., Sausen R., Grassl H., and Fiebig M. Comment on 'Control of fossil-fuel particulate black carbon and organic matter, possibly the most effective method of slowing global warming' by M. Z. Jacobson. *J. Geophys. Res.*, 2003, **108**(S24), ACH10/1–ACH10/2; ACH11/1–ACH11/4.
49. Feingold G. Modeling of the first indirect effect: Analysis of measurement requirements. *Geophys. Res. Lett.*, 2003, **30**(19), ASC7/1–ASC7/4.
50. Feingold G., Eberhard W. L., Veron D. E., and Previdi M. First measurements of the Twomey indirect effect using ground–based remote sensors. *Geophys. Res. Lett.*, 2003, **30**(6), 20/1–20/4.
51. Fiebig M., Petzold A., Wandlinger U., Wendisch M., Kiemle C., Stifter A., Ebert M., Rother T., and Leiterer U. Optical closure for an aerosol column: Method, accuracy, and inferable properties applied to a biomass-burning aerosol and its radiative forcing. *J. Geophys. Res.*, 2002, **107**(D21), LAC11/1–LAC11/13.
52. Folland C. K., Karl T. R., and Salinger M. J. Observed climate variability and change. *Weather*, 2002, **57**(3), 269–278.
53. Fortmann M. Zum Einfluss troposphärischer Aerosole auf das Klima der Arktis. *Ber. Polar- und Meeresforsch.*, 2004, 486, I–II, 1–142.
54. Fridling A. M., Ackerman A. S., Jensen E. J., Heymsfield A. J., Poellot M. R., Stevens D. E., Wang D., Miloshevich L. M., Baumgardner D., Lawson R. P., *et al.* Evidence for the predominance of mid-tropospheric aerosols as subtropical anvil cloud nuclei. *Science*, 2004, **304**, 718–722.
55. Garrett T. J., Russel L.M., Ramaswamy V., Maria S. F., and Huebert B. J. Microphysical and radiative evolution of aerosol plumes over the tropical North Atlantic Ocean. *J. Geophys. Res.*, 2003, **108**(D1), 11/1–11/16.
56. Giorgi F., Bi X., and Qian Y. Indirect and direct effects of anthropogenic sulfate on the climate of East Asia as simulated with a regional coupled climate-chemistry/aerosol model. *Climatic Change*, 2003, **58**, 345–376.

57. Gong S. L. and Barrie L. A. Simulating the impact of sea salt on global nss sulphate aerosol. *J. Geophys. Res.*, 2003, **108**(D16), AAC4/1–AAC4/18.
58. Gong S. L., Zhang X. Y., Zhao T. L., and Barrie L. A. Sensitivity of Asian dust storm to natural and anthropogenic factors. *Geophys. Res. Lett.*, 2004, **31**(7), LO7210/1–LO7210/4.
59. Goody R. Observing and thinking about the atmosphere. *Annu. Rev. Energy Environ.*, 2002, **27**, 1–20.
60. Griggs D. J. and Noguer M. Climate Change 2001: The scientific Basis. Contribution of Working Group 1 to the Third Assessment Report of the Intergovernmental Panel on Climate Change. *Weather*, 2002, **57**(8), 267–269.
61. Haigh J. D. Climate variability and the influence of the Sun. *Science*, 2001, **294**(5549), 2109–2111.
62. Haigh J. D. Radiative forcing of climate change. *Weather*, 2002, **57**(8), 278–283.
63. Hansen J., Sato M., and Ruedy R. Radiative forcing and climate response. *J. Geophys. Res.*, 1997, **102**, 6831–6864.
64. Hansen J., Sato M., Nazarenko L., Ruedy R., Laws A., Koch D., Tegen I., Hall T., Shindell D., Santer B., *et al.* Climate forcings in Goddard Institute for Space Studies S1 2000 simulations. *J. Geophys. Res.*, 2002, **107**(D18), ACL2/1–ACL2/37.
65. Hass H., van Loon M., Kessler C., Stern R., Matthijsen J., Sauter F., Zlatev Z., Langner J., Foltescu V., and Schaap M. *Aerosol Modelling: Results and Intercomparison from European Regional-Scale Modelling Systems.* FURUTRAC-2 ISS, Munich, Germany, 2003, 77 pp.
66. Henning S., Bojinski S., Diehl K., Ghan S., Nyeki S., Weingartner E., Wurzler S., and Baltenspreger U. Aerosol partitioning in natural-phase clouds. *Geophys. Res. Lett.*, 2004, **31**(6), LO6101/1–LO6101/4.
67. Hoffer A., Kiss G., Blazcó M., and Gelancsér A. Chemical characterization of humic-like substances (HULIS) formed from a lignin-type precursor in model cloud water. *Geophys. Res. Lett.*, 2004, **31**(6), LO6115/1–LO6115/4.
68. Horvath H., Arboledas L., Alados O. F. J., Jovano V. O., Gangl M., Kaller W., Sánchez C., Saurzepf H., and Seidt S. Optical characteristics of the aerosol in Spain and Austria and its effect on radiative forcing. *J. Geophys. Res.*, 2002, **107**(D19), AAC9/1–AAC9/18.
69. Hoskins B. J. Climate change at cruising altitude? *Science*, 2003, **301**, 469–470.
70. Hubbard K. G. and Lin X. Realtime data filtering models for air temperature measurements. *Geophys. Res. Lett.*, 2002, **29**(10), 67/1–67/4.
71. Huebert B. J., Bates T., Russel P. B., Shi G., Kim Y. J., Kawamura K., Carmichel G. and Nakajima T. An overview of ACE–Asia: Strategies for quantifying the relationships between Asian aerosols and their climatic impacts. *J. Geophys. Res.*, 2003, **108**(D23), ACE1/1–ACE1/20.
72. Hughes T. P., Baird A. H., Bellwood D. R., Card M., Connoly S. R., Folke C., Grosberg R., Hoegh-Gulberg O., Jackson J. B. C., Kleypas J., *et al.* Climate change, human impacts, and the resilience of coral reefs. *Science*, 2003, **301**, 529–533.
73. *IPCC Third Assessment Report* (Volume 1). *Climate Change 2001. The Scientific Basis.* Cambridge University Press, UK, 2001, 881 pp.
74. Jacobson M. Z. Control of fossil-fuel particulate black carbon and organic matter, possibly the most effective method of slowing global warming. *J. Geophys. Res.*, 2002, **107**(D19), 4410, doi: 10.1029/2001JD002044.

75. Jacobson M. Z. Reply to comment by J. F. Penner on 'Control of fossil-fuel particulate black carbon and organic matter, possibly the most effective method of slowing global warming'. *J. Geophys. Res.*, 2003, **108**(D24), ACH15/1–ACH15/4.
76. Jacobson M. Z. Reply to comment by J. Feichter *et al.* on 'Control of fossil-fuel particulate black carbon and organic matter, possibly the most effective method of slowing global warming'. *J. Geophys. Res.*, 2003, **108**(D24), ACH11/1–ACH11/4.
77. Jacobson M. Z. Reply to comment by D. P. Chock *et al.* on 'Control of fossil-fuel particulate black carbon and organic matter, possibly the most effective method of slowing global warming'. *J. Geophys. Res.*, 2003, **108**(D24), ACH13/1–ACH13/4.
78. Jaworowski Z. The global warming folly. *21st Century Science and Technology*, 1999, **12**(4), 64–75.
79. Johnson B. T., Shine K. P., and Forster P.M. The semi-direct aerosol effect: Impact of absorbing aerosols on marine stratocumulus. *Quart. J. Roy. Meteorol. Soc.*, 2004, **130**B(599), 1407–1422.
80. Kahn R. A., Ogren J. A., Ackerman T. P., Rosenberg J., Charlson R. J., Diner D. H., Holben B. N., Menzies R. T., Miller M. A., and Seinfeld J. H. Aerosol data sources and their roles within PARAGON. *Bull. Amer. Meteorol. Soc.*, 2004, **85**(10), 1511–1522.
81. Kim D.-H., Sohn B.-J., Nakajima T., Takemura T., Choi B.-C., and Yoon S.-C. Aerosol optical properties over east Asia determined from ground-based sky radiation measurements. *J. Geophys. Res.*, 2004, **109**(D2), DO2209, doi: 10.1029/2003JD003387.
82. Kim E. and Hopke P. K. Improving source identification of fine particles in a rural northeastern US area utilizing temperature resolved carbon fractions. *J. Geophys. Res.*, 2004, **109**(D9), DO9204/1–DO9204/13.
83. Kondratyev K. Ya. *Key Problems of Global Ecology*. All-Russian Institute for Scientific and Technical Information Publ., Moscow, 1990, 454 pp. [in Russian].
84. Kondratyev K. Ya. *Global Climate*. Nauka Publ., St Petersburg, 1992, 359 pp. [in Russian].
85. Kondratyev K. Ya. and Galindo I. *Volcanic Activity and Climate*. A. Deepak Publ., Hampton, VA, 1997, 382 pp.
86. Kondratyev K. Ya. *Multidimensional Global Change*. Wiley–Praxis, Chichester, UK, 1998, 761 pp.
87. Kondratyev K. Ya. *Ecodynamics and Geopolicy* (Volume 1). *Global Problems*. St Petersburg Branch of RAS Publ., 1999, 1036 pp. [in Russian].
88. Kondratyev K. Ya. *Climatic Effects of Aerosols and Clouds*. Springer–Praxis, Chichester, UK, 1999, 264 pp.
89. Kondratyev K. Ya. and Cracknell A. P. *Observing Global Climate Change*. London, Taylor & Francis, 1999, 562 pp.
90. Kondratyev K. Ya. and Varotsos C. A. *Atmospheric Ozone Variability: Implications for Climate Change, Human Health, and Ecosystems*. Springer–Praxis, Chichester, UK, 2000, 614 pp.
91. Kondratyev K. Ya. and Demirchian K. S. Global climate changes and carbon cycle. *Proc. Russ. Geogr. Soc.*, 2001, **132**(4), 1–20 [in Russian].
92. Kondratyev K. Ya. Key issues of global change at the end of the second millennium. Our fragile world: Challenges and opportunities for sustainable development. *EOLSS Vorrunner*, 2001, **1**, 147–165.
93. Kondratyev K. Ya. Global climate change: Facts, assumptions, and perspectives of research. *Optics of the Atmos. and Ocean*, 2002, **15**(10), 1–16 [in Russian].

94. Kondratyev K. Ya. Global climate change and the Kyoto Protocol. *Időjárás*, 2002, **106**(2), 1–37.
95. Kondratyev K. Ya. Radiative forcing due to aerosol. *Optics of the Atmos. and Ocean*, 2003, **16**(1), 5–18 [in Russian].
96. Kondratyev K. Ya. and Krapivin V. F. Global changes: Real and potential in future. *Research of the Earth from Space*, 2003, **4**, 1–10 [in Russian].
97. Kondratyev K. Ya., Krapivin V. F., and Savinykh V. P. *Perspectivess of Civilization Development. Multi-Dimensional Analysis*. Logos Publ., Moscow, 2003, 575 pp. [in Russian].
98. Kondratyev K. Ya., Krapivin V. F., and Varotsos C. A. *Global Carbon Cycle and Climate Change*. Springer–Praxis, Chichester, UK, 2003, 278 pp.
99. Kondratyev K. Ya. Global climate change: Observational data and numerical modelling results. *Research of the Earth from Space*, 2004, **2**, 61–96 [in Russian].
100. Krishnan K. and Ramanathan V. Evidence of surface cooling from absorbing aerosols. *Geophys. Res. Lett.*, April 29, 2002, 10.1029/2001 GL014687.
101. Kristjánsson J. E. Studies of the aerosol indirect effect from sulfate and black carbon aerosols. *J. Geophys. Res.*, 2002, **107**(D15), AAC1/1–AAC1/19.
102. Kukla G. *Comments on the CCSO Strategic Plan*. Manuscript, 24 September 2003, 6 pp.
102a. Lan K.-M., Kim K.-M., and Hsu N.-C. Observational evidence of effects of absorbing aerosols on seasonal-to-interannual anomalies of the Asian Monsoon. *Exchanges*, 2005, **10**(3), 7–9.
103. Lesins D. S. and Lohmann U. GCM aerosol radiative effects using geographically varying aerosol sizes deduced from AERONET measurements. *J. Atmos. Sci.*, 2003, **60**(22), 2747–2764.
104. Levis S., Wiedinmyer C., Bonan G. B., and Guenther A. Simulating biogenic volatile organic compounds emissions in the Community Climate System Model. *J. Geophys. Res.*, 2003, **108**(D21), ACH2/1–ACH2/9.
105. Levitus S., Antonov J. I., Wang J., Delworth T. L., Dixon K. W., and Broccoli A. J. Anthropogenic warming of Earth's climate system. *Science*, 2001, **292**(5515), 267–270.
106. Lindsey R. and Simmon R. Escape from the Amazon. *The Earth Observer*, 2003, **15**(2), 8–13.
107. Lindzen R. S., Chou M.-D., and Hou A. Y. Does the Earth have an adaptive infrared iris? *Bull. Amer. Meteorol. Soc.*, 2001, **82**, 417–432.
108. Lindzen R. S. and Gianitsis C. Reconciling observations of global temperature change. *Geophys. Res. Lett.*, 2003, **29**(10), X1–X3.
109. Liu G., Shao H., Coakley J. A., Jr., Curry J. A., Haggerty J. A., and Tschudi M. A. Retrieval of cloud droplet size from visible and microwave radiometric measurements during INDOEX: Implication to aerosols' indirect radiative effect. *J. Geophys. Res.*, 2003, **108**(D1), 2/1–2/10.
110. Loeb N. G. and Kato S. Top-of-atmosphere direct radiative effect of aerosols over the tropical oceans from the Clouds and the Earth's Radiant Energy System (CERES) satellite instrument. *J. Climate*, 2002, **15**, 1474–1484.
111. Loeschler H. W., Bentz J. A., Oberbauer S. F., Ghosh T. K., Tompson R. V., and Loyalka S. K. Characterization and dry deposition of carbonaceous aerosols in a wet tropical forest canopy. *J. Geophys. Res.*, 2004, **109**(D2), DO2309/1–DO2309/12.
112. Loginov V. F. and Mikutski V. S. Assessment of the anthropogenic signal in the climate of cities. *Proc. Russ. Geogr. Soc.*, 2000, **132**(1), 23–31 [in Russian].

113. Lohmann U. and Lesins G. Stronger constraints on the anthropogenic indirect aerosol effect. *Science*, 2002, **297**(5595), 1012–1015.
114. Lohmann U. Can anthropogenic aerosols decrease the snowfall rate? *J. Atmos. Sci.*, 2004, **61**(10), 2457–2468.
115. Majorovicz H., Safanda J., and Skinner W. East to west retardation in the onset of the recent warming across Canada inferred from inversions of temperature logs. *J. Geophys. Res.*, 2002, **107**(B10), ETG6/11–ETG6/12.
116. Manzini E., Steil B., Brühl C., Giorgetta M. A., and Krüger K. A new interactive chemistry–climate model. 2. Sensitivity of the middle atmosphere to ozone depletion and increase in greenhouse gases and implications for recent stratospheric cooling. *J. Geophys. Res.*, 2003, **108**(D14), ACL10/1–ACL10/22.
117. Maria S. F., Russel L. M., Gilles M. K., and Myneni S. C. B. Organic aerosol growth mechanisms and their climate-forcing implications. *Science*, 2004, **306**(5703), 1921–1924.
118. Markowicz K. M., Flatau P. J., Quinn P. K., Carrico C. M., Flatau M. K., Vogelmann A. M., Bates D., Liu M., and Rood M. J. Influence of relative humidity on aerosol radiative forcing: An ACE-Asia experiment perspective. *J. Geophys. Res.*, 2003, **108**(D23), ACE30/1–ACE30/12.
119. Markowicz K. M., Flatau P. J., Vogelmann A. M., Quinn P. K., and Welton E. J. Clear-sky infrared aerosol radiative forcing at the surface and the top of the atmosphere. *Quart. J. Roy. Meteorol. Soc.*, 2003, **129**, 2927–2947.
120. McKitrick R. Trends in data on air temperature obtained with internal correlations taken into account. *Proc. Russ. Geogr. Soc.*, 2002, **134**(3), 16–24 [in Russian].
121. Mebust M. R., Eder B. K., Binkowski F. S., and Roselle S. J. Models-3 Community Multiscale Air Quality (CMAQ) model aerosol component. 2. Model evaluation. *J. Geophys. Res.*, 2003, **108**(D6), AAC4/1–AAC4/18.
122. Melnikova I. N. and Vasilyev A. V. *Short-Wave Solar Radiation in the Earth's Atmosphere*. Springer Verlag, Berlin, Heidelberg, 2004, 303 pp.
123. Menon S., Hansen J., Nazarenko L., and Luo Y. Climate effects of black carbon aerosols in China and India. *Science*, 2002, **297**, 2250–2253.
124. Menon S., Brenguier J.-L., Boucher O., Davison P., del Genio A. D., Feichter J., Ghan S., Guibert S., Liu X., Lohman U., *et al.* Evaluating aerosol/cloud/radiation process parameterizations with single-column models and Second Aerosol Characterization Experiment (ACE-2) cloudy column observations. *J. Geophys. Res.*, 2003, **108**(D24), AAC2/1–AAC2/19.
125. Mitra A. P. Indian Ocean Experiment (INDOEX): An overview. *Indian J. Mar. J. Sci.*, 2004, **33**(1), 30–39.
126. Mohr T. and Bridge J. The evolution of the integrated global Earth observing system. *Studying the Earth from Space*, 2003, No. 1, 64–73.
127. Moritz R. E., Bitz C. M., and Steig E. J. Dynamics of recent climate change in the Arctic. *Science*, 2002, **297**, 1497–1502.
128. Munoz-Alpizar R., Blanchet J.-P., and Quintanar A. I. Application of the NARCM model to high-resolution aerosol simulations: Case study of Mexico City basin during the Investigacιœn Sobre Materia Particulada y Deterioro Atmosférico – aerosol and Visibility Research measurements campaign. *J. Geophys. Res.*, 2003, **108**(D15), AAC7/1–AAC7/14.
129. Myhre G., Grini A., Heywood J. M., Stordal F., Chatenet B., Tanré D., Sundet J. K., and Isaksen I. S. A. Modeling the radiative impact of mineral dust during the

Saharan Dust Experiment (SHADE) campaign. *J. Geophys. Res.*, 2003, **108**(D18), SAH6/1–SAH6/13.

130. Myhre G., Stordal F., Berglen T. F., Sundet J. T., and Isaksen I. S. A. Uncertainties in the radiative forcing due to sulfate aerosols. *J. Atmos. Sci.*, 2004, **61**(5), 485–498.
131. Myhre G., Stordal F., Johmsrud M., Ignatov A., Mischenko M. I., Geogdzhaev I. V., Tanré D., Denzé J.-L., Goloub P., Nakajima T., *et al.* Intercomparison of satellite retrieved aerosol optical depth over the ocean. *J. Atmos. Sci.*, 2004, **61**, 499–513.
132. Nakajima T., Sekiguchi M., Tekemura T., Uno I., Higurashi A., Kim D., Sohn B. J., Oh S.-N., Nakajima T. Y., Ohta S., *et al.* Significance of direct and indirect radiative forcing of aerosols in the East China Sea region. *J. Geophys. Res.*, 2003, **108**(D23), ACE26/1–ACE26/16.
133. Noone K. J., Targino A., Olivares G., Glantz P., and Jansson J. Aerosols and their role in the Earth's energy balance. *Global Change Newsletter*, 2004, No. 59, 7–10.
134. Novakov T., Ramanathan V., Hansen J. E., Kirchstetter T. W., Sato M., Sinton J. E., and Sathaye J. A. Large historical changes of fossil-fuel black carbon aerosols. *Geophys. Res. Lett.*, March 20, 2003, **30**(6), 1324, doi: 10.1029/2002 GL016345.
135. Osborne S. R., Haywood J. M., Francis P. N., and Dubovik O. Short-wave radiative effects of biomass burning aerosol during SAFARI 2000. *Quart. J. Roy. Meteorol. Soc.*, 2004, **130B**(599), 1423–1447.
136. Pavolonis M. J. and Key J. R. Antarctic cloud radiative forcing at the surface estimated from the AVHRR Polar Pathfinder and ISCCPDI data sets, 1985–1993. *J. Appl. Meteorol.*, 2003, **42**, 827–840.
137. Peng Y. and Lohmann U. Sensitivity study of the spectral dispersion of the cloud droplet size distribution on the indirect aerosol effect. *Geophys. Res. Lett.*, 2003, **30**(10), 14/1–14/4.
138. Penner J. E. Comment on 'Control of fossil-fuel particulate black carbon and organic matter, possibly the most effective method of slowing global warming' by M. Z. Jacobson. *J. Geophys. Res.*, 2003, **108**(D21), ACH14/1–ACH14/5; ACH5/1–ACH5/4.
139. Penner J. E., Zhang S. Y., and Chuang C. C. Soot and smoke aerosol may not warm climate. *J. Geophys. Res.*, 2003, **108**(D21), AAC1/1–AAC1/9.
140. Penner J. E., Dong X., and Chen Y. Observational evidence of a change in radiative forcing due to the indirect aerosol effect. *Nature*, 2004, **247**, 231–234.
141. Pirjola L., Lehtinen K. E. J., Hansson H.-C., and Külmálá M. How important is nucleation in regional/global modelling? *Geophys. Res. Lett.*, 2004, **31**(12), L12109/1–L12109/4.
142. Pitman A. J. The evolution of, and revolution in, land surface schemes designed for climate models. *Int. J. Climatol.*, 2003, **23**(5), 479–510.
143. Podgorny I. A. and Ramanathan V. A modeling study of the direct effect of aerosols over the tropical Indian Ocean. *J. Geophys. Res.*, 2001, **104**(20), 24097–24104.
144. Poore R. Z., Qwinn T. M., and Verardo S. Century-scale movement of the Atlantic Intertropical Convergence Zone linked to solar variability. *Geophys. Res. Lett.*, 2004, **31**(12), L12214/1–L12214/4.
145. Potter G. L. and Cess R. D. Testing the impact of clouds on the radiation budget of 19 atmospheric general circulation models. *J. Geophys. Res.*, 2004, **109**(D2), D02106/1–D02106/9.
146. Power H. C. and Goyal A. Comparison of aerosol and climate variability over Germany and South Africa. *Int. J. Climatol.*, 2003, **23**, 921–941.
147. Prospero J. M. and Lamb P. J. African droughts and dust transport to the Caribbean: Climate change implications. *Science*, 2003, **302**(5647), 1024–1027.

148. Quaas J., Bouher O., Dufresne J.-L., and Le Treut H. Impacts of greenhouse gases and aerosol direct and indirect effects on clouds and radiation in atmospheric GCM simulations of the 1930–1989 period. *Climate Dynamics*, 2004, **23**, 779–789.
149. Randel W. J., Wu F., and Rios W. R. Thermal variability of the tropical tropopause region derived from GPS/MET observations. *J. Geophys. Res.*, 2003, **108**(D1),7/1–7/12.
150. Reid J. S., Prins E. M., Westphal D. L., Schmidt C. C., Richardson K. A., Cristopher S. A., Eck T. F., Reid E. A., Curdis C. A., and Hoffman J. P. Real-time monitoring of South American smoke particle emissions and transport using a coupled remote sensing/box-mode approach. *Geophys. Res. Lett.*, 2004, **31**(6), L06107/1–L06107/5.
151. Rodriguez M. A. and Dabdub D. IMAGES-SCAPE 2: A modeling study of size- and chemically resolved aerosol thermodynamics in a global chemical transport model. *J. Geophys. Res.*, 2004, **109**(D2), D02203/1 –D02203/27.
152. Rossow W. B. Workshop on climate system feedbacks. *GEWEX News.*, 2003, **13**(1), 12–14.
153. Santer B. D., Sausen R., Wigley T. M. L., Boyle J. S., Achuta Rao K., Doutriaux C., Hansen J. E., Meehl G. A., Roeckner E., Ruedy R., *et al.* Behavior of tropopause height and atmospheric temperature in models, re-analysis, and observations: Decadal changes. *J. Geophys. Res.*, 2003, **108**(D1), 1/1–1/22.
153a. Sateesh S. K., Ramanathan V., Holben B. N., Moorthy K. K., Loeb N.-G., Maring H., Prospero J. M., and Savoie D. Chemical, microphysical, and radiative effects of Indian Ocean aerosols. *J. Geophys. Res.*, 2002, **107**(D23), AAC20/1–AAC20/13.
154. Science and Technology for Sustainable Development, a G8 Action Plan, June 2, 2003.
155. Seinfeld J. H., Carmichael G. R., Arimoto R., Conant W. C., Brechtel F. J., Bates T. S., Cahill T. A., Clarke A. D., Doherty S. J., Flatau P. J., *et al.* ACE-Asia. Regional climatic and atmospheric chemical effects of Asian dust and pollution. *Bull. Amer. Meteorol. Soc.*, 2004, **85**(3), 367–380.
156. Seinfeld J. H., Kahn R. A., Anderson T. L., Charlon R. J., Davies R., Diner D. J., Ogren J. A., Schwartz S. E., and Wielicki B. A. Scientific objectives, measurement needs, and challenges motivating the PARAGON aerosol initiative. *Bull. Amer. Meteorol. Soc.*, 2004, **85**(10), 1503–1509.
157. Shamir N. J. and Veizer J. Celestial driver of Phanerozoic climate? *GSA Today*, 2003, **13**(7), 4–10.
158. Sekiguchi M., Nakajima T., Suzuki K., Kawamoto K., Higurashi A., Rosenfeld D., Sano I., and Mukai S. A study of the direct and indirect effects of aerosols using global satellite data sets of aerosol and cloud parameters. *J. Geophys. Res.*, 2003, **108**(D22), AAC4/1–AAC4/15.
159. Soon W., Baliunas S., Idso C., Idso S., and Legates D. R. Reconstructing climatic and environmental changes of the past 1000 years: A re-appraisal. *Energy and Environment*, 2003, **14**(2–3), 233–296.
160. Stone R. S., Dutton E. G., Harris S. M., and Longenecker D. Earlier spring snowmelt in northern Alaska as an indicator of climate change. *J. Geophys. Res.*, 2002, **107**(D10), ACL10/1–ACL10/15.
161. *Strategic Plan for the Climate Change Science Program*. Washington, D.C., 2003, 202 pp.
162. Tabazadeh A., Yokelson R. J., Singh H. B., Hobbs P. V., Crawford J. H., and Iraci L. T. Heterogeneous chemistry involving methanol in tropospheric clouds. *Geophys. Res. Lett.*, 2004, **31**(6), L06114/1–L06114/4.

163. Takemura T., Nakajima T., Higurashi A., Ohta S., and Sugimoto N. Aerosol distributions and radiative forcing over the Asian Pasific region simulated by Spectral Radiation-transport Model for Aerosol Species (SPRINTARS). *J. Geophys. Res.*, 2003, **108**(D23), ACE27/1–ACE27/10.
164. Takemura T., Nozawa T., Emori S., Nakajima T. Y., and Nakajima T. Simulation of climate response to aerosol direct and indirect effects with aerosol transport–radiation model. *J. Geophys. Res.*, 2005, **110**(D02202), doi: 10.1025/2004JD005029.
165. Taubman B. F., Marufu L. T., Vant-Hull B. L., Piety C. A., Doddridge B. G., Dickerson R. R., and Li Z. Smoke over haze: Aircraft observations of chemical and optical properties and the effects on heating rates and stability. *J. Geophys. Res.*, 2004, **109**(D2), D022061/1–D022061/16.
166. The U.S. Climate Change Science Program. *Vision for the Program and Highlights of the Science Strategic Plan*. A Report by the Climate Change Science Program and the Subcommittee on Global Change Research. Washington, D.C., July 2003, 34 pp.
167. Timmerman A., Jin F.-F., and Collins M. Intensification of the annual cycle in the tropical Pacific due to greenhouse warming. *Geophys. Res. Lett.*, 2004, **31**(12), L12208/1–L12208/4.
168. Toth F. L. Climate policy in light of climate science: The ICCIPS Project. *Clim. Change*, 2003, **56**(1–2), 7–76.
169. Toole D. A. and Siegel D. A. Light-driven cycling of dimethylsulfide (DMS) in the Sargasso Sea: Closing the loop. *Geophys. Res. Lett.*, May 6, 2004, **31**(L09308), doi: 10.1029/2004 GL019581.
170. Tsukernik M., Chase T. N., Serreze M. C., Barry R. G., Pielke R., Sr., Herman B., and Zeng X. On the regulation of minimum mid-tropospheric temperatures in the Arctic. *Geophys. Res. Lett.*, 2004, **31**(6), L06112/1 –L06112/4.
171. Vasilyev A. V. and Melnikova I. N. *Short-Wave Solar Radiation in the Earth's Atmosphere. Calculations. Measurements. Interpretation.* St. Petersburg Sci. Centre, RAS, St Petersburg, 2002, 388 pp. (in Russian)
172. Vetrov A. A. and Romankevich E. A. *Carbon Cycle in the Russian Arctic*. Springer-Verlag, Heidelberg, 2004, 332 pp.
173. Victor D. G. *The Collapse of the Kyoto Protocol and the Struggle to Slow Global Warming*. Princeton Univ. Press, 2001, 178 pp.
174. Vogelmann A. M., Flatau P. J., Szcordrak M., Markowicz K. M., and Minnett P. J. Observations of large aerosol infrared forcing at the surface. *Geophys. Res. Lett.*, 2003, **30**(12), 1655, doi: 10.1029/20022002GL016829.
175. Wallace J. M. and Thompson D. W. J. Annual modes and climate prediction. *Phys. Today*, 2002, **55**(2), 28–33.
176. Wang P.-H., Minnis P., Wielicki B. A., Wong T., Cess R. D., Zhang M., Vann L. B., and Kent G. S. Characteristics of the 1997/1998 El Niño cloud distributions from SAGE-II observations. *J. Geophys. Res.*, 2003, **108**(D1), 5/1–5/11.
177. Wang T.-J., Min J.-Z., Xu Y.-F., and Lam K.-S. Seasonal variations of anthropogenic sulfate aerosol and direct radiative forcing over China. *Meteorol. Atmos. Phys.*, 2003, **84**, 185–198.
178. Waple A. M. and Lawrimore J. H. (eds). State of the climate in 2002. *Bull. Amer. Meteorol. Soc.*, 2003, **84**(6), S1–S68.
179. Watterson I. G. and Dix M. R. Effective sensitivity and heat capacity in the response of climate models to greenhouse gas and aerosol forcings. *Quart. J. Roy. Meteorol. Soc.*, 2005, **131A**(605), 259–280.

180. Weaver C. P. Efficiency of storm tracks an important climate parameter? The role of cloud radiative forcing in poleward heat transport. *J. Geophys. Res.*, 2003, **108**(D1), 5/1–5/6.
181. Wendisch M., Keil A., Müller D., Wandinger U., Wendling P., Stifter A., Petzold A., Fiebig M., Wiegner M., Freudenthaler V., *et al.* Aerosol–radiation interaction in the cloudless atmosphere during LACE–98. 1. Measured and calculated broad-band solar and spectral surface insolations. *J. Geophys. Res.*, 2002, **107**(D21), LAC6/1–LAC6/20.
182. Wendling P., Stifter A., Mayer B., Fiebig M., Kiemle C., Flentje H., Wendisch M., Armbruster W., Leiterer U., von Hoyningen-Huene W., *et al.* Aerosolradiation interaction in the cloudless atmosphere during LACE-98. 2. Aerosol-induced solar irradiance changes determined from airborne pyranometer measurements and calculations. *J. Geophys. Res.*, 2003, **108**(D21), LAC12/1–LAC12/15.
183. Xu J., Bergin M. H., Greenwald R., and Russel P. B. Direct aerosol radiative forcing in the Yangtze delta region of China: Observation and model estimation. *J. Geophys. Res.*, 2003, **108**(D2), AAC7/1–AAC7/12.
184. Yabe T., Höller R., Tohno S., and Kadahara M. An aerosol climatology at Kyoto: Observed local radiative forcing and columnar optical properties. *J. Appl. Meteorol.*, 2003, **42**, 841–850.
185. Zender C. S., Bian H., and Newman D., Mineral dust entrainment and deposition (DEAD) model: Description and 1990s dust climatology. *J. Geophys. Res.*, 2003, **108**(D14), AAC8/1–AAC8/19.
186. Zhang H., Henderson-Sellers A., and McGuffie K. The compounding effects of tropical deforestation and greenhouse warming of climate. *Clim. Change*, 2001, **49**, 309–338.
187. Zhang J. and Christopher S. A. Long-wave radiative forcing of Saharan dust aerosol estimated from MODIS, MISR, and CERES observations on Terra. *Geophys. Res. Lett.*, 2003, **30**(23), ASC3/1–ASC3/4.
188. Zhao T. X.-P., Dubovik O., Smirnov A., Holben B. N., Sapper J., Pietras C., Voss K. J., and Fronin R. Regional evaluation of an advanced very high resolution radiometer (AVHRR) two-channel aerosol retrieval algorithm. *J. Geophys. Res.*, 2004, **109**(D2), D02204/1–D02204/13.

Index

absorption 15, 18, 21, 134, 224, 292
advection 116
acid 106, 121, 166, 195, 233, 251
actinometric
  microwave sensing 5
  radiosoundings 48
advection 32
aerological observations 11
aerosol
  absorption 136
  atmospheric 31, 138, 167, 188, 227, 278
  basic component 29
  chemical
    composition 28, 71, 136
    matrices 155
  cloud 11
  concentration 10, 169, 253
  dry 24
  dust 6, 68, 142, 218, 242
  energy characteristics 36
  formation 26, 37, 187, 190, 289
  haze 147
  large particle 30
  marine 68, 246
  matter 31
  monodisperse aerosol dynamics 227
  natural 67
  oceanic 287
  optical characteristics 25
  optics 50
  organic 28, 189, 202, 230, 277
  particles 15, 31, 190, 201, 233, 286
  physico-chemical properties 142, 274
  photochemical reactions 26
  pollution 8
  real 36
  Saharan
    dust 108
    layer 34
  size distribution 141
  solid particles 9
  stratospheric 43, 291
  structure 25
  substance 66
  tropospheric 32, 112, 175
  urban/industrial 68, 228
Afganistan 52
Africa 34, 145, 225
agricultural lands 27
air
  humidity 9
  pollution 8, 287
  temperature 11
airborne sensors 250
aircraft 137
albedo 5, 19, 68, 191, 246
alcanoic acid 71
algorithm 67, 108
altitude 18
aluminium 162, 167, 207, 285

Amazon River 105
America 61, 105, 135
ammonium 74, 190, 277
anthropogenic
  aerosol 41, 187, 238, 283
  emissions 166, 213
  forcings 74
  particles 26
  pollution 248
  sources of aerosol 25
apatite 52
approximation 36
Arabian Sea 147
Arctic
  haze 120, 238
  troposphere 120
Asia 142, 153, 241, 267
Atlanta 74
Atlantic Ocean 26, 34, 170, 207, 225
atmosphere 16, 75, 191, 223, 274, 288
atmospheric
  aerosol 24, 195, 274
  aerosol experiments 3
  boundary layer 8
  chemistry 288
  composition 8
  correction 73
  haze 8
  pollution 7, 246, 281
  radiative regime 27, 32
  thickness 19
  top level 134
  turbidity 52

backscattering 24, 72, 245
backward trajectories 152
balance
  energy 4
  heat 4
balloon dust sonde 44
Barents Sea 41
Bering Sea 50
biofuel 150, 279
biogeochemical cycles 277
biogenic
  emissions 71, 108
  precursors 276
  sources 71
biomass burning 68, 143, 150, 225, 241, 277
biosmoke 279
boundary layer 7, 9, 33, 244, 278

calcium 76, 165, 233
calibration 69
Canada 71, 225, 252
carbon
  aerosol 222
  black 70, 188, 265, 278
  -containing substances 70
  dioxide 8
  elemental 71, 74
  monoxide 8, 10, 284
  organic 76, 142, 188, 222, 226, 278
  soot 207
  -sulphate particles 111
carbonates 71
Caspian Sea 14
chemical
  analysis 12, 44, 107
  balance 277
  composition 66, 117, 167, 197, 281
  reactions 274
  transformation 67
China 143, 220, 242
circulation 38, 170
climate
  global 7, 10
  urban 7
cloud
  albedo 21
  condensation nuclei 142
  conditions 9
  convective 37
  droplets 24, 191, 197
  -free atmosphere 67, 193
  ice 218, 275
  layer 21
  microphysical characteristics 27
  optical thickness 24
  properties 4
  stratocumulus 72
  top 24
cloudiness 20, 29, 69, 152
coagulation 37
coal combustion products 219
coastal
  atmosphere 220
  zone 122

coefficient
  absorption 36, 70, 77, 112
  aerosol scattering 151
  Angström 72, 139, 146
  attenuation 36
  backscattering 148
  enrichment 157
  extinction 77, 245
  scattering 36, 77
condensation 37, 188, 243, 276
correlation
  analysis 91, 94
  coefficient 99, 161, 253
cyclone 41, 153
decomposition 190, 276
density 113
deposition 140, 141, 144, 212, 243, 275
desert 19, 26, 53, 220, 254
diurnal variations 7, 29, 286
divergence 16
droplet size distribution 24
droplets' modal radius 21
dry sedimentation 250
dust 51, 82, 135, 158, 188, 278
dust-free layer 36

earthquake 103
electron microscope 44, 83, 139, 274
environmental monitoring 71
equivalent optical constants 31
eruption 271
Europe 41, 109, 123, 193, 253, 278
evaporation 66, 188, 276
evolution 49
extrapolation 16

field experiment 20, 191, 199
flying laboratory 48, 133, 151, 192, 247
forest
  canopy 108
  tropical 107, 283
fossil fuel burning 119, 150
fractal aggregates 49
France 51
free atmosphere 10, 13, 40
Germany 51
gigantic particles 28
global
  atmosphere 223
  budget of tropospheric ozone 279
  ecosystem 231
  monitoring 63
  warming 61
gravimetric mass 113
greenhouse gases 70

heat 5, 33
heavy metals 158, 190
hemisphere 72, 142, 213
heterogeneous
  reactions 121, 206, 271, 277, 289
  transformation 275
homogeneous surface 32
human
  activity 7, 37
  health 144
humic polymer matter 277
humidity 29, 112, 147, 195, 246, 288
humus-like substances 107

India 147, 245
Indian Ocean 146, 244
Industrial
  emissions 190
  region 279
intra-tropical convergence zone 35
ionization 66
Iran 52
iron 159, 233

Japan 268

Kamchatka 153, 159, 233
Kara-Bogaz-Gol 14
Kara-Kum desert 27
Kazakhstan 27, 154, 164, 174
Kirghizstan 153

laser 11, 253
lidar 110, 245
lithogenous elements 89
lithosphere 158
long-range transport 141, 191, 242
long-term trend 45
lower stratosphere 6, 193

macroelements 91, 93
marine/coastal air masses 122
mass
  balance 141
  spectrometer 75, 77
mass-spectroscopic analysis 66
measurements
  actinometric 24
  airborne 12
  aircraft 7, 39
    pyranometric 26
  lidar 73
  photoacoustic 70
  spectral 45
Mediterranean basin 140
metals 76, 233, 292
meteorological
  conditions 29
  factors 10
  parameters 278
  regime 11
  situations 115
methane 231
method
  Monte Carlo 111, 281
Mexico 80, 252, 284
microanalysis 110
microclimate 10
microelements 95, 158, 167
mineralogical composition 52
minor gas components 8, 265
mode
  log-normal 122
  nucleation 122, 195
model
  air quality 281
  atmospheric 110
  global 278
  numerical 32, 281
  regional numerical 67
  semiempirical 69
moisture 5
molecular absorption 24, 81
momentum 7
morphological
  classification of aerosol 48
  structure 48
  types 50

natrium 144, 161, 175
near-water-surface layer 27
nitrate 66, 74, 163, 175, 188, 206
nitrogen
  dioxide 10
  oxides 8, 190, 281
northward transport 145
nucleation 70, 199, 201, 233, 276
numerical
  modelling 10, 109, 136, 192, 200, 288
  simulation 40, 221, 227, 233, 267

observational
  data 5, 50, 71, 137, 149
  experiment 106, 114
  network 76
observations 10, 35, 137, 220, 252
ocean 117, 139, 231, 292
optical thickness 21, 62, 281
organic
  compounds 145
  gases 26
  matter 28, 148, 248, 276
oscilation 232
oxidized organics 66
oxygen 21, 222, 250
ozone 47, 192, 198, 265, 292

Pacific Ocean 68, 142, 216, 246
parameters
  meteorological 7, 13
parametrization 39
particles 84, 196, 239
perspective 64
petroleum burning product 81
photochemical
  processes 38, 250
  reactions 26, 81, 286
photodegradation 224
photoelectric counter 103
photolysis 115, 265, 271, 280
photometer 110, 138, 150, 245, 252
physical
  characteristics 20
  factors 65
physico-chemical processes 37
phytoplankton 222
planetary cooling 19

platform 21
plume 111, 135, 145, 173, 194
polycyclic aromatic hydrocarbons 149
polluted atmosphere 139, 281
pollution 12, 167, 210
potassium 144, 175, 206
precipitation 10, 29, 83, 137, 189, 212
precursors 282
probability function 17

quartz particles 29

radar 5
radiation
  absorbed by clouds 24
  balances 40
  direct solar 72
  downward flux 22
  flux divergence 13, 23
  flux 7, 20, 280
  incoming 13, 24
  long-wave 6
  scattered 72
  short-wave 6, 12, 22, 62
  solar 12, 198, 227, 248, 286
  thermal transport 55
  transfer 12, 195
  upward 40
radiative
  data 20
  effects 46, 209
  flux 13
  forcing 247
  heat flux 3
  regime 9
  temperature 12, 47, 134
radiometer 65, 72, 134, 252
radiometric noises 69
refraction index 32, 38, 40, 68, 82
regularization 64
rainwater 152
remote sensing 38, 61, 65, 108, 252, 280
retrieval algorithm 69
rocks 76, 233, 250
roughness 5, 68
Russia 154, 252

Saharan dust 34, 108, 138, 219
satellite 108, 252
scanning spectro-radiometer 280
scattering 10, 39, 64, 139
sea
  salt 67, 140, 192, 219, 235
  sprays 37
  surface 26, 191, 211
season
  dry 98
  rainy 98
sedimentation 275
Siberia 154, 159, 167, 231
silicon 157, 175, 233
simulation 115
sodium 76
soil 141, 158, 220, 249, 276
solar zenith angle 14, 26
solubility 282
spatial
  resolution 63
  structure 10
special meteorological regime 122
spectral
  brightness 13
  distribution 13, 17, 23
  optical measurements 81
spectrometer 14, 62, 83, 136
spectrophotometer 6
spherical particles 138
steppe 26
stereophotogrammetry 9
storm dynamics 71
stratosphere 232
strontium 100
submicron particles 75
subtropical anticyclones 145
sulphate 31, 74, 148, 188, 215, 235
sulphur
  compounds 67, 215
  dioxide 8,10, 166, 189, 214
sun photometers 68, 252
surface
  albedo 10, 24, 72
  layer 102, 175
  observations 109
  ocean 249
surface–atmosphere interaction 32

synoptic
  conditions 21
  maps 33
  processes 42
system, surface–atmosphere 35

Tadjikistan 51
temperature 7, 33, 47, 196, 293
thermodynamic equilibrium 79
trace
  elements 74
  metals 152
transmission 21, 24
transparency 10, 30, 40
tropopause 45, 233
troposphere 4, 13, 110, 148, 158, 203
turbulent
  diffusion 9
  heat flux 7, 32
Turkmenistan 154

urban
  atmosphere 12
  climatology 8
  environment 199

variability 21, 77, 80, 146, 187, 223
vegetation 71, 190, 240, 267
vertical
  gradient 249
  motions 200
  profile 19, 24, 72, 148
  structure 119
volatile organic compounds 118
volcanic
  aerosol 188, 231, 251
  ash 85
  eruption 96, 231
  plume 145
volcano
  Colima 92, 101, 104, 287
  Fuego 104
  Kilauea 233
  Kudriavy 231
  Merapi 231
  Popocatepetl 90
Volga 41

water
  rain 98
  vapour 21, 61, 233
wavelength 13, 18, 119, 198, 247
weather conditions 82
White Sea 41
wind 12, 29, 53, 147, 188, 200
winter monsoon period 151

Zaporozhye 27
zirconium 88

Printing: Mercedes-Druck, Berlin
Binding: Stein+Lehmann, Berlin